LINEAR CIRCUIT ANALYSIS

S. MADHU
Professor of Electrical Engineering
Rochester Institute of Technology

With contributions from
R. Unnikrishnan
Professor of Electrical Engineering
Rochester Institute of Technology

LINEAR CIRCUIT ANALYSIS

Prentice Hall
Englewood Cliffs, New Jersey 07632

Library of Congress Cataloging-in-Publication Data

MADHU, SWAMINATHAN.
   Linear circuit analysis

   Bibliography: p.
   Includes index.
     1. Electric circuits, Linear. 2. Electric networks.
I. Unnikrishnan, R. II. Title.
TK454.M224 1988   621.319′2   87-14398
ISBN 0-13-536673-9

Editorial/production supervision: Mary Jo Stanley
Interior design: Kenny Beck
Cover and chapter opening design: Christine Gehring-Wolf
Cover photo: FPG/Chris Simpson
Manufacturing buyer: Cindy Grant

PSpice™ is a registered trademark of
MicroSim Corporation, Laguna Hills, California

© 1988 by Prentice-Hall, Inc.
A Division of Simon & Schuster
Englewood Cliffs, New Jersey 07632

All rights reserved. No part of this book may be
reproduced, in any form or by any means,
without permission in writing from the publisher.

Printed in the United States of America
10 9 8 7 6 5 4 3 2

ISBN 0-13-536673-9

Prentice-Hall International (UK) Limited, *London*
Prentice-Hall of Australia Pty. Limited, *Sydney*
Prentice-Hall Canada Inc., *Toronto*
Prentice-Hall Hispanoamericana, S.A., *Mexico*
Prentice-Hall of India Private Limited, *New Delhi*
Prentice-Hall of Japan, Inc., *Tokyo*
Prentice-Hall of Southeast Asia Pte. Ltd., *Singapore*
Editora Prentice-Hall do Brasil, Ltda., *Rio de Janeiro*

To
my children
and
my brother

CONTENTS

Preface XIX
A Note on Notation for Voltages and Currents XXIII

CHAPTER 1
BASIC CONCEPTS, COMPONENTS, AND LAWS OF ELECTRIC CIRCUITS 1

1-1 Basic Units and Definitions 2
    1-1-1 Potential Difference (Voltage) 2
    1-1-2 Electric Current 2
    1-1-3 Electrical Power and Energy 4
1-2 Voltage 7
    1-2-1 Voltage Polarity Notation 8
    1-2-2 Ideal Voltage Sources 9
1-3 Current 10
    1-3-1 Ideal Current Sources 10
1-4 Power Delivered and Power Received 11
1-5 Classification of Circuits 13
    1-5-1 Lumped and Distributed Circuits 13
    1-5-2 Linear and Nonlinear Circuits 14
    1-5-3 Time-variant and Time-invariant Circuits 17

1-6 Resistance 20
    1-6-1   Ohm's Law 20
    1-6-2   Power Dissipated in a Resistor 21
    1-6-3   Nonlinear Resistors 21
    1-6-4   Effect of Temperature 23
1-7 Circuit Diagram Conventions 23
1-8 Kirchhoff's Laws 23
    1-8-1   Kirchhoff's Voltage Law (KVL) 24
    1-8-2   Kirchhoff's Current Law (KCL) 28
1-9 Voltage and Current Sources 32
    1-9-1   Determination of the Voltage Source Model 34
    1-9-2   Determination of the Current Source Model 36
    1-9-3   Conversion between the Models of a Source 38
    1-9-4   Dependent Sources 41
1-10 An Overview of Circuit Analysis 42
1-11 Summary of Chapter 43
Answers to Exercises 44
Problems 45

CHAPTER 2
SERIES AND PARALLEL CIRCUITS 56

2-1 The Series Circuit 56
    2-1-1   Voltage Dividers 59
    2-1-2   Series Circuits with Multiple Sources 62
2-2 The Parallel Circuit 65
    2-2-1   Current Dividers 68
    2-2-2   Parallel Circuits with Multiple Sources 71
2-3 Duality 74
2-4 Series and Parallel Circuits with Dependent Sources 74
    2-4-1   Model of an Amplifier 76
2-5 Series-Parallel and Ladder Networks 78
    2-5-1   Series-Parallel Reduction Method 78
    2-5-2   Ladder Network Procedure 82
2-6 Circuit with A Nonlinear Component 88
2-7 Summary of Chapter 90
Answers to Exercises 91
Problems 93

CHAPTER 3
NODAL ANALYSIS 103

3-1 Network Topology—Choice of Independent Voltages 104
3-2 Principles of Nodal Analysis 106
    3-2-1   Resistors in Nodal Analysis 107
    3-2-2   Nodal Equations 108
    3-2-3   Standard Form of Nodal Equations 113

        3-2-4   Nodal Equations for Circuits with Voltage Sources   118
        3-2-5   Circuits with Dependent Sources   125
3-3   Algebraic Discussion of Nodal Equations   130
        3-3-1   Nodal Conductance Matrix   130
        3-3-2   Cramer's Rule and the Solution of Nodal Equations   133
        3-3-3   Circuits with Dependent Current Sources   136
        3-3-4   Transfer and Driving Point Resistances   139
3-4   Operational Amplifiers   144
        3-4-1   Symbol of an Op Amp   145
        3-4-2   Analysis of the Inverting Op Amp Circuit   146
        3-4-3   Approximate Model of an Op Amp   148
3-5   The Noninverting Op Amp Circuit   149
3-6   The Summing Op-Amp Circuit   151
3-7   Summary of Chapter   153
Answers to Exercises   154
Problems   157

CHAPTER 4
METHODS OF BRANCH CURRENTS, LOOP, AND MESH ANALYSIS   168

4-1   The Method of Branch Currents   168
        4-1-1   DC Analysis of a Transistor Circuit   173
        4-1-2   Operational Amplifiers   176
        4-1-3   Circuits Containing Current Sources   177
4-2   The Concept of Loop Currents   180
4-3   Network Topology—Choice of Independent Loop Currents   181
        4-3-1   Number of Independent Currents   182
        4-3-2   Selection of Independent Closed Paths   183
        4-3-3   Relationship of Branch Currents to Loop Currents   183
4-4   Principles of Loop Analysis   184
4-5   Mesh Analysis   189
4-6   Standard Form of Loop and Mesh Equations   191
4-7   Circuits with Linear Dependent Voltage Sources   195
4-8   Circuits Containing Current Sources   197
        4-8-1   Circuits with Linear Dependent Current Sources   199
        4-8-2   Mesh Analysis of Circuits with Current Sources   201
4-9   Duality   203
4-10   Mesh Resistance Matrix   205
        4-10-1   Application of Cramer's Rule to the Solution of Mesh Equations   207
        4-10-2   Circuits Containing Dependent Voltage Sources   209
        4-10-3   Loop Resistance Matrix   211
        4-10-4   Driving Point and Transfer Conductances   213
4-11   Summary of Chapter   216
Answers to Exercises   216
Problems   220

CHAPTER 5
NETWORK THEOREMS 230

5-1 Thevenin's Theorem 231
    5-1-1 Procedure for Finding a Thevenin Equivalent Circuit 232
5-2 Norton's Theorem 238
    5-2-1 Conversion between Thevenin and Norton Equivalent Circuits 242
5-3 Maximum Power Transfer Theorem 244
5-4 The Superposition Theorem 247
5-5 Reciprocity Theorem 250
5-6 Proofs of Theorems 253
    5-6-1 Proof of Thevenin's Theorem 253
    5-6-2 Proof of the Superposition Theorem 255
    5-6-3 Proof of Reciprocity Theorem 256
5-7 Summary of Chapter 256
Answers to Exercises 257
Problems 257

CHAPTER 6
CIRCUITS CONTAINING CAPACITORS AND INDUCTORS 264

6-1 Properties and Relationships of a Capacitor 264
    6-1-1 Current through a Capacitor 265
    6-1-2 Energy Stored in a Capacitor 267
    6-1-3 Voltage Continuity Condition 267
6-2 Properties and Relationships of an Inductor 267
    6-2-1 Duality between Capacitance and Inductance 268
    6-2-2 Energy Stored in an Inductor 269
    6-2-3 Current Continuity Condition 269
6-3 Equations of Simple RC and RL Circuits 270
    6-3-1 Solution of a First-Order Linear Differential Equation 271
    6-3-2 The Step Function 271
    6-3-3 Step Responses of a Series RC Circuit 273
    6-3-4 Discussion of the Step Response of the Series RC Circuit 275
    6-3-5 Time Constant and Rise Time 277
    6-3-6 Step-by-Step Procedure for the Solution of RC Circuits with a Step Input 278
    6-3-7 Discussion of the Pulse Response of an RC Circuit 281
    6-3-8 Change of Conditions in an RC Circuit 283
6-4 Step Response of a Parallel RC Circuit 286
6-5 Step Response of RL Circuits 288
    6-5-1 Discussion of the Step Response of an RL Circuit 289
    6-5-2 Time Constant of a Series RL Circuit 290

6-6 Zero Input and Zero State Response of First-Order Circuits   295
6-7 Step Response of a Second-Order Circuit   297
    6-7-1 Zero Input Response of a Series RLC Circuit   298
    6-7-2 Zero State Response of an RLC Circuit (Step Input)   305
6-8 Integrating Op Amp Circuit   307
6-9 Summary of Chapter   308
Answers to Exercises   309
Problems   311

CHAPTER 7
SINUSOIDAL STEADY STATE–TIME DOMAIN ANALYSIS   318

7-1 Importance of Sinusoids in Circuit Analysis   318
7-2 Basic Definitions and Relationships of Sinusoidal Functions   319
    7-2-1 Phase Angle   320
    7-2-2 Phase Difference between Two Signals   322
    7-2-3 Decomposition of a General Sinusoidal Signal   323
    7-2-4 Combination of Sinusoidal Functions   324
7-3 Voltage, Current, and Power in Single Components   326
7-4 The Series RL Circuit–Time Domain Solution   329
7-5 The Series RC Circuit–Time Domain Solution   331
7-6 The Parallel GLC Circuit–Time Domain Solution   332
7-7 Instantaneous and Average Power in AC Circuits   334
    7-7-1 Average Power   335
7-8 RMS Values of Time-Varying Signals   338
    7-8-1 RMS Value of a Sinusoidal Voltage   339
7-9 Summary of Chapter   340
Answers to Exercises   341
Problems   342

CHAPTER 8
PHASORS, IMPEDANCE, AND ADMITTANCE   347

8-1 Complex Exponential Functions   347
8-2 The Phasor Concept   350
    8-2-1 Phasors and Sinusoidal Forcing Functions   352
8-3 The Series RL Circuit under Complex Exponential Excitation   354
8-4 Analysis of Circuits in the Sinusoidal Steady State: Procedure Using Phasors   356

8-5   Impedance, Resistance, and Reactance   360
    8-5-1   Impedance of Single Elements   362
    8-5-2   Series Circuits   363
    8-5-3   Impedance Triangles and Phasor Diagrams   367
8-6   Admittance, Conductance, and Susceptance   369
    8-6-1   Admittance of Single Elements   371
    8-6-2   Parallel Circuits   372
    8-6-3   Admittance Triangles and Phasor Diagrams   375
    8-6-4   Relationships between Impedance and Admittance Components   377
8-7   Simple Equivalent Circuits   378
8-8   An Impedance Bridge   380
8-9   Power in the Sinusoidal Steady State   381
    8-9-1   Power Factor   382
    8-9-2   Apparent Power and Reactive Power   384
    8-9-3   Modification of Power Factor   386
    8-9-4   Complex Power   389
8-10   Summary of Chapter   391
Answers to Exercises   392
Problems   395

CHAPTER 9
NETWORKS IN THE SINUSOIDAL STEADY STATE   402

9-1   Networks with Single Sources   402
9-2   Networks with Multiple Sources   408
9-3   Nodal Analysis   408
    9-3-1   Nodal Analysis in the Presence of Voltage Sources   412
9-4   Loop and Mesh Analysis   417
    9-4-1   Loop and Mesh Analysis in the Presence of Current Sources   422
9-5   Transfer Functions of a Network   426
    9-5-1   Frequency Response Characteristics of a Transfer Function   428
    9-5-2   Complex Exponential Driving Functions and the Concept of Negative Frequency   430
9-6   Transfer Functions from Mesh Impedance Matrices   433
9-7   Network Theorems   437
    9-7-1   Thevenin's Theorem   437
    9-7-2   Norton's Theorem   440
    9-7-3   Maximum Power Transfer Theorem   443
    9-7-4   The Superposition Theorem   447
    9-7-5   The Reciprocity Theorem   450
9-8   Summary of Chapter   453
Answers to Exercises   454
Problems   456

CHAPTER 10
MAGNETIC CIRCUITS, COUPLED COILS AND THREE-PHASE CIRCUITS   472

10-1 Basic Relationships of Magnetic Fields   472
10-2 A Basic Magnetic Circuit   475
    10-2-1 Reluctance of a Magnetic Circuit   477
10-3 Principles of Analysis of Magnetic Circuits   478
    10-3-1 Determination of *mmf* for Specified Flux   479
    10-3-2 Determination of Flux for Specified *mmf*   481
10-4 Magnetically Coupled Coils   485
    10-4-1 Polarity of Mutually Induced Voltages   486
    10-4-2 Coefficient of Coupling   487
    10-4-3 Voltage Components in Coupled Coils   487
    10-4-4 Dot Polarity Notation   489
    10-4-5 Coupled Coils in Series   491
    10-4-6 Coupled Coils in Parallel   491
10-5 Analysis of Circuits with Coupled Coils   492
10-6 Energy Stored in Coupled Coils   494
10-7 The T Equivalent Circuit for Coupled Coils   497
10-8 Transformers   499
    10-8-1 The Ideal Transformer   499
    10-8-2 Impedance Transformation Using Ideal Transformers   504
10-9 Three-Phase Systems   505
    10-9-1 Double Subscript Notation   505
    10-9-2 Balanced Three-Phase Sources   507
    10-9-3 Y Connected Load   509
    10-9-4 Delta Connected Load   512
10-10 Power in Three-Phase Circuits   515
    10-10-1 Use of Wattmeters in Measuring Power   517
10-11 Summary of Chapter   520
Answers to Exercises   521
Problems   523

CHAPTER 11
RESONANT CIRCUITS   528

11-1 Resonance in a Parallel GLC Circuit   528
    11-1-1 Properties of the Parallel Resonant Circuit   532
    11-1-2 Energy Considerations   534
11-2 Bandwidth of the Parallel GLC Circuit   539
    11-2-1 Definition of Bandwidth   539
11-3 The Two-Branch Parallel RLC Circuit   544
    11-3-1 Application of Tuned Circuits   549

11-4 Resonance in a Series RLC Circuit   551
    11-4-1 Energy Considerations in a Series RLC Circuit   554
    11-4-2 Resonant Q of a Series RLC Circuit   555
    11-4-3 Bandwidth of a Series Resonant Circuit   557
11-5 Reactance/Susceptance Curves of Lossless Networks   561
11-6 Summary of Chapter   566
Answers to Exercises   566
Problems   568

CHAPTER 12
COMPLEX FREQUENCY AND NETWORK FUNCTIONS   573

12-1 The Concept of Complex Frequency   573
12-2 Impedance and Admittance in the Complex Frequency Domain   575
    12-2-1 Analysis of Circuits in the Complex Frequency Domain   576
12-3 Critical Frequencies of a Network Function   578
    12-3-1 Poles and Zeros of Network Functions   579
12-4 Frequency Response and Bode Diagrams   582
    12-4-1 Amplitude Response Diagrams   583
    12-4-2 Phase Response Diagrams   589
    12-4-3 Bode Diagrams for $H(s)$ with a Single Zero   590
    12-4-4 Bode Diagrams of a Network   592
12-5 Complex Frequency Viewpoint of Resonance   598
    12-5-1 Magnitude and Angle of a Driving Point Function   598
    12-5-2 Response of a Tuned Circuit   600
    12-5-3 Movement of Zeros of $Y(s)$ as a Function of Q_{CO} 601
    12-5-4 Variation of $Y(s)$ as a function of s 602
12-6 Summary of Chapter   606
Answers to Exercises   607
Problems   608

CHAPTER 13
TWO-PORTS   614

13-1 Basic Notation and Definitions   615
13-2 Open-Circuit Impedance Parameters (Formal Derivation)   616
13-3 Open-Circuit Impedance Parameters of a Two-Port   619
13-4 Use of the z Parameters in the Analysis of a Two-Port System   623
13-5 Models of a Two-Port Using z Parameters   624
    13-5-1 Model of a Reciprocal Two-Port   624
    13-5-2 Models of a Nonreciprocal Two-Port   626
13-6 Short-Circuit Admittance Parameters of a Two-Port   629

13-7   Models of a Two-Port Using y Parameters   632
13-8   Relationship between z and y Parameters   635
13-9   The Hybrid Parameters   636
  13-9-1   Determination of the h Parameters of a Two-Port   637
13-10 Model of a Two-Port Using h Parameters   642
  13-10-1 Analysis of a Two-Port System Using h Parameters   643
13-11 Relationship between h and z Parameters   644
13-12 Transmission or *ABCD* Parameters   646
13-13 Interconnection of Two-Port Networks   646
  13-13-1 Parallel Connection of Two-Ports   646
  13-13-2 Cascaded Connection of Two-Ports   649
13-14 Balanced and Unbalanced Two-Ports   653
13-15 Summary of Chapter   653
Answers to Exercises   654
Problems   656

CHAPTER 14
NONSINUSOIDAL SIGNALS–FOURIER METHODS   661

14-1   Fourier Series Representation of a Periodic Signal   661
  14-1-1   Determination of the Fourier Series Coefficients   663
  14-1-2   Effect of Waveform Symmetry on Fourier Series   668
14-2   Steady State Response of a Circuit with Periodic Input   673
  14-2-1   Average Power in a Circuit with Periodic Input   675
  14-2-2   RMS Value of a Periodic Signal   678
14-3   Exponential Fourier Series   679
  14-3-1   Determination of the Coefficients c_k 679
  14-3-2   Relationships between Exponential and Trigonometric Forms of Fourier Series   681
14-4   Steady State Response of a Circuit Using Exponential Fourier Series   683
14-5   Frequency Spectra of Periodic Signals   685
  14-5-1   Amplitude and Phase Spectra   685
  14-5-2   Power Spectrum   686
14-6   Minimum Mean Square Error Property of Fourier Series   689
14-7   Nonperiodic Signals: The Fourier Transform   690
  14-7-1   Energy Density Spectrum of a Signal   693
  14-7-2   Properties of the Fourier Transform   695
14-8   The Inverse Fourier Transform   699
14-9   Response of Linear Networks to Nonperiodic Signals   701
14-10 Response of a Linear Network to Arbitrary Signals   702
  14-10-1 Representation of an Arbitrary Signal by Means of Impulse Functions   703
  14-10-2 The Impulse Function   705

14-11 The Convolution Integral 708
  14-11-1 Evaluation of the Convolution Integral 711
  14-11-2 Fourier Transform of Convolution 717
14-12 Response of a Linear System 720
14-13 Limitations of the Fourier Transform 721
14-14 Summary of Chapter 721
Answers to Exercises 723
Problems 725

CHAPTER 15
LAPLACE TRANSFORMATION 734

15-1 The Laplace Transform 735
  15-1-1 Region of Convergence for the Laplace Transform 735
  15-1-2 Causal Signals and Systems 738
15-2 Use of Laplace Transforms in Analysis of Linear Networks 741
  15-2-1 Some Properties of the Laplace Transform 741
  15-2-2 Current Voltage Relationships in Basic Circuit Components 743
  15-2-3 Analysis of a Linear Network 745
15-3 Determination of the Inverse Laplace Transform 748
  15-3-1 Procedure for Finding the Partial Fraction Expansion of $F(s)$ 750
15-4 Step-by-Step Procedure for Analysis of Networks Using Laplace Transforms 759
15-5 Initial Value and Final Value Theorems 768
15-6 Convolution Theorem 769
15-7 Role of Critical Frequencies in the Response of Networks 769
15-8 An Overview of Network Analysis 770
15-9 Summary of Chapter 772
Answers to Exercises 772
Problems 774

APPENDIX A   Computer-Based Problems 781

APPENDIX B   Determinants and Matrices 785
  B-1 Determinants 785
  B-2 Minors and Cofactors of a Determinant 786
  B-3 Laplace's Expansion of a Determinant 787
  B-4 Cramer's Rule 787
  B-5 Matrices 789

APPENDIX C   Complex Numbers and Complex Algebra 791

APPENDIX D SPICE Program for Circuit Analysis 797

- D-1   Introduction   797
- D-2   Types of Analysis   798
- D-3   Input Format   798
- D-4   Circuit Description   799
- D-5   Title, Comment, and .END Statements   799
- D-6   Element Statements   800
- D-7   Linear Dependent Sources   801
- D-8   Independent Source Functions   802
- D-9   Control Statements   804
- D-10 Examples of SPICE applications   808

APPENDIX D SPICE Program for Circuit Analysis 797

- D-1   Introduction   797
- D-2   Types of Analysis   798
- D-3   Input Format   798
- D-4   Circuit Description   799
- D-5   Title, Comment, and .END Statements   799
- D-6   Element Statements   800
- D-7   Linear Dependent Sources   801
- D-8   Independent Source Functions   802
- D-9   Control Statements   804
- D-10 Examples of SPICE applications   808

BIBLIOGRAPHY   811

ANSWERS TO EVEN-NUMBERED PROBLEMS   813

INDEX   827

PREFACE

The analysis of electric circuits and networks is a fundamental tool of the electrical engineering student, since it provides the foundation for the study of linear systems, electronic circuits, power systems, control systems, and communication systems. A strong background in methodical circuit analysis is essential for intelligent and successful design, as well as for using the computer for the analysis of complex networks. The aim of this text is to present a reasonably rigorous treatment of the methods of circuit analysis and their applications, it is intended for use by sophomores and juniors in electrical engineering.

With the growing need for accommodating a wide variety of important technical areas in an electrical engineering program, there has been an unavoidable trend toward curtailing the time allotted to teaching circuit analysis. As a result, faculty usually find it difficult to devote enough time in the classroom to working a sufficient number of examples and drilling the student in the various topics in circuit analysis. The difficulty has been exacerbated in many colleges and universities by large class sizes. I have tried to alleviate the problem by including numerous worked examples, as well as a large number of exercise problems in each chapter. *A section at the end of each chapter provides not only the final answers to the exercise problems, but also appropriate intermediate steps to help students check their work.* The intermediate steps should help reassure students whose final answers are incorrect even though they have applied the principles correctly. Each chapter also includes 30 to 50 homework problems organized by sections within the chapter; the final answers to roughly 50 percent of these problems are provided at the end of the book. A combination of the worked examples, exercise problems with intermediate steps and answers, and the large number of chapter-end problems should help students develop a strong working background in the principles and procedures of circuit analysis.

The general mathematical background expected of the student is a year of differential and integral calculus. There are some optional sections in Chapters 3 and 4 and some later chapters requiring a knowledge of the theory of determinants (especially Cramer's rule) and basic matrix notation. Students lacking a background in the use of determinants and basic principles of matrix notation will find it helpful to study the summary of the relevant information provided in Appendix B. First- and second-order linear differential equations are encountered in Chapter 6, and the necessary steps for solving them (by using an assumed solution) are presented in that chapter. A knowledge of complex numbers and complex algebra is essential for Chapter 8 and the chapters that follow. A summary of the basic concepts and rules of complex numbers and complex algebra is presented in Appendix C.

Motivation is an important problem in teaching circuit analysis. Since circuit analysis is usually taught when the students have not been exposed to other electrical engineering subjects (except probably digital systems and microprocessors), they are overwhelmed by the steady flow of information with very little clue about what to do with it. I have tried to address this problem by providing examples of circuits from electronics (biasing calculations, small signal analysis, and pulse response, as well as operational amplifiers) and communications (filters and bandwidth considerations of communication systems). The op amp, with its simple approximate model, provides an attractive and convenient tool for illustrating the use of circuit analysis in practical electronic circuits. It is first introduced in Chapter 3 (on nodal analysis) and revisited periodically in later chapters, where its linear applications are discussed.

The degree and manner in which students use the computer at this point in their program vary widely. At this early level, it is more important for the student to grasp the principles and methods of circuit analysis by focusing on circuits of moderate complexity rather than simply obtaining the final answers by means of a computer. Nevertheless, it is desirable to give the student the opportunity to write programs in FORTRAN or BASIC as an adjunct of a course (or courses) in circuit analysis. With this end in view, a set of computer-based problems is provided in Appendix A.

When the student has progressed sufficiently in circuit analysis, it becomes appropriate to use commercially available software packages to study complex circuits of practical importance. A summary of SPICE is provided in Appendix D for this purpose. Since there is a proliferation of personal computers on campuses, an adaptation of SPICE, called *PSpice*,™ created by MicroSim Corporation for use with the personal computer, should prove attractive to many students and professors. Prentice Hall also publishes *A Guide to Circuit Simulation and Analysis Using PSpice*™ by Paul Tuinenga. The hardcover edition includes a student version of the PSpice program. A paperback edition of the text only is also available.

The order of treatment of the topics of circuit analysis follows classical lines. The treatment of circuits containing only resistors and sources is presented first, since such circuits need only the algebra of real numbers and variables and permit the student to concentrate on the principles and methods of circuit analysis without worrying about more involved mathematical manipulations. Chapter 1 deals with basic definitions important in the study of lumped, linear, and time-invariant circuits, Ohm's law, and Kirchhoff's laws. Simple circuit configurations are discussed in that chapter. Chapters

2, 3, 4, and 5 develop the methods of circuit analysis for networks containing resistors and dependent and independent sources. Chapter 6 discusses the properties and relationships of capacitors and inductors and the time-domain determination of the step response of first- and second-order circuits. The classical (time domain) approach to the sinusoidal steady state response of circuits is covered in Chapter 7. Chapter 8 introduces the concepts of phasors, impedance, and admittance, and deals with series and parallel configurations.

The analysis of more general circuits in the sinusoidal steady state is discussed in Chapter 9 by modifying the information already presented in Chapters 3, 4, and 5. Chapter 10 discusses the principles of magnetic circuits, magnetically coupled coils, and balanced three-phase circuits. Parallel and series resonance is studied in Chapter 11 by using impedance and admittance concepts. A discussion of reactance/susceptance curves of lossless networks is also included in that chapter. Complex frequency is introduced on an ad hoc basis in Chapter 12 as an important tool in the study of the frequency response characteristics of driving point and transfer functions of networks. Bode diagrams of circuits with simple poles and zeros are dealt with in detail in that chapter. Resonance is revisited by using vectors in the complex frequency plane to illustrate the link between sinusoidal steady state response, natural response, and the use of complex frequency.

Chapter 13 discusses two-port networks and their analysis by using z, y, and h parameters. The *ABCD* parameters are introduced by means of exercise problems in that chapter. The appropriate characterization of some of the interconnections of two-ports are also presented in Chapter 13. Chapters 14 and 15 extend the analysis of networks to other than sinusoidal signals. The discussion in Chapter 14 starts with the use of Fourier series for periodic signals and is then extended to the use of Fourier transforms in the case of nonperiodic signals of finite duration. The use of convolution for finding the time domain response of a network to an arbitrary signal is also presented in that chapter. Laplace transform is presented in Chapter 15 as an extension of the Fourier transform by using complex frequency to provide convergence of the transform. The use of Laplace transform in the analysis of circuits is also studied in that chapter. It concludes with an overview of circuit analysis as presented in this text.

The material in this book could be covered comfortably in a one-year sequence in circuit analysis without any omissions. In my school, we find it possible to cover Chapters 1 through 13 in two academic quarters by skipping a small portion of the material in those chapters. The material in Chapters 1 through 9 would fit the one-semester course in circuit analysis taught in a number of colleges and universities. Such coverage will provide the student with the essential background in circuit analysis needed for courses in electronics and linear systems that normally follow the course in circuit analysis.

A solution manual for use by faculty is available from the publisher. The solution manual provides complete solutions to the problems in the 15 chapters, the computer-based problems in Appendix A, and also several sample problems solved by using SPICE (or PSpice).

The writing of this book has been a long project, and I have been helped along the way by a substantial number of people. Even though I cannot thank every one of them individually, I do wish to express my sincere appreciation to the following reviewers, whose constructive and careful criticisms have contributed significantly to the final manuscript: Professors Elizabeth Koster, Arthur Dickerson, Charles Nelson, Peter Scott, and Sidney Shapiro. My colleague, Unnikrishnan, was instrumental in the starting of this project, and the material in Chapter 10 is based on his contributions. I also wish to thank another colleague, David Perlman, whose notes on SPICE have been useful in writing Appendix D. I have had the good fortune of working with two of the most patient and understanding editors: Tim Bozik and Bernard Goodwin. Their continued faith in this book has been a strong driving force. Finally, I cannot fully express how grateful I am to my wife Janice, who had serious doubts that this book would ever be finished, patiently put up with portions of the manuscript strewn over various parts of the house for months, and accepted my excuses for not doing a number of things.

S. Madhu
Rochester, New York

A NOTE ON NOTATION FOR VOLTAGES AND CURRENTS

The following notation is used to represent voltages and currents:

Lowercase letters explicitly written as functions of time, $v(t)$ and $i(t)$, denote time-varying voltages and currents, respectively. Often the explicit dependence is not shown, and the functions are written simply as v or i.

Uppercase letters V and I denote parameters that are constant in time-varying voltages and currents. Examples are the constant value of a voltage or current in the case of dc (Chapters 1 through 6), and the amplitude of a sinusoidal voltage or current in the case of ac (Chapter 7 and those that follow).

In Chapters 1 through 5, the networks are purely resistive, and their analysis is not affected by whether the voltages and currents are constants or time-varying. In these chapters, lower- and uppercase letters are used interchangeably.

Lowercase letters in boldface, $\mathbf{v}(t)$ and $\mathbf{i}(t)$, denote *complex time-varying functions*, usually complex exponential functions (Chapter 8 and those that follow).

Uppercase letters in boldface, $\mathbf{V}$ and $\mathbf{I}$, are used for *phasors* representing sinusoidal voltages and currents (Chapter 8 and those that follow). The boldface is sometimes omitted (especially in the later chapters), however, when the context is sufficient to indicate that the voltages and currents are phasors (with magnitude and angle). The same comment applies to letters used to represent impedances and admittances: boldface is not always used when the context indicates that the quantity has both magnitude and angle.

LINEAR CIRCUIT ANALYSIS

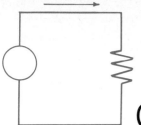

CHAPTER 1
BASIC CONCEPTS, COMPONENTS, AND LAWS OF ELECTRIC CIRCUITS

Electrical components (e.g., resistors, coils, capacitors, and batteries) are familiar to us from a study of electricity and magnetism in physics. An *electric circuit* is an interconnection of electrical components for performing some function. The term *electric network* usually refers to a rather complex electric circuit, but the terms *circuit* and *network* are used interchangeably. As one progresses through the study of electric circuits and networks, a point is reached where the interest is no longer on the structural details of the network but on how it behaves when excited by a particular source of electrical energy. In such cases, the term *electrical system* is used.

The aim of *circuit analysis* is the evaluation of quantities, such as voltages, currents, and power, or ratios of such quantities in a given circuit. *Synthesis* of networks, on the other hand, deals with the methodical design of networks to satisfy specified input-output characteristics. A strong understanding of analysis forms the foundation for intelligent and efficient synthesis.

Procedures used in circuit analysis are built on basic laws rooted in certain physical principles. These principles govern the behavior of the components and the constraints imposed by their interconnections. The underlying laws of circuit analysis are only a few in number; they are fairly simple to state and understand. The actual application of the laws to the methodical analysis of circuits, however, leads to a variety of systematic techniques to be presented in this text.

The purpose of this chapter is to introduce the basic concepts of electric circuits, some of the electrical components, and some of the basic laws of circuit analysis.

1-1 BASIC UNITS AND DEFINITIONS

The three basic physical quantities of a physical system are *mass* (measured in *kilograms,* abbreviated *kg*), *length* (measured in *meters,* abbreviated *m*) and *time* (measured in *seconds,* abbreviated *s*). A fourth basic quantity, the *electric charge,* is needed when the physical system also contains electrical components. The symbol for electric charge is *q,* and it is measured in *coulombs* (abbreviated *C*). Special names are given to the units of various quantities in electrical systems (usually in honor of pioneering scientists); but all these units can be expressed in terms of the four basic units: kilograms, meters, seconds, and coulombs.

1-1-1 Potential Difference (Voltage)

An electric field exists in the neighborhood of an electric charge or a collection of electric charges. Since the electric field exerts a force on any electric charge placed in it, work has to be performed (that is, energy has to be expended) to move the electric charge from one point to another in the electric field. The work needed to move a *positive* electric charge of *1 coulomb* (+1 C) from a point A to a point B in an electric field is defined as the *potential difference* between those two points.

> **The potential difference represents the *change in potential energy* of a unit positive charge between two points.**

The unit of potential difference is the *volt* (abbreviated *V*). It follows from the above definition that *1 volt equals 1 joule/coulomb,* where energy is measured in joules. Since the unit for potential difference is volts, the phrase *voltage between two points* is commonly used to denote the potential difference between two points.

Example 1-1 The potential energy of an electron is found to change by 1.2×10^{-16} J when it moves between two points. Calculate the voltage between the two points.

Solution Magnitude of electronic charge = 1.6×10^{-19} C
Work done in moving the electron = 1.2×10^{-16} J
Therefore, the voltage between the two points is $1.2 \times 10^{-16}/1.6 \times 10^{-19} = 750$ V ∎

The term *electromotive force* (abbreviated *emf*) is often used when the voltage is due to a source of energy such as a battery or a generator.

In referring to the voltage between two points A and B, it is often convenient to use one of the points, say A, as a *reference*. This permits the comparison of potentials at different points in a circuit using the potential at point A as a common basis. In such cases, the term *voltage at point B* is used to represent the *potential difference between the point B and the reference point A.*

It is important to note that a voltage is *always* measured *between two points.* Even when the phrase "voltage *at* a point" is used, it actually means the voltage between that point and some reference point.

1-1-2 Electric Current

Current flow in a material is due to the motion of free electric charges under the influence of an external force acting on the material. Current carriers can be negatively charged (e.g., *free electrons* in metals and semiconductors) or positively charged (e.g., *holes* in semiconductors). When the flow of free charges is due to an external electric field, an electron moves in a direction opposite to the vector representing the electric field, whereas a hole moves in the same direction as that vector.

The *conventional* direction of *positive current* is always taken as that in which a *free positive charge* would flow under the influence of an external force. Therefore, when there is a motion of electrons, the direction of positive current is *opposite* to the direction of motion of the electrons.

The current through a cross section of a material is defined by the *time-rate of change of electric charge* through that section. That is, if q is the electric charge, which varies with time t, then the resulting current i is given by

$$i = \frac{dq}{dt} \qquad (1\text{-}1)$$

The unit of current is the *ampere* (abbreviated A). One ampere represents a rate of flow of *1 coulomb/second*.

Example 1-2 The electric charge in a cross section of a material is found to vary with time as shown in Fig. 1-1(a). Determine and sketch the current $i(t)$.

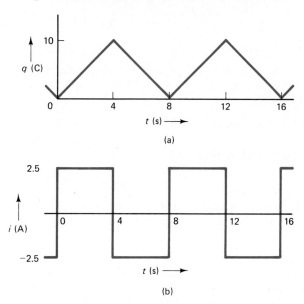

Fig. 1-1 Charge and current waveforms for Example 1-2.

Solution Since the graph of $q(t)$ is in the form of straight-line segments in this example, dq/dt is evaluated from the *slopes* of the different segments. The segments in Fig. 1-1(a) have slopes of $+2.5$ and -2.5 alternately. Therefore,

$$i(t) = \begin{cases} 2.5 \text{ A when } q(t) \text{ has a positive slope} \\ -2.5 \text{ A when } q(t) \text{ has a negative slope} \end{cases}$$

The waveform of the current $i(t)$ is shown in Fig. 1-1(b). ■

Example 1-3 The electric charge in a cross section of a material varies according to the equation

$$q(t) = Q_o \cos Bt \text{ C}$$

where Q_o and B are constants. Determine the current $i(t)$ through that section.

Solution

$$i = \frac{dq}{dt}$$
$$= -Q_o B \sin Bt \text{ A}$$

Exercise 1-1* The variation of electric charge in a cross section of a material is of the form shown in Fig. 1-2. Determine and sketch $i(t)$.

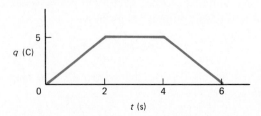

Fig. 1-2 Charge waveform for Exercise 1-1.

Exercise 1-2 The electric charge in a cross section of a material is found to vary according to the equation

$$q(t) = Q_o e^{-pt} \text{ C}$$

where Q_o and p are positive constants. Determine and sketch $i(t)$.

1-1-3 Electrical Power and Energy

Power is the rate at which work is done. Consider an elementary charge dq moving from point A to point B in a time interval dt, and let the potential difference between A and B be v. Then the work done on the charge dq is

$$dw = v\, dq = v(i\, dt) \tag{1-2}$$

since $i = dq/dt$. Then the power is given by

$$p = \frac{dw}{dt} = vi \tag{1-3}$$

The unit of power is the *watt* (abbreviated W). One watt equals *1 joule/second*. It can be verified that the product (volts × amperes) has the dimensions of joules/second.

In general, the voltage v and the current i vary with time. The quantity p given by Eq. (1-3) represents the *instantaneous power* in the circuit and varies as a function of time. It is, therefore, preferable to write Eq. (1-3) in the form

$$p(t) = v(t)i(t) \tag{1-4}$$

explicitly showing the dependence on time.

Example 1-4 The voltage across an electrical component is a constant at 15 V, and the current through it is a constant at 6 A. Calculate the power in the component.

---

*Answers to the exercises, along with some intermediate steps where appropriate, will be found at the end of the chapter.

Solution Since the voltage and current are constant, the power is also a constant and equals $(15 \times 6) = 90$ W.

Example 1-5 The voltage across a circuit is given by

$$v(t) = 12 \cos 100t \text{ V}$$

and the current through it is

$$i(t) = 5 \cos 100t \text{ A}$$

Determine $p(t)$. Sketch $v(t)$, and $p(t)$.

Solution
$$p(t) = v(t)i(t) = 60 \cos^2 100t$$
$$= 30 + 30 \cos 200t$$

where the formula $\cos^2 X = (1 + \cos 2X)/2$ has been used.

The sketches of $v(t)$, $i(t)$, and $p(t)$ are shown in Fig. 1-3. It is seen that $p(t)$ varies continuously between a minimum of zero and a maximum of 60 W. The average value of $p(t)$ is 30 W.

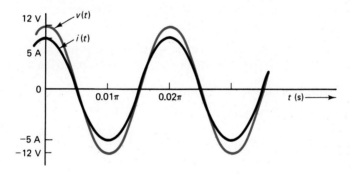

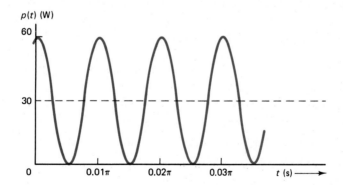

Fig. 1-3 Sinusoidal current, voltage, and power waveforms.

Exercise 1-3 The voltage across an electric circuit is a constant at 12 V, and the current through it is given by

$$i(t) = (10 + 5 \cos 20t) \text{ A}$$

Determine and sketch $p(t)$.

Exercise 1-4 The voltage across a component is

$$v(t) = 10 \cos 50t \text{ V}$$

The current through the component is equal to dv/dt. Obtain the expression for $p(t)$.

Energy in electric circuits is the work done in moving a charge between two points with a potential difference between them. Energy is either delivered to or received by a component. Since power equals the derivative of energy, *energy equals the integral of power*. The energy delivered or received by a component at time t is therefore given by

$$w(t) = \int p(t)dt \qquad (1\text{-}5)$$

The unit of energy is *joules* (abbreviated J).

The energy delivered to or received by a component in the time interval $(0, t)$ is given by

$$w(0,t) = \int_0^t p(u)du \qquad (1\text{-}6)$$

where u is a dummy variable of integration. The energy in a component at any time t can then be considered as the sum of two parts:

Energy at time t = (Initial energy in the component at $t = 0$)
+ [Energy delivered or received in the time interval $(0,t)$]

The integral in Eq. (1-5) can then be written as

$$w(t) = w(0) + \int_0^t p(u)du \qquad (1\text{-}7)$$

where $w(0)$ is the initial energy in the component.

Example 1-6 Determine the energy delivered at time t to the component of Example 1-4. Assume that $w(0) = 0$.

Solution Since power p is a constant at 90 W,

$$w(t) = w(0) + \int_0^t 90 du = 90t \text{ J}$$

The energy delivered increases linearly with time. ■

Example 1-7 The voltage across and the current through a component are given by

$$v(t) = V_o e^{-pt} \text{ V} \quad \text{and} \quad i(t) = I_o e^{-pt} \text{ A}$$

where V_o, I_o, and p are constants. Obtain an expression for the energy delivered to the component in the interval $(0,t)$.

Solution First obtain the expression for the power $p(t)$.

$$p(t) = v(t)i(t) = V_o I_o e^{-2pt} \text{ W}$$

The energy delivered in the interval $(0,t)$:

Basic Concepts, Components, and Laws of Electric Circuits

$$w(0,t) = \int_0^t p(u)du = \int_0^t V_o I_o e^{-2pu} du$$
$$= -(1/2p)V_o I_o (e^{-2pt} - 1) \quad \blacksquare$$

Exercise 1-5 The voltage across a component is a constant at 12 V, and the current through it is a constant at 7.5 A. Calculate the energy delivered to the component at $t = 3$ s, assuming the energy is zero at $t = 0$.

Exercise 1-6 Power companies measure energy in kilowatt-hours (*kWh*), which is the product of power in kilowatts and the number of hours over which the power is used. Calculate the equivalent of 1 kWh in joules.

Exercise 1-7 The voltage across and the current in a component are given by $v(t) = (10t - 8)$ V and $i(t) = 15$ A (constant), respectively. Determine and sketch the instantaneous power $p(t)$. Calculate the energy consumed in the intervals (a) (0,8 s) and (b) (8 s,12 s).

1-2 VOLTAGE

Voltages in an electric circuit can be constant at all times (or at least during the interval of observation), or vary as a function of time, as shown in Fig. 1-4.

Voltages that are constant at all times are referred to as *direct-current* or *dc* voltages. Time-varying voltages are classified according to whether they have repetitive or nonrepetitive waveforms, or whether their waveforms are deterministic or nondeterministic. Such classifications of time-varying voltages will be discussed in Chapters 7 and 14.

Voltages in an electric circuit can be observed on an oscilloscope or measured by a voltmeter. The so-called *dc voltmeter* measures the *average value* of the voltage taken over a certain time interval. When the voltage remains constant, its average value equals the constant value. In the case of time-varying voltages, the average value gives only one important parameter of the voltage. Another important parameter is its *root-mean-square,* or *rms,* value. The rms value of a time-varying voltage is the square *root* of the

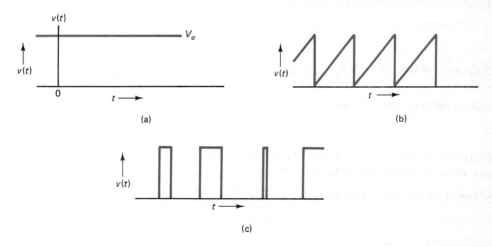

Fig. 1-4 (a) Constant and (b, c) time-varying voltages.

average (or *mean*) value of the *square* of the voltage. This may appear at first to be a complicated way of measuring voltage, but it actually is the natural form of response of certain voltmeter mechanisms. The *110 V* specification of the household and laboratory power supply in the United States is the rms value of the available voltage.

1-2-1 Voltage Polarity Notation

The potential energy of a unit positive charge situated in an electric field varies, in general, from one point to another. Given two points in an electric field, therefore, the potential at one of those points is, in general, higher than the other. (Whenever potentials at different points are compared with one another, *all potentials* are measured with respect to the *same reference point*.) In circuit diagrams, the fact that one terminal of a component is at a higher potential than the other is indicated by placing *plus* and *minus* polarity marks at the component's terminals, as illustrated in Fig. 1-5(a).

Fig. 1-5 Reference polarity marks for a voltage. (a) $v_1 = v_A - v_B$. (b) Circuit for Example 1-8.

The polarity marks in Fig. 1-5(a) mean that the voltage v_1 across the component is given by

$$v_1 = v_A - v_B \qquad (1\text{-}8)$$

where v_A and v_B are the potentials at points A and B, respectively. Voltage v_1 can be positive, zero, or negative, depending on whether the potential at terminal A, respectively, is greater than, equal to, or less than the potential at terminal B.

In circuit problems, it is usually not known beforehand which terminal of a component is actually at a higher potential than the other. It is customary to arbitrarily choose one terminal of the component as the positive reference and the other becomes the negative reference.

> Then the voltage across the component is equal to potential at the terminal chosen as the positive reference minus the potential at the terminal chosen as the negative reference.

The value obtained from the above relationship can be positive, zero, or negative.

Example 1-8 The component shown in Fig. 1-5(b) is part of a circuit. For each of the following pairs of values of potential at P and Q, calculate the voltage v_x.
(a) $v_P = 8$ V; $v_Q = 5$ V. (b) $v_P = -5$ V; $v_Q = 1$ V.
(c) $v_P = 2$ V; $v_Q = 2$ V. (d) $v_P = -7$ V; $v_Q = 5$ V.

Solution In all cases, v_x is calculated from the equation:

$$v_x = v_P - v_Q$$

(a) $v_x = 3$ V (point P is 3 V higher than point Q)
(b) $v_x = -6$ V (P is 6 V lower than Q)
(c) $v_x = 0$ V (P and Q are at the same potential)
(d) $v_x = -12$ V (P is 12 V lower than Q)

Fig. 1-6 Components for Example 1-9.

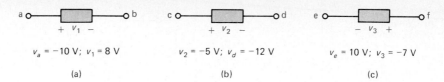

Example 1-9 Circuit components are shown in Fig. 1-6. The reference polarity marks, the voltage across the component, and the potential at one terminal are given in each case. Calculate the potential at the other terminal.

Solution Resist all temptation to guess. Write an equation for each case and solve it.
(a) $v_1 = v_a - v_b$. $v_1 = 8$ V; $v_a = -10$ V. Therefore, $v_b = -18$ V.
(b) $v_2 = v_c - v_d$. $v_2 = -5$ V; $v_d = -12$ V. Therefore, $v_c = -17$ V.
(c) $v_3 = v_f - v_e$. $v_3 = -7$ V; $v_e = 10$ V. Therefore, $v_f = 3$ V. ∎

Exercise 1-8 Calculate the voltage v_x in each of the components shown in Fig. 1-7. Values of potential at the terminals are shown.

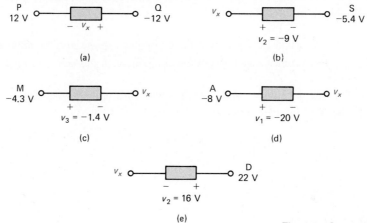

Fig. 1-7 Components for Exercise 1-8.

1-2-2 Ideal Voltage Sources

A source of electrical energy (for example, a battery) is used to supply power to a component or an arbitrary circuit connected to it, that is, to a *load*.

> When the energy source has the property that the voltage across its terminals is *independent* of the current supplied by it to the load, it is called an *ideal voltage source*.

The symbol for an ideal voltage source is shown in Fig. 1-8(a), where v_s denotes the voltage available across the terminals of the source. An ideal voltage source is shown connected to a load in Fig. 1-8(b). The voltage v_L across the load is not affected by how large or small the load current i_L is.

If the voltage v_s has a constant value at all times, then a plot of v_s as a function of the current i_L is a horizontal straight line, as shown in Fig. 1-8(c).

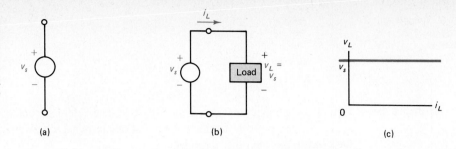

Fig. 1-8 Ideal voltage source. (a) Symbol for the ideal voltage source. (b) Voltage v_s is not affected by the load current i_L. (c) Case of a constant (or dc) voltage source.

As a practical matter, however, the voltage across the terminals of an energy source does not remain independent of the current supplied by it. In circuit analysis, nonideal voltage sources are represented by models, which will be considered later in this chapter.

1-3 CURRENT

The current through a component in an electric circuit is indicated by an arrow, representing the *assumed direction* of current flow, as shown in Fig. 1-9. There also exists a voltage with reference voltage polarities (plus and minus), leading to the two situations shown in the figure:

1. Current flows through the component in the *plus-to-minus* direction
2. Current flows through the component in the *minus-to-plus* direction

The term *associated reference polarities* is used when the current direction is chosen to be in the *plus-to-minus* direction of reference voltage polarities.

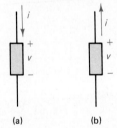

Fig. 1-9 Current direction in a component. Associated reference polarities are shown in part (a).

1-3-1 Ideal Current Sources

An energy source with the property that the current supplied by it is independent of the voltage across its terminals is called an *ideal current source*.

The symbol for an ideal current source is shown in Fig. 1-10(a), where i_s represents the current supplied by the source. An ideal current source is shown connected to a load in Fig. 1-10(b). The current i_L in the load is not affected by how large or small the voltage v_L across the load is.

If the current i_s supplied by the source has a constant value at all times, then a plot of i_L as a function of the load voltage v_L is a horizontal straight line, as shown in Fig. 1-10(c).

As a practical matter, however, the current supplied by an energy source is not independent of the voltage across the load connected to it. In circuit analysis, nonideal current sources are represented by models, which will be considered later in this chapter.

Fig. 1-10 Ideal current source. (a) Symbol for the ideal current source. (b) Current i_s is not affected by the load voltage v_L. (c) Case of a constant (or dc) current source.

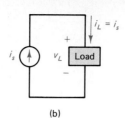

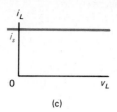

(a)   (b)   (c)

1-4 POWER DELIVERED AND POWER RECEIVED

The instantaneous power associated with a component is given by Eq. (1-9):

$$p(t) = v(t)i(t) \tag{1-9}$$

Since there is an assignment of *voltage reference polarities* and *direction* for the current through a component, two distinct cases are possible as shown in Fig. 1-9:

1. Current direction is from the plus to the minus terminal of voltage polarities (associated reference polarities)
2. Current direction is from the minus to the plus terminal of voltage polarities

Consider first the situation in Fig. 1-9(a), where the current is chosen flowing from the positive reference terminal to the negative reference terminal (associated reference polarities). Let both the voltage and current be positive. Then, positive current carriers are flowing from a higher to lower potential while negative current carriers are flowing from a lower to higher potential. That is, carriers are moving *in the direction of the force* due to an electric field set up in the component by an external agent (such as an electric generator). Therefore, work is being done by the external agent, and the component is *receiving* electrical energy from the external agent.

If, in a component, the *current direction is from the plus terminal to the minus terminal* (that is, associated reference polarities are used), then the product vi at any instant of time t represents the *power received* by the component at that instant of time.

Consider now the situation in Fig. 1-9(b), where the current is flowing from the negative reference terminal to the positive reference terminal. This is exactly the opposite of the previous case. That is, current carriers are moved through the component in *opposition to the force* due to an electric field. Therefore, work must be done by the component itself, and it is *delivering* energy to some other component or circuit.

If, in a component, the current direction is *from the minus terminal to the plus terminal*, then the product *vi* at any instant of time *t* represents the *power delivered* by the component at that instant of time.

The two cases can be summarized:

- vi = power *received* by a component if the direction of current is from *the plus to the minus* terminal of voltage polarities (associated reference polarities).
- vi = power *delivered* by a component if the direction of current is from *the minus to the plus* terminal of voltage polarities.

Example 1-10   In each of the cases shown in Fig. 1-11, calculate the power delivered or received by the component.

Fig. 1-11 Circuits for the power calculations of Example 1-10.

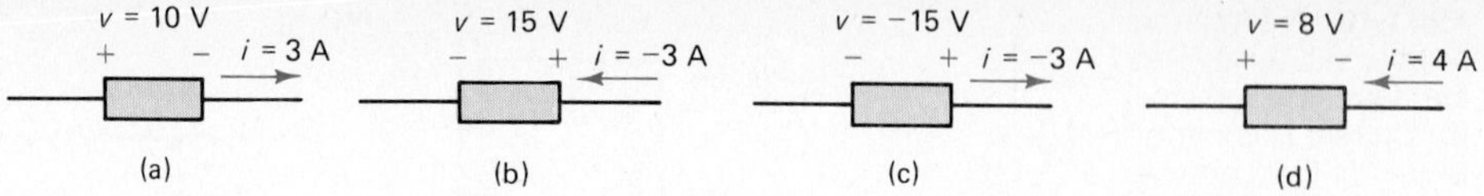

(a) $v = 10$ V, $i = 3$ A
(b) $v = 15$ V, $i = -3$ A
(c) $v = -15$ V, $i = -3$ A
(d) $v = 8$ V, $i = 4$ A

Solution In Fig. 1-11(a) and (b), the current direction is from plus to minus voltage polarities. Therefore, the product vi represents the power received in these two cases.
(a) $vi = 30$ W. The component is *receiving 30 W*.
(b) $vi = (15)(-3) = -45$ W. The component is *receiving -45 W*.

In Fig. 1-11(c) and (d), the current direction is from minus to plus voltage polarities. Therefore, the product vi represents the power delivered in these two cases.
(c) $vi = (-15)(-3) = 45$ W. The component is *delivering 45 W*.
(d) $vi = 32$ W. The component is *delivering 32 W*. ∎

The last example shows that the product vi is negative in some instances. Negative values of vi should be interpreted as follows: *receiving $-X$ watts* is equivalent to *delivering $+X$ watts;* and *delivering $-X$ watts* is equivalent to *receiving $+X$ watts*.

Example 1-11 The voltage across a component is $v = 12$V (constant at all times), and the current (whose direction is from the plus to the minus terminal of voltage polarities) is given by

$$i(t) = (8 + 10 \sin 4\pi t) \text{ A}$$

(a) Determine the instantaneous power $p(t)$ received by the component. (b) Sketch $p(t)$ as a function of time for $0 < t < 0.75$ s. (c) Calculate the maximum and minimum values of power received by the component. (d) Make use of the symmetry of the plot of $p(t)$ to estimate the average value of the power received by the component.

Solution (a) Instantaneous power:

$$p(t) = v(t)i(t) = (96 + 120 \sin 4\pi t) \text{ W}$$

Since the given current direction is from plus to minus voltage polarities, $p(t)$ represents the power *received* by the component at time t.

(b) The plot of $p(t)$ for the interval 0 to 0.75 s is shown in Fig. 1-12. Note that the power received by the component is positive for part of the interval and negative for part of the interval.

Fig. 1-12 Waveform for the power received in Example 1-11.

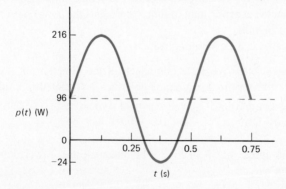

Basic Concepts, Components, and Laws of Electric Circuits

(c) The power received has a maximum value of 96 + 120 = 216 W and a minimum value of 96 − 120 = −24 W.

(d) From the symmetry of the waveform, the average value of power received (the dashed line) is 96 W. ∎

Exercise 1-9 Determine for each case of Fig. 1-13 whether the component is receiving or delivering power, and calculate the value of the power.

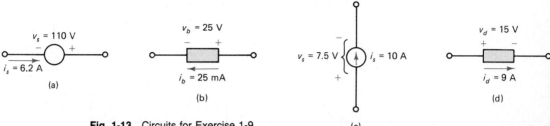

Fig. 1-13 Circuits for Exercise 1-9.

Exercise 1-10 It is given that each of the components of Fig. 1-14 is *delivering* +20 W of power. Calculate the value of the missing quantity (voltage or current) for each case, including the proper sign for each answer.

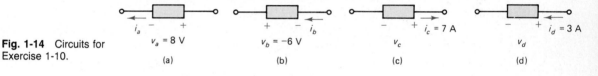

Fig. 1-14 Circuits for Exercise 1-10.

1-5 CLASSIFICATION OF CIRCUITS

Circuits are classified into different categories: lumped or distributed, linear or nonlinear, and time-variant or time-invariant. The analysis techniques to be presented in this text will be confined to *lumped, linear, time-invariant circuits*. Such a group covers a large number of networks of practical importance and interest. Even though the methods developed in this and the following chapters do not directly apply to other classes of networks, they nevertheless serve as an essential foundation for developing analysis methods for other types of networks.

1-5-1 Lumped and Distributed Circuits

Consider a taut string with one end fastened to a fixed point and the other end excited by some force. The resulting disturbance travels along the string in the form of a wave with an associated wavelength. The displacement and velocity at any point at a given instant of time depends on the position of the point on the line. In an analogous manner, electrical energy propagates in the form of electromagnetic waves (with the velocity of light). For example, the current and voltage on a power transmission line at any point at a given instant of time depends on the position of the point on the line. The current entering one end of the line is not necessarily equal to the current leaving at the other end, and the voltage can be measured at an infinite number of points along the line. Such components are called *distributed components*. Networks partly or wholly made up of distributed components are called *distributed networks*. The analysis of distributed networks requires the use of the theory of electromagnetic waves and is outside the scope of this book.

In principle, all circuit components are distributed components, since electrical energy takes a finite nonzero time to propagate through any component. From a practical viewpoint, however, when the physical dimensions of a component are much smaller than a *quarter-wavelength* of the waves propagating through it, the following approximating assumptions are made. The dependence of current and voltage on distance are neglected. A single value of current in the component and a single value of voltage across the component are defined for the entire component at a given instant of time. This is analogous to a body of small dimension in a mechanical system, where a single value of displacement and a single value of velocity is used for the entire body at a given instant of time.

A component in which it is possible to specify a single value of current and a single value of voltage for the whole component at a given instant of time is called a *lumped component*; a circuit made up *entirely* of lumped components is called a *lumped circuit*.

The criterion for a lumped component approximation is that the physical dimensions be much less than $\lambda/4$, where λ is the wavelength, given by $\lambda = c/f$, where c is the velocity of light (3×10^8 m/s), and f is the frequency of the electrical energy.

A component of length 2 cm, for example, can be treated as a lumped component when 2 cm $\ll \lambda/4$. *Much less than* is often taken to mean *10 percent or smaller*. In the present case, then, we take $\lambda/4 = 20$ cm $= 0.2$ m, or $\lambda = 0.8$ m. This leads to a frequency $f = c/\lambda = 3.75 \times 10^8$ Hz, which is a very high frequency indeed. Note that even the 2-cm-long component has to be treated as a distributed component at frequencies above 3.75×10^8 Hz.

Exercise 1-11 The frequency of electrical energy in power systems in the United States is 60 Hz. Estimate the maximum length of a transmission line that can be treated as a lumped circuit.

All components mentioned in the preceding sections were (tacitly) assumed to be lumped components, and the discussion in this textbook is confined to lumped circuits. Lumped circuits are analyzed by means of Kirchhoff's laws (to be introduced later in this chapter), without having to solve electromagnetic wave equations.

1-5-2 Linear and Nonlinear Circuits

Scaling Property

An electrical component is said to have the *scaling property* if it satisfies the following condition.

If forcing function $x_1(t) \rightarrow$ response $y_1(t)$,

then forcing function $kx_1(t) \rightarrow$ response $ky_1(t)$ where k is any constant.

The forcing function is either a voltage or current applied to the element (by a source), and the response is either a voltage or current caused by the forcing function.

For example, suppose

$$50 \cos 100t \text{ V} \to 12 \cos 100t \text{ A}$$

is a component with the scaling property. Then

$$5 \cos 100t \text{ V} \to 1.2 \cos 100t \text{ A}$$

Additive Property

An electrical component is said to have the *additive property* if it satisfies the following condition.

If forcing function $x_1 \to$ response $y_1(t)$,

and forcing function $x_2(t) \to$ response $y_2(t)$,

then $[x_1(t) + x_2(t)] \to [y_1(t) + y_2(t)]$

The forcing functions can be voltages or currents, and the response functions can also be voltages or currents.

For example, suppose

$$\text{current } 10e^{-5t} \text{ A} \to \text{voltage } 2.5e^{-5t} \text{ V}$$

and

$$\text{current } 3 \cos 377t \text{ A} \to \text{voltage } 12 \cos 377t \text{ V}$$

in a component with the additivity property. Then

$$\text{current } (10e^{-5t} + 3 \cos 377t) \to \text{voltage } (2.5e^{-5t} + 12 \cos 377t)$$

A component that possesses *both the scaling and additive properties* is said to be *linear*. A circuit made up entirely of linear components is a *linear circuit*.

Example 1-12 The voltage-current relationship of an electrical component is

$$v = \frac{di}{dt} \tag{1-10}$$

Determine if the component is linear.

Solution *Scaling Condition* Suppose a current i_1 in the component produces a voltage v_1. Then, from Eq. (1-10),

$$v_1 = \frac{di_1}{dt} \tag{1-11}$$

Now, change the current to ki_1. Then, using Eq. (1-10),

$$v = \frac{d(ki_1)}{dt} = k\left(\frac{di_1}{dt}\right)$$

or, from Eq. (1-11),

$$v = kv_1$$

It is seen that if $i_1 \to v_1$, then $ki_1 \to kv_1$ for the given component. Therefore, the scaling condition is satisfied.

1-5 Classfication of Circuits

Additive Condition Suppose a current i_a produces a voltage v_a and a current i_b produces a voltage v_b in the component. Then from Eq. (1-10),

$$v_a = \frac{di_a}{dt} \quad \text{and} \quad v_b = \frac{di_b}{dt} \tag{1-12}$$

Now let the current in the component be $i_a + i_b$. Then Eq. (1-10) leads to

$$v = \frac{d(i_a + i_b)}{dt}$$

$$= \frac{di_a}{dt} + \frac{di_b}{dt}$$

or, from Eq. (1-12),

$$v = v_a + v_b$$

Thus, if $i_a \rightarrow v_a$ and $i_b \rightarrow v_b$,

then $i_a + i_b \rightarrow v_a + v_b$

Therefore, the additive condition is satisfied by the component.
Since both the scaling and additive conditions are satisfied, the component is linear. ∎

Example 1-13 Test the linearity of a device whose voltage-current relationship is given by

$$i = av + bv^2 \tag{1-13}$$

where a and b are constants.

Solution *Scaling Condition* Let $v_1 \rightarrow i_1$. Then, from Eq. (1-13),

$$i_1 = av_1 + bv_1^2 \tag{1-14}$$

Now change the voltage to kv_1. Then, from Eq. (1-13),

$$i = a(kv_1) + b(kv_1)^2 \tag{1-15}$$
$$= kav_1 + k^2 bv_1^2$$

It is seen, from Eqs. (1-14) and (1-15), that

$$i \neq ki_1$$

Therefore, the scaling condition is not satisfied by the given device, and the device is *not linear*. Note that once the scaling condition fails, it is not necessary to check the additive condition to prove that the device is nonlinear. ∎

Exercise 1-12 The voltage current relationship of a component is given by $v = \int i\, dt$. Test the component for linearity.

Exercise 1-13 The voltage-current relationship of a component is given by $i = v(dv/dt)$. Test the component for linearity.

Principle of Superposition

The two aspects of linearity, scaling and additivity, are usually combined into a single statement as follows. If the response of a component due to the weighted sum of a number of forcing functions is the weighted sum of the responses due to each forcing function acting alone on the component, then the component is linear. That is,

$$\text{if } x_1 \to y_1, x_2 \to y_2, \ldots, x_n \to y_n$$

where the x's represent the forcing functions and the y's the response functions, and

$$a_1 x_1 + a_2 x_2 + \ldots + a_n x_n \to a_1 y_1 + a_2 y_2 + \ldots + a_n y_n$$

where the a's are all constants, then the component is linear.

Conversely,

if a component is linear, then its response to a weighted sum of several forcing functions is a weighted sum of the responses due to each forcing function acting alone on the component.

This last statement is known as the *principle of superposition* and plays a pivotal role in the analysis of linear circuits and systems.

The principle of superposition leads to straightforward methods of analysis of linear circuits. The methods of circuit analysis presented in this text will be confined to linear circuits (except for the discussion of the graphical analysis of a simple nonlinear circuit). As a practical matter, all electrical components and devices exhibit a certain degree of nonlinearity. Since the analysis of nonlinear circuits and systems is complicated, time-consuming, and often impractical, it becomes necessary to use idealized linear models in circuit analysis. It should be remembered that such idealization frequently leads to results that do not match laboratory measurements with 100 percent accuracy (especially if the models have not been developed carefully for a given device), and some adjustments and refinements are necessary in practical design.

1-5-3 Time-variant and Time-invariant Circuits

If the result of shifting the forcing function applied to a component by a time interval is to simply shift the response by the same time interval, then the component is said to be *time-invariant*. That is,

if forcing function $x(t) \to$ response $y(t)$,

then forcing function $x(t - t_o) \to$ response $y(t - t_o)$

in a time-invariant component, as shown in Fig. 1-15.

A circuit made up entirely of time-invariant components is called a *time-invariant circuit*.

Example 1-14 The voltage-current relationship of a component is given by

$$v(t) = L\left(\frac{di(t)}{dt}\right) \tag{1-16}$$

where L is a constant. Show that it is time-invariant.

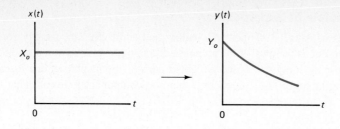

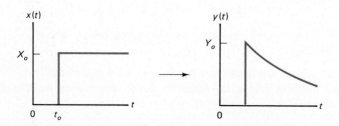

Fig. 1-15 Time invariance. A shift in forcing function simply shifts the response.

Solution Let a current $i_1(t)$ produce a response $v_1(t)$. Then

$$v_1(t) = L\left(\frac{di_1(t)}{dt}\right) \tag{1-17}$$

Now consider a current $i_2(t)$ obtained by shifting $i_1(t)$ by t_o. That is,

$$i_2(t) = i_1(t - t_o)$$

and let the resulting response be $v_2(t)$. Then, from Eq. (1-16),

$$\begin{aligned} v_2(t) &= L\left(\frac{di_2(t)}{dt}\right) \\ &= L\left(\frac{di_1(t - t_o)}{dt}\right) \end{aligned} \tag{1-18}$$

Using the substitution $t - t_o = u$ and $dt = du$, the right-hand side of Eq. (1-18) becomes

$$L\left(\frac{di_1(u)}{du}\right)$$

But, from Eq. (1-17)

$$\begin{aligned} L\left(\frac{di_1(u)}{du}\right) &= v_1(u) \\ &= v_1(t - t_o) \end{aligned} \tag{1-19}$$

Comparing Eqs. (1-18) and (1-19),

$$v_2(t) = v_1(t - t_o)$$

Therefore, the only effect of a shift in the forcing function by t_o is a shift in the response by t_o also. The component is time-invariant. ∎

Example 1-15 Suppose the voltage-current relationship of a component is given by

$$i(t) = t\left(\frac{dv(t)}{dt}\right) \quad (1\text{-}20)$$

Is the component time-invariant?

Solution Let voltage $v_1(t)$ produce a response $i_1(t)$. Then from Eq. (1-20),

$$i_1(t) = t\left(\frac{dv_1(t)}{dt}\right) \quad (1\text{-}21)$$

Now consider a voltage $v_2(t) = v_1(t - t_o)$, which produces a response $i_2(t)$. Then, from Eq. (1-20),

$$i_2(t) = t\left(\frac{dv_2(t)}{dt}\right)$$

$$= t\left(\frac{dv_1(t - t_o)}{dt}\right) \quad (1\text{-}22)$$

Substituting $u = (t - t_o)$ and $du = dt$, the right-hand side of Eq. (1-22) becomes

$$(u + t_o)\left(\frac{dv_1(u)}{du}\right) = u\left(\frac{dv_1(u)}{du}\right) + t_o\left(\frac{dv_1(u)}{du}\right)$$

The first term in the above expression gives $i_1(u) = i_1(t - t_o)$ from Eq. (1-21). Therefore,

$$i_2(t) = i_1(t - t_o) + t_o\left(\frac{dv_1(t - t_o)}{dt}\right)$$

Since $i_2(t)$ is not equal to $i_1(t - t_o)$, the component is not time-invariant. ∎

Exercise 1-14 If a component's voltage-current relationship is given by

$$v(t) = K\int i(t)\, dt$$

where K is a constant, is the component time-invariant?

It is important to note that the terms *time-variant* and *time-invariant* refer to properties of *components and circuits*. They should not be confused with *time-varying signals* and *constant signals*. The response of a *time-invariant component* to a *time-varying signal* will, in general, vary with time. When the applied signal is shifted by t_o, however, the *shape* of the response does not change, but the response is *shifted* by t_o.

Equations governing the behavior of lumped, linear, time-invariant circuits are, in general, ordinary linear differential equations with constant coefficients. The solution of such equations is straightforward. The methods of basic circuit analysis are, therefore, confined to lumped, linear, time-invariant circuits.

1-6 RESISTANCE

The current flowing in a conductor depends on the voltage across its terminals. If v is the voltage (in volts) and i is the current (in amperes), then the ratio $R = v/i$ is called the *resistance* of the conductor. The unit of resistance is the *ohm* (denoted by the symbol Ω). *One ohm equals 1 volt/ampere.* A conductor with resistance is called a *resistor*, and its symbol is shown in Fig. 1-16.

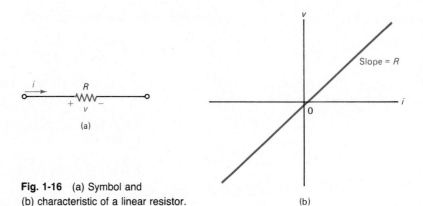

Fig. 1-16 (a) Symbol and (b) characteristic of a linear resistor.

1-6-1 Ohm's Law

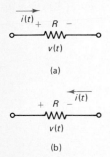

Fig. 1-17 Ohm's law: (a) $v = Ri$ when the current direction is assigned from plus to minus, and (b) $v = -Ri$ when the current direction is assigned from minus to plus.

The plot of current versus voltage in a *linear time-invariant resistor* is a straight line passing through the origin as shown in Fig. 1-16(b). The resistance R of a linear time-invariant resistor is a constant independent of the current through it or the voltage across its terminals. (Unless otherwise stated, linear time-invariant resistors are assumed in the following discussion.)

Current flows from a *higher* to a *lower* potential in a conductor, and the normal current direction is *from the plus* terminal *to the minus* terminal of voltage polarities in a resistor, as indicated in Fig. 1-17(a). With such a choice, the voltage and current in a resistor are related through *Ohm's law*:

$$v(t) = Ri(t) \quad \text{Ohm's law} \qquad (1\text{-}23)$$

In some analysis problems, however, the current in a resistor may be assigned a direction *from the minus terminal to the plus terminal* of voltage polarities, as indicated in Fig. 1-17(b). (Such a situation occurs fairly routinely in electronic circuit analysis.) Noting that a current $i(t)$ flowing in one direction is equivalent to a current $-i(t)$ in the opposite direction, Ohm's law for the situation shown in Fig. 1-17(b) becomes

$$v(t) = -Ri(t) \qquad (1\text{-}24)$$

The distinction between the two situations in Fig. 1-17 is very important and must be carefully observed in writing equations for resistors.

Example 1-16 Calculate the voltage or current (whichever is not given) for each resistor shown in Fig. 1-18.

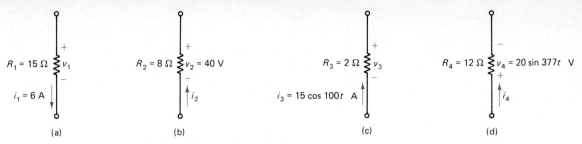

Fig. 1-18 Resistors for Example 1-16.

Solution First, decide which of the two equations, Eq. (1-23) or (1-24), applies in each case.
(a) Current from plus to minus. $v_1 = R_1 i_1 = 90$ V.
(b) Current from minus to plus. $v_2 = -R_2 i_2$. Therefore, $i_2 = -5$ A.
(c) $v_3 = -R_3 i_3 = -30 \cos 100 t$ V.
(d) $v_4 = R_4 i_4$, and $i_4 = 1.67 \sin 377t$ A.

Exercise 1-15 Calculate the voltage or current (whichever is not given) for each resistor shown in Fig. 1-19.

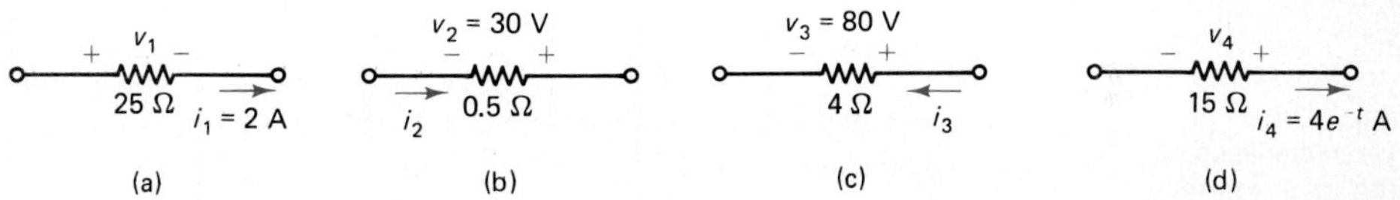

Fig. 1-19 Resistors for Exercise 1-15.

1-6-2 Power Dissipated in a Resistor

A resistor receives power from a source of electrical energy and dissipates it in the form of heat. It does not store energy, nor can it deliver power to another component. Since $p(t) = v(t)i(t)$ [Eq. (1-9)], the power dissipated in a resistor is given by

$$p(t) = R[i(t)]^2 = [v(t)]^2/R \qquad (1\text{-}25)$$

Example 1-17 Calculate the power dissipated in each of the resistors of Example 1-16.

Solution Using Eq. (1-25), (a) $p = 540$ W, (b) $p = 200$ W, (c) $p = 450 \cos^2 100t$ W, and (d) $p = 33.3 \sin^2 377t$ W.

Exercise 1-16 Calculate the power dissipated in each of the resistors of Exercise 1-15.

1-6-3 Nonlinear Resistors

A resistor whose current-voltage relationship does not satisfy Ohm's law is a *nonlinear resistor*. Typical graphs of the characteristics of nonlinear resistors are shown in Fig. 1-20.

The resistance of a nonlinear resistor varies as a function of current (or voltage). The device whose characteristic is shown in Fig. 1-20(b) has the additional property that its

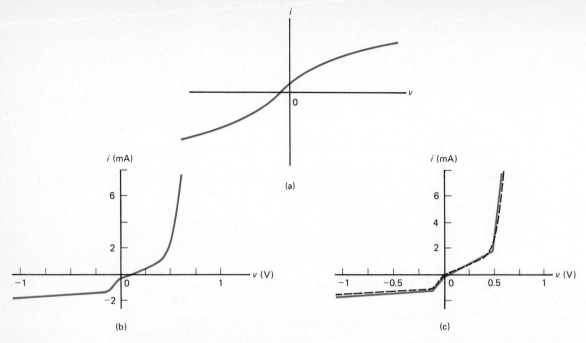

Fig. 1-20 Characteristics of nonlinear resistors.

resistance values are low (in the order of tens of ohms) when $v > 0$ (first quadrant of the graph) and very high (in the order of several thousand ohms) when $v < 0$ (third quadrant). Such a device is said to be *nonohmic* or *rectifying*.

In order to use linear circuit analysis methods on a nonlinear resistor, it is necessary to replace it by one or more linear models. A common approach is to approximate the current-voltage characteristic by means of straight-line segments, as indicated in Fig. 1-20(c). Note that the resistance of each straight-line segment is a constant and valid for a particular range of voltage values.

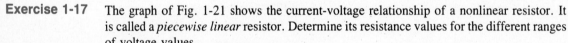

Exercise 1-17 The graph of Fig. 1-21 shows the current-voltage relationship of a nonlinear resistor. It is called a *piecewise linear* resistor. Determine its resistance values for the different ranges of voltage values.

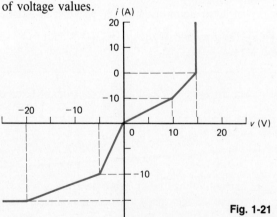

Fig. 1-21 Piecewise approximation of a nonlinear resistor characteristic for Exercise 1-17.

1-6-4 Effect of Temperature

In any resistor, the power dissipation due to the flow of current causes the temperature to increase (in proportion to i^2), which leads to a change in the resistance, except in the case of resistors made up of certain special alloys. In some materials, the resistance decreases as temperature increases, and in others, the resistance increases as temperature increases. Thus, the resistance of any resistor is affected by the flow of current. Most resistors are, therefore, nonlinear. When linear resistors are used in circuit analysis, they are approximate models of real resistors. The error introduced by such an approximation will depend on the material of which the resistor is made. For example, the resistance of molded-carbon composition resistors used in laboratory circuits varies by no more than a specified percentage (5 or 10 percent) from the nominal value if the power rating is not exceeded. Treating such resistors as linear usually leads to results of acceptable error margins.

Two other components commonly used in electric circuits are *capacitors* and *inductors*, which are discussed in Chapter 6.

1-7 CIRCUIT DIAGRAM CONVENTIONS

Components such as resistors and energy sources are shown by specific symbols in a circuit diagram, and connecting wires (assumed to have zero resistance) are shown by lines. A dark dot at the junction of two or more conductors denotes the presence of an electrical connection between the conductors at the junction, as shown in Fig. 1-22(a). The absence of an electrical connection at a crossover is usually indicated by the absence of a dark dot, as shown in Fig. 1-22(b), but other conventions are also used, as shown in Fig. 1-22(c).

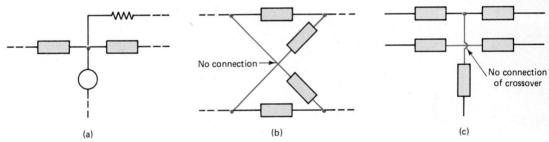

Fig. 1-22 (a) Dark dot at the intersection of two lines indicates an electrical connection. (b) Absence of a dark dot indicates no connection. (c) Absence of an electrical connection is also shown by a crossover.

1-8 KIRCHHOFF'S LAWS

Circuit analysis involves the determination of the values of the currents and voltages in a circuit set up by a forcing function or by some initial stored energy or a combination of the two. These values are governed by a number of constraints, some of which are due to the manner in which the components are interconnected in a circuit. Two fundamental laws of circuit analysis, *Kirchhoff's voltage law* (KVL) and *Kirchhoff's current law* (KCL), derived from the properties of electric fields dictate the constraints imposed by the interconnection of the circuit components. Kirchhoff's laws are valid for *any lumped circuit* whether linear or nonlinear, time-variant or time-invariant. The discussion in this and the following chapters, however, are confined to *lumped, linear, and time-invariant* circuits with only an occasional discussion of some simple nonlinear circuit.

1-8-1 Kirchhoff's Voltage Law (KVL)

An electric field has the *conservative property*, namely the work in moving an electric charge around a closed path in the field is zero. Kirchhoff's voltage law states the above property in terms of voltages in a circuit:

the algebraic sum of the voltages around a closed path in an electric circuit is zero.

The two key terms in the statement of KVL are *closed path* and *algebraic sum of voltages*.

A *closed path* in an electric circuit is any path that starts at some point in the circuit and returns to that point after traversing a number of components. A closed path is indicated by a *looping arrow*, as shown in Fig. 1-23(a). Note that a closed path can jump across several components or even across an open pair of terminals (Path 3 in Fig. 1-23(a)).

Now, suppose a unit positive charge is moved clockwise around the closed path shown in Fig. 1-23(b). When the charge is moved from the higher-potential terminal to the lower-potential terminal of a component (that is, *from the reference positive to the reference negative* terminal), it *loses* a potential energy equal to the voltage across the component. For example, a unit charge moving across component #1 in the direction of the path in Fig. 1-23(b) would lose v_1 joules. On the other hand, when a unit positive charge is moved from the lower-potential terminal to the higher-potential terminal of a component (that is, *from the reference negative to the reference positive terminal*), it *gains* a potential energy equal to the voltage across the component. For example, a unit charge moving across component #2 in the direction of the path in Fig. 1-23(b) would gain v_2 joules. The losses and gains in the potential energy of a unit charge moving across the components in a closed path must be subtracted and added, respectively, to determine the net change in its potential energy. The resulting *algebraic sum* of the voltages in the closed path must be zero, since the work done in moving a charge through a closed path in an electric field is zero (conservative property of the electric field).

The following *sign convention* is used in writing KVL equations.

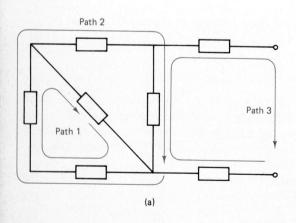
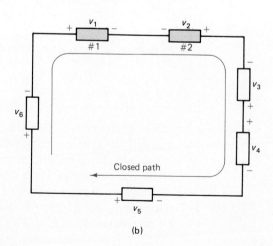

Fig. 1-23 Closed paths in a circuit.

- If the direction of path is *from plus* to minus across a component, that voltage is *added* in the KVL equation.
- If the direction of path is *from minus* to plus across a component, that voltage is *subtracted* in the KVL equation.

The algebraic sum of the voltages in a closed path, obtained by using the above convention, is then equated to zero.

For the closed path in Fig. 1-23(b), the direction of the path is from + to − in v_1, v_4, and v_6, and from − to + in v_2, v_3, and v_5. Therefore, the KVL equation becomes

$$v_1 - v_2 - v_3 + v_4 - v_5 + v_6 = 0$$

Example 1-18 Write the KVL equations for the three closed paths in the circuit of Fig. 1-24.

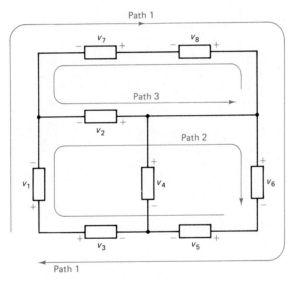

Fig. 1-24 Closed paths for the KVL equations for Example 1-18.

Solution *Path 1:*
$$v_1 - v_7 - v_8 + v_6 + v_5 - v_3 = 0$$

Path 2:
$$v_1 - v_2 + v_6 + v_5 - v_3 = 0$$

Path 3: (Note the *counterclockwise* direction of this path)
$$-v_2 + v_7 + v_8 = 0$$

Example 1-19 Calculate the voltage v_8 in the circuit of Fig. 1-25(a).

Solution Two closed paths are available for this calculation, as shown separately in Figs. 1-25(b) and 1-25(c). (Both should give the same answer for v_8.)

The KVL equation of the closed path in Fig. 1-25(b) is

$$-v_1 - v_2 - v_5 + v_6 + v_8 - v_7 + v_4 = 0$$

which gives (after substituting the numerical values)

$$v_8 = -100 \text{ V}$$

1-8 Kirchhoff's Laws

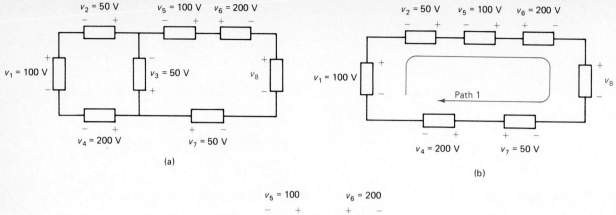

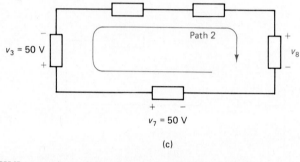

Fig. 1-25 Circuit for Example 1-19. Voltage v_8 can be calculated by using either the path in part (b) or in (c).

The KVL equation of the closed path in Fig. 1-25(c) is

$$v_3 - v_5 + v_6 + v_8 - v_7 = 0$$

which also gives $v_8 = -100$ V. ∎

Since the actual polarities of voltages across the components in a circuit are not always known beforehand, they are usually assigned arbitrarily. If the assigned polarities match the actual polarities in a component, then the answer will be a positive voltage. If the assigned polarities do not match the actual polarities, then the answer will be a negative voltage, as in the preceding example. It is not necessary, however, to go back and reassign polarities in such cases; the answer is left as a negative value.

Exercise 1-18 Write the KVL equation for the closed paths shown in Fig. 1-26(a).

Exercise 1-19 Obtain an expression for the voltage marked v_x in terms of $v_a, \ldots, v_e$ in Fig. 1-26(b).

Exercise 1-20 In each of the circuits shown in Figs. 1-26(c) and (d), evaluate the voltage marked v_x.

Open Circuit

A pair of terminals is said to be *open circuited* when there is an open gap between them, which represents an *infinite* resistance. *Current through an open-circuited pair of terminals is zero* (except when an *infinite* voltage is applied across the terminals). The *voltage across an open circuit is, in general, not zero.*

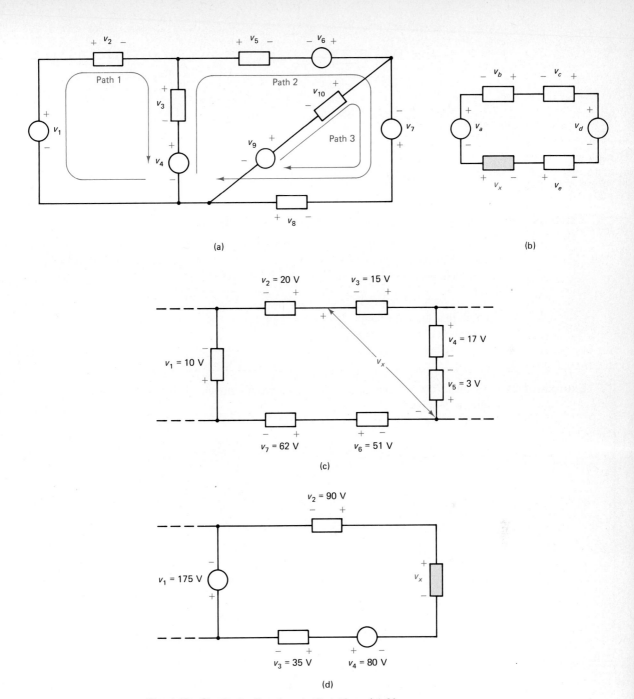

Fig. 1-26 Circuits for Exercises 1-18, 1-19, and 1-20.

1-8 Kirchhoff's Laws

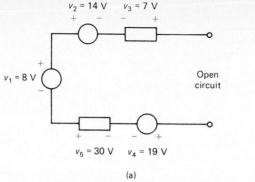
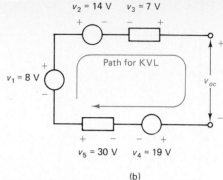

Fig. 1-27 Calculation of the open-circuit voltage for Example 1-20.

(a) (b)

Example 1-20 Calculate the voltage across the open circuit in Fig. 1-27(a).

Solution First, label the voltage across the open circuit as v_{oc} and assign reference polarities (arbitrarily) as shown in Fig. 1-27(b). Next write a KVL equation around the closed path that includes the open circuit:

$$-v_1 + v_2 - v_3 + v_{oc} + v_4 - v_5 = 0$$

(Note that the path jumps across the open circuit and v_{oc} is included to account for the voltage due to the jump.)

Therefore, $v_{oc} = 12$ V. ∎

Exercise 1-21 Calculate the voltage across each open circuit shown in Fig. 1-28. Use the polarities already assigned.

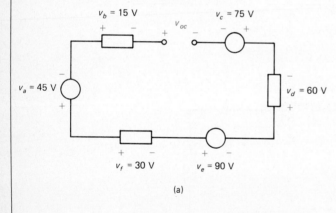
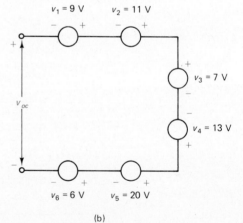

Fig. 1-28 Circuits for Exercise 1-21.

Note that the only restriction in using KVL is that the circuits be lumped circuits; no restriction is imposed about the linearity or time invariance of the components.

1-8-2 Kirchhoff's Current Law (KCL)

The law of conservation of electric charge states that throughout any process occurring within a system, the algebraic sum of the electric charges within the system does not

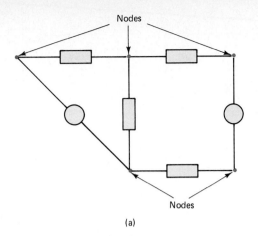

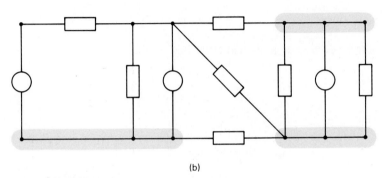
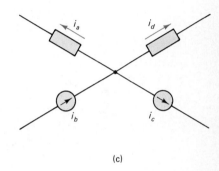

Fig. 1-29 Nodes in a circuit. A node is often shown spread out as in part (b) so as to avoid clutter.

change. Since currents involve flow of charges, Kirchhoff's current law restates the charge conservation principle in terms of currents in a circuit:

The algebraic sum of the currents at a node in an electric circuit is zero.

The key terms in the statement of KCL are *node* and *algebraic sum of the currents*.

The terminals of a branch in a circuit are called *nodes*, as shown in Fig. 1-29(a), and KCL applies to any node in a circuit. When many branches meet at a common node, a single node is often shown spread out in order to avoid clutter in the circuit diagram, as shown in Fig. 1-29(b).

Consider the portion of a circuit shown in Fig. 1-29(c), with currents assigned to the branches, where some currents are entering and others are leaving the node. A distinction must be made between them by using appropriate *algebraic signs* and this leads to the "algebraic sum" referred to in the statement of KCL.

The following *sign convention* will be used in writing KCL equations.

- A current *leaving* a node is written with a *positive* sign.
- A current *entering* a node is written with a *negative* sign.

The algebraic sum of the currents at a node, obtained by using the above convention, is then equated to zero.

The KCL equation of the node in Fig. 1-29(c) is

$$i_a - i_b + i_c + i_d = 0$$

1-8   Kirchhoff's Laws

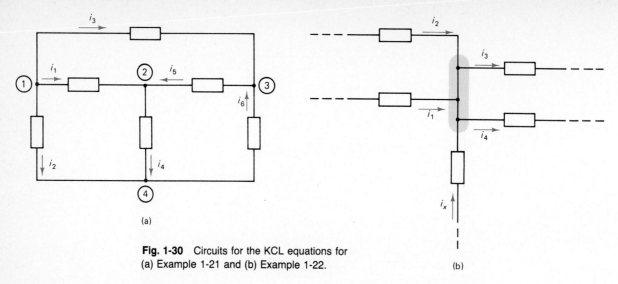

Fig. 1-30 Circuits for the KCL equations for (a) Example 1-21 and (b) Example 1-22.

Example 1-21 Write a KCL equation for each node in the circuit of Fig. 1-30(a).

Solution Node 1: $\quad i_2 + i_1 + i_3 = 0$

Node 2: $\quad -i_1 + i_4 - i_5 = 0$

Node 3: $\quad i_5 - i_6 - i_3 = 0$

Node 4: $\quad -i_2 - i_4 + i_6 = 0$

It is found that the equation for *node 4* is equal to the *negative sum* of the other three equations. That is, the equation for node 4 is a *linear combination* of the other three equations, and node 4 is, consequently, not an independent node. A formal discussion of the independence of nodes in a circuit is presented in Chapter 3. ∎

Example 1-22 Determine the current i_x in terms of the other currents in Fig. 1-30(b).

Solution The KCL equation for the given node is

$$-i_1 - i_2 + i_3 - i_x + i_4 = 0$$

Therefore,

$$i_x = -i_1 - i_2 + i_3 + i_4$$

∎

Exercise 1-22 Write the KCL equations of nodes 1, 2, and 3 in the circuit of Fig. 1-31(a). Verify that the KCL equation of node 4 is equal to a linear combination of the equations of the other three nodes.

Exercise 1-23 Obtain an expression for current i_x in the circuit of Fig. 1-31(b) in terms of the other currents.

Exercise 1-24 Evaluate current i_x in the circuit of Fig. 1-31(c).

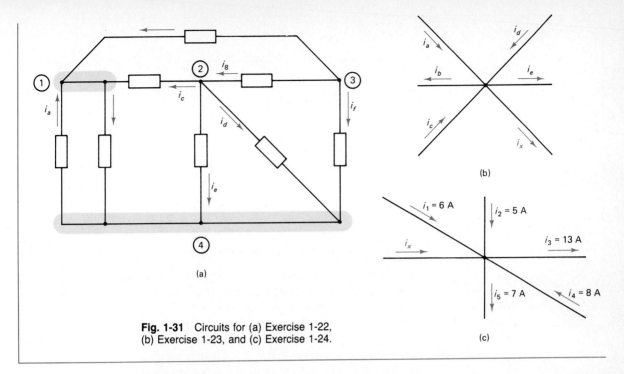

Fig. 1-31 Circuits for (a) Exercise 1-22, (b) Exercise 1-23, and (c) Exercise 1-24.

Short Circuit

A *short circuit* is said to exist across a pair of terminals connected by a wire of zero resistance. *The voltage across a short circuit is zero* (unless an infinite current is forced through it). Even though the voltage is zero across a short circuit, the current through it is, in general, not zero.

Example 1-23 Calculate the current through the short circuit between points A and B in the circuit of Fig. 1-32.

Solution The KCL equation at node A:

$$-i_1 - i_2 - i_3 + i_{sc} + i_4 = 0$$

which leads to

$$i_{sc} = -2.5 \text{ A}$$

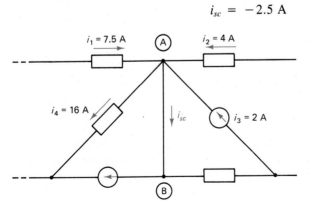

1-8 Kirchhoff's Laws

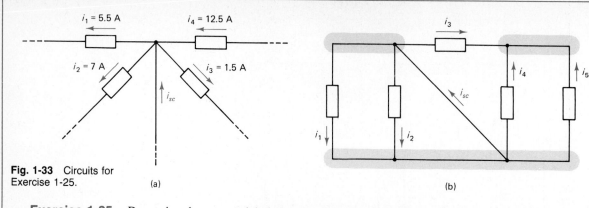

Fig. 1-33 Circuits for Exercise 1-25.

Exercise 1-25 Determine the current labeled i_{sc} in each circuit of Fig. 1-33.

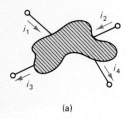

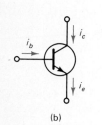

Fig. 1-34 KCL is also applicable to the currents at the terminals of a circuit or a device.

Since the actual directions of currents in the branches of a circuit are not always known beforehand, they are usually assigned arbitrarily. If the assigned direction matches the actual direction in a branch, then the answer will be a positive current. If the assigned direction does not match the actual direction, then the answer will be a negative current, as in the previous example. It is not necessary, however, to go back and reassign current directions in such cases; the answer is left as a negative value.

In the preceding examples and exercises, KCL has been applied to junctions of individual branches in a circuit. KCL can also be applied to more general situations, as indicated in Fig. 1-34.

In Fig. 1-34(a), the shape may represent a portion of a more complex circuit or a multiterminal device. If the four paths in the diagram are the only external connections to the shape, then the KCL equation

$$-i_1 - i_2 + i_3 + i_4 = 0$$

is valid. An example of such a situation occurs in bipolar junction transistors and other electronic devices. Figure 1-34(b) shows the symbol of a bipolar transistor with the relevant currents, and the three currents are related by the KCL equation:

$$i_e - i_b - i_c = 0$$

Note that the only restriction in applying KCL in a circuit is that it should be a lumped circuit; no restriction is imposed about the time invariance or linearity of the individual components.

The following sections will discuss the use of Kirchhoff's laws in the development of models of nonideal sources.

1-9 VOLTAGE AND CURRENT SOURCES

Ideal voltage and current sources were discussed in Sections 1-2 and 1-3, respectively. The voltage across the terminals of an ideal voltage source is independent of the load current drawn from it, and the current through the terminals of an ideal current source is independent of the voltage across a load connected to it. Sources available in practice do not, however, behave like ideal voltage or ideal current sources: that is, the voltage and current available at their terminals vary with the load. For example, the voltage of the battery in an automobile decreases noticeably when the engine is being started.

Concept of Equivalent Circuits

The analysis of a circuit is often facilitated by the use of *equivalent circuits* (or *models* made up of circuit components). That is, a device (such as a transistor), a component (such as a nonideal source), or even a subnetwork (i.e., a portion of a network) is replaced by a circuit that exhibits the *same terminal characteristics* as the original device, component, or subnetwork. If the equivalent circuits are properly developed for a given situation and carefully used, then the results obtained by utilizing them will fit the original circuit. Equivalent circuits and models are encountered at various points in this and the following chapters.

Care should be exercised in using an equivalent circuit, since it is usually developed or specified with respect to a set of terminals:

> **The equivalent circuit in any situation is valid only up to a set of terminals and ceases to be valid beyond those terminals.**

This point is further elaborated in Subsection 1-9-3 for the case of models of nonideal sources.

Models of Nonideal Sources

Consider a nonideal source represented by a black box with a variable load resistance, as shown in Fig. 1-35(a). The arrow through the resistance denotes a variable component. Voltage v_{ab} is the *terminal voltage* and i_a is the *terminal current* of the source. If, as shown in Fig. 1-35(b), the terminal voltage *varies linearly* with the terminal current as the load is varied, then there are two possible models for the source. The validity of

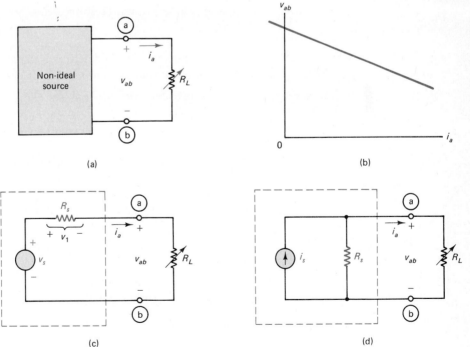

Fig. 1-35 Models of a nonideal source. (a) Nonideal source with a load. (b) *v-i* characteristic of the nonideal source. (c) Voltage source (Thevenin) model. (d) Current source (Norton) model.

these models is based on Thevenin's and Norton's theorems, which are discussed in Chapter 5.

1. *Voltage Source Model.* In the voltage source model, an ideal voltage source v_s is in *series* with resistor R_s, as shown in Fig. 1-35(c). In a series connection, the components are connected in such a way that the same current flows through them.
2. *Current Source Model.* In the current source model, an ideal current source i_s is in *parallel* with a resistor R_s, as shown in Fig. 1-35(d). In a parallel connection, the components are connected in such a way that the same voltage appears across them.

(The models in Figs. 1-35(c) and (d) are often referred to as the *Thevenin model* and the *Norton model,* respectively.)

The resistor R_s represents the *internal resistance* of the source. Internal resistance is just a concept used to model the given source; the source may not actually contain an internal component identifiable as a resistor. Note that if $R_s = 0$, the voltage source model becomes an ideal voltage source and $v_{ab} = v_s$ independent of the load current. If $R_s = \infty$, then the current source model becomes an ideal current source and $i_a = i_s$ independent of the load voltage. Thus, ideal sources are special cases of nonideal sources. In the following discussion, it will be assumed that a given source has a *finite, nonzero* internal resistance.

1-9-1 Determination of the Voltage Source Model

The KVL equation relating the different voltages in the model of Fig. 1-35(c) is

$$-v_s + v_1 + v_{ab} = 0 \quad (1\text{-}26)$$

But, $v_1 = R_s i_a$ (Ohm's law). Therefore, Eq. (1-26) becomes

$$v_{ab} = v_s - R_s i_a \quad (1\text{-}27)$$

where v_s and R_s are constants. Equation (1-27) describes a straight line on a graph with v_{ab} as the ordinate and i_a as the abscissa, as shown in Fig. 1-36(a).

> The intersections of the straight line with the two axes are v_s on the vertical axis, obtained by putting $i_a = 0$ in Eq. (1-27), and v_s/R_s on the horizontal axis, obtained by putting $v_{ab} = 0$ in Eq. (1-27).

Fig. 1-36 Measurements for the determination of the voltage source model.

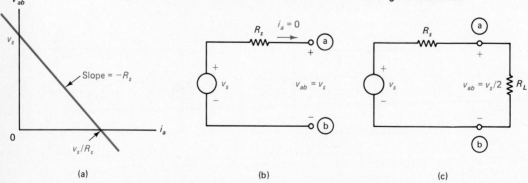

Now, consider *open circuiting the terminals a–b* of the source, as shown in Fig. 1-36(b). Then,

$$i_a = 0 \quad \text{due to the open circuit}$$

which makes the voltage across $R_s = 0$. Therefore,

$$v_{ab} = v_s \quad \text{under the open circuit condition}$$

Thus,

v_s in the voltage source model = open-circuit voltage measured across the load terminals (1-28a)

The basic principle in finding the resistance R_s is that if the voltages across R_s and R_L are made equal (by a proper choice of R_L), then R_s and R_L should be equal also (by Ohm's law). When the above condition occurs, the voltage v_s is split equally between R_s and R_L. Therefore, connect a load resistance R_L to terminals a–b and vary its value until the voltage v_{ab} becomes exactly *one-half of the open-circuit voltage* as shown in Fig. 1-36(c). When this condition occurs, Eq. (1-27) shows that

$$R_L = R_s$$

Therefore,

R_s = the value of R_L needed to make the load voltage across a–b one-half of the open-circuit voltage (1-28b)

The voltage source model of a nonideal source is thus given by Eqs. (1-28a) and (1-28b).

Example 1-24 The terminal voltage of a battery (with an open-circuit voltage of 12 V) reduces to 6 V when a resistance of 5 Ω is connected to its terminals. (a) Obtain the voltage source model of the battery. (b) Calculate the current flowing through a short circuit placed between the battery terminals. (c) Sketch the graph of the terminal voltage as a function of the terminal current.

Solution (a) The model consists of $v_s = 12$ V in series with $R_s = 5$ Ω, as shown in Fig. 1-37(a).

Fig. 1-37 Model and load line for Example 1-24.

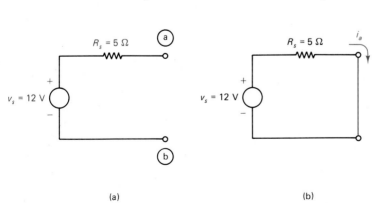
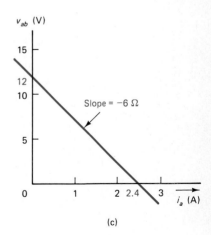

1-9 Voltage and Current Sources

(b) When a–b are short circuited, as shown in Fig. 1-37(b), $i_a = v_s/R_s = 2.4$ A.
(c) The graph is a straight line drawn by joining the point at $v_{ab} = v_s = 12$ V on the vertical axis to the point at $i_a = v_s/R_s = 2.4$ A, as shown in Fig. 1-37(c). ∎

Exercise 1-26 A load resistance of 120 Ω reduces the terminal voltage to 12 V of a source whose open-circuit voltage is 24 V. (a) Draw the model of the source. (b) Sketch the graph of the terminal voltage against the terminal current. (c) Calculate the current when the load terminals are short circuited.

1-9-2 Determination of the Current Source Model

The KCL equation of the model in Fig. 1-38(a) is given by

$$-i_s + i_1 + i_a = 0 \quad (1\text{-}29\text{a})$$

But $i_1 = v_{ab}/R_s$ by Ohm's law, and Eq. (1-29a) becomes

$$i_a = i_s - \frac{v_{ab}}{R_s} \quad (1\text{-}29\text{b})$$

which describes a straight line on a graph with i_a as the ordinate and v_{ab} as the abscissa, as shown in Fig. 1-38(b).

The points of intersection of the straight line with the two axes are: i_s on the vertical axis [obtained by putting $v_{ab} = 0$ in Eq. (1-29b)] and $R_s i_s$ on the horizontal axis [obtained by putting $i_a = 0$ in Eq. (1-29b)].

Fig. 1-38 Measurements for the determination of the current source model.

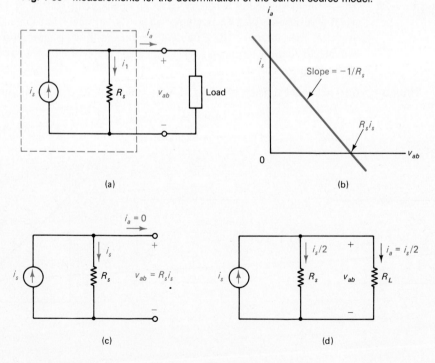

If terminals a–b are open circuited, as shown in Fig. 1-38(c), then $i_a = 0$ and Eq. (1-29b) leads to

$$v_{ab} = R_s i_s \quad \text{open-circuit voltage} \quad (1\text{-}30)$$

The current source model for a given source is obtained by using the same two measurements used for the voltage source model. The open-circuit voltage of the given source is measured. Then the load resistance is adjusted until the voltage across it is exactly one-half of the open-circuit voltage. Under this condition, Eq. (1-30) leads to

$$v_{ab} = R_s i_s / 2$$

which means that the current through R_s is $i_s/2$ and the current through R_L is also $i_s/2$, as is shown in Fig. 1-38(d). Therefore, $R_L = R_s$ when the load voltage is one-half the open-circuit voltage.

Then, Eq. (1-30) gives

$$i_s \text{ in the current source model} = (\text{open-circuit voltage of the source})/R_s$$

Example 1-25 The open-circuit voltage of a source is 18 V. The terminal voltage reduces to 9 V when the load resistance is 5 kΩ. (a) Obtain the current source model. (b) Sketch the graph of the terminal current as a function of the terminal voltage. (c) Calculate the current that would flow in a short circuit across the terminals of the given source.

Solution (a) The internal resistance of the source is 5 kΩ. Since the open circuit voltage is 18 V, $i_s = 18/(5 \times 10^{-3}) = 3.6$ mA. The model is shown in Fig. 1-39 (a).
(b) The graph is a straight line joining the point $i_s = 3.6$ mA on the vertical axis with the point $R_s i_s = 18$ V on the horizontal axis, as shown in Fig. 1-39 (b).
(c) A short circuit from a to b, as shown in Fig. 1-39 (c), makes the voltage $v_{ab} = 0$. Since the voltage across R_s is zero, no current flows in R_s. Therefore, the current in the short circuit $= i_s = 3.6$ mA. ∎

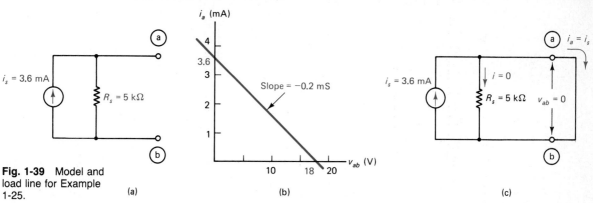

Fig. 1-39 Model and load line for Example 1-25.

Exercise 1-27 The open-circuit voltage of a source is 24 V and a load resistance of 470 Ω reduces the load voltage to 12 V. (a) Obtain the current source model. (b) Sketch the graph of the terminal current as a function of the terminal voltage.

An examination of the two models of a nonideal source shows that the *current flowing in a short circuit* across the source terminals equals i_s in the current source model and

v_s/R_s in the voltage source model. Therefore, the following two relationships are also useful in setting up the models.

$$i_s \text{ in the current source model} = \text{short-circuit current of the given source}$$

and

$$R_s \text{ in either model} = \text{(open-circuit voltage)/(short-circuit current)}$$

Making a short-circuit measurement on a given source is usually not feasible, since it may permanently damage the source. The preferred experimental procedure is to adjust R_L to give one-half the open-circuit voltage.

Exercise 1-28 The open-circuit voltage and short-circuit current of a source are given as 12 V and 0.3 A, respectively. Obtain the two models of the source.

1-9-3 Conversion Between the Models of a Source

Depending on a given problem, one model is found to be more convenient than the other. Conversion of one model to the other is therefore necessary in some problems.

Consider the two models of a nonideal source, as shown in Figs. 1-40(a) and (b). The value of the internal resistance is the same in both cases, but it is connected in *series with the voltage source* and *in parallel with the current source*.

Now consider an arbitrary load resistance R_L connected to the two models, as shown in Figs. 1-40(c) and (d).

For the voltage source model, the KVL equation is

$$-v_s + v_1 + v_{ab} = 0$$

which leads to, after using Ohm's law,

$$(R_s + R_L)i_a = v_s$$

and

$$v_{ab} = R_L i_a = [R_L/(R_L + R_s)]v_s \qquad (1\text{-}31)$$

For the current source model, the KCL equation is

$$-i_s + i_1 + i_a = 0$$

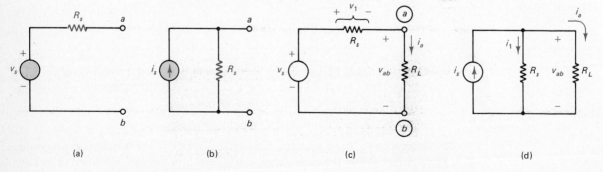

Fig. 1-40 Conversion between the models of a source. Models in parts (a) and (b) are equivalent if $v_s = R_s i_s$.

which leads to, after using $i_1 = v_{ab}/R_s$ and $i_a = v_{ab}/R_L$,

$$v_{ab} = \left(\frac{R_L R_s}{(R_s + R_L)}\right) i_s \tag{1-32}$$

Equating Eqs. (1-31) and (1-32), it is seen that

$$v_s = R_s i_s \quad \text{and} \quad i_s = v_s/R_s \tag{1-33}$$

Equation (1-33) permits the determination of i_s if v_s is known and *vice versa*.

Example 1-26 Convert each of the voltage source models in Figs. 1-41(a) and (b) to an equivalent current source model.

Solution (a) The internal resistance is $R_s = 15\ \Omega$. The source current is $v_s/R_s = 5$ A. The equivalent current source model of Fig. 1-41(a) is shown in Fig. 1-41(c).
(b) The internal resistance is $R_s = 10\ \Omega$. The source current is $v_s/R_s = 12$ A. The equivalent current source model of Fig. 1-41(b) is shown in Fig. 1-41(d). ∎

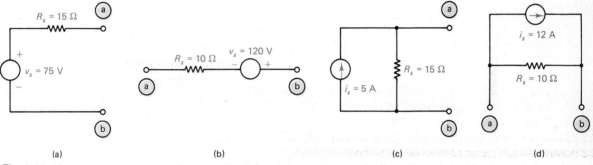

Fig. 1-41 Source conversions for Example 1-26.

The direction of the current in the current source must be carefully determined and clearly indicated in the diagram of the converted model. The direction of the current in the equivalent current source is determined by *matching* the directions of the current in a short circuit placed across the terminals a–b in the two models.

Example 1-27 Convert each of the current source models in Figs. 1-42(a) and (b) to an equivalent voltage model.

Solution (a) The internal resistance is $R_s = 0.1\ \Omega$. The source voltage is $R_s i_s = 2.5$ V. The voltage source model of Fig. 1-42(a) is shown in Fig. 1-42(c).
(b) The internal resistance is $R_s = 11\ \Omega$. The source voltage is $R_s i_s = 110$ V. The voltage source model of Fig. 1-42(b) is shown in Fig. 1-42(d). ∎

Fig. 1-42 Source conversions for Example 1-27.

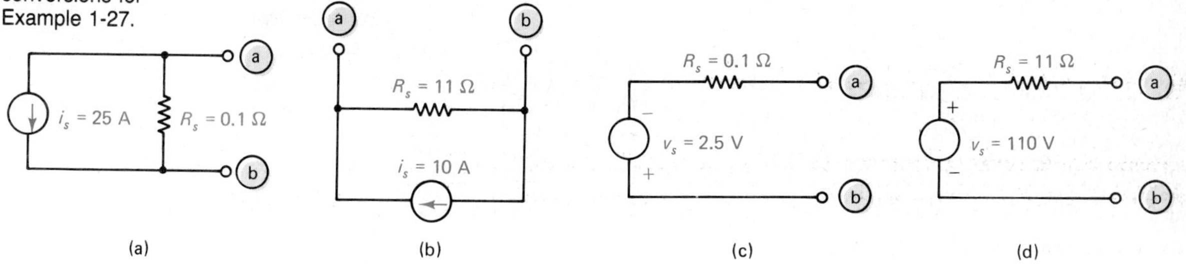

The polarity marks on the voltage source must be carefully determined and clearly indicated in the diagram of the voltage source model. The polarities are determined by *matching* the voltage polarities on an open circuit across the terminals a–b in the two models.

Exercise 1-29 Convert each source model in Fig. 1-43 to the other form.

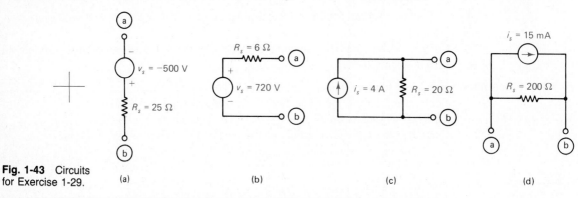

Fig. 1-43 Circuits for Exercise 1-29.

Important Points to Note in the Conversion Procedure

1. The conversion procedure depends on the presence of a *finite nonzero* internal resistance R_s. If $R_s = 0$ in the voltage source model (the case for an *ideal* voltage source), or $R_s = \infty$ in the current source model (the case for an *ideal* current source), the conversion procedure will lead to *indeterminate* results. Therefore, if the given source is ideal, then it *cannot* be converted to the other form.

2. Equivalence between the two source models is valid only *outside* the terminals a–b of the source. That is, the two models yield identical results for any and all calculations *outside* the dashed lines in Fig. 1-44. For any calculations inside the dashed lines, the two models will, in general, yield different results. For example, suppose both models are open-circuited so as to make $i_a = 0$. Then the power delivered by v_s [as shown in Fig. 1-44(a)] is zero since it supplies no current, whereas the power supplied by i_s [as shown in Fig. 1-44(b)] is $R_s i_s$, since the current i_s would flow through R_s, and the two results are seen to be quite different.

Care should, therefore, be taken to use equivalent circuits only within valid circuit boundaries.

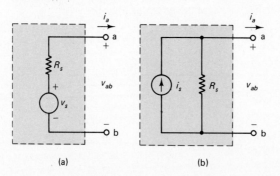

Fig. 1-44 The two models yield the same results for any connections made to the terminals a–b. The equivalence is not valid inside the dashed lines.

Basic Concepts, Components, and Laws of Electric Circuits

1-9-4 Dependent Sources

If the voltage of an ideal voltage source in a circuit is independent of the voltage or current in another branch, then it is called an *independent* voltage source. A similar definition is valid for an independent current source. Ideal source components in the models of batteries and generators are examples of independent sources.

A *dependent* (or *controlled*) source, on the other hand, has the property that its voltage or current is dependent on the voltage or current in another branch of the circuit.

Dependent sources arise in the models of active devices such as transistors and electronic circuits. For example, a bipolar junction transistor has three terminals (emitter, base, and collector) with a current associated with each terminal, and the collector current can be expressed as a specific fraction of the emitter current. This dependence of the collector current on the emitter current leads to a dependent current source in the model of a bipolar junction transistor.

A dependent source is represented by a *diamond* shaped symbol, as shown in Fig. 1-45. As shown in the diagrams, a dependent source is a voltage source or a current source, and the quantity that controls the voltage or current of a dependent source is a voltage or current in another branch. When the voltage or current in a dependent source is related to the controlling voltage or current by a constant that does not vary with time, which is the case in all four circuits of Fig. 1-45, the source is a *linear, time-invariant dependent source*, which is the only kind that will be discussed in this and the following chapters.

Fig. 1-45 Linear dependent sources. (a) Dependent current source controlled by the voltage in a branch. (b) Dependent current source controlled by the current in a branch. (c) Dependent voltage source controlled by the voltage in a branch. (d) Dependent voltage source controlled by the current in a branch.

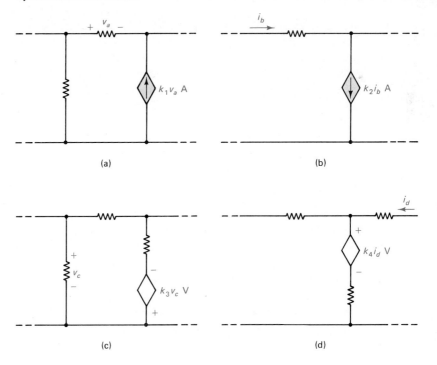

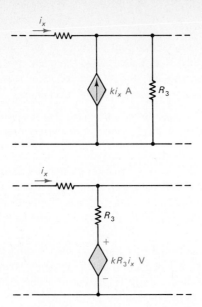

Fig. 1-46 Conversion between dependent current and voltage source models.

Except for the fact that their voltage or current is controlled by what is happening somewhere else in a circuit, dependent sources are analyzed in the same manner as independent sources. For example, the conversion procedure of source models applies to dependent sources also. An example of such a conversion is shown in Fig. 1-46.

1-10 AN OVERVIEW OF CIRCUIT ANALYSIS

The primary aim of circuit analysis is to determine the response of a circuit to a given forcing function. The forcing function is a voltage or current applied to a circuit, and the response is a voltage or current at some part of the circuit. But circuit analysis serves other important purposes as well. It forms the basis of *network theory,* which is a study of the behavior of general networks. Some basic concepts of network theory are presented in later chapters. The ultimate aim of circuit analysis can be considered to be *network synthesis, that is, the design of a network* to meet a set of specifications about the input and output. Methods of synthesis will not, however, be considered in this book.

Even though it would be desirable to develop methods of circuit analysis applicable to all types of circuits, the mathematical complexity of analyzing distributed, nonlinear, or time-variant networks makes it necessary to restrict our attention here to lumped, linear, time-invariant circuits.

Since it is usually convenient to proceed from simple cases to difficult cases in a gradual manner, we start with the simplest type of circuit problems: circuits made up of resistors and driven by sources of constant voltage or current. Algebra is the only mathematical tool needed for the analysis of such circuits, and we can concentrate on the basic principles and procedures of circuit analysis without complicated mathematics.

The addition of energy storage elements such as inductors and capacitors leads to ordinary linear differential equations with constant coefficients describing the circuits. Again, we start with the simple case of such circuits driven by sources of constant voltage or current before considering time-varying forcing functions.

Among the wide variety of time-varying forcing functions, the sinusoidal function (that is, a voltage or current described by a sine or cosine function of time) occupies a most important position. Because of its importance, a great deal of attention will be devoted to the analysis of circuits driven by sinusoidal sources. The response of circuits to sinusoidal forcing functions are governed by ordinary linear differential equations with constant coefficients. Since the solution of such equations tends to be tedious, an alternative method is developed in which differential equations are transformed to algebraic equations with complex coefficients.

From sinuosoidal forcing functions progress is made to more general forcing functions, where a periodic function can be expressed as the sum of sinusoidal components and this sum is known as a Fourier series. Fourier analysis in conjunction with the principle of superposition permits the analysis of circuits driven by periodic nonsinusoidal functions.

When forcing functions are not periodic, several methods of analysis are available. One is convolution, which is a statement of the principle of superposition in the form of an integral. Convolution permits the analysis of a circuit in the time domain and is applicable even when the forcing function cannot be expressed in a convenient mathematical form. In the case of forcing functions that are of finite duration, circuits can also be analyzed by Fourier transforms that are modifications of the Fourier methods for periodic functions. It is found, however, that Fourier transforms have certain limitations and are not applicable to all forcing functions of interest. A more versatile method of attack is provided by Laplace transforms, which permit the analysis of circuits under a wide variety of forcing functions in both the time domain and the frequency domain.

The principle of superposition will be seen to form the foundation on which the structure of linear circuit analysis is built. In fact, the difficulty in analyzing nonlinear circuits directly is their failure to obey the principle of superposition.

1-11 SUMMARY OF CHAPTER

Potential difference (or voltage) between two points in an electric field is the work required to move a positive unit charge between the two points. The unit of potential difference is the *volt:* $1 \text{ V} = 1$ joule/coulomb. Voltage reference polarity marks ($+$ and $-$) are necessary to indicate the reference positive and reference negative terminals of a component.

Electric current is defined as the rate of change of electric charge across a section of a material. The direction of current is taken as that in which free positive charges would flow under the influence of an electric field.

Power is the rate of doing work. If v is the voltage across and i the current through a component, then $p = vi$ represents the instantaneous power. When p varies as a function of time, the root-mean-square (rms) value is usually specified. A component can deliver or receive power.

The product vi represents te power received by a component if the current is flowing from the reference positive to the reference negative terminals of a component. (Such an assignment of voltage polarities and current direction is referred to as associated reference polarities.)

The product vi represents the power delivered by a component if the current is flowing from the reference negative to the reference positive terminals of a component.

An ideal voltage source has the property that its terminal voltage is independent of the current supplied by it to a load. An ideal current has the property that its terminal current is independent of the voltage across a load connected to it. Energy sources in practice are nonideal and their models are considered in Chapter 2.

Components and circuits can be classified in different ways. The treatment in this book will be limited to lumped, linear, time-invariant circuits. Circuits with physical dimensions much smaller than a quarter wavelength of the signals are considered lumped. Linearity is defined by the principle of superposition: the response of the circuit to a weighted sum of forcing functions is given by the weighted sum of the responses to the individual forcing functions. Time invariance implies that a shift of the forcing function in time causes a shift in the response by the same interval without affecting the shape or form of the response. Kirchhoff's laws are used to analyze lumped, linear, time-invariant circuits. Other types of circuits usually required more sophisticated techniques that are outside the scope of this text.

The voltage across and a current through a conductor are related by Ohm's law: $v = Ri$. A linear resistor has the property that its resistance R is a constant independent of the current through it. Almost all practical resistors are nonlinear. Linear approximations are used, however, in order to apply linear circuit analysis techniques.

The two basic laws on which circuit analysis is founded are Kirchhoff's voltage law (KVL) and Kirchhoff's current law (KCL). KVL states that the algebraic sum of the voltages around a closed path in an electric circuit is zero. KCL states that the algebraic sum of the currents at a node in an electric circuit is zero.

The sign convention used for KVL is add a voltage if the direction of the path is from the positive reference polarity to the reference negative polarity, and subtract a voltage if the direction of the path is from the negative reference polarity to the positive reference polarity. The algebraic sum of all the voltages in any closed path obtained by using the above convention is equated to zero.

The sign convention used for KCL is add a current if the current is leaving the node, and subtract a current if it is entering the node. The algebraic sum of all the currents at any node obtained by using the above convention is equated to zero.

A nonideal source, in which the terminal voltage varies linearly with the terminal current (as the load is varied) may be represented by one of two models: the Thevenin model, consisting of an ideal voltage source in series with a resistance, or a Norton model, consisting of an ideal current source in parallel with a resistance. The two models are determined by making two measurements: an open-circuit measurement and finding the load resistance that reduces the load voltage to one-half the open-circuit voltage. Either model is converted to the other by using some simple conversion formulas.

A dependent source is one whose voltage or current is controlled by the voltage or current in another branch of the circuit. Dependent sources occur in models of active devices such as transistors.

Answers to Exercises

[*Wherever appropriate, some intermediate steps are given below to help you check your work.*]

1-1 $i = 2.5$ A $(0 < t < 2)$; $i = 0$ A $(2 < t < 4)$; $i = -2.5$ A $(4 < t < 6)$; $i = 0$ A $(t > 6)$.

1-2 $i = -(pQ_o)e^{-pt}$.

1-3 $p = 120 + 60 \cos 20t$ W.

1-4 $i = -500 \sin 50t$ A; $p = -5000 \sin 50t \cos 50t$ W.

1-5 $w = 270$ J.

1-6 3.6×10^6 J.

1-7 (a) 3840 J and (b) 5520 J.

1-8 (a) -24 V; (b) -14.4 V; (c) -2.9 V; (d) 12 V; (e) 6 V.

1-9 (a) Delivering 682 W; (b) receiving 625 mW; (c) receiving 75 W; (d) receiving 135W.

1-10 (a) $v_a i_a = -20$ W, $i_a = -2.5$ A; (b) $v_b i_b = +20$ W, $i_b = -3.33$ A; (c) $v_c i_c = +20$ W, $v_c = 2.86$ V; (d) $v_d i_d = -20$ W, $v_d = -6.67$ V.

1-11 $\lambda/4 = 1.25 \times 10^6$ m. Maximum length $= 1.25 \times 10^5$ m.

1-12 The element is linear since both scaling and additive properties are satisfied.

1-13 The element is nonlinear since it fails the scaling test.

1-14 The element is time-invariant since $i(t - t_o)$ results in $v(t - t_o)$.

1-15 (a) 50 V; (b) -60 A; (c) 20 A; (d) $-60e^{-t}$ A.

1-16 (a) 100 W; (b) 1800 W; (c) 1600 W; (d) $240\,e^{-2t}$ W.

1-17 Resistance $= 1/\text{slope}$. $R = \infty$ ($v < -20$ V); $R = 3\,\Omega$ ($-20 < v < -5$ V); $R = 0.5\,\Omega$ ($-5 < v < 0$ V); $R = 2\,\Omega$ ($0 < v < 10$ V); $R = 1\,\Omega$ ($10 < v < 15$ V); $R = 0$ ($v > 15$ V).

1-18 Path 1: $-v_1 + v_2 + v_3 + v_4 = 0$. Path 2: $-v_4 - v_3 + v_5 - v_6 - v_7 - v_8 = 0$. Path 3: $-v_9 - v_{10} - v_7 - v_8 = 0$.

1-19 $v_x = -v_a - v_b - v_c + v_d - v_e$.

1-20 (a) $v_x = -v_1 + v_2 + v_6 - v_7 = -1$ V or $v_x = -v_3 + v_4 - v_5 = -1$ V; (b) $v_x = -v_1 + v_2 + v_4 - v_3 = -40$ V.

1-21 (a) $v_{oc} = -v_a - v_b + v_c + v_d + v_e + v_f = 195$ V; (b) $v_{oc} = -v_1 - v_2 + v_3 - v_4 + v_5 + v_6 = 0$.

1-22 Node 1: $-i_a + i_b - i_c - i_h = 0$. Node 2: $i_c + i_e + i_d - i_g = 0$. Node 3: $i_h + i_f + i_g = 0$. (Node 4: $i_a - i_b - i_e - i_d - i_f = 0$, which is also the negative sum of the equations of nodes 1, 2, 3.)

1-23 $i_x = i_a - i_b + i_c - i_e + i_d$.

1-24 $i_x = -i_1 - i_2 + i_3 - i_4 + i_5 = 1$ A.

1-25 (a) 1.5 A and (b) $i_1 + i_2 + i_3$ or $i_1 + i_2 - i_4 - i_5$.

1-26 (a) Model: 24 V in series with 120 Ω. (b) Straight line cuts the horizontal axis at 0.2 A and the vertical axis at 24 V. (c) $i = v_s/R_s = 0.2$ A.

1-27 (a) Model: 51.1-mA source in parallel with 470 Ω. (b) Straight line cuts the horizontal axis at 24 V and the vertical axis at 51.1 mA.

1-28 Voltage source model: 12-V source in series with 40 Ω. Current source model: 0.3-A source in parallel with 40 Ω.

1-29 (a) 20-A source in parallel with 25 Ω. Current in the source directed from terminal a to terminal b. (b) 120-A source in parallel with 6 Ω. Current in the source directed from terminal b to terminal a. (c) 80-V source in series with 20 Ω. Positive polarity at terminal a. (d) 3-V source in series with 200 Ω. Positive polarity at terminal b.

PROBLEMS

Sec. 1-1 Basic Units and Definitions

1-1 The work needed to move a positive charge of 0.25 C from point A to point B in an electric field is 20 J. (a) Calculate the potential difference between the two points. (b) Calculate the work done in moving an electron between the two points. (Magnitude of electronic charge $= 1.6 \times 10^{-19}$ C.)

1-2 The potential difference between two points A and B is 250 V with A at the higher potential. (a) Calculate the change in potential energy of an electron as it goes from B to A. (b) Does its potential energy increase or decrease?

1-3 The flow of charge in a material is given by $q(t) = te^{-t}$ C for $t > 0$. (a) Obtain an expression for the current $i(t)$ for $t > 0$. (b) Draw a neat sketch of $q(t)$ and $i(t)$ on the same graph.

1-4 The variation of the electric charge in a section of a material is as shown in Fig. 1-47. Determine and carefully sketch the current $i(t)$.

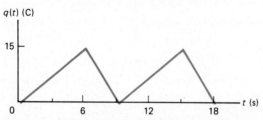

Fig. 1-47 Variation of electric charge for Problem 1-4.

1-5 A piece of material has a uniform distribution of free charges moving at a constant (average) velocity. Let ρ denote the charge density (i.e., charge per unit volume in C/m³) and u the velocity. (a) Consider an elementary section of length Δx in the direction of motion of the charges and assume that the charge in that section takes Δt to move out of the section. Show that by letting Δx and $\Delta t \to 0$, the current in the material is given by $i = A\rho u$, where A is the area of cross section of the material. The above expression is of particular importance in the study of semiconductor devices. (b) If the number of free electrons in a circular copper wire of diameter 2×10^{-3} m is $8.5 \times 10^{31}/\text{m}^3$, calculate the velocity of the electrons for $i = 1$ A. How does this velocity compare with that of light at which electromagnetic energy travels?

1-6 The voltage across a component is a constant at 12 V. The current in it is $i(t) = 15 - 15 \sin 2\pi t$ A. (a) Determine the instantaneous power $p(t)$ and sketch it in the interval (0, 1 s). (b) Calculate the energy received by the component in the interval (0.5, 1 s).

1-7 The voltage across a component is $v(t) = K_1 t$ V and the current in it is $i(t) = K_2 e^{-t}$ A, where K_1 and K_2 are constants. Obtain the expression for the energy $w(t)$ assuming $w(0) = 0$.

1-8 The pulse train produced by a transmitter has a maximum power level P_o, as shown in Fig. 1-48. The duration of each pulse is 5 µs and the period $T = 0.1$ ms. (1 µs = 10^{-6} s and 1 ms = 10^{-3} s.) The energy contained in each pulse is 2 J. (a) Calculate the value of P_o. (b) Calculate the average power of the pulse train, where the average power is defined as the area of each pulse divided by the period T.

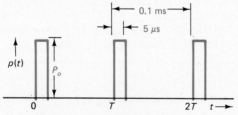

Fig. 1-48 Pulse train for Problem 1-8.

1-9 The voltage $v(t)$ and current $i(t)$ in a component are as shown in Fig. 1-49. (a) Sketch the power $p(t)$. (b) Determine the total energy received by the component.

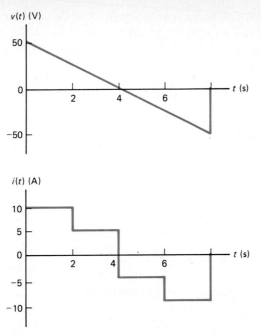

Fig. 1-49 Voltage and current in a component for Problem 1-9.

Sec. 1-2 Voltage

1-10 The voltages v_A, v_B, and v_C in Fig. 1-50 are measured with reference to a common ground. (a) Write expressions for v_1 and v_2. (b) If $v_A = 15$ V, $v_C = 15$ V, and $v_2 = -34$ V, calculate v_1 and v_B.

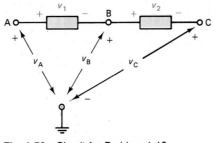

Fig. 1-50 Circuit for Problem 1-10.

1-11 Calculate the voltage across each of the components in Fig. 1-51. Use the reference polarities already assigned. The voltages at different terminals are shown.

Fig. 1-51 Components for Problem 1-11.

Problems

Fig. 1-52 Circuit for Problem 1-12.

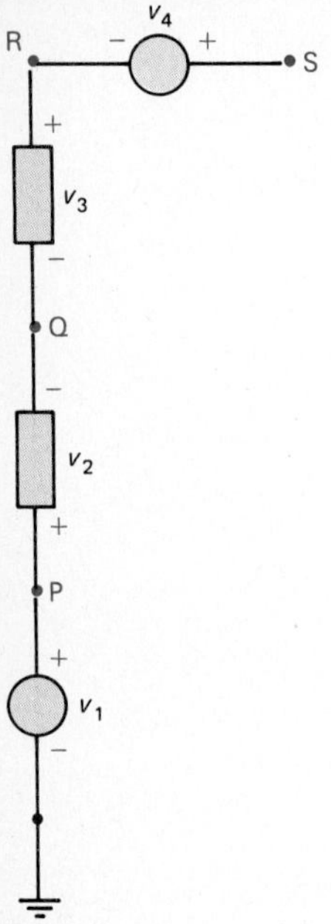

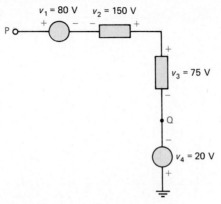

1-12 Obtain expressions for the voltages at points P, Q, R, and S (with reference to ground) in terms of voltages v_1, v_2, v_3, and v_4 in the circuit shown in Fig. 1-52.

1-13 Calculate the work done in moving an electron from point P to point Q in the circuit for Fig. 1-53.

Fig. 1-53 Circuit for Problem 1-13.

Sec. 1-4 Power Delivered and Power Received

1-14 Calculate the power received or delivered by each component in Fig. 1-54. State in each case whether power is received or delivered.

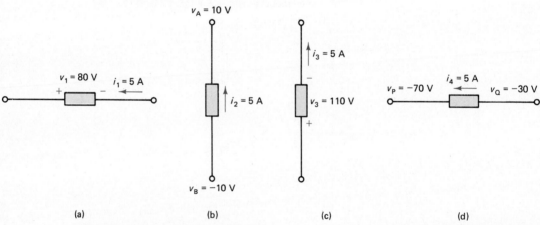

Fig. 1-54 Components for Problem 1-14.

1-15 If a current of 16 A flows from terminal P to ground through the various components in the circuit for Problem 1-13 (Fig. 1-53), calculate the power in each component and state if it is received or delivered.

1-16 The voltages and currents in the circuit of Fig. 1-55 are $v_1 = 100$ V, $v_2 = -50$ V, $v_3 = 75$ V, $v_4 = 225$ V; $i_1 = -7$ A, $i_2 = 13$ A, $i_3 = -6$ A. (a) Calculate the power delivered or received by each component and state in each case if power is delivered or received. (b) Verify that the total power received equals the total power delivered in the circuit. (This result is valid only for *complete* circuits and not for portions of a circuit. A formal statement of the above result is known as *Tellegen's theorem*.)

1-17 The voltage and current across a component (with *associated* reference polarities assigned) are given by $v(t) = 50 \cos 20\pi t$ V and $i(t) = 3 \sin 20\pi t$ A. Determine and carefully sketch the power received by the component in the interval (0, 0.2 s). What is the average value of the power?

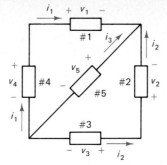

Fig. 1-55 Circuit for Problem 1-16.

Sec. 1-5 Classification of Circuits

1-18 The conductor ribbons used for interconnecting various printed circuit boards in a personal computer have lengths in the order of 0.1 m. Determine the maximum frequency for which they can be treated as lumped elements.

1-19 Test the linearity of the component whose current-voltage relationship is given by

$$v = \frac{d^2 i}{dt^2}$$

1-20 The voltage-current relationship of a component is given by

$$v = V_o + Ri$$

Where V_o and r are constants. Test the component for linearity. Draw a graph of the voltage as a function of the current. Note that even though the graph is a straight line, the component does not satisfy the linearity property. (When such components are analyzed, a model is used in which the constant term V_o is represented by a constant voltage source and the Ri term by a linear resistor.)

1-21 It was (tacitly) assumed that in the component of Exercise 1-12 the initial voltage is zero. Suppose there is an initial voltage V_o, so that the voltage-current relationship becomes

$$v = \int_0^t i(u)\,du + V_o$$

Test the linearity of the component. (In analysis, this nonzero initial condition is treated by using a model similar to the one mentioned in the previous problem.)

1-22 The forcing voltage functions and the corresponding current responses of a certain linear circuit are

$$v(t) = 10 \text{ V} \rightarrow i(t) = 20 \text{ A}$$
$$v(t) = \cos 10t \rightarrow i(t) = 2 \cos 10t - 5 \sin 10t$$

Determine the response $i(t)$ by using the principle of superposition when the forcing function is (a) $v(t) = 13.2 + 5 \cos 10t$ V; (b) $v(t) = -8 + 2 \cos 10t$ V; and (c) $v(t) = -11 - 2.56 \cos 10t$.

1-23 Test the time-invariance of the components whose voltage-current relationships were given in Problems 1-20 and 1-21.

1-24 The forcing-current function on a linear time-invariant circuit is given by

$$i(t) = \begin{cases} 0 & t < 0 \\ 5 \text{ A} & t > 0 \end{cases}$$

The resulting voltage is

$$v(t) = \begin{cases} 0 & t < 0 \\ 12e^{-3t} \text{ V} & t > 0 \end{cases}$$

Determine the response of the circuit for a forcing-current function

$$i(t) = \begin{cases} 0 & t < -4 \text{ s} \\ -10 \text{ A} & t > -4 \text{ S} \end{cases}$$

Sec. 1-6 Ohm's Law

1-25 Calculate the current or voltage, whichever is unknown, in each of the resistors shown in Fig. 1-56. Include the appropriate signs in the answers.

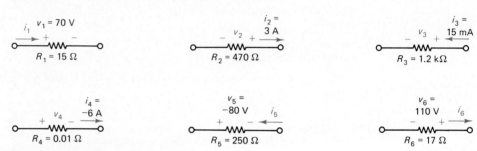

Fig. 1-56 Resistors for Problem 1-25.

1-26 The current through a conductor of resistance 470 Ω is 1.2 A. Calculate the change in potential energy of an electron moving between the two terminals of the conductor.

1-27 A resistance R consumes 100 W of power from a battery of voltage V_{bb}. (a) If the resistance is increased to $4R$ but the battery voltage is unchanged, evaluate the power consumed by the new resistance. (b) If the battery voltage is doubled and the resistance is $4R$, evaluate the power consumed by the resistance. (c) If the battery voltage V_{bb} delivers 625 W to a resistance R', find the ratio R'/R.

1-28 The piecewise linear model of a nonlinear resistor is shown in the graph of Fig. 1-57. Write an equation for each straight-line segment. For each segment, the nonlinear resistor can be modeled

Fig. 1-57 Graph for Problem 1-28.

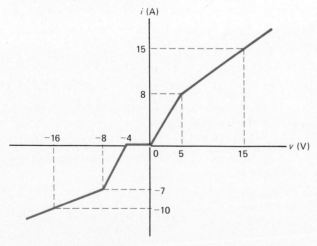

Basic Concepts, Components, and Laws of Electric Circuits

as a series combination of an ideal voltage source and a linear resistor. Set up the models and indicate the range of current values for which each model is valid.

Sec. 1-7 Kirchhoff's Voltage and Current Laws

1-29 Write the KVL equations for the closed paths indicated in each of the circuits of Fig. 1-58.

Fig. 1-58 Circuits for Problem 1-29.

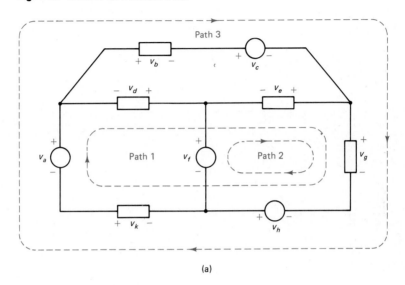

(a)

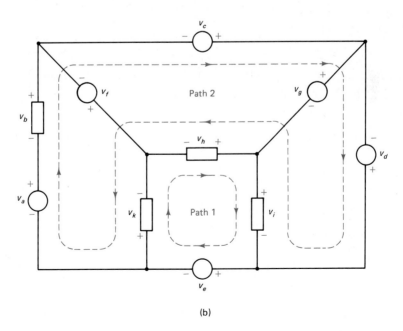

(b)

Problems 51

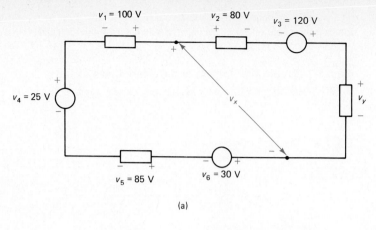

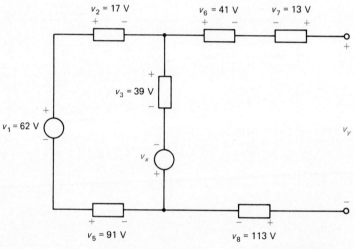

Fig. 1-59 Circuits for Problem 1-30.

1-30 Evaluate the voltages labeled v_x and v_y in the circuits shown in Fig. 1-59.

1-31 In the network of Fig. 1-60, (a) obtain an expression relating the voltages v_a, v_b, v_c, and v_d; (b) if $v_a = 100$ V, $v_d = 100$ V, and $v_{in} = 300$ V, calculate v_o, v_b, and v_c.

Fig. 1-60 Network for Problem 1-31.

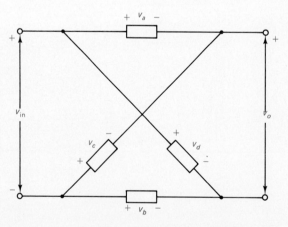

52   Basic Concepts, Components, and Laws of Electric Circuits

1-32 Using suitable closed paths, calculate the voltage across and the power *delivered* by each of the three current sources in Fig. 1-61.

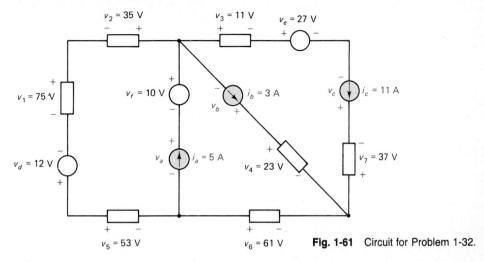

Fig. 1-61 Circuit for Problem 1-32.

1-33 (a) Write the KCL equations for nodes 1, 2, and 3 in the circuit of Fig. 1-62. (b) Show that the equation for the remaining node can be expressed as a linear combination of the three equations of part (a).

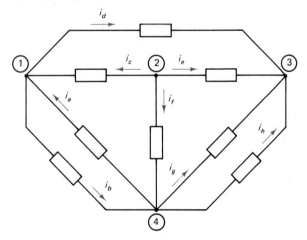

Fig. 1-62 Circuit for Problem 1-33.

1-34 In each of the cases shown in Fig. 1-63, determine the value of current i_x.

Fig. 1-63 Circuits for Problem 1-34.

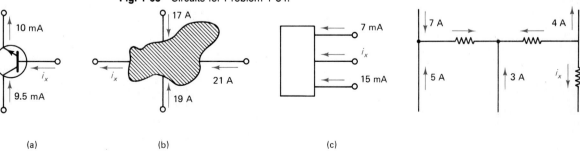

(a)　　　　　(b)　　　　　(c)

Problems

1-35 Evaluate currents i_1, i_2, i_3, and i_4 in the circuit of Fig. 1-64.

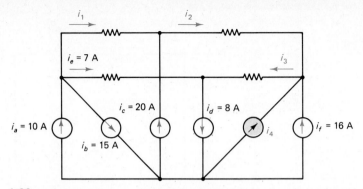

Fig. 1-64 Circuit for Problem 1-35.

1-36 Determine the current in and the power supplied by each of the three voltage sources in the circuit of Fig. 1-65.

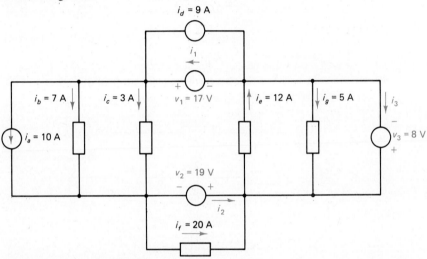

Fig. 1-65 Circuit for Problem 1-36.

Sec. 1-8 Voltage and Current Sources

1-37 The open-circuit voltage of a nonideal source is 14 V. It is found that a load resistance of 24 Ω reduces the terminal voltage to one-fourth of the open-circuit voltage. Obtain (a) the voltage source model and (b) the current source model.

1-38 For each of the source models shown in Fig. 1-66, obtain the equivalent source model of the other type. Include diagrams of the models showing the source voltage polarities or current directions as appropriate.

Fig. 1-66 Source models for Problem 1-38.

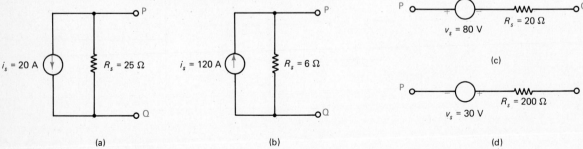

54 Basic Concepts, Components, and Laws of Electric Circuits

1-39 The graphs in Fig. 1-67 show the results of measuring the terminal voltage versus the terminal current for two circuits. The voltage polarities and current direction used in the measurements are as shown in the figure. For each case, obtain the two source models, including diagrams showing the voltage polarity marks and current directions.

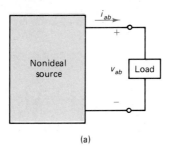

Fig. 1-67 Terminal voltage versus terminal current for two circuits of Problem 1-39.

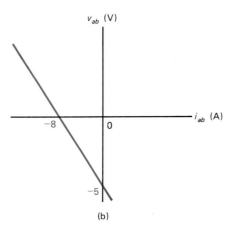

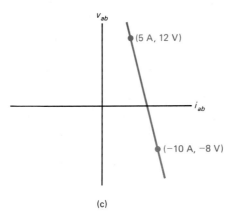

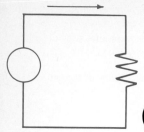

CHAPTER 2
SERIES AND PARALLEL CIRCUITS

The two fundamental laws of circuit analysis, Kirchhoff's voltage law (KVL) and Kirchhoff's current law (KCL) were discussed in Chapter 1. As stated earlier, Kirchhoff's laws are valid for *any lumped circuit* whether linear or nonlinear, time-variant or time-invariant. The discussion in this and the following chapters is, however, confined to *lumped, linear, and time-invariant* circuits, with only an occasional discussion of some simple nonlinear circuits. The application of Kirchhoff's laws to circuits containing certain configurations of resistors are presented in this chapter. The analysis of circuits studied does not involve the solution of simultaneous equations; and the absence of algebraic manipulations makes it possible to concentrate on the fundamental properties of electric circuits.

As an illustration of the application of the analysis techniques to some practical engineering situations, the model of an electronic amplifier and its analysis are also discussed in this chapter. The model is used as the basis of introducing the operational amplifier, which is a versatile and commonly used circuit in electronic design. The operational amplifier is revisited in Chapters 3, 4 and 6.

The simplest circuit that can be solved by applying KVL, KCL, and Ohm's law is one made up of resistors in series or in parallel. The solution of series circuits is discussed first.

2-1 THE SERIES CIRCUIT

Two branches are *in series* if they are physically connected so that the *same* current flows through them regardless of what makes up those two branches.

A series circuit contains a single closed path made up of components in series and a KVL equation is written for that closed path. A series circuit containing only resistors and

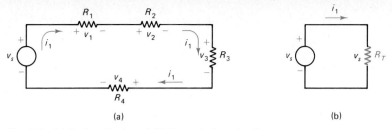

Fig. 2-1 (a) Series circuit and (b) its equivalent circuit.

voltage sources is analyzed by using Ohm's law in combination with the KVL equation.

Consider a series circuit driven by a single voltage source v_s, as shown in Fig. 2-1(a). There is only one current i_1 in the circuit. Voltage polarities are assigned to the resistors so that i_1 flows from the higher potential (plus) terminal to the lower potential (minus) terminal in each resistor. (This is the "associated reference polarities" defined in Chapter 1.)

As stated in Chapter 1, the convention for writing the KVL equation for a closed path is

Add a voltage if the path is from *plus to minus* and subtract if the path is from *minus to plus*.

Following the direction of the current in Fig. 2-1(a), the KVL equation is

$$-v_s + v_1 + v_2 + v_3 + v_4 = 0$$

which leads to

$$v_1 + v_2 + v_3 + v_4 = v_s \tag{2-1}$$

Each voltage term on the left side of Eq. (2-1) is replaced by an Ri product by using Ohm's law.

$$R_1 i_1 + R_2 i_1 + R_3 i_1 + R_4 i_1 = v_s$$

or

$$i_1 = \frac{v_s}{(R_1 + R_2 + R_3 + R_4)} \tag{2-2}$$

which is solved for i_1.

It is possible to set up a simple circuit, as shown in Fig. 2-1(b), so that the current in it $i_1 = v_s/R_T$ is the same as in the original circuit shown in Fig. 2-1(a). Then the circuit of Fig. 2-1(b) is said to be *equivalent* to that of Fig. 2-1(a). It follows from Eq. (2-2) that the relationship between Figs. 2-1(a) and (b) is

$$R_T = R_1 + R_2 + R_3 + R_4 \tag{2-3}$$

Therefore,

the total resistance of two or more resistors in series is equal to the sum of the individual resistances,

and

the current in a series circuit is the ratio of the applied voltage to the total resistance.

2-1   The Series Circuit

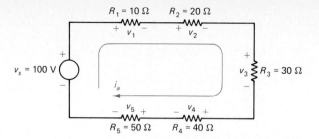

Fig. 2-2 Circuit for Example 2-1.

Example 2-1 Calculate the current i_a and the voltage across each resistor in the circuit of Fig. 2-2.

Solution The total resistance of the circuit is

$$R_T = R_1 + R_2 + R_3 + R_4 + R_5 = 150 \text{ }\Omega$$

Therefore,

$$i_a = v_s/R_T = 100/150 = 0.667 \text{ A}$$

The voltages across the resistors are (using Ohm's law) $v_1 = R_1 i_a = 6.67$ V; $v_2 = 13.34$ V; $v_3 = 20.01$ V; $v_4 = 26.68$ V; and $v_5 = 33.35$ V. ∎

If the five voltage values are added, the sum should equal v_s. The actual sum is 100.05 (instead of 100.00) and the discrepancy is due to the *round-off error* in calculating i_a. This situation occurs in almost all problems, of course. Trying to minimize this error by retaining a large number of significant digits in numerical answers is absurd, however, since such accuracy cannot be matched in the usual laboratory measurements. As a general rule, *three* significant digits are retained in numerical answers in this text, letting the round-off error fall where it may.

Exercise 2-1 Calculate the current and the voltage across each resistor in each of the circuits of Fig. 2-3.

Fig. 2-3 Circuits for Exercise 2-1.

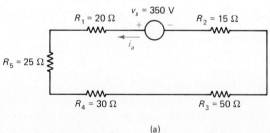

(a)

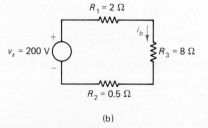

(b)

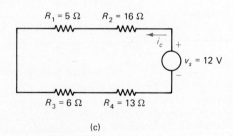

(c)

58   Series and Parallel Circuits

Example 2-2 Given that the 5-Ω resistor in the circuit of Fig. 2-4 consumes a power of 80 W (constant), determine the voltage v_s of the source, the power dissipated in each of the other resistors, and the power supplied by the source.

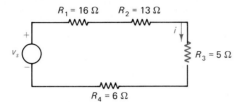

Fig. 2-4 Circuit for Example 2-2.

Solution The power consumed by a resistor is given by

$$p = i^2 R$$

Since $p = 80$ W and $R = 5$ Ω, $i = 4$ A.
 The total resistance of the circuit is 40 Ω. Therefore, $v_s = 160$ V.
 The power dissipated by the resistors are 256 W (in R_1), 208 W (in R_2), and 96 W (in R_4).
 The power supplied by the voltage source is

$$p_s = v_s i = 640 \text{ W}$$

This should equal the sum of the powers dissipated by the four resistors in the circuit. ■

Exercise 2-2 If the 5-Ω resistor of Fig. 2-4 is to dissipate 160 W, recalculate the results of the previous example.

Exercise 2-3 The power supplied by the voltage source in the circuit of Fig. 2-5 is 600 W. Determine the value of R_x.

Exercise 2-4 If the value of each resistor in the circuit of Fig. 2-5 is multiplied by a factor of N, what is the effect on the power supplied by v_s?

Exercise 2-5 If the source voltage in Fig. 2-5 is multiplied by a factor of N, what will be the effect on the power dissipated in each resistor compared with the results of Exercise 2-3?

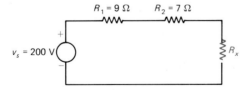

Fig. 2-5 Circuit for Exercises 2-3 to 2-5.

2-1-1 Voltage Dividers

A special case of a series circuit is two resistors in series. Such a circuit divides the applied voltage in proportion to the two resistances and is known as a *voltage* divider. For the voltage divider shown in Fig. 2-6, the current in the circuit is

$$i = v_s/(R_1 + R_2)$$

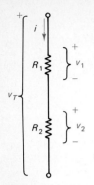

Fig. 2-6 Voltage divider.

The voltage across R_1 is

$$v_1 = R_1 i = \left(\frac{R_1}{R_1 + R_2}\right) v_s \qquad (2\text{-}4a)$$

and the voltage across R_2 is

$$v_2 = R_2 i = \left(\frac{R_2}{R_1 + R_2}\right) v_s \qquad (2\text{-}4b)$$

Equations (2-4a) and (2-4b) are known as *voltage divider* formulas. The voltage divider formula states that the voltage across each resistor equals the ratio of that resistance to the total resistance.

Also, the ratio of the voltages across the resistors in Fig. 2-6 is given by

$$\frac{v_1}{v_2} = \frac{R_1}{R_2}$$

and, for unequal resistances, the *larger voltage* appears across the *larger resistance*.

Example 2-3 Two resistors, $R_1 = 100 \, \Omega$ and $R_2 = 400 \, \Omega$, are in series with a total voltage $v_T = 50$ V across them. Calculate the voltage across R_1.

Solution $v_1 = [R_1/(R_1 + R_2)] v_s = (100/150) \times 50 = 10$ V ∎

The following example illustrates the use of a voltage divider in the measurement of voltages using a certain kind of voltmeter.

Example 2-4 The basic mechanism of an *analog dc voltmeter* is called the *D'Arsonval movement*, which has a resistance R_m, as shown in Fig. 2-7(a). It is quite delicate and the maximum value of v_m that can be safely applied to the mechanism is of the order of tens of millivolts. (1 mV = 10^{-3} V.) When a voltage $v_T > v_m$ is to be measured, the voltage across the mechanism is limited by placing a resistance R_1 in series with it, as shown in Fig. 2-7(b). Obtain an expression for R_1 in terms of R_m, v_m, and v_T. Calculate R_1 when $R_m = 50$ Ω, $v_m = 50$ mV, and $v_T = 100$ V.

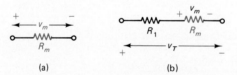

Fig. 2-7 Voltmeter for Example 2-4.

Solution: From the voltage divider formula, Eq. (2-4),

$$v_m = \left(\frac{R_m}{R_m + R_1}\right) v_T$$

Therefore,

$$R_1 = R_m[(v_T/v_m) - 1]$$

When $R_m = 50 \, \Omega$, $v_m = 50 \times 10^{-3}$ V, and $v_T = 100$ V, $R_1 = 10^5 = 100$ kΩ. (1 k$\Omega = 10^3 \, \Omega$.) ∎

Exercise 2-6 Calculate the voltage v_2 in each of the circuits in Fig. 2-8.

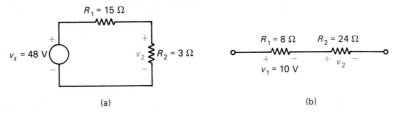

(a)   (b)

Fig. 2-8 Circuits for Exercise 2-6.

Exercise 2-7 If the D'Arsonval movement of Example 2-4 is placed in series with a 5-kΩ resistor, what is the maximum voltage v_T that can be measured by the voltmeter?

Useful Approximations

When one of the resistances is *much larger* than the other in a voltage divider, the *total resistance* can be taken as approximately equal to the *larger* resistance and the smaller resistance neglected. That is, if in Fig. 2-6,

$$R_1 \gg R_2, \text{ then } R_T = R_1 + R_2 \approx R_1$$

where the symbol $\approx$ means *approximately equal to*. Also, the voltage across the larger resistance $\approx$ the total voltage.

The voltage across the smaller resistance is not zero, however; it is simply much smaller than the total voltage. The above approximation is useful for making rough calculations (especially in electronic circuits). The question is, of course, how to define "much larger than." This depends upon the *allowable margin of error*, which may be as high as 10 percent in calculations involving electronic devices. In such cases, approximations are permitted if one resistor is ten or more times larger than the other. In circuit analysis problems, however, the allowable margin of error varies with the instructor.

Example 2-5 Consider the results obtained in the voltmeter example, Example 2-4. Calculate the total resistance, the voltage across each resistor when the smaller resistance is neglected, and the error (percent) introduced by the approximation.

Solution Neglecting the 50 Ω due to R_m,

- Total resistance $R_T \approx R_1 = 100$ kΩ (100.05 kΩ is the exact value)
- Voltage across $R_1 \approx v_T = 100$ V (99.95 V is the exact value from $[R_1/(R_1 + R_m)]v_T$)
- Voltage across R_m: $v_m = (R_m/R_T)v_t \approx 0.05$ V (0.04998 V is the exact value)
- Percent error = 100 × (error/exact value) = 0.05 percent for all three cases ∎

Exercise 2-8 Consider the voltmeter circuit of Exercise 2-7. Calculate the total resistance, the voltage across each resistor, and the error introduced when the smaller resistance is neglected.

2-1 The Series Circuit

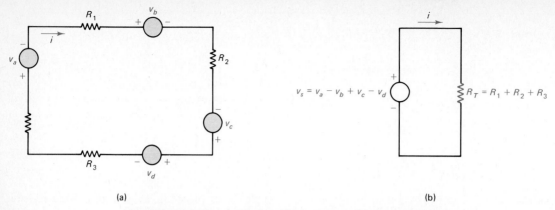

Fig. 2-9 (a) Series circuit with multiple sources. (b) Equivalent circuit.

2-1-2 Series Circuits with Multiple Sources

Consider a series circuit with more than one voltage source, as shown in Fig. 2-9(a). Choosing a direction for the current i as indicated, the KVL equation of the closed path in the circuit is

$$v_a + R_1 i + v_b + R_2 i - v_c + v_d + R_3 i = 0$$

The above equation is rewritten by keeping the Ri products on the left side and moving the source terms to the right side (since the source voltages are usually specified quantities).

$$(R_1 + R_2 + R_3)i = -v_a - v_b + v_c - v_d \tag{2-5}$$

It is seen that the resistances can be summed to a single resistance. The source voltages are added algebraically to obtain a net source voltage acting on the circuit, as shown in Fig. 2-9(b). The rest of the procedure is exactly the same as for a series circuit with a single voltage source.

Example 2-6 (a) Calculate the current in the circuit of Fig. 2-10(a). (b) Calculate the voltage between points A and B.

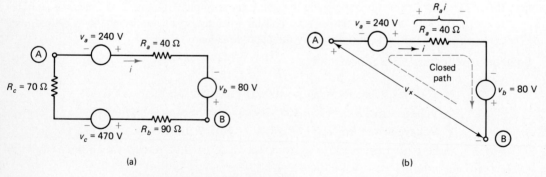

Fig. 2-10 Circuit for Example 2-6.

Solution (a) The KVL equation is

$$-v_a + R_a i - v_b + R_b i + v_c + R_c i = 0$$

62 Series and Parallel Circuits

which gives

$$(R_a + R_b + R_c)i = v_a + v_b - v_c$$

Therefore, $i = -150/200 = -0.75$ A.

(b) Label the voltage between points A and B as v_x and assign reference polarities. Write a KVL equation for a closed path obtained by starting from B, going over to A across v_x and returning to B through the components of the circuit. One such path is shown in Fig. 2-10(b), and its KVL equation is

$$-v_x - v_a + R_a i - v_b = 0$$

which gives

$$v_x = -240 + (40)(-0.75) - 80 = -350 \text{ V}$$

∎

Exercise 2-9 Verify the value of v_x obtained in Example 2-6 by using the other available closed path.

Example 2-7 Determine the voltage v_{AB} in the circuit of Fig. 2-11(a).

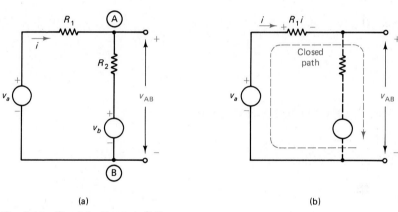

(a)   (b)

Fig. 2-11 Circuit for Example 2-7.

Solution First solve for the current i.
The KVL equation is

$$-v_a + R_1 i + R_2 i + v_b = 0$$

which gives

$$i = (v_a - v_b)/(R_1 + R_2) \qquad (2\text{-}6)$$

There are two possible closed paths for calculating v_{AB}. One choice is shown in Fig. 2-11(b) and its KVL equation is

$$-v_a + R_1 i + v_{AB} = 0$$

which leads to, after substituting for i from Eq. (2-6),

$$v_{AB} = v_a - R_1 i = (R_2 v_a + R_1 v_b)/R_T$$

where $R_T = R_1 + R_2$.

∎

2-1 The Series Circuit

Exercise 2-10 Verify the expression for v_{AB} obtained in the previous example by using the other available closed path.

Exercise 2-11 Calculate the current in each of the circuits shown in Fig. 2-12.

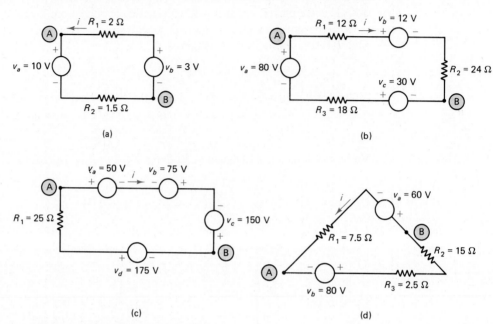

Fig. 2-12 Circuits for Exercise 2-11.

Exercise 2-12 Calculate the voltage between points A and B in each of the circuits of Exercise 2-11, with A chosen as the positive reference.

Exercise 2-13 Determine the value of the source voltage v_x in Fig. 2-13 so as to make $v_{50} = 150$ V.

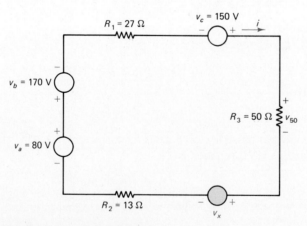

Fig. 2-13 Circuit for Exercise 2-13.

2-2 THE PARALLEL CIRCUIT

Two branches are *in parallel* if they are physically connected so that the *same* voltage appears across them regardless of what makes up those two branches.

A circuit in which all the components are connected in parallel across one pair of nodes is called a *parallel circuit*. The number of *independent* nodes in a circuit is *one less than the total number of nodes* (a statement that will be established in Chapter 3). Thus, there is only one independent node in a parallel circuit and a KCL equation is written for one of the two nodes. A parallel circuit containing only resistors and current sources can be solved by using Ohm's law in combination with the KCL equation.

Consider a parallel circuit driven by a single current source i_s, as shown in Fig. 2-14(a). The voltage v_1 is common to all the branches. A current is assigned to each resistor flowing from the higher to the lower potential terminal of v_1 (associated reference polarities).

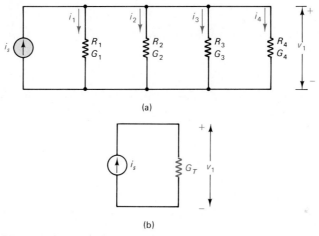

Fig. 2-14 (a) Parallel circuit and (b) its equivalent circuit.

As discussed in Chapter 1, the KCL equation for the upper node is written using the convention: *add* if a current *leaves* the node and *subtract* if a current *enters* the node.

$$-i_s + i_1 + i_2 + i_3 + i_4 = 0$$

which leads to

$$i_1 + i_2 + i_3 + i_4 = i_s \qquad (2\text{-}7)$$

Each current on the left side of Eq. (2-7) is replaced by a ratio of the form v/R by using Ohm's law. Using ratios such as v/R tends to make the algebra somewhat messy. It is therefore convenient to introduce the parameter *conductance*.

Conductance **is defined as the reciprocal of resistance and denoted by the symbol** G.

The unit of conductance is the *siemens* (abbreviated S). (The old unit for conductance was "mho," which can still be found in some textbooks.) Thus,

$$G = \frac{1}{R} \qquad (2\text{-}8)$$

and if R is in ohms, then G is in siemens. Ohm's law then assumes the alternative form

$$v = (1/G)i \quad \text{or} \quad i = Gv \tag{2-9}$$

Equations (2-9) and (2-7) lead to

$$G_1v_1 + G_2v_1 + G_3v_1 + G_4v_1 = i_s$$

where $G_k = 1/R_k$ (k = 1, 2, 3, 4).
Therefore,

$$v_1 = \frac{i_s}{(G_1 + G_2 + G_3 + G_4)}$$

which is solved for v_1.

The given circuit can be replaced by a simple circuit, as shown in Fig. 2-14(b), provided the voltage $v_1 = i_s/G_T$ in it is the same as in the original circuit of Fig. 2-14(a). That is, the circuit of Fig. 2-14(b) is *equivalent* to the original circuit if

$$G_T = G_1 + G_2 + G_3 + G_4 \tag{2-10}$$

The *total conductance of two or more resistors in parallel* equals the sum of the individual conductances,

and

the voltage across a parallel circuit is the ratio of the applied current to the total conductance.

Example 2-8 Calculate the voltage v_1 and the current in each resistor in the circuit of Fig. 2-15.

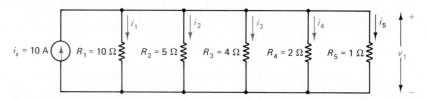

Fig. 2-15 Circuit for Example 2-8.

Solution The individual conductances are $G_1 = 0.1$ S, $G_2 = 0.2$ S, $G_3 = 0.25$ S, $G_4 = 0.5$ S, and $G_5 = 1$ S.

The total conductance is

$$G_T = G_1 + G_2 + G_3 + G_4 + G_5 = 2.05 \text{ S}$$

Therefore,

$$v_1 = i_s/G_T = 4.88 \text{ V}$$

The currents through the individual resistors are (by using Ohm's law) $i_1 = v_1G_1 = 0.488$ A, $i_2 = 0.976$ A, $i_3 = 1.22$ A, $i_4 = 2.44$ A, and $i_5 = 4.88$ A. ∎

The five currents should add up to $i_s = 10$ A (within the limits of round-off error).

Exercise 2-14 Calculate the voltage across and current in each resistor in each of the circuits of Fig. 2-16.

Series and Parallel Circuits

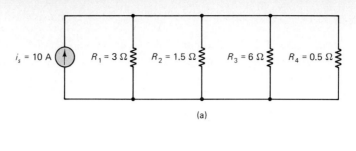
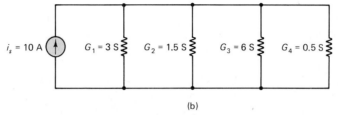

Fig. 2-16 Circuits for Exercise 2-14.

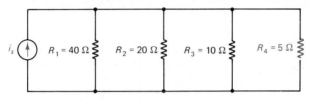

Fig. 2-17 Circuit for Example 2-9.

Example 2-9 Given that the 5-Ω resistor in the circuit of Fig. 2-17 consumes a power of 80 W (constant), determine the current i_s supplied by the source, the power dissipated by each resistor, and the power supplied by the source.

Solution The power consumed by a resistor is given by

$$p = v^2/R = v^2 G$$

Since $p = 80$ W and $R = 5\ \Omega$, $v = 20$ V.

The total conductance of the circuit $= 0.375$ S. Therefore, $i_s = vG_T = 7.5$ A.

The power dissipated by the resistors are (by using $p = v^2 G$): 80 W (for R_1), 40 W (for R_2), 20 W (for R_3), and 10 W (for R_4). The power supplied by the source is $vi_s = 150$ W.

Exercise 2-15 If the 5-Ω resistor of Fig. 2-17 is to dissipate 160 W, recalculate the results of Example 2-9.

Exercise 2-16 The power supplied by the current source in Fig. 2-18 is 600 W. Determine the value of R_x.

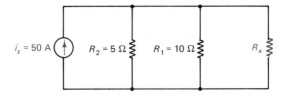

Fig. 2-18 Circuit for Exercise 2-16.

2-2 The Parallel Circuit

2-2-1 Current Dividers

A special case of the parallel circuit is two resistors in parallel. Such a circuit divides an applied current into two parts in inverse proportion to the two resistances and is called a *current divider* (analogous to the voltage divider). Even though a current divider is a special case of a number of conductances in parallel, it is usually more convenient to analyze it in terms of *resistances* rather than conductances. For the circuit shown in Fig. 2-19(a), the total conductance is

$$G_T = G_1 + G_2$$

By using $R = 1/G$, the above equation is rewritten as

$$1/R_T = 1/R_1 + 1/R_2$$

which leads to

$$R_T = \frac{R_1 R_2}{R_1 + R_2} \tag{2-11}$$

The total resistance of two resistances in parallel equals the ratio of the product to the sum of the two resistances.

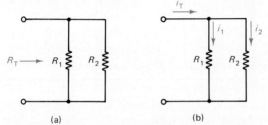

Fig. 2-19 (a) Current divider. (b) Current flow through the current divider.

Remember that the (product/sum) rule applies only to *two* resistors in parallel. It does not apply to three or more resistors in parallel.

It is important to keep in mind that in any parallel combination of resistors, the total resistance is always *less than the smallest* of the individual resistances.

The KCL equation of the circuit of Fig. 2-19(b) is

$$i_T = i_1 + i_2 \tag{2-12}$$

The voltages across R_1 and R_2 must be equal to each other (due to the parallel connection). Therefore,

$$R_1 i_1 = R_2 i_2 \tag{2-13}$$

Equations (2-12) and (2-13) lead to

$$i_1 = \left(\frac{R_2}{R_1 + R_2}\right) i_T \tag{2-14a}$$

and

$$i_2 = \left(\frac{R_1}{R_1 + R_2}\right) i_T \tag{2-14b}$$

Equations (2-14a) and (2-14b) are known as *current divider formulas*. Note that the *denominator is the same* $(R_1 + R_2)$ *in both formulas*. Also, the current through either branch is *proportional to the* resistance of the other *branch*.

For unequal resistors, the *larger current* flows through the *smaller resistor*. (This is to be expected since the Ri products of the two branches must be equal.)

Example 2-10 Calculate (a) the total resistance and (b) the current in each resistor of Fig. 2-20.

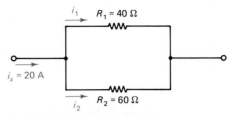

Fig. 2-20 Circuit for Example 2-10.

Solution Using Eqs. (2-11), (2-14a), and (2-14b),

$$R_T = R_1 R_2/(R_1 + R_2) = 24 \ \Omega$$
$$i_1 = [R_2/(R_1 + R_2)]i_s = 12 \ \text{A}$$
$$i_2 = [R_1/(R_1 + R_2)]i_s = 8 \ \text{A}$$

Note that the two currents add up to $i_s = 20$ A and the voltage across each resistor is 480 V. ∎

Example 2-11 Determine the value of R_x in Fig. 2-21 so as to make the total resistance 100 Ω. Calculate the current in each resistor if a current of 16 A is fed to the circuit.

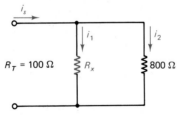

Fig. 2-21 Circuit for Example 2-11.

Solution From Eqs. (2-15),

$$100 = 800 R_x/(800 + R_x)$$

which, when solved, gives $R_x = 114 \ \Omega$.

Using the current divider formulas and a total current of 16 A,

$$\text{current in 800-}\Omega \text{ resistor} = [R_x/(R_x + 800)](16) = 2 \ \text{A}$$
$$\text{current in } R_x = [800/(R_x + 800)](16) = 14 \ \text{A}$$

(An alternative procedure is to first find the voltage across the circuit, $16 \times 100 = 1600$ V, and then use Ohm's law for each resistor.) ∎

2-2 The Parallel Circuit

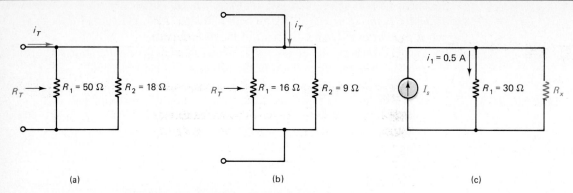

Fig. 2-22 Circuits for Exercises 2-17 to 2-19.

Exercise 2-17 Calculate the total resistance of the circuits of Figs. 2-22(a) and (b).

Exercise 2-18 Calculate the current in each resistor of the circuits of Fig. 2-22(a) and (b) when the total current is 100 A.

Exercise 2-19 If the current source in Fig. 2-22(c) delivers 50 W of power to the circuit, evaluate R_x.

The current divider principle is used in the measurement of currents using certain types of ammeters, as discussed in the following example.

Example 2-12 The D'Arsonval movement (introduced in Example 2-4) is also the basic mechanism of *dc ammeters*. The movement can withstand a current i_m which is of the order of a few mA. If a current $i_T > i_m$ is to be measured, then it is necessary to limit the current in the movement by providing a parallel resistor (known as a *shunt* resistor) R_1 as shown in Fig. 2-23. Obtain an expression for R_1 in terms of R_m, i_m, and i_T. Calculate R_1 when $R_m = 50\ \Omega$, $i_m = 2$ mA, and $i_T = 10$ A.

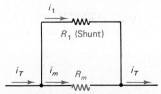

Fig. 2-23 Ammeter for Example 2-12.

Solution Using the current divider formula,

$$i_m = [R_1/(R_1 + R_m)]i_T$$

or

$$R_1 = R_m i_m/(i_T - i_m)$$

For the given numerical values, $R_1 = 0.01\ \Omega$.

Note that the current through the meter movement is only 2 mA (or 0.002 A) compared with the current of 9.998 A in R_1. Since the current in R_1 is much greater than that in R_m, it is to be expected that R_1 is very small in comparison with R_m. ∎

Exercise 2-20 If the D'Arsonval movement of Example 2-12 is placed in shunt with a 5×10^{-3}-Ω resistor, calculate the maximum current that can be measured by the ammeter.

Useful Approximations

When one of the resistances is *much larger* than the other in a current divider, the *total resistance* can be taken as approximately equal to the *smaller* resistance and the larger resistance ignored. That is, if in Fig. 2-19,

$$R_1 \gg R_2, \text{ then } R_T \approx R_2$$

Also, the current in the *smaller* resistance $\approx$ the total current.

The current in the larger resistance is not zero, however; it is simply much smaller than the total current. The above approximation is useful in making rough calculations. The "much larger than" is normally taken as "ten or more times larger than" in electronic circuits, which leads to a maximum error of 10 percent.

Example 2-13 For the ammeter of Example 2-12, calculate the total resistance and the current in each resistor when the larger resistor is ignored, and calculate the error (percent) introduced by the approximation.

Solution Neglecting the 50 Ω due to R_m,

- Total resistance $\approx R_1 = 0.01$ Ω (0.009998 Ω is the exact value)
- Current in R_1: $i_1 \approx i_T = 10$ A (9.998 A is the exact value)
- Current in R_m: $i_m = [R_1/(R_1 + R_m)]i_T = (R_1/R_m)i_T = 0.002$ A (or 2 mA) (1.9996 mA is the exact value)
- Error = 0.02 percent in all three cases

Exercise 2-21 Repeat the calculations of Example 2-13 for the ammeter circuit of Exercise 2-20.

2-2-2 Parallel Circuits with Multiple Sources

Consider a parallel circuit with more than one current source as shown in Fig. 2-24(a). Assign polarities to the voltage v across the circuit. The KCL equation of node 1 is

$$i_a + G_1 v + i_b + G_2 v - i_c + G_3 v + i_d = 0$$

Keeping the Gv products on the left side and moving the source terms to the right side, the equation becomes

$$(G_1 + G_2 + G_3)v = -i_a - i_b + i_c - i_d$$

It is seen that the conductances can be added to give a single conductance. The source currents are added algebraically to obtain a net current feeding the circuit, as shown in Fig. 2-24(b). The rest of the procedure is exactly the same as for a parallel circuit with a single current source.

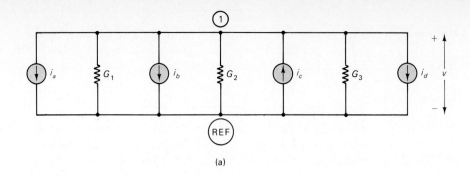

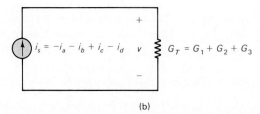

Fig. 2-24 (a) Parallel circuit with multiple sources and its (b) equivalent circuit.

Example 2-14 (a) Calculate the voltage across the circuit of Fig. 2-25(a). (b) Calculate the current i_p.

Solution (a) The KCL Equation is

$$-i_s + G_1v - i_a + G_2v + i_b + G_3v = 0$$

or

$$(G_1 + G_2 + G_3)v = i_s + i_a - i_b$$

Therefore,

$$v = 35/6.5 = 5.38 \text{ V}$$

Fig. 2-25 Circuit for Example 2-14.

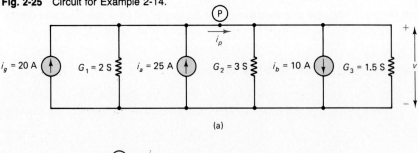

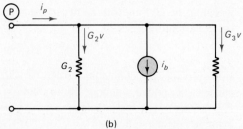

Series and Parallel Circuits

(b) Write a KCL equation using the branches to the right of point P. Figure 2-25(b) shows the currents in those branches.

$$-i_p + G_2 v + i_b + G_3 v = 0$$

Therefore, $i_p = 34.2$ A. ∎

Exercise 2-22 Recalculate the current i_p in Example 2-14 by using branches to the left of point P.

Example 2-15 Given that the current in G_3 is $i_3 = 60$ A in the circuit of Fig. 2-26, evaluate the current i_x.

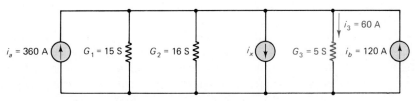

Fig. 2-26 Circuit for Example 2-15.

Solution The voltage across G_3 (and all the other branches as well) is $v = i_3/G_3 = 12$ V. The KCL equation is

$$-i_a + G_1 v + G_2 v + i_x + G_3 v - i_b = 0$$

Therefore, $i_x = 48$ A. ∎

Exercise 2-23 Calculate the voltage across each of the circuits shown in Fig. 2-27.

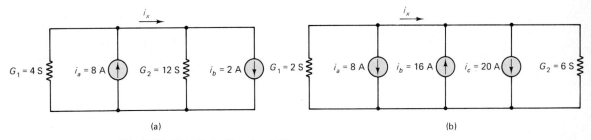

(a) (b)

Fig. 2-27 Circuits for Exercise 2-23.

Exercise 2-24 In each circuit of Exercise 2-23, evaluate the current i_x.

Exercise 2-25 Determine the value of the conductance G_x in Fig. 2-28 so as to make $i_q = 15$ A.

Fig. 2-28 Circuit for Exercise 2-25.

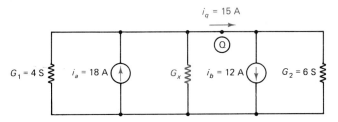

2-2 The Parallel Circuit

2-3 DUALITY

A careful review of the discussions of series circuits in Section 2-1 and parallel circuits in Section 2-2 shows a strong similarity between the various relationships and properties of the two types of circuits.

For example, the total resistance of a number of *resistances in series* is equal to the *sum of the resistances,* whereas the total conductance of a number of *conductances in parallel* is equal to the *sum of the conductances.* In fact, if the statements and relationships of Section 2-1 on series circuits were rewritten by replacing *voltage* by *current, current* by *voltage,* and *resistance* by *conductance,* then the resulting paragraphs would lead to the corresponding statements and relationships in Section 2-2 on parallel circuits. The problems at the end of the chapter covering Sections 2-1 and 2-2 also show this similarity.

That is, even though the geometrical structure and tools of analysis of series circuits and parallel circuits are quite different, the relationships and properties of one connection can be readily transformed to those of the other by methodically interchanging voltage with current, resistance with conductance, and series connection with parallel connection. This correspondence between the two types of circuit configuration is called *duality.* Quantities such as voltage and current form dual pairs. The following is a summary of dual pairs:

$$\text{voltage} \leftrightarrow \text{current}$$
$$\text{resistance} \leftrightarrow \text{conductance}$$
$$\text{series connection} \leftrightarrow \text{parallel connection}$$
$$\text{open circuit} \leftrightarrow \text{short circuit}$$

The above is only a partial list and some other dual pairs are encountered in later chapters.

Duality not only exhibits a symmetry in circuit theory but is a useful concept as well. Theorems derived for a series configuration, for example, lead to theorems that are valid for a parallel circuit, and the results obtained from the analysis of some general series circuits can be used in the study of dual parallel circuits. Such applications of the principle of duality are presented in Chapters 4, 5, 6, 8, 9, 11 and 12.

2-4 SERIES AND PARALLEL CIRCUITS WITH DEPENDENT SOURCES

Dependent sources were introduced in Chapter 1; they have the property that their voltage or current is controlled by a voltage or current in a branch in another part of the circuit. Models of electronic devices include dependent sources so as to account for the terminal properties of the devices. The analysis of circuits containing dependent sources uses the same principles and procedures as those containing only independent sources, as illustrated by the following examples.

Example 2-16 Determine the current in the circuit of Fig. 2-29.

Solution It is advisable to first set up the relationship for the voltage controlling the dependent source: v_1 across resistor R_1.

$$v_1 = R_1 i_1 \qquad (2\text{-}15)$$

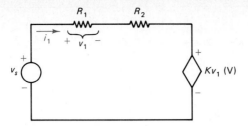

Fig. 2-29 Series circuit with a dependent source for Example 2-16.

The KVL equation is

$$-v_s + R_1 i_1 + R_2 i_1 + K v_1 = 0 \qquad (2\text{-}16)$$

Substitute for v_1 from Eq. (2-16) in Eq. (2-15).

$$[R_1 (1 + K) + R_2] i_1 = v_s$$

Therefore,

$$i_1 = \frac{v_s}{R_1 (1 + K) + R_2}$$ ∎

Example 2-17 Find the voltage across the dependent source in the circuit of Fig. 2-30.

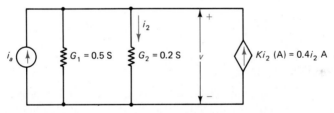

Fig. 2-30 Parallel circuit with a dependent source for Example 2-17.

Solution Set up the relationship for the current controlling the dependent source.

$$i_2 = G_2 v = 0.2v \text{ A}$$

The KCL equation is

$$-i_a + G_1 v + G_2 v - K i_2 = 0$$

Therefore,

$$G_1 v + G_2 v - K(0.2v) = i_a$$

which gives

$$0.62v = 10$$

or

$$v = 16.1 \text{ V}$$ ∎

2-4 Series and Parallel Circuits with Dependent Sources

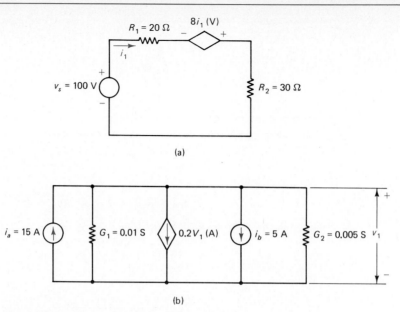

Fig. 2-31 (a) Circuit for Exercise 2-26. (b) Circuit for Exercise 2-27.

Exercise 2-26 Determine the current in the circuit of Fig. 2-31(a).

Exercise 2-27 Determine the voltage in the circuit of Fig. 2-31(b).

2-4-1 Model of an Amplifier

An amplifier is a circuit that receives an input voltage and produces an output voltage of a larger amplitude than the input. For our purposes, it is sufficient to think of an amplifier as a "black box" with a pair of input terminals and a pair of output terminals, as shown in Fig. 2-32(a). v_1 and i_1 are the input voltage and current, and v_2 and i_2 are the output voltage and current (across a load not shown in the diagram). One of the models used to represent an amplifier is shown in Fig. 2-32(b). R_i represents the *input resistance* of the amplifier. (The model used in the discussion is usually referred to as the *voltage controlled voltage source model*. Other models are also used for an amplifier in electronics textbooks.) The electronic devices in the amplifier circuit cause the input voltage v_1 to be multiplied by some factor A and this is represented by the dependent voltage source Av_1. The resistor R_o represents the *output resistance* of the amplifier.

Now connect a signal source v_s in series with a resistor R_s to the input terminals of the amplifier and a load resistance R_L to the output terminals, as shown in Fig. 2-32(c). Note that the circuit is made up of two *separate* series paths; the only link between them is through the dependent source. The amplifier circuit is analyzed by writing the KVL equations of the two closed paths.

Input Side: $\qquad -v_s + R_s i_1 + R_i i_1 = 0$

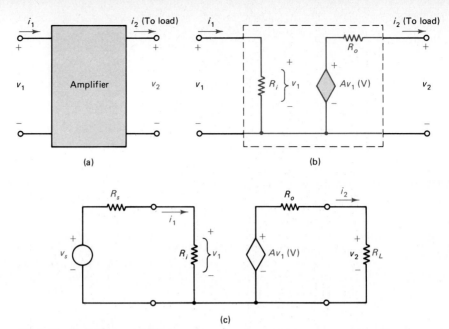

Fig. 2-32 Model of an amplifier.

which leads to

$$i_1 = \frac{v_s}{R_s + R_i} \qquad (2\text{-}17)$$

Output Side: The voltage controlling the dependent source is v_1, which is given by

$$v_1 = R_i i_1 \qquad (2\text{-}18)$$

where i_1 is given by Eq. (2-17). The KVL equation for the output side is

$$-Av_1 + R_o i_2 + R_L i_2 = 0 \qquad (2\text{-}19)$$

which, in conjunction with Eq. (2-17), gives

$$i_2 = AR_i i_1/(R_o + R_L)$$

$$= \frac{AR_i v_s}{(R_s + R_i)(R_o + R_L)} \qquad (2\text{-}20)$$

$$v_2 = R_L i_2 \qquad (2\text{-}21)$$

Equations (2-17) through (2-21) are used to solve for any desired quantities in the amplifier circuit.

Exercise 2-28 Use the equations derived above to find the voltage gain v_2/v_s of the amplifier.

Exercise 2-29 Exercise 2-28 can be solved more efficiently by using the voltage divider formulas on the input and output sides of Fig. 2-32(c). Show the solution.

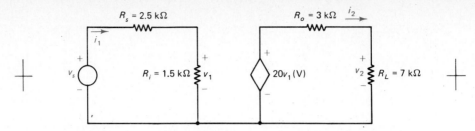

Fig. 2-33 Circuit for Exercise 2-30.

Exercise 2-30 Analyze the amplifier circuit shown in Fig. 2-33. Evaluate the following: (a) v_2/v_s and (b) i_2/i_1.

We will return to the model of the amplifier at various points in the text. It is important to note that even though the active devices and the circuits in actual amplifiers are generally complicated, the circuit set up by using appropriate linear models are solvable by using the principles of linear circuit analysis.

2-5 SERIES-PARALLEL AND LADDER NETWORKS

Progressing from strictly series and strictly parallel circuits, we next consider circuits with resistors connected in series and parallel driven by a single source. Two methods are available for their analysis.

1. The circuit is reduced by starting at the branch farthest from the source and using successive series and parallel combinations of the resistors. The reduction leads to a simple equivalent circuit (one source and one resistor). This circuit is solved and the result is then used to evaluate the currents and voltages in the original circuit by reversing the steps used in the reduction. This procedure is known as *series-parallel reduction* of a network.

2. The analysis is started by assuming the current to be 1 A (or the voltage to be 1 V) in the branch farthest from the source. The currents and voltages in the other branches conforming to the above assumption are then calculated by using KVL, KCL, and Ohm's law and moving toward the source, one branch at a time. The principle of linearity is then invoked to find the actual currents and voltages in the circuit. This method is known as the *ladder* network procedure.

2-5-1 Series-Parallel Reduction Method

The procedure of series-parallel reduction is best illustrated through an example. Consider the circuit of Fig. 2-34(a).

Starting with the branch farthest from the source, resistors R_1 and R_2 are in series. These are added to a single resistance $R_a = R_1 + R_2$ and the circuit reduces to that shown in Fig. 2-34(b). R_a is now in parallel with R_3 and they are combined to a single resistance R_b leading to the circuit of Fig. 2-34(c). The succeeding steps with the relevant resistances are shown in Figs. 2-34(d) to (f). The final circuit contains a resistor R_e in series with the source. R_e is the total resistance seen by the source. The current supplied by the source is $i_s = v_s/R_e$.

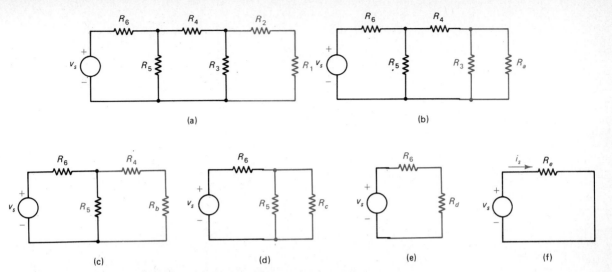

Fig. 2-34 Series-parallel reduction of a circuit.

Example 2-18 Determine the total resistance seen by the source and the voltage across it in the circuit of Fig. 2-35(a).

Solution The successive reductions are shown in Figs. 2-35(b) through (e). The total resistance seen by the source is 7.5 Ω and the voltage across it is $10 \times 7.5 = 75$ V. ∎

Fig. 2-35 Circuit for Example 2-18.

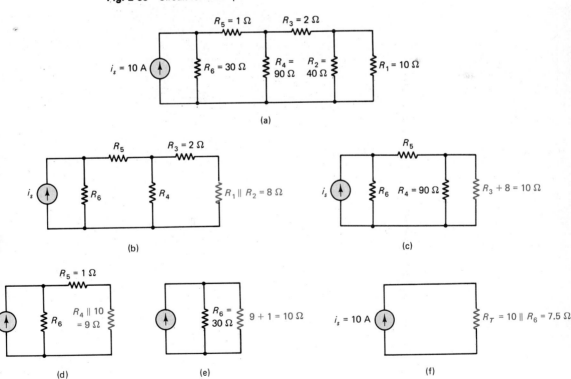

2-5 Series-Parallel and Ladder Networks

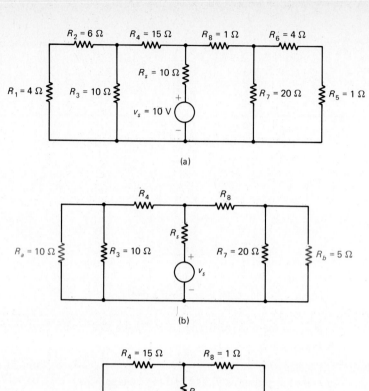

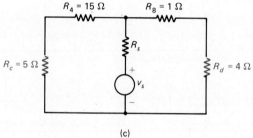

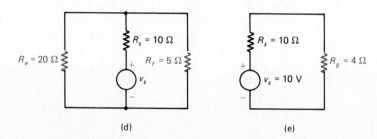

Fig. 2-36 Circuit for Example 2-19.

Example 2-19 Determine the power supplied by the 10-V source in Fig. 2-36(a).

Solution This circuit has two sections, one on either side of the source. Each section is reduced separately at first as shown in the diagrams of Figs. 2-36(b) to (c). Then the two resistors R_e and R_f in Fig. 2-36(d) are in parallel and combined into R_g as shown in Fig. 2-36(e). R_g is in series with R_s.

- Total resistance: $R_T = R_s + R_g = 14\ \Omega$
- Current from the source: $i_s = v_s/R_T = 0.714$ A
- Power delivered: $v_s i_s = 7.14$ W

Series and Parallel Circuits

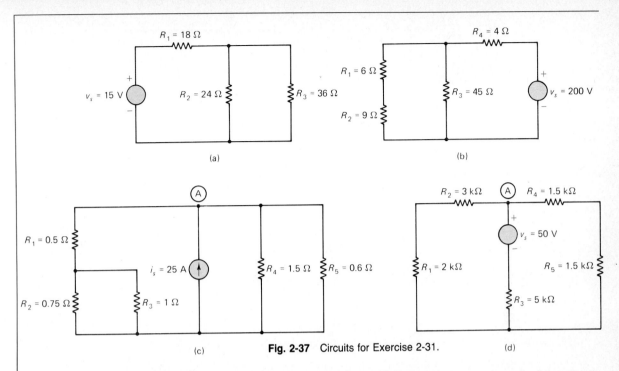

Fig. 2-37 Circuits for Exercise 2-31.

Exercise 2-31 Calculate the total resistance and the power supplied by the source in each of the circuits of Fig. 2-37.

Determination of the Branch Currents

If the aim of the analysis is to find the currents in some or all the branches of the circuit, then it is necessary to retrace a path through the intermediate steps of reduction. The branch currents (and voltages) are usually calculated by using the current and voltage division equations.

As an example, Fig. 2-38 shows the calculation of branch currents in the circuit of Fig. 2-34, which was reduced earlier by series-parallel combinations. Current i_s splits into i_5 and i_c given by

$$i_5 = \left(\frac{R_c}{R_5 + R_c}\right) i_s$$

$$i_c = \left(\frac{R_5}{R_5 + R_c}\right) i_s$$

Current i_c splits into i_3 and i_a given by

$$i_3 = \left(\frac{R_a}{R_3 + R_a}\right) i_c$$

$$i_a = \left(\frac{R_3}{R_3 + R_a}\right) i_c$$

All the branch currents in the circuit of Fig. 2-34 are now known.

2-5 Series-Parallel and Ladder Networks

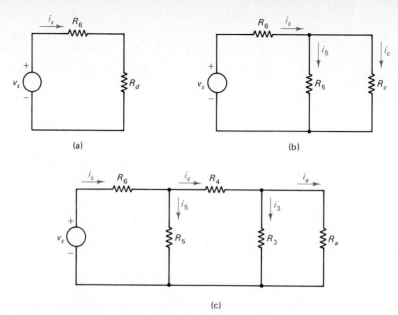

Fig. 2-38 Calculation of branch currents by current division.

Example 2-20 Determine the currents in all the branches of the circuit of Fig. 2-39(a).

Solution Figures 2-39(b) through (e) show the reduction of the circuit to a single resistance. Figures 2-39(e) through (h) show the calculations of the branch currents by retracing the reduction steps. Current division has been used in each step. Figure 2-39(h) shows the original circuit with the values of the branch currents. ■

Exercise 2-32 Evaluate the branch currents in the circuits of Fig. 2-37.

In applying series-parallel reductions, it is important to start from a point farthest from the source and work *towards* the source. The source branch remains unaffected throughout the reduction. In some circuits, the resistance seen from a pair of terminals is to be calculated and the source cannot be explicitly shown. In such cases, the required resistance is what would be seen by a source if it were connected to the terminal pair.

The series-parallel reduction procedure is efficient when the quantities to be determined are those associated with the source itself: the resistance seen by the source, the current or voltage of the source, and the power supplied by the source. When the currents in individual branches are to be determined, the procedure tends to be inefficient and tedious. The ladder network procedure is more convenient in such cases.

2-5-2 Ladder Network Procedure

Networks in which the configuration of branches is alternately vertical and horizontal resemble a ladder and are called *ladder networks*.

The principle underlying the ladder network procedure is as follows. Suppose the currents and voltages in a network driven by a source v'_s have been determined. What is the effect on those values if the source v'_s is replaced by one with voltage $v_s = Kv'_s$?

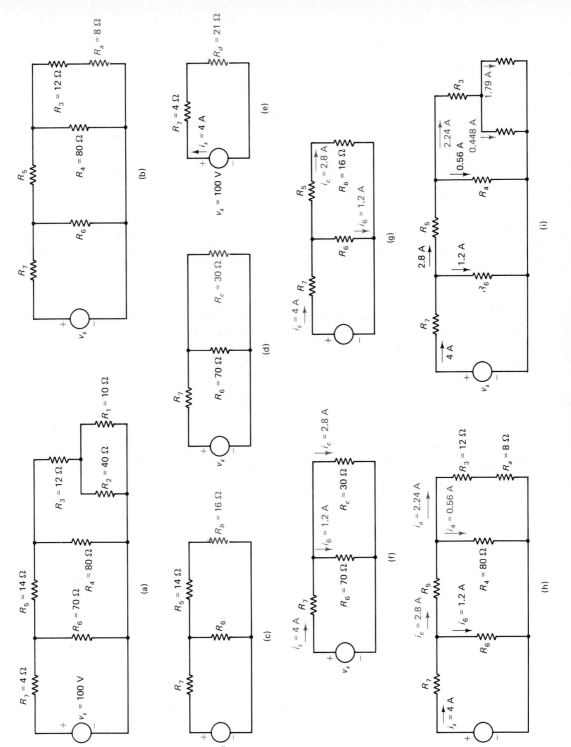

Fig. 2-39 Circuit for Example 2-20.

2-5 Series-Parallel and Ladder Networks

Since the network is linear, the current and voltage in each branch change by a factor $K = v_s/v'_s$.

Exercise 2-33 Return to the circuits of Fig. 2-37 and replace the source in Fig. 2-37(a) by a 115-V source and the source in Fig. 2-37(c) by a 15-mA source. Determine the new values of branch currents and voltages by using the multiplication factor K. Verify the new results by direct calculations on the circuits.

The ladder network method starts by *supposing* that the value of the current (or voltage) in the branch farthest from the source as 1 A (or 1 V). Since the given source v_s cannot conform to the above supposition (except of course by accident), an unknown source v'_s is used in place of v_s. Starting with the branch where a unit current or voltage was assumed, KCL, KVL, and Ohm's law are used as appropriate and we move step by step towards the source. This leads to a set of values of branch currents i'_n and voltages v'_n (where $n = 1, 2, 3, \ldots$ denotes the different branches) and a source voltage v'_s. This means that

if **the circuit were driven by a source v'_s, *then* the branch currents and voltages are given by i'_n and v'_n obtained in the above calculations.**

If, on the other hand, the circuit were driven by the original source v_s, then the branch currents and voltages i'_n and v'_n should be multiplied by the factor

$$K = v_s/v'_s$$

Even though the above statements have been made in connection with a voltage source driving the circuit, the principle is equally valid when the given circuit is driven by a current source. The factor is then i_s/i'_s.

The ladder network procedure is illustrated by the following example.

Example 2-21 Determine the voltages and currents in the circuit of Fig. 2-40(a).

Solution The circuit is shown redrawn in Fig. 2-40(b) after assigning currents and voltages. An unknown source voltage v'_s has replaced the given voltage source. Assume $i'_1 = 1$ A.

Voltages v'_1 and v'_2 are calculated by using Ohm's law:

$$v'_1 = R_1 i'_1 = 1 \text{ V} \quad \text{and} \quad v'_2 = R_2 i'_1 = 2 \text{ V}$$

The voltage across R_3 is (from KVL) the sum of the voltages across R_1 and R_2:

$$v'_3 = v'_1 + v'_2 = 3 \text{ V}$$

The current in R_3 is obtained by Ohm's law:

$$i'_3 = v'_3/R_3 = 0.75 \text{ A}$$

The current in R_4 is (from KCL) the sum of the currents in R_3 and R_2:

$$i'_4 = i'_1 + i'_3 = 1.75 \text{ A}$$

The procedure continues in the above manner. The voltage or current in a branch is calculated at each step by using Ohm's law, KVL, or KCL, depending on the available information at that point.

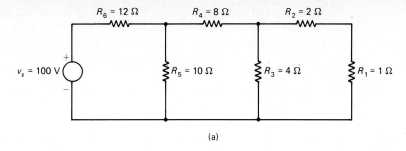

(a)

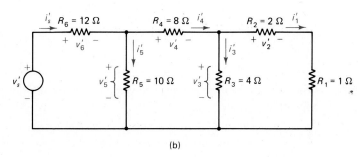

(b)

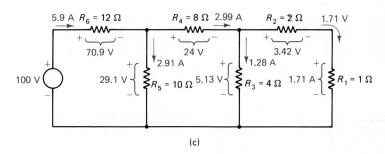

(c)

Fig. 2-40 Ladder network of Example 2-21.

$$v'_4 = R_4 i'_4 = 14 \text{ V}$$
$$v'_5 = v'_4 + v'_3 = 17 \text{ V (KVL)}$$
$$i'_5 = v'_5/R_5 = 1.7 \text{ A}$$
$$i'_s = i'_4 + i'_5 = 3.45 \text{ A (KCL)}$$
$$v'_6 = R_6 i'_s = 41.4 \text{ V}$$
$$v'_s = v'_6 + v'_5 = 58.4 \text{ V (KVL)}$$

If $v'_s = 58.4$ V, then the voltages and currents in the circuit are those calculated above. But the given voltage source is $v_s = 100$ V. Therefore, the branch voltages and currents must be multiplied by the factor

$$K = v_s/v'_s = 100/58.4 = 1.71$$

Figure 2-40(C) shows the given circuit with the currents and voltages when driven by the 100-V source. The total resistance seen by the source can also be calculated.

$$R_T = v_s/i_s = 100/5.90 = 16.9 \text{ }\Omega$$

There is no need to retrace the reduction steps in the ladder network procedure (in contrast to the series-parallel reduction procedure). Also, it is usually not necessary to write each intermediate step explicitly as was done in the last example. The calculations can be done directly on the circuit diagram and the problem solved without writing numerous equations.

Even though an assumption of unit *current* in a branch was the starting point of the last example, it is equally valid to start by assuming a unit *voltage* in a branch, which may be more convenient in some problems.

In some problems, the source voltage or current is not specified and the objective is to determine the *ratio* of the voltage or current in some branch to the source voltage or source current. Such a ratio does not depend on the actual value of the source voltage or current. For instance, it can be verified that the ratios i_1'/v_s' and i_1/v_s are equal to each other in Example 2-21.

Exercise 2-34 Determine the branch currents and voltages in the circuit of Fig. 2-41 by using the ladder network procedure.

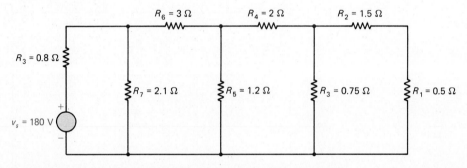

Fig. 2-41 Circuit for Exercise 2-41.

Exercise 2-35 Use the ladder network procedure in the circuit of Fig. 2-39(a).

Example 2-22 Use the ladder network procedure in the circuit of Fig. 2-42(a) and determine v_o/v_s.

Solution The circuit extends to both sides of the source. It is necessary to start by assuming a current (or voltage) in both resistors R_d and R_4 (farthest from the source). If both the currents (or voltages) are assumed to have a value of unity, the result is two different values of v_s' (one from the left half and the other from the right half), which would lead to a conflict. It is therefore necessary to assume a current of 1 A (or voltage of 1 V) in one of the two resistors and an *unknown* current of x A (or voltage of x V) in the other. Since v_o/v_s is asked for in the present problem, start by assuming

$$v_o' = 1 \text{ V} \quad \text{and} \quad i_d' = x \text{ A}$$

The ladder network procedure is used separately on the two sections of the network until point P (which is where the two halves of the circuit meet) is reached.

Right Half:
From Fig. 2-42(b): $v_o' = 1$ V $i_1' = 2$ A $v_3' = R_3 i_1' = 3$ V
$v_2' = v_3' + v_o' = 4$ V $i_2' = v_2'/R_2 = 8$ A
$i_3' = i_2' + i_1' = 10$ A $v_1' = R_1 i_3' = 2.0$ V
$v_P' = v_1' + v_2' = 6.0$ V

Series and Parallel Circuits

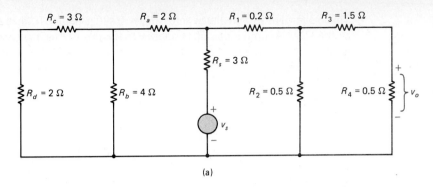

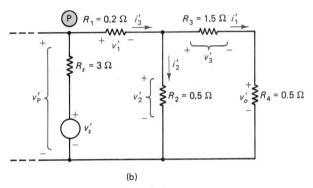

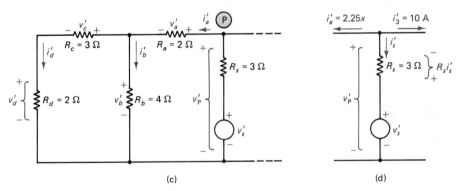

Fig. 2-42 Circuit for Example 2-22.

Left Half:
From Fig. 2-42(c): $i'_d = x$ A $\qquad v'_d = R_d i'_d = 2x$ V
$\qquad\qquad\qquad v'_c = R_c i'_d = 3x$ V $\qquad v'_b = v'_c + v'_d = 5x$ V
$\qquad\qquad\qquad i'_b = v'_b/R_b = 1.25x$ A
$\qquad\qquad\qquad i'_a = i'_b + i'_d = 2.25x$ A
$\qquad\qquad\qquad v'_a = R_a i'_a = 4.5x$ V
$\qquad\qquad\qquad v'_P = v'_b + v'_a = 9.5x$ V

The two values of v'_P, 6.0 V and 9.5x V, must be equal to each other since KVL dictates the voltage at P to have a unique value for a given source. Therefore,

$$9.5x = 6.0 \quad \text{or} \quad x = 0.632 \text{ A}$$

2-5 Series-Parallel and Ladder Networks

This means that if the voltage across R_4 is 1 V (as assumed), then the current in R_d must be 0.632 A. A similar comment applies to the other branch currents and voltages in the left half of the circuit.

Now consider the branch containing the source, as shown in Fig. 2-42(d):

$$i'_s = i'_a + i'_3 = 2.25x + 10 = 11.4 \text{ A} \qquad \text{since } x = 0.632$$
$$v'_s = v'_P + R_s i'_s = 40.3 \text{ V}$$

Therefore, $v'_s = 40.3$ V when $v'_o = 1$ V, or

$$v_o/v_s = v'_o/v'_s = 1/40.3 = 2.48 \times 10^{-2} \qquad \blacksquare$$

Exercise 2-36 Determine the branch currents and voltages of the circuits in Fig. 2-37(c) and (d) by using the ladder network procedure.

If the series-parallel reduction procedure is compared with the ladder network procedure, the former is efficient when only the quantities associated with the source terminals are of interest. When the aim of the analysis is to find the currents and voltages elsewhere in the circuit, the ladder network procedure is far more efficient since it leads to the desired results in a single series of calculations.

2-6 CIRCUIT WITH A NONLINEAR COMPONENT

A nonlinear series circuit is considered in this section to show that its solution requires a different approach from the linear series circuit.

The device labeled NL in Fig. 2-43(a) is nonlinear, that is, the current *versus* voltage for the device is not a straight line passing through the origin. The relationship between its voltage v_n and current i_n is given by the *characteristic curve* of Fig. 2-43(b). Suppose it is desired to solve for the values of v_n and i_n when v_s and R_s are given.

Any solution to the circuit must satisfy:

1. The *device constraint;* that is, any pair of values (v_n, i_n) must give a point on the device characteristic in order to be a valid solution
2. The *KVL equation* of the closed path is

$$-v_s + R_s i_n + v_n = 0$$

which is rewritten as

$$i_n = \frac{v_s}{R_s} - \frac{v_n}{R_s} \qquad (2\text{-}22)$$

Equation (2-22) describes a straight line, called the *load line,* as shown in Fig. 2-43(c), on a set of axes with i_n as ordinate and v_n as abscissa. It has a slope $= -1/R_s$.

The load line joins point v_s/R_s on the vertical axis [obtained by putting $v_n = 0$ in Eq. (2-22)] to point v_s on the horizontal axis [obtained by putting $i_n = 0$ in Eq. (2-22)]. The load line is a *graphical representation* of the constraint due to KVL in the given circuit.

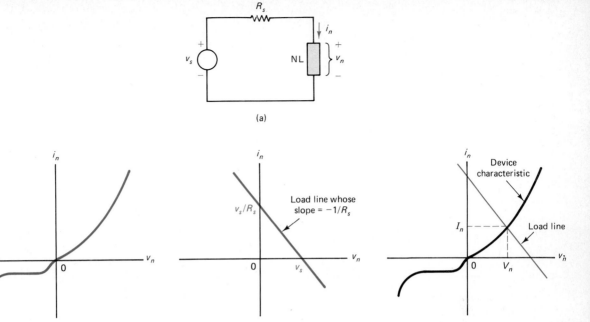

Fig. 2-43 (a) Circuit with a nonlinear component (NL). (b) Characteristic of the nonlinear device. (c) Load line. (d) Intersection of load line and device characteristic.

The *point of intersection* of the load line with the device characteristic, as shown in Fig. 2-43(d), satisfies the device constraint since it lies on the device characteristic and the KVL constraint since it lies on the load line. Therefore,

the coordinates of the point of intersection give the values of v_n and i_n for the circuit.

Example 2-23 Find the voltage across the nonlinear device and the current in the circuit of Fig. 2-44(a) for the following values of v_s: (a) 6 V, (b) 2 V, and (c) −4 V. The device characteristic is shown in Fig. 2-44(b).

Solution Three load lines are needed since three different values of v_s are given. Draw each load line by joining the point at v_s on the horizontal axis to the point at $v_s/R_s = 0.05v_s$ on the vertical axis, as shown in Fig. 2-44(c). Find v_n at the intersection of each load line with the device characteristic.
(a) $v_s = 6$ V and $0.05v_s = 0.3$ A. The point of intersection is at $v_n = 3.5$ V and $i_n = 0.125$ A.
(b) $v_s = 2$ V and $0.05v_s = 0.1$ A. $v_n = 1.5$ V and $i_n = 0.025$ A.
(c) $v_s = -4$ V and $0.05v_s = -0.2$ A. $v_n = -2.5$ V and $i_n = -0.075$ A. ∎

Exercise 2-37 Suppose v_s in the circuit of the last example is kept fixed at 3 V. For each of the following values of R_s, determine the voltage across the nonlinear device and the current in the circuit: (a) 10 Ω, (b) 15 Ω, and (c) 25 Ω.

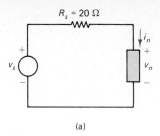

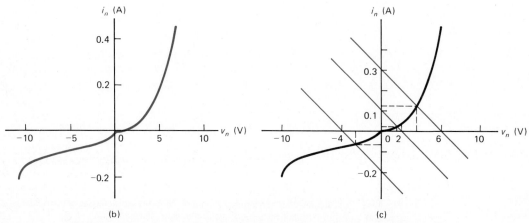

Fig. 2-44 (a) Circuit, (b) device characteristic, and (c) load lines for Example 2-23.

2-7 SUMMARY OF CHAPTER

In a series combination of resistors, the total resistance equals the sum of the individual resistances, and the current is obtained by dividing the voltage applied to the circuit by the total resistance. A voltage divider is a special case of a series circuit made up of two resistors in series. The voltage across each resistance equals the ratio of that resistance to the total resistance multiplied by the applied voltage. When one resistance is much larger than the other, the total resistance can be taken as approximately equal to the larger resistance and the applied voltage is essentially across the larger resistance.

In a parallel combination of resistors, the total conductance equals the sum of the individual conductances, and the voltage is obtained by dividing the current applied to the circuit by the total conductance. A current divider is a special case of a parallel circuit made up of two resistors in parallel. The current in each resistance is proportional to the resistance of the other branch. The larger current flows through the smaller resistance. When one resistance is much larger than the other, the total resistance can be taken as approximately equal to the smaller resistance and the applied current flows essentially through the smaller resistance.

Duality is the correspondence between parameters and relationships in circuits. The following are dual pairs: current and voltage, resistance and conductance, series connection and parallel connection, open circuit and short circuit. Duality is useful in extending the results obtained for one circuit configuration to the dual configuration.

Circuits with dependent sources are analyzed in the same manner as those with only independent sources. Dependent sources occur in the model of electronic devices and circuits. One possible model of a basic amplifier uses an input resistance and a dependent voltage source in series with an output resistance. When a signal source is connected to the input and a load resistance is connected to the output, the amplifier circuit consists of two series circuits.

A circuit made up of both series and parallel combinations of resistors and driven by a single source can be analyzed either by a series-parallel reduction procedure or by a ladder network procedure. In the series-parallel reduction procedure, the resistances are combined alternately in series and in parallel until the circuit reduces to a single source connected to a single resistance. The determination of the individual branch currents and voltages requires the retracing of the steps used in the reduction process and applying the current division formula. In the ladder network procedure, an initial assumption is made about the voltage or current in the branch farthest from the source. KCL, KVL, and Ohm's law are used repeatedly to obtain the values of the currents and voltages in all the branches consistent with the initial assumption. Then the concept of linearity is invoked to find the branch currents and voltages for the specific value of the source voltage or current driving the given network. The ladder network has the advantage that the retracing of steps (needed in the reduction procedure) is not required.

Answers to Exercises

2-1 $R_T = 140\ \Omega$ and $i_a = 2.5$ A. Voltages across the resistors are 50 V, 37.5 V, 125 V, 75 V, and 62.5 V. (b) $R_T = 10.5\ \Omega$ and $i_b = 19.05$ A. Voltages across the resistors are 38.1 V, 152.4 V, and 9.52 V. (c) $R_T = 40\ \Omega$ and $i_c = 0.30$ A. Voltages across the resistors are 1.5 V, 4.8 V, 1.8 V, and 3.9 V.

2-2 $i = 5.66$ A; $R_T = 40\ \Omega$; $v_s = 226$ V. Power dissipated by the resistors are 513 W, 416 W, 160 W, and 192 W. Power delivered by $v_s = 1279$ W.

2-3 $i_s = 3$ A; $R_T = 66.7\ \Omega$; $R_x = 50.7\ \Omega$.

2-4 Current decreases by a factor N. Power supplied by the source also decreases by a factor N.

2-5 Current increases by a factor N. Power increases by a factor N^2.

2-6 (a) $v_2 = 8$ V and (b) $v_2 = 30$ V.

2-7 $v_T = 5.05$ V.

2-8 $R_T \approx R_1 = 5000\ \Omega$. $v_T \approx R_1 i \approx 5$ V. Error is 0.99 percent.

2-9 $v_x = -R_c i - v_c - R_b i = -350$ V.

2-10 $v_{AB} = (R_2 v_a + R_1 v_b)/(R_1 + R_2)$.

2-11 (a) $i = (v_b - v_a)/R_T = -2$ A. (b) $i = (v_a - v_b + v_c)/R_T = 1.82$ A. (c) $i = (-v_a + v_b + v_c + v_d)/R_T = 14$ A. (d) $i = (v_b - v_a)/R_T = 0.8$ A.

2-12 (a) $v_{AB} = v_a + R_2 i = v_b - R_1 i = 7$ V. (b) $v_{AB} = R_1 i + v_b + R_2 i = v_a - R_3 i + v_c = 77.5$ V. (c) $v_{AB} = v_a - v_b - v_c = v_d - R_1 i = -175$ V. (d) $v_{AB} = +R_2 i + R_3 i - v_b = -v_a - R_1 i = -66$ V.

2-13 $i = 3$ A. $v_x = v_a - v_b + v_c - R_T i = -210$ V.

2-14 (a) $v = 3.16$ V. Currents are 1.05 A, 2.10 A, 0.526 A, and 6.32 A. (b) $v = 0.909$ V. Currents are 2.73 A, 1.36 A, 5.45 A, and 0.455 A.

2-15 $v = 28.3$ V and $i_s = 10.6$ A. Powers dissipated are 20 W, 40 W, 80 W, and 160 W.

2-16 $v = 12$ V; $G_T = 4.17$ S; $R_x = 0.258\ \Omega$.

2-17 (a) 13.2 Ω and (b) 5.76 Ω.

2-18 (a) 26.5 A and 73.5 A.   (b) 36 A and 64 A.

2-19 $i_s = 3.33$ A. $R_x/(R_x + R_1) = 0.5/3.33$. $R_x = 5.30\ \Omega$.

2-20 $v_m = 100$ mV and $i_T = 20.002$ A.

2-21 $R_T \approx R_1 = 5 \times 10^{-3}\ \Omega$ (exact value is $4.9995 \times 10^{-3}\ \Omega$). $i_T \approx i_1 = 20$ A (exact value is 20.002 A). Error is 0.01 percent.

2-22 $i_p = i_s - G_1v + i_a = 34.2$ A.

2-23 (a) 0.375 V and (b) -1.5 V.

2-24 (a) $i_x = i_a - G_1v = i_b + G_2v = 6.5$ A.
(b) $i_x = -G_1v - i_a = -i_b + i_c + G_2v = -5$ A.

2-25 $v = (i_q - i_b)/G_2 = 0.5$ V. $G_xv = -i_q + i_a - G_1v$. $G_x = 2$ S.

2-26 $i_1 = v_s/(R_1 + R_2 - 8) = 2.38$ A.

2-27 $v_1 = (-i_b + i_a)/(G_1 + G_2 + 0.2) = 46.5$ V.

2-28 Voltage gain $= AR_iR_L/(R_s + R_i)(R_o + R_L)$.

2-29 $v_2 = A[R_L/(R_L + R_o)][R_i/(R_i + R_s)]v_s$, which gives the same expression for voltage gain as in Exercise 2-41.

2-30 $i_1 = 0.25 \times 10^{-3}\ v_s$; $v_i = 0.375\ v_s$; $v_2 = 5.25\ v_s$; $i_2 = 0.75 \times 10^{-3}\ v_s$. Voltage gain $= 5.25$. Current gain $= 3.0$.

2-31 (a) $R_T = 32.4\ \Omega$; $i_s = 0.463$ A; $P = 6.94$ W.   (b) $R_T = 15.25\ \Omega$; $i_s = 13.1$ A; $P = 2623$ W.   (c) $R_T = 0.293\ \Omega$; $v_s = 7.32$ V; $P = 183$ W.   (d) $R_T = 6.88$ kΩ; $i_s = 7.27$ mA; $P = 0.364$ W.

2-32 (The subscripts of the current symbols here match those of the resistors in Fig. 2-37.)
(a) $i_s = i_1 = 0.463$ A; $i_2 = 0.278$ A; $i_3 = 0.185$ A.   (b) $i_s = i_4 = 13.1$ A; $i_1 = i_2 = 9.82$ A; $i_3 = 3.28$ A.   (c) $i_1 = 7.90$ A; $i_2 = 4.51$ A; $i_3 = 3.39$ A; $i_4 = 4.89$ A; $i_5 = 12.2$ A.
(d) $i_s = i_3 = 7.27$ mA; $i_2 = 2.73$ mA; $i_4 = 4.54$ mA.

2-33 (a) Multiply all currents and voltages in the answers to Exercise 2-32(a) by the factor $115/15 = 7.67$.   (b) Multiply all currents and voltages in the answers to Exercise 2-32(c) by the factor $15 \times 10^{-3}/25 = 6 \times 10^{-4}$.

2-34 The same notation scheme as in Example 2-21 is used in the answers to this and the next two Exercises. Assume $i'_1 = 1$ A. Then $v'_1 = 0.5$ V; $v'_2 = 1.5$ V; $v'_3 = 2$ V; $i'_3 = 2.67$ A; $i'_4 = 3.67$ A; $v'_4 = 7.33$ V; $v'_5 = 9.33$ V; $i'_5 = 7.78$ A; $i'_6 = 11.4$ A; $v'_6 = 34.3$ V; $v'_7 = 43.6$; $i'_7 = 20.8$ A; $i'_s = 32.2$ A; voltage across $R_s = 25.8$ V; $v'_s = 69.4$ V. Multiply all the above voltage and current values by the ratio $= 180/69.4 = 2.59$ to get the results for the given 180-V source.

2-35 Assume $i'_1 = 1$ A. Then $v'_1 = v'_2 = 10$ V; $i'_2 = 0.25$ A; $i'_3 = 1.25$ A; $v'_3 = 15$ V; $v'_4 = 25$ V; $i'_4 = 0.312$ A; $i'_5 = 1.56$ A; $v'_5 = 21.9$ V; $v'_6 = 46.9$ V; $i'_6 = 0.670$ A; $i'_7 = 2.23$ A; $v'_7 = 8.93$ V; $v'_s = 55.8$ V. Multiply all the above voltage and current values by the ratio $100/55.8 = 1.79$ to get the results for the given 100-V source.

2-36 (a) *Right Half:* Assume $i'_5 = 1$ A. Then $v'_4 = v'_5 = 0.6$ V; $i'_4 = 0.4$ A; $i'_a = 1.4$ A; $v'_P = 0.6$ V. *Left Half:* Assume $i'_2 = x$. Then $v'_2 = v'_3 = 0.75x$ V; $i'_3 = 0.75x$ A; $i'_1 = 1.75x$ A; $v'_1 = 0.875x$ V; $v'_P = 1.625x$ V. $x = 0.369$; $i'_1 = 0.646$ A; $i'_s = 2.05$ A. In the above answers, use $x = 0.369$ where appropriate, and also multiply all the voltage and current values by the ratio $25/2.05 = 12.2$ to get the results for the given 25-A source.   (b) *Right Half:* Assume $i'_5 = 1$ mA. Milliamperes are used since the product of 1 mA and 1 kΩ gives 1 V, which is rather convenient. $v'_5 = 1.5$ V; $v'_4 = 1.5$ V; $v'_P = 3$ V. *Left Half:* Assume $i'_1 = 1$ mA. $v'_2 = 2x$ V; $v'_2 = 3x$ V; $v'_P = 5x$. $x = 0.6$; $i'_1 = 0.6$ mA; $i'_s = 1.6$ mA; $v'_3 = 8$ V; $v'_s = 11$ V. Use $x = 0.6$

where appropriate and also multiply all the above voltage and current values by the ratio $50/11 = 4.54$ to get the results for the given 50-V source.

2-37 (a) $v_n = 2.5$ V and $i_n = 0.06$ A.   (b) $v_n = 2.3$ V and $i_n = 0.05$ A.   (c) $v_n = 2.1$ V and $i_n = 0.04$ A.

PROBLEMS

Sec. 2-1 The Series Circuit

2-1 Determine (a) the total resistance, (b) the current supplied by the source, and (c) the voltage labeled v_x in each of the circuits shown in Fig. 2-45.

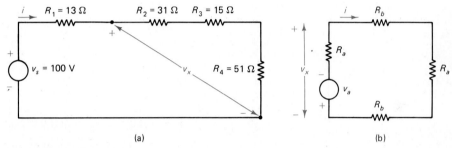

Fig. 2-45 Circuits for Problem 2-1.

2-2 If the power supplied by the voltage source in the circuit of Fig. 2-46 is 75 W, determine (a) the value of R_x and (b) the power dissipated by each of the resistors.

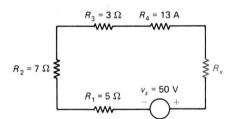

Fig. 2-46 Circuit for Problem 2-2.

2-3 If the voltage across the resistor R_3 in the circuit of Fig. 2-47 is 3.6 V, determine the voltage across each of the other resistors and the voltage v_s.

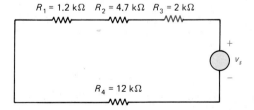

Fig. 2-47 Circuit for Problem 2-3.

2-4 A source with $v_s = 110$ V is available. It is necessary to use a voltage divider to provide a voltage of 6.3 V to some load resistance. Design the voltage divider. Assume that the maximum power that can be dissipated by available resistors is 0.5 W.

2-5 A voltmeter with a multiple range of voltages to be measured can be designed by using a D'Arsonval movement and several resistors as indicated in Fig. 2-48. The value placed next to each contact position in the diagram represents the maximum voltage to be measured for that position. If the resistance of the movement is 40 Ω and the maximum voltage it can safely withstand is 80 mV, determine the values of the resistors.

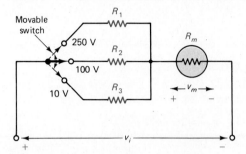

Fig. 2-48 Circuit for Problem 2-5.

2-6 The voltage across two resistors R_1 and R_2 in series is 12 V. For each of the following values of the ratio R_2/R_1, find the error introduced in the value of the voltage across R_2 if approximate calculations are made by neglecting the smaller resistance: (a) 0.01, (b) 0.2, (c) 2, and (d) 2000.

2-7 Consider the circuit shown in Fig. 2-49, where the load resistance R_L is variable, and the source voltage v_s and the resistance R_s remain fixed. (a) Obtain an expression for the power P_L dissipated in the load resistance in terms of v_s, R_s, and R_L. (b) By evaluating the derivative dP_L/dR_L and equating it to zero, obtain an expression for the value of R_L at which P_L becomes a maximum, and (c) obtain an expression for the maximum value of P_L.

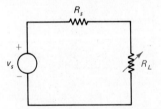

Fig. 2-49 Circuit for Problem 2-7.

2-8 Obtain expressions for the circuit current and the voltage across each resistor in the circuit of Fig. 2-50.

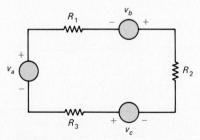

Fig. 2-50 Circuit for Problem 2-8.

2-9 Determine the voltage v_x in each of the circuits shown in Fig. 2-51.

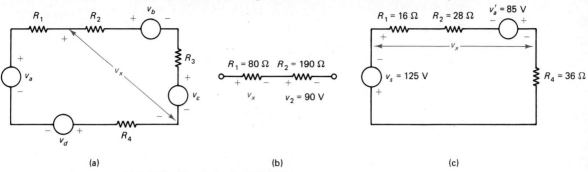

Fig. 2-51 Circuits for Problem 2-9.

2-10 If the voltage v_{oc} in the circuit of Fig. 2-52 is 100 V, calculate the value of the source voltage v_b.

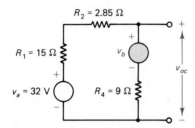

Fig. 2-52 Circuit for Problem 2-10.

Sec. 2-2 The Parallel Circuit

2-11 Determine (a) the total conductance, (b) the voltage across the source, and (c) the current i_x in each of the circuits shown in Fig. 2-53.

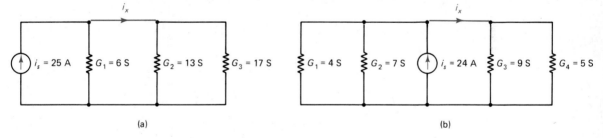

Fig. 2-53 Circuits for Problem 2-11.

2-12 If the power supplied by the current source in the circuit of Fig. 2-54 is 75 W, determine (a) the value of G_x and (b) the power dissipated in each resistor.

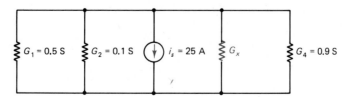

Fig. 2-54 Circuit for Problem 2-12.

Problems 95

2-13 If the current in the conductance G_3 in the circuit of Fig. 2-55 is 3.6 A, determine the current in each of the other conductances and the source current i_s.

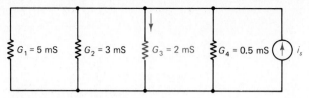

Fig. 2-55 Circuit for Problem 2-13.

2-14 A current source with $i_s = 20$ mA is available. It is necessary to use a current divider to provide a current of 5 mA to some load resistance. Design the current divider. Assume that the maximum power that can be dissipated by the available resistors is 0.5 W.

2-15 An ammeter with a multiple range of currents to be measured can be designed by using a D'Arsonval movement and several resistors, as shown in Fig. 2-56. The value next to each contact position in the diagram represents the maximum current to be measured for that position. If the resistance of the movement is 40 Ω and the maximum current it can safely withstand is 5 mA, determine the values of the resistors.

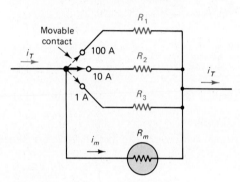

Fig. 2-56 Circuit for Problem 2-15.

2-16 The current through a parallel combination of two resistors R_1 and R_2 is 8 A. For each of the following values of the ratio R_2/R_1, find the error introduced in the value of the current in R_2 if approximate calculations are made by neglecting the larger resistance: (a) 0.01, (b) 0.2, and (c) 2000.

2-17 Consider the circuit shown in Fig. 2-57, where the conductance G_L is variable, and the source current i_s and the conductance G_s remain fixed. (a) Obtain an expression for the power P_L dissipated in G_L in terms of i_s, G_s, and G_L. (b) By equating the derivative dP_L/dG_L to zero, obtain an expression for the value of G_L at which P_L becomes a maximum, and (c) an expression for the maximum value of P_L.

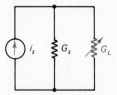

Fig. 2-57 Circuit for Problem 2-17.

Series and Parallel Circuits

2-18 Obtain expressions for the voltage across the circuit and the current in each conductance for the circuit of Fig. 2-58.

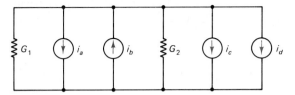

Fig. 2-58 Circuit for Problem 2-18.

2-19 Determine the current i_x in each of the circuits shown in Fig. 2-59.

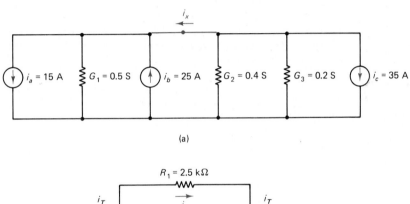

Fig. 2-59 Circuits for Problem 2-19.

2-20 If the current i_{sc} for the circuit of Fig. 2-60 is to be 30 A, calculate the value of the source current i_b.

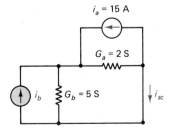

Fig. 2-60 Circuit for Problem 2-20.

Problems

2-21 Each circuit shown in Fig. 2-61 can be converted to a series circuit by changing the current source in parallel with a resistance into an equivalent voltage source in series with a resistance. (a) Determine the current i_1 after performing such a conversion. (b) Determine the power supplied by the two voltage sources v_b and v_c. (c) Return to the original circuit and determine the voltage across the current source and the power supplied by it.

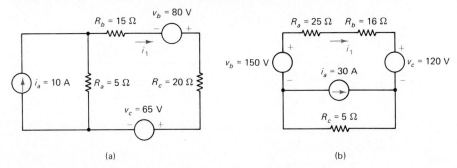

Fig. 2-61 Circuits for Problem 2-21.

2-22 Each circuit shown in Fig. 2-62 can be converted to a parallel circuit by changing the voltage source in series with a resistance into an equivalent current source in parallel with a resistance. (a) Determine the voltage v_1. (b) Determine the power supplied by the two current sources i_a and i_b. (c) Return to the original circuit and determine the current in the voltage source and the power supplied by it.

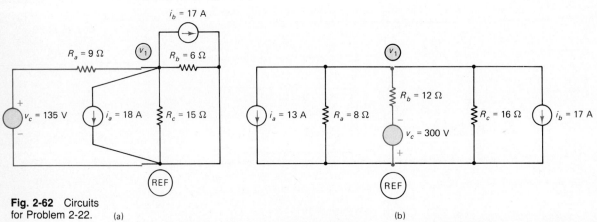

Fig. 2-62 Circuits for Problem 2-22.

Sec. 2-3 Circuits with Dependent Sources

2-23 For each of the circuits shown in Fig. 2-63, determine (a) the current in the circuit and (b) the power delivered by the dependent source.

Fig. 2-63 Circuits for Problem 2-23.

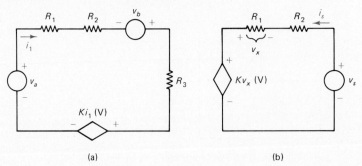

Series and Parallel Circuits

2-24 For each of the circuits shown in Fig. 2-64, determine (a) the voltage across the circuit and (b) the power delivered by the dependent source.

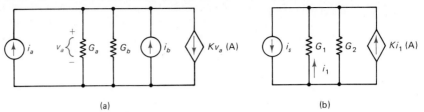

Fig. 2-64 Circuits for Problem 2-24.

2-25 Each circuit shown in Fig. 2-65 can be converted to a series circuit by changing the dependent current source in parallel with a resistance into an equivalent dependent voltage source in series with a resistance. Determine the current in each circuit after performing such a conversion.

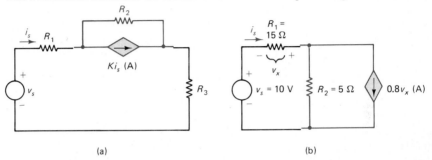

Fig. 2-65 Circuits for Problem 2-25.

2-26 The circuit shown in Fig. 2-66 can be converted to a parallel circuit by changing the dependent voltage source in series with a resistance into an equivalent dependent current source in parallel with a resistance. Determine the voltage across the circuit after performing such a conversion.

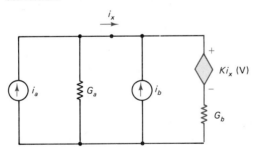

Fig. 2-66 Circuit for Problem 2-26.

2-27 The circuit of Fig. 2-67 shows a model of a transistor amplifier, slightly more complicated than the one discussed in the text. (This model is the same as the hybrid parameter model to be introduced in Chapter 13). Note that the input side can be analyzed as a series circuit and the output side as a parallel circuit. Obtain an expression for (a) the current i_1, (b) the voltage v_2, and (c) the ratio v_2/v_s.

Fig. 2-67 Circuit for Problem 2-27.

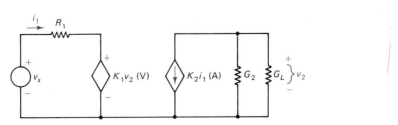

Problems

99

Sec. 2-4 Series-Parallel and Ladder Networks

2-28 Calculate the total resistance seen by the source and the current supplied by it in each of the circuits shown in Fig. 2-68.

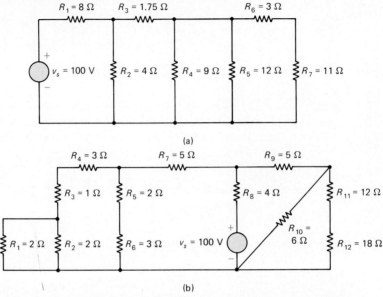

Fig. 2-68 Circuits for Problem 2-28.

2-29 Determine all the branch currents in the circuits of Problem 2-28.

2-30 Use the ladder network procedure in each of the circuits of Problem 2-28 to determine the branch currents.

2-31 Determine the branch currents and voltages in the circuit of Fig. 2-69 when it is driven by a voltage source $v_s = 100$ V. What are the answers if the given circuit is driven by a current source $i_s = 100$ A?

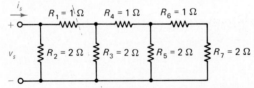

Fig. 2-69 Circuit for Problem 2-31.

2-32 Start by assuming that $v_{10} = 1$ V and $v_9 = y$ V in the circuit of Fig. 2-70, work toward the source, and calculate all the branch currents and voltages in the circuit. Be sure to determine the value of y and use it in the appropriate voltages and currents.

Fig. 2-70 Circuit for Problem 2-32.

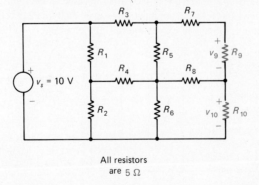

All resistors are 5 Ω

2-33 A slide-wire resistance, or *rheostat* (shown in Fig. 2-71), is used as a voltage divider in laboratory experiments. It consists of a coil of wire wound uniformly over an insulating cylinder with a sliding contact C on the coil. The resistance between any two points on the coil is assumed to be proportional to the distance between those two points. Assume that the total resistance between points A and B is 20 kΩ. If v_{AB} = 110 V and the contact C is one-fourth of the way down from A, determine the voltage between C and B for the following two cases: (a) no load is connected and (b) R_L = 5 kΩ. Also, calculate (c) the position of the contact so as to have 55 V across a 5-kΩ load.

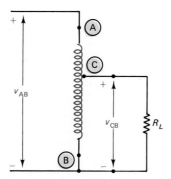

Fig. 2-71 Circuit for Problem 2-33.

2-34 For the circuit of Fig. 2-72(a), obtain an expression for the voltage v_L across R_L.

If the voltage across R_L is to be measured by a voltmeter [connected in parallel with R_L, as shown in Fig. 2-72(b)], the finite resistance R_v of the voltmeter introduces an error in the measurement, since it alters the given circuit. Obtain an expression for the voltage v'_L measured across the load with the voltmeter connected.

Define *percent error* by 100 $[(v_L - v'_L)/v_L]$. Obtain an expression for the percent error.

Suppose an error of 10 percent in the measurement of the voltage v_L is permissible. Obtain an expression for R_v in terms of R_L and R_s for a 10 percent error.

If R_s = 10 kΩ and R_L = 15 kΩ, calculate R_v for a 10-percent error.

Fig. 2-72 Circuits for Problem 2-34.

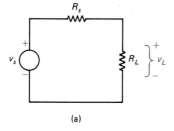

(a)

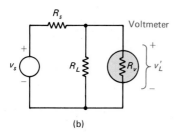

(b)

2-35 For the circuit of Fig. 2-73, points A and B are required to be at the same voltage (with respect to a common reference node) so that if a resistance were connected between the two points, no current would flow in it. Obtain a relationship between the four resistances R_1, R_2, R_3, and R_4 to satisfy the above condition.

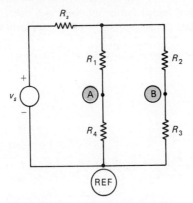

Fig. 2-73 Circuit for Problem 2-35.

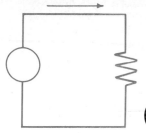

CHAPTER 3
NODAL ANALYSIS

The analysis of circuits, in general, requires writing and solving a set of simultaneous equations. These equations are mathematical expressions of Kirchhoff's laws and voltage-current relationships applied to the circuit and its components. Except in special configurations, such as those considered in Chapter 2, which did not require a set of simultaneous equations for their solution, systematic procedures are necessary to select the proper set of independent variables and write the equations of a given circuit. Two such systematic procedures are provided by nodal analysis (discussed in this chapter) and loop and mesh analysis (presented in Chapter 4).

In a circuit where the number of branches is B, there are $2B$ unknown quantities: B branch currents and B branch voltages. This does not mean, however, that the circuit needs a set of $2B$ simultaneous equations in $2B$ unknowns because the voltages and currents are not all independent of one another. For example, if the current in a resistor is known, the voltage across it is also known through Ohm's law. Also, if the voltages across all but one of the components in a closed path are known, then the voltage across the remaining component can be calculated by using Kirchhoff's voltage law. Similarly, if the currents in all but one of the branches meeting at a node are known, the current in the remaining branch can be calculated by using Kirchhoff's current law. Thus, the number of *independent* voltage and current variables in a circuit with B branches is not necessarily equal to $2B$.

A result from the theory of determinants states that a system of *n equations in n unknowns* gives a unique set of values of the unknowns if and only if the n unknowns form *an independent set*. Thus, the choice of an independent set of variables in a circuit forms the first important step in analysis.

One choice of independent variables for a circuit uses the voltages at the nodes of a circuit measured with respect to a common reference node. Such a choice leads to *nodal analysis*.

3-1 NETWORK TOPOLOGY— CHOICE OF INDEPENDENT VOLTAGES

The number of independent branch voltages and currents in a circuit depends on the manner in which the various components are interconnected to form the circuit and not on what are the individual components. For example, in a series circuit, there is only a single independent current variable regardless of what the actual components in the circuit are. Therefore, it is important to concentrate on the *geometry of interconnection* of the components or *topology* of the network for selecting a set of independent variables in a network.

The topological aspects of a circuit are displayed by the graph of the circuit in which the branches are shown simply as lines connected in the same configuration as in the given circuit (without showing the actual component in each branch).

An example of a circuit and its graph is shown in Fig. 3-1.

Let

$$B = \text{number of branches}$$

and

$$N = \text{number of nodes}$$

in a given circuit. For example, $B = 9$ and $N = 6$ in the graph of Fig. 3-1(b).

A *subset* of the branches of the graph, called a *tree*, is first chosen, so as to satisfy the following two conditions:

(1) the tree should not have any closed paths, and (2) it should be possible to go from any node in the graph to any other node by striking a path through the branches of the tree.

It is possible to draw several different trees for a given graph. Figures 3-1(c) and (d) show two possible trees for the graph of Fig. 3-1(b). Any branch chosen to form a tree is referred to as a *tree-branch*.

Fig. 3-1 A network, its graph, and two possible trees.

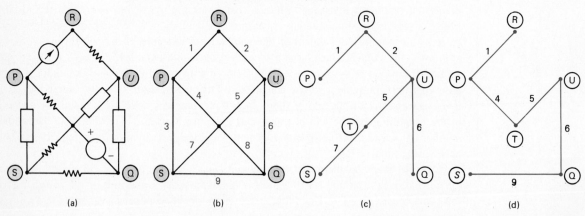

To form a tree, it takes one branch to connect the first two nodes and then an additional branch to connect each successive node. Therefore, it takes $N - 1$ branches to form a tree; that is, if T = number of tree-branches in a graph, then

$$T = N - 1$$

Exercise 3-1 For each circuit shown in Fig. 3-2, draw the graph. For each graph, draw several trees.

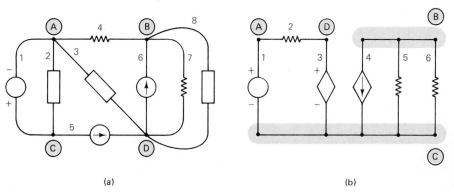

Fig. 3-2 Circuit for Exercise 3-1.

Kirchhoff's voltage law makes it possible to express the voltage between any two nodes (called a *node-pair* voltage) in a graph as the algebraic sum of voltages across the set of tree-branches between the two nodes. For example, the voltage between nodes P and Q in Fig. 3-1(c) is given by the algebraic sum of voltages across the tree-branches numbered 1, 2, and 6. Similarly, the voltage between nodes R and S in Fig. 3-1(d) is given by the algebraic sum of voltages across the tree-branches 1, 4, 5, 6, and 9.

Since a tree is set up so that there is a path made up of tree-branches from any one node to any other node, it follows that *all the node-pair voltages of a circuit are known if the voltages across all the tree-branches are known.*

Exercise 3-2 For each tree set up in Exercise 3-1, list the branch voltages that should be (algebraically) added to express the voltage between nodes A and B indicated in the given circuits.

Let

$$n = \text{number of } \textit{independent} \text{ node-pair voltages}$$

in a circuit. Imagine a short circuit placed across a tree-branch so as to make the voltages at the two terminals of that branch equal. This would reduce the number of independent node-pair voltages in the circuit by one to $n - 1$. If the short circuiting were continued to the other tree-branches, each short circuit would reduce the number of independent node-pair voltages by one at a time. Eventually, if short circuits were placed across *all* the tree-branches, the number of independent node-pair voltages would be reduced to zero, since all the nodes would now be at the same voltage.

Therefore, the number of independent node-pair voltages in a circuit equals the number of tree-branches, or

$$n = T = N - 1$$

The above conclusion can also be reached by the following alternative argument. Suppose the number of independent node-pair voltages n is *less than* T. Then it would be possible to reduce the number of independent node-pair voltages to zero by short circuiting just n branches. But this is not possible since $T - n$ tree-branches would remain without short circuits across them and hence with nonzero node-pair voltages. On the other hand, assume the number of independent node-pair voltages n to be *greater than* T. This would imply that even after *all* the tree-branches were short circuited, there would still be $n - T$ nonzero node-pair voltages. This is clearly impossible since the circuit has been reduced to nothing by shorting all the tree-branches. Therefore, $n = T = N - 1$.

Exercise 3-3 Consider the circuit of Fig. 3-3 with five nodes. Let v_{jk} denote the voltage between nodes j and k (where $j, k = 1, 2, 3, 4,$ and 5) with j taken as the positive reference. Verify that there are only four independent node-pair voltages by expressing v_{12} as an algebraic sum of $v_{23}, v_{34}, v_{45},$ and v_{51}.

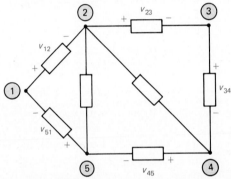

Fig. 3-3 Circuit for Exercise 3-3.

The above discussion shows that one approach to the analysis of a circuit is to write a set of $n = N - 1$ equations with n node-pair voltages as the unknowns to be solved.

In nodal analysis, one node of the circuit is selected as a *reference* or *datum* node. Then the voltages at all the other nodes measured with respect to the datum node become a set of independent node-pair voltages.

Exercise 3-4 In the circuit of Exercise 3-3, choose node 3 as the reference node. Denote the node-to-datum voltages by $v_1, v_2, v_4,$ and v_5. Express the following node-pair voltages in terms of the above node-to-datum voltages: (a) v_{14}, (b) v_{25}, (c) v_{51}, and (d) v_{32}.

The topology of a circuit is discussed again in Chapter 4 for choosing a set of independent current variables in a circuit. The present discussion is devoted to nodal analysis.

3-2 PRINCIPLES OF NODAL ANALYSIS

The variables used in nodal analysis are node-to-datum voltages (called simply *node voltages*), which are voltages at the nodes of a circuit measured with respect to a common reference point. The reference point is usually chosen as a node in the given circuit, but it can also be taken outside the circuit. However, *the number of independent node voltages in a circuit is always one less than the total number of nodes*, regardless of where the reference is selected.

In some circuits, the reference node may already be specified, as indicated by a ground symbol or the label *REF*. If a reference node is not prescribed, any one of the nodes in the given circuit can be chosen as the reference.

When the reference node is chosen within the circuit, its voltage becomes the zero base, and nodal equations are written for all the other nodes.

Kirchhoff's current law is the primary tool in nodal analysis.

The currents at each independent node are algebraically added and equated to zero. The currents in each equation are then separated into two groups:

1. Those due to sources
2. Those due to components with known current-voltage relationships

Since the discussion in this chapter is limited to circuits with only resistors (besides sources), the current-voltage relationships of the second group are Ohm's law equations, thus introducing node voltages into the equations.

The above procedure leads to a set of n equations with n unknown node voltages. The equations are then solved for the node voltages in terms of the known source quantities.

3-2-1 Resistors in Nodal Analysis

First consider Ohm's law as it should appear in nodal analysis. Suppose the voltages at the terminals of a resistor R_1 are v_a and v_b (measured with respect to some reference node) as shown in Fig. 3-4(a). Let the current in R_1 be i_1 taken in the direction a to b and the voltage be v_1 [Fig. 3-4(b), where *associated reference polarities** have been used]. The Ohm's law equation is

$$i_1 = v_1/R_1 \qquad (3\text{-}1)$$

Algebraic manipulations in nodal analysis are found to be more manageable if conductances are used instead of resistances. Using $G_1 = 1/R_1$, Eq. (3-1) becomes

$$i_1 = G_1 v_1 \qquad (3\text{-}2)$$

But, in Fig. 3-4(b),

$$v_1 = v_a - v_b$$

Fig. 3-4 Resistor in nodal analysis: $i_1 = G_1(v_a - v_b)$.

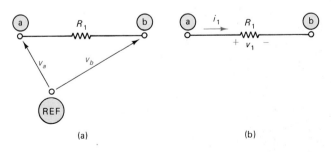

*Associated reference polarities (defined in Chapter 1) assign the direction of current in a component from the reference positive to negative voltage polarities.

and Eq. (3-2) becomes

$$i_1 = G_1(v_a - v_b) \tag{3-3}$$

Exercise 3-5 Choose the current in R_1 in the above discussion as i_1' in the direction b to a. Choose voltage v_1' using *associated* reference polarities. Obtain an expression for i_1' similar to Eq. (3-3).

Equation (3-3) and the equation obtained in Exercise 3-5 show that

if the two terminals of a resistor have node voltages v_j and v_k, then

current in the direction j to $k = G(v_j - v_k)$

where G is the conductance of the resistor.

Exercise 3-6 Write the expressions for the currents in the various branches shown in Fig. 3-5. The voltage at the node labeled REF should be treated as zero.

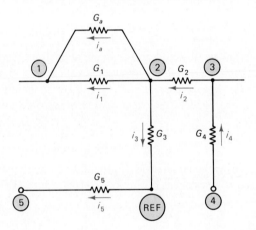

Fig. 3-5 Circuit for Exercise 3-6.

3-2-2 Nodal Equations

Consider the circuit shown in Fig. 3-6. The reference node has already been picked, and the voltages at the other nodes are denoted by v_j, where $j = 1, 2, 3$ is the number labeling each node. All node voltages are measured with respect to the REF node.

Assign currents to the resistors with directions chosen arbitrarily as shown in Fig. 3-6(b). Write KCL equations at the three nodes. (Recall the convention introduced in Chapter 1:

minus sign for currents *entering* and plus sign for currents *leaving* a node.)

Node 1: $\qquad -i_a + i_1 + i_2 - i_3 + i_b = 0 \tag{3-4}$

Node 2: $\qquad i_3 - i_b + i_4 + i_5 - i_c = 0 \tag{3-5}$

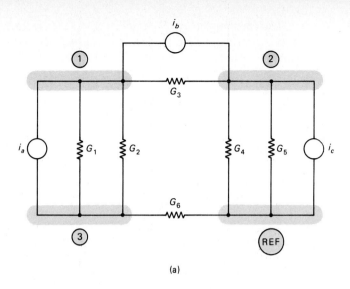

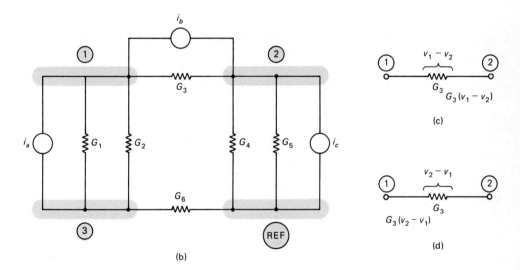

Fig. 3-6 Development of nodal analysis: (a) Given circuit, (b) arbitrary assignment of branch currents, (c) node 1: current leaves node 1 through branch G_3, and (d) node 2: current leaves node 2 through branch G_3.

Node 3: $\qquad i_a - i_1 - i_2 - i_6 = 0 \qquad$ (3-6)

In order to work toward a *standard form* of nodal equations (to be defined shortly), move the terms due to the *current sources* in the equations to the *right-hand side*.

Node 1: $\qquad i_1 + i_2 - i_3 = i_a - i_b \qquad$ (3-7)

Node 2: $\qquad i_3 + i_4 + i_5 = i_b + i_c \qquad$ (3-8)

Node 3: $\qquad -i_1 - i_2 - i_6 = -i_a \qquad$ (3-9)

3-2 **Principles of Nodal Analysis**

Each term on the *left-hand side* of an equation represents the *current through a resistor* and is rewritten as the product of a conductance and the difference between two node voltages. That is,

$$i_1 = G_1(v_1 - v_3) \qquad i_2 = G_2(v_1 - v_3)$$
$$i_3 = G_3(v_2 - v_1) \qquad i_4 = G_4(v_2 - 0)$$
$$i_5 = G_5(v_2 - 0) \qquad i_6 = G_6(0 - v_3)$$

Note (once again) that the voltage at the reference node is treated as zero.

Substitution of the above expressions into Eqs. (3-7), (3-8), and (3-9) and some rearrangement leads to the following equations:

Node 1: $\qquad G_1(v_1 - v_3) + G_2(v_1 - v_3) + G_3(v_1 - v_2) = i_a - i_b \qquad$ (3-10)

Node 2: $\qquad G_3(v_2 - v_1) + G_4(v_2 - 0) + G_5(v_2 - 0) = i_b + i_c \qquad$ (3-11)

Node 3: $\qquad G_1(v_3 - v_1) + G_2(v_3 - v_1) + G_6(v_3 - 0) = -i_a \qquad$ (3-12)

The last three equations are the nodal equations of the given circuit. These equations have the following features:

1. In each equation, the right-hand side terms together represent the *net current entering that node* from all the *current source* branches connected to that node.
2. Each term on the left-hand side of an equation represents a current *leaving that node* through a conductance branch connected to that node.

For example, consider the current in the conductance G_3 connected between nodes 1 and 2. In Eq. (3-10) for node 1, the current in G_3 appears as $G_3(v_1 - v_2)$, that is, as a current *leaving node 1* through G_3 [Fig. 3-6(c)]; but in Eq. (3-11) for node 2, the current in G_3 appears as $G_3(v_2 - v_1)$, that is, as a current *leaving node 2* through G_3 [Fig. 3-6(d)].

Thus each nodal equation is of the form

Sum of currents *leaving the node* through the different conductances connected to that node = Net current *entering the node* due to sources connected to that node

Therefore, the general procedure for writing the *equation of node j* is as follows.

Right-hand side: Form the algebraic sum of currents due to the sources connected to node j, using a plus sign for a source current entering the node and a minus sign for a source current leaving the node.

Left-hand side: Form the sum of currents *leaving* node j through the resistors connected to node j. Each term will be of the form $G_{jk}(v_j - v_k)$, where k denotes any other node linked to node j by the conductance G_{jk}.

Example 3-1 Write the nodal equations of the circuit in Fig. 3-7.

Solution Node 1: For the right-hand side, the current is $i_a - i_b$ because source current i_a enters node 1 and i_b leaves node 1.

For the left-hand side, the current leaving node 1 through G_1 is $G_1(v_1 - 0) = G_1v_1$; through G_2 is $G_2(v_1 - 0) = G_2v_1$; and through G_3 is $G_3(v_1 - v_2)$.

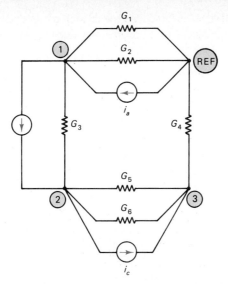

Fig. 3-7 Circuit for Example 3-1.

The equation for node 1 is, therefore,

$$G_1 v_1 + G_2 v_1 + G_3(v_1 - v_2) = i_a - i_b \tag{3-13}$$

By a similar argument, equations for nodes 2 and 3 are

$$G_3(v_2 - v_1) + G_5(v_2 - v_3) + G_6(v_2 - v_3) = i_b - i_c \tag{3-14}$$

$$G_4 v_3 + G_5(v_3 - v_2) + G_6(v_3 - v_2) = i_c \tag{3-15}$$

Equations (3-13) through (3-15) form the set of nodal equations of the given circuit.

Note that every term on the left-hand side of each equation has the *physical significance* of a *current leaving that node*. Consider G_5 connected between nodes 2 and 3, for example. The current *leaving node 2* through G_5 is $G_5(v_2 - v_3)$ and this is the term due to G_5 in the equation for *node 2*. The current *leaving node 3* through G_5 is $G_5(v_3 - v_2)$ and this is the term in the equation due to G_5 for *node 3*. Similar statements are valid for the terms due to G_3 and G_6. ∎

Example 3-2 Determine the power supplied by each of the current sources in the circuit of Fig. 3-8.

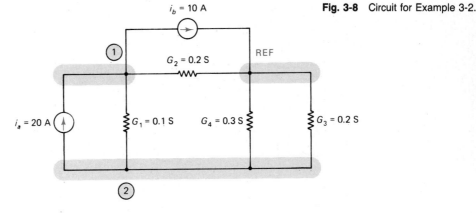

Fig. 3-8 Circuit for Example 3-2.

3-2 Principles of Nodal Analysis

Solution In order to calculate the power associated with each current source, it is first necessary to determine the voltage across it. Write the nodal equations.

Node 1: $$G_1(v_1 - v_2) + G_2 v_1 = i_a - i_b$$

or
$$0.3 v_1 - 0.1 v_2 = 10 \tag{3-16}$$

Node 2: $$G_1(v_2 - v_1) + G_3 v_2 + G_4 v_2 = -i_a$$

or
$$-0.1 v_1 + 0.6 v_2 = -20 \tag{3-17}$$

Equations (3-16) and (3-17) form the set of nodal equations of the given circuit. Solving the two equations,

$$v_1 = 23.5 \text{ V} \quad \text{and} \quad v_2 = -29.4 \text{ V}$$

- Voltage across the i_a source $= v_1 - v_2 = 52.9$ V
- Power delivered by the i_a source $= i_a(v_1 - v_2) = 1058$ W
- Voltage across the i_b source $= v_1 = 23.5$ V
- Power received by the i_b source $= i_b v_1 = 235$ W ∎

Exercise 3-7 Calculate the current and the power dissipated in each resistor of Example 3-2. Verify that the *algebraic* sum of all the powers (including the sources and the resistors) in the circuit is zero.

Exercise 3-8 Write the nodal equations for the circuit of Fig. 3-9(a).

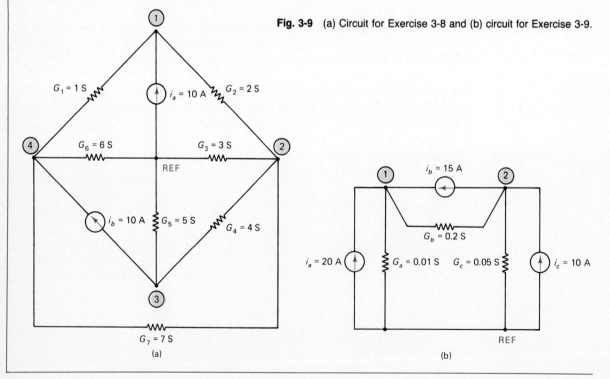

Fig. 3-9 (a) Circuit for Exercise 3-8 and (b) circuit for Exercise 3-9.

Checking Your Work:

It is possible to verify the correctness of answers to a problem in nodal analysis by evaluating all the branch currents (even when they are not specifically asked for) and checking whether KCL is satisfied at all the nodes.

Exercise 3-9 Use nodal analysis and determine the currents in all the branches for the circuit of Fig. 3-9(b).

Exercise 3-10 Write the nodal equations and calculate the voltages for the circuit of Fig. 3-10.

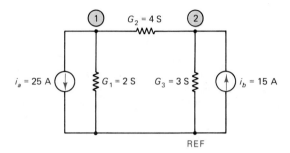

Fig. 3-10 Circuit for Exercise 3-10.

3-2-3 Standard Form of Nodal Equations

It was seen in Example 3-2 and Exercises 3-7 and 3-9 that after setting up the nodal equations by means of the procedure outlined in the previous section, it was necessary to collect the coefficients of the different voltages in each equation before solving the set of equations.

When the coefficients of the different voltages are collected (and simplified where possible) and the voltage terms on the left-hand side are written in the same order in all the equations (that is, v_1 term followed by v_2 term followed by v_3 term . . .), the nodal equations are said to be in *standard form*.

For example, the nodal equations in Eqs. (3-10) through (3-12) for the circuit of Fig. 3-6 have the standard form:

Node 1: $\qquad (G_1 + G_2 + G_3)v_1 - G_3v_2 - (G_1 + G_2)v_3 = i_a - i_b \qquad$ (3-16b)

Node 2: $\qquad -G_3v_1 + (G_3 + G_4 + G_5)v_2 = i_b + i_c \qquad$ (3-17b)

Node 3: $\qquad -(G_1 + G_2)v_1 + (G_1 + G_2 + G_6)v_3 = -i_a \qquad$ (3-18)

Exercise 3-11 Write the nodal equations of Example 3-1 and Exercise 3-8 in standard form.

It is possible to write the standard form of the nodal equations of a circuit *directly* (without going through the intermediate steps outlined in Subsection 3-2-2) when the sources contained in the circuit are all *independent current sources*.

For this, first examine Eqs. (3-16) to (3-18).

An examination of the equations shows that the *coefficient of v_j in the equation for node j* (where $j = 1, 2, 3$) is the *positive sum of all the conductances meeting at node j*.

For example, the coefficient of v_2 in the equation for node 2 is $G_3 + G_4 + G_5$, the positive sum of all the conductances meeting at node 2.

Equations (3-16) through (3-18) also have the property that the *coefficient of v_k* (where $k = 1, 2, 3$) in the *equation for node $j \neq k$*) is the *negative sum of the conductances connected between nodes j and k*.

For example, the coefficient of v_3 in the equation for node 1 is $-(G_1 + G_2)$, which is the negative sum of the conductances connected between nodes 1 and 3.

The general validity of the above features of the coefficients in the standard form of nodal equations can be established as follows.

Let G_{jk} be a conductance connected between nodes j and k in a circuit. Then the term due to G_{jk} in the equation for node j is

$$G_{jk}(v_j - v_k) = G_{jk}v_j - G_{jk}v_k \qquad (3\text{-}19)$$

Therefore, the coefficient of v_j in the equation for node j due to the presence of G_{jk} is $+G_{jk}$. Every conductance connected to node j appears as a (positive) coefficient of v_j in the equation for node j.

Equation (3-19) also shows that the coefficient of v_k in the equation for node j due to the presence of G_{jk} connected between the two nodes is $-G_{jk}$. Every conductance connected between nodes j and k similarly appear as a (negative) coefficient of v_k in the equation for node j.

The rules for writing the standard form of nodal equations are

- The coefficient of v_j in the equation for node j = the positive sum of all conductances connected to node j.
- The coefficient of v_k in the equation for node j ($k \neq j$) = the negative sum of all conductances directly connected between nodes j and k.

If there is no conductance directly connected between nodes j and k, then the coefficient of v_k in the equation for node j is *zero*. For example, in Eq. (3-17) for node 2, the coefficient of v_3 is zero because there is no conductance directly connected between nodes 2 and 3 in Fig. 3-6.

The above rules permit the writing of the standard form directly in circuits where all the sources are *independent* current sources. Writing the standard form of nodal equations of circuits with other types of sources (e.g., dependent current sources) usually needs some steps in addition to the application of the above rules.

Example 3-3 Write the nodal equations for the circuit of Fig. 3-11 in standard form.

Solution The right-hand side of each equation is the *net current entering that node*.

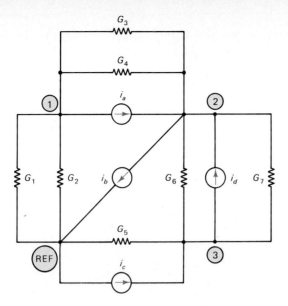

Fig. 3-11 Circuit for Example 3-3.

The left-hand side is set up by using the rules stated above for the coefficients of the voltages.

Node 1 Equation: Source current i_a is *leaving* node 1. Therefore,

$$\text{right-hand side} = -i_a$$

The coefficient of v_1 is the positive sum of all the conductances (G_1, G_2, G_3, and G_4) connected to node 1. The coefficient of v_2 is the negative sum of the conductances (G_3 and G_4) connected between nodes 1 and 2. The coefficient of $v_3 = 0$ because there is no conductance directly connected between nodes 1 and 3.

The equation for node 1 is, therefore, given by

$$(G_1 + G_2 + G_3 + G_4)v_1 - (G_3 + G_4)v_2 = -i_a \tag{3-20}$$

The equations of the other two nodes are written by using a similar argument.

Node 2: $\quad -(G_3 + G_4)v_1 + (G_3 + G_4 + G_6 + G_7)v_2 - (G_6 + G_7)v_3$
$$= i_a - i_b + i_d \tag{3-21}$$

Node 3: $\quad -(G_6 + G_7)v_2 + (G_5 + G_6 + G_7)v_3 = i_c - i_d \tag{3-22}$

Equations (3-20), (3-21), and (3-22) are the nodal equations of the given circuit in standard form. ∎

Nodal equations will be expected to be put in standard form beyond this point even when not so stated explicitly.

3-2 Principles of Nodal Analysis

Exercise 3-12 Write the nodal equations for each of the circuits shown in Fig. 3-12.

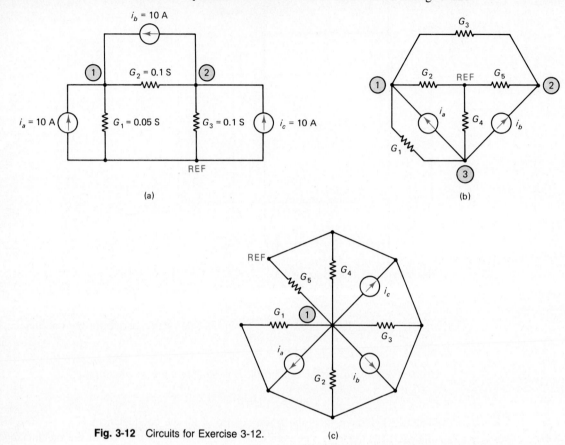

Fig. 3-12 Circuits for Exercise 3-12.

Example 3-4 Determine the currents in and the power supplied or received by each of the branches for the circuit shown in Fig. 3-13.

Fig. 3-13 Circuit for Example 3-4.

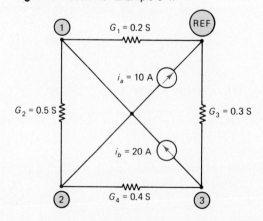

Solution The nodal equations are

Node 1:
$$(G_1 + G_2)v_1 - G_2v_2 = i_b$$
or
$$0.7v_1 - 0.5v_2 = 20 \tag{3-23}$$

Node 2:
$$-G_2v_1 + (G_2 + G_4)v_2 - G_4v_3 = -i_a$$
or
$$-0.5v_1 + 0.9v_2 - 0.4v_3 = -10 \tag{3-24}$$

Node 3:
$$-G_4v_2 + (G_4 + G_3)v_3 = -i_b$$
or
$$-0.4v_2 + 0.7v_3 = -20 \tag{3-25}$$

Equations (3-23) through (3.25) are solved for the voltages and the solution is

$$v_1 = 12.3 \text{ V} \qquad v_2 = -22.7 \text{ V} \qquad v_3 = -41.6 \text{ V}$$

(The negative values of v_2 and v_3 indicate that the voltages at nodes 2 and 3 are *lower* than the voltage at the reference node by 22.7 and 41.6 V, respectively.)

$$\text{Voltage across each conductance} = \textit{difference} \text{ between the voltages at its two terminals}$$

Figure 3-14 shows the numerical values of branch voltages (with reference polarities). Then the currents are calculated by using $i = Gv$ (Ohm's law). The values of the currents are also shown in Fig. 3-14.

The power dissipated by each conductance is calculated by using any of the following relationships: v^2G, i^2/G, or vi. The values of power dissipated are

- 30.2 W in G_1, 612 W in G_2, 519 W in G_3, and 143 W in G_4.
- The i_b source delivers $(v_1 - v_3)i_b = 1078$ W.
- The i_a source delivers $(0 - v_2)i_a = 227$ W.

Since this is a complete circuit, the total power delivered (by the two sources) must equal the total power dissipated by the four resistors (within round-off error). ∎

Fig. 3-14 Circuit for Example 3-4 showing the branch voltages and branch currents.

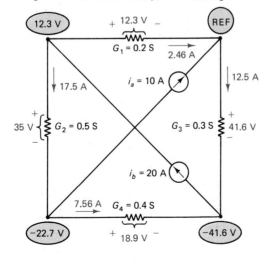

Exercise 3-13 Use nodal analysis and determine the power delivered or received by each component in the circuit of Fig. 3-15(a).

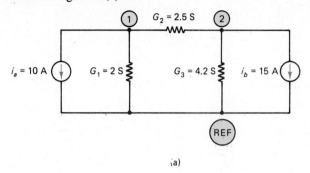

(a)

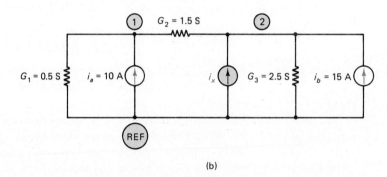

(b)

Fig. 3-15 (a) Circuit for Exercise 3-13 and (b) circuit for Exercise 3-14.

Exercise 3-14 If the voltage at node 1 in Fig. 3-15(b) is given as 20 V, calculate the source current i_x.

3-2-4 Nodal Equations for Circuits with Voltage Sources

The circuits used in the discussion of nodal analysis up to this point contained only current sources. The currents due to such sources appear directly in the nodal equations. When ideal voltage sources are present in a circuit, they introduce a constraint between the voltages at a pair of nodes. The nodal analysis of circuits with voltage sources, therefore, requires some adjustment of the procedure discussed in the previous subsection.

The voltage across an ideal voltage source is usually specified, and it provides a relationship between the voltages at the two terminals of the source. Such an equation is a statement of the *dependence* between the two voltages and leads to a reduction of the number of unknown independent node voltages. On the other hand, since the current supplied by a voltage source is a *variable* depending on the circuit connected to it, the current through a voltage source becomes an additional *unknown* quantity.

Consider the circuit shown in Fig. 3-16. One terminal of the ideal voltage source has been chosen as the reference node. (The wisdom of such a choice will become apparent soon.) Note that the junction of the two branches v_a and R_a is chosen as a node. There are three independent nodes as indicated.

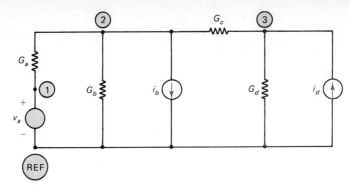

Fig. 3-16 Circuit with voltage sources in nodal analysis.

The presence of the ideal voltage source v_a gives a relationship between the voltage at node 1 and the reference node voltage: $v_1 - 0 = v_a$. That is,

$$v_1 = v_a$$

Since the source voltage v_a is usually known, v_1 also becomes a known quantity. This reduces the number of *unknown* independent voltages to two (that is, v_2 and v_3). There is no need to write an equation for node 1, and equations are written only for nodes 2 and 3.

This saving of labor has resulted from choosing the *reference node* at *one terminal of the ideal voltage source*.

Equations for nodes 2 and 3 are written by using the procedure already discussed in the previous subsection.

Node 2: $\qquad -G_a v_1 + (G_a + G_b + G_c) v_2 - G_c v_3 = -i_b$

Since $v_1 = v_a$, a known quantity, the $G_a v_a$ term is moved to the right-hand side.

$$(G_a + G_b + G_c) v_2 - G_c v_3 = G_a v_a - i_b \qquad (3\text{-}26)$$

Node 3: $\qquad -G_c v_2 + (G_c + G_d) v_3 = i_d \qquad (3\text{-}27)$

Equations (3-26) and (3-27) are solved for the voltages v_2 and v_3 in terms of v_a, i_a, and i_d.

Example 3-5 Determine (a) the current supplied by the voltage source v_b and (b) the power delivered by the current source i_c in the circuit of Fig. 3-17(a).

Solution Choose the reference node at one terminal of the source v_b and number the other nodes as indicated in Fig. 3-17(b). Note that the bottom line is node 4.
Then

$$v_1 = 0 - v_b = -v_b = -150 \text{ V}$$

Write equations for nodes 2, 3, and 4.

Node 2: $\qquad -G_b v_1 + (G_a + G_b + G_c) v_2 - G_c v_3 - G_a v_4 = i_a$

3-2 Principles of Nodal Analysis

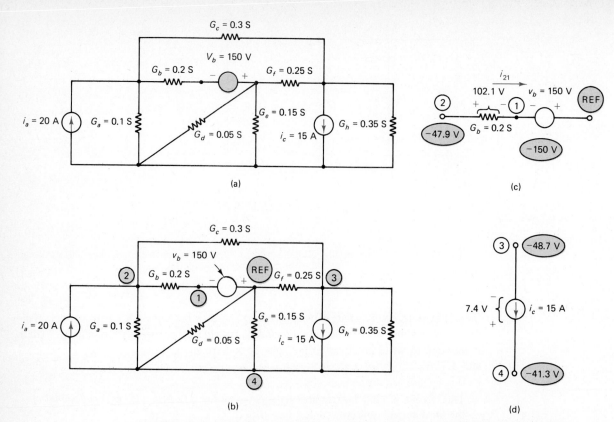

Fig. 3-17 Circuit for Example 3-5: (a) Given circuit, (b) assignment of reference node and node voltages, (c) calculation of current in the v_b source, and (d) calculation of voltage across the i_c source.

which, after substituting $v_1 = -150$ V and the numerical values of the conductances, leads to

$$0.6v_2 - 0.3v_3 - 0.1v_4 = -10 \tag{3-28}$$

Node 3:
$$-G_c v_2 + (G_c + G_f + G_h)v_3 - G_h v_4 = -i_c$$

or
$$-0.3v_2 + 0.9v_3 - 0.35v_4 = -5 \tag{3-29}$$

Node 4:
$$-G_a v_2 - G_h v_3 + (G_a + G_d + G_e + G_h)v_4 = -i_a + i_c$$

or
$$-0.1v_2 - 0.35v_3 + 0.65v_4 = -5 \tag{3-30}$$

Solving Eqs. (3-28) through (3-30), the voltages are

$$v_2 = -47.9 \text{ V} \quad v_3 = -48.7 \text{ V} \quad v_4 = -41.3 \text{ V}$$

(a) Referring to Fig. 3-17(c), the current supplied by the source v_b is i_{21}, which is also the current in G_b.

$$i_{21} = G_b(v_2 - v_1) = 0.2(-47.9 + 150) = 20.4 \text{ A}$$

(b) Referring to Fig. 3-17(d), the power delivered by the current source i_c is

$$v_c i_c = (v_4 - v_3)i_c = 111 \text{ W}$$
∎

Exercise 3-15 Calculate the power delivered by the current source i_a and the power dissipated in G_c in the circuit of Example 3-5.

Exercise 3-16 Use nodal analysis and determine the currents in all the branches of the circuit shown in Fig. 3-18(a).

Exercise 3-17 Choose the terminal common to the sources v_a and v_b as the reference node in Fig. 3-18(b). Find the power delivered by each of the three sources.

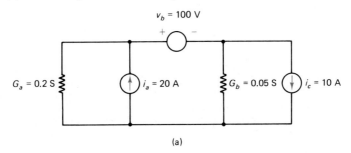

(a)

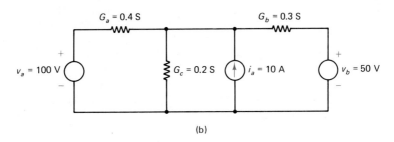

(b)

Fig. 3-18 (a) Circuit for Exercise 3-16 and (b) circuit for Exercise 3-17.

Consider now the circuit of Fig. 3-19. The reference node has been chosen at one terminal of one of the ideal voltage sources, and the other nodes are numbered as shown. The voltage source v_a places a constraint on the node voltage v_1:

$$v_1 = (v_a - 0) = v_a \tag{3-31}$$

The voltage source v_c places a constraint on the node voltages v_3 and v_4:

$$v_4 - v_3 = v_c$$

or

$$v_4 = v_3 + v_c \tag{3-32}$$

Thus, even though there are four independent node voltages in the given circuit, Eqs. (3-31) and (3-32) reduce the number of *unknown* independent node voltages to two.

On the other hand, it becomes necessary to introduce a current i_x in the voltage source branch in order to apply KCL at nodes 3 and 4. (Note that a KCL equation cannot be written for either node 3 or node 4 without introducing the current i_x in the voltage

3-2 Principles of Nodal Analysis

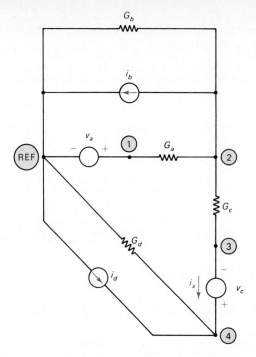

Fig. 3-19 Circuit with voltage sources in nodal analysis. The current i_x in the v_c source must be included in the equations for nodes 3 and 4.

source.) The current i_x is an additional unknown quantity in the circuit. The direction of the current i_x is chosen *arbitrarily*, as indicated in Fig. 3-19.

Write equations for nodes 2, 3, and 4.

Node 2:
$$-G_a v_1 + (G_a + G_b + G_c)v_2 - G_c v_3 = -i_b$$

Using $v_1 = v_a$ [Eq. (3-31)] and moving the v_a term to the right-hand side, the equation for node 2 becomes

$$(G_a + G_b + G_c)v_2 - G_c v_3 = -i_b + G_a v_a \tag{3-33}$$

Node 3:
$$-G_c v_2 + G_c v_3 = -i_x$$

Since i_x is an unknown quantity, it is moved to the left-hand side, and the equation for node 3 becomes

$$-G_c v_2 + G_c v_3 + i_x = 0 \tag{3-34}$$

Node 4:
$$G_d v_4 = i_x + i_d$$

Using Eq. (3-31), the last equation becomes

$$G_d v_3 + G_d v_c = i_x + i_d$$

The $G_d v_c$ term is moved to the right because v_c is a known quantity, and the i_x term is moved to the left because i_x is unknown. The equation for node 4 becomes

$$G_d v_3 - i_x = i_d - G_d v_c \tag{3-35}$$

Equations (3-33) through (3-35) are solved for v_2, v_3, and i_x. These values, in conjunction with v_a and v_c are then used to determine any branch current or voltage in the given circuit.

Example 3-6 Determine the current supplied by the voltage source v_b in the circuit of Fig. 3-20 through nodal analysis *without* changing the reference node assignment already given.

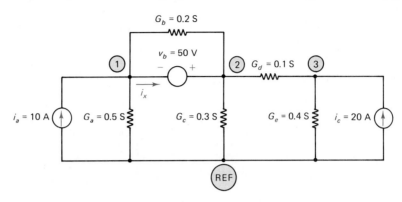

Fig. 3-20 Circuit for Example 3-6.

Solution Constraint due to the ideal voltage source v_b:

$$v_2 - v_1 = v_b \quad \text{or} \quad v_2 = v_1 + 50 \quad (3\text{-}36)$$

Choose a current i_x through v_b. Equations for nodes 1, 2, and 3 are written by referring to Fig. 3-20.

Node 1:
$$(G_a + G_b)v_1 - G_b v_2 = i_a - i_x$$
That is,
$$0.7v_1 - 0.2v_2 = 10 - i_x$$

The last equation leads to (after using Eq. (3-36) and some manipulation)

$$0.5v_1 + i_x = 20 \quad (3\text{-}37)$$

Node 2:
$$-G_b v_1 + (G_b + G_c + G_d)v_2 - G_d v_3 = i_x$$

Using Eq. (3-36) and rearranging the terms,

$$0.4v_1 - 0.1v_3 - i_x = -30 \quad (3\text{-}38)$$

Node 3:
$$-G_d v_2 + (G_d + G_e)v_3 = i_c$$

which gives, after using Eq. (3-36),

$$-0.1v_1 + 0.5v_3 = 25 \quad (3\text{-}39)$$

Equations (3-37) through (3-39) are the set of nodal equations. Their solution gives

$$i_x = 22.8 \text{ A}$$

which is the current in the v_b source. ∎

Exercise 3-18 Determine all the branch currents in the circuit in Example 3-6.

Exercise 3-19 Redo Example 3-6 and Exercise 3-17 by shifting the reference node to the positive terminal of the v_b source.

Exercise 3-20 Determine the node voltages, branch currents, and the power delivered by the sources for the circuit of Figs. 3-21(a) and (b). Note that the values given are *resistances* and not conductances.

Exercise 3-21 Using the reference node already assigned, determine the node voltages and the branch currents for the circuit of Fig. 3-21(c).

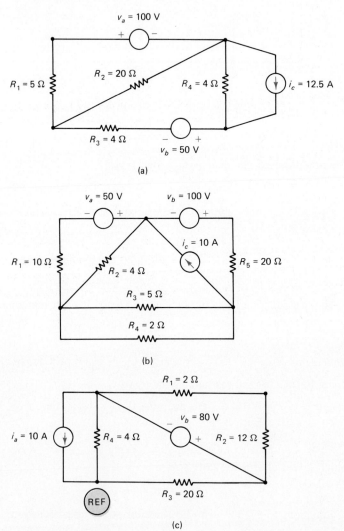

Fig. 3-21 (a) and (b) Circuits for Exercise 3-20 and (c) circuit for Exercise 3-21.

Nodal Analysis

Summary of Procedure for Nodal Analysis of Circuits with Voltage Sources:

Unless some restrictions are imposed by the statement of the problem, choose the reference node at one terminal of an ideal voltage source.

Write relationships between node voltages at the terminals of ideal voltage sources in terms of the source voltages. These constraints are used to reduce the number of unknown independent node voltages.

If it is necessary to write the equation for a node at the terminal of an ideal voltage source, assign a current through that source as a new unknown quantity and treat this current in the same manner as a source current when writing the nodal equation.

Use the constraints due to the ideal voltage sources and reduce the number of equations to be solved to equal the number of unknown independent node voltages plus the number of unknown currents through the ideal voltage sources.

Alternative Methods:

There are two other procedures besides the one discussed above that are also useful in the nodal analysis of circuits with voltage sources.

In one procedure, *ideal voltage sources with series resistors are replaced by ideal current sources with parallel resistors* by using the conversion rules discussed in Chapter 1. (This procedure is discussed in Problem 3-20 at the end of the chapter.) The conversion procedure is inapplicable, however, to any ideal voltage source not in series with a resistor.

The other procedure uses the concept of the so-called "supernode," which will not be discussed in this text. It is described in the literature.

3-2-5 Circuits with Dependent Sources

The concept of dependent voltage and dependent current sources, which appear in the models of active devices such as transistors, was introduced in Chapter 2. The value of the voltage of a dependent voltage source (or the current in a dependent current source) is determined by the voltage or current in another branch. In a *linear* dependent voltage source, there is a *linear relationship* between the voltage of the source and the parameter that controls the voltage. Similarly, in a *linear* dependent current source, the source current and the parameter controlling it are related linearly. The procedures of nodal analysis discussed in the preceding sections can be directly extended to circuits containing linear dependent sources.

In writing nodal equations, terms due to dependent sources are treated in the same manner as those due to independent sources.

The following examples illustrate the analysis of circuits with dependent sources.

Example 3-7 The circuit of Fig. 3-22 has a dependent source whose current Kv_1 is controlled by the voltage at node 1. (K is a constant.) Write the nodal equations.

Solution Node 1:
$$(G_1 + G_2)v_1 - G_2v_2 = i_s \tag{3-40}$$

Node 2:
$$-G_2v_1 + (G_2 + G_3)v_2 = -Kv_1 \tag{3-41}$$

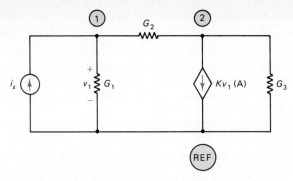

Fig. 3-22 Circuit with a dependent source for Example 3-7.

The v_1 terms on both sides of Eq. (3-41) are combined and placed on the left-hand side.

$$(K - G_2)v_1 + (G_2 + G_3)v_2 = 0 \tag{3-42}$$

The nodal equations of the given circuit are given by Eqs. (3-40) and (3-42).

Example 3-8 The circuit of Fig. 3-23 is the model of an amplifier (known as the common-collector bipolar transistor amplifier). The portion of the circuit shown within the dashed lines is an approximate model of the transistor, whereas the remainder of the circuit shows connections made to the terminals of the transistor.

Determine (a) the voltage gain (v_2/v_1), and (b) the current gain (i_o/i_s).

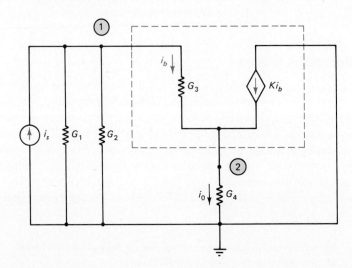

Fig. 3-23 Circuit for Example 3-8.

Solution The model contains a dependent source whose current is controlled by the current i_b in the G_3 branch. Since i_b is not an independently specified parameter of the circuit, first write an equation for i_b using Ohm's law and the voltages at the two terminals of G_3.

$$i_b = G_3(v_1 - v_2) \tag{3-43}$$

Node 1: $\qquad (G_1 + G_2 + G_3)v_1 - G_3 v_2 = i_s \tag{3-44}$

Node 2: $\qquad\qquad\qquad\qquad\qquad = Ki_b \tag{3-45}$

126 Nodal Analysis

Since i_b is not an independently specified quantity, use Eq. (3-43) in Eq. (3-45) to eliminate i_b.

$$-G_3v_1 + (G_3 + G_4)v_2 = KG_3(v_1 - v_2)$$

Collecting coefficients, the equation for node 2 becomes

$$-(K + 1)G_3v_1 + [(K + 1)G_3 + G_4]v_2 = 0 \qquad (3\text{-}46)$$

The nodal equations of the given circuit are given by Eqs. (3-44) and (3-46), which are repeated:

$$(G_1 + G_2 + G_3)v_1 - G_3v_2 = i_s$$
$$-(K + 1)G_3v_1 + [(K + 1)G_3 + G_4]v_2 = 0$$

(a) *Voltage Gain:* Solve the nodal equations for v_1 and v_2 to evaluate the ratio v_2/v_1. The determinant Δ of the system of Eqs. (3-44) and (3-46) is given by

$$\Delta = (G_1 + G_2 + G_3)[(K + 1)G_3 + G_4] - (K + 1)G_3^2$$
$$= (G_1 + G_2)[(K + 1)G_3 + G_4] + G_3G_4 \qquad (3\text{-}47)$$

The node voltage v_1 is given by

$$v_1 = \frac{1}{\Delta} \begin{vmatrix} i_s & -G_3 \\ 0 & (K + 1)G_3 + G_4 \end{vmatrix}$$

Therefore,

$$v_1 = (1/\Delta)[(K + 1)G_3 + G_4]i_s \qquad (3\text{-}48)$$

where Δ is the determinant given by Eq. (3-47).
Similarly, the node voltage v_2 is given by

$$v_2 = (1/\Delta)(K + 1)G_3 i_s \qquad (3\text{-}49)$$

From Eqs. (3-48) and (3-49),

$$\frac{v_2}{v_1} = \frac{(K + 1)G_3}{[(K + 1)G_3 + G_4]}$$

which is the voltage gain of the amplifier.

(b) *Current Gain:* The current i_o in G_4 is given by

$$i_o = G_4 v_2$$
$$= (1/\Delta)(K + 1)G_3 G_4 i_s$$

Therefore,

$$i_o/i_s = (1/\Delta)(K + 1)G_3 G_4$$
$$= \frac{(K + 1)G_3 G_4}{(G_1 + G_2)[(K + 1)G_3 + G_4] + G_3 G_4}$$

which is the current gain of the amplifier. ∎

Exercise 3-22 Obtain an expression for the current i_b, which controls the dependent source current, in the circuit of Example 3-8.

Exercise 3-23 Write the nodal equations for the circuit of Fig. 3-24(a) and determine the voltage v_2.

Exercise 3-24 The circuit of Fig. 3-24(b) is the model of an amplifier known as the common-emitter bipolar transistor amplifier. Note that the only link between the input side and the output side is through the dependent source. Obtain an expression for the voltage gain v_2/v_1.

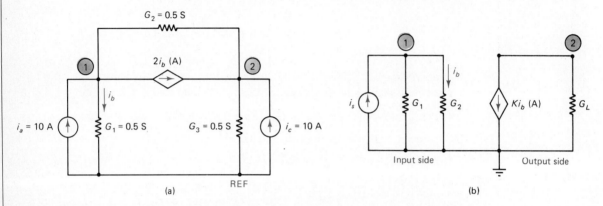

Fig. 3-24 (a) Circuit for Exercise 3-23 and (b) circuit for Exercise 3-24.

Example 3-9 Determine the resistance seen by the source i_s in the circuit of Fig. 3-25.

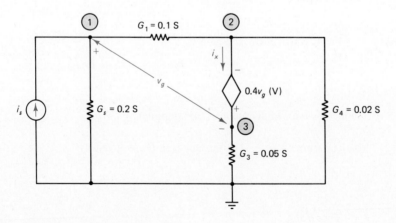

Fig. 3-25 Circuit for Example 3-9.

Solution The resistance seen by source i_s is given by v_1/i_s. The voltage v_1 is obtained here by nodal analysis.

There are three independent nodes in the circuit, as shown in the figure, and the dependent voltage source places a constraint between the voltages at nodes 2 and 3.

$$v_3 - v_2 = 0.4v_g \tag{3-50}$$

But the voltage v_g controlling the dependent source voltage is given by

$$v_g = v_1 - v_3 \tag{5-51}$$

so that Eq. (3-50) becomes

$$v_3 - v_2 = 0.4(v_1 - v_3)$$

or

$$-0.4v_1 - v_2 + 1.4v_3 = 0 \tag{3-52}$$

Equation (3-52) is a relationship between the three voltages due to the presence of the dependent source.

Equation for Node 1: $\quad (G_s + G_1)v_1 - G_1v_2 = i_s$

which gives (after substitution of numerical values)

$$0.3v_1 - 0.1v_2 = i_s \tag{3-53}$$

To write equations for nodes 2 and 3, it is necessary to introduce a current i_x in the $0.4v_g$ source, as shown in Fig. 3-25.

Equation for Node 2: $\quad -G_1v_1 + (G_1 + G_4)v_2 = -i_x$

Substituting numerical values and moving i_x term to the left-hand side, we have

$$-0.1v_1 + 0.12v_2 + i_x = 0 \tag{3-54}$$

Equation for Node 3: $\quad G_3v_3 = i_x$

which becomes

$$0.05v_3 - i_x = 0 \tag{3-55}$$

There are four unknowns, the three node voltages and the current i_x, and four equations, Eqs. (3-52) through (3-55), which are repeated:

$$-0.4v_1 - v_2 + 1.4v_3 = 0$$
$$0.3v_1 - 0.1v_2 = i_s$$
$$-0.1v_1 + 0.12v_2 + i_x = 0$$
$$0.05v_3 - i_x = 0$$

The solution of the above equations gives

$$v_1 = 4.08i_s$$

The resistance seen by the current source is, therefore,

$$v_1/i_s = 4.08 \ \Omega$$

■

3-2 Principles of Nodal Analysis

Exercise 3-25 Determine the ratio v_3/i_s for the circuit shown in Fig. 3-26(a).

Exercise 3-26 Determine the resistance seen by the current source shown in Fig. 3-26(b).

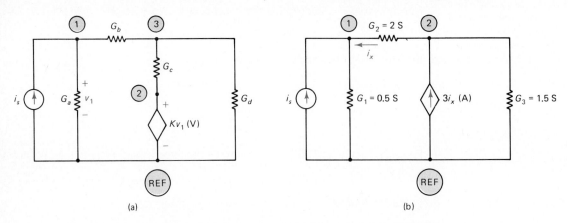

Fig. 3-26 (a) Circuit for Exercise 3-25 and (b) circuit for Exercise 3-26.

3-3 ALGEBRAIC DISCUSSION OF NODAL EQUATIONS*

The application of the theory of determinants to nodal analysis leads to some general results of circuit theory. Some aspects of such an application are discussed in this section. The principles developed here are useful in the proof of several important theorems of circuit theory in later chapters. A knowledge of matrix notation and the basic theory of determinants is assumed in this section. A brief review of the theory of determinants is presented in Appendix C.

3-3-1 Nodal Conductance Matrix

First, consider circuits in which all the sources are *independent current sources*. For such circuits, the coefficient of v_k in the equation for node k is the positive sum of all the conductances meeting at node k, and the coefficient of v_j in the equation for node k is the negative sum of the conductances linking nodes j and k.

The *nodal conductance matrix* of the circuit is the ($n \times n$) array of the set of coefficients of the voltages on the left-hand side of the nodal equations (in standard form). The n rows of the matrix correspond to the n nodal equations and the n columns to the n voltages.

The nodal conductance matrix of the circuit of Example 3-3, whose nodal equations were given by Eqs. (3-20) through (3-22), is

$$[g] = \begin{bmatrix} G_1 + G_2 + G_3 + G_4 & -(G_3 + G_4) & 0 \\ -(G_3 + G_4) & (G_3 + G_4 + G_6 + G_7) & -(G_6 + G_7) \\ 0 & -(G_6 + G_7) & (G_5 + G_6 + G_7) \end{bmatrix}$$

*Section 3-3 can be omitted without loss of continuity.

Exercise 3-27 Write the nodal conductance matrices for the circuits of Exercises 3-11 and 3-12.

The nodal conductance matrix of a circuit with n independent nodes is of the general form:

$$[g] = \begin{bmatrix} g_{11} & g_{12} & \cdots & g_{1j} & \cdots & g_{1n} \\ g_{21} & g_{22} & \cdots & g_{2j} & \cdots & g_{2n} \\ \cdot & \cdot & & \cdot & & \cdot \\ g_{k1} & g_{k2} & \cdots & g_{kj} & & g_{kn} \\ \cdot & \cdot & & \cdot & & \cdot \\ g_{n1} & g_{n2} & \cdots & g_{nj} & \cdots & g_{nn} \end{bmatrix}$$

The first subscript of each element in the matrix denotes the row and the second subscript the column of the matrix in which that element is found. That is, g_{kj} (where k and j are any integers 1, 2, 3, . . . , n) is the element in the kth row and the jth column of the matrix.

The elements in the kth *row* ($k = 1, 2, 3, \ldots, n$) are the coefficients of the voltages in the equation for node k. The elements in the jth *column* ($j = 1, 2, 3, \ldots, n$) are the coefficients of the voltage v_j in the different nodal equations.

Consider, for example, the following nodal conductance matrix:

$$[g] = \begin{bmatrix} G_1 + G_2 + G_3 & -G_3 & -(G_1 + G_2) \\ -G_3 & G_3 + G_6 & 0 \\ -(G_1 + G_2) & 0 & G_1 + G_2 + G_5 \end{bmatrix}$$

The elements in the first row, $g_{11} = G_1 + G_2 + G_3$, $g_{12} = -G_3$, $g_{13} = -(G_1 + G_2)$ are, respectively, coefficients of the voltages v_1, v_2, and v_3 in the equation for node 1.

The elements in the third column of the matrix, $g_{13} = -(G_1 + G_2)$, $g_{23} = 0$, $g_{33} = G_1 + G_2 + G_5$) are coefficients of the voltage v_3 in the equations for nodes 1, 2, and 3, respectively.

Exercise 3-28 For the nodal conductance matrices obtained in Exercise 3-27, (a) list the following elements: g_{11}, g_{22}, g_{23}, and g_{31}; (b) list the coefficients of the voltages in the equation for node 3; (c) list the coefficients of v_2 for the different nodes.

The elements of the nodal conductance matrix of a circuit containing only independent current sources and conductances are seen to have the following properties.

1. The element g_{kk} (where $k = 1, 2, 3, \ldots, n$) is the positive sum of all the conductances meeting at node k. This follows from the rule established in Subsec. 3-2-3: the coefficient of v_k in the equation for node k is the positive sum of all the conductances meeting at node k.

2. The element g_{kj} (where k and j can assume any integer values between 1 and n, but $k \neq j$) is the negative sum of the conductances linking nodes k and j. This follows from the rule established in Subsec. 3-2-3: the coefficient of the voltage v_j in the equation for node k is the negative sum of the conductances linking nodes j and k.

3-3 Algebraic Discussion of Nodal Equations

3. The elements g_{jk} and g_{kj} ($j \neq k$) are equal. That is, the element in the *j*th row and *k*th column is the same as that in the *k*th row and *j*th column. This follows from the fact that the coefficient of v_j in the equation for node k and the coefficient of v_k in the equation for node j are both equal to the negative sum of the conductances linking nodes j and k.

Passive and Active Networks:

A resistor is said to be *passive* if the power dissipated by it, the product *vi*, is always positive. That is, a passive resistor cannot deliver power to another component. The resistance (*v/i*) of a passive resistor is always positive. In Chapter 6, two other passive components, the passive capacitor and passive inductor are introduced. A circuit made up entirely of passive components is called a *passive circuit*.

If the ratio *v/i* for an element were negative, indicating a *negative resistance*, then the power $vi = Ri^2$ received by it would also be negative. That is, a negative resistance would *deliver* power to other components. A component with a negative value of resistance is called an *active* resistor. Besides negative resistances, ideal voltage and current sources are *active* components also since they can deliver power to other components. Circuits containing one or more active components are called *active* circuits.

The circuits discussed in the present section belong to the class of passive circuits driven by independent current sources. We have seen that, in such circuits, the nodal conductance matrix has the property:

$$g_{jk} = g_{kj}$$

The nodal conductance matrix of a passive circuit (driven by independent current sources) is symmetric about the principal diagonal.

As will be seen later in this chapter, *dependent* sources can introduce voltage coefficients in nodal equations, which violate the property $g_{jk} = g_{kj}$. That is, the nodal conductance matrix of an active circuit may not be symmetric.

The nodal conductance matrix of passive circuits in which all the sources are independent current sources is written directly by inspection of the circuit by using the three properties discussed earlier.

Example 3-10 Write the nodal conductance matrix of the circuit in Fig. 3-27.

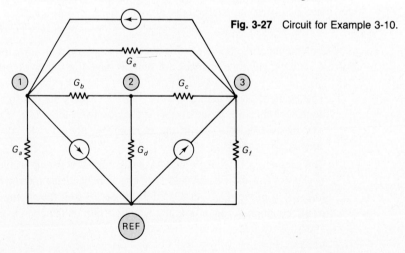

Fig. 3-27 Circuit for Example 3-10.

Solution: Since the given network contains only independent current sources and conductances, the nodal conductance matrix is written by inspection. These are three independent nodes in the circuit and, consequently, the conductance matrix has three rows and three columns. The elements of the first row are

$$g_{11} = \text{positive sum of all conductances meeting at node 1}$$
$$= G_a + G_b + G_e$$
$$g_{12} = \text{negative sum of the conductances linking nodes 1 and 2}$$
$$= -G_b$$
$$g_{13} = \text{negative sum of the conductances linking nodes 1 and 3}$$
$$= -G_e$$

The other elements of the matrix are obtained in a similar manner. The matrix is given by

$$[g] = \begin{bmatrix} G_a + G_b + G_e & -G_b & -G_e \\ -G_b & G_b + G_c + G_d & -G_c \\ -G_e & -G_c & G_c + G_e + G_f \end{bmatrix}$$

The matrix is seen to be symmetric. For example, $g_{12} = g_{21} = -G_b$. ∎

Exercise 3-29 Write the nodal conductance matrix of each of the circuits shown in Fig. 3-28.

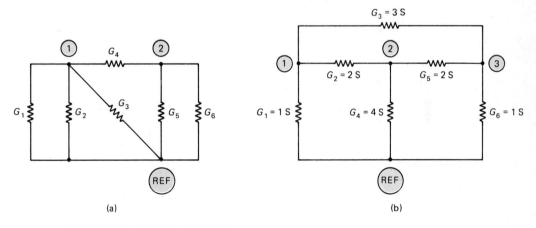

Fig. 3-28 Circuit for Exercise 3-29.

3-3-2 Cramer's Rule and the Solution of Nodal Equations

Cramer's rule from the theory of determinants sets up a procedure for the systematic solution of a system of n independent equations in n unknowns. Apart from its use in the systematic solution of a set of n simultaneous equations, Cramer's rule plays a more important part in circuit theory: it facilitates the derivation of a number of general results and theorems.

Consider a system of nodal equations given by

$$g_{11}v_1 + g_{12}v_2 + \ldots + g_{1n}v_n = i_1$$
$$g_{21}v_1 + g_{22}v_2 + \ldots + g_{2n}v_n = i_2$$
$$\ldots\ldots\ldots\ldots\ldots\ldots\ldots\ldots\ldots\ldots$$
$$g_{n1}v_1 + g_{n2}v_2 + \ldots + g_{nn}v_n = i_n$$

The solution of such a system of equations by using Cramer's rule involves the following steps:

1. Evaluate the determinant of the nodal conductance matrix: $\|g\|$.

$$\|g\| = \begin{vmatrix} g_{11} & g_{12} & \cdots & g_{1j} & \cdots & g_{1n} \\ g_{21} & g_{22} & \cdots & g_{2j} & \cdots & g_{2n} \\ \cdot & \cdot & & \cdot & & \cdot \\ g_{k1} & g_{k2} & \cdots & g_{kj} & \cdots & g_{kn} \\ \cdot & \cdot & & \cdot & & \cdot \\ g_{n1} & g_{n2} & \cdots & g_{nj} & \cdots & g_{nn} \end{vmatrix}$$

2. To find the value of the voltage v_j (where j is any integer between 1 and n), the jth column of the nodal conductance matrix is replaced by the currents $i_1, i_2, \ldots, i_n$ from the right-hand sides of the equations and the determinant of the matrix so obtained is divided by $\|g\|$.

That is,

$$\qquad\qquad\qquad\qquad\qquad j\text{th column}$$

$$v_j = \frac{1}{\|g\|} \begin{vmatrix} g_{11} & g_{12} & \cdots & i_1 & \cdots & g_{1n} \\ g_{21} & g_{22} & \cdots & i_2 & \cdots & g_{2n} \\ \cdot & \cdot & & \cdot & & \cdot \\ g_{k1} & g_{k2} & \cdots & i_k & \cdots & g_{kn} \\ \cdot & \cdot & & \cdot & & \cdot \\ g_{n1} & g_{n2} & \cdots & i_n & \cdots & g_{nn} \end{vmatrix} \qquad (3\text{-}56)$$

Let

$$A_{kj} = \text{cofactor of the } k\text{th row and } j\text{th column}$$

That is, A_{kj} is the determinant obtained after deleting the kth row and jth column of the nodal conductance matrix. Using Laplace's expansion about the jth column of the numerator determinant in Eq. (3-56) (and letting $k = 1, 2, \ldots, n$), the expression for the voltage at node j becomes

$$v_j = (1/\|g\|)(A_{1j}i_1 + A_{2j}i_2 + \ldots + A_{jj}i_j + \ldots + A_{nj}i_n) \qquad (3\text{-}57)$$

or

$$v_j = \frac{1}{\|g\|} \sum_{k=1}^{n} A_{kj} i_k \qquad (3\text{-}58)$$

Equation (3-58) is the general expression for the solution of the nodal equations of a circuit with n independent nodes. In Eq. (3-58),

$\|g\|$ = determinant of the nodal conductance matrix of the given circuit
A_{kj} = cofactor of the kth row and jth column of the nodal conductance matrix
i_k = the net current entering the kth node due to independent current sources connected to that node

Example 3-11 Determine the nodal voltages in the circuit of Fig. 3-29 by using Cramer's rule in the form of Eq. (3-58).

Fig. 3-29 Circuit for Example 3-11.

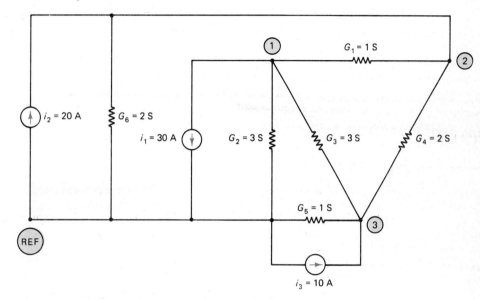

Solution The nodal equations of the circuit are

$$7v_1 - v_2 - 3v_3 = -30$$
$$-v_1 + 5v_2 - 2v_3 = 20$$
$$-3v_1 - 2v_2 + 6v_3 = 10$$

The determinant $\|g\|$ is given by

$$\|g\| = \begin{vmatrix} 7 & -1 & -3 \\ -1 & 5 & -2 \\ -3 & -2 & 6 \end{vmatrix} = 119$$

Using Eq. (3-58) with $n = 3$, the node voltages are given by

$$v_j = (1/\|g\|) \sum_{k=1}^{3} A_{kj} i_k$$
$$= (1/\|g\|)(A_{1j} i_1 + A_{2j} i_2 + A_{3j} i_3)$$

The voltage at node 1 is given by

$$v_1 = (1/\|g\|)(A_{11} i_1 + A_{21} i_2 + A_{31} i_3) \tag{3-59}$$

3-3 Algebraic Discussion of Nodal Equations

The cofactors A_{11}, A_{21}, and A_{31} are given by

$$A_{11} = \begin{vmatrix} 5 & -2 \\ -2 & 6 \end{vmatrix} \qquad A_{21} = (-1)\begin{vmatrix} -1 & -3 \\ -2 & 6 \end{vmatrix} \qquad A_{31} = \begin{vmatrix} -1 & -3 \\ 5 & -2 \end{vmatrix}$$

Therefore,

$$A_{11} = 26 \qquad A_{21} = 12 \qquad A_{31} = 17$$

Equation (3-59) then gives

$$v_1 = -3.11 \text{ V}$$

The voltages at the other two nodes are found by a similar procedure.

$$v_2 = 3.95 \text{ V} \qquad \text{and} \qquad v_3 = 1.43 \text{ V} \qquad \blacksquare$$

Exercise 3-30 The nodal equations of a circuit are given by

$$16v_1 - 6v_2 - 4v_3 = 10$$
$$-6v_1 + 12v_2 = -10$$
$$-4v_1 + 8v_3 = 0$$

Evaluate the three nodal voltages by using Cramer's rule.

3-3-3 Circuits with Dependent Current Sources

The preceding discussion (of a nodal conductance matrix and the use of Cramer's rule for solving nodal equations) is readily extended to cover circuits with dependent current sources, that is, to active circuits. The only difference is that the nodal conductance matrix of such circuits cannot always be written directly by inspection. It is usually necessary to first write the nodal equations and manipulate them to obtain the standard form before writing the nodal conductance matrix.

Example 3-12 Obtain the nodal conductance matrix of the circuit shown in Fig. 3-30.

Solution The given circuit has a dependent source. The current i_x controlling its current is given by

$$i_x = G_4(v_1 - v_2) \tag{3-60}$$

Fig. 3-30 Circuit for Example 3-12.

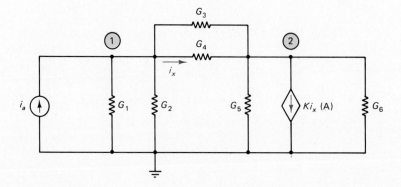

136 Nodal Analysis

The nodal equations are given by

Node 1: $(G_1 + G_2 + G_3 + G_4)v_1 - (G_3 + G_4)v_2 = i_a$ (3-61)

Node 2: $-(G_3 + G_4)v_1 + (G_3 + G_4 + G_5 + G_6)v_2 = -Ki_x$

Substituting for i_x from Eq. (3-60), the last equation becomes

$$-[G_3 + G_4(1 - K)]v_1 + [G_3 + G_4(1 - K) + G_5 + G_6]v_2 = 0 \quad (3\text{-}62)$$

Equations (3-61) and (3-62) are the nodal equations of the given circuit. The nodal conductance matrix is

$$[g] = \begin{bmatrix} G_1 + G_2 + G_3 + G_4 & -(G_3 + G_4) \\ -[G_3 + G_4(1 - K)] & G_3 + G_4(1 - K) + G_5 + G_6 \end{bmatrix}$$

It is seen that $g_{12} = -(G_3 + G_4)$, and $g_{21} = -[G_3 + G_4(1 - K)]$; the matrix is not symmetric.

Exercise 3-31 Obtain the nodal conductance matrix of each of the circuits shown in Fig. 3-31.

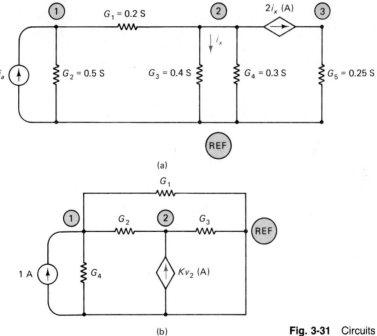

Fig. 3-31 Circuits for Exercise 3-31.

Except for the lack of symmetry in the nodal conductance matrix of an active circuit, the treatment of the matrix of such a circuit follows exactly the same procedure as that of passive circuits. For example, Cramer's rule [as given by Eq. (3-58)] is useful for solving the nodal equations.

It is necessary, however, to pay particular attention to the *order of the subscripts* of the cofactors because $A_{jk} \neq A_{kj}$, in general.

3-3 Algebraic Discussion of Nodal Equations

Fig. 3-32 Circuit for Example 3-13.

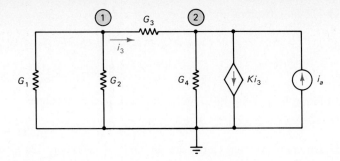

Example 3-13 Determine (a) the nodal voltages in the circuit of Fig. 3-32 by using Cramer's rule and (b) the resistance seen by the current source i_a.

Solution The current controlling the dependent source is given by

$$i_3 = G_2(v_1 - v_2) \tag{3-63}$$

Node 1 Equation: $\quad (G_1 + G_2 + G_3)v_1 - G_3 v_2 = 0 \tag{3-64}$

Node 2 Equation: $\quad -G_3 v_1 + (G_3 + G_4)v_2 = -Ki_3 + i_a$

Substituting for i_3 from Eq. (3-63), the equation for node 2 becomes

$$(K - 1)G_3 v_1 + [G_3(1 - K) + G_4]v_2 = i_a \tag{3-65}$$

Equations (3-64) and (3-65) are the nodal equations of the given circuit. The nodal conductance matrix is

$$[g] = \begin{bmatrix} G_1 + G_2 + G_3 & -G_3 \\ -G_3(1 - K) & G_3(1 - K) + G_4 \end{bmatrix}$$

(a) The determinant of the nodal conductance matrix is

$$\|g\| = (G_1 + G_2)[G_3(1 - K) + G_4] + G_3 G_4 \tag{3-66}$$

Using Eq. (3-58),

$$v_j = (1/\|g\|)(A_{1j}i_1 + A_{2j}i_2)$$

where $i_1 = 0$ and $i_2 = i_a$ for the given circuit. Therefore,

$$v_1 = A_{21}i_a/\|g\|$$
$$= G_3 i_a/\|g\|$$

and

$$v_2 = A_{22}i_a$$
$$= (G_1 + G_2 + G_3)i_a/\|g\|$$

(b) The resistance seen by the current source i_a is

$$\frac{v_2}{i_a} = \frac{(G_1 + G_2 + G_3)}{\|g\|}$$

where $\|g\|$ is given by Eq. (3-66). The determinant $\|g\|$ depends on the factor $(1 - K)$ and $\|g\|$ can be negative for a sufficiently large K. Then the ratio v_2/i_a becomes negative.

That is, the resistance seen by the source becomes *negative*. A situation like this was also encountered in Exercise 3-26. It is, therefore, possible to "create" a negative resistance by using an active device (whose model contains a dependent source) in a circuit. ∎

Exercise 3-32 Determine the voltage ratio v_2/v_1 in each of the circuits of Exercise 3-31 (see Fig. 3-31).

The emphasis on the solution of nodal equations by using Cramer's rule in the preceding discussion should not lead the reader to conclude that Cramer's rule is the preferred method for solving all nodal equations. In fact, the equations of the previous example could be readily solved by noting that $v_2 = 2v_1$ from Eq. (3-64) and using this fact in Eq. (3-65). The purpose of stressing Cramer's rule, on the other hand, is to facilitate the theoretical developments in the following section and the proof of theorems to be presented in Chapter 5.

3-3-4 Transfer and Driving Point Resistances

Consider a linear circuit (which may contain linear dependent sources) with n independent nodes, which contains no *independent* sources. Such a network is called a *relaxed*, or *dead*, network. Assume a relaxed linear network contained in the black box of Fig. 3-33(a), where leads are shown attached to the nodes of the circuit for the purposes of connecting external independent sources to make the circuit spring it into action.

Suppose a single independent current source is connected between node r and the reference node as indicated in Fig. 3-33(b). The leads to the other nodes are left open as shown. Under the single-source connection, the *net source current* entering any node (except node r) is zero. That is, in the equation for the voltage v_j in the Cramer's rule formula of Eq. (3-58), repeated here,

$$v_j = (1/\|g\|) \sum_k A_{kj} i_k$$

$$i_k = \begin{cases} 0 & \text{when } k \neq r \\ i_r & \text{when } k = r \end{cases}$$

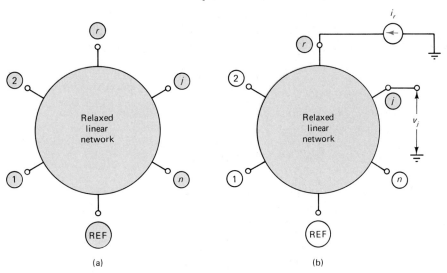

Fig. 3-33 (a) Relaxed network with multiple nodes. (b) Network fed by a single current source.

Therefore, Eq. (3-58) reduces to

$$v_j = (1/\|g\|)(A_{rj}i_r) \qquad j = 1, 2, \ldots, n \qquad (3\text{-}65)\text{--}(3\text{-}58a)$$

where A_{rj} is the cofactor of the rth row and jth column of the nodal conductance matrix. Equation (3-58a) gives the voltage at any node (with respect to the reference node) when the circuit is driven by a single current source connected between any node r and the reference node. The ratio of the voltage at any node to the source current at node r leads to two situations: the transfer resistance and the driving point resistance.

Transfer Resistance:

The ratio of the voltage at any node j to the source current i_r is obtained from Eq. (3-58a)

$$\frac{v_j}{i_r} = \frac{A_{rj}}{\|g\|} \qquad (3\text{-}66)\text{--}(3\text{-}65a)$$

The ratio v_j/i_r is called the *transfer resistance* from node r to node j.

The transfer resistance is a measure of the effect at some node j due to a current source connected to some other node r of a circuit. Transfer resistances are of importance in the study of amplifiers and communication circuits. For example, suppose it is required to find the output voltage of an amplifier when an input current signal is applied. Equation (3-65a) shows that such a determination requires only a knowledge of the nodal conductance matrix of the circuit.

Example 3-14 Write the nodal conductance matrix of the circuit in Fig. 3-34. Determine the transfer resistance (a) from node 2 to node 3 and (b) from node 1 to node 2.

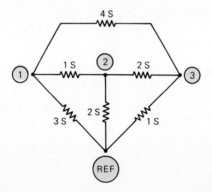

Fig. 3-34 Circuit for Example 3-14.

Solution The nodal conductance matrix is written by inspection.

$$[g] = \begin{bmatrix} 8 & -1 & -4 \\ -1 & 5 & -2 \\ -4 & -2 & 7 \end{bmatrix}$$

The determinant $\|g\|$ is found to be

$$\|g\| = 145$$

(a) The transfer resistance from node 2 to node 3 is given by

$$v_3/i_2 = (A_{23}/\|g\|)$$

where

$$A_{23} = -1 \begin{vmatrix} 8 & -1 \\ -4 & -2 \end{vmatrix} = 20$$

Therefore, $v_3/i_2 = 20/145 = 0.138\Omega$.

(b) The transfer resistance from node 1 to node 2 is given by

$$v_2/i_1 = A_{12}/\|g\|$$

where

$$A_{12} = -1 \begin{vmatrix} -1 & -2 \\ -4 & 7 \end{vmatrix} = 15$$

Therefore, $v_2/i_1 = 15/145 = 0.103\Omega$. ■

Exercise 3-33 Determine the other four transfer resistances of the circuit in Example 3-14. Note that the symmetry of the matrix $[g]$ leads to equality between pairs of transfer resistances.

Example 3-15 Obtain the nodal conductance matrix of the circuit shown in Fig. 3-35, and determine (a) the transfer resistance from node 2 to node 1 and (b) from node 1 to node 2.

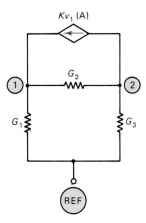

Fig. 3-35 Circuit for Example 3-15.

Solution Since there is a dependent source in the circuit, we start by writing the nodal equations:

Node 1: $(G_1 + G_2)v_1 - G_2v_2 = Kv_1$

or $(G_1 + G_2 - K)v_1 - G_2v_2 = 0$ \hfill (3-67)

Node 2: $-G_2v_1 + (G_2 + G_3)v_2 = -Kv_1$

or $-(G_2 - K)v_1 + (G_2 + G_3)v_2 = 0$ \hfill (3-68)

3-3 Algebraic Discussion of Nodal Equations

From Eqs. (3-67) and (3-68), the nodal conductance matrix is

$$[g] = \begin{bmatrix} G_1 + G_2 - K & -G_2 \\ -(G_2 - K) & G_2 + G_3 \end{bmatrix}$$

The determinant of the matrix is found to be

$$\|g\| = G_1 G_2 + G_1 G_3 + G_2 G_3 - K G_3 \qquad (3\text{-}69)$$

(a) The transfer resistance from node 2 to node 1 is given by

$$v_1/i_2 = A_{21}/\|g\|$$
$$= G_2/\|g\|$$

where $\|g\|$ is given by Eq. (3-69).

(b) The transfer resistance from node 1 to node 2 is given by

$$v_2/i_1 = A_{12}/\|g\|$$
$$= (G_2 - K)/\|g\|$$

where $\|g\|$ is given by Eq. (3-69).

Note that since the nodal conductance matrix is not symmetric, the transfer resistance from node 1 to node 2 is not the same as from node 2 to node 1. ∎

Exercise 3-34: Obtain the nodal conductance matrix of each of the circuits shown in Fig. 3-36 and determine all possible transfer resistances.

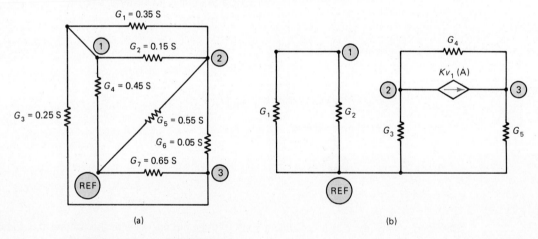

Fig. 3-36 Circuits for Exercise 3-34.

When the conductance matrix of a circuit is *symmetric* (the case of passive circuits),

$$A_{rj} = A_{jr}$$

and

transfer resistance from node r to node j = transfer resistance from node j to node r

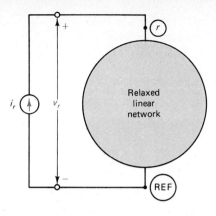

Fig. 3-37 Driving point resistance at node $r = v_r/i_r$.

Driving Point Resistance:

Consider the case where $j = r$ in Eq. (3-66). That is, the voltage is measured at the same node as the one driven by the current source [Fig. 3-37]. Then Eq. (3-66) becomes

$$\frac{v_r}{i_r} = \frac{A_{rr}}{\|g\|} \tag{3-70}$$

The ratio v_r/i_r is the effective resistance seen by the current source connected between node r and the reference node. It is, therefore, called the *driving point* resistance of the circuit with respect to node r.

When a current source is applied to a terminal designated as the *input terminal*, the driving point resistance seen by the source is referred to as the *input resistance*. When a current source is applied to a terminal designated as the *output terminal* of the circuit, the driving point resistance seen by the source is referred to as the *output resistance*. In Chapter 5, the driving point resistance seen from a designated pair of terminals of a circuit will be referred to as the *Thevenin resistance*.

Example 3-16 Determine the three driving point resistances of the circuit of Example 3-14.

Solution The driving point resistance seen from node 1 is

$$v_1/i_1 = A_{11}/\|g\|$$

where

$$A_{11} = \begin{vmatrix} 5 & -2 \\ -2 & 7 \end{vmatrix} = 31$$

so that

$$v_1/i_1 = 31/145 = 0.214 \, \Omega$$

The driving point resistance seen from node 2 is

$$v_2/i_2 = A_{22}/\|g\| = 40/145 = 0.276 \, \Omega$$

The driving point resistance seen from node 3 is

$$v_3/i_3 = A_{33}/\|g\| = 39/145 = 0.269 \, \Omega$$

That is, if the given circuit were viewed from node 1, it would be equivalent to a single resistance of 0.214 Ω. Similarly if viewed from node 2, it would be equivalent to a single resistance of 0.276 Ω; from node 3, a single resistance of 0.269 Ω. ∎

Example 3-17 Determine the input and output resistances of the circuit of Example 3-15.

Solution The input resistance is the driving point resistance seen from node 1, and is given by

$$\frac{A_{11}}{\|g\|} = \frac{(G_2 + G_3)}{\|g\|}$$

The output resistance is the driving point resistance seen from node 2, and is given by

$$\frac{A_{22}}{\|g\|} = \frac{(G_1 + G_2 - K)}{\|g\|}$$

From Eq. (3-69),

$$\|g\| = G_1 G_2 + G_1 G_3 + G_2 G_3 - K G_3$$

∎

Exercise 3-35 Obtain the nodal conductance matrix of the circuit shown in Fig. 3-38, which is the model of a common-emitter transistor amplifier. (a) Determine all the transfer resistances. (b) Determine the input and output resistances of the amplifier.

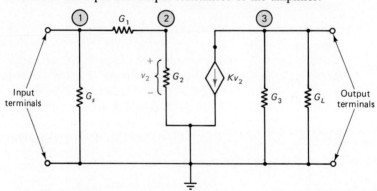

Fig. 3-38 Circuit for Exercise 3-35.

The preceding discussion shows that the driving point and transfer resistances of a relaxed linear circuit can be determined once the nodal conductance matrix of the circuit is known. The procedure involves a calculation of the appropriate cofactor and dividing it by the determinant of the conductance matrix.

The nodal conductance matrix and the use of Cramer's rule thus provide a systematic and efficient approach to the determination of driving point and transfer resistances of a circuit.

3-4 OPERATIONAL AMPLIFIERS

It is worthwhile at this point to consider some circuits of practical importance and apply nodal analysis to them. In order to apply the analysis techniques developed to this point, however, we have to restrict ourselves to circuits containing only resistors and sources.

Even though such a restriction may appear to be severe, it so happens that the models of many circuits in electronics need only resistors and dependent sources. This is illustrated by the model of an amplifier discussed in Chapter 2 and in the examples presented earlier in this chapter. This particular class of amplifier is known as an *operational amplifier,* usually abbreviated *op amp*.

The model of an amplifier (used in Sec. 2-4) contains three components: the *input resistance* R_i, the *output resistance* R_o, and a dependent voltage source Av_i, where A is the *voltage gain* of the amplifier. The three parameters R_i, R_o, and A can be either determined experimentally for a given amplifier or obtained from theoretical considerations of devices such as transistors. It should be stressed that electronic devices are inherently nonlinear and the models used in analysis are only *linear approximations* to the characteristics of actual electronic devices or circuits. Also, there are several *levels of approximation* (and corresponding linear models), depending upon the type of signal applied to the amplifier, the degree of accuracy desired, and the type of computational facilities employed.

Consider a resistance R_F connected between the output side and the input side of the model of an amplifier, as shown in Fig. 3-39(a).

R_F serves as a link between the output and input sides and *feeds back* information from the output side to the input side. Hence, it is called a *feedback resistance*.

The circuit of Fig. 3-39(a) is thus a model of a *feedback amplifier* circuit. Specially designed feedback amplifiers are commercially available in which the gain A is made extremely large (of the order 10^6), the input resistance R_i is also made extremely large (of the order 10^6 Ω and much higher in some cases), whereas the output resistance R_o is made extremely small (of the order 100 Ω). Such amplifiers are called *operational amplifiers* because they were originally designed for performing mathematical operations like addition, subtraction, and integration. With the advent of integrated circuits technology, op amps have become a standard component in a wide variety of electronic circuits.

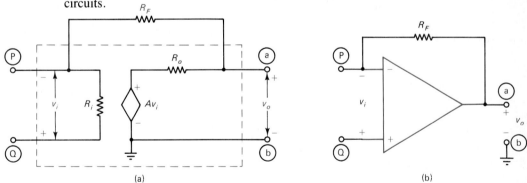

Fig. 3-39 (a) Amplifier model with feedback. (b) Symbol of an op amp with feedback.

3-4-1 Symbol of an Op Amp

The portion of the op amp enclosed by the dashed lines in Fig. 3-39(a) is symbolically shown as a large triangle with two input terminals and one output terminal, as shown in Fig. 3-39(b). The feedback resistor R_F is shown externally connected to the terminals of the op-amp symbol.

The input terminal marked with a minus sign is called the *inverting input* terminal and the one marked with a plus sign is called the *noninverting input* terminal.

In order to see the basis for this labeling, let P and Q denote the inverting and noninverting terminals, respectively. Then, the input voltage to the op amp is defined by

$$v_i = v_Q - v_P \tag{3-71}$$

and the output in Fig. 3-39(b) is given by

$$v_o = Av_i = A(v_Q - v_P)$$
$$= -Av_P + Av_Q$$

It is seen in the above equation, that the term due to a voltage applied to P is multiplied by a minus sign, that is, it is inverted by the amplifier. On the other hand, the term due to a voltage applied to Q is multiplied by a plus sign, that is, it is not inverted by the amplifier. Consequently, P is called the *inverting* input terminal and Q the *noninverting* input terminal.

The minus and plus signs on the op-amp symbol do not mean that only a negative signal should be connected to terminal P or only a positive signal connected to terminal Q. In actual practice, any signal can be connected to one or the other or both terminals, depending upon the particular application.

3-4-2 Analysis of the Inverting Op Amp Circuit

Consider a voltage signal v_s in series with a resistance R_1 connected to the inverting input terminal of the op amp, while a noninverting input terminal is connected to ground, as shown in Fig. 3-40(a). The circuit configuration in Fig. 3-40(a) is called the *inverting op-amp* circuit.

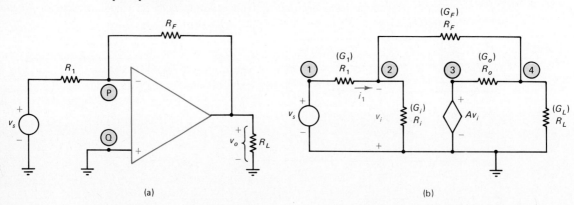

Fig. 3-40 (a) Inverting op-amp circuit. (b) Inverting op-amp circuit with the op amp replaced by a model.

In order to find the voltage gain v_o/v_s and the input resistance of the amplifier, replace the op amp by its model, as shown in Fig. 3-40(b). The nodal equations are

Node 1:
$$v_1 = v_s \tag{3-72}$$

Node 2:
$$-G_1 v_1 + (G_1 + G_i + G_F)v_2 - G_F v_4 = 0$$

which, in combination with Eq. (3-72), leads to

$$(G_1 + G_i + G_F)v_2 - G_F v_4 = G_1 v_s \tag{3-73}$$

Node 3:
$$v_3 = Av_i = -Av_2 \tag{3-74}$$

Node 4:
$$-G_F v_2 - G_o v_3 + (G_F + G_o + G_L) v_4 = 0$$

which, in combination with Eq. (3-74), leads to

$$(AG_o - G_F)v_2 + (G_F + G_o + G_L) v_4 = 0 \tag{3-75}$$

Equations (3-73) and (3-75) are to be solved for v_2 and v_4. The determinant of the system of two equations, Eqs. (3-73) and (3-75), is, after some manipulation,

$$\|g\| = (G_1 + G_i)(G_F + G_o + G_L) + G_F[(A + 1)G_o + G_L] \tag{3-76}$$

The voltage v_2 is given by

$$v_2 = G_1(G_F + G_o + G_L)v_s/\|g\| \tag{3-77}$$

and the output voltage v_o is given by

$$v_o = v_4 = -G_1(AG_o - G_F)v_s/\|g\| \tag{3-78}$$

The voltage gain is given by

$$\frac{v_o}{v_s} = -\frac{G_1(AG_o - G_F)}{\|g\|} \tag{3-80}$$

The input resistance of the amplifier is defined by

$$R_{in} = v_2/i_1 \tag{3-81}$$

where i_1 is the current supplied by v_s. Since i_1 is the current through G_1, the expression for i_1 is

$$i_1 = (v_s - v_2)G_1 \tag{3-82}$$

Equations (3-80) to (3-82) lead to

$$R_{in} = \frac{(G_F + G_o + G_L)}{G_i(G_F + G_o + G_L) + G_F[(A + 1)G_o + G_L]} \tag{3-83}$$

Exercise 3-36 An op amp has the following specifications: $G_i = 10^{-6}$ S, $G_o = 10^{-2}$ S, and $A = 2 \times 10^5$. Calculate the voltage gain and the input resistance of the inverting op-amp circuit when $G_1 = 10^{-3}$ S, $G_F = 10^{-4}$ S, and $G_L = 10^{-3}$ S.

Exercise 3-37 In typical op-amp circuits, $G_i \ll G_1$, $G_o \gg G_F$, and $G_o \gg G_L$. Use these criteria and obtain approximate expressions for the determinant $\|g\|$, the voltage gain, and the input resistance of the inverting op-amp circuit. Use the numerical values of Exercise 3-36 in the approximate expressions, and compare the results with those calculated in Exercise 3-36.

Exercise 3-38 The μA741 is a commercially available op amp very popular with electronic designers. Its input resistance is given by the manufacturer as 2 MΩ and its output resistance as 75 Ω. Using a factor of 100 to define "much greater than," determine the *maximum* value of R_1 and the *minimum* values of R_F and R_L to satisfy the conditions stated in Exercise 3-37.

3-4-3 Approximate Model of an Op Amp

The last two exercises show that certain valid approximations can be made in practical op-amp circuits.

1. The value of G_i of a practical op amp is small enough to be neglected in comparison with G_s. Therefore, G_i can be taken as zero or $R_i = \infty$ in the op-amp model of Fig. 3-39.
2. The value of G_o of a practical op amp is very large compared with G_F or G_L. Therefore, G_o can be taken as infinity or $R_o = 0$ in the op-amp model of Fig. 3-39.

The op-amp model resulting from the above two approximations is shown in Fig. 3-41(a). The op amp in the approximate model presents an *open circuit* at its input terminals. This means that

the input currents (shown as i_1 and i_1' in the diagram) to the op amp are zero. The op amp draws no input current.

It is also seen that, in the approximate model, the output voltage v_o of the op amp is *identically* equal to Av_i.

Thus the output voltage depends only upon the gain A and the input voltage v_i; it is independent of the load resistance R_L.

Because of this lack of dependence of the output voltage on the load resistance, it is customary to omit the load resistance in op-amp circuit diagrams.

Fig. 3-41 (a) Approximate model of an op amp. (b) Inverting op-amp circuit with the op amp replaced by an approximate model.

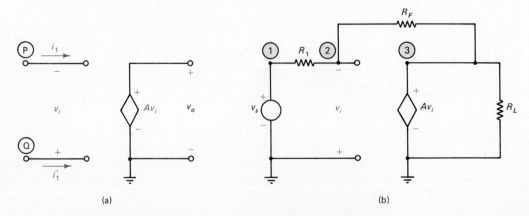

Nodal Analysis

(A third approximation, namely, the *gain A* of the op amp is *infinite,* is also frequently used in electronics. This leads to the so called *ideal op amp*. We will, however, use the approximate model of Fig. 3-39 with a finite gain A.)

Analysis of the Inverting Op-Amp Circuit Using the Approximate Model:

When the approximate op-amp model is used in the inverting op-amp circuit of Fig. 3-40, the resulting circuit is as shown in Fig. 3-41(b). Nodal analysis of that circuit leads to the following expressions (the same as those obtained in Exercise 3-37):

$$\frac{v_o}{v_s} = -\frac{AG_1}{G_1 + G_F(1 + A)} \tag{3-84}$$

$$R_{in} = \frac{R_F}{(1 + A)} \tag{3-85}$$

Exercise 3-39 Write the nodal equations of the circuit of Fig. 3-41(b) and derive Eqs. (3-84) and (3-85).

Exercise 3-40 In practical op amps, the gain A is of the order of 10^5 or higher, so that $A \gg 1$ and $AG_F \gg G_1$. What is the effect of this approximation on Eqs. (3-84) and (3-85)?

The approximate model of the op amp is used in the following discussion because it saves a great deal of work and the results obtained are sufficiently accurate for all practical purposes.

3-5 THE NONINVERTING OP AMP CIRCUIT

A signal can be applied to the noninverting input terminal of an op amp instead of the inverting input, as shown in Fig. 3-42(a). Such a configuration is known as the *noninverting op-amp circuit*. Note that the feedback resistor R_F is still connected between the inverting input and the output terminals. Since the op amp is assumed to have infinite input resistance and draws no input current, it is necessary to have resistor R_1 connected between the inverting input and ground so as to provide a path for current flow through R_F. Replacing the op amp by its approximate model leads to the equivalent circuit shown in Fig. 3-42(b).

The nodal equations for the circuit of Fig. 3-42(b) are:

Node 1: $$(G_1 + G_F)v_1 - G_F v_2 = 0 \tag{3-86}$$

Node 2: $$v_2 = Av_i \tag{3-87}$$

Node 3: $$v_3 = v_s \tag{3-88}$$

Also,

$$v_i = v_3 - v_1 = v_s - v_1 \tag{3-89}$$

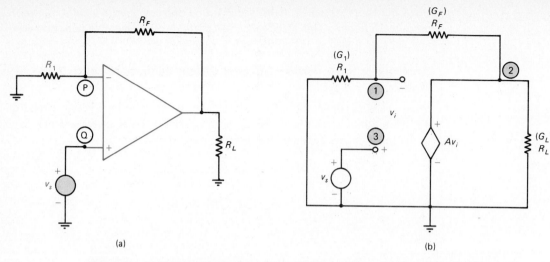

Fig. 3-42 (a) Noninverting op-amp circuit. (b) Noninverting op-amp circuit with the op amp replaced by an approximate model.

Equations (3-86) through (3-89) lead to

$$[G_1 + G_F(A + 1)]v_1 = AG_F v_s$$

or

$$v_1 = AG_F v_s/[G_1 + G_F(A + 1)] \qquad (3\text{-}90)$$

From Eqs. (3-89) and (3-90),

$$v_i = v_s - v_1$$
$$= (G_1 + G_F)v_s/[G_1 + G_F(A + 1)] \qquad (3\text{-}91)$$

The voltage gain is given by

$$v_o/v_s = Av_i/v_s$$
$$= \frac{A(G_1 + G_F)}{[G_1 + G_F(A + 1)]} \qquad (3\text{-}92)$$

The input resistance of the noninverting op amp is infinite because the current supplied by the signal source is zero.

Exercise 3-41 Calculate the voltage gain of the noninverting op-amp circuit when $G_1 = 10^{-3}$ S, $G_F = 10^{-4}$ S, and $G_L = 10^{-3}$ S. Assume $A = 2 \times 10^5$.

Exercise 3-42 What is the effect on Eq. (3-92) when the gain A becomes sufficiently large as in Exercise 3-40?

A comparison of the voltage gains of the inverting and noninverting op-amp circuits leads to the following observations:

Nodal Analysis

1. The voltage gain of the *inverting* op-amp circuit is *negative*, whereas that of the *noninverting* op-amp circuit is *positive*. That is, the polarity of the output voltage is the same as the signal v_s in the noninverting op-amp circuit, but opposite to the input signal in the inverting op-amp circuit.
2. The voltage gain of the *inverting* op-amp circuit can have any magnitude between *zero and infinity*, whereas the voltage gain of the *noninverting* op-amp circuit has a *minimum value of 1*.

Exercise 3-43 Establish the above-mentioned property about the range of voltage-gain magnitudes of the two op-amp circuits.

Exercise 3-44 For each set of values given, calculate (a) the voltage gain and (b) the input resistance for an inverting op-amp circuit and a noninverting op-amp circuit. (i) $G_1 = 10^{-3}$ S, $G_F = 9 \times 10^{-3}$ S, $A = 10^3$. (ii) $G_1 = 9 \times 10^{-3}$ S, $G_F = 10^{-3}$ S, $A = 10^3$. (iii) $G_1 = G_F = 10^{-4}$ S, $A = 10^3$.

3-6 THE SUMMING OP-AMP CIRCUIT

It was mentioned earlier that the op amp was originally designed to perform certain mathematical operations such as addition, subtraction, differentiation, and integration. We will consider here the addition of two or more signals through the use of a summing op-amp circuit. Differentiation and integration are discussed in Chapter 6. Consider three voltage signals, v_{i1}, v_{i2}, and v_{i3}, applied to the inverting input terminal of an op amp through resistors R_1, R_2, and R_3, respectively, as shown in Fig. 3-43(a). The output of the circuit is equal to the *negative weighted sum* of the inputs. To prove this, replace the op amp by its approximate model, leading to the equivalent circuit of Fig. 3-43(b).

The equation for node 1 is

$$-G_1 v_{i1} - G_2 v_{i2} - G_3 v_{i3} + (G_1 + G_2 + G_3 + G_F)v_1 - G_F v_2 = 0$$

Using $v_i = -v_1$ and $v_2 = Av_i = -Av_1$, the equation for node 1 is rewritten as

$$[G_F(A + 1) + G_1 + G_2 + G_3]v_1 = G_1 v_{i1} + G_2 v_{i2} + G_3 v_{i3}$$

Fig. 3-43 (a) Summing op-amp circuit. (b) Summing op-amp circuit with the op amp replaced by an approximate model.

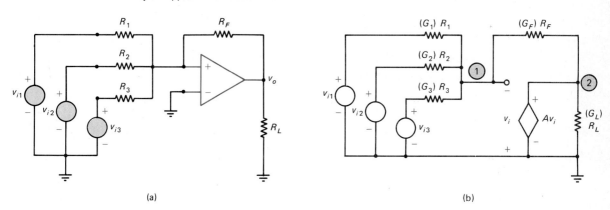

or

$$v_1 = \frac{G_1 v_{i1} + G_2 v_{i2} + G_3 v_{i3}}{G_F(A + 1) + G_1 + G_2 + G_3}$$

and the output voltage is given by

$$v_o = v_2 = -A v_1$$

$$= \frac{-A(G_1 v_{i1} + G_2 v_{i2} + G_3 v_{i3})}{G_F(A + 1) + G_1 + G_2 + G_3}$$

The expression for the output voltage is of the form

$$v_o = -(K_1 v_{i1} + K_2 v_{i2} + K_3 v_{i3}) \tag{3-93}$$

where K_1, K_2, and K_3 are constants given by

$$K_1 = AG_1/[G_F(A + 1) + G_1 + G_2 + G_3]$$
$$K_2 = AG_2/[G_F(A + 1) + G_1 + G_2 + G_3]$$
$$K_3 = AG_3/[G_F(A + 1) + G_1 + G_2 + G_3]$$

The output voltage of the given op-amp circuit is, therefore, the negative weighted sum of the three inputs.

Proceeding in a similar manner, the output voltage of a summing op-amp circuit with k input signals, $v_{i1}, v_{i2}, \ldots, v_{ik}$ applied to the inverting input terminal through resistors $R_1, R_2, \ldots, R_k$, respectively, is given by the expression

$$v_o = -(K_1 v_{i1} + K_2 v_{i2} + \ldots + K_k v_{ik})$$

where

$$K_j = \frac{AG_j}{G_F(A + 1) + G_1 + G_2 + \ldots + G_k}$$

Exercise 3-45 Determine the output voltage in terms of the input voltages in a summing op-amp circuit if $R_1 = 10$ kΩ, $R_2 = 2$ kΩ, $R_3 = 20$ kΩ and $R_F = 20$ kΩ. Assume $A = 2 \times 10^5$.

Exercise 3-46 Suppose the three inputs to a summing op-amp circuit are given as $v_{i1} = 0.08$ V, $v_{i2} = 0.02$ V, and $v_{i3} = -0.12$ V. $R_F = 100$ kΩ. Given that $R_1 = 5$ kΩ and $R_2 = 10$ kΩ, find the value of R_3 so as to make the output voltage equal zero. Assume $A = 2 \times 10^5$.

If the resistors in a summing op-amp circuit are chosen so that $R_1 = R_2 = \ldots = R_k$, then the output becomes

$$v_o = -K(v_{i1} + v_{i2} + \ldots + v_{ik})$$

where

$$K = \frac{AG_1}{G_1 + G_F(A + 1)}$$

The output then equals the negative sum of the inputs multiplied by a scale factor K.

Exercise 3-47 Obtain a relationship for G_F in terms of G_1 and A in order to make the scale factor $K = 1$. Modify the expression for the case where A is sufficiently large.

The above analysis shows that it is possible to choose the feedback resistor so as to obtain the negative sum of two or more input signals applied to a summing op-amp circuit. If it is desired to obtain the *positive* sum of the input signals, then the output of the summing amplifier is applied to an inverting op-amp circuit with a voltage gain of -1 as indicated in Fig. 3-44. The output v_o is then the positive sum of the input signals.

The op amp finds numerous applications in electronic circuits where signals have to be added or subtracted, or the weighted sum of a set of signals is desired. For example, threshold logic circuits can be designed so that the output can be made to switch from a low value to a high value (or *vice versa*), depending upon the weighted sum of several inputs. The problems at the end of the chapter consider several op-amp circuits. Op-amp circuits are also discussed in Chapters 4 and 6.

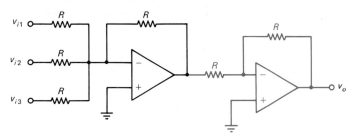

Fig. 3-44 Obtaining an output equal to the *positive* sum of the inputs by using a second op amp with a gain of -1.

3-7 SUMMARY OF CHAPTER

The first step in the analysis of a circuit is the choice of a complete set of independent variables, which may be a set of branch currents or voltages. The topology of a network is the tool for selecting a set of independent voltage variables or currents. In studying the topology of a network, the geometry of connections of its branches is first represented in the form of a graph. A tree is a subset of the graph, providing a path from any node to any other node through its branches but with no closed paths. Tree-branches correspond to a set of independent node-pair voltages that can be used for writing a set of independent equations. The number of independent node-pair voltages in a circuit is exactly equal to the number of tree-branches, which is one less than the total number of nodes.

Nodal analysis uses a set of independent node-to-datum voltages (that is, voltages between the nodes of the circuit and a common reference node) as the variable. One of the nodes in the circuit is chosen as the reference node and KCL equations are written for the remaining nodes. Nodal equations are expressions of Kirchhoff's current law, where the current through each resistive branch is written as the product of a conductance and the difference between the node voltages at the terminals of the branch. Solution of the nodal equations gives the node voltages. Once the voltages at all the nodes are known, it is possible to determine the current in any branch and other quantities of interest in the given circuit.

The left-hand side of the standard form of the equation for node k contains the terms involving the node voltages: the coefficient of voltage v_k is the sum of all the conductances meeting at node k; the coefficient of v_j ($j \neq k$) is the negative of the sum of the conductances linking nodes j and k. The right-hand side of the equation is the algebraic sum of the currents due to sources meeting at node k: a source current entering the node is added and a source current leaving the node is subtracted. If dependent sources are present, they are treated as if they were ordinary current sources in setting up the equations. Some manipulations are then necessary to obtain the standard form of the equations.

When a circuit contains voltage sources, a judicious choice of the reference node often leads to a significant reduction in the computations.

The application of the theory of determinants to nodal equations leads to the concept of the nodal conductance matrix whose elements are the coefficients of the voltages in the form of a matrix. The nodal conductance matrix of networks containing resistors and independent sources (but no dependent sources) can be written by inspection of the circuit. The matrices of such networks are symmetric about the principal diagonal. For networks containing dependent sources, it is necessary to first write some of the equations explicitly and rearrange terms before writing the matrix. The matrices of such networks are, in general, not symmetric.

The application of Cramer's rule to the nodal conductance matrix leads to the concepts of driving point resistance and transfer resistance. The driving point resistance represents the total resistance seen looking in from any specified node of the network. The transfer resistance is a measure of the effect of a current source connected to a node to the voltage at a different node. The determination of the driving point and transfer resistances of a circuit is systematic once the nodal conductance matrix of the circuit is known.

An operational amplifier is a specially designed, commercially available amplifier with certain special properties: extremely high input resistance, extremely low output resistance, and extremely high gain. The inverting op-amp circuit has an output opposite in polarity to the input and a gain whose magnitude lies in the range zero to infinity. The noninverting op-amp circuit has an output of the same polarity as the input and a gain of magnitude greater than 1. The summing op-amp circuit has an output equal to the (negative) weighted sum of the input signals, with the weights being controlled by resistors in series with the signal sources.

Answers to Exercises

3-1 See Fig. 3-45. The trees shown here may not all be the same as the ones you set up.

3-2 (a) *Tree 1:* branches 1, 5, and 8. *Tree 2:* branch 4. *Tree 3:* branch 4. *Tree 4:* branches 3 and 7. (b) *Tree 1:* branches 1 and 4. *Tree 2:* branches 1 and 5. *Tree 3:* branches 2, 3, and 6.

3-3 $v_{12} = -v_{23} - v_{34} - v_{45} - v_{51}$.

3-4 (a) $v_{14} = v_1 - v_4$; (b) $v_{25} = v_2 - v_5$; (c) $v_{51} = v_5 - v_1$; (d) $v_{32} = -v_2$.

3-5 $i_1' = G_1(v_b - v_a)$.

3-6 $i_a = G_a(v_2 - v_1)$; $i_1 = G_1(v_2 - v_1)$; $i_2 = G_2(v_3 - v_2)$; $i_3 = G_3 v_2$; $i_4 = G_4(v_4 - v_3)$; and $i_5 = -G_5 v_5$.

3-7 279.8 W in G_1; 110.4 W in G_2; 172.9 W in G_3; and 259.3 W in G_4.

3-8 Node 1: $3v_1 - 2v_2 - v_4 = 10$. Node 2: $-2v_1 + 16v_2 - 4v_3 - 7v_4 = 0$. Node 3: $-4v_2 + 9v_3 = -10$. Node 4: $-v_1 - 7v_2 + 14v_4 = 10$.

3-9 Node 1: $0.21v_1 - 0.2v_2 = 35$. Node 2: $-0.2v_1 + 0.25v_2 = -10$. $v_1 = 540$ V and $v_2 = 392$ V. Branch currents are: 29.6 A in 0.2 S; 5.4 A in 0.01 S; and 19.6 A in 0.05 S.

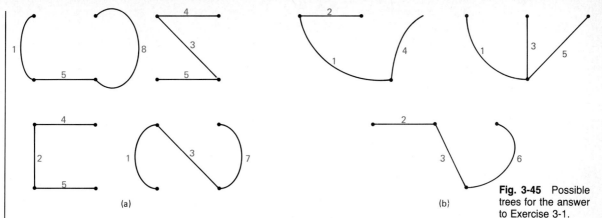

Fig. 3-45 Possible trees for the answer to Exercise 3-1.

3-10 Node 1: $6v_1 - 4v_2 = -25$.
Node 2: $-4v_1 + 7v_2 = 15$. $v_1 = -4.42$ V and $v_2 = -0.385$ V.

3-11 (a) Node 1: $(G_1 + G_2 + G_3)v_1 - G_3v_2 = i_a - i_b$.
Node 2: $-G_3v_1 + (G_3 + G_5 + G_6)v_2 - (G_5 + G_6)v_3 = i_b - i_c$.
Node 3: $-(G_5 + G_6)v_2 + (G_4 + G_5 + G_6)v_3 = i_c$. (b) See answer to Exercise 3-8.

3-12 (a) $0.15v_1 - 0.1v_2 = 20$; $-0.1v_1 + 0.2v_2 = 0$. (b) $(G_1 + G_2 + G_3)v_1 - G_3v_2 - G_1v_3 = i_a$; $-G_3v_1 + (G_3 + G_5)v_2 = i_b$; $-G_1v_1 + (G_1 + G_4)v_3 = -i_a - i_b$.
(c) $(G_1 + G_5 + G_4 + G_3 + G_2)v_1 = -i_a - i_b - i_c$.

3-13 Node equations: $4.5v_1 - 2.5v_2 = -10$; $-2.5v_1 + 6.7v_2 = -15$.
$v_1 = -4.37$ V and $v_2 = -3.87$ V. i_a delivers 43.7 W; i_b delivers 58.0 W. Powers dissipated: 38.2 W in G_1; 0.625 W in G_2; and 62.9 W in G_3.

3-14 $i_x = 35$ A.

3-15 Voltage across the i_a source = 6.6 V (bottom terminal positive). Power received = 132 W. Voltage across G_c = 0.8 V (left side positive). Power dissipated = 0.192 W.

3-16 $v_1 = 100$ V and $v_2 = 40$ V. Branch currents: 12.0 A in G_a; 8.0 A in the 100-V source; and 2.0 A in G_b.

3-17 $v_1 = 100$ V; $v_2 = 50$ V; and $v_3 = 72.2$ V. Current through the v_a source = 11.1 A; it delivers 1110 W. Current through the v_b source = 6.66 A; it receives 333 W. i_a source delivers 722 W.

3-18 2.8 A in G_a; 10 A in G_b; 13.3 A in G_c; 0.52 A in G_d; and 19.8 A in G_e.

3-19 (a) $v_1 = -50$ V. Node 2 (bottom): $1.2v_2 - 0.4v_3 = -55$.
Node 3 (extreme right): $-0.4v_2 + 0.5v_3 = 20$. $v_2 = -44.3$ V; $v_3 = 4.54$ V; and $i_x = 22.8$ A.
(b) Node 1 (bottom): $v_1 = -50$ V. Node 2 (extreme left): $v_2 = 50$ V and $v_3 = 22.2$ V. Other answers are the same as for Exercise 3-17.

3-20 (a) With the bottom node as reference, node 1 at the left of the v_b source and numbering the remaining nodes counterclockwise, $v_1 = -50$. Node 2: $0.5v_2 - 0.25v_3 = -17.5$.
Node 3: $-0.2v_2 + 0.2v_3 - i_a = 0$. Node 4: $-0.05v_2 + 0.3v_3 + i_a = 17.5$ A, where i_a is the current in the v_a source (right to left). v_a delivers 933 W; v_b delivers 334 W; i_c delivers 958 W. Power dissipated: 435 W in G_1; 143 W in G_2; 178 W in G_3; and 1475 W in G_4. (b) Choose the node common to v_a and v_b as reference. Node 1 is at the negative terminal of v_a; node 2 is at the positive terminal of v_b; node 3 is at the bottom left; and node 4 is at the bottom right. $v_1 = -50$ V and $v_2 = 100$ V. Node 3: $1.05v_3 - 0.7v_4 = -5$. Node 4: $-0.7v_3 + 0.75v_4 = -5$. $v_3 = -24.4$ V and $v_4 = -29.4$ V. v_a delivers 128 W; i_c delivers 294 W; and v_b delivers 647 W. Power dissipated: 65.5 W in G_1; 149 W in G_2; 5 W in G_3; 12.5 W in G_4; and 837 W in G_5.

Answers to Exercises

3-21 Number the nodes clockwise starting at top left. $v_3 = (v_1 + 80)$. Node 1: $0.75v_1 - 0.5v_2 + i_b = -10$. Node 2: $-0.583v_1 + 0.583v_2 = 6.66$. Node 3: $0.133v_1 - 0.0833v_2 - i_b = -10.7$, where i_b is the current in the v_b source (from top to bottom). $v_1 = -46.7$ V; $v_2 = -35.2$ V; $v_3 = 33.3$ V; $i_b = 7.42$ A. Branch currents: 11.7 A in G_4; 5.75 A in G_1; and 1.66 A in G_3.

3-22 $i_b = (G_3 G_4 / \|g\|) i_s$.

3-23 Node 1: $2v_1 - 0.5v_2 = 10$. Node 2: $-1.5v_1 + v_2 = 10$. $v_2 = 28$ V.

3-24 Node 1: $(G_1 + G_2) v_1 = i_s$. Node 2: $G_L v_2 = -K i_b$. $v_2/v_1 = -K(G_2/G_L)$

3-25 Node 1: $(G_a + G_b) v_1 - G_b v_3 = i_s$. Node 2: $v_2 = K v_1$. Node 3: $- G_b v_1 - G_c v_2 + (G_b + G_c + G_d) v_3 = 0$. $v_3/i_s = (G_b + K G_c)/\|g\|$, where $\|g\| = (G_a + G_b)(G_b + G_c + G_d) - G_b(G_b + K G_c)$.

3-26 $i_x = 2(v_2 - v_1)$. Node 1: $2.5v_1 - 2v_2 = i_s$. Node 2: $4v_1 - 2.5v_2 = 0$. $v_1 = -1.43 i_s$. Input resistance $= v_1/i_s = -1.43 \Omega$. Yes, the resistance is *negative*. This is a possibility in circuits with dependent sources.

3-27 Use the answers given above for Exercises 3-11 and 3-12.

3-28 N/A below denotes "not applicable." (a) $g_{11} = G_1 + G_2 + G_3$; 3; 0.15; $G_1 + G_2 + G_3$; $G_1 + G_2 + G_3 + G_4 + G_5$. $g_{22} = G_3 + G_5 + G_6$; 16; 0.2; $G_3 + G_5$; N/A. $g_{23} = -(G_5 + G_6)$; -4; N/A; 0; N/A. $g_{31} = 0$; 0; N/A; $-G_1$; N/A.
(b) $0v_1, -(G_5 + G_6)v_2, (G_4 + G_5 + G_6)v_3$; $0v_1, -4v_2, 9v_3, 0v_4$; N/A; $-G_1 v_1, 0v_2, (G_1 + G_4) v_3$. (c) $-G_3, (G_3 + G_5 + G_6), -(G_5 + G_6)$; $-2, 16, -4, -7$; $-0.1, 0.2$; $-G_3, G_3 + G_5, 0$; N/A.

3-29 (a) $\begin{bmatrix} G_1 + G_2 + G_3 + G_4 & -G_4 \\ -G_4 & G_4 + G_5 + G_6 \end{bmatrix}$ (b) $\begin{bmatrix} 6 & -2 & -3 \\ -2 & 8 & -2 \\ -3 & -2 & 6 \end{bmatrix}$

3-30 $v_1 = 0.455$ V; $v_2 = -0.606$ V; and $v_3 = 0.227$ V

3-31 (a) $\begin{bmatrix} 0.7 & -0.2 & 0 \\ -0.2 & 1.7 & 0 \\ 0 & -0.8 & 0.25 \end{bmatrix}$ (b) $\begin{bmatrix} G_1 + G_2 + G_4 & -G_2 \\ -G_2 & G_2 + G_3 - K \end{bmatrix}$

3-32 (a) $v_1 = 1.48 i_a$; $v_2 = 0.174 i_a$; $v_3 = 0.557 i_a$; $v_2/v_1 = 0.118$.
(b) $v_1 = (G_2 + G_3 - K)/\|g\|$; $v_2 = G_2/\|g\|$; $v_2/v_1 = [G_2/(G_2 + G_3 - K)]$.

3-33 $v_3/i_1 = 0.152\ \Omega$. $v_1/i_2 = 0.103\ \Omega$. $v_1/i_3 = 0.152\ \Omega$. $v_2/i_3 = 0.138\ \Omega$.

3-34 (a) $\begin{bmatrix} 1.2 & -0.5 & -0.25 \\ -0.5 & 1.1 & -0.05 \\ -0.25 & -0.05 & 0.95 \end{bmatrix}$

Transfer resistances: 1 to 2 (or 2 to 1) $0.524\ \Omega$; 1 to 3 (or 3 to 1) $0.322\ \Omega$; and 2 to 3 (or 3 to 2) $0.198\ \Omega$.

(b) $\begin{bmatrix} G_1 + G_2 & 0 & 0 \\ K & G_3 + G_4 & -G_4 \\ -K & -G_4 & G_4 + G_5 \end{bmatrix}$

$\|g\| = (G_1 + G_2)[(G_3 + G_4)(G_4 + G_5) - G_4^2]$.
Transfer resistances: $v_2/i_1 = -KG_5/\|g\|$; $v_3/i_1 = KG_3/\|g\|$; $v_1/i_2 = 0$; $v_3/i_2 = G_4(G_1 + G_2)/\|g\|$; $v_1/i_3 = 0$; $v_2/i_3 = G_4(G_1 + G_2)/\|g\|$.

3-35 $\begin{bmatrix} G_s + G_1 & -G_1 & 0 \\ -G_1 & G_1 + G_2 & 0 \\ 0 & K & G_3 + G_2 \end{bmatrix}$

$\|g\| = (G_s + G_1)[(G_1 + G_2)(G_3 + G_L) - G_1^2]$.

Transfer resistances: $v_1/i_2 = v_2/i_1 = G_1(G_3 + G_L)/\|g\|$; $v_1/i_3 = 0$; $v_2/i_3 = 0$; $v_3/i_1 = -KG_1/\|g\|$; $v_3/i_2 = -K(G_s + G_1)/\|g\|$. Driving point resistances: $R_{in} = (G_1 + G_2)(G_3 + G_L)/\|g\|$; $R_{out} = 1/(G_3 + G_2)$.

3-36 Gain $= -10$ and $R_{in} = 0.056$ Ω.

3-37 $\|g\| = G_o[G_1 + (A + 1)G_F]$. $v_o/v_s \approx -AG_1/[G_1 + (A + 1)G_F]$. $R_{in} \approx 1/G_F(A + 1)$. Gain $= -10$; $R_{in} = 0.05$ Ω.

3-38 $R_1 < 20$ kΩ. $R_F > 7.5$ kΩ; and $R_L > 7.5$ kΩ.

3-39 Nodal equation $(G_1 + G_F)v_2 + AG_Fv_2 = G_1v_s$ leads to the desired results.

3-40 $v_o/v_s = -(G_1/G_F)$. $R_{in} = 0$.

3-41 Gain $= 11$.

3-42 Gain $= (G_1 + G_F)/G_F$.

3-43 Gain can be made zero by making $G_1 = 0$ in the inverting op-amp circuit. As G_1 increases to large values, the gain approaches $-A$. In the noninverting op amp, the gain is 1 when $G_1 = 0$. As G_1 increases to large values, the gain approaches $+A$.

3-44 (i) Gain $= -0.111$ and $R_{in} = 0.111$ Ω for the inverting connection; gain $= 1.11$ and $R_{in} = \infty$ for the noninverting connection. (ii) Gain $= -8.91$ and $R_{in} = 0.999$ Ω for the inverting connection; gain $= 9.90$ and $R_{in} = \infty$ for the noninverting connection. (iii) Gain $= -0.998$ and $R_{in} = 9.99$ Ω for the inverting connection; gain $= 2.00$; and $R_{in} = \infty$ for the noninverting connection.

3-45 $v_o = -(2v_{i1} + 10v_{i2} + v_{i3})$.

3-46 $R_3 = 6.67$ kΩ.

3-47 $G_F = G_1$.

PROBLEMS

Sec. 3-1 Network Topology—Choice of Independent Voltages

3-1 For each of the circuits shown in Fig. 3-46, (a) draw the graph, and (b) draw three trees.

3-2 For the graph of the circuit in Fig. 3.46(a), (a) draw a tree containing the branches labeled A, C, E, G, and H; (b) make a list of the tree-branches whose voltages should be algebraically added to obtain the node-pair voltages: (i) 1 and 4; (ii) 2 and 6.

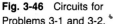

Fig. 3-46 Circuits for Problems 3-1 and 3-2.

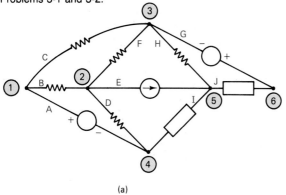

(a)

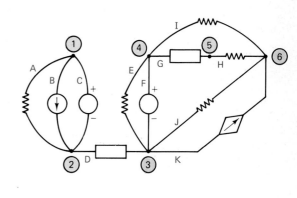

(b)

3-3 In the graph shown in Fig. 3-47, denote the voltage across each branch by v_j, where j = A, B, ..., L are the letters associated with the branches in the diagram. The reference voltage polarities are as shown in the diagram. Set up a tree consisting of branches labeled A, B, D, E, F, H, and J. Write an expression in the form of an algebraic sum of tree-branch voltages for each of the following node-pair voltages: (a) Nodes 1 and 4, with node 1 taken as reference positive; (b) nodes 5 and 7, with node 5 taken as reference positive; (c) nodes 1 and 8, with node 1 taken as reference positive.

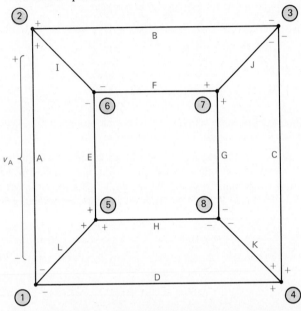

Fig. 3-47 Graph for Problems 3-3, 3-4, and 3-5.

3-4 Repeat the work of the previous problem for the graph in that problem using a tree comprising branches labeled I, J, K, L, B, C, and D.

3-5 In the graph of Fig. 3-47, choose node 8 as the datum (reference) node. Write the voltages at the other nodes with respect to the datum node in terms of the tree-branch voltages in the tree set up in Problem 3-3.

3-6 Repeat the work of Problem 3-5 using the tree set up in Problem 3-4.

Sec. 3-2 Principles of Nodal Analysis

3-7 The numerical value shown next to a node in each of the resistors shown in Fig. 3-48 is the voltage at that node with respect to a reference node. Calculate the current in each resistor.

Fig. 3-48 Circuit for Problem 3-7.

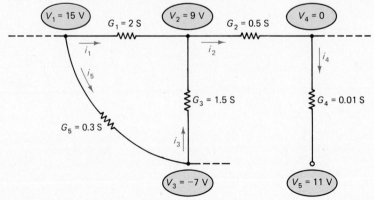

158   Nodal Analysis

3-8 The diagrams in Fig. 3-49 show several partial circuits. The connections to node 1 are complete, however. For each circuit shown, assign a current to each conductance branch. Write a KCL equation at node 1. Replace each current in the equation (except the source current) by the product of the relevant conductance and the difference between node voltages, and obtain an expression relating the source current to the node voltages and conductances.

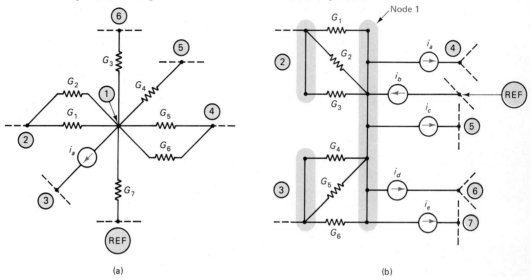

Fig. 3-49 Circuits for Problem 3-8.

3-9 Write the nodal equations for each of the circuits shown in Fig. 3-50. (It is assumed that the equations are to be put in standard form in this and future problems.)

Fig. 3-50 Circuits for Problem 3-9.

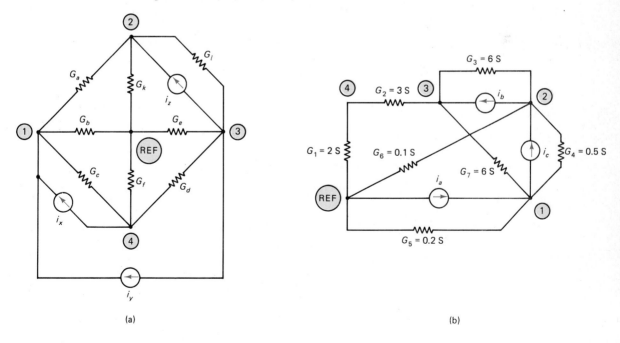

Problems 159

3-10 Use nodal analysis and determine (a) the currents in all the branches and (b) the power delivered or received by each component in the circuit shown in Fig. 3-51. (c) Verify that the total power received equals the total power delivered in the circuit.

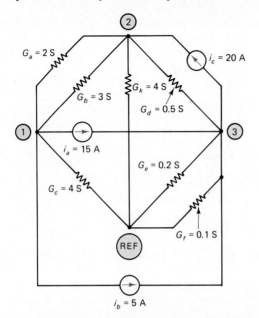

Fig. 3-51 Circuit for Problems 3-10, 3-11, and 3-12.

3-11 Suppose the reference node in the circuit of Problem 3-10 is chosen at what was originally designated node 1. What will be the new values of the node voltages, the branch currents, and the power quantities obtained in the previous Problem 3-10? Arrive at your answers without redoing the problem.

3-12 The circuit of Fig. 3-51 has been set up in the laboratory. The voltage at the node designated as reference (in Problem 3-10) is measured with respect to a metal chassis and found to be 13 V. What will be the values of the voltages at the other nodes (with respect to the metal chassis)? What will be the values of the branch currents and the power quantities calculated in Problem 3-10? Arrive at your answers without redoing the problem.

3-13 Determine the nodal voltages in the circuit of Fig. 3-52. The result should be put in the form of a ratio of polynomials involving the conductances.

Fig. 3-52 Circuit for Problem 3-13.

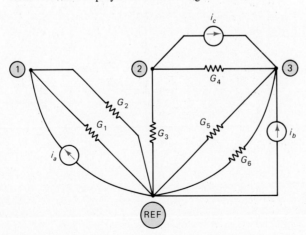

160 Nodal Analysis

3-14 Use nodal analysis and determine the ratio v_2/i_a in the circuit shown in Fig. 3-53.

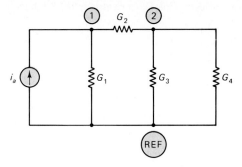

Fig. 3-53 Circuit for Problem 3-14.

3-15 It is known that the i_b current source in the circuit of Fig. 3-54 *delivers* a power of 300 W to the circuit. Use nodal analysis and determine all the branch currents.

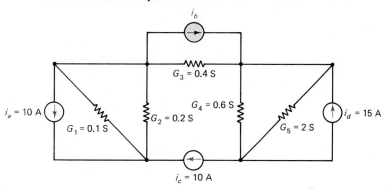

Fig. 3-54 Circuit for Problems 3-15 and 3-16.

3-16 Suppose the power delivered by the i_b current source in the circuit of Problem 3-15 is zero. Use nodal analysis and determine all the branch currents.

3-17 The conductance G_a in the circuit of Fig. 3-55 is known to dissipate a power of 100 W. Evaluate the current i_a (which may be positive or negative) and the node voltages. If there is more than one correct solution, find all solutions.

Fig. 3-55 Circuit for Problem 3-17.

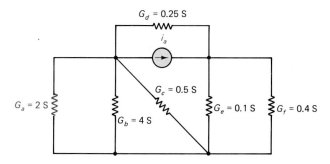

Problems

Subsec. 3-2-4 Circuits Containing Voltage Sources

3-18 Use nodal analysis and determine the current supplied by the voltage source in the circuit of Fig. 3-56. Choose a reference node so as to reduce the amount of computations.

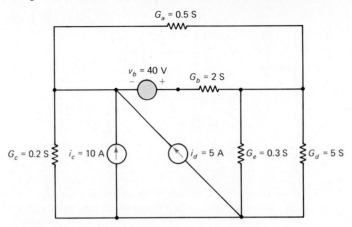

Fig. 3-56 Circuit for Problems 3-18, 3-19, and 3-21.

3-19 In the circuit of Problem 3-18, choose the node at the bottom as the reference node. Write the new set of nodal equations and recalculate the result asked for in Problem 3-18.

3-20 Use nodal analysis and find the power delivered by each of the sources in the circuit of Fig. 3-57.

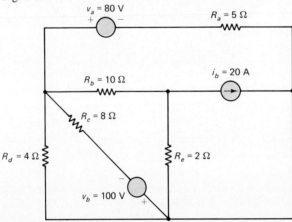

Fig. 3-57 Circuit for Problems 3-20 and 3-22.

3-21 It was mentioned in the text that an alternative procedure is available for the nodal analysis of circuits containing voltage sources. In Chapter 2, the conversion of a voltage source in series with a resistance by a current source in parallel with a resistance was discussed. Using that conversion procedure, it is possible to change a circuit containing voltage sources (and current sources) to one that contains current sources only. Do this for the circuit of Problem 3-18. Choose the node at the bottom as the reference node and write the equations for the various nodes. Solve the equations for the node voltages. Return to the *original* diagram (Fig. 3-56) and calculate the current supplied by the voltage source. Note that this procedure is not fully applicable if the circuit contains one or more voltage sources without a series resistance.

3-22 Repeat the work of Problem 3-20 using the procedure outlined in the previous problem.

Subsec. 3-2-5 Circuits Containing Dependent Sources

3-23 Write the nodal equations of each of the circuits shown in Fig. 3-58.

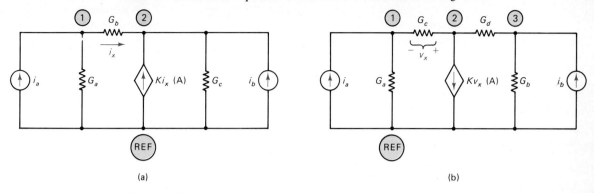

Fig. 3-58 Circuits for Problem 3-23.

3-24 Use nodal analysis and determine the node voltages and branch currents in the circuit shown in Fig. 3-59.

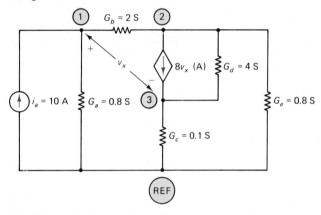

Fig. 3-59 Circuit for Problem 3-24.

3-25 Use nodal analysis and determine the conductance i_s/v_1 seen by the current source i_s in each of the circuits shown in Fig. 3-60.

Fig. 3-60 Circuits for Problem 3-25.

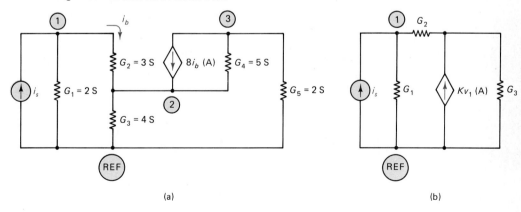

Problems

3-26 The circuit shown in Fig. 3-61 is the model of an amplifier (called the common-base bipolar transistor amplifier). Use nodal analysis and determine (a) the voltage ratio v_2/v_1, (b) the current ratio i_2/i_s, and (c) the resistance v_1/i_s seen by the current source. (d) Remove the current source and reconnect it between node 2 and the ground node. Determine the resistance v_2/i_s seen by the current source in the new position.

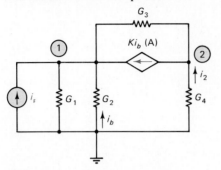

Fig. 3-61 Circuit for Problem 3-26.

3-27 Repeat the work of Problem 3-26 on the circuit shown in Fig. 3-62.

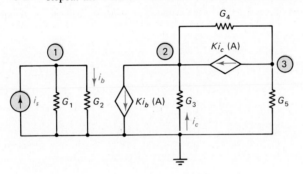

Fig. 3-62 Circuit for Problem 3-27.

Sec. 3-3 Algebraic Discussion of Nodal Analysis

3-28 Write the nodal conductance matrix of each of the circuits shown in Fig. 3-63.

Fig. 3-63 Circuits for Problem 3-28.

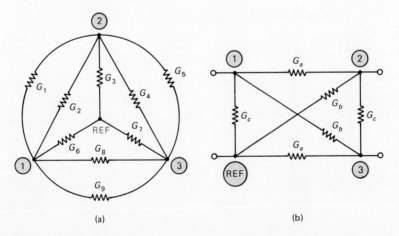

Nodal Analysis

3-29 The nodal conductance matrices of two circuits are given. Set up a circuit that will match each of the given matrices.

(a) $\begin{bmatrix} 10 & -4 & -2 \\ -4 & 9 & -1 \\ -2 & -1 & 3 \end{bmatrix}$ (b) $\begin{bmatrix} G_1 + G_2 & 0 & 0 \\ 0 & G_3 + G_4 & -G_3 \\ 0 & -G_3 & G_4 + G_5 \end{bmatrix}$

3-30 Use Cramer's rule and determine the node voltages for the circuits whose conductance matrices are given in Problem 3-29, when the following current sources are connected: $i_1 = 10$ A (between node 1 and ground) and $i_2 = 15$ A (between node 2 and ground).

3-31 Obtain the nodal conductance matrix of each of the circuits shown in Fig. 3-64.

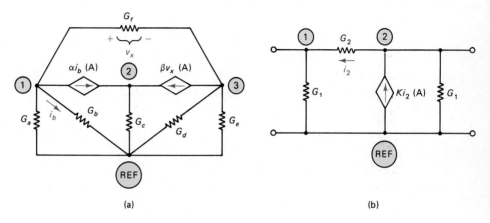

Fig. 3-64 Circuits for Problem 3-31.

3-32 The nodal conductance matrices of two circuits are given. In each case, set up a circuit diagram that can give rise to that matrix.

(a) $\begin{bmatrix} G_1 + G_2 & -G_2 \\ (K - G_2) & G_2 + G_3 \end{bmatrix}$ (b) $\begin{bmatrix} 10 & -4 & -2 \\ 0 & 6 & -3 \\ -2 & -3 & 7 \end{bmatrix}$

3-33 For the circuit shown in Fig. 3-65, (a) set up the nodal conductance matrix. Use Cramer's rule to evaluate (b) the node voltage v_2 and (c) the driving point resistance at node 1 when a current source i_1 is connected between node 1 and the reference node.

Fig. 3-65 Circuit for Problem 3-33.

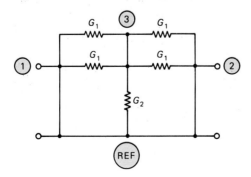

3-34 Repeat the work of Problem 3-33 for the circuit shown in Fig. 3-66.

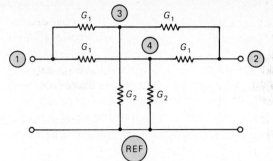

Fig. 3-66 Circuit for Problem 3-34.

3-35 Determine the transfer resistance (a) v_3/i_2 from node 2 to node 3 and (b) v_2/i_3 from node 3 to node 2 in the circuits whose conductance matrices were given in Problem 3-29.

3-36 Determine the transfer resistance (a) v_1/i_2 from node 2 to node 1 and (b) v_2/i_1 from node 1 to node 2 in the circuits whose conductance matrices were given in Problem 3-32.

3-37 Determine the ratio v_2/i_1 in the circuit shown in Fig. 3-67.

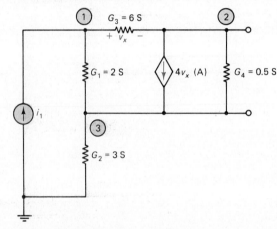

Fig. 3-67 Circuit for Problem 3-37.

3-38 Determine the three driving point resistances of each of the circuits in Problem 3-29.

3-39 Determine all possible driving point resistances of each of the circuits in Problem 3-32.

3-40 Determine the input and output resistances of each of the circuits shown in Fig. 3-68.

Fig. 3-68 Circuits for Problem 3-40.

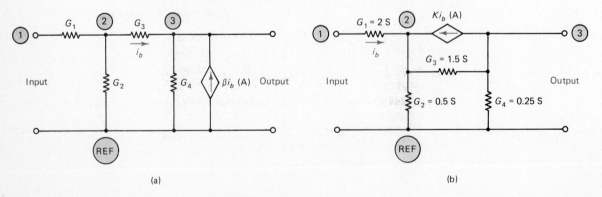

(a)   (b)

Nodal Analysis

Sec. 3-4 Operational Amplifiers

(The approximate model of the op amp should be used in solving the following problems.)

3-41 The gain of an op amp is $A = 10^6$. For each of the following sets of values, calculate the voltage gain and the input resistance of the inverting op-amp circuit: (a) $G_1 = 0.01$ S, $G_F = 10^{-5}$ S, and $G_L = 0.1$ S; (b) $G_1 = 0.01$ S, $G_F = 10^{-2}$ S, and $G_L = 10$ S.

3-42 In the circuit of Fig. 3-40, replace v_s by a short circuit. Remove R_L and connect a current source i_T in its place. Determine the resistance seen by the current source. This is the output resistance of the inverting op-amp circuit.

3-43 A signal source that generates 20 mV and has an internal resistance of 500 kΩ is connected to the inverting input of an op amp (with the noninverting input grounded). $A = 10^6$. Determine the value of the feedback resistor R_F needed to produce an output of 5 V.

Sec. 3-5 The Noninverting Op-Amp Circuit

3-44 Repeat the work of Problem 3-41 for the noninverting op-amp circuit.

3-45 Repeat the work of Problem 3-42 for the circuit of Fig. 3-42.

3-46 Repeat the work of Problem 3-43 if the signal source is connected to the noninverting input terminal of the op amp (with the inverting terminal connected to ground through a resistance R_1 = the resistance of the source).

3-47 Obtain an expression for the output voltage of the op-amp circuit of Fig. 3-69.

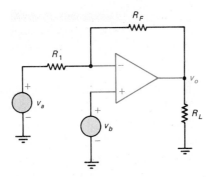

Fig. 3-69 Circuit for Problems 3-47 and 3-48.

3-48 Repeat Problem 3-47 by using the following procedure: first calculate the component of v_o if only v_a is present (with v_b replaced by a short circuit), then calculate the component of v_o if only v_b is present (with v_a replaced by a short circuit), and add the two components so obtained to get the total output voltage. (This is an application of the principle of superposition.)

Sec. 3-6 The Summing Op-Amp Circuit

3-49 In a summing op-amp circuit (using an op amp with $A = 10^5$), $R_F = 100$ kΩ. Determine the values of the resistors R_1, R_2, and R_3 so that the output is $v_o = -(1.25v_{i1} + v_{i2} + 0.75v_{i3})$.

3-50 Consider the three signal sources (along with the resistors) of Fig. 3-43 connected to the *noninverting* input terminal of the op amp instead of the inverting input. Assume that the inverting input is connected to ground through resistor R_4. Obtain an expression for the output voltage of the op-amp circuit so obtained.

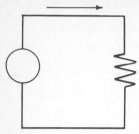

CHAPTER 4
METHODS OF BRANCH CURRENTS, LOOP, AND MESH ANALYSIS

The method of nodal analysis discussed in Chapter 3 uses a set of independent node-pair (or node-to-datum) voltages in a circuit as the unknown quantities and the equations are based on Kirchhoff's current law. An alternative procedure is to choose a set of branch currents as the unknown quantities and write equations based on Kirchhoff's voltage law. In the *method of branch currents,* a separate current is assigned to each branch of the circuit and then the necessary set of independent equations is written using Kirchhoff's voltage law around a set of independent closed paths. The method of branch currents is effective for simple circuits, but it loses its effectiveness in even moderately complex circuits. The reason is the lack of a methodical choice of independent closed paths or branch currents in the circuit. *Loop analysis,* on the other hand, is based on a methodical choice of independent currents around closed paths (or loops) in a circuit. *Mesh analysis* is a special form of loop analysis, in which the meshes of the network form the closed paths of currents. Both loop analysis and mesh analysis provide systematic approaches to circuit analysis in the same manner as nodal analysis. The methods of branch currents, loop analysis, and mesh analysis are presented in this chapter.

4-1 THE METHOD OF BRANCH CURRENTS

It is possible to assign an independent set of branch currents in simple circuits through a step-by-step application of Kirchhoff's current law at successive nodes. KVL equations are then written around the requisite number of closed paths. The equations are then solved for the currents in the various branches of the circuit. Such a procedure is called the *method of branch currents.* Even though it is not as systematic as loop analysis and

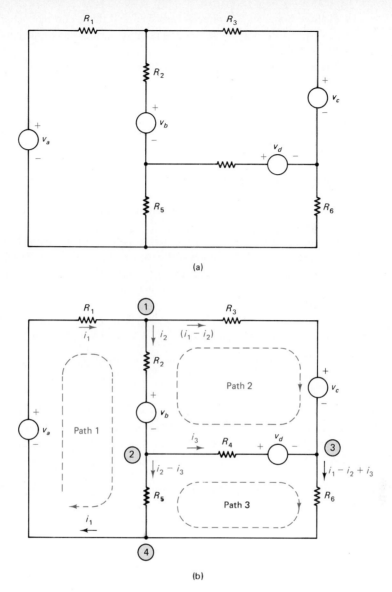

Fig. 4-1 Assignment of branch currents and closed paths for the method of branch currents.

mesh analysis, it is found to be particularly useful in a number of problems of practical interest such as, for example, dc analysis of transistor circuits.

Consider the circuit in Fig. 4-1(a). Start at any branch (say R_1), and assign a current i_1 to it. Follow i_1 to node 1, where two new branches are encountered. Assign a current i_2 to one branch (say R_2). With the use of KCL at node 1, the current in the other branch (R_3) then becomes ($i_1 - i_2$). Follow the current i_2 to node 2, where two new branches are encountered: a current i_3 is assigned to one branch (R_4) and the current in the other branch is written as $i_2 - i_3$ by using KCL at node 2. At node 3, there are three branches and the currents in two of these are already specified: i_3 in one and $i_1 - i_2$ in the other. KCL at node 3 gives the current in R_6 as $i_1 - i_2 + i_3$. Finally, at node 4, the currents

4-1 The Method of Branch Currents

in all three branches are already specified and these currents satisfy KCL at that node. Node 4, therefore, merely serves as a point of verification of the currents already assigned in the circuit.

The assignment of branch currents in the given circuit is shown in Fig. 4-1(b). Since no new current symbols were introduced except when necessary, and the current in a branch was expressed in terms of those already assigned whenever possible, the set of branch currents chosen by the above procedure forms an independent set. Since there are three independent branch currents, i_1, i_2, and i_3, in the circuit, a system of three equations is needed. These equations are obtained by using KVL around three closed paths in the circuit.

The closed paths used for writing the KVL equations should be independent of one another so as to have a system of independent equations. The following procedure is used in the selection of an independent set of closed paths*.

The first closed path is chosen arbitrarily. Each successive closed path after that must include at least one branch that is not a member of the closed paths already selected. Also, the total number of closed paths must be exactly equal to the number of independent branch currents determined earlier.

A set of three closed paths satisfying the above criteria for the present example is shown in Fig. 4-1(b).

In writing the KVL equation for a closed path, the following convention (already introduced in Chapter 2) is used:

If the direction of the path is *from plus* to minus across a component, *add* the voltage in the KVL equation.

If the direction of the path is *from minus* to plus across a component, *subtract* the voltage in the KVL equation.

The current direction in a resistor is from the positive to the negative voltage reference polarities (associated reference polarities). Therefore, if the direction of the path *coincides* with the direction of current in a resistor, then the path goes from plus to minus and the "*RI*" product due to that resistor must be *added;* if the direction of the path *opposes* the direction of current in a resistor, then the "*RI*" produce due to that resistor must be subtracted.

The KVL equations for the paths in Fig. 4-1(b) are given by

Path 1: $$-v_a + R_1 i_1 + R_2 i_2 + v_b + R_5(i_2 - i_3) = 0 \qquad (4\text{-}1)$$

Path 2: $$-v_b - R_2 i_2 + R_3(i_1 - i_2) + v_c - v_d - R_4 i_3 = 0 \qquad (4\text{-}2)$$

Path 3: $$R_4 i_3 + v_d + R_6(i_1 - i_2 - i_3) - R_5(i_2 - i_3) = 0 \qquad (4\text{-}3)$$

Equations (4-1) to (4-3) form a complete set of branch current equations of the given circuit and can be solved for i_1, i_2, and i_3. Once these three currents are known, it is possible to calculate the currents and voltages in any branch of the circuit.

The choice of branch currents or closed paths is clearly not unique and different sets of independent branch currents can be assigned and different sets of closed paths selected for a circuit. This is illustrated in the following exercise.

---

*Refer to E. A. Guillemin, *Introductory Circuit Theory* (Chapter 1, Section 7) (New York: John Wiley, 1953).

Exercise 4-1 In the circuit of Fig. 4-1(a), assign branch currents i_a (left to right) in R_1, i_b (left to right) in R_3, and i_c (downward) in R_5. Label the currents in the other branches in terms of the above three currents. Choose closed path 1 through v_a, R_1, R_3, v_c, R_6, and back to v_a. The other two paths are the same as those indicated in Fig. 4-1(b). Write the set of branch current equations for this set of closed paths and assignment of branch currents.

Exercise 4-2 Three branch currents have already been specified for the circuit shown in Fig. 4-2. Assign appropriate currents to the other branches, and write a complete set of branch current equations.

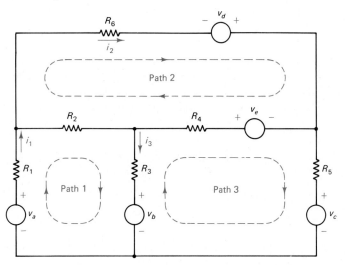

Fig. 4-2 Circuit for Exercise 4-2.

Example 4-1 Determine (a) the current in each resistor and (b) the power received or delivered by each voltage source in the circuit shown in Fig. 4-3(a) on page 172.

Solution The assignment of branch currents is shown in Fig. 4-3(b). There are three independent branch currents. Three closed paths are selected, as shown in Fig. 4-3(b).

Equation of Path 1:

$$R_1 i_1 - v_a + R_4(i_1 - i_2) - v_b + v_c + R_2(i_1 - i_2 - i_3) = 0$$

By substituting numerical values, collecting coefficients, and moving the voltage source terms to the right-hand side, the above equation becomes

$$45i_1 - 35i_2 - 20i_3 = 200 \tag{4-4}$$

Equation of Path 2:

$$v_b - R_4(i_1 - i_2) + R_6 i_2 - R_5 i_3 = 0$$

which leads to

$$-15i_1 + 45i_2 - 25i_3 = -150 \tag{4-5}$$

4-1 The Method of Branch Currents

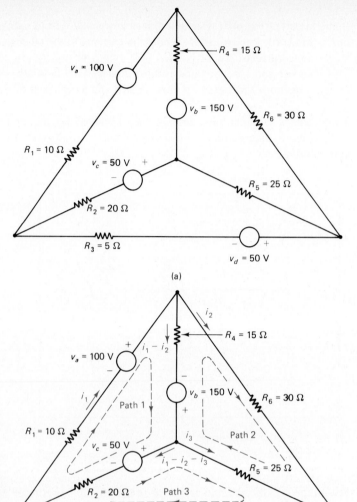

Fig. 4-3 Circuit for Example 4-1. (a) Given circuit. (b) Assignment of branch currents and closed paths.

Equation of Path 3:

$$-R_2(i_1 - i_2 - i_3) - v_c + R_5 i_3 + v_d + R_3(i_2 + i_3) = 0$$

which leads to

$$-20i_1 + 25i_2 + 50i_3 = 0 \qquad (4\text{-}6)$$

Equations (4-4) to (4-6) are solved for i_1, i_2, and i_3. It is found that

$$i_1 = 5.17 \text{ A} \qquad i_2 = -0.366 \text{ A} \qquad i_3 = 2.25 \text{ A}$$

The currents in the various resistors are $i_1 = 5.17$ A in R_1; $i_1 - i_2 - i_3 = 3.29$ A in R_2; $i_2 + i_3 = 1.88$ A in R_3; $i_1 - i_2 = 5.54$ A in R_4; $i_3 = 2.25$ A in R_5; and $i_2 = -0.366$ A in R_6.

The power associated with the voltage sources are $v_a i_1 = 517$ W (delivered by v_a); $v_b(i_1 - i_2) = 831$ W (delivered by v_b); $v_c(i_1 - i_2 - i_3) = 164$ W (received by v_c); and $v_d(i_2 + i_3) = 94$ W (received by v_d). ∎

Exercise 4-3 Write a set of branch current equations for the closed paths indicated in the circuit of Fig. 4-4(a). Note that currents have already been assigned to two of the branches.

Exercise 4-4 Determine the currents i_1 and i_2 in the circuit of Fig. 4-4(b).

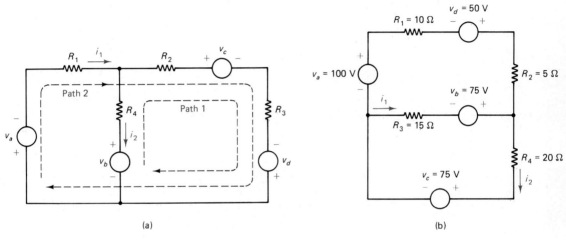

Fig. 4-4 Circuits for Exercises 4-3 and 4-4.

4-1-1 DC Analysis of a Transistor Circuit

The method of branch currents is particularly suitable for the determination of the dc voltages and currents in a transistor circuit, as illustrated by the following example.

Example 4-2 For the transistor in the circuit of Fig. 4-5(a), assume (as a first-level approximation) that the current I_c is related to the current I_b through a constant β:

$$I_c = \beta I_b \tag{4-7}$$

Assume that the dc level of the voltage V_{be} is given. The three currents I_b, I_c, and I_e satisfy KCL:

$$I_e - I_b - I_c = 0 \tag{4-8}$$

since the transistor can be considered as a node. [Refer to Fig. 2-13(b) in Chapter 2.] The voltages V_{cb} and V_{ce} and the current I_c are to be determined.

(a) Assign a set of branch currents to the circuit. (b) Choose the necessary number of closed paths and write the KVL equations to solve for the branch currents. Avoid paths involving V_{cb} and V_{ce} since they are not known. (c) Solve the equations and obtain an expression for I_b. (d) Determine the voltages V_{cb} and V_{ce} for the following numerical values: $R_1 = 10$ kΩ, $R_2 = 90$ kΩ, $R_e = 1$ kΩ, $R_c = 4$ kΩ, β = 50; $V_{CC} = 15$ V, and $V_{be} = 0.7$ V.

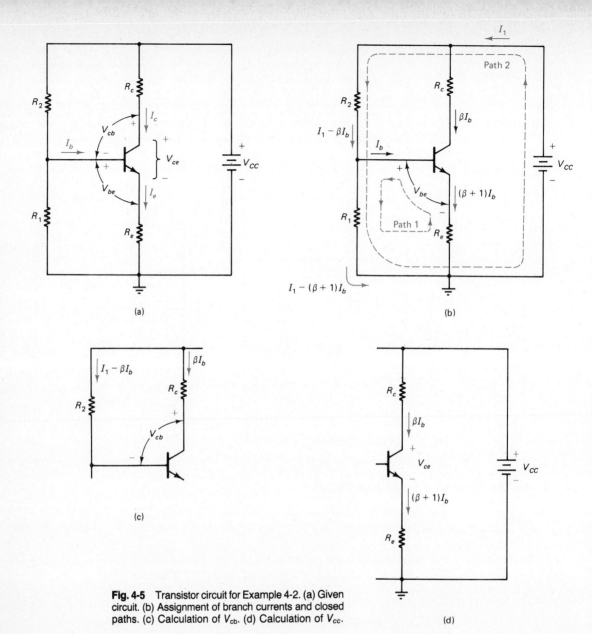

Fig. 4-5 Transistor circuit for Example 4-2. (a) Given circuit. (b) Assignment of branch currents and closed paths. (c) Calculation of V_{cb}. (d) Calculation of V_{cc}.

Solution (a) The three currents I_b, I_c, and I_e are related through Eqs. (4-7) and (4-8). From Eq. (4-7),

$$I_c = \beta I_b$$

Equation (4-7) combined with Eq. (4-8) leads to

$$I_e = (\beta + 1)I_b \qquad (4\text{-}9)$$

Thus, I_e and I_c can both be expressed in terms of I_b, and there is only one independent current among the three associated with the transistor.

Methods of Branch Currents, Loop, and Mesh Analysis

Let I_1 be the current supplied by the battery V_{CC}. The currents in all the branches of the given circuit are expressed in terms of I_1 and I_b, as shown in Fig. 4-5(b). There are only two independent currents in the circuit.

(b) KVL equations for two closed paths are necessary to solve for the two currents I_1 and I_b. Neither of these closed paths should involve the voltages V_{cb} and V_{ce}, which are unknown. Two closed paths are shown in Fig. 4-5(b), one of which includes the voltage V_{be} (whose value is assumed to be given).

(c) The KVL equations of the two closed paths are

Path 1:
$$R_1[I_1 - (\beta + 1)I_b] - R_e(\beta + 1)I_b - V_{be} = 0$$

which leads to

$$R_1 I_1 - [(\beta + 1)(R_1 + R_e)]I_b = V_{be} \tag{4-10}$$

Path 2:
$$-V_{CC} + R_2(I_1 - \beta I_b) + R_1[I_1 - (\beta + 1)I_b] = 0$$

which leads to

$$(R_1 + R_2)I_1 - [\beta R_2 + (\beta + 1)R_1]I_b = V_{CC} \tag{4-11}$$

Equations (4-10) and (4-11) are solved for I_b. It is found that

$$I_b = \frac{R_1 V_{CC} - (R_1 + R_2)V_{be}}{R_1 R_2 + (\beta + 1)R_e(R_1 + R_2)}$$

(d) Using the numerical values in the expression for I_b,

$$I_b = 1.33 \times 10^{-5} \text{ A}$$

or 13.3 microamperes. By substituting the above value of I_b and the numerical values in Eq. (4-11), it is found that

$$I_1 = 8.18 \times 10^{-4} \text{ A}$$

or 0.818 mA.

To find V_{cb}, choose a closed path that includes that voltage. Such a path is shown in Fig. 4-5(c), and the KVL equation for that path is

$$V_{cb} - R_2(I_1 - \beta I_b) + R_e(\beta I_b) = 0$$

which yields

$$V_{cb} = 11.0 \text{ V}$$

To find V_{ce}, choose a closed path that includes that voltage. Such a path is shown in Fig. 4-5(d), and the KVL equation for that path is

$$V_{ce} + R_e(\beta + 1)I_b - V_{CC} + R_e(\beta I_b) = 0$$

which yields

$$V_{ce} = 11.7 \text{ V}$$

This last value could also have been obtained by going around the transistor and writing the KVL equation

$$V_{be} + V_{cb} - V_{ce} = 0$$

4-1 The Method of Branch Currents

Exercise 4-5 In the circuit of Example 4-2, calculate the power supplied by the battery and the power dissipated in each resistor. Calculate the power dissipated by the transistor, which is the difference between the power supplied by the battery and the total power dissipated by the resistors.

Exercise 4-6 The circuit of Fig. 4-6 shows a slightly modified version of that in Example 4-2. Using the numerical values given earlier, calculate the current I_b and the voltages V_{cb} and V_{cc}.

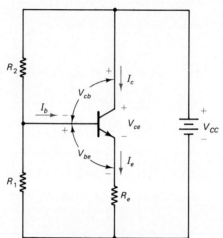

Fig. 4-6. Circuit for Exercise 4-6.

4-1-2 Operational Amplifiers

Operational amplifier circuits were analyzed in Chapter 3 using nodal analysis. The following example illustrates the use of the method of branch currents in the analysis of op-amp circuits.

Example 4-3 The inverting op-amp circuit is shown in Fig. 4-7(a). Obtain the expression for the voltage ratio v_o/v_s.

Solution Branch currents i_1 and i_L are assigned to the circuit, as shown in Fig. 4-7(b). By writing a KVL equation for the closed path consisting of v_s, R_1, and v_i, the expression for v_i,

Fig. 4-7 (a) Inverting op-amp circuit for Example 4-3 and (b) its equivalent circuit.

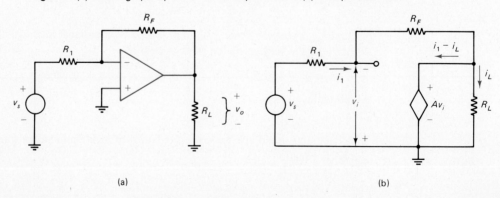

(a) (b)

176 Methods of Branch Currents, Loop, and Mesh Analysis

which controls the voltage of the dependent source, is

$$v_i = R_1 i_1 - v_s \tag{4-12}$$

The KVL equation for the path v_s, R_1, R_F, and Av_i is

$$R_1 i_1 + R_F i_1 = v_s - Av_i$$

which becomes, after using Eq. (4-12),

$$[R_1(1 + A) + R_F]i_1 = v_s(1 + A)$$

Therefore,

$$i_1 = \frac{v_s(1 + A)}{R_1(1 + A) + R_F} \tag{4-13}$$

The output voltage is given by

$$v_o = R_L i_L$$

and the KVL equation for the path Av_i, R_L gives

$$v_o = R_L i_L = Av_i$$

By using Eqs. (4-12) and (4-13), it is found that

$$\frac{v_o}{v_s} = -\frac{AR_F}{R_1(1 + A) + R_F} \tag{4-14}$$

which is the desired result.

When the gain A is sufficiently large, Eq. (4-14) reduces to

$$\frac{v_o}{v_s} = -\frac{R_F}{R_1}$$

which is a commonly used formula for inverting op-amp circuits. ∎

Exercise 4-7 Using $G = 1/R$, show that the equation for the voltage ratio obtained in Chapter 3 for the inverting op-amp circuit is identical to Eq. (4-14).

4-1-3 Circuits Containing Current Sources

When a circuit contains a current source, the current supplied by it is a known quantity and consequently chosen as one of the branch currents. It is important to remember that the voltage across a current source is *not a known quantity*. Therefore, it is advisable to avoid including the branch with a current source as part of a closed path whenever possible. The following example illustrates the procedure.

Example 4-4 Determine the current supplied by the v_b source in the circuit of Fig. 4-8(a).

Solution The circuit contains a dependent current source. The current due to the dependent source and the current controlling it are already specified. By using KCL at node 1, the current in R_1 is $(K + 1)i_2$. By denoting the current in R_4 by i_4, KCL at node 2 gives the current through R_3 as $i_4 - Ki_2$. The circuit of Fig. 4-8(b) shows the assignment of branch currents.

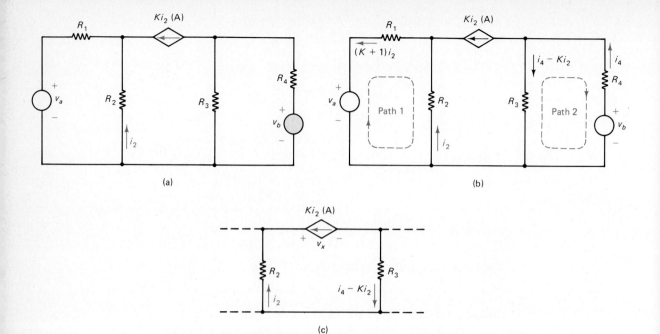

Fig. 4-8 Circuit for Example 4-4. (a) Given circuit. (b) Assignment of branch currents and closed paths. (c) Calculation of voltage across the current source.

Since there are two independent current variables, i_2 and i_4, KVL equations for two closed paths are needed. The two closed paths chosen are indicated in Fig. 4-8(b) and the resulting KVL equations are

Path 1:
$$-v_a - R_1(K+1)i_2 - R_2 i_2 = 0$$

Collecting coefficients and moving v_a to the right-hand side, the above equation becomes

$$-[(K+1)R_1 + R_2]i_2 = v_a \qquad (4\text{-}15)$$

Path 2:
$$-R_3(i_4 - Ki_2) + v_b - R_4 i_4 = 0$$

Collecting coefficients and moving v_b to the right-hand side, the above equation becomes

$$KR_3 i_2 - (R_3 + R_4)i_4 = -v_b \qquad (4\text{-}16)$$

Equations (4-15) and (4-16) are the two equations to be solved. From Eq. (4-15),

$$i_2 = v_a/[(K+1)R_1 + R_2] \qquad (4\text{-}17)$$

By combining Eqs. (4-16) and (4-17), it is found after some manipulation that

$$i_4 = \frac{-KR_3 v_a + [(K+1)R_1 + R_2]v_b}{(R_3 + R_4)[(K+1)R_1 + R_2]}$$

which is the current supplied by the v_b source. ∎

Suppose that in the above circuit it is required to find the voltage across the Ki_2 source. Then a voltage, say v_x, is assigned to that branch and KVL is written around a closed path that includes that source, as indicated in Fig. 4-8(c). The KVL equation for the closed path is

$$v_x + R_3(i_4 - Ki_2) + R_2 i_2 = 0$$

which is solved for v_x.

Exercise 4-8 Obtain the expression for v_x using the equation given above.

Exercise 4-9 Use the method of branch currents to determine the resistance v_T/i_T seen by the voltage source v_T in each of the circuits of Fig. 4-9.

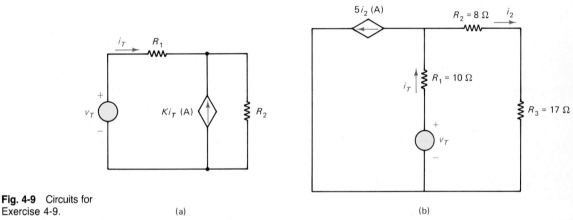

Fig. 4-9 Circuits for Exercise 4-9.

Summary of the Method of Branch Currents

Start with a current i_1 in some branch, proceed to each of the nodes in succession, and split the current at each node. Use KCL as appropriate at each node to express a branch current in terms of those already specified, and introduce a new current symbol only when necessary. By a careful use of this procedure, it is possible to arrive at a set of independent branch currents.

Choose a set of closed paths, equal in number to the number of independent branch currents. In selecting closed paths, each should include at least one branch not included in those already selected.

In writing the KVL equation for a closed path, *add* the voltage across a component if the path goes from *plus to minus*. *Subtract* the voltage if the path goes from *minus to plus*. In the case of resistors, *add* the RI product if the path direction *coincides* with the current direction, and *subtract* if the path direction *opposes* the current direction.

Take advantage of the presence of a current source by using its current as one of the branch currents. It should be kept in mind that the voltage across a current source is an unknown quantity when writing the KVL equation of a path that includes the current source.

The disadvantage in the method of branch currents is the lack of a systematic approach to the assignment of independent branch currents and the selection of independent closed paths to fit the set of branch currents assigned to the circuit. There is a need to make a number of judgments both in the assignment of branch currents and the selection of closed paths, which becomes cumbersome in even moderately complex circuits. As will be seen in the following discussion, the methods of loop analysis and mesh analysis provide a systematic approach to the assignment of currents and selection of closed paths in a circuit.

4-2 THE CONCEPT OF LOOP CURRENTS

Consider the circuit shown in Fig. 4-10(a) with branch currents assigned as shown. Consider the current through the R_b branch. Even though there is only a *single* current through R_b denoted by $i_1 - i_2$, it is possible to imagine that there are two *separate current components* i_1 and i_2 in R_b flowing in *opposite* directions [as indicated in Fig. 4-10(b)], so that the *net* current in R_b is $i_1 - i_2$. It then becomes possible to use two *circulatory or loop currents* i_1 and i_2 in the circuit, as indicated in Fig. 4-10(c). It is apparent that the current in a branch is not actually made up of two (or more) physically distinguishable loop current components. But, such a separation is conceptually feasible and leads to a systematic procedure for assigning currents to a circuit.

The current in any branch of the circuit of Fig. 4-10(c) can be expressed in terms of the two loop currents i_1 and i_2: current in $R_a = i_1$, current in $R_b = i_1 - i_2$, and current in $R_c = i_2$. The above idea is readily extended to more complicated circuits. That is, in a given circuit, it is possible to choose a set of loop currents in such a manner that the current in any branch can be expressed as the algebraic sum of loop currents.

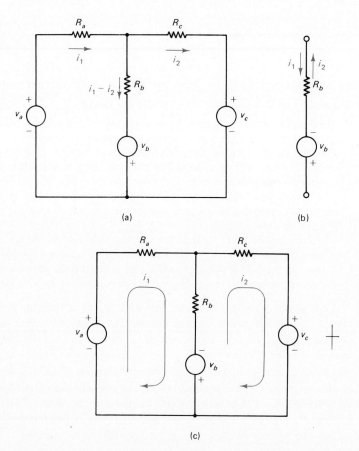

Fig. 4-10 Concept of loop currents. The current $i_1 - i_2$ in R_b in part (a) is treated as the difference between two component currents i_1 and i_2 in part (b), leading to the two loop currents in part (c).

Exercise 4-10 Several circuits are shown in Fig. 4-11 with loop currents already specified. For each circuit, express the current in each branch as the algebraic sum of the loop currents.

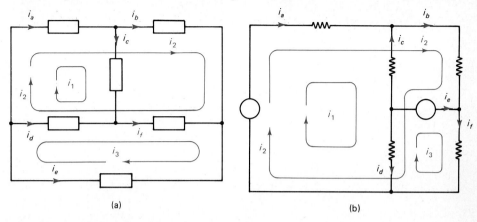

Fig. 4-11 Circuits for Exercise 4-10.

4-3 NETWORK TOPOLOGY—CHOICE OF INDEPENDENT LOOP CURRENTS

The selection of a set of independent loop currents is based on the concepts of network topology introduced in Chapter 3, where the *graph* of a network and the selection of a *tree* were defined.

The graph simply shows the interconnection of the branches of a circuit but not the actual components in each branch. A *tree* is a subgraph satisfying two conditions: (a) there should be no closed path in a tree, and (b) it must be possible to strike a path from any node to any other node through the tree-branches.

The tree forms the basis for establishing the independence of node-pair voltages in a circuit. The graph of a circuit and one possible tree are shown in Figs. 4-12(a) and (b).

Links in a Graph:

Now, consider the branches of a graph not used in the formation of a tree as, for example, branch numbers 3, 4, 8, and 9 in Fig. 4-12(a).

Branches not used in setting up a tree are called *links*.

Suppose, in a given graph,

N = total number of nodes
B = total number of branches
T = number of tree-branches
L = number of links

Then the number of tree-branches is given by

$$T = (N - 1)$$

4-3 Network Topology—Choice of Independent Loop Currents 181

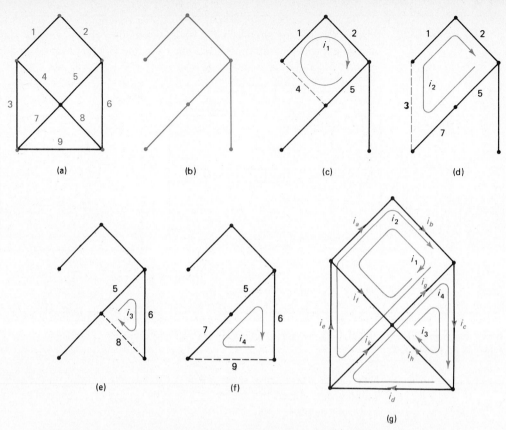

Fig. 4-12 Choice of independent loop currents. (a) Graph of a circuit. (b) A possible tree. (c) to (f) Loop currents set up by individual links. (g) Complete set of loop currents.

as was shown in Chapter 3. Of the B branches, T have been used in the tree, and the remaining branches must be the links. Therefore,

$$L = (B - T) = (B - N + 1) \tag{4-18}$$

If any of the links from the graph of a circuit were added to the tree, it would immediately create a closed path and consequently facilitate the flow of a current (a loop current).

This is illustrated by Figs. 4-12(c) through (f), where each circuit shows the closed path and the loop current set up by adding a link to the tree of Fig. 4-12(b).

Exercise 4-11 Calculate the number of links in the graph of each circuit of Exercise 4-10.

4-3-1 Number of Independent Currents

The number of independent currents in a circuit equals the number of links in its graph. The proof of the above statement is as follows:

First, note that the addition of each link to a tree sets up a closed path and facilitates the flow of a current independent of the presence of any other link. Consequently, there must be *at least* as many independent currents in a circuit as there are links in its graph.

Next, assume that the number of independent currents is greater than the number of links in the graph. This would imply that even after removing all the links, at least one more current would still flow in the circuit. This leads to a contradiction because the removal of all the links leaves behind a tree with no closed paths, making it impossible for a current to flow. Therefore, the number of independent loop currents *cannot be greater than* the number of links in the graph.

Since there must be at least as many independent currents as the number of links, and the number of independent currents cannot be greater than the number of links, it follows that

the number of independent currents in a circuit is exactly equal to the number of links in its graph.

[The above rule verifies (in retrospect) the correctness of the number of independent branch currents in the circuits analyzed by the method of branch currents as discussed in Sec. 4-1.]

4-3-2 Selection of Independent Closed Paths

The loop current established by placing a link on a tree depends for its existence only upon that link and not any other link: remove that link and the loop disappears. Therefore, the procedure of setting up a tree and placing links on the tree, one at a time, also establishes a set of *independent loops* for the currents besides giving the correct number of independent current variables. The paths described by these loops form a set of independent closed paths for writing KVL equations.

The complete set of independent loop currents for the graph of Fig. 4-12 [using the tree of Fig. 4-11(b)] is shown in Fig. 4-12(g).

The selection of a complete set of independent loop currents for a given circuit is not unique because it is possible, in general, to set up two or more trees for a given graph. Each tree leads to a different set of loop currents for the circuit. It should be remembered that the *number* of independent loop currents remains the same for a given circuit, however, and is given by Eq. (4-18).

Exercise 4-12 Consider the circuit of Fig. 3-7 (in Chapter 3). Calculate the number of independent loop currents. Set up several different trees. For each tree, show the paths of the independent loop currents.

4-3-3 Relationship of Branch Currents to Loop Currents

Since the loop currents chosen by the procedure described above form a complete set of independent currents, it must be possible to relate the current in each branch to the loop currents. Given a tree for a graph, each link sets up its own loop current and there is no other loop current involving that link. Consequently, the branch current in any link is equal to the loop current set up by that link. On the other hand, a tree-branch is, *in general,* common to the closed paths set up by different links. Therefore, there are two or more loop currents in a tree-branch, and the *net current in each tree-branch is the algebraic sum of the loop currents flowing through it.*

The relationships between the branch currents in the circuit of Fig. 4-12 and the loop currents are given by

$$i_a = i_1 + i_2 \qquad i_b = i_1 + i_2 \qquad i_c = i_3 + i_4$$
$$i_d = i_4 \qquad i_e = i_2 \qquad i_f = -i_1$$
$$i_g = -i_1 - i_2 + i_3 + i_4 \qquad i_h = i_3 \qquad i_k = -i_2 + i_4$$

Exercise 4-13 For each set of loop currents obtained in Exercise 4-12, write the relationships between the branch currents and the loop currents.

Exercise 4-14 Show that the currents given in each circuit of Exercise 4-10 (Fig. 4-11) form a complete set of independent loop currents by setting up the tree from which those loop currents were established and adding the links one at a time.

4-4 PRINCIPLES OF LOOP ANALYSIS

The discussion in the previous section shows that the procedure for selecting a set of independent loop currents is

1. Draw the graph of the circuit.
2. Set up a tree.
3. Add one link at a time to the tree and delineate the corresponding loop current.

The loop currents chosen in the above manner have the following properties:

1. They form a complete set of independent current variables for the circuit.
2. The closed paths described by the loops are independent of one another.
3. The current in any branch equals the (algebraic) sum of the loop currents.

Because of the above properties, it is possible to determine the currents in all the branches of a circuit by solving a set of KVL equations around the loops with the loop currents as the current variables. The resulting method of analysis is called *loop analysis*.

The basic principles for writing the KVL equation for a loop are the same as those for the method of branch currents.

Traverse a closed path by following the loop current that delineates each loop, and set the algebraic sum of the voltages around the loop equal to zero.

Add the voltage across a component if the direction of the path is from the plus to minus voltage polarities of the component. Subtract the voltage across a component if the direction of the path is from the minus to plus voltage polarities of the component. Equate the algebraic sum to zero.

As for the resistors in a loop, there can be more than one loop current through a resistor and the net current is the algebraic sum of the loop currents through it. The following convention leads to a consistent rule for adding the voltages across resistors in a loop: *If the net current is always evaluated in the direction of the loop for which the equation is being written, then the path direction coincides with the direction of the* net current. Then the voltage in a resistance, given by the RI product, (resistance times net current), is always added *in the KVL equation*.

Example 4-5 For the circuit shown in Fig. 4-13(a), a set of loop currents has already been chosen using the tree (shown in solid lines) and links (shown in dashed lines) in Fig. 4-13(b). Write the loop equations.

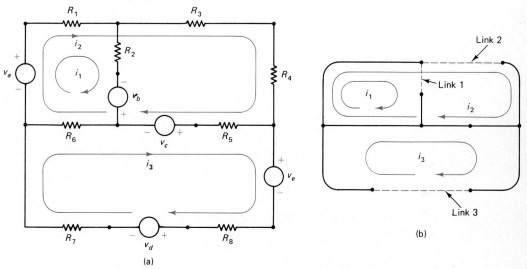

Fig. 4-13 (a) Circuit for Example 4-5. (b) Tree and links to fit the given set of loop currents.

Solution *Loop 1* is in the direction of i_1. First consider the resistors.

Resistor R_1 has two loop currents, both in the direction of Loop 1. Therefore,

$$\text{Net current in } R_1 \text{ in the direction of Loop 1} = i_1 + i_2$$

and

$$\text{Voltage term for } R_1 = +R_1(i_1 + i_2)$$

Resistor R_2 has only one loop current, and

$$\text{Voltage term for } R_2 = +R_2 i_1$$

Resistor R_6 has three loop currents, i_1 and i_2 in the direction of Loop 1 and i_3 in the opposite direction. Therefore,

$$\text{Net current in } R_6 \text{ in the direction of Loop 1} = i_1 + i_2 - i_3$$

and

$$\text{Voltage term for } R_6 = +R_6(i_1 + i_2 - i_3)$$

The KVL equation for Loop 1 is, therefore, given by

$$-v_a + R_1(i_1 + i_2) + R_2 i_1 - v_b + R_6(i_1 + i_2 - i_3) = 0$$

The above equation is rewritten by *moving the voltage source terms to the right-hand side*:

$$R_1(i_1 + i_2) + R_2 i_1 + R_6(i_1 + i_2 - i_3) = v_a + v_b \qquad (4\text{-}19)$$

4-4 Principles of Loop Analysis

Loop 2 is in the direction of i_2.

Net current in R_1 in the direction of Loop 2 $= i_2 + i_1$ and the voltage term for
$$R_1 = +R_1(i_2 + i_1)$$

Current in $R_3 = i_2$ and the voltage term for $R_3 = +R_3 i_2$.
Current in $R_4 = i_2$ and the voltage term for $R_4 = +R_4 i_2$.
Net current in R_5 in the direction of Loop 2 $= i_2 - i_3$ and voltage term for
$$R_5 = +R_5(i_2 - i_3)$$

Net current in R_6 in the direction of Loop 2 $= i_2 + i_1 - i_3$ and voltage term for
$$R_6 = +R_6(i_2 + i_1 - i_3)$$

The KVL equation for Loop 2 is, therefore, given by
$$-v_a + R_1(i_2 + i_1) + R_3 i_2 + R_4 i_2 + R_5(i_2 - i_3) + v_c + R_6(i_2 + i_1 - i_3) = 0$$
which is rewritten (by moving the voltage source terms to the right-hand side) as
$$R_1(i_2 + i_1) + R_3 i_2 + R_4 i_2 + R_5(i_2 - i_3) + R_6(i_2 + i_1 - i_3) = v_a - v_c \quad (4\text{-}20)$$

Loop 3 is in the direction of i_3. Using the same argument as for the other two loops, the KVL equation for Loop 3 is given by
$$R_6(i_3 - i_1 - i_2) - v_c + R_5(i_3 - i_2) + v_e + R_8 i_3 + v_d + R_7 i_3 = 0$$
which is rewritten as
$$R_6(i_3 - i_1 - i_2) + R_5(i_3 - i_2) + R_8 i_3 + R_7 i_3 = v_c - v_e - v_d \quad (4\text{-}21)$$

Equations (4-19) to (4-21) are the loop equations of the given circuit. ■

The loop equations of a circuit have the following features:

1. In the equation for each loop, the right-hand side is the algebraic sum of terms due to source voltages and represents the *net increase in potential* due to the sources in the loop.
2. Each term on the left-hand side represents the voltage across a resistor due to the net current in the resistor in the direction of the loop. Thus, the left-hand side represents the total *decrease in potential* due to the resistors.

Thus, each loop equation states that the total decrease in potential in the loop due to all the resistors in it equals the total increase in potential in the loop due to the sources in it.

The above discussion is readily extended to the case of a general network, leading to the following procedure for writing the loop equation for loop j in a network.

Right-Hand Side: Form the algebraic sum of voltages due to the sources present in loop j. Use a *plus sign* if the loop goes from *minus to plus through the source* and a minus sign otherwise.

Left-Hand Side: Add the *RI* product due to all the resistors present in the loop. For each resistor, the current is taken as the net current through the resistor *in the direction of loop* j.

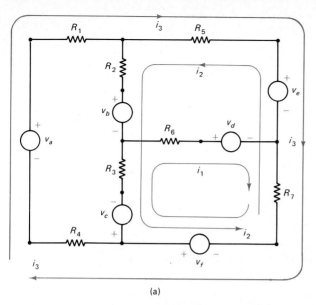
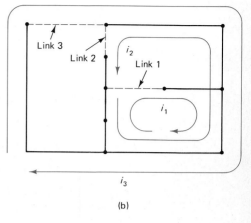

Fig. 4-14 (a) Circuit for Example 4-6. (b) Tree and links to fit the given set of loop currents.

Example 4-6

For the circuit shown in Fig. 4-14(a), a set of loop currents has been selected by using the tree and links shown in Fig. 4-14(b). Write the loop equations.

Solution Using the procedure outlined earlier, the loop equations are

Loop 1:
$$R_3(i_1 - i_2) + R_6 i_1 + R_7(i_1 - i_2 + i_3) = -v_c - v_d + v_f$$

Loop 2:
$$R_3(i_2 - i_1) + R_7(i_2 - i_1 - i_3) + R_5(i_2 - i_3) + R_2 i_2 = v_c - v_f + v_e - v_b$$

Loop 3:
$$R_1 i_3 + R_5(i_3 - i_2) + R_7(i_3 - i_2 + i_1) + R_4 i_3 = v_a - v_e + v_f$$ ∎

Exercise 4-15 Consider the circuit of Fig. 4-13 (of Example 4-5). Set up a tree so that the *links* are provided by the R_1, R_4, and R_5 branches. Write a set of loop equations for the resulting configuration of loop currents.

Exercise 4-16 Consider the circuit of Fig. 4-14 (of Example 4-6). Set up a tree so that the *links* are provided by the R_3, R_5, and R_7 branches. Write a set of loop equations for the resulting configuration of loop currents.

Example 4-7 Determine the power supplied by each of the sources in the circuit of Fig. 4-15. Loop currents have already been assigned.

Solution The loop equations are given by

Loop 1:
$$R_1(i_1 + i_2) + R_2 i_1 = v_a - v_b$$

4-4 Principles of Loop Analysis

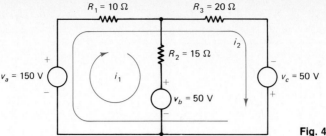

Fig. 4-15 Circuit for Example 4-7.

which (after collecting coefficients) leads to

$$(R_1 + R_2)i_1 + R_1 i_2 = v_a - v_b$$

which (after substituting numerical values) gives

$$25i_1 + 10i_2 = 100 \tag{4-22}$$

Loop 2:
$$R_1(i_1 + i_2) + R_3 i_2 = v_a + v_c$$

which leads to

$$10i_1 + 30i_2 = 200 \tag{4-23}$$

Solving Eqs. (4-22) and (4-23),

$$i_1 = 1.54 \text{ A} \quad \text{and} \quad i_2 = 6.15 \text{ A}$$

The current in the v_a source is $i_1 + i_2$, and the power supplied by it is $v_a(i_1 + i_2) = 1154$ W.

The current in the v_b source is i_1, and the power received by it is $v_b i_1 = 77$ W.
The current in the v_c source is i_2, and the power supplied by it is $v_c i_2 = 308$ W.

Checking Your Answers:

Choose a closed path that was not used as one of the loops in solving the problem and see if KVL is satisfied. ■

Exercise 4-17 Calculate the power dissipated by each resistor in the circuit of Example 4-7. Verify that the total power dissipated in the circuit equals the total power supplied to it.

Exercise 4-18 For the circuit shown in Fig. 4-16, verify by setting up a tree that the set of loop equations shown is a complete independent set. Use loop analysis and determine the power supplied by each source and dissipated by each resistor.

Fig. 4-16 Circuit for Exercise 4-18.

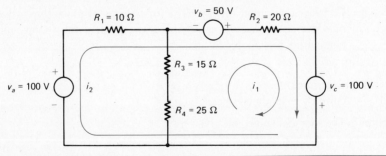

Methods of Branch Currents, Loop, and Mesh Analysis

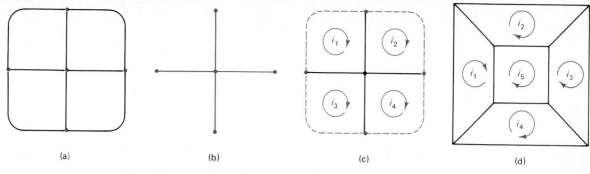

Fig. 4-17 Mesh currents.

4-5 MESH ANALYSIS

When a network has a graph that is mappable on the surface of a sphere, there exists an alternative approach to the selection of independent current variables and closed paths. If the term *network* is taken as analogous to a net (such as a fishing net), then the analogy can be extended by using the term *meshes* for the *open spaces* in the network.

Consider the graph of Fig. 4-17(a). By choosing the tree of Fig. 4-17(b), the resulting loops coincide with the meshes, as shown in Fig. 4-17(c). The set of mesh currents is simply one choice of loop currents in this circuit. Therefore, the mesh currents form a complete set of independent current variables for the circuit of Fig. 4-17(a).

On the other hand, consider the mesh currents in the graph of Fig. 4-17(d). They also form a complete set of independent current variables even though we cannot find a tree that will lead to a set of loops coincident with the meshes of the network.

Circuits whose graphs can be drawn on a plane (without any crossovers) or on the surface of a sphere have the property that the mesh currents form a complete set of independent current variables.

The above property is based on the fact that "a sufficient (though not necessary) procedure to insure the independence of closed paths is to select them successively in such a way that each additional path involves at least one branch that is not part of any of the previously selected paths."* It can be seen that the above condition was satisfied by the mesh current assignments of Fig. 4-17(d).

Therefore, it is possible to choose the meshes as closed paths (which is only possible if the graph is mappable on a plane or on the surface of a sphere) and the mesh currents as the current variables in the analysis of some circuits. The procedure of analyzing a circuit by writing KVL equations for the meshes with mesh currents as the variables is known as *mesh analysis*.

Since the only difference between loop analysis and mesh analysis is the method of choosing the current variables and closed paths, the steps in mesh analysis are the same as those in loop analysis. There is a minor difference:

there are only two mesh current components in any branch common to two meshes, whereas in loop analysis it is possible to have even three or more loop current components in a branch common to two loops.

---

*Refer to E. A. Guillemin, *Introductory Circuit Theory* (Chapter 1, Section 7) (New York: John Wiley, 1953).

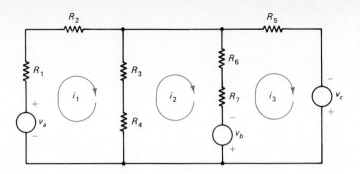

Fig. 4-18 Circuit for Example 4-8.

Example 4-8 Write the mesh equations for the circuit of Fig. 4-18.

Solution Using the same procedure as in loop analysis, the mesh equations are

Mesh 1: $\quad R_1 i_1 + R_2 i_1 + R_3(i_1 - i_2) + R_4(i_1 - i_2) = v_a$

Mesh 2: $\quad R_3(i_2 - i_1) + R_4(i_2 - i_1) + R_6(i_2 - i_3) + R_7(i_2 - i_3) = v_b$

Mesh 3: $\quad R_7(i_3 - i_2) + R_6(i_3 - i_2) + R_5 i_3 = v_c - v_b$ ∎

Example 4-9 Determine the power supplied by each of the voltage sources in the circuit of Fig. 4-19. (Pay careful attention to the directions of the mesh currents.)

Solution The mesh equations are

Mesh 1: $\quad R_1 i_1 + R_2(i_1 + i_2) = -v_a + v_b$

or

$$25 i_1 + 15 i_2 = -50 \qquad (4\text{-}24)$$

Mesh 2: $\quad R_2(i_2 + i_1) + R_3 i_2 + R_4(i_2 - i_3) = v_b$

or

$$15 i_1 + 70 i_2 - 25 i_3 = 150 \qquad (4\text{-}25)$$

Mesh 3: $\quad R_4(i_3 - i_2) + R_5 i_3 = -v_c$

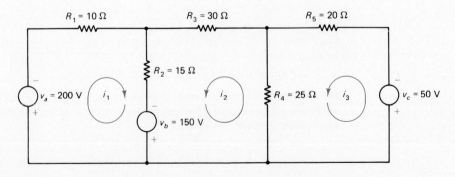

Fig. 4-19 Circuit for Example 4-9.

or

$$-25i_2 + 45i_3 = -50 \qquad (4\text{-}26)$$

Solving Eqs. (4-24) to (4-26),

$$i_1 = -3.94 \text{ A} \qquad i_2 = 3.23 \text{ A} \qquad i_3 = 0.683 \text{ A}$$

- Power *delivered* by v_a is $v_a(-i_1) = 788$ W.
- Power *delivered* by v_b is $v_b(i_1 + i_2) = -106$ W.
- Power *delivered* by v_c is $v_c(-i_3) = -34.2$ W.

Exercise 4-19 Write the mesh equations of the circuits in Figs. 4-13 and 4-14 (refer to Exercises 4-14 and 4-15).

Exercise 4-20 Write the mesh equations of the circuit in Fig. 4-20(a), solve them, and find the power delivered or received by each component in the circuit.

Exercise 4-21 Determine the ratio v_o/v_s in the circuit of Fig. 4-20(b) by using (a) mesh analysis and (b) the ladder network approach.

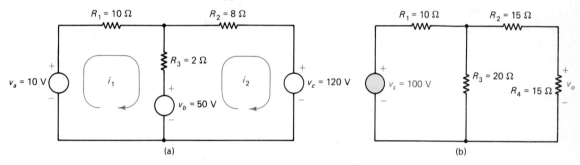

Fig. 4-20 (a) Circuit for Exercise 4-20 and (b) circuit for Exercise 4-21.

4-6 STANDARD FORM OF LOOP AND MESH EQUATIONS

The standard form of loop and mesh equations are obtained by (a) collecting the coefficients of the different currents in the equation (and simplifying where possible) and (b) writing the current terms on the left-hand side in the same order in all the equations (that is, i_1 term followed by i_2 term followed by i_3 term . . .).

Example 4-10 Write the standard form of (i) the loop equations for the loop currents indicated in the circuit shown in Fig. 4-21(a) and (ii) the mesh equations of the circuit shown in Fig. 4-21(b) on page 192.

Solution (i) The loop equations are

Loop 1: $\qquad R_1 i_1 + R_2(i_1 - i_2) + R_3(i_1 - i_2 - i_3) = v_a - v_b$

or

$$(R_1 + R_2 + R_3)i_1 - (R_2 + R_3)i_2 - R_3 i_3 = v_a - v_b \qquad (4\text{-}27)$$

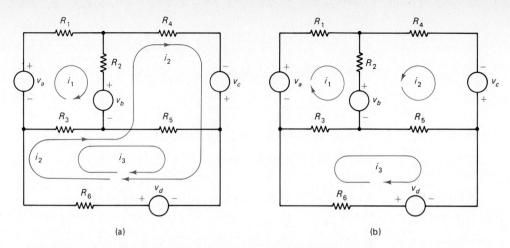

Fig. 4-21 Circuit for Example 4-10. (a) Loop analysis. (b) Mesh analysis.

Loop 2:

$$R_2(i_2 - i_1) + R_4 i_2 + R_6(i_2 + i_3) + R_3(i_2 + i_3 - i_1) = v_b + v_c + v_d$$

or

$$-(R_2 + R_3)i_1 + (R_2 + R_4 + R_3 + R_6)i_2 + (R_3 + R_6)i_3 = v_b + v_c + v_d \quad (4\text{-}28)$$

Loop 3:

$$R_3(i_3 + i_2 - i_1) + R_5 i_3 + R_6(i_2 + i_3) = v_d$$

or

$$-R_3 i_1 + (R_3 + R_6)i_2 + (R_3 + R_5 + R_6)i_3 = v_d \quad (4\text{-}29)$$

Equations (4-27) to (4-29) are the loop equations of the circuit in standard form.

(ii) The mesh equations are

Mesh 1:

$$R_1 i_1 + R_2(i_1 + i_2) + R_3(i_1 - i_3) = v_a - v_b$$

or

$$(R_1 + R_2 + R_3)i_1 + R_2 i_2 - R_3 i_3 = v_a - v_b \quad (4\text{-}30)$$

Mesh 2:

$$R_4 i_2 + R_2(i_2 + i_1) + R_5(i_2 + i_3) = -v_c - v_b$$

or

$$R_2 i_1 + (R_4 + R_2 + R_5)i_2 + R_5 i_3 = -v_c - v_b \quad (4\text{-}31)$$

Mesh 3:

$$R_6 i_3 + R_3(i_3 + i_1) + R_5(i_3 + i_2) = v_d$$

or

$$-R_3 i_1 + R_5 i_2 + (R_3 + R_5 + R_6)i_3 = v_d \quad (4\text{-}32)$$

Equations (4-30) to (4-32) are the mesh equations of the given circuit in standard form. ∎

The standard form of the loop and mesh equations of the circuit in the preceding example have the following features. Even though the term *loop* is used in the following statements, they are equally valid for mesh equations also.

1. Coefficient of i_j in the equation for loop j is the positive sum of all the resistances present in loop j.

   For example, the coefficient of i_1 in the equation for loop 1 is the sum of the resistances present in loop 1.

2. Coefficient of i_k in the equation for loop j ($k \neq j$) is the sum of the resistances in the branches common to loops j and k.

   The sign of the coefficient is *positive* if loop currents i_j and i_k flow in the *same* direction through the common branch.

   The sign of the coefficient is *negative* if loop currents i_j and i_k flow in opposite directions through the common branch.

   For example, the coefficient of i_2 in the equation for loop 1 is the positive sum of the resistances common to loops 1 and 2 because i_1 and i_2 are flowing in the same direction, whereas the coefficient of i_3 in the equation for loop 1 is the negative of the resistance common to loops 1 and 3 because i_1 and i_3 are flowing in opposite directions.

The above observations are valid for any network *containing only independent voltage sources and resistors*, as seen from the following argument. Let R_{jk} be a resistance common to loops j and k. Then the RI product due to R_{jk} in the equation for loop j is

$$R_{jk}(i_j + i_k) = R_{jk}i_j + R_{jk}i_k \tag{4-33a}$$

if i_j and i_k are in the same direction through R_{jk}; and

$$R_{jk}(i_j - i_k) = R_{jk}i_j - R_{jk}i_k \tag{4-33b}$$

if i_j and i_k are in opposite directions through R_{jk}.

Therefore, the coefficient of i_j in the equation for loop j due to the presence of R_{jk} is $+R_{jk}$. Every resistance present in loop j appears as a positive coefficient of i_j in the equation for loop j. Thus, the coefficient of i_j in the equation for loop j is the positive sum of all the resistances present in loop j.

Equation (4-33a) also shows that when the currents i_j and i_k are in the same direction through R_{jk}, the coefficient of i_k in the equation for loop j due to R_{jk} is $+R_{jk}$. Thus, if i_k and i_j are in the same direction in the branch(es) common to loops j and k, the coefficient of i_k in the equation for loop j is the positive sum of all resistances common to loops j and k.

Similarly, Eq. (4-33b) shows that if i_j and i_k are in opposite directions in the branch(es) common to loops j and k, then the coefficient of i_k in the equation for loop j is the negative sum of all the resistances common to loops j and k.

In the case of *mesh analysis*, it is generally convenient to choose all currents circulating clockwise (or all counterclockwise). Then the coefficient of i_k in the equation for mesh j is *always* the negative sum of the resistances common to meshes j and k.

The above statements facilitate the writing of the standard form of the loop equations of a circuit directly, when the circuit contains only independent voltage sources and resistors. For circuits containing current sources or dependent sources, some additional rearrangement of terms is usually needed.

It will be assumed from this point onward that all loop and mesh equations arc to be put in standard form whether or not explicitly stated.

Exercise 4-22 Write the loop equations for Exercises 4-15 and 4-16 in standard form.

Exercise 4-23 In the circuits of Exercises 4-15 and 4-16, choose mesh currents (all clockwise), and write the mesh equations in standard form.

Example 4-11 For the circuit shown in Fig. 4-22(a), (a) find the values of the loop currents, and (b) determine the voltage between points A and B.

Solution (a) The loop equations are

Loop 1: $\qquad (R_1 + R_2 + R_3)i_1 - R_3 i_2 + R_1 i_3 = v_a - v_b$

or

$$6i_1 - 3i_2 + i_3 = 50 \qquad (4\text{-}34)$$

Fig. 4-22 (a) Circuit for Example 4-11. (b) Calculation of v_{AB}.

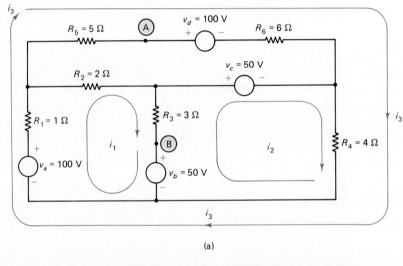

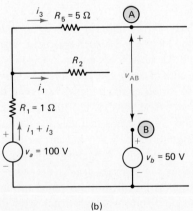

(b)

Loop 2:
$$-R_3 i_1 + (R_3 + R_4) i_2 + R_4 i_3 = v_b - v_c$$

or

$$-3 i_1 + 7 i_2 + 4 i_3 = 0 \tag{4-35}$$

Loop 3:
$$R_1 i_1 + R_4 i_2 + (R_1 + R_5 + R_6 + R_4) i_3 = v_a - v_c$$

or

$$i_1 + 4 i_2 + 16 i_3 = 0 \tag{4-36}$$

Solving Eqs. (4-34) to (4-36), it is found that

$$i_1 = 12.0 \text{ A} \qquad i_2 = 6.48 \text{ A} \qquad i_3 = -2.37 \text{ A}$$

(b) To find the voltage between points A and B, consider a portion of the circuit that includes the two points and a set of branches between them, as shown in Fig. 4-22(b), and write a KVL equation:

$$-v_{AB} - R_5 i_3 - R_1 (i_1 + i_3) + v_a - v_b = 0$$

Therefore, $v_{AB} = 52.2$ V. ∎

Exercise 4-24 Solve Example 4-11 by using mesh analysis.

Exercise 4-25 Determine the power supplied by each of the voltage sources in the circuit of Fig. 4-23(a).

Exercise 4-26 If it is given that $i_1 = 20$ A in the circuit of Fig. 4-23(b), determine the value of v_x.

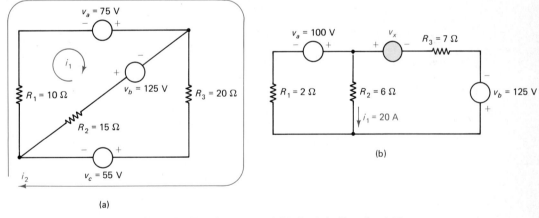

Fig. 4-23 (a) Circuit for Exercise 4-25 and (b) circuit for Exercise 4-26.

4-7 CIRCUITS WITH LINEAR DEPENDENT VOLTAGE SOURCES

When a circuit contains a linear dependent voltage source, the procedure involves the same steps as those used in the previous discussion.

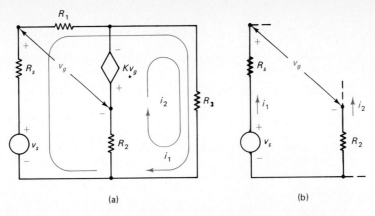

Fig. 4-24 (a) Circuit for Example 4-12. (b) Calculation of v_g controlling the dependent source.

Example 4-12 Determine the currents i_1 and i_2 in the circuit of Fig. 4-24(a).

Solution It is first necessary to obtain a relationship for the parameter controlling the dependent voltage source, which is v_g. Consider a portion of the circuit containing v_g and relevant branches, so that a KVL equation can be written. Such a portion is shown in Fig. 4-24(b). Writing a KVL equation for the closed path B to A through v_g, R_s, v_s, and R_2,

$$-v_s + R_s i_1 + v_g - R_2 i_2 = 0$$

Therefore,

$$v_g = v_s - R_s i_1 + R_2 i_2 \tag{4-37}$$

The loop equations are

Loop 1:
$$R_s i_1 + R_1 i_1 + R_3 (i_1 + i_2) = v_s$$

which (after collecting coefficients) gives

$$(R_s + R_1 + R_3) i_1 + R_3 i_2 = v_s \tag{4-38}$$

Loop 2:
$$R_3 (i_1 + i_2) + R_2 i_2 = -K v_g \tag{4-39}$$

Substituting for v_g from Eq. (4-37) into Eq. (4-38), and rearranging terms,

$$(R_3 - K R_s) i_1 + [R_3 + R_2 (1 + K)] i_2 = -K v_s \tag{4-40}$$

Equations (4-39) and (4-40) are solved for i_1 and i_2. It is found that

$$i_1 = \frac{(R_3 + R_2)(1 + K) v_s}{\| r \|}$$

and

$$i_2 = -\frac{[R_3 (1 + K) + K R_1] v_s}{\| r \|}$$

where

$$\| r \| = (R_s + R_1 + R_3)[R_3 + R_2 (1 + K)] - R_3 (R_3 - K R_s)$$

∎

Exercise 4-27 If the voltage across R_3 in the circuit of Example 4-12 is denoted by v_o, determine the voltage ratio v_o/v_s in that circuit.

Exercise 4-28 Write the loop equations for the circuit of Fig. 4-25(a) using the R_1 and R_2 branches as links. Determine the power supplied by the Kv_2 source.

Exercise 4-29 Determine the input resistance $R_{in} = v_s/i_1$ in the circuit of Fig. 4-25(b).

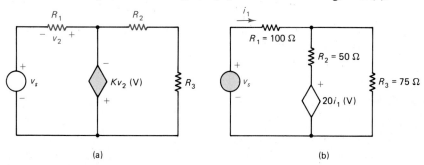

Fig. 4-25 (a) Circuit for Exercise 4-28 and (b) circuit for Exercise 4-29.

4-8 CIRCUITS CONTAINING CURRENT SOURCES

Consider a circuit containing one or more current sources. *The presence of a current source in a branch makes the current in that branch a known quantity. Thus, each current source reduces the number of unknown independent current variables by one. It is possible to exploit this feature by choosing the current source branch as a* link *in assigning a set of loop currents to the circuit.* Then the loop current defined by that link becomes a known quantity and it is therefore not necessary to write an equation for that loop. As a simple example of the use of the above idea, consider the circuit shown in Fig. 4-26(a). By choosing the current source branch as a link, as shown in Fig. 4-26(b), the resulting set of independent loop currents for the circuit is obtained, as shown in Fig. 4-26(c).

Then i_1, being equal to i_b, is a known quantity and it is not necessary to write an equation for loop 1. The equation for loop 2 is

$$R_a i_1 + (R_a + R_b)i_2 = v_a \quad (4\text{-}41)$$

Using $i_1 = i_b$, Eq. (4-35) can be solved for i_2:

$$i_2 = (v_a - R_a i_b)/(R_a + R_b)$$

Fig. 4-26 (a) Circuit with a current source in loop analysis. (b) Choosing the current source branch as a link reduces computational effort. (c) Resulting set of independent loop currents.

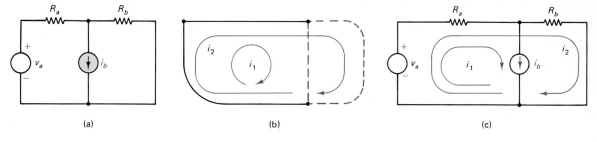

4-8 Circuits Containing Current Sources

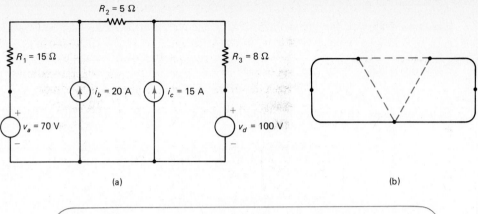

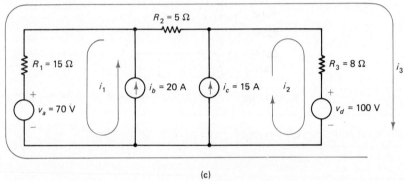

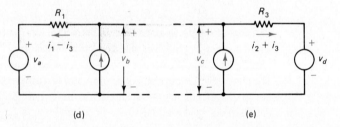

Fig. 4-27 (a) Circuit for Example 4-13. (b) Current source branches chosen as links. (c) Loop currents. (d) and (e) Calculation of voltages across the current sources.

Example 4-13 Determine the power supplied by each of the sources in the circuit of Fig. 4-27(a).

Solution A tree is set up, as shown in Fig. 4-27(b), so as to make the two current source branches act as links, and the resulting set of loop currents is shown in Fig. 4-27(c). Then,

$$i_1 = i_b = 20 \text{ A}$$

and

$$i_2 = i_c = 15 \text{ A}$$

There remains only one unknown independent current variable, i_3. The equation for loop 3 is

$$-R_1 i_1 + R_3 i_2 + (R_1 + R_2 + R_3)i_3 = v_a - v_d$$

which leads to, after using $i_1 = 20$ A, $i_2 = 15$ A, and the other numerical values,

$$28 i_3 = 150$$

or $i_3 = 5.36$ A.

198   Methods of Branch Currents, Loop, and Mesh Analysis

Power supplied by the v_a source $= v_a(i_3 - i_1) = -1025$ W.
Power supplied by the v_d source $= v_d(-i_3 - i_2) = -2036$ W.
(The negative values of power mean that the v_a and v_d sources are actually *receiving* 1025 W and 2036 W, respectively, from the other sources.)

In order to calculate the power supplied by the i_b source, it is necessary to calculate the voltage across it. For this, label the voltage across the current source v_b, and choose a closed path, including the current source and appropriate branches, as shown in Fig. 4-27(d). The KVL equation for the closed path is

$$-v_b + R_1(i_1 - i_3) + v_a = 0$$

which gives $v_b = 290$ V. Therefore,

$$\text{Power supplied by the } i_b \text{ source} = v_b i_b = 5792 \text{ W}$$

A similar procedure with the i_c source [Fig. 4-27(e)] gives

$$\text{Power supplied by the } i_c \text{ source} = v_c i_c = 3943 \text{ W}$$

■

Exercise 4-30 Determine the value of v_c using KVL in the closed path shown in Fig. 4-27(e), and verify the value obtained for the power supplied by the i_c source.

Exercise 4-31 The voltage across the i_b source in the circuit of Fig. 4-27(c) can also be determined by the alternative closed path R_2, R_3, v_d, and i_b. Verify the results obtained earlier using this closed path.

Exercise 4-32 Calculate the current in the resistor R_L in the circuit of Fig. 4-28 by using loop analysis. Only one loop equation is needed.

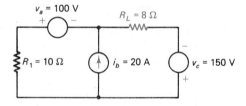

Fig. 4-28 Circuit for Exercises 4-32 and 4-33.

Exercise 4-33 Determine the power supplied by the current source in the circuit of Exercise 4-32.

4-8-1 Circuits with Linear Dependent Current Sources

The analysis of a circuit with linear *dependent* current sources follows exactly the same steps as the previous example.

Example 4-14 Determine the ratio v_o/v_s in the circuit of Fig. 4-29(a).

Solution A careful selection of links leads to significant simplification in the analysis of the given circuit. By choosing the dependent source branch as a link, the current in that source becomes one of the loop currents.

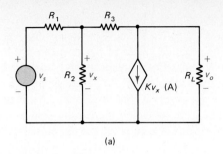

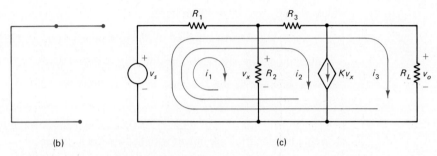

Fig. 4-29 (a) Circuit for Example 4-14. (b) Selected tree. (c) Loop currents.

The voltage v_x across R_2 controls the current source. Choosing the branch with R_2 as a link makes v_x a function of a single loop current, which provides some computational advantage.

The voltage v_o is defined across the resistor R_L. Choosing the branch with R_L as a link makes v_o a function of a single loop current, which also provides some computational advantage.

The tree that conforms to the above choice of links is shown in Fig. 4-29(b), and the circuit with the resulting set of loop currents is shown in Fig. 4-29(c).

$$i_2 = Kv_x$$

and, since $v_x = R_2 i_1$,

$$i_2 = KR_2 i_1 \qquad (4\text{-}42)$$

Since i_2 is related to i_1 by Eq. (4-42), there are only two unknown independent currents, i_1 and i_3, and equations are written for loops 1 and 3.

Loop 1: $\qquad (R_1 + R_2)i_1 + R_1 i_2 + R_1 i_3 = v_s$

which, on using Eq. (4-36), becomes

$$(R_1 + R_2 + KR_1 R_2)i_1 + R_1 i_3 = v_s \qquad (4\text{-}43)$$

Loop 3: $\qquad R_1 i_1 + (R_1 + R_3)i_2 + (R_1 + R_3 + R_L)i_3 = v_s$

which, on using Eq. (4-43), becomes

$$[R_1 + KR_2(R_1 + R_3)]i_1 + (R_1 + R_3 + R_L)i_3 = v_s \qquad (4\text{-}44)$$

Equations (4-43) and (4-44) are the loop equations (in standard form) of the given circuit. Solving them for i_3,

$$i_3 = \frac{R_2(1 - KR_3)v_s}{(R_1 + R_2)(R_3 + R_L) + R_1R_2(1 + KR_L)}$$

and, since $v_o = R_L i_3$, the ratio v_o/v_s is given by

$$\frac{v_o}{v_s} = \frac{R_L R_2 (1 - KR_3)}{(R_1 + R_2)(R_3 + R_L) + R_1R_2(1 + KR_L)}$$

■

Exercise 4-34 Determine the ratio v_2/v_s in the circuit of Fig. 4-30(a).

Exercise 4-35 Determine (a) the input resistance v_s/i_1 and (b) the voltage ratio v_o/v_s in the circuit of Fig. 4-30(b).

Exercise 4-36 Choose links carefully so as to minimize the amount of computation and determine the power delivered by the dependent current source in Fig. 4-30(c).

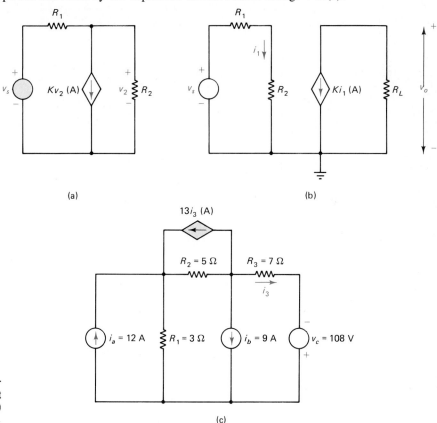

Fig. 4-30 (a) Circuit for Exercise 4-34, (b) circuit for Exercise 4-35, and (c) circuit for Exercise 4-36.

4-8-2 Mesh Analysis of Circuits with Current Sources

Consider now the case when a branch with a current source has not been chosen as a link in a circuit with one or more current sources. This happens especially when one routinely uses mesh analysis and the current source is in a branch common to two meshes.

Such a procedure makes the analysis of a circuit significantly more complex by increasing the number of equations to be solved. The following example and exercises illustrate the computational disadvantage of using mesh analysis in circuits with current sources.

Example 4-15 Determine the voltage across the resistor R_4 in the circuit shown in Fig. 4-31(a) by using mesh analysis.

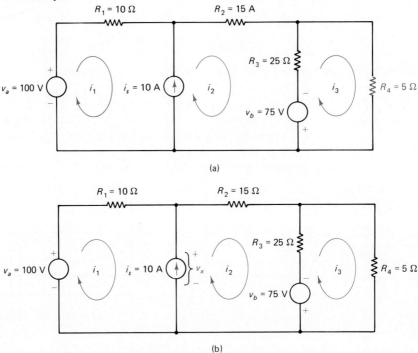

Fig. 4-31 (a) Mesh analysis of a circuit with a current source. The voltage v_x across the current source in part (b) must be included in the equations for meshes 1 and 2.

Solution The current i_s in the current source leads to the following relationship between the mesh currents i_1 and i_2.

$$i_2 - i_1 = i_s = 10 \text{ A}$$

or

$$i_2 = (i_1 + 10) \text{ A} \tag{4-45}$$

The above equation does reduce the number of unknown independent current variables by one. But the voltage across a current source is not a known quantity and, consequently, introduces a new unknown denoted by the symbol v_x in Fig. 4-31(b). The voltage v_x must be included in the equations for meshes 1 and 2. Thus, there are three unknown independent variables in the circuit, i_1, i_3, and v_x, and a system of three equations is needed.

(Note that with a careful choice of *loop currents* instead of mesh currents, the problem needs only two loop equations. The complexity of solving a set of simultaneous equations increases *geometrically* with the number of equations and it is advisable to use an approach that requires the smallest number of equations to be solved.)

The mesh equations of the circuit are as follows:

Mesh 1: Treat v_x like the voltage across a source in writing the equation.
$$R_1 i_1 = v_a - v_x$$

Move the unknown quantity v_x to the left-hand side. Using the given numerical values, the above equation becomes

$$10 i_1 + v_x = 100 \qquad (4\text{-}46)$$

Mesh 2:
$$(R_2 + R_3) i_2 - R_3 i_3 = v_x + v_b$$

Move the unknown quantity v_x to the left-hand side and use Eq. (4-45) to replace i_2 by $i_1 + 10$. Using the given numerical values, Eq. (4-46) then becomes

$$40 i_1 - 25 i_3 - v_x = -325 \qquad (4\text{-}47)$$

Mesh 3:
$$-R_3 i_2 + (R_3 + R_4) i_3 = -v_b$$

which gives, after using Eq. (4-45) and the numerical values,

$$-25 i_1 + 30 i_3 = 175 \qquad (4\text{-}48)$$

Solving Eqs. (4-46) to (4-48), it is found that

$$i_3 = 3.57 \text{ A}$$

Therefore, the voltage across $R_4 = R_4 i_3 = 17.9$ V. ∎

Exercise 4-37 In Example 4-15, determine the power delivered by the current source.

Exercise 4-38 Choose a set of loop currents for the circuit of Example 4-15 so that only two loop equations are needed for solving the circuit. Solve the equations and find the voltage across R_4.

Exercise 4-39 Repeat Exercise 4-34 by using mesh analysis.

Exercise 4-40 Repeat Exercise 4-36 by using mesh analysis.

4-9 DUALITY

The concept of duality was introduced in Chapter 2 (Section 2-5-3) and the following dual pairs were discussed in connection with series and parallel circuits.

current ↔ voltage
resistance ↔ conductance
series connection ↔ parallel connection
short circuit ↔ open circuit

Duality also applies to analysis methods based on nodal, loop, and mesh equations. For example, note that Kirchhoff's voltage law equates the algebraic sum of *voltages* around a *closed path* to zero, whereas Kirchhoff's current law equates the algebraic sum of *currents at a node* to zero. Thus,

KVL and *KCL* are a dual pair

Also, since KVL is written for a closed path and KCL for a node,

closed path is the dual of a *node*

Consider a system of nodal equations for a set of independent nodes for some circuit. If the voltage variables in the equations are replaced by currents and conductance symbols by resistances, the resulting system of equations should correspond to the loop or mesh equations of a circuit that is the dual of the given circuit. Therefore, it would appear that the dual of a system of nodal equations is a system of loop or mesh equations.

When nodal equations are written with *node-to-datum voltages* as the independent variables (as in Chapter 3), then their dual is a set of mesh equations of a circuit with all mesh currents circulating in the same direction (all clockwise or all counterclockwise).

The validity of the above statement is established by comparing the features of nodal equations and mesh equations. In nodal analysis, the coefficient of voltage v_j in the equation for node j is the sum of all the conductances connected to node j. The coefficient of the voltage v_k in the equation for node j is the negative sum of the conductances common to nodes j and k. In mesh analysis, the coefficient of current i_j in the equation for mesh j is the sum of all the resistances connected in mesh j. The coefficient of the current i_k in the equation for mesh j is the negative sum of the resistances common to meshes j and k (provided that all currents are circulating in the same direction). Therefore, if the nodal equations of a circuit are transformed by using duality, the resulting equations are the mesh equations of the dual circuit. The converse is also true: if the mesh equations of a circuit are transformed by using duality, the resulting equations are the nodal equations of the dual circuit. Note that the dual of a node is a mesh rather than an arbitrary closed path. An example of a dual pair of circuits is shown in Fig. 4-32.

Fig. 4-32 Dual pair of circuits. Mesh equations of part (a) are the duals of the nodal equations of part (b).

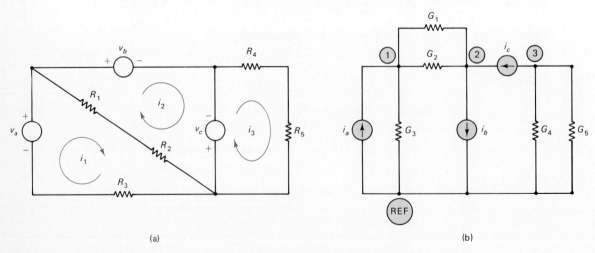

Exercise 4-41 Show that the two circuits of Fig. 4-32 are a dual pair by writing the equations.

Exercise 4-42 Consider the nodal equations given by Eqs. (3-13) to (3-15) in Chapter 3. Replace v by i, i by v, and G by R in the equations. Draw the diagram of the circuit described by the resulting mesh equations.

In the general case of loop analysis, on the other hand, the coefficient of current i_k in the equation for loop j can be *positive or negative* even if all loop currents are circulating in the same direction. (Refer to the equation for loop 2 in the circuit of Example 4-9.) Therefore, if the nodal equations of a circuit are transformed by using duality, the resulting equations need not represent the general loop equations of a dual circuit (except that mesh equations are a special case of loop equations). Thus, a system of loop equations and a system of nodal equations are not necessarily duals of each other. It so happens, however, that the dual of a general system of loop equations is a set of KCL equations written by using *node-pair voltages* (that is, voltages defined between pairs of nodes without reference to a datum node). Such a procedure was not discussed in this chapter and the interested student should refer to the literature on network theory.

Based on the above discussion (and Chapter 2), the following table of dual pairs is obtained.

$$\text{current} \leftrightarrow \text{voltage}$$
$$\text{resistance} \leftrightarrow \text{conductance}$$
$$\text{node-to-datum voltage} \leftrightarrow \text{mesh current}$$
$$\text{node-pair voltage} \leftrightarrow \text{loop current}$$
$$\text{series connection} \leftrightarrow \text{parallel connection}$$
$$\text{short circuit} \leftrightarrow \text{open circuit}$$

It should be stressed that each of the above pairs represents *mutual duals:* each item in a pair is the dual of the other. Equations, expressions, and general results of a linear network are applicable to its dual by using the transformations listed above. This approach is used at various points in this text for proving certain properties and theorems.

4-10 MESH RESISTANCE MATRIX

When duality is applied to the algebraic discussion of nodal analysis in Section 3-3 (Chapter 3), the corresponding properties and relationships of mesh analysis are obtained, as shown in the following summary:

> The *mesh resistance matrix* of a circuit is the arrangement of the coefficient of the mesh currents in the mesh equations of the circuit (in standard form) in the form of an $n \times n$ matrix, with n rows corresponding to the n mesh equations and n columns corresponding to the n independent mesh currents.

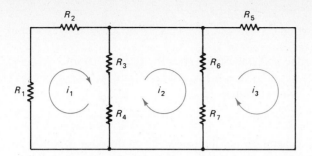

Fig. 4-33 Illustrative circuit for a mesh resistance matrix.

The mesh resistance matrix of the circuit in Fig. 4-33 is given by

$$\begin{bmatrix} R_1 + R_2 + R_3 + R_4 & -(R_3 + R_4) & 0 \\ -(R_3 + R_4) & R_3 + R_4 + R_6 + R_7 & -(R_6 + R_7) \\ 0 & -(R_6 + R_7) & R_5 + R_6 + R_7 \end{bmatrix}$$ ■

Exercise 4-43 Write the mesh resistance matrices of the circuits of Exercise 4-22.

For a circuit with n independent mesh currents and containing *only independent voltage sources and resistances* (that is, a passive circuit driven by independent voltage sources), the mesh resistance matrix is of the general form

$$\begin{bmatrix} r_{11} & r_{12} & \cdots & r_{1j} & \cdots & r_{1n} \\ r_{21} & r_{22} & \cdots & r_{2j} & \cdots & r_{2n} \\ \cdot & \cdot & & \cdot & & \cdot \\ r_{k1} & r_{k2} & \cdots & r_{kj} & \cdots & r_{kn} \\ \cdot & \cdot & & \cdot & & \cdot \\ r_{n1} & r_{n2} & \cdots & r_{nj} & \cdots & r_{nn} \end{bmatrix}$$

The elements in the kth *row* ($k = 1, 2, 3, \ldots, n$) are the coefficients of the currents in the equation for mesh k. The elements in the jth *column* ($j = 1, 2, 3, \ldots, n$) are the coefficients of the current i_j in the different mesh equations.

Exercise 4-44 In the mesh resistance matrices obtained in Exercise 4-43, (a) list the following elements: r_{11}, r_{22}, r_{23}, and r_{31}; (b) list the coefficients for the currents in the equation for mesh 3; and (c) list the coefficients for i_2 for the different meshes.

The elements of the mesh resistance matrix of a *passive* circuit driven by *independent voltage sources* have the following properties:

1. The element r_{kk} ($k = 1, 2, 3, \ldots, n$) is the positive sum of all the resistances in mesh k.

The element r_{kj} (k and j can assume any integer values between 1 and n, but $k \neq j$) is the negative sum of the resistances common to meshes k and j. Note that all mesh currents must circulate in the same direction for this property to be valid. 3.

$r_{kj} = r_{jk}$; that is, the matrix is *symmetric* about the principal diagonal.

The above properties of the mesh resistance matrix of a passive circuit driven by independent voltage sources make it possible to write the mesh resistance matrix by inspection.

Write the mesh resistance matrix for the circuit of Fig. 4-34.

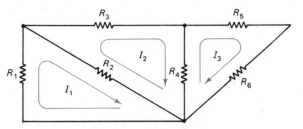

Fig. 4-34 Circuit for Example 4-16.

Solution The mesh resistance matrix is given by

$$\begin{bmatrix} R_1 + R_2 & -R_2 & 0 \\ -R_2 & R_2 + R_3 + R_4 & -R_4 \\ 0 & -R_4 & R_4 + R_5 + R_6 \end{bmatrix}$$

Exercise 4-45 Write the mesh resistance matrix for each of the circuits shown in Fig. 4-35.

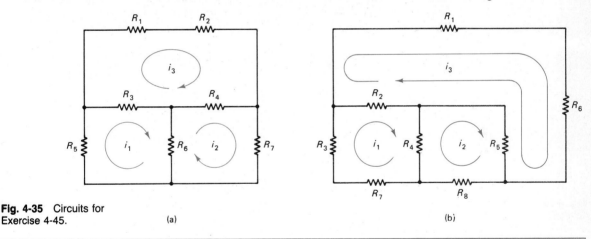

Fig. 4-35 Circuits for Exercise 4-45.

4-10-1 Application of Cramer's Rule to the Solution of Mesh Equations

Consider a system of mesh equations given by

$$r_{11}i_1 + r_{12}i_2 + \ldots + r_{1j}i_j + \ldots + r_{1n}i_n = v_1$$
$$r_{21}i_1 + r_{22}i_2 + \ldots + r_{2j}i_j + \ldots + r_{2n}i_n = v_2$$

$$\vdots$$

$$r_{k1}i_1 + r_{k2}i_2 + \ldots + r_{kj}i_j + \ldots + r_{kn}i_n = v_k$$

$$\vdots$$

$$r_{n1}i_1 + r_{n2}i_2 + \ldots + r_{nj}i_j + \ldots + r_{nn}i_n = v_n$$

4-10 Mesh Resistance Matrix

where the voltage term on the right-hand side of each equation represents the net increase in potential difference around that mesh due to voltage sources.

Let $\|r\|$ denote the determinant of the mesh resistance matrix $[r]$. Then the mesh current i_j ($j = 1, 2, 3, \ldots, n$) is given by

$$i_j = (1/\|r\|)[A_{1j}v_1 + A_{2j}v_2 + \ldots + A_{jj}v_j + \ldots + A_{nj}v_j] \qquad (4\text{-}49)$$

$$= \frac{\sum_{k=1}^{n} A_{kj}v_k}{\|r\|}$$

where

A_{kj} = cofactor of the kth row and jth column of the mesh resistance matrix.

v_k = the net increase in potential difference in mesh k due to the voltage sources in it.

$\|r\|$ = the determinant of the mesh resistance matrix.

Example 4-17 Use Cramer's rule to determine the mesh currents in the circuit of Fig. 4-36.

Solution The mesh equations are

Mesh 1: $\qquad (R_1 + R_2 + R_3)i_1 - R_2 i_2 - R_3 i_3 = v_a - v_b$

or

$$21i_1 - 7i_2 - 9i_3 = 50 \qquad (4\text{-}50)$$

Mesh 2: $\qquad -R_2 i_1 + (R_2 + R_5 + R_4)i_2 - R_5 i_3 = v_c$

or

$$-7i_1 + 21i_2 - 11i_3 = 125 \qquad (4\text{-}51)$$

Mesh 3: $\qquad -R_3 i_1 - R_5 i_2 + (R_5 + R_3 + R_6)i_3 = -v_c + v_b$

Fig. 4-36 Circuit for Example 4-17.

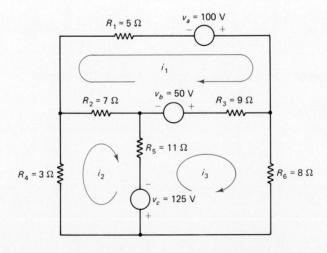

Methods of Branch Currents, Loop, and Mesh Analysis

or
$$-9i_1 - 11i_2 + 28i_3 = -75 \qquad (4\text{-}52)$$

The mesh resistance matrix is given by

$$[r] = \begin{bmatrix} 21 & -7 & -9 \\ -7 & 21 & -11 \\ -9 & -11 & 28 \end{bmatrix}$$

The determinant of the matrix $[r]$ is found to be

$$\|r\| = 5348$$

The different cofactors of the determinant are

$$A_{11} = \begin{vmatrix} 21 & -11 \\ -11 & 28 \end{vmatrix} = 467 \qquad A_{12} = -\begin{vmatrix} -7 & -11 \\ -9 & 28 \end{vmatrix} = 295$$

$$A_{13} = \begin{vmatrix} -7 & 21 \\ -9 & -11 \end{vmatrix} = 266 \qquad A_{21} = A_{12} = 295$$

$$A_{22} = \begin{vmatrix} 21 & -9 \\ -9 & 28 \end{vmatrix} = 507 \qquad A_{23} = -\begin{vmatrix} 21 & -7 \\ -9 & -11 \end{vmatrix} = 294$$

$$A_{31} = A_{13} = 266 \qquad A_{32} = A_{23} = 294$$

$$A_{33} = \begin{vmatrix} 21 & -7 \\ -7 & 21 \end{vmatrix} = 392$$

Using Cramer's rule [Eq. (4-49)],

$$i_1 = (1/\|r\|)(A_{11}v_1 + A_{21}v_2 + A_{31}v_3)$$

where $v_1 = 50$ V, $v_2 = 125$ V, and $v_3 = -75$ V from the right-hand sides of Eqs. (4-50), (4-51), and (4-52).

$$i_1 = 7.53 \text{ A}$$
$$i_2 = (1/\|r\|)(A_{12}v_1 + A_{22}v_2 + A_{32}v_3)$$
$$= 10.5 \text{ A}$$
$$i_3 = (1/\|r\|)(A_{13}v_1 + A_{23}v_2 + A_{33}v_3)$$
$$= 3.86 \text{ A}$$

Exercise 4-46 Use Cramer's rule and determine the mesh currents in the circuit for Exercise 4-24.

4-10-2 Circuits Containing Dependent Voltage Sources

The use of the mesh resistance matrix and Cramer's rule in the analysis of a circuit applies also to circuits containing dependent voltage sources. The main difference, however, is that the mesh resistance of such circuits cannot be written completely by inspection. It is necessary to write the mesh equations and then manipulate them into the standard form before writing the mesh resistance matrix.

The mesh resistance matrix of an active network (that is, a network containing dependent sources) is, in general, *not symmetric* about the principal axis; that is,

$$r_{jk} \neq r_{kj}$$

The lack of symmetry makes the cofactors A_{jk} and A_{kj} different from each other, and special care must be taken about the *order of subscripts* in the cofactor terms of Eq. (4-43) when using Cramer's rule.

Except for the additional steps needed for obtaining the mesh resistance matrix and the possible lack of symmetry in such a matrix, the treatment of the mesh resistance matrix of an active circuit follows exactly the same procedure as passive circuits.

Example 4-18 For the circuit shown in Fig. 4-37, (a) obtain the mesh resistance matrix, and (b) determine the voltage ratio v_o/v_s, where v_o is the voltage across R_L.

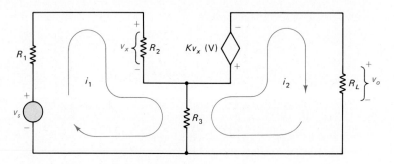

Fig. 4-37 Circuit for Example 4-18.

Solution (a) Since the circuit contains a dependent source, we start by writing the mesh equations. The parameter controlling the voltage of the dependent source is given by

$$v_x = R_2 i_1 \tag{4-53}$$

The mesh equations are

Mesh 1:
$$(R_1 + R_2 + R_3)i_1 - R_3 i_2 = v_s \tag{4-54}$$

Mesh 2:
$$-R_3 i_1 + (R_3 + R_L)i_2 = -K v_x$$

which becomes, after using Eq. (4-53) for v_x and rearranging terms,

$$(KR_2 - R_3)i_1 + (R_3 + R_L)i_2 = 0 \tag{4-55}$$

The mesh resistance matrix is obtained from Eqs. (4-54) and (4-55):

$$\begin{bmatrix} R_1 + R_2 + R_3 & -R_3 \\ KR_2 - R_3 & R_3 + R_L \end{bmatrix}$$

(b) The voltage v_o is given by

$$v_o = R_L i_2$$

and i_2 is obtained by using Cramer's rule.

The determinant of the mesh resistance matrix is

$$\|r\| = (R_1 + R_2 + R_3)(R_3 + R_L) + R_3(KR_2 - R_3) \quad (4\text{-}56)$$
$$= (R_1 + R_2)(R_3 + R_L) + R_3(R_L + KR_2)$$

Using Eq. (4-43) for Cramer's rule, the current i_2 is given by

$$i_2 = (1/\|r\|)(A_{12}v_1 + A_{22}v_2)$$

where $v_1 = v_s$ and $v_2 = 0$ from the right-hand sides of Eqs. (4-54) and (4-55). The cofactors are

$$A_{12} = -(KR_2 - R_3) \quad \text{and} \quad A_{22} = R_1 + R_2 + R_3$$

so that

$$i_2 = (R_3 - KR_2)v_s/\|r\|$$

The voltage ratio v_o/v_s is given by

$$\frac{V_o}{V_s} = \frac{R_L(R_3 - KR_2)}{(R_1 + R_2)(R_3 + R_L) + R_3(R_L + KR_2)}$$

■

Exercise 4-47 Find the current supplied by the source v_s in the circuit of Fig. 4-38(a).

Exercise 4-48 Use mesh analysis and determine the voltage ratio v_2/v_s in the circuit of Fig. 4-38(b).

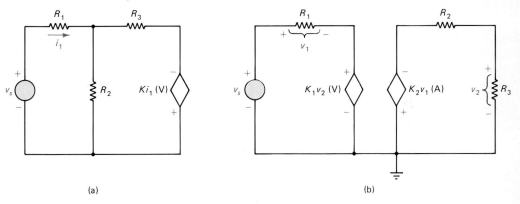

Fig. 4-38 (a) Circuit for Exercise 4-47 and (b) circuit for Exercise 4-48.

4-10-3 Loop Resistance Matrix

Attention was focused on the mesh resistance matrix in the above discussion in order to make use of the duality between nodal conductance matrices and mesh resistance matrices and transfer the results obtained in Chapter 3 to this chapter. It is also possible to write a *loop resistance matrix* for a network and apply the relationships and discussion of the preceding sections to the analysis of a network using loop resistance matrices.

Example 4-19 For the system of loop currents shown in the circuit of Fig. 4-39, (a) obtain the loop resistance matrix, and (b) determine the ratio v_s/i_1 when $R_1 = 2\,\Omega$, $R_2 = 4\,\Omega$, $R_3 = 8\,\Omega$, $R_4 = 7\,\Omega$, and $R_5 = 9\,\Omega$.

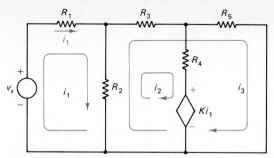

Fig. 4-39 Circuit for Example 4-19.

Solution (a) The given circuit contains a dependent source. The parameter controlling the source is i_1.

The loop equations are

Loop 1: $\quad (R_1 + R_2)i_1 - R_2i_2 - R_2i_3 = v_s \quad$ (4-57)

Loop 2: $\quad -R_2i_1 + (R_2 + R_3 + R_4)i_2 + (R_2 + R_3)i_3 = -Ki_1$

which, after moving the $-Ki_1$ term to the left, leads to

$$(K - R_2)i_1 + (R_2 + R_3 + R_4)i_2 + (R_2 + R_3)i_3 = 0 \quad (4\text{-}58)$$

Loop 3: $\quad -R_2i_1 + (R_2 + R_3)i_2 + (R_2 + R_3 + R_5)i_3 = 0 \quad$ (4-59)

The loop resistance matrix is obtained from Eqs. (4-57) to (4-59):

$$[r] = \begin{bmatrix} R_1 + R_2 & -R_2 & -R_2 \\ K - R_2 & R_2 + R_3 + R_4 & R_2 + R_3 \\ -R_2 & R_2 + R_3 & R_2 + R_3 + R_5 \end{bmatrix}$$

(b) On using the numerical values, the loop resistance matrix becomes

$$[r] = \begin{bmatrix} 6 & -4 & -4 \\ K & -419 & 12 \\ -4 & 12 & 21 \end{bmatrix}$$

The determinant is

$$\|r\| = 36K + 1274$$

The current i_1 is given by [from Cramer's rule, Eq. (4-49)]

$$i_1 = (1/\|r\|)(A_{11}v_1 + A_{21}v_2 + A_{31}v_3)$$

where $v_1 = v_s$ and $v_2 = v_3 = 0$ from the right-hand sides of Eqs. (4-57) to (4-59). Therefore,

$$i_1 = \frac{A_{11}v_s}{\|r\|} = \frac{255\, v_s}{36K + 1274}$$

and

$$v_s/i_1 = 0.141k + 4.997$$

∎

Exercise 4-49 Write the loop resistance matrix of the circuit shown in Fig. 4-40(a), and determine the power delivered by each of the two voltage sources.

Exercise 4-50 Write the loop resistance matrix of the circuit shown in Fig. 4-40(b), and determine the current i_1.

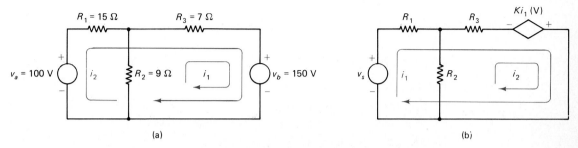

Fig. 4-40 (a) Circuit for Exercise 4-49 and (b) circuit for Exercise 4-50.

4-10-4 Driving Point and Transfer Conductances

Consider a relaxed linear network (that is, one containing no independent sources) with m independent loop currents and a loop resistance matrix $[r]$ inside a black box, as indicated in Fig. 4-41(a). (The following discussion is also valid for mesh currents and the mesh resistance matrix.) Two branches, labeled p and j, have been extended outside the box in the diagram for purposes of external connections and measurements. An opening is made in branch p and a voltage source v_p is connected, as shown in Fig. 4-41(b).

Since there is only one independent voltage source, v_p in loop p, the net voltage due to the voltage sources in loop k from Cramer's rule, Eq. (4-49) is given by

$$v_k = 0 \quad \text{except when } k = p$$

Then the current in loop j is, from Eq. (4-49),

$$i_j = (A_{pj}v_p)/\|r\| \tag{4-61}$$

where A_{pj} is the cofactor of the pth row and rth column of $[r]$.

Equation (4-61) gives the current in loop j due to the presence of a voltage source in loop p in a circuit.

Fig. 4-41 Driving point and transfer conductances of a relaxed network.

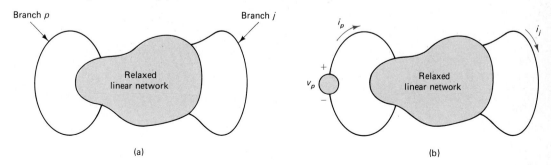

Transfer Conductance:

The ratio of the current i_j in loop j due to a voltage source v_p in loop p is called the *transfer conductance* from loop p to loop j.

From Eq. (4-61),

$$\frac{i_j}{v_p} = \frac{A_{pj}}{\|r\|} \qquad (4\text{-}62)$$

The transfer conductance from one loop to another is a measure of the effect in a loop due to a voltage source placed in a different loop. The transfer conductance is a useful tool in the analysis and design of electronics and communication circuits.

Driving Point Conductance:

The ratio of the current i_j in loop j due to a voltage source placed in the *same* loop is called the *driving point* conductance with respect to loop j.

By making $p = j$ in Eq. (4-56), it is seen that

$$\frac{i_j}{v_j} = \frac{A_{jj}}{\|r\|} \qquad (4\text{-}63)$$

The driving point conductance with respect to any loop is the effective conductance seen by a voltage source placed in that loop.

Equations (4-62) and (4-63) permit the determination of the transfer and driving point conductances of a circuit whose loop (or mesh) resistance matrix is known.

Example 4-20 For the circuit shown in Fig. 4-42, determine (a) the loop resistance matrix for the set of loop currents shown, (b) the driving point conductance with respect to loop 1 and the transfer conductance from loop 1 to loop 3, (c) the driving point conductance with respect to loop 3 and the transfer conductance from loop 3 to loop 1.

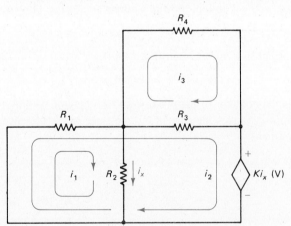

Fig. 4-42 Circuit for Example 4-20.

Solution (a) Since the circuit contains a dependent source, we start by writing the loop equations to obtain the loop resistance matrix.

The current controlling the dependent source is

$$i_x = i_1$$

Loop 1:
$$(R_1 + R_2)i_1 + R_1 i_2 = 0 \quad (4\text{-}64)$$

Loop 2:
$$R_1 i_1 + (R_1 + R_3)i_2 - R_3 i_3 = -K i_x$$

which becomes, after using $i_x = i_1$

$$(R_1 + K)i_1 + (R_1 + R_3)i_2 - R_3 i_3 = 0 \quad (4\text{-}65)$$

Loop 3:
$$-R_3 i_2 + (R_3 + R_4)i_3 = 0 \quad (4\text{-}66)$$

The loop resistance matrix is, from Eqs. (4-64) to (4-66),

$$[r] = \begin{bmatrix} R_1 + R_2 & R_1 & 0 \\ R_1 + K & R_1 + R_3 & -R_3 \\ 0 & -R_3 & R_3 + R_4 \end{bmatrix}$$

The determinant of the loop resistance matrix is found to be

$$\|r\| = R_1 R_3 (R_2 + R_4 - K) + R_1 R_4 (R_2 - K) + R_2 R_3 R_4 \quad (4\text{-}67)$$

(b) The driving point conductance with respect to loop 1 is

$$\frac{i_1}{v_1} = \frac{A_{11}}{\|r\|} = \frac{R_1 R_3 + R_1 R_4 + R_3 R_4}{\|r\|}$$

The transfer conductance from loop 1 to loop 3 is

$$\frac{i_3}{v_1} = \frac{A_{13}}{\|r\|} = -\frac{R_3 (R_1 + K)}{\|r\|}$$

(c) The driving point conductance with respect to loop 3 is

$$\frac{i_3}{v_3} = \frac{A_{33}}{\|r\|} = \frac{R_1 (R_3 - K) + R_1 R_2 + R_2 R_3}{\|r\|}$$

The transfer conductance from loop 3 to loop 1 is

$$\frac{i_1}{v_3} = \frac{A_{31}}{\|r\|} = -\frac{R_1 R_3}{\|r\|}$$

Exercise 4-51 For each of the circuits shown in Fig. 4-43, set up the loop resistance matrices and determine all possible driving point and transfer conductances.

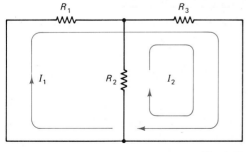
(a)

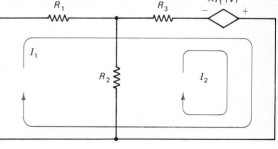
(b)

Fig. 4-43 Circuits for Exercise 4-51.

4-10 Mesh Resistance Matrix

4-11 SUMMARY OF CHAPTER

Two methods of analysis of circuits were developed and discussed in this chapter.

The method of branch currents applies KVL and KCL directly to the analysis of a circuit. A set of independent branch currents is chosen by proceeding to each node in succession and deciding at each node whether a new current variable has to be introduced. A set of independent closed paths is chosen by including a new branch in each successive path. KVL equations are written for the closed paths and these can be solved to determine all the branch currents. Even though the method of branch currents is not as systematic as the methods of nodal and loop (or mesh) analysis, it is a useful and effective procedure for simple circuits. It is found to be particularly effective in the dc analysis of electronic circuits.

Loop analysis is based on the concept of circulating currents: the current through a branch can be split into two or more component currents for the purposes of analysis, with each current component circulating through a loop. Network topology provides a systematic method of obtaining a set of independent loop currents. Mesh analysis uses the meshes of a network that is mappable on a plane or on the surface of a sphere as the independent loops. In either case, KVL equations are written for the loops (meshes) and their solution leads to the determination of the currents in the branches of the network. When a network contains current sources, a judicious choice of loops permits a considerable savings in computation.

Mesh analysis is the dual of nodal analysis. The results obtained by using the theory of determinants in nodal analysis can be directly converted to mesh analysis by using duality. Driving point and transfer conductances can be defined either on a mesh analysis basis or a loop analysis basis, and they can be determined from a knowledge of the mesh resistance or loop resistance matrix of the network.

One of the questions that arises when solving a given circuit problem is what is the "best" method of attack? Even though there is no precise answer to the above question, it is important to examine which method, if any, requires fewer equations to solve, since the computational complexity of a system of equations increases geometrically with the number of equations.

Answers to Exercises

4-1 $-v_a + R_1 i_a + R_3 i_b + v_c + R_6(i_a - i_c) = 0.$ $-v_b - R_2(i_a + R_3 i_b + v_c - v_d - R_4(i_a - i_b - i_c) = 0.$ $R_4(i_a - i_b - i_c) + v_d R_6(i_a - i_c) - R_5 i_c = 0.$

4-2 $-v_a + R_1 i_1 + R_2(i_1 - i_2) + R_3 i_3 + v_b = 0.$ $R_6 i_2 - v_d - v_e - R_4(i_1 - i_2 - i_3) - R_2(i_1 - i_2) = 0.$ $-v_b - R_3 + R_4(i_1 - i_2 - i_3) + v_e + R_5(i_1 - i_3) + v_c = 0.$

4-3 $(R_2 + R_3)i_1 - (R_4 + R_2 + R_3)i_2 = v_b - v_c + v_d.$
$(R_1 + R_2 + R_3)i_1 - (R_2 + R_3)i_2 = -v_a - v_c + v_d.$

4-4 $-30i_1 + 15i_2 = 75$ and $15i_1 + 20i_2 = 0.$ $i_1 = -1.82$ A and $i_2 = 1.36$ A.

4-5 R_1: 1.4×10^{-4} A, 1.96×10^{-4} W; R_2: 1.53×10^{-4} A, 2.11×10^{-3} W: R_c: 6.65×10^{-4} A, 1.77×10^{-3} W; and R_e: 6.78×10^{-4} A, 4.60×10^{-4} W. Power supplied by $V_{CC} = 12.3 \times 10^{-3}$ W. Power dissipated by the transistor $= 7.76 \times 10^{-3}$ W.

4-6 $V_{ce} = 14.3$ V and $V_{cb} = 13.6$ V.

4-7 Verification.

4-8 $v_x = \dfrac{(KR_3R_4 - R_2R_3 - R_2R_4)v_a - [(K + 1)R_1 + R_2]R_3 v_b}{(R_3 + R_4)[(K + 1)R_1 + R_2]}$

4-9 (a) $R_{in} = R_1 + (K + 1)R_2$ and (b) 14.2 Ω.

4-10 (a) $i_a = i_1 + i_2$; $i_b = i_2$; $i_c = i_1$; $i_d = -i_1 - i_2 + i_3$; $i_e = -i_3$; $i_f = -i_2 + i_3$.
(b) $i_a = i_1 + i_2$; $i_b = i_2$; $i_c = -i_1$; $i_d = i_1 + i_2 - i_3$; $i_e = -i_2 + i_3$; $i_f = i_3$.

4-11 (a) $N = 4$; $B = 6$; $T = 3$; $L = 3$.   (b) $N = 5$; $B = 7$; $T = 4$; $L = 3$.

4-12 Six independent loop currents. See Fig. 4-44.

Fig. 4-44 Answers to Exercises 4-12 and 4-42.

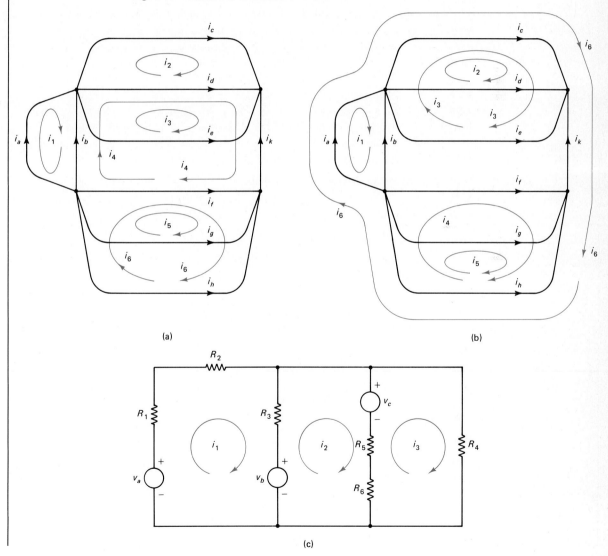

4-13 Answers will vary depending on your answers to Exercise 4-12.

4-14 Verification.

4-15 Loop 1: $(R_1 + R_2 + R_6)i_1 - (R_2 + R_6)i_2 - R_6i_3 = v_a + v_b$.
Loop 2: $- (R_2 + R_6)i_1 + (R_2 + R_3 + R_4 + R_8 + R_7 + R_6)i_2 - (R_6 + R_8 + R_7)i_3$
$= -v_b - v_e - v_d$.
Loop 3: $-R_6i_1 - (R_6 + R_8 + R_7)i_2 + (R_5 + R_8 + R_7 + R_6)i_3 = v_c - v_e - v_d$.

4-16 Loop 1: $(R_1 + R_2 + R_3 + R_4)i_1 - R_2i_2 + (R_1 + R_2 + R_4)i_3 = v_a - v_b + v_c$.
Loop 2: $- R_2i_1 + (R_2 + R_5 + R_6)i_2 - (R_2 + R_6)i_3 = v_b - v_e + v_d$.
Loop 3: $(R_1 + R_2 + R_4)i_1 - (R_2 + R_6)i_2 + (R_1 + R_2 + R_6 + R_7 + R_4)i_3$
$= v_a - v_b - v_d + v_f$.

4-17 592 W in R_1; 35.6 W in R_2; and 756 W in R_3.

4-18 Loop 1: $60i_1 + 20i_2 = 150$. Loop 2: $20i_1 + 30i_2 = 250$.
$i_1 = 0.357$ A and $i_2 = 8.57$ A. v_a delivers 857 W;
v_b, 411 W; v_c, 821 W. Dissipated: 734 W in R_1; 1349 W in R_2; 1.91 W in R_3; 3.19 W in R_4.

4-19 (a) $(R_1 + R_2 + R_6)i_1 - R_5i_2 - R_6i_3 = v_a + v_b$. $-R_2i_1 + (R_2 + R_3 + R_4 + R_5)i_2$
$- R_5i_3 = v_b - v_c$. $-R_6i_1 - R_5i_2 + (R_6 + R_5 + R_8 + R_7)i_3 = c_c - v_e - v_d$.
(b) $(R_1 + R_2 + R_3 + R_4)i_1 - R_2i_2 - R_3i_3 = v_a - v_b + v_c$.
$-R_2i_1 + (R_2 + R_5 + R_6)i_2 - R_6i_3 = v_b - v_e + v_d$.
$-R_3i_1 - R_6i_2 + (R_3 + R_6 + R_7)i_3 = -v_c - v_d + v_f$.

4-20 $12i_1 - 2i_2 = -40$; $-2i_1 + 10i_2 = -70$; $i_1 = -4.66$ A; $i_2 = -7.93$ A; v_a receives 46.6 W; v_b receives 164 W; v_c delivers 952 W. Dissipated: 217 W in R_1; 503 W in R_2; 21.5 W in R_3.

4-21 $30i_1 - 20i_2 = 100$; $-20i_1 + 50i_2 = 0$. $v_o/v_s = 0.273$.

4-22 See answers to Exercises 4-15 and 4-16.

4-23 (a) $(R_1 + R_2 + R_6)i_1 - R_2i_2 - R_6i_3 = v_a + v_b$. $-R_2i_1 + - R_5i_3 = -v_b - v_c$.
$-R_6i_1 - R_5i_2 + (R_6 + R_5 + R_7 + R_8)i_3 = v_c - v_e - v_d$.
(b) $(R_1 + R_2 + R_3 + R_4)i_1 - R_2i_2 - R_3i_3 = v_a - v_b + v_c$.
$-R_2i_1 + (R_2 + R_5 + R_6)i_2 - R_6i_3 = v_b - v_e + v_d$.
$-R_3i_1 - R_6i_2 + (R_3 + R_6 + R_7)i_3 = -v_c - v_d + v_f$.

4-24 $6i_1 - 3i_2 - 2i_3 = 50$; $-3i_1 + 7i_2 = 0$; $-2i_1 + 13i_3 = -50$. $i_1 = 9.60$ A; $i_2 = 4.12$ A; $i_3 = -2.37$ A; $v_{AB} = 52.2$ V.

4-25 $25i_1 + 10i_2 = 200$; $10i_1 + 30i_2 = 20$. $i_1 = 8.92$ A; $i_2 = -2.31$ A. 496 W delivered by v_a; 1115 W delivered by v_b; 126 W delivered by v_c.

4-26 Assign i_2 to R_3. $8i_1 + 2i_2 = 100$; $2i_1 + 9i_2 = (225 - v_x)$. $v_x = 455$ V.

4-27 $v_o/v_s = R_3[R_2(1 + K) - KR_1]/\|r\|$.

4-28 $i_1 = v_s/[R_1(1 + K)]$. $i_2 = Kv_s/[(1 + K)(R_2 + R_3)]$. Power supplied by Kv_2 is
$- v_s^2 K(R_2 + R_3 + KR_1)/[(1 + K)^2 R_1(R_2 + R_3)]$.

4-29 $170i_1 - 50i_2 = v_s$; $-70i_1 + 125i_2 = 0$. $i_1 = 7.04 v_s$ mA. $R_{in} = 142$ Ω.

4-30 $-v_c + R_3(i_2 + i_3) + v_d = 0$. $v_c = 263$ V. Power = 3943 W.

4-31 $-v_b + R_2i_3 + R_3(i_2 + i_3) + v_d = 0$. $v_b = 290$ V.

4-32 $i_1 = -i_b = -20$ A; $(R_1 + R_L)i_2 = -v_a + v_c + R_1i_b$. $i_2 = 13.9$ A.

4-33 Voltage across i_b source = 38.9 V. Power received = 778 W.

4-34 $i_1 = Kv_2 = KR_2i_2$; $R_1i_1 + (R_1 + R_2)i_2 = v_s$. $v_2/v_s = R_2/(R_1 + R_2 + KR_1R_2)$.

4-35 $R_{in} = R_1 + R_2$; $v_o/v_s = -KR_L/(R_1 + R_2)$.

4-36 Choose the three branches with current sources as links. $i_1 = i_a = 12$ A; $i_2 = i_b = 9$ A; $i_4 = -13i_3$. Loop 3 Equation: $-R_1i_1 + (R_1 + R_2)i_2 + (R_1 + R_2 + R_3)i_3 - R_2i_4 = v_c$. Voltage across current source $= R_2(i_3 + i_2 - i_4) = 108$ V. Power delivered $= 1264$ W.

4-37 Voltage across the current source $= 127$ V. Power delivered $= 1272$ W.

4-38 $i_1 = -i_s = -10$ A. Loop 2: $50i_2 - 25i_3 = 275$. Loop 3: $-25i_2 + 30i_3 = -75$. Voltage across $R_4 = 17.9$ V.

4-39 Voltage across current source; $v_x = v_2$. Mesh 1: $R_1i_1 + v_x = v_s$. Mesh 2: $R_2i_2 = v_x$. Expression for v_2/v_s is the same as in Exercise 4-34.

4-40 Denote the voltage across the i_b source by v_b. Mesh 1: $i_1 = i_a = 12$ A. Mesh 2: $73i_3 + v_b = -36$. Mesh 3: $7i_3 - v_b = 108$. $i_3 = 0.900$ A. Same answer for power as in Exercise 4-36.

4-41 Equations show duality.

4-42 See Fig. 4-44(c).

4-43
(a) $\begin{bmatrix} R_1 + R_2 + R_6 & -R_2 & -R_6 \\ -R_2 & R_2 + R_3 + R_4 + R_5 & -R_5 \\ -R_6 & -R_5 & R_6 + R_5 + R_7 + R_8 \end{bmatrix}$

(b) $\begin{bmatrix} R_1 + R_2 + R_3 + R_4 & -R_2 & -R_3 \\ -R_2 & R_2 + R_5 + R_6 & -R_6 \\ -R_3 & -R_6 & R_3 + R_6 + R_7 \end{bmatrix}$

4-44 (a) $r_{11} = (R_1 + R_2 + R_6), (R_1 + R_2 + R_3 + R_4)$; $r_{22} = (R_2 + R_3 + R_4 + R_5), (R_2 + R_5 + R_6)$; $r_{23} = -R_5, -R_6$; $r_{31} = -R_6, -R_3$. (b) $-R_6, -R_5, (R_6 + R_5 + R_7 + R_8)$; $-R_3, -R_6, (R_3 + R_6 + R_7)$. (c) $-R_2, (R_2 + R_3 + R_4 + R_5), -R_5$; $-R_2, (R_2 + R_5 + R_6), -R_6$.

4-45
(a) $\begin{bmatrix} R_5 + R_3 + R_6 & -R_6 & -R_3 \\ -R_6 & R_6 + R_4 + R_7 & -R_4 \\ -R_3 & -R_4 & R_1 + R_2 + R_3 + R_4 \end{bmatrix}$

(b) $\begin{bmatrix} R_3 + R_2 + R_4 + R_7 & -R_4 & -R_2 \\ -R_4 & R_4 + R_5 + R_8 & -R_5 \\ -R_2 & -R_5 & R_1 + R_6 + R_5 + R_2 \end{bmatrix}$

4-46 $\|r\| = 401$. Cofactors of column 1 are 91, 39, and 14. $i_1 = 9.60$ A. Cofactors of column 2 are 39, 74, and 6. $i_2 = 4.12$ A. Cofactors of column 3 are 14, 6, and 33. $i_3 = -2.37$ A.

4-47 $\|r\| = R_1R_2 + R_1R_3 + R_2(R_3 - K)$. $i_1 = (R_2 + R_3)/\|r\|$.

4-48 $R_1i_1 + K_1R_3i_2 = v_s$; $K_2R_1i_1 + (R_2 + R_3)i_2 = 0$. $\|r\| = R_1(R_2 + R_3) - K_1K_2R_1R_3$. $v_2/v_s = -K_2R_1R_3/\|r\|$.

4-49 $\begin{bmatrix} 16 & 7 \\ 7 & 22 \end{bmatrix}$.

$\|r\| = 303$. $i_1 = -9.74$ A. $i_2 = 0.825$ A. Power delivered: 82.5 W by v_a; 1337 W by v_b.

4-50 $\|r\| = (R_1 + R_3 - K)(R_2 + R_3) - R_3(R_3 - K)$; $i_1 = (R_2 + R_3)v_s/\|r\|$.

4-51 (a) $\|r\| = R_1R_2 + R_1R_3 + R_2R_3$. $(R_2 + R_3)/\|r\|$. $(R_1 + R_3)/\|r\|$. $-R_3/\|r\|$.
(b) $\|r\| = R_1R_2 + R_1R_3 + R_2(R_3 - K)$. $(R_2 + R_3)/\|r\|$. $(R_1 + R_3 - K)/\|r\|$. $-R_3/\|r\|$. $(K - R_3)/\|r\|$.

PROBLEMS

Sec. 4-1 The Method of Branch Currents

4-1 In the circuits of Fig. 4-45, some branch currents have already been assigned. Label the remaining branch currents in terms of those already given, and write a set of branch current equations for each circuit.

Fig. 4-45 Circuits for Problem 4-1.

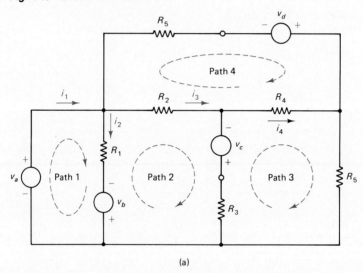

(a)

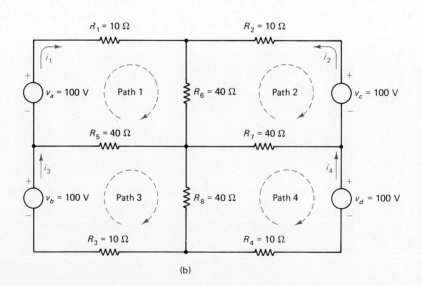

(b)

4-2 For each of the circuits shown in Fig. 4-46, determine (a) all the branch currents, (b) the power delivered by each voltage source, and (c) the voltage between the points A and B (with A as reference positive).

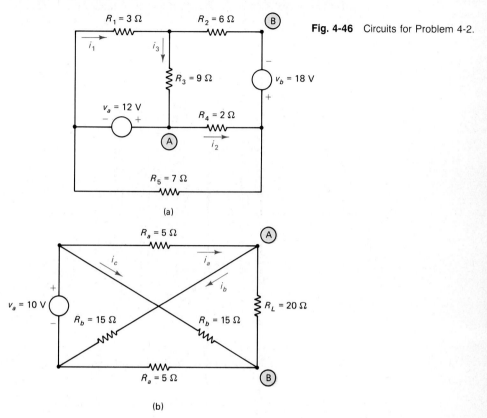

Fig. 4-46 Circuits for Problem 4-2.

4-3 Determine the current I_b and the voltages marked V_{cb} and V_{ce} in the transistor circuit shown in Fig. 4-47. Equations (4-7) and (4-8) in the text (Example 4-2) are valid for the transistor.

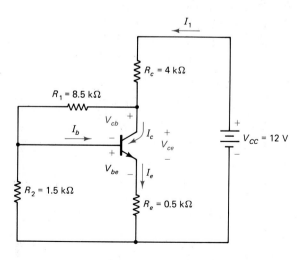

Fig. 4-47 Circuit for Problem 4-3.

Problems

4-4 The circuit shown in Fig. 4-48 is used as part of differential amplifiers. The box marked v_D can be treated as some device whose voltage is a known quantity. Again, Eqs. (4-7) and (4-8) in the text are valid for the transistor. Obtain an expression for i_c in terms of β, the resistances, and the voltages labeled v_{be} and v_D. If $\beta \gg 1$, it is possible to make

$$i_c = V_{EE}R_2/[R_3(R_2 + R_1)]$$

Determine the relationship that must be satisfied by v_D and v_{be} for the above equation for i_c to be valid.

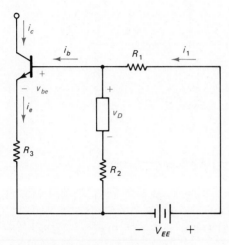

Fig. 4-48 Circuit for Problem 4-4.

4-5 Analyze the noninverting op-amp circuit discussed in Sec. 3-5, Chapter 3, using the method of branch currents.

4-6 Analyze the summing op-amp circuit discussed in Sec. 3-6, Chapter 3, using the method of branch currents.

4-7 Determine the value of the current i_L in the circuit of Fig. 4-49.

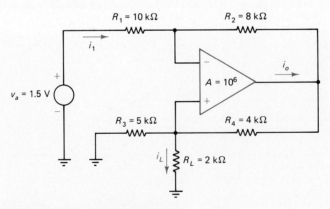

Fig. 4-49 Circuit for Problem 4-7.

4-8 Use the method of branch currents in the circuit of Fig. 4-50, and determine the power delivered or received by each of the components.

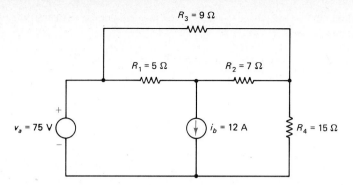

Fig. 4-50 Circuit for Problem 4-8.

4-9 For the circuit of Fig. 4-51, it is required to make the voltage between points A and B equal zero. Determine the value of the current supplied by the i_c source. What effect does the resistance R_c have on the circuit?

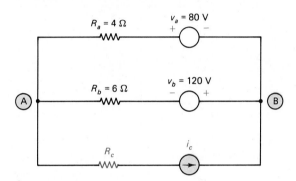

Fig. 4-51 Circuit for Problem 4-9.

4-10 For the circuit of Fig. 4-52, determine the power delivered or received by each component. Determine the range of values of R_L in terms of v_b and i_a for which the power delivered by the voltage source becomes negative; that is, it receives positive power.

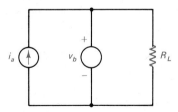

Fig. 4-52 Circuit for Problem 4-10.

Secs. 4-2 and 4-3 The Concept of Loop Currents, Network Topology—Choice of Independent Loop Currents

4-11 The relationships between a set of independent loop currents and the currents in the individual branches of a network can be set up in the form of a table, called the *tie set schedule*. In the tie set schedule, a row is set up for each loop current ($i_1, i_2, \ldots$) and a column for each branch current ($i_a, i_b, \ldots$). Each element in the table is a $+1$, -1, or 0, depending on whether the loop current is going in the same direction as, opposing, or not related to the particular branch current.

Problems

For example, the tie set schedule of the loop current assignment in Fig. 4-12(g) in the text is as shown below.

	i_a	i_b	i_c	i_d	i_e	i_f	i_g	i_h	i_k
i_1	+1	+1	0	0	0	−1	−1	0	0
i_2	+1	+1	0	0	+1	0	−1	0	−1
i_3	0	0	+1	0	0	0	+1	+1	0
i_4	0	0	+1	+1	0	0	+1	0	+1

Each row of the tie set schedule shows the relationship of a loop current to the various branch currents, whereas each column shows the relationship of a branch current to the various loop currents. The tie set schedule can also be used to test the independence of the loop currents chosen for a given circuit. If a row in the tie set schedule can be expressed as a linear combination of the other rows, then the loops given by the tie set schedule do not form an independent set.

For the circuit shown in Fig. 4-53, set up the tie set schedule for each of the following cases: (a) Branches d, e, f, g, and h chosen as links. (b) Branches a, d, e, g, and i chosen as links. (c) Mesh currents chosen as the independent variables.

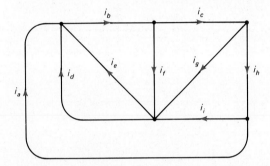

Fig. 4-53 Circuit for Problem 4-11.

4-12 A set of loop currents has been assigned in the graph of Fig. 4-54. Write the tie set schedule and show that the row for the loop current i_1 can be written as an algebraic sum of the elements in the other three rows. The given choice of loop currents is, therefore, not an independent set even though the number of loop currents is correct.

Fig. 4-54 Circuit for Problem 4-12.

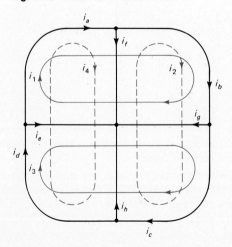

Methods of Branch Currents, Loop, and Mesh Analysis

Sec. 4-4 Principles of Loop Analysis

4-13 For the circuit shown in Fig. 4-55, write a set of loop equations for each of the following cases: (a) R_2, R_d, R_b, and R_c are chosen as links; (b) R_a, R_b, R_d, and R_c are chosen as links; (c) R_1, R_1, R_c, and R_b are chosen as links.

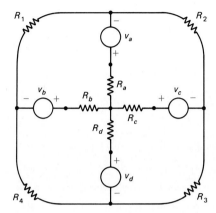

Fig. 4-55 See Problems 4-13, 4-17, and 4-30.

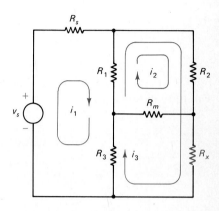

Fig. 4-56 Wheatstone's bridge circuit for Problems 4-14 and 4-18.

4-14 The circuit in Fig. 4-56 is known as *Wheatstone's bridge*, which used to measure the value of an unknown resistor (R_x in the circuit). When the resistors R_1, R_2, and R_3 are adjusted until there is no current through the resistor R_m, the bridge is said to be *balanced*. Determine the relationship between the resistors R_1, R_2, R_3, and R_x when the bridge is balanced, and obtain an expression for R_x in terms of the other three resistors. How does the value of R_s or R_m affect the relationship obtained?

4-15 For the circuit of Fig. 4-57, use loop analysis and (a) determine the power supplied by each voltage source, (b) the voltage between points A and B (with A as reference positive), and (c) the voltage between points B and C (with B as reference positive).

Fig. 4-57 Circuit for Problems 4-15 and 4-19.

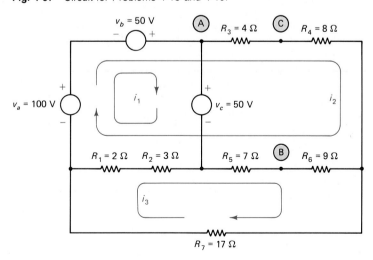

Problems

4-16 Determine the current i_s in the circuit of Fig. 4-58 when (a) the terminals A to B are open-circuited and (b) when the terminals A to B are short-circuited. Also, in part (a), determine the open-circuit voltage across A to B, and in part (b), determine the current in the short circuit from A to B.

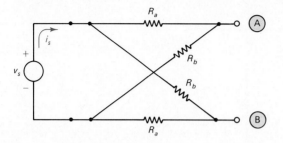

Fig. 4-58 Circuit for Problem 4-16.

4-17 Write the mesh equations for the circuit in Problem 4-13.

4-18 Solve the Wheatstone bridge problem (Problem 4-14) by using mesh analysis.

4-19 Solve the circuit given in Problem 4-15 by using mesh analysis.

4-20 The resistor configurations shown in the two circuits of Fig. 4-59 are equivalent to each other in the sense that the current supplied by each of the three voltage sources is the same in both circuits. The configuration in Fig. 4-59(a) is known as a *delta* network, whereas that in Fig. 4-59(b) is known as a *wye* (or *star*) network. Obtain the relationships that permit the determination of the resistances in the equivalent wye network for a given delta network and vice versa.

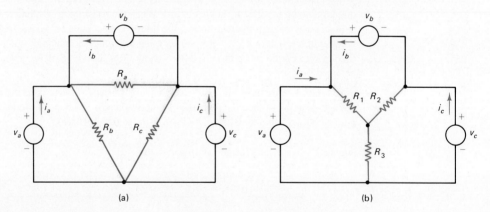

Fig. 4-59 Delta-wye transformation for Problem 4-20.

4-21 Determine the value of i_x in the circuit of Fig. 4-60.

Fig. 4-60 Circuit for Problem 4-21.

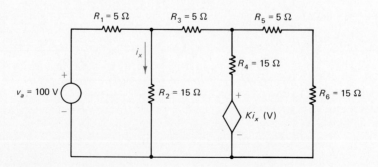

226   Methods of Branch Currents, Loop, and Mesh Analysis

4-22 Determine the voltage v_3 in the circuit of Fig. 4-61.

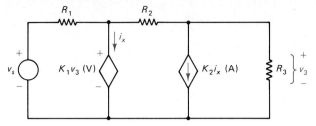

Fig. 4-61 Circuit for Problems 4-22 and 4-31.

4-23 (a) Determine the ratio v_o/v_s in the circuit shown in Fig. 4-62. (b) Modify the given circuit by replacing the source v_s by a short circuit and the resistor R_3 by a voltage source v_T. Determine the resistance seen by v_T in the modified circuit.

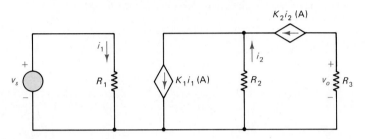

Fig. 4-62 Circuit for Problem 4-23.

4-24 Determine the power delivered or received by each component in the circuit of Fig. 4-63. Determine the range of values of R_b in terms of v_a and i_c for which the power delivered by the current source is negative, that is, the current source receives positive power.

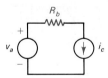

Fig. 4-63 Circuit for Problem 4-24.

4-25 Determine the value of i_x due to the current source so that the voltage across R_1 is zero in the circuit of Fig. 4-64.

Fig. 4-64 Circuit for Problem 4-25.

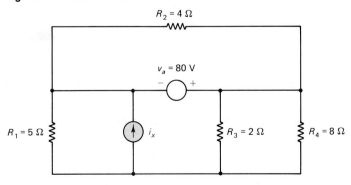

Problems

4-26 Determine the voltage labeled v_o in the circuit of Fig. 4-65.

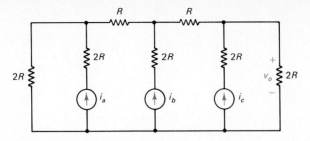

Fig. 4-65 Circuit for Problem 4-26.

4-27 Determine the power delivered by each of the four sources in the circuit of Fig. 4-66 by using (a) the method of branch currents, (b) nodal analysis, (c) loop analysis in which the current source branches are chosen as links, and (d) mesh analysis.

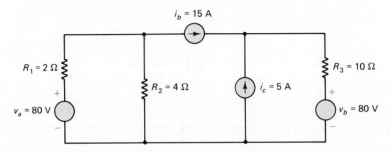

Fig. 4-66 Circuit for Problem 4-27.

Sec. 4-9 Duality

4-28 Set up the dual networks for the networks analyzed (using nodal analysis) in Problem 3-9 of Chapter 3.

4-29 Set up the dual network for the one analyzed (using nodal analysis) in Example 3-5 of Chapter 3.

4-30 Set up the dual network for the one analyzed (using mesh analysis) in Problem 4-13.

4-31 Set up the dual network for the one analyzed (using mesh analysis) in Problem 4-22.

Sec. 4-10 Mesh Resistance Matrix

4-32 Write the mesh resistance matrix of the circuit shown in Fig. 4-67. Use Cramer's rule and determine the values of the mesh currents when each resistance is 10Ω and a 100-V source is placed in series with R_1.

4-33 Write the mesh resistance matrix of the circuit for Problem 3-28 in Chapter 3.

4-34 Obtain the mesh resistance matrix of the circuit shown in Fig. 4-68.

4-35 In the circuit of Problem 4-32, assume that each resistor is $10\ \Omega$. Make an opening in the R_5 branch and insert a voltage source v_3. Determine the driving point conductance and the transfer conductances between meshes 3 and 1 and between meshes 3 and 2.

4-36 For the circuit of Problem 4-35, insert a voltage source v_1 in series with R_1 (after removing the source in the R_5 branch). Determine the driving point and the two possible transfer conductances.

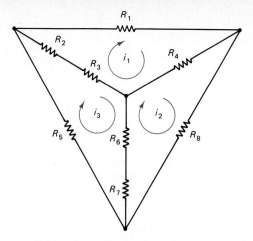

Fig. 4-67 Circuit for Problems 4-32, 4-35, and 4-36.

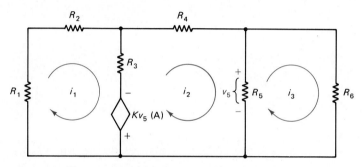

Fig. 4-68 Circuit for Problems 4-34 and 4-37.

4-37 In the circuit of Problem 4-34, use the following numerical values: $R_1 = R_3 = R_5 = R_6 = 15\ \Omega$ and $R_2 = R_4 = 20\ \Omega$. (a) Insert a voltage source v_1 in series with R_1 and determine the driving point conductance and the transfer conductance i_3/v_1. (b) Remove the v_1 source. Insert a voltage source v_3 in series with R_6 and determine the driving point conductance and the transfer conductance i_1/v_3.

4-38 Write the mesh resistance matrix of the circuit of Fig. 4-69. Determine the driving point conductance with respect to mesh 1, the driving point conductance with respect to mesh 2, the transfer conductance from mesh 1 to mesh 2, and the transfer conductance from mesh 2 to mesh 1.

Fig. 4-69 Circuit for Problem 4-38.

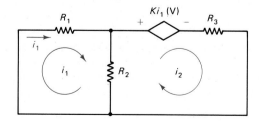

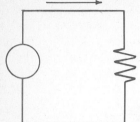

CHAPTER 5
NETWORK THEOREMS

The methods of nodal and loop analysis discussed in the previous chapters provide a standard and routine approach to the analysis of networks. There are, however, many situations where the analysis of a network is more effectively and efficiently done by using theorems derived from the general properties of linear networks. They serve as an attractive alternative (and a more efficient approach) to the analysis of networks in certain special situations. Network theorems also provide a special insight into the response of networks and lead to a better understanding of network theory. The discussion in this chapter addresses theorems used most frequently in the solution of circuit problems: Thevenin's and Norton's theorems, maximum power transfer theorem, superposition theorem, and reciprocity theorem. The discussion in this chapter (as in previous chapters) is restricted to resistive networks containing independent and linear dependent sources. The use of theorems in other types of networks is discussed in Chapters 6 and 9.

Cases occur in which one portion of a network remains fixed while the other portion is variable, and the aim of the problem is to determine the currents and voltages in the variable portion of the circuit. A typical example is a circuit in which only the load resistance is varied. In such cases, it is possible to replace the fixed portion of the circuit by a simple equivalent circuit by using *Thevenin's* or *Norton's* theorem. The use of the simple equivalent circuit greatly reduces the quantity and complexity of computations needed to analyze the given circuit. A question related to the above situation is to maximize the power being transferred from the fixed portion of the circuit to the load and the answer is provided by the *maximum power transfer* theorem.

The currents and voltages in the branches of a circuit being driven by two or more sources can be determined by using the *superposition theorem*. This theorem facilitates the solution by evaluating the response of the circuit under the influence of each of the sources acting individually and then suitably adding the responses so obtained. Superposition is an indispensable tool for analyzing circuits driven by a mixture of sources (as, for example, sources of different frequencies). Superposition is also the basis for the derivation of a general approach (called *convolution*) to the analysis of networks subjected to arbitrary signals. Convolution is discussed in Chapter 14.

It is possible to transfer the results obtained for the response of one branch in a circuit when driven by a source in another branch to the converse situation where the response branch and source branch are interchanged. Such a transfer is made possible by means of the *reciprocity theorem*.

The use of the above theorems is restricted to lumped linear networks. In addition to the above constraint, the reciprocity theorem requires the additional condition that the circuit be passive.

The early sections of this chapter concentrate on understanding the implications of the various theorems and their applications in circuit analysis. The *proofs* of the theorems based on the algebra of loop resistance matrices are presented at the end of this chapter (Section 5-6).

5-1 THEVENIN'S THEOREM

Consider a network in which the currents and voltage in a particular portion are of interest. This is indicated in Fig. 5-1(a), where the "load circuit" denotes the portion of the network under observation and the "fixed circuit" inside the black box represents the remainder of the network. It is assumed that the fixed circuit is a lumped linear circuit made up of resistors, linear dependent sources, and independent sources. No such constraint is placed on the load circuit. The objective is to replace the fixed circuit with a simple equivalent circuit consisting of a voltage source v_{Th} in series with a resistor R_{Th} as shown in Fig. 5-1(b).

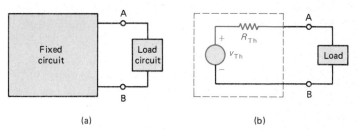

Fig. 5-1 (a) Fixed linear circuit and (b) its Thevenin equivalent.

> **Equivalence here implies that the current and voltage in the load circuit are the same whether the load is driven by the original fixed circuit or by the equivalent circuit.**

For example, it is possible to replace the portion of the specific network shown enclosed by dashed lines in Figure 5-2(a) by the simple series equivalent circuit within the dashed lines in Fig. 5-2(b). The currents and voltage in the load circuit are the same whether they are determined by using the original network of Fig. 5-2(a) or the equivalent network of Fig. 5-2(b). The circuits within the dashed lines in either diagram are *equivalent* as far as any *external* connections made to the terminal pair A–B are concerned. The equivalence ceases to be valid once we go inside the dashed lines.

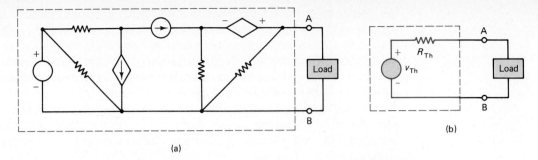

Fig. 5-2 Illustrating the concept of a Thevenin equivalent circuit. (a) Original network. (b) Thevenin equivalent.

Statement of Thevenin's Theorem

Given a network of linear elements and sources, the voltages and currents in any general load connected to a terminal pair A–B of the network can be determined by replacing the network by an equivalent circuit consisting of a voltage source in series with a resistance.

The voltage v_{Th} of the source in the equivalent circuit is equal to the voltage that would appear across the terminal pair A–B if those terminals were *open-circuited*.

The resistance R_{Th} in the equivalent circuit is the total resistance seen *looking into* the network from the terminal pair A–B, when the *independent* sources in the network have been made inactive. (Such a network is called a *relaxed* network.)

Certain important points should be noted in the above statement:

1. The lumped linearity restriction affects only the portion of the network being replaced by a Thevenin equivalent. The restriction does not extend to the load. As a practical matter, however, if the load is not a lumped linear circuit, then the solution of the problem (after replacing the fixed circuit by its Thevenin equivalent) is, in general, difficult.
2. A Thevenin equivalent is defined with respect to a *specified pair of terminals* of a network. The equivalent circuits obtained by looking in from different terminal pairs of a network are different from one another.

5-1-1 Procedure for Finding a Thevenin Equivalent Circuit

The statement of Thevenin's theorem and the steps in its proof (presented in Section 5-6) lead to the following procedure for finding the Thevenin equivalent circuit as seen from a terminal pair of a (lumped linear) network.

The load is removed and set aside. In some cases, the load may already be absent from the given circuit.

The Thevenin voltage is defined as the voltage across the open-circuited terminal pair A–B. The terminals A–B are therefore open-circuited and the voltage across them is calculated by appropriate analysis.

The Thevenin resistance is defined as the total resistance looking in from the terminal pair A–B of the relaxed circuit. A relaxed circuit (also called somewhat morbidly a "dead circuit") should contain no *independent* sources. Therefore, the given circuit is first made *relaxed* by disabling all the *independent* sources. An independent *voltage source* is disabled by replacing it with a *short* circuit so as to make the voltage zero. An independent current source is disabled by replacing it with an *open* circuit so as to make the current zero. *Dependent sources, if any, are left untouched.* The total resistance looking in from the terminals A–B of the relaxed circuit gives the Thevenin resistance.

In circuits that do not contain any dependent sources, the Thevenin resistance is usually calculated by series and parallel combinations of resistances.

In circuits that contain dependent sources, however, the presence of such sources precludes the calculation of Thevenin resistance by series and parallel combinations. In such cases, the usual procedure is to connect a test source, as indicated in Fig. 5-3, and determine the ratio v_T/i_T to find the Thevenin resistance. *The Thevenin resistance of networks containing dependent sources may be negative in some cases.* It is, therefore, critically important to keep track of the polarities of voltages and directions of current in the determination of the Thevenin resistance.

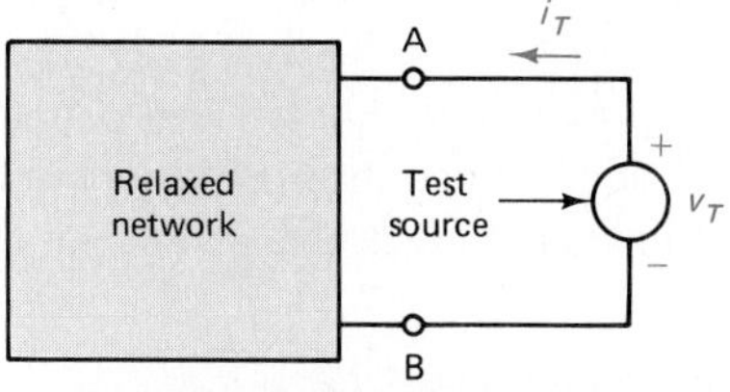

Fig. 5-3 Using a test source to find R_{Th} of a relaxed circuit with dependent sources.

Note that the direction of the current i_T in the test source in Fig. 5-3 is from the *negative to the positive* terminals of the source.

The following examples illustrate the determination of the Thevenin equivalent circuit.

Example 5-1 Obtain the Thevenin equivalent as seen from the terminals A–B of the circuit in Fig. 5-4(a).

Fig. 5-4 (a) Circuit for Example 5-1. (b) Calculation of v_{Th}. (c) Calculation of R_{Th}.

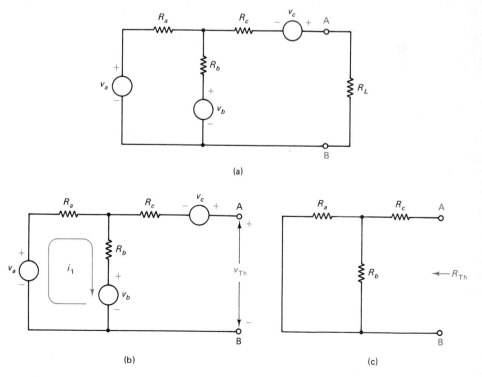

5-1   Thevenin's Theorem

Solution The load resistor R_L is first removed and set aside.

Determination of v_{Th}: Figure 5-4(b) shows the circuit with terminals A–B open-circuited. There is only a single current i_1, given by

$$i_1 = (v_a - v_b)/(R_a + R_b) \tag{5-1}$$

In order to find the voltage v_{Th}, choose a closed path that includes the terminals A–B and write the KVL equation. (Note that the voltage across $R_c = 0$, since there is no current through it.)

$$-v_{Th} + v_c + R_b i_1 + v_b = 0$$

which leads to, after using Eq. (5-1),

$$v_{Th} = v_c + \frac{R_b v_a + R_a v_b}{R_a + R_b} \tag{5-2}$$

Determination of R_{Th}: Figure 5-4(c) shows the relaxed circuit (obtained by replacing the voltage sources by short circuits). Since the circuit does not contain any dependent sources, the resistance looking in from A–B is obtained by series-parallel combinations, and

$$R_{Th} = R_c + (R_a \parallel R_b) \tag{5-3}$$
$$= R_c + \frac{R_a R_b}{R_a + R_b}$$

Equations (5-2) and (5-3) define the Thevenin equivalent as seen from the terminals A–B of the given circuit. ∎

Exercise 5-1 Obtain the Thevenin equivalent of the circuit of Fig. 5-5 as seen from the terminal pair A–B.

Exercise 5-2 For the circuit shown in Fig. 5-5, obtain the Thevenin equivalent as seen from the terminal pair C–D.

Exercise 5-3 For the circuit shown in Fig. 5-5, obtain the Thevenin equivalent as seen from the terminal pair E–F.

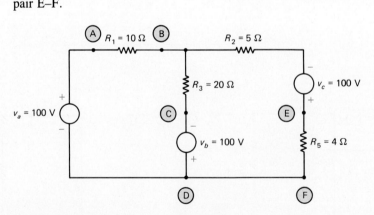

Fig. 5-5 Circuit for Exercises 5-1, 5-2, and 5-3.

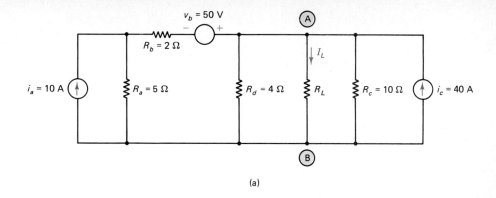

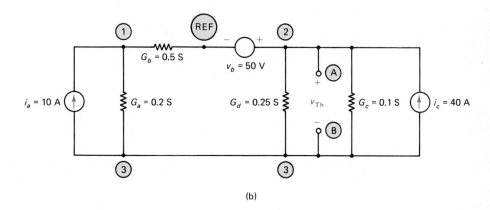

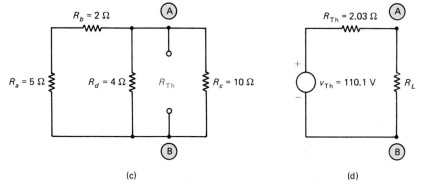

Fig. 5-6 (a) Circuit for Example 5-2. (b) Calculation of v_{Th}. (c) Calculation of R_{Th}. (d) Calculation of i_L.

Example 5-2 Calculate the current i_L in the circuit of Fig. 5-6(a) for the following values of R_L: (a) 2 Ω, (b) 1 Ω, and (c) 5 Ω.

Solution Since only one branch in the circuit is varied, an efficient approach is to first replace the rest of the network by a Thevenin equivalent as seen from the terminals of R_L.

Remove the resistor R_L and set it aside.

Determination of v_{Th}: Figure 5-6(b) shows the circuit with terminals A–B open-circuited. By choosing the negative terminal of v_b as the reference and numbering the other

5-1 Thevenin's Theorem 235

nodes as indicated, the voltage v_{Th} across terminals A–B is

$$v_{Th} = v_2 - v_3 \tag{5-4}$$

The nodal equations of the circuit are

Node 1: $$(G_a + G_b)v_1 - G_a v_3 = i_a$$

or $$0.7v_1 - 0.2v_3 = 10 \tag{5-5}$$

Node 2: $$v_2 = 50 \text{ V}$$

Node 3: $$-G_a v_1 + (G_d + G_e)v_2 + (G_a + G_d + G_e)v_3 = -i_a - i_c$$

which becomes, after using $v_2 = 50$ V and substituting numerical values,

$$-0.2v_1 + 0.55v_3 = -32.5 \tag{5-6}$$

Solving Eqs. (5-5) and (5-6) for v_3,

$$v_3 = -60.1 \text{ V}$$

Therefore, from Eq. (5-4),

$$v_{Th} = v_2 - v_3 = 110.1 \text{ V}$$

Determination of R_{Th}: The relaxed circuit [Fig. 5-6(c)] is obtained by replacing the i_a and i_c sources by open circuits and the v_b source by a short circuit. The Thevenin resistance is the resistance seen from the terminals A–B (obtained by series-parallel combination):

$$R_{Th} = R_c \parallel [R_d \parallel (R_a + R_b)] = 2.03 \text{ }\Omega$$

The Thevenin equivalent circuit is shown in Fig. 5-6(d) with the load resistance R_L restored to its original position. The current i_L is given by

$$i_L = v_{Th}/(R_{Th} + R_L)$$

(a) For $R_L = 2 \text{ }\Omega : i_L = 27.3$ A,   (b) for $R_L = 1 \text{ }\Omega : i_L = 36.3$ A, and   (c) for $R_L = 5 \text{ }\Omega : i_L = 15.7$ A.   ∎

Exercise 5-4 In the circuit of Example 5-2, use a value of $R_L = 8 \text{ }\Omega$ and obtain the Thevenin equivalent as seen from the terminals of R_a.

Example 5-3 Obtain the Thevenin equivalent as seen from the terminals A–B in the circuit of Fig. 5-7(a).

Solution *Determination of v_{Th}:* The terminals A–B are already open. Nodal analysis can be used for finding v_{Th}. Labeling the nodes, as shown in Fig. 5-7(b), the Thevenin voltage is

$$v_{Th} = v_3$$

The voltage controlling the dependent source is

$$v_x = v_2 - v_3 \tag{5-7}$$

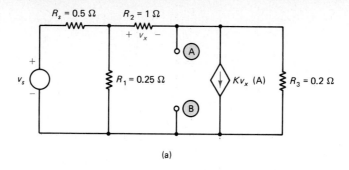

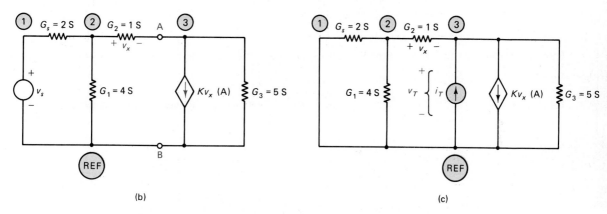

Fig. 5-7 (a) Circuit for Example 5-3. (b) Calculation of v_{Th}. (c) Calculation of R_{Th}.

The nodal equations are

Node 1: $\qquad v_1 = v_s$

Node 2: $\qquad -G_s v_1 + (G_1 + G_2 + G_s)v_2 - G_2 v_3 = 0$

or $\qquad 7v_2 - v_3 = 2v_s \qquad (5\text{-}8)$

Node 3: $\qquad -G_2 v_2 + (G_2 + G_3)v_3 = -Kv_x$

which leads to, after using Eq. (5-7),

$$(K - 1)v_2 + (6 - K)v_3 = 0 \qquad (5\text{-}9)$$

Solving Eqs. (5-8) and (5-9),

$$v_{Th} = v_3 = \frac{2(K - 1)v_s}{(6K - 41)} \qquad (5\text{-}10)$$

Determination of R_{Th}: Note that the *dependent source is retained* in forming the relaxed version of the circuit [Fig. 5-7(c)]. The presence of the dependent source makes it impossible to find R_{Th} by using only series and parallel combinations. A test source is inserted at terminals A–B as indicated and

$$R_{Th} = v_3/i_T \qquad (5\text{-}11)$$

5-1 Thevenin's Theorem

It is seen that $v_1 = 0$ in Fig. 5-7(c) and the equations for the other two nodes are

Node 2:
$$7v_2 - v_3 = 0 \tag{5-12}$$

Node 3:
$$(K - 1)v_2 + (6 - K)v_3 = i_T \tag{5-13}$$

Solving Eqs. (5-12) and (5-13),

$$R_{Th} = \frac{v_3}{i_T} = \frac{7}{41 - 6K} \; \Omega \tag{5-14}$$

The Thevenin equivalent as seen from the terminals A–B of the given circuit is defined by Eqs. (5-10) and (5-14). ∎

Exercise 5-5 For the circuit shown in Fig. 5-8(a), determine the Thevenin equivalent as seen from the terminals A–B.

Exercise 5-6 For the circuit shown in Fig. 5-8(b), determine the Thevenin equivalent as seen from the terminals C–D.

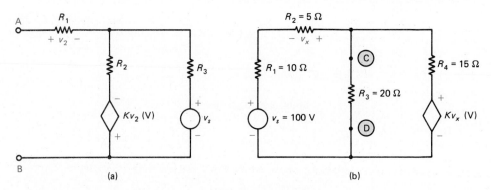

Fig. 5-8 (a) Circuit for Exercises 5-5 and (b) Circuit for Exercise 5-6.

5-2 NORTON'S THEOREM

Norton's theorem is the *dual* of Thevenin's theorem. By replacing *voltage* by *current*, *resistance* by *conductance*, and *open circuit* by *short circuit* in the statement of Thevenin's theorem of Section 5-1, the following statement is obtained:

> Given a network made up of linear elements and sources, the voltages and currents in any general load connected to a terminal pair A–B of the network can be determined by replacing the network by an equivalent circuit consisting of a current source in parallel with a conductance.

The current i_N of the source in the equivalent circuit is equal to the current that would appear across the terminal pair A–B if those terminals were *short-circuited*. The conductance G_N in the equivalent circuit is the total conductance seen *looking into* the network from the terminal pair A–B when the *independent* sources in the network have been made inactive.

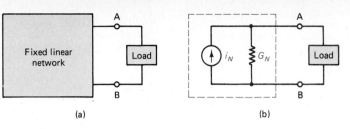

Fig. 5-9 (a) Fixed linear circuit and (b) its Norton equivalent.

The Norton equivalent is illustrated in Fig. 5-9. Current i_N in Fig. 5-9(b) is the current that flows from A to B if a short circuit were placed across the terminals A–B in Fig. 5-9(a). Conductance G_N is the conductance seen from the terminals A–B when the *independent* sources in the fixed network of Fig. 5-9(a) have been made inactive.

The comments made earlier immediately after the statement of Thevenin's theorem are valid for Norton's theorem also.

Procedure for Finding a Norton Equivalent Circuit

Given a network for which it is desired to find the Norton equivalent as seen from a specified terminal pair A–B, remove whatever is connected between A and B and set it aside.

Determination of i_N: Place a short circuit across the terminals A–B and determine the current flowing *from A to B* through the short circuit. This gives i_N in the Norton equivalent.

Determination of G_N: Make the network relaxed by replacing all independent *voltage sources* by *short circuits* and all independent *current sources* by *open circuits*. Determine the conductance seen looking in from terminals A–B (remember to remove the short circuit across A–B). This gives G_N in the Norton equivalent. In the case of circuits with no dependent sources, the Norton conductance is usually found by series and parallel combinations of resistors, but when the circuit contains dependent sources, it is necessary to use a test source at the terminals A–B and calculate the resistance seen by the test source by nodal or loop analysis.

A comparison of the steps to determine the Norton conductance with those for the Thevenin resistance shows that

the Norton conductance G_N and the Thevenin resistance R_{Th} are simply *reciprocals* of each other.

That is,

$$G_N = 1/R_{Th}$$

Example 5-4 Determine the Norton equivalent of the circuit of Fig. 5-10(a) as seen from the terminals A–B.

Solution *Determination of i_N:* R_L is removed and a short circuit is placed across the terminals A–B leading to the circuit shown in Fig. 5-10(b). The current in the short circuit is the sum of the current in G_3 ($= G_3 v_s$) and the current through the dependent source. That is,

$$i_N = G_3 v_2 + K v_x$$

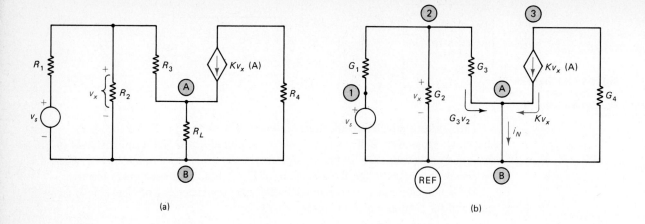

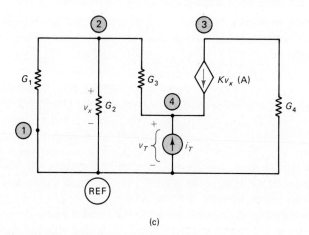

Fig. 5-10 (a) Circuit for Example 5-4. (b) Calculation of i_N. (c) Calculation of G_N.

but, referring to Fig. 5-10(b),

$$v_x = v_2$$

Therefore,

$$i_N = (G_3 + K)v_2$$

and v_2 has to be expressed in terms of v_s so as to obtain an expression for i_N in terms of v_s.

The nodal equations of the circuit are

Node 1: $\qquad v_1 = v_s$

Node 2: $\qquad -G_1 v_1 + (G_1 + G_2 + G_3)v_2 = 0$
or $\qquad (G_1 + G_2 + G_3)v_2 = G_1 v_s \qquad (5\text{-}15)$

That is,

$$v_2 = G_1 v_s/(G_1 + G_2 + G_3) \qquad (5\text{-}16)$$

240 Network Theorems

and Eq. (5-16) leads to

$$i_N = \frac{G_1(K + G_3)v_s}{G_1 + G_2 + G_3} \tag{5-17}$$

Determination of G_N: The relaxed circuit is shown in Fig. 5-10(c). A test source has been inserted between terminals A–B because the relaxed circuit contains a dependent source. The Norton conductance G_N is given by

$$G_N = i_T/v_4 \tag{5-18}$$

Note that

$$v_x = v_2 \tag{5-19}$$

as before.

The equations for nodes 2 and 4 are

Node 2: $$(G_1 + G_2 + G_3)v_2 - G_3v_4 = 0 \tag{5-20}$$

Node 4: $$-G_3v_2 + G_3v_4 = i_T + Kv_x$$

which becomes, after using $v_x = v_2$ from Eq. (5-19),

$$-(K + G_3)v_2 + G_3v_4 = i_T \tag{5-21}$$

Solving Eqs. (5-20) and (5-21) for v_4,

$$v_4 = \frac{(G_1 + G_2 + G_3)i_T}{G_3(G_1 + G_2 - K)}$$

and, from Eq. (5-18),

$$G_N = \frac{G_3(G_1 + G_2 - K)}{G_1 + G_2 + G_3} \tag{5-22}$$

The desired Norton equivalent is defined by Eqs. (5-17) and (5-22). ∎

Exercise 5-7 Determine the Norton equivalent of the circuit of Fig. 5-11 as seen from the terminals A–B.

Fig. 5-11 Circuit for Exercise 5-7.

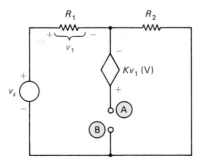

Exercise 5-8 For the circuits specified in Exercises 5-1 through 5-3, determine *directly* the Norton equivalents.

Exercise 5-9 Repeat the work of Exercise 5-8 for the circuits specified in Exercises 5-5 and 5-6.

5-2-1 Conversion between Thevenin and Norton Equivalent Circuits

Since the two theorems, Thevenin's and Norton's, are related to each other by duality, it is possible to obtain the Thevenin equivalent from the Norton equivalent and vice versa. It was already pointed out that the Norton conductance and Thevenin resistance are mutual reciprocals, that is,

$$G_N R_{Th} = 1 \qquad (5\text{-}23)$$

Consider a fixed network [Fig. 5-12(a)] replaced by its Thevenin equivalent as seen from the terminal pair A–B [Fig. 5-12(b)]. The two systems are equivalent as far as the terminal pair A–B is concerned. Therefore, the Norton source current can be found by placing a short circuit across A–B in either Fig. 5-12(a) or in Fig. 5-12(b). If a short circuit is placed across the terminals A–B in Fig. 5-12(b), the current in the short circuit, i_N, is given by

$$i_N = v_{Th}/R_{Th} \qquad (5\text{-}24)$$

Equations (5-23) and (5-24) permit the conversion of one equivalent circuit to the other and the conversion procedure is shown diagrammatically in Fig. 5-13.

The conversion between Thevenin and Norton equivalents (Fig. 5-13) is identical to the conversion between the voltage source and current source models discussed in Chapter 2 (Subsection 2-6-1). Thus, the Thevenin–Norton theorems provide the theoretical basis

Fig. 5-12 Short circuit current in terms of the Thevenin equivalent of a fixed linear network.

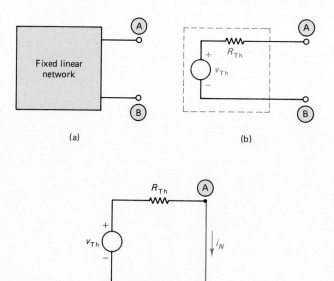

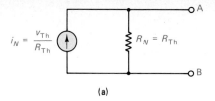

(a)

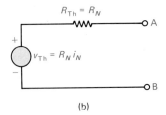

(b)

Fig. 5-13 Conversion between Thevenin and Norton equivalent circuits.

for the source conversion presented in Chapter 2 and also explains the names "Thevenin model" and "Norton model," used for the voltage source and current source models, respectively.

From Eq. (5-24),

$$R_{Th} = \frac{v_{Th}}{i_N} \qquad (5\text{-}25)$$

That is, the

Thevenin resistance of a network as seen from a terminal pair A–B is the ratio of the open-circuit voltage across A–B to the short-circuit current from A–B.

This relationship provides an alternative approach to the determination of the Thevenin resistance (or Norton conductance) of a network.

Exercise 5-10 Use the results of Exercises 5-1, 5-2, 5-3, 5-5, 5-6, 5-8, and 5-9 to verify the above statements regarding the conversion between Thevenin and Norton equivalent circuits.

In using Thevenin's and Norton's theorems, it should be remembered that the equivalent circuits are useful only for calculations pertaining to the load network. They are of no use in determining the currents and voltages in the branches of the fixed network that they have replaced.

Thevenin's (or Norton's) theorem is an obvious labor-saving device in networks where a branch (or a portion of the circuit) is variable. There are also situations that may not involve variable networks, where Thevenin's and Norton's theorems are useful in letting us concentrate on one portion of the network while essentially ignoring the remainder of the network. In any circuit, it is always worth asking whether Thevenin's or Norton's theorem can either simplify the calculations or give some insight into the behavior of those portions of particular interest in the given circuit. When properly used, these two theorems are extremely useful tools in the study of a wide variety of network problems.

As an example, consider the circuit containing a nonlinear device, as shown in Fig. 5-14(a). If it is required to find the voltage across (or the current in) the nonlinear device,

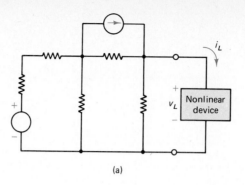

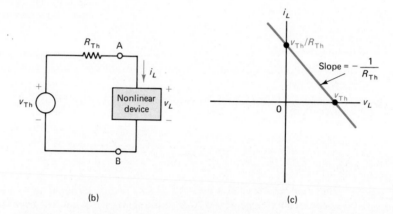

Fig. 5-14 Application of Thevenin's theorem to a circuit with a nonlinear component.

then a Thevenin equivalent is first obtained for the portion of the circuit excluding the nonlinear device, as indicated in Fig. 5-14(b). As discussed in Chapter 2, the circuit is then solved by drawing a load line: the load line intersects the v_L axis at v_{Th} and the i_L axis at v_{Th}/R_{Th} and has a slope of $-1/R_{Th}$ [Fig. 5-14(c)]. The coordinates of the point of intersection of the load line with the device characteristic give the voltage across and the current in the nonlinear device.

Exercise 5-11 Suppose a Norton equivalent is used to replace the circuit in Fig. 5-14(a) except for the nonlinear device. Draw the relevant load line by using the parameters of the Norton equivalent.

5-3 MAXIMUM POWER TRANSFER THEOREM

One of the most important considerations in electronics and communication systems is to maximize the power transferred from a given network (or system) to a load. As an example, the resistance of a loudspeaker to be connected to an amplifier should be selected so as to transfer the maximum power from the amplifier to the speaker. In the case of loads connected to lumped linear networks, the problem of maximum power transfer is attacked conveniently by replacing the network by a Thevenin equivalent.

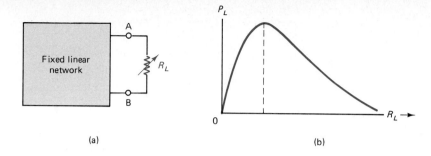

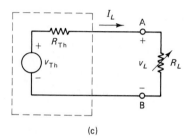

Fig. 5-15 Maximum power transfer from a fixed linear network to a variable load.

In the circuit of Fig. 5-15(a), let the black box contain a lumped linear circuit that is *fixed in its composition*, while the load R_L is a variable resistance. As R_L varies, the power delivered to it by the fixed circuit varies, as shown in Fig. 5-15(b): the power delivered is zero when R_L is zero, increases at first as R_L increases, reaches a maximum for some value of R_L, and then decreases toward zero as R_L becomes infinite. The value of R_L for which the power delivered to the load is a maximum is given by the *maximum power transfer theorem*.

The *maximum power transfer theorem for resistive networks* states that the power transferred to a variable load resistance by a fixed lumped linear network is a maximum when the load resistance is equal to the Thevenin resistance as seen from the terminals of the load resistor.

The maximum power transferred to the load is $v_{Th}^2/4R_{Th}$, where v_{Th} and R_{Th} make up the Thevenin equivalent circuit of the fixed circuit.

To prove the above theorem, replace the fixed circuit by its Thevenin equivalent, as shown in Fig. 5-15(c). The power delivered to R_L is then

$$P_L = v_L^2/R_L \qquad (5\text{-}26)$$
$$= \frac{v_{Th}^2 R_L}{(R_L + R_{Th})^2}$$

The condition to be satisfied for P_L to be a maximum when R_L is the variable is

$$dP_L/dR_L = 0$$

Differentiating Eq. (5-26) with respect to R_L, it is found that $dP_L/dR_L = 0$ when

$$R_L = R_{Th} \qquad (5\text{-}27a)$$

Substitution of R_{Th} for R_L in Eq. (5-26) leads to

$$P_{L(\max)} = v_{Th}^2/4R_L \qquad (5\text{-}27b)$$

5-3   Maximum Power Transfer Theorem

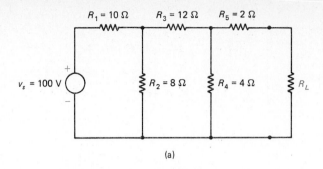

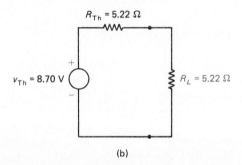

Fig. 5-16 (a) Circuit for Example 5-5. (b) Thevenin equivalent.

Example 5-5 Determine the value of the load resistance to be connected to the circuit of Fig. 5-16 so as to transfer maximum power from the circuit to the load. Also calculate the value of the maximum power transferred.

Solution The Thevenin equivalent of the given circuit as seen from the terminal pair A–B is shown in Fig. 5-16(b). Therefore,

$$R_L = 5.22 \ \Omega$$

is the load resistance needed for maximum power transfer.

When $R_L = 5.22 \ \Omega$, the power delivered to it is 3.62 W.

When the load resistance is made equal to the Thevenin resistance seen from the load terminals, the load is said to be *matched* to the network. Matching is an essential condition whenever maximum power is to be transferred from a *fixed network* to a variable load resistance.

Under matched conditions, the two resistances R_{Th} and R_L are equal and consume equal amounts of power. That is, one-half of the total power generated by v_{Th} is consumed by R_{Th}, leaving only the other half to the load. Thus, the efficiency of power transfer is only 50 percent under matched conditions. This low efficiency is the price to pay when the aim is to transfer as much power as possible from a given circuit to the load and the given circuit is *fixed* and cannot be casually altered by the user. Matching the load to the circuit is an important consideration in electronics and communication circuits, but is not an economical practice in the case of power systems. In power systems, an attempt is made to reduce the *Thevenin resistance* to a much smaller value than the load resistance so as to increase the efficiency of power transfer. ∎

Exercise 5-12 Determine the value of R_L for maximum power transfer in the circuit of Fig. 5-17, and calculate the maximum power dissipated in the load.

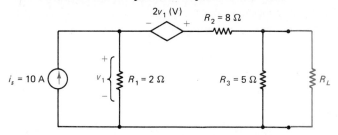

Fig. 5-17 Circuit for Exercise 5-12.

Exercise 5-13 Restate the maximum power transfer theorem in terms of the parameters of a Norton equivalent circuit.

Exercise 5-14 A black box is known to contain a fixed resistive network. A load R_L is connected to the terminals of the black box and adjusted so as to maximize the power dissipated in it. The maximum power dissipated is 400 W. When R_L is adjusted to one-half of its value needed for maximum power transfer, the *voltage* across it is 10 V. Determine the Thevenin and Norton models of the circuit.

5-4 THE SUPERPOSITION THEOREM

The principle of superposition is the cornerstone of the theory of linear networks and systems. In fact, as stated in Chapter 1, the testing of the linearity of a network or system is through the principle of superposition. Besides serving as a computational tool in circuit analysis, the theorem also leads to some general results of importance in the analysis and design of linear systems. Superposition may not always be the most efficient tool for purely resistive circuits driven by constant voltage (or current) sources. On the other hand, it is an indispensable tool in circuits driven by different types of sources (such as, for example, sinusoidal sources of different frequencies). The *superposition theorem* is applicable to any linear network (including those containing linear dependent sources).

The superposition theorem states that

the current or voltage associated with a branch in a linear network equals the sum of the current or voltage components set up in that branch due to each of the *independent* sources acting one at a time on the circuit.

For example, the current i_L in R_L in the circuit of Fig. 5-18 can be written as the sum

$$i_L = i_{La} + i_{Lb} + i_{Lc}$$

Fig. 5-18 Illustration of the superposition theorem.

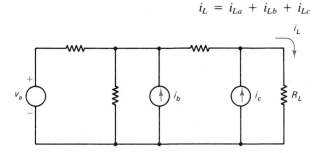

where i_{La} is the current component in R_L when v_a alone is present in the circuit, i_{Lb} is the current component in R_L when i_b alone is present in the circuit, and i_{Lc} is the current component in R_L when i_c alone is present in the circuit.

Remember that in making a source inactive, a voltage source is replaced by a short circuit and a current source by an open circuit.

Example 5-6 Determine the voltage v_1 in the circuit of Fig. 5-19(a) by using the superposition theorem.

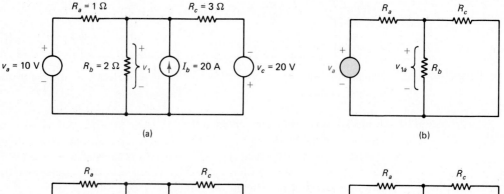

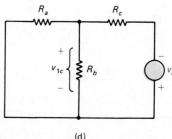

Fig. 5-19 (a) Circuit for Example 5-6. (b), (c), and (d) Calculation of components of v_1.

Solution The components of the voltage v_1 due to each of the three sources acting individually are first calculated and then added (algebraically) to obtain the total voltage v_1.

Case 1: Only v_a is active while i_b is replaced by an open circuit and v_c by a short circuit, as shown in Fig. 5-19(b). The voltage v_{1a} across R_b is obtained by using voltage division.

$$v_{1a} = \frac{v_a(R_b \| R_c)}{R_a + (R_b \| R_c)} = 5.46 \text{ V}$$

Case 2: Only i_b is active while v_a and v_c are replaced by short circuits, as shown in Fig. 5-19(c). The voltage v_{1b} across R_b is given by

$$v_{1b} = i_b/(G_a + G_b + G_c) = 10.9 \text{ V}$$

Case 3: Only v_c is active while v_a is replaced by a short circuit and i_b by an open circuit, as shown in Fig. 5-19(d). The voltage v_{1c} across R_b is given by

$$v_{1c} = \frac{-v_c(R_a \| R_b) \, v_c}{R_c + (R_a \| R_b)} = -3.63 \text{ V}$$

The total voltage v_1 across R_b is, therefore,

$$v_1 = v_{1a} = v_{1b} + v_{1c} = 12.73 \text{ V} \qquad \blacksquare$$

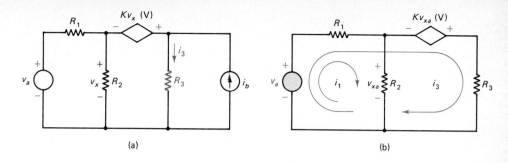

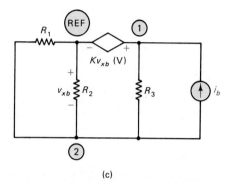

Fig. 5-20 (a) Circuit for Example 5-7. (b) and (c) Calculation of components of i_3.

Example 5-7 Find the current i_3 in R_3 in the circuit of Fig. 5-20(a) by using superposition.

Solution *Case 1:* Voltage v_a is acting alone. Current i_b is replaced by an open circuit. The dependent source is left in place. Referring to Fig. 5-20(b),

$$v_{xa} = R_2 i_1 \quad (5\text{-}28)$$

Loop 1:
$$(R_1 + R_2)i_1 + R_1 i_{3a} = v_a \quad (5\text{-}29)$$

Loop 2:
$$R_1 i_1 + (R_1 + R_3)i_{3a} = v_a + K v_{xa}$$

which becomes, after using Eq. (5-28),

$$(R_1 - KR_2)i_1 + (R_1 + R_3)i_{3a} = v_a \quad (5\text{-}30)$$

Solving Eqs. (5-29) and (5-30),

$$i_{3a} = \frac{R_2(K + 1)v_a}{R_1 R_3 + R_2 R_3 + R_1 R_2(K + 1)} \quad (5\text{-}31)$$

Case 2: Current i_b is acting alone. Voltage v_a is replaced by a short circuit. The dependent source is retained. Referring to Fig. 5-20(c) and using nodal analysis,

Node 1:
$$v_1 = K v_{xb}$$

But, $v_{xb} = -v_2$ and the above equation becomes

$$v_1 = -K v_2 \quad (5\text{-}32)$$

Node 2:
$$-G_3 v_1 + (G_1 + G_2 + G_3)v_2 = -i_b$$

5-4 The Superposition Theorem

which leads to, after using Eq. (5-32),

$$v_2 = \frac{-i_b}{G_1 + G_2 + G_3(K + 1)}$$

The current i_{3b} in R_3 is given by

$$i_{3b} = G_3(v_1 - v_2)$$

Since, from Eq. (5-32), $v_1 = -Kv_2$,

$$\begin{aligned}i_{3b} &= -G_3(K + 1)v_2 \\ &= \frac{G_3(K + 1)i_b}{G_1 + G_2 + G_3(K + 1)} \\ &= \frac{R_1R_2(K + 1)i_b}{R_1R_3 + R_2R_3 + R_1R_2(K + 1)}\end{aligned} \qquad (5\text{-}33)$$

By combining Eqs. (5-32) and (5-33), the total current i_3 in R_3 of the given circuit is

$$i_3 = \frac{R_2(K + 1)(v_a + R_1 i_b)}{R_1R_3 + R_2R_3 + R_1R_2(K + 1)} \qquad \blacksquare$$

Exercise 5-15 In the circuit of Example 5-6, calculate the current i_c through R_c by superposition.

Exercise 5-16 Use superposition to determine the voltage v_o in each of the circuits shown in Fig. 5-21.

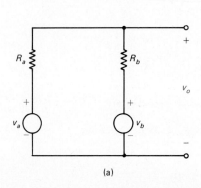
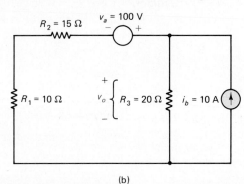

Fig. 5-21 Circuits for Exercise 5-16.

5-5 Reciprocity Theorem

The *reciprocity theorem* is valid for passive linear networks (that is, linear networks that do not contain dependent sources) and permits the transfer of the result obtained for one branch of a circuit due to a source in another branch to the reciprocal situation when the response branch and source branch are interchanged. Consider the two situations shown in Fig. 5-22, where both black boxes contain the same relaxed passive linear network. In Fig. 5-22(a), the voltage at node k due to a source i_j connected between node j and the reference node is measured. In Fig. 5-22(b), the voltage at node j due to a source i_k

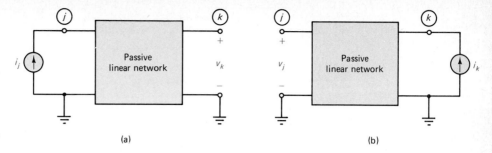

Fig. 5-22 Reciprocity theorem: $i_j/v_k = i_k/v_j$.

connected between node k and the reference node is measured. The reciprocity theorem states that

$$i_j/v_k = i_k/v_j$$

Similarly, in the situations shown in Figs. 5-23(a) and (b), the reciprocity theorem states that

$$v_j/i_k = v_k/i_j$$

An important restriction of the reciprocity theorem is that it applies *only* to the *ratio of a voltage to a current* or the *ratio of a current to a voltage*. It does not apply to the ratio of two voltages or the ratio of two currents.

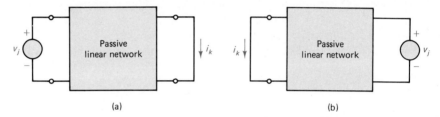

Fig. 5-23 Reciprocity theorem: $v_j/i_k = v_k/i_j$.

Example 5-8 Verify that the ratio of the voltage v_3 to the source current i_1 in the circuit of Fig. 5-24(a) equals the ratio of the voltage v_1 to the source current i_3 in the circuit of Fig. 5-24(b).

Solution The ladder network procedure is used to evaluate v_3/i_1 in Fig. 5-24(a) and v_1/i_3 in Fig. 5-24(b), and the details are shown in the diagrams. It is seen that both ratios equal 0.238. ∎

A combination of reciprocity and superposition often provides a useful tool in circuit analysis and measurements, as illustrated by the following example.

Fig. 5-24 Circuits for Example 5-8. (a) Calculation of v_3/i_1. (b) Calculation of v_1/i_3.

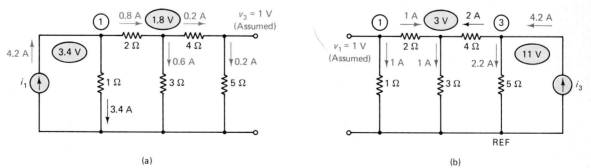

5-5 Reciprocity Theorem

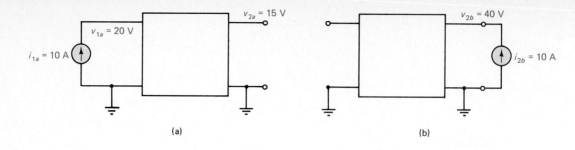

Fig. 5-25 Combination of reciprocity and superposition for Example 5-9.

Example 5-9 Two sets of measurements are made on a passive linear network, as indicated in Figs. 5-25(a) and (b). Calculate the voltages v_1 and v_2 in the arrangement of Fig. 5-25(c).

Solution *Use of Reciprocity:* From Fig. 5-25(a), $v_{2a}/i_{1a} = 1.5$. Therefore, in Fig. 5-25(b), $v_{1b}/i_{2b} = 1.5$ by reciprocity.

We now have the following information:

When a current source i_1 is connected to node 1 (with no source at node 2), $v_{1a}/i_a = 20/10 = 2$, and $v_{2a}/i_a = 15/10 = 1.5$. That is,

$$v_{1a} = 2i_a \quad \text{and} \quad v_{2a} = 1.5i_a$$

When a current source i_2 is connected to node 2 (with no source at node 1), $v_{1b}/i_b = 1.5$ and $v_{2b}/i_b = 40/10 = 4$. That is,

$$v_{1b} = 1.5i_b \quad \text{and} \quad v_{2b} = 4i_b$$

Use of Superposition: For the situation shown in Fig. 5-25(c), the voltages are obtained by adding the components when each current source is acting alone.

$$v_1 = v_{1a} + v_{1b} = 2i_a + 1.5i_b = 32.5 \text{ V}$$
$$v_2 = v_{2a} + v_{2b} = 1.5i_a + 4i_b = 67.5 \text{ V}$$

∎

Exercise 5-17 Show that the reciprocity theorem is not valid for the circuit shown in Fig. 5-26(a).

Fig. 5-26 (a) Circuit for Exercise 5-17, and (b) circuit for Exercise 5-18 and 5-19.

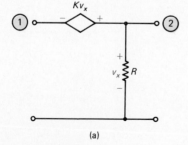

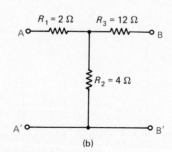

Network Theorems

Exercise 5-18 Verify the reciprocity theorem for the circuit of Fig. 5-26(b) by (a) connecting a voltage source between the terminals A–A' and calculating the current in a short circuit across B–B' and (b) connecting a voltage source between the terminals B–B' and calculating the current in a short circuit across A–A'.

Exercise 5-19 Use the results of Exercise 5-18 and superposition to find the branch currents in the circuit of Fig. 5-26(b) when a 100-V source is connected between A–A' and a 75-V source is connected between B–B'.

5-6 PROOFS OF THEOREMS

The proofs presented here assume familiarity with the algebraic treatment of nodal and loop analyses in Chapters 3 and 4.

5-6-1 Proof of Thevenin's Theorem

Consider an arbitrary load connected to a fixed linear network, as shown in Fig. 5-27(a). The equivalent situation (using Thevenin's theorem) is shown in Fig. 5-27(b). The objective is to determine the expressions that relate v_{Th} and R_{Th} to the parameters of the fixed linear network in Fig. 5-27(a).

For the system in Fig. 5-27(b),

$$i_L = (v_{Th} - v_L)/R_{Th} \tag{5-34}$$

Let the system in Fig. 5-27(a) be described by a set of n loop equations with currents $i_1, i_2, \ldots, i_n$, and let

$$i_n = i_L \tag{5-35}$$

Fig. 5-27 (a) Fixed linear network and (b) its Thevenin equivalent. (c) Load replaced by a source v_L.

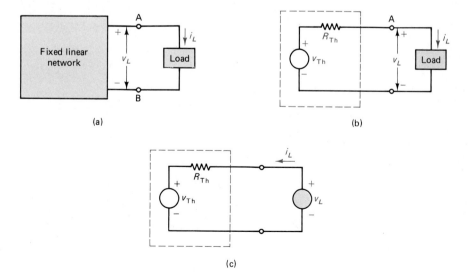

that is, the load branch is taken as the link defining the loop current i_n. Then the loop equations of the system in Fig. 5-27(a) will be of the form

$$r_{11}i_1 + r_{12}i_2 + \ldots + r_{1n}i_n = v_1$$
$$r_{21}i_2 + r_{22}i_2 + \ldots + r_{2n}i_n = v_2 \tag{5-36}$$
$$\vdots$$
$$r_{n1}i_1 + r_{n2}i_2 + \ldots + r_{nn}i_n = v_n - v_L$$

where the load voltage v_L is directly included in the last equation because the relationship between v_L and i_L ($= i_n$) is not specified. It is important to note that the load network is fully covered by the term v_L in the last equation, and, therefore, none of the coefficients on the left-hand sides of the above set of equations depend on the load network.

The current i_n is given by (using Cramer's rule):

$$i_n = \frac{1}{\|r\|} \begin{vmatrix} r_{11} & r_{12} & \ldots & r_{1(n-1)} & v_1 \\ r_{21} & r_{22} & \ldots & r_{(n-1)} & v_2 \\ \vdots & & & & \\ r_{n1} & r_{n2} & \ldots & r_{n(n-1)} & v_n - v_L \end{vmatrix}$$

Laplace's expansion of the numerator determinant of the above expression about the nth column leads to

$$i_n = (1/\|r\|)[A_{1n}v_1 + A_{2n}v_2 + \ldots + A_{nn}(v_n - v_L)] \tag{5-37}$$

where A_{jk} is the cofactor of the jth row and the kth column and $\|r\|$ is the determinant of the coefficients of the currents in Eq. (5-36). Equation (5-37) can be rewritten in the form (after some manipulations)

$$i_n = \frac{\left[\sum_{j=1}^{n} (A_{jn}/A_{nn}) v_j \right] - v_l}{\|r\|/A_{nn}} \tag{5-38}$$

Since $i_n = i_L$, a comparison of Eqs. (5-34) and (5-38) shows that

$$v_{\text{Th}} = \sum_{j=1}^{n} (A_{jn}/A_{nn})v_j \tag{5-39}$$

and

$$R_{\text{Th}} = \frac{\|r\|}{A_{nn}} \tag{5-40}$$

First consider the expression for R_{Th}. Neither $\|r\|$ nor A_{nn} in Eq. (5-40) includes any term due to the load network. Therefore, R_{Th} does not depend on the load network and it must be removed before calculating R_{Th}. The expression $A_{nn}/\|r\|$ is (from the discussion in Chapter 4) the driving point conductance seen from the nth link of the network when the network contains no *independent* sources. Therefore, $\|r\|/A_{nn}$ is the driving point resistance seen looking into the fixed network of Fig. 5-27(a) when the fixed network does not contain any independent sources. Thus, R_{Th} is the driving point resistance seen from the terminals of the load network after removing the load network and making the fixed network relaxed.

Now consider Eq. (5-39) for v_{Th}. It is necessary to show that this is the voltage appearing across the terminals A–B if they are open-circuited. Since the load network is completely arbitrary, it is convenient to represent it by a fictitious source of voltage V_L delivering a current i_L, with the relationship between v_L and i_L being arbitrarily dictated by the original load network. This is only a *conceptual* source and does not alter the voltages and currents in the given system. The modified system is as shown in Fig. 5-27(c). Now let the voltage v_L be varied until the current i_L become identically equal to zero so as to *stimulate* an open-circuit conditional cross A–B. The voltage v_L needed to make $i_L = 0$ must equal the open-circuit voltage across A–B. Making $i_n = i_L = 0$ in Eq. (5-38), the open-circuit voltage across A–B is

$$v_L = \sum_{j=1}^{n} (A_{jn}/A_{nn})v_j \tag{5-41}$$

which is the same as the expression for v_{Th} in Eq. (5-39). Thus, the voltage of the Thevenin source is equal to the open-circuit voltage appearing across the load network terminals when those terminals are open-circuited.

Note that the only assumption made in the above proof is that the fixed network be linear. It may contain linear dependent sources.

5-6-2 Proof of the Superposition Theorem

Consider a general linear network with n independent node voltages $v_1, v_2, \ldots, v_n$, driven by current sources $i_1, i_2, \ldots, i_n$. It was shown in Chapter 3 that the voltage v_k at the kth node is given by

$$v_k = \frac{\sum_{j=1}^{n} A_{jk}i_j}{\|g\|} \tag{5-42}$$

where $\|g\|$ is the determinant of the nodal conductance matrix, and A_{jk} is the cofactor of the jth row and kth column of that matrix.

Suppose only one current source i_r is left active while all the others are disabled (by replacing them with open circuits). Let v_{kr} be the component of the voltage of node k when only i_r is active. Then, from Eq. (5-42),

$$v_{kr} = \frac{A_{rk}i_r}{\|g\|} \tag{5-43}$$

If the process of keeping only one current source active at a time is repeated by making $r = 1, 2, 3, \ldots$ in succession, then the components of v_k due to the single-source connections are given by Eq. (5-43), with $r = 1, 2, 3, \ldots$. A comparison of Eqs. (5-42) and (5-43) shows that the total voltage at node k when *all* the sources are simultaneously active is given by

$$v_k = v_{k1} + v_{k2} + \ldots + v_{kn}$$

which is the statement of the superposition theorem: the total voltage at a node is the sum of the components of that node voltage due to each source acting independently on the network.

The above argument is readily extended to a network containing voltage sources of current sources or a combination of both types.

It should be noted that the only assumption made in the above proof about the network is that it should be linear. It may contain linear dependent sources.

5-6-3 Proof of Reciprocity Theorem

It was seen in Chapter 3 that the nodal conductance matrix of a *passive linear network* is symmetric about the principal diagonal. That is,

$$g_{jk} = g_{kj} \qquad (j \neq k)$$

Therefore, the cofactors A_{jk} and A_{kj} are also equal. Now, if the network is driven by a current source i_j connected between node j and the reference node and the resulting voltage at node k is v_k, then

$$v_k/i_j = A_{jk}/\|g\| \qquad (5\text{-}44)$$

where $\|g\|$ is the determinant of the nodal conductance matrix. On the other hand, if the network is driven by a current source i_k connected between node k and the reference node, the resulting voltage v_j at node j is given by

$$v_j/i_k = A_{kj}/\|g\| \qquad (5\text{-}45)$$

For a *passive* network, $A_{jk} = A_{kj}$ and the two ratios v_k/i_j and v_j/i_k given by Eqs. (5-44) and (5-45) are equal.

Note that both passivity and linearity are required for reciprocity to be valid.

5-7 SUMMARY OF CHAPTER

It is possible to replace a linear resistive network as seen from a specified pair of terminals by a Thevenin equivalent circuit consisting of a voltage source in series with a resistance. The given circuit may contain linear dependent sources. The voltage of the Thevenin source is equal to the open-circuit voltage across the specified terminal pair and the Thevenin resistance is the resistance seen looking into the network from the terminal pair. For networks without dependent sources, the Thevenin resistance is usually determined by series/parallel combinations of resistances for networks. When linear dependent sources are present, a test source is applied to the terminal pair and the resistance presented to the source is calculated by some suitable analysis procedure.

The Norton equivalent circuit is the dual of the Thevenin equivalent. The current source in a Norton equivalent equals the current through a short circuit placed across the specified terminal pair. The Norton conductance is simply the reciprocal of the Thevenin resistance.

In the case of a variable load resistance attached to a network that is fixed in its composition, maximum power is transferred from the network to the load when the load resistance is equal to the Thevenin resistance seen from the load terminals. Then the load resistance is said to match the fixed network. The maximum power transferred to the matched load is $v_{Th}^2/4R_{Th}$.

The total current or voltage associated with a branch in a linear circuit can be determined by superimposing the contributions from the individual sources acting on the circuit one at a time. Superposition is an important and fundamental concept in the study of linear networks and systems.

The ratio of the voltage at a node to the current fed to another node is unaffected by the interchange of the source node and the response node. A similar statement is valid for the ratio of the current in a branch due to a voltage source in another branch. Reciprocity applies only to passive linear networks (that is, networks without any dependent sources).

Answers to Exercises

5-1 $v_{Th} = 200$ V and $R_{Th} = 6.21\ \Omega$.

5-2 $v_{Th} = -5.27$ V and $R_{Th} = 24.7\ \Omega$. (The negative sign in the value of v_{Th} means that terminal D is actually at a higher voltage than terminal C.)

5-3 $v_{Th} = 133$ V and $R_{Th} = 11.7\ \Omega$.

5-4 $v_{Th} = 75.2$ V and $R_{Th} = 4.10\ \Omega$.

5-5 $v_{Th} = v_s R_2/(R_2 + R_3)$. $R_{Th} = [R_1 R_2 + R_1 R_3 + R_3(R_2 - KR_1)]/(R_2 + R_3)$.

5-6 $v_{Th} = (1500 - 500K)/(30 - 5K)$. $R_{Th} = 225/(30 - 5K)\ \Omega$.

5-7 $i_N = v_s(R_2 + KR_1)/R_1 R_2(1 - K)$. $G_N = (R_1 + R_2)/R_1 R_2(1 - K)$.

5-8 The Norton source currents are $i_N = 32.2$ A, -0.213 A, and 11.4 A. The Norton conductance values are the reciprocals of the Thevenin resistances already calculated.

5-9 The Norton source currents are $i_N = v_s R_2/[R_1 R_2 + R_2 R_3 + R_1 R_3(1 - K)]$ and $(1500 - 500K)/225$. G_N is the reciprocal of R_{Th} already determined.

5-10 Source conversion formulas lead to the desired answers.

5-11 The load line intersects the i_L axis at i_N and the v_L axis at i_N/G_N. Its slope is $-G_N$.

5-12 $v_{Th} = 15.8$ V and $R_{Th} = 3.68\ \Omega$. $R_L = 3.68\ \Omega$. Power dissipated in $R_L = 17.0$ W.

5-13 Use duality on the statement of the maximum power transfer theorem in the text.

5-14 $v_{Th} = 30$ V and $R_{Th} = 0.5625\ \Omega$.

5-15 Current components are 1.82 A, 3.64 A, and 5.45 A. $i_c = 10.9$ A.

5-16 (a) $v_o = (R_a v_b + R_b v_a)/(R_a + R_b)$. (b) Components are 44.4 V and 111 V. $v_o = 156$ V.

5-17 $v_2/i_1 = R$ and $v_1/i_2 = (1 - K)R$. (Note that if $K = 0$, then reciprocity is present.)

5-18 $i_2/v_1 = i_1/v_2 = 0.05$.

5-19 16.2 A in R_1, 16.9 A in R_2, and 0.625 A in R_3.

PROBLEMS

Sec. 5-1 Thevenin's Theorem

5-1 Obtain the Thevenin equivalent of the network in Fig. 5-28 with reference to each of the following pairs of terminals: (a) A–B, (b) C–D, and (c) E–F.

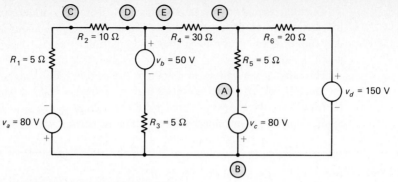

Fig. 5-28 Circuit for Problem 5-1.

5-2 Obtain the Thevenin equivalent of the network in Fig. 5-29 as seen from the terminals A–B.

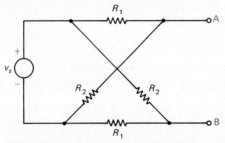

Fig. 5-29 Circuit for Problems 5-2 and 5-6.

5-3 (a) Obtain the Thevenin equivalent of the network in Fig. 5-30 as seen by R_L and use the equivalent circuit to determine the ratio v_o/v_s. (b) Verify your result by using nodal analysis on the given network.

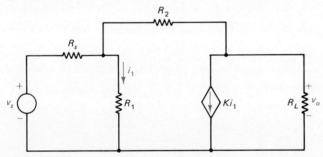

Fig. 5-30 Circuit for Problems 5-3 and 5-6.

5-4 Determine the input resistance R_{in} of the network in Fig. 5-31.

Fig. 5-31 Circuit for Problem 5-4.

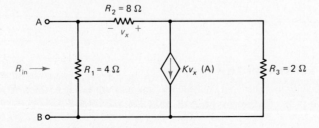

Network Theorems

5-5 Obtain the Thevenin equivalent of the op-amp circuit shown in Fig. 5-32 as seen from the output terminal and ground. Use the approximate model for the op amp.

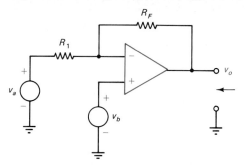

Fig. 5-32 Circuit for Problem 5-5.

Sec. 5-2 Norton's Theorem

5-6 Determine the Norton source current in the circuits of Problems 5-1, 5-2, and 5-3 *directly*. Verify that $R_{Th} = v_{Th}/i_N$.

5-7 Determine the Norton equivalent of the circuit shown in Fig. 5-33.

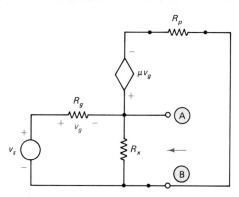

Fig. 5-33 Circuit for Problem 5-7.

5-8 For the circuit of Fig. 5-34, (a) determine the open-circuit voltage across A–B, (b) determine the short-circuit current from A to B, and (c) obtain the Thevenin and Norton equivalent circuits as seen from A–B.

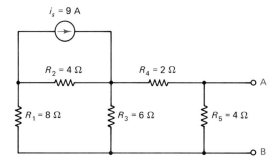

Fig. 5-34 Circuit for Problem 5-8.

5-9 When the terminals A–B of a black box containing a resistive network are short-circuited, the current form A to B is 6 A. The short circuit is removed and a source of 80 V in series with a resistance of 20 Ω is connected between A and B (with the positive terminal at A). The voltage across A–B is then 100 V. Determine the Thevenin and Norton equivalents of the network in the black box.

Problems 259

5-10 The component labeled NL in the circuit of Fig. 5-35 is a nonlinear element whose voltage-current relationship is given by $v_L = i_L^2$. Determine the values of v_L and i_L.

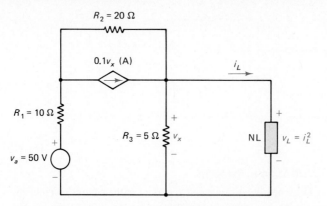

Fig. 5-35 Circuit for Problem 5-10.

Sec. 5-3 Maximum Power Transfer Theorem

5-11 In the circuit of Fig. 5-36, determine the value of R_L for maximum power transfer and calculate the maximum power dissipated in R_L.

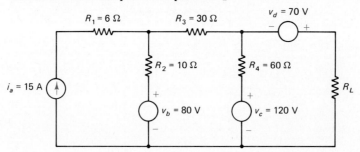

Fig. 5-36 Circuit for Problem 5-11.

5-12 In each of the circuits of Fig. 5-37, the variable resistor is indicated by an arrow through the resistor symbol. In each case, determine the value of the variable resistor so that maximum power is dissipated in the *load resistor* R_L.

Fig. 5-37 Circuits for Problem 5-12.

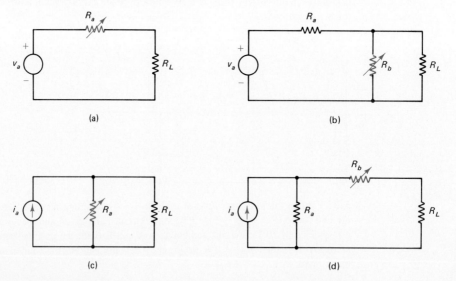

260 Network Theorems

5-13 In the circuit of Fig. 5-38, the source (v_s in series with R_s) is to be matched to the circuit from the input side. At the same time, the load resistance R_L is to be matched to the circuit from the output side. Obtain expressions for R_s and R_L in terms of R. Determine the power supplied by v_s and the power dissipated in R_L under the above conditions.

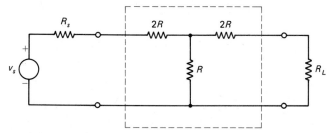

Fig. 5-38 Circuit for Problem 5-13.

5-14 Consider a source v_s in series with R_s (with both v_s and R_s fixed in their values) feeding a variable load R_L. Write an expression for the efficiency given by the ratio P_L/P_s, where P_L is the power dissipated in the load, and P_s is the power supplied by v_s. Draw a graph showing the variation of P_s, P_L, and P_L/P_s as a function of R_L (with all three plots on the same set of axes). What is the value of R_L for which each of the three quantities becomes a maximum?

Sec. 5-4 Superposition Theorem

5-15 Determine the voltage marked v_{oc} in the circuit of Fig. 5-39 by using superposition.

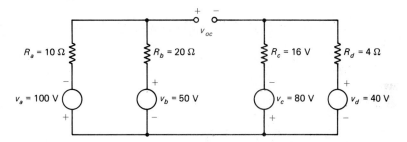

Fig. 5-39 Circuit for Problem 5-15.

5-16 Determine the short-circuit current marked i_{sc} in the circuit of Fig. 5-40 by using superposition.

Fig. 5-40 Circuit for Problem 5-16.

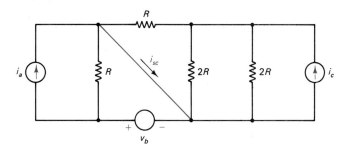

Problems 261

5-17 Determine the voltage v_o in the circuit of Fig. 5-41 by using superposition.

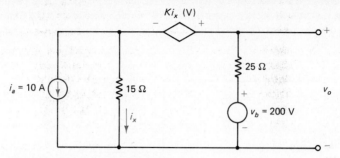

Fig. 5-41 Circuit for Problem 5-17.

5-18 Obtain a relationship for the source voltage v_b in terms of i_a, i_c, K, and R so that the voltage v_2 is zero in the circuit of Fig. 5-42.

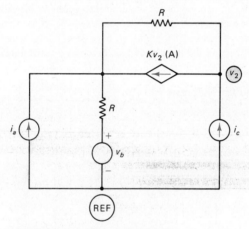

Fig. 5-42 Circuit for Problem 5-18.

Sec. 5-5 Reciprocity Theorem

5-19 Verify the reciprocity theorem for the network shown in Fig. 5-43 by finding (a) the ratio v_3/i_1 when a current source i_1 is connected to node 1, and (b) the ratio v_1/i_3 when a current source i_3 is connected to node 3.

Fig. 5-43 Circuit for Problems 5-19 and 5-20.

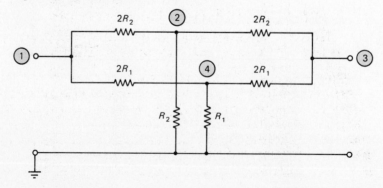

5-20 Verify the reciprocity theorem for the network of Problem 5-19 by finding (a) the ratio i_1/v_3 when a voltage source v_3 is connected between node 3 and ground and i_1 is the current through a short circuit between node 1 and ground, and (b) the ratio i_3/v_1 when a voltage source v_1 is connected between node 1 and ground and i_3 is the current through a short circuit between node 3 and ground.

5-21 Four terminals are available for a passive linear network, as indicated in Fig. 5-44. Measurements are made by connecting a current source of $i_s = 1$ A to a single node at a time (with the other terminal of the current source grounded). The following data are obtained. *Current source at node 1:* $v_1 = 90$ V, $v_2 = 30$ V, and $v_3 = -10$ V. *Current source at node 2:* $v_2 = 60$ V, and $v_3 = 15$ V. *Current source at node 3:* $v_3 = 40$ V. Determine the three node voltages when three current sources are connected to the network (simultaneously): $i_1 = 10$ A to node 1, $i_2 = -10$ A to node 2, and $i_3 = 15$ A to node 3.

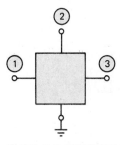

Fig. 5-44 Circuit for Problem 5-21.

5-22 A three-port network is shown in Fig. 5-45. The network can be described by the set of three equations:

$$v_1 = r_{11}i_1 + r_{12}i_2 + r_{13}i_3$$
$$v_2 = r_{21}i_1 + r_{22}i_2 + r_{23}i_3$$
$$v_3 = r_{31}i_1 + r_{32}i_2 + r_{33}i_3$$

where the parameters r_{jk} need to be determined.

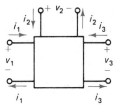

Fig. 5-45 Circuit for Problem 5-22.

Measurements are made yielding the following data: With $v_1 = 100$ V and $v_2 = v_3 = 0$, it is found that $i_1 = 10$ A, $i_2 = -10$ A, and $i_3 = -4$ A. When $v_2 = 100$ V and $v_1 = v_3 = 0$, $i_2 = 12$ A and $i_3 = 7$ A. When $v_3 = 100$ V and $v_1 = v_2 = 0$, $i_3 = 9$ A.

For a passive linear network, determine the values of the parameters r_{jk}.

If the network is linear but not passive, can the parameters be determined from the given information? Explain.

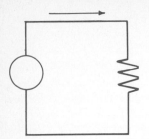

CHAPTER 6
CIRCUITS CONTAINING CAPACITORS AND INDUCTORS

Capacitors and inductors have the ability to store energy in the electric and magnetic fields, respectively, and this property makes them useful components in electrical circuits and systems. The voltage and current in inductors and capacitors are related through derivatives (or integrals). Consequently, the KVL and KCL equations of circuits that contain inductors or capacitors or both are *differential* equations rather than algebraic equations. When the capacitors and inductors are linear and their values are time-invariant, the differential equations are found to be ordinary linear differential equations with constant coefficients. The determination of the response of circuits containing inductors or capacitors or both, therefore, requires the solution of ordinary linear differential equations with constant coefficients.

After a brief discussion of capacitors and inductors (which the student is assumed to have studied in a physics course), this chapter concentrates on circuits containing a single capacitor and resistor (first-order RC circuits) subjected to a constant voltage or current source. The procedure used for the RC circuits is then extended to the analysis of circuits containing a single inductor and resistor (first-order RL circuits) subjected to a constant current or voltage source. The chapter concludes with a treatment of circuits containing a capacitor, an inductor, and a resistor (second-order RLC circuits).

6-1 PROPERTIES AND RELATIONSHIPS OF A CAPACITOR

A capacitor consists of two electrodes separated by an insulating medium called the dielectric.

In a *linear* capacitor, the electric charge q on the electrodes is proportional to the voltage v across the capacitor;

that is

$$q = Cv \qquad (6\text{-}1)$$

where C is the capacitance. The unit of capacitance is *farads* (abbreviated F): 1 farad equals 1 coulomb/volt. Practical capacitors have capacitances in the order of microfarads (1 μF = 10^{-6} F) or picofarads (1 pF = 10^{-12} F).

6-1-1 Current through a Capacitor

Current is the rate of change of an electric charge and, from Eq. (6-1), the current through a capacitor is given by

$$i(t) = dq/dt = d(Cv/dt) = C(dv/dt) \qquad (6\text{-}2)$$

where C is assumed to be a time-invariant constant. For Eq. (6-2) to be valid, the reference polarities of the voltage v and the direction of the current i must be as shown in Fig. 6-1 (associated reference polarities were introduced in Chapter 1).

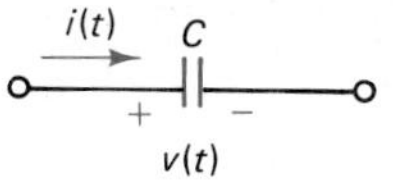

Fig. 6-1 Voltage-current relationship of a capacitor.

Note that when the voltage across a capacitor is a constant, $dv/dt = 0$ and $i = 0$. Therefore *the current through a capacitor is zero when the voltage across it is a constant.*

Given the current $i(t)$ in a capacitor, the voltage across it is given by

$$v(t) = (1/C) \int i(t)dt \qquad (6\text{-}3)$$

which follows from Eq. (6-2). The use of the indefinite integral leads to an arbitrary constant whose value can be found if the voltage across the capacitor is known at a specified instant of time. If the voltage at some instant of time $t = t_0$ is known to be $v(t_0)$, then Eq. (6-3) leads to

$$v(t) = (1/C) \int_{t_0}^{t} i(u)du + v(t_0) \qquad (6\text{-}4)$$

where u is a dummy variable. Frequently, the voltage across the capacitor is known at the beginning of the time of observing its response—that is, at $t = 0$. Then Eq. (6-4) leads to

$$v(t) = (1/C) \int_{0}^{t} i(u)du + v(0) \qquad (6\text{-}5a)$$

If the capacitor is initially uncharged, then $v(0) = 0$, and

$$v(t) = (1/C) \int_{0}^{t} i(u)du \qquad \text{[initially uncharged capacitor]} \qquad (6\text{-}5b)$$

Example 6-1 The voltage waveform $v(t)$ across a capacitor (C = 250 pF) is shown in Fig. 6-2(a). Determine and sketch $i(t)$.

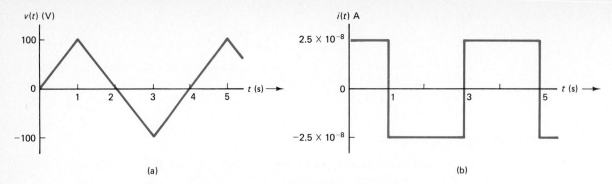

Fig. 6-2 Voltage and current waveforms for Example 6-1.

Solution Since $v(t)$ is made up of straight line segments, the derivative dv/dt is readily determined by using their slopes in this example. The current waveform $i(t)$ obtained by multiplying the slopes by C is shown in Fig. 6-2(b). ∎

Example 6-2 The current in a capacitor is given by $i(t) = I_0 e^{-pt} (t > 0)$, where I_0 and p are constants. Obtain an expression for the capacitor voltage $v(t)$ for $t > 0$, given that $v(t) = v_0$ at $t = 0$.

Solution From Eq. (6-5a)

$$v(t) = (1/C) \int_0^t i(u)\,du + v(0)$$
$$= (1/C) \int_0^t I_0 e^{-pu}\,du + v_0$$
$$= (I_0/pC)(1 - e^{-pt}) + v_0 \qquad ∎$$

Exercise 6-1 If the voltage across a capacitor is given by $v(t) = At + B$, where A and B are constants, determine and sketch the current waveform.

Exercise 6-2 The voltage across a 30 pF capacitor is as shown in Fig. 6-3. Sketch $i(t)$.

Fig. 6-3 Waveform for Exercise 6-2.

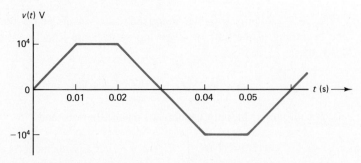

Exercise 6-3 The current in a capacitor ($C = 500$ pF) is given by $i(t) = 20 \cos 10^3 t$ A. Determine the voltage across the capacitor if it is given that $v(t) = 50$ V at $t = 0$.

6-1-2 Energy Stored in a Capacitor

The charge on the electrodes of a capacitor sets up an electric field, and the work done in charging a capacitor is stored as energy in its electric field. In order to obtain a relationship for the energy stored, consider the power relationship:

$$p(t) = v(t)i(t) \tag{6-6}$$
$$= v(t)C\frac{dv(t)}{dt}$$

Now

$$\frac{d[v(t)]^2}{dt} = 2\,v(t)\,\frac{dv(t)}{dt}$$

Therefore, Eq. (6-6) becomes

$$p(t) = \frac{d}{dt}\left[\frac{1}{2}C[v(t)]^2\right] \tag{6-7}$$

The energy $w(t)$ stored in the capacitor is related to power through the equation

$$p(t) = \frac{dw(t)}{dt} \tag{6-8}$$

Comparing Eqs. (6-7) and (6-8), the energy stored in a capacitor at any instant of time t is given by

$$w(t) = (1/2)C[v(t)]^2 \tag{6-9}$$

Exercise 6-4 Obtain an expression for the energy stored in a capacitor (C = 150 pF) if the voltage across it is given by $v(t) = (10t - 5)$ V. Evaluate the energy at $t = 0$, and at $t = 15$ s. Calculate the instant of time at which the energy is zero. Calculate the energy gained or lost in the interval (0 to 1 s).

6-1-3 Voltage Continuity Condition

Consider the energy $w(t)$ in a physical system as a function of time. If there were discontinuities in the graph of $w(t)$ as a function of time, then the derivative $dw/dt = p(t)$, the power, at those points would be infinite. A discontinuity in energy is therefore possible if and only if an infinite amount of power is available. Therefore, in any physical system where infinite power is not available, energy must be a *continuous* function of time. Since the energy stored in a capacitor is directly proportional to the square of the voltage, the continuity of energy leads to the *voltage continuity condition* for a capacitor:

The voltage across a capacitor must be a continuous function of time (except when an infinite amount of power is available).

6-2 PROPERTIES AND RELATIONSHIPS OF AN INDUCTOR

Current flowing through a conductor sets up a magnetic field in the region surrounding the conductor, and there is a magnetic flux linking the conductor. If the current varies with time, then the magnetic flux also varies with time, and a voltage is induced in the

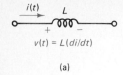

Fig. 6-4 Voltage-current relationships of an inductor: (a) Using associated reference polarities. (b) Using the Lenz's law viewpoint. Note that the two are equivalent.

conductor proportional to the rate of change of the magnetic flux. This is the phenomenon of *electromagnetic induction*. Devices (usually in the form of coils) specifically designed to exploit electromagnetic induction are called *inductors* and are denoted by a coil as shown in Fig. 6-4.

In a *linear* inductor, the magnetic flux linkage is proportional to the current; that is

$$\psi = Li$$

where L is called the *coefficient of self-induction,* or simply the *inductance*. The unit of inductance is *henries* (abbreviated H). The voltage due to electromagnetic self-induction is then given by

$$v(t) = d\Psi/dt = L[di(t)/dt] \qquad (6\text{-}10)$$

where it has been assumed that L is a time-invariant constant. *Associated reference polarities as shown in Fig. 6-4(a) must be used in order for Eq. (6-10) to be valid*. Note that the voltage across an inductor is zero when the current through it is a constant.

In physics textbooks, the relationship between self-induced voltage and the current in an inductor is usually given in the form of *Lenz's law:*

$$v(t) = -L di(t)/dt$$

with a minus sign present in the equation. The Lenz's law equation corresponds to the inductance acting as a *source delivering* power vi, with the generated voltage $v(t)$ opposing the change in the current in the coil; see Fig. 6-4(b). When associated reference polarities are used, as shown in Fig. 6-4(a), the inductance is *receiving* power vi. The change from "delivering power" to "receiving power" is taken into account by changing the minus sign in Lenz's law to the plus sign in Eq. (6-10).

6-2-1 Duality between Capacitance and Inductance

A comparison of the voltage-current relationship of a capacitor as in Eq. (6-2), and that of an inductor, as in Eq. (6-10), shows that

capacitance and inductance are duals

of each other, since replacing v by i and i by v in one equation leads to the other. The duality between inductance and capacitance will be used here to extend the discussion of the capacitor to the case of the inductor.

Duals of Eqs. (6-3), (6-4), (6-5a), and (6-5b) lead to relationships useful in finding the current in an inductor when the voltage $v(t)$ is given.

$$i(t) = (1/L) \int v(t) dt \qquad (6\text{-}11a)$$

$$= (1/L) \int_{t_0}^{t} v(u) du + i(t_0) \qquad (6\text{-}11b)$$

when the value of the current is known at some instant of time t_0. When the initial value of current (at $t = 0$) is known, the relationship becomes

$$i(t) = (1/L) \int_{0}^{t} v(u) du + i(0) \qquad (6\text{-}11c)$$

and when the initial current is zero

$$i(t) = (1/L)\int_0^t v(u)du \qquad \text{[initial zero current]} \qquad (6\text{-}11\text{d})$$

Exercise 6-5 The current in an inductance ($L = 5$ mH) is as shown in Fig. 6-5. Determine and sketch the voltage across the inductance.

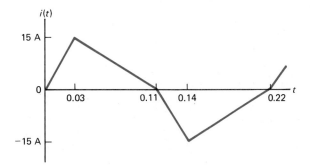

Fig. 6-5 Current for Exercise 6-5.

Exercise 6-6 The voltage across an inductance L is given by $v(t) = V_0 e^{-pt}$ where V_0 and p are constants. Determine the current in the inductance if it is given that $i(t) = 0$ at $t = 0$.

6-2-2 Energy Stored in an Inductor

The work done in setting up the magnetic flux in an inductor is stored as energy in its magnetic field. The energy stored in the magnetic field of an inductor is given by

$$w(t) = (1/2)L[i(t)]^2 \qquad \text{dual of Eq. (6-9)} \qquad (6\text{-}12)$$

Exercise 6-7 Derive Eq. (6-12) for the energy stored in an inductor.

Exercise 6-8 Obtain an expression for the energy stored in an inductor ($L = 0.1$ H) if the current in it is given by $i(t) = 10e^{-2t}$ A. Evaluate the energy at $t = 0$, and at $t = 0.5$ s. Calculate the energy gained or lost in the interval (0 to 1 s).

6-2-3 Current Continuity Condition

The energy stored in an inductance is a *continuous* function of time (unless an infinite amount of power is available). The energy stored in an inductor is proportional to the square of the current, and the continuity of energy leads to the *current continuity condition for an inductor:*

The current in an inductor must be a *continuous* function of time (unless an infinite amount of power is available).

This condition is the dual of the voltage continuity condition of a capacitor.

6-3 Equations for Simple RC and RL Circuits

In general, electric circuits are made up of resistances, capacitances, and inductances, and their responses are determined by using Kirchhoff's laws and the different methods of analysis discussed in earlier chapters. We start with simple circuits containing a resistance and a capacitance, or a resistance and an inductance.

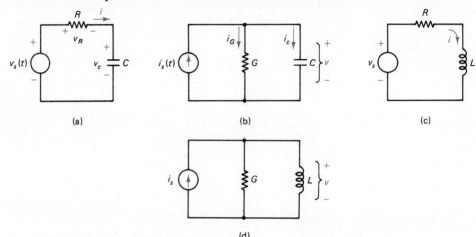

Fig. 6-6 Simple RC and RL circuit configurations. Circuits in (a) and (d) are a dual pair. Circuits in (b) and (c) are a dual pair.

The KVL equation of a series connection of a resistance R and a capacitance C driven by a voltage source $v_s(t)$, shown in Fig. 6-6(a), is given by

$$v_c(t) + v_R(t) = v_s(t)$$

which becomes, after using Eq. (6-3) and Ohm's law

$$(1/C) \int i(t)dt + Ri(t) = v_s(t) \qquad (6\text{-}13a)$$

It is convenient to differentiate once through the equation to get rid of the integral. Equation (6-13a) then becomes

$$i(t) + RC \frac{di(t)}{dt} = C \frac{dv_s(t)}{dt} \qquad (6\text{-}13b)$$

For the parallel circuit shown in Fig. 6-6(b), the KCL equation is

$$i_c(t) + i_G(t) = i_s(t)$$

which, after using Eq. (6-1) and Ohm's law, leads to

$$\frac{dv(t)}{dt} + \frac{G}{C}v(t) = \frac{1}{C} i_s(t) \qquad (6\text{-}14)$$

The equations of the series RL and parallel GL circuits in Figs. 6-6(c) and (d) are written by using duality:

$$\frac{di(t)}{dt} + \frac{R}{L}i(t) = \frac{1}{L}v_s(t) \qquad \text{dual of Eq. (6-14)}$$

$$v(t) + GL \frac{dv(t)}{dt} = L\frac{di_s(t)}{dt} \qquad \text{dual of Eq. (6-13)}$$

The four equations obtained above belong to the category of ordinary linear differential equations with constant coefficients. Since the highest derivative in each equation is of the first order, they are differential equations of the first order, and the four circuits described by them are called *first-order circuits*.

6-3-1 Solution of a First-Order Linear Differential Equation

The detailed discussion of the solution of first-order linear differential equations is presented in standard mathematics texts on differential equations. A brief summary of the steps is presented here for the special case of a constant forcing function. The general procedure is to assume a suitable solution, with some coefficients whose values must be determined.

Given the differential equation

$$dx/dt + px(t) = F \qquad (6\text{-}15)$$

where p and F are known constants, the general solution is assumed to be of the form

$$x(t) = K_1 e^{-pt} + K_2 \qquad (6\text{-}16\text{a})$$

where K_1 and K_2 are constant coefficients to be determined.

To determine the value of K_2, the expression for $x(t)$ in Eq. (6-16a) and dx/dt obtained by differentiating Eq. (6-16a) are substituted in the given differential equation, Eq. (6-15). This leads to

$$K_2 = F$$

so that Eq. (6-16a) now reads

$$x(t) = K_1 e^{-pt} + F \qquad (6\text{-}16\text{b})$$

The constant K_1 is then evaluated by using a known initial condition, usually the value of $x(t)$ at $t = 0$, in Eq. (6-16b). Suppose $x(t) = X_0$ at $t = 0$. Then putting $t = 0$ and $x(t) = X_0$ in Eq. (6-16b)

$$X_0 = K_1 + F$$

or

$$K_1 = (X_0 - F)$$

and the final expression for $x(t)$ becomes

$$x(t) = (X_0 - F)e^{-pt} + F \qquad (6\text{-}16\text{c})$$

which is the solution of the given differential equation.

Example 6-3 Find the solution of the differential equation

$$dx/dt + 0.5x(t) = 20$$

given the initial condition $x(0) = 6$.

Solution Since the steps involved in the solution of a first-order differential equation are straightforward, it is convenient to proceed step by step, rather than plug the given numbers into a formula. The general solution is

$$x(t) = K_1 e^{-0.5t} + K_2$$

and
$$dx/dt = -0.5K_1 e^{-0.5t}$$

Substitution of $x(t)$ and dx/dt in the original differential equation gives
$$K_2 = 20/0.5 = 40$$
so that
$$x(t) = K_1 e^{-0.5t} + 40$$

Putting $t = 0$ in the last expression and using the given initial condition $x(0) = 6$, it is found that
$$K_1 = [x(0) - 40] = -34$$

Therefore, the solution of the given equation is
$$x(t) = -34 e^{-0.5t} + 40 \quad \blacksquare$$

Exercise 6-9 Find the solution of the differential equation
$$di/dt + 2i(t) = 40$$
if $i(0) = -4$ A.

6-3-2 The Step Function

As we noted earlier, the discussion in this chapter will be confined to the behavior of circuits subjected to a constant forcing function. Consider a circuit in a black box, as shown in Fig. 6-7(a), to which is applied a constant voltage source V_B. The source begins to act upon the circuit at $t = 0$ by means of a switch that closes at $t = 0$. Then the input voltage v_i to the circuit jumps from 0 to V_B instantaneously at $t = 0$, as indicated in Fig. 6-7(b). Such a function is called a *step function*. A *unit step function*, $u(t)$, shown in Fig. 6-7(c), is defined by

$$u(t) = \begin{cases} 0 & t < 0 \\ 1 & t > 0 \end{cases}$$

Note the presence of the discontinuity at $t = 0$.

A step function that is zero until $t = 0$ and V_B for $t > 0$ is written as $V_B u(t)$:

$$V_B u(t) = \begin{cases} 0 & t < 0 \\ V_B & t > 0 \end{cases}$$

Fig. 6-7 (a) Circuit with constant voltage source. (b) A step function. (c) A unit step function.

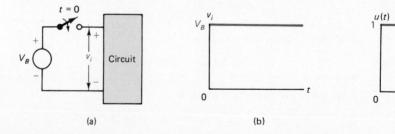

which is the function in Fig. 6-7(b). For practical purposes, a step function is simply a representation of a constant source made active at $t = 0$.

6-3-3 Step Responses of a Series RC Circuit

The differential equation of a series RC circuit driven by a voltage step $V_B u(t)$, as shown in Fig. 6-8(a), is

$$Ri(t) + (1/C) \int i(t)dt = V_B u(t) \tag{6-17}$$

The responses of interest are the current $i(t)$ in the circuit and the voltage $v_c(t)$ across the capacitor. Suppose we wish first to find the response $v_c(t)$. The voltage $v_c(t)$ and the current $i(t)$ are related by these equations:

$$v_c(t) = (1/C) \int i(t)dt \quad \text{and} \quad i(t) = C[dv_c(t)/dt] \tag{6-18}$$

After using Eqs. (6-18) and some manipulations, Eq. (6-17) becomes

$$\frac{dv_c(t)}{dt} + \frac{1}{RC}v_c(t) = \frac{1}{RC}V_B u(t) \tag{6-19}$$

Consider first the interval $t > 0$:

$$t > 0: \quad u(t) = 1$$

and Eq. (6-19) becomes

$$\frac{dv_c(t)}{dt} + \frac{1}{RC}v_c(t) = \frac{1}{RC}V_B \tag{6-20}$$

which is of the form of Eq. (6-15), with $p = (1/RC)$ and $F = (1/RC)V_B$. Its general solution is, from Eq. (6-16a)

$$v_c(t) = K_1 e^{-(1/RC)t} + K_2 \tag{6-21}$$

Substituting Eq. (6-21) in Eq. (6-20), we find that

$$K_2 = V_B$$

so that the solution becomes

$$v_c(t) = K_1 e^{-(t/RC)} + V_B \tag{6-22}$$

Initial Condition

The constant K_1 is determined by making use of a known initial condition. Assume that the capacitor had no charge before $t = 0$. That is

$$v_c(t) = 0 \quad \text{when} \quad t < 0 \tag{6-23}$$

At $t = 0$, the applied voltage to the circuit jumps to V_B and activates the circuit. But the *voltage continuity condition* for the capacitor demands that the voltage $v_c(t)$ remain continuous and not undergo any instantaneous changes. Therefore, immediately after $t = 0$, an instant that may be called $t = 0+$, the voltage $v_c(t)$ will still be zero. That is

$$v_c(t) = 0 \quad \text{at} \quad t = 0+ \tag{6-24}$$

Equation (6-24) gives the initial condition to be used for evaluating K_1 in Eq. (6-22).

Putting $t = 0+$ and $v_c(0+) = 0$ in Eq. (6-22), we find that

$$K_1 = -V_B$$

Eq. (6-22) then becomes

$$v_c(t) = V_B - V_B e^{-(t/RC)} \qquad (t > 0) \qquad (6\text{-}25)$$

which is the voltage across the capacitor in the interval $t > 0$.
Therefore, the voltage across the capacitor is given by

$$v_c(t) = \begin{cases} 0 & t < 0 \\ (V_B - V_B e^{-t/RC}) & t > 0 \end{cases}$$

which may be written more compactly as

$$v_c(t) = V_B(1 - e^{-t/RC})u(t) \qquad (6\text{-}26)$$

The current $i(t)$ in the RC circuit is obtained by using the equation

$$i(t) = C[dv_c(t)/dt]$$

Therefore

$$i(t) = \begin{cases} 0 & t < 0 \\ (V_B/R)e^{-t/RC} & t > 0 \end{cases} \qquad (6\text{-}27a)$$

or

$$i(t) = (V_B/R)e^{-t/RC}u(t) \qquad (6\text{-}27b)$$

The graphs of $v_c(t)$ and $i(t)$ are shown in Fig. 6-8(b).

Example 6-4 A step voltage $v_s(t) = 150\,u(t)$ is applied to a series RC circuit with $R = 40\ k\Omega$ and $C = 50\ \mu F$. Assume that the initial voltage on the capacitor is zero. (a) Write the differential equation of the circuit in terms of the capacitor voltage $v_c(t)$. (b) Obtain expressions for $v_c(t)$ and $i(t)$. (c) Evaluate $v_c(t)$ and $i(t)$ at $t = 2\ s$. (d) Determine the time taken for the capacitor to charge up to 140 V.

Solution (a) The differential equation in terms of the current $i(t)$ is

$$Ri(t) + (1/C) \int i(t)dt = v_s(t)$$

or

$$40 \times 10^3 i(t) + 2 \times 10^4 \int i(t)dt = 150 u(t)$$

Using $i(t) = C[dv_c(t)/dt]$, the differential equation becomes

$$dv_c(t)/dt + 0.5 v_c(t) = 75 \qquad (t > 0) \qquad (6\text{-}28)$$

(b) The general solution is of the form

$$v_c(t) = K_1 e^{-0.5t} + K_2 \qquad (6\text{-}29)$$

Using Eq. (6-29) in (6-28), $K_2 = 150$, so that $v_c(t) = K_1 e^{-0.5t} + 150$.
Using the initial condition $v_c(0) = 0$ in the last equation, $K_1 = -150$.
Therefore

$$v_c(t) = [150 - 150 e^{-0.5t}]u(t) \quad V \qquad (6\text{-}30)$$

and

$$i(t) = C[dv_c(t)/dt] = 3.75 \times 10^{-3} e^{-0.5t} \text{ A} \qquad (t > 0) \qquad (6\text{-}31)$$

(c) The values of $v_c(t)$ and $i(t)$ at $t = 2$ are obtained by substituting $t = 2$ in Eqs. (6-30) and (6-31): $v_c = 94.8$ V and $i = 1.38 \times 10^{-3}$ A

(d) If $v_c = 140$ V at $t = t_1$, then Eq. (6-30) becomes

$$140 = 150 - 150 e^{-0.5 t_1}$$

which gives $t_1 = -2 \ln(10/150) = 5.42$ s. ∎

Exercise 6-10 A 5000 pF capacitor in series with a 470 kΩ resistor is driven by a source $v_s(t) = 120 u(t)$ V. If the capacitor has no initial charge, obtain the expressions for $v_c(t)$ and $i(t)$ for $t > 0$.

Exercise 6-11 If the initial voltage $v_c(0) = 50$ V (instead of zero) in the circuit of the previous exercise, obtain the expressions for $v_c(t)$ and $i(t)$.

Exercise 6-12 A series RC circuit is excited by a source $v_s = 25 u(t)$ V. The capacitor has no initial charge. If it is given that $i(t) = 6.25$ A at $t = 0+$ and that $v_c(t) = 12.5$ V at $t = 100$ ms, determine the values of R and C.

6-3-4 Discussion of the Step Response of the Series RC Circuit

Voltage Buildup in a Capacitor

Let us examine the response of a series RC circuit to a step voltage from a physical point of view; see Fig. 6-8(a). The voltage across the capacitor is assumed to be zero when $t < 0$. At $t = 0$, the voltage V_B is applied to the circuit.

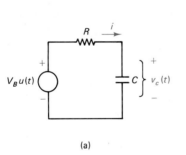
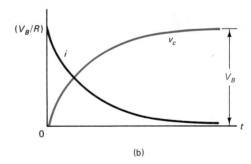

Fig. 6-8 (a) Series RC circuit under step input. (b) Step response of a series RC circuit.

The voltage continuity condition of the capacitor requires that $v_c = 0$ at $t = 0+$. The condition of the circuit at $t = 0+$ is as shown in Fig. 6-9(a). Using KVL around the closed path at $t = 0+$

$$v_R = V_B \quad \text{and} \quad i = (v_R/R) = (V_B/R) \quad \text{at } t = 0+$$

The current jumps up instantaneously from zero at $t < 0$ to (V_B/R) at $t = 0+$. (Note that the continuity condition need not be met by the *current through a capacitor*.)

The current (V_B/R) flows through the capacitor and starts charging it. As the charge on the capacitor builds up, $v_c(t)$ increases and $v_R(t)$ decreases, since $v_R + v_c = V_B$, a

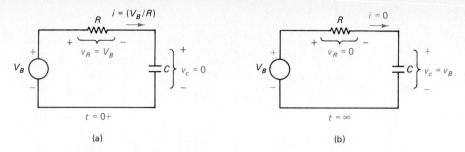

Fig. 6-9 Snapshots of a series RC circuit at (a) $t = 0+$, and (b) $t = \infty$.

constant. The decrease in v_R is accompanied by a smaller charging current $i(t)$. The smaller charging current slows down the rate of charge buildup in the capacitor, as can be seen by the gradual reduction of the slope of the $v_c(t)$ curve in Fig. 6-8(b). The variation of the capacitor voltage $v_c(t)$ and $i(t)$ depends upon the exponential term $e^{-t/RC}$ in Eqs. (6-26) and (6-27).

As the capacitor voltage increases and eventually becomes equal to the applied voltage V_B, as shown in Fig. 6-9(b), the voltage v_R across the resistor becomes zero. Since $v_R = 0$, $i = (v_R/R) = 0$, and no further charging takes place. The capacitor voltage stays constant at V_B.

The state of affairs where the capacitor voltage has reached a constant value and no further charging occurs is referred to as the steady state of the circuit.

The steady state occurs when the exponential term $e^{-t/RC}$ becomes zero, which occurs theoretically at $t = \infty$. From a practical point of view, however, the exponential term becomes negligibly small when $t = 5RC$ ($e^{-5} < 0.01$), and *the steady state may therefore be assumed to start at $t = 5RC$.*

The exponential term $-V_B e^{-t/RC}$ appearing in the expression for $v_c(t)$ in Eq. (6-26) has a value of $-V_B$ at $t = 0$ and approaches zero exponentially as t increases. Because of its temporary existence, this term is called the *transient component* of the capacitor voltage. The voltage across the capacitor for $t > 0$, given by Eq. (6-26), is thus seen to be the sum of two components: a transient component $-V_B e^{-t/RC}$ which goes to zero as $t \to \infty$, and a steady state component V_B which is a constant. Similarly, the current in the circuit is a sum of a transient component ($V_B/R(e^{-t/RC}$, and a steady state component which is zero in the present circuit.

The transient component is also called the *natural response* of the circuit. It represents the manner in which the passive components (R and C) would respond if left to themselves. Because of the energy storage and voltage continuity properties of the capacitor and the energy dissipation in a resistor, the circuit dictates the manner in which the voltage and current will vary in it, and this gives rise to the natural response term. The shape and form of the natural response depend only upon R and C. The applied voltage source affects only the magnitude of the natural response term at $t = 0$: V_B for the voltage and (V_B/R) for the current.

The steady state components of the capacitor voltage and current represent the response forced upon the circuit by the applied voltage source: the steady state components are constant, since the applied voltage is a constant. The steady state component represents the *forced response* of the circuit.

6-3-5 Time Constant and Rise Time

By expressing the resistance R and the capacitance C in terms of the basic dimensions (charge, mass, length, and time), it is possible to show that the *product RC has the dimensions of time*. It is called the *time constant* of the RC circuit and denoted by the symbol τ. Then the exponential term may be written as $e^{-t/\tau}$, where $\tau = RC$ s.

Consider the expression for the voltage across the capacitor:

$$v_c(t) = V_B(1 - e^{-t/\tau})$$

At $t = \tau$, the value of v_c becomes

$$v_c = V_B(1 - e^{-1}) = 0.632\, V_B$$

The time constant is therefore the time taken by the capacitor in a series RC circuit driven by a step voltage to reach *63.2 percent of the final (steady state) value*.

As we mentioned earlier, the circuit is assumed to have reached the steady state after *5 time constants*.

Figure 6-10(a) shows the variation of $v_c(t)$ for two circuits with the same steady state voltage, but different time constants: $\tau_1 > \tau_2$. The circuit with the *larger time constant* takes *longer* to reach steady state.

The speed with which the capacitor reaches steady state thus depends upon the time constant: A larger time constant leads to a slower response.

When the capacitor voltage in an RC circuit is observed in the laboratory through an oscilloscope, it is not easy to pinpoint the instants of time where the capacitor voltage starts from zero and where it has reached steady state. The measurement of the time constant by using the time at which the capacitor voltage reaches 63.2 percent of is final value becomes difficult. A more convenient parameter is the *10 to 90 percent rise time* (usually referred to simply as *rise time*), t_R, as shown in Fig. 6-10(b).

Rise time is defined as the time taken for the capacitor voltage to increase from $0.1V_B$ to $0.9V_B$.

By evaluating the instants of time at which $v_c = 0.1V_B$ and $v_c = 0.9V_B$ in Eq. (6-26), the formula for rise time is found to be as follows:

$$t_R = 2.2RC \tag{6-32}$$

Fig. 6-10 Speed of charging of a capacitor: (a) Larger time constant means slower charging. (b) Definition of rise time.

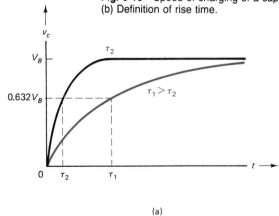
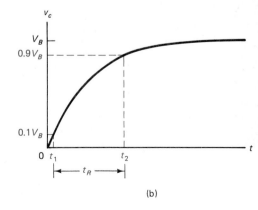

(a) (b)

That is, the *rise time equals 2.2 time constants*. The time constant is known once the rise time is known.

Exercise 6-13 Using Eq. (6-26), evaluate the time t_1 at which $v_c(t) = 0.1V_B$ and the time t_2 at which $v_c(t) = 0.9V_B$. Determine the rise time $t_R = (t_2 - t_1)$ and verify Eq. (6-32).

Exercise 6-14 Determine the rise time for an RC circuit with $R = 100$ kΩ and $C = 50$ pF.

Exercise 6-15 An RC circuit is to reach steady state in 0.025 s. Find its time constant and rise time.

6-3-6 Step-by-Step Procedure for the Solution of RC Circuits with a Step Input

The discussion about the transient component, steady state component, and time constant presented in the previous section shows that when a series RC circuit is driven by a step voltage, the response may be obtained by using the following steps:

1. Write an expression of the form: (transient term + steady state term) = $K_1 e^{-t/\tau} + K_2$ for the complete response, where the time constant $\tau = RC$.
2. Evaluate the steady state term K_2 by examining the circuit when the capacitor voltage has reached a constant value and the current through the capacitor is zero.
3. Use the initial condition in the expression for the *complete response* to find K_1. The initial condition is obtained by inspecting the circuit at $t = 0+$ and applying the voltage continuity condition to the capacitor voltage. Remember that there is no continuity condition for the capacitor current.
4. In the case of circuits containing a single capacitor but not in the form of a series RC circuit, first obtain the Thevenin equivalent, as seen from the capacitor terminals, and then use the steps above.

The following examples illustrate the application of this procedure to RC circuits.

Example 6-5 The circuit shown in Fig. 6-11(a) contains a voltage source V_a which remains constant at 150 V for all t ($t > -\infty$) and a step voltage source $v_B(t) = 200\, u(t)$ V which becomes active at $t = 0$. Determine the voltage $v_c(t)$.

Solution Note that the condition of the circuit changes at $t = 0$ by the activation of the source v_B. Therefore, the response is considered in two parts: the interval $t < 0$ when only the V_a source is active, and the interval $t > 0$ when both sources are active.

Interval $t < 0$: In this interval, $v_B(t) = 0$, and the circuit is as shown in Fig. 6-11(b). Replace the circuit by a Thevenin equivalent, as seen from the capacitor terminals; see Fig. 6-11(c). Since the V_a source has been active in the circuit since $t = -\infty$, the circuit must have reached a steady state by the time $t = 0$. Therefore, the capacitor voltage must equal the source voltage:

$$v_c(0-) = 112.5 \text{ V}$$

Next consider $t > 0$. The circuit is now as shown in Fig. 6-11(d). Again, replace the circuit by a Thevenin equivalent, as seen from the capacitor terminals in Fig. 6-11(e).

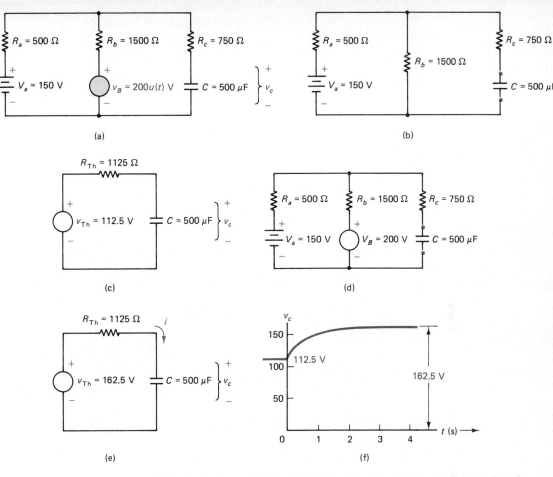

Fig. 6-11 (a) Circuits for Example 6-11. (b) and (c): Interval $t < 0$. (d) and (e): Interval $t > 0$. (f) Capacitor voltage.

The time constant of the circuit is $\tau = RC = (1125 \times 500 \times 10^{-6}) = 0.5625$ s. Therefore, the response $v_c(t)$ is of the form

$$v_c(t) = K_1 e^{-t/0.5625} + K_2 \quad (6\text{-}33)$$

Evaluate the constant K_2 by using the steady state condition:

$$\text{steady state value of } v_c(t) \text{ as } t \to +\infty = 162.5 \text{ V}$$

Therefore $K_2 = 162.5$ V, and Eq. (6-33) becomes

$$v_c(t) = K_1 e^{-1.78t} + 162.5 \quad (6\text{-}34)$$

K_1 is evaluated by using the initial condition. The initial value of v_c is $v_c(0+) = v_c(0-) = 112.5$ V (due to voltage continuity). Putting $t = 0$ and $v_c = 112.5$ in Eq. (6-34), $112.5 = K_1 + 162.5$, or $K_1 = -50$.

Therefore, Eq. (6-34) becomes

$$v_c(t) = -50e^{-1.78t} + 162.5 \quad (t > 0)$$

6-3 Equations for Simple RC and RL Circuits

The voltage across the capacitor in the given circuit is given by:

$$V_c(t) = \begin{cases} 112.5 \text{ V} & t < 0 \\ -50e^{-1.78t} + 162.5 \text{ V} & t > 0 \end{cases}$$

The variation of the capacitor voltage as a function of time is shown in Fig. 6-11(f). ∎

Exercise 6-16 Use the expressions for $v_c(t)$ obtained in the example above and do the following: (a) Obtain the expressions for the current in R_c. (b) Use the results of part (a) to obtain the expressions for the voltage at the top node of the circuit in Fig. 6-11(a). (c) Use the results of part (b) to obtain the expressions for the voltages across R_a and R_b. (d) Use the results of part (c) to obtain the expressions for the currents supplied by the two voltage sources.

Exercise 6-17 (Discharge of a capacitor) The voltage applied to a series RC circuit is given by $V_B u(-t)$ where

$$u(-t) = \begin{cases} 1 & t < 0 \\ 0 & t > 0 \end{cases}$$

Obtain the expressions for the capacitor voltage and capacitor current for all t, and sketch the two responses.

Example 6-6 Pulse Response of a Series RC Circuit

The response of an RC circuit to a rectangular pulse is important in electronics and communication systems, since a large number of signals of practical interest are in the form of rectangular pulses. Consider the rectangular pulse shown in Fig. 6-12(a) applied to the RC circuit of Fig. 6-12(b). Assuming that $v_c(0-) = 0$, determine the voltage and current response of the RC circuit.

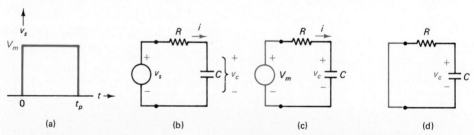

(a) (b) (c) (d)

Fig. 6-12 Pulse response of an RC circuit. The rectangular pulse in (a) is applied to the circuit in (b). (c) The circuit in the interval $0 < t < t_p$. (d) The circuit in the interval $t > t_p$.

Solution Since there is a change of conditions in the circuit at $t = 0$ and at $t = t_p$, the response is considered in the two intervals: $0 < t < t_p$ and $t_p < t < \infty$.

Interval $0 < t < t_p$: The situation during the interval $0 < t < t_p$ is as shown in Fig. 6-12(c): a step voltage is applied to a series RC circuit. From Subsection 6-3-3

$$v_c(t) = V_m(1 - e^{-t/RC}) \qquad (0 < t < t_p) \qquad (6\text{-}35)$$

and

$$i(t) = (V_m/R)e^{-t/RC} \qquad (0 < t < t_p)$$

Note that the circuit has no way of knowing beforehand that the input will be reduced to zero at $t = t_p$, and therefore it responds as if the voltage will remain at V_m for all $t > 0$.

Interval $t_p < t < \infty$: The situation for the interval $t > t_p$ is as shown in Fig. 6-12(d). The capacitor starts with an initial voltage given by Eq. (6-35) evaluated at $t = t_p$, and discharges toward zero. Since the steady state value is zero, the expression for $v_c(t)$ in this interval is of the form

$$v_c(t) = K_1 e^{-t/RC} \tag{6-36}$$

where the value of K_1 is determined by using the value of v_c at $t = t_p+$ (which is the initial condition for this interval). Due to the voltage continuity condition

$$v_c(t_p+) = v_c(t_p-)$$

and from Eq. (6-35)

$$v_c(t_p-) = V_m(1 - e^{-t_p/RC}) \tag{6-37}$$

Using Eq. (6-37) in Eq. (6-36) with $t = t_p$,

$$K_1 e^{-t_p/RC} = V_m(1 - e^{-t_p/RC})$$

which leads to

$$K_1 = V_m(e^{t_p/RC} - 1)$$

so that Eq. (6-36) becomes

$$v_c(t) = V_m(e^{t_p/RC} - 1)e^{-t/RC} \qquad (t > t_p) \tag{6-38}$$

The capacitor current in the interval $t > t_p$ is obtained from Eq. (6-38)

$$\begin{aligned} i(t) &= C[dv_c/dt] \\ &= -(V_m/R)(e^{-t_p/RC} - 1)e^{-t/RC} \qquad (t > t_p) \end{aligned} \tag{6-39}$$

Collecting the various expressions obtained, the pulse response of the series RC circuit is given by

$$v_c(t) = \begin{cases} V_m(1 - e^{-t/RC}) & 0 < t < t_p \\ V_m(e^{t_p/RC} - 1)e^{-t/RC} & t > t_p \end{cases}$$

$$i(t) = \begin{cases} (V_m/R)e^{-t/RC} & 0 < t < t_p \\ -(V_m/R)(e^{t_p/RC} - 1)e^{-t/RC} & t > t_p \end{cases}$$

∎

6-3-7 Discussion of the Pulse Response of an RC Circuit

Electronic amplifiers and communication networks have the same effect on an input pulse as an *RC* circuit. Consequently, the output pulse of such systems is of the same form as the voltages in an *RC* circuit subjected to a pulse input. The discussion of the preceding example can be used to determine how well the output of an amplifier resembles the input pulse.

The diagrams of Fig. 6-13 show the output pulse (voltage across the capacitor) in a series *RC* circuit for three different cases:

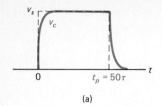 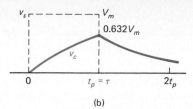 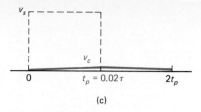

Fig. 6-13 Effect of time constant on pulse response: (a) Slight rounding off occurs when the time constant is very small compared with pulse duration. (b) The pulse is recognizable when the time constant = pulse duration. (c) The pulse is indistinguishable when the time constant is very large compared with pulse duration.

1. Pulse duration is much larger than the time constant: $t_p = 50\tau$.
2. Pulse duration = time constant.
3. Pulse duration is much smaller than the time constant: $t_p = (\tau/50)$.

In all cases, the output voltage starts from zero at $t = 0$, charges up to a certain value at the end of the input pulse, and then discharges toward zero.

When the input pulse duration is large compared with the time constant, as shown in Fig. 6-13(a), the output reaches a constant value (equal to the amplitude V_m of the input pulse) long before the input pulse disappears. When the input pulse returns to zero, the output voltage decreases from V_m to zero. *The output pulse is a rounded-off version of the input pulse.* As the duration of the input pulse is made even longer in comparison with the time constant RC, the rounding off of the output pulse is less pronounced, and it begins to resemble the input pulse more closely. The voltage continuity condition of the capacitor makes it impossible for the output to resemble the input pulse exactly (with its discontinuities at $t = 0$ and t_p), but it is possible to make the output voltage reasonably faithful to the input when the input pulse is of extremely long duration compared with the time constant.

When the duration of the input pulse equals the time constant, as shown in Fig. 6-13(b), the output voltage reaches a value of $0.632V_m$ at the end of the input pulse. The output does not resemble the input, but it is possible to detect the presence of a pulse. *Choosing a time constant equal to the duration of the input pulse is therefore used as an acceptable rule of thumb in some RC circuits through which pulses are transmitted.*

When the duration of the input pulse is much smaller than the time constant, as Fig. 6-13(c) shows, the output voltage reaches only a small fraction of V_m at $t = t_p$ and then discharges from this small value toward zero. The presence of a pulse in the output is hardly noticeable in this case. In fact, if several pulses were applied in quick succession to the RC circuit, the output would be in the form of a small ripple instead of a series of distinguishable pulses.

The discussion shows that when pulses are transmitted through a capacitive network with the output taken across the capacitor, it is necessary to make the *pulse duration much larger than the time constant* of the circuit to produce an output that is a reasonable facsimile of the input pulse.

The time constant of an RC circuit will also be encountered later in the discussion of the frequency response of RC networks; it will be seen to be closely related to the bandwidth of the RC circuit.

Exercise 6-18 Use the expressions for the current $i(t)$ in the RC circuit driven by a pulse input to find the voltage v_R across the resistor. Sketch $v_R(t)$ for the three cases: (a) $t_p = 50\tau$; (b) $t_p = \tau$; (c) $t_p = 0.02\tau$. Discuss the resemblance between v_R and the input pulse in the three cases.

6-3-8 Change of Conditions in an RC Circuit

Circuit conditions may change before the steady state is reached in an RC circuit. An example is the RC circuit with an input pulse whose duration is comparable to the time constant. The principles of analysis of such circuits are exactly similar to those used earlier in this section. *It is, however, necessary to write a different expression for each interval and evaluate the constants in the expressions by using the steady state value and initial condition appropriate to each interval.*

Example 6-7 Obtain the expression for the voltage $v_c(t)$ in the circuit of Fig. 6-14(a) if the switch S remains closed until $t = 0.5$ s and then opened. Assume that $v_c = 0$ at $t = 0-$.

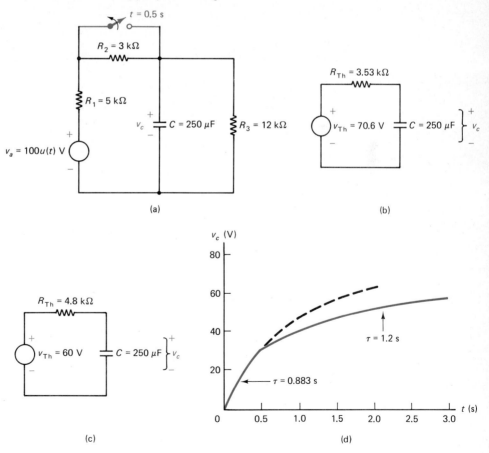

Fig. 6-14 (a) Circuits for Example 6-7. (b) Thevenin equivalent for $0 < t < 0.5$ s. (c) Thevenin equivalent for $t > 0.5$ s. (d) Capacitor voltage.

6-3 Equations for Simple RC and RL Circuits

Solution There are two intervals of interest: $0 < t < 0.5$ s and $t > 0.5$ s.

Interval $0 < t < 0.5$ s: The switch shorts out R_2. Replace the circuit by a Thevenin equivalent, as seen from the capacitor terminals:

$$R_{Th} = (R_1 \| R_3) = 3.53 \text{ k}\Omega$$

and

$$v_{Th} = v_a[R_3/(R_1 + R_3)] = 70.6 \text{ V}$$

The resulting Thevenin equivalent is shown in Fig. 6-14(b). For this circuit

$$\text{time constant } \tau = 0.883 \text{ s}$$

and the response is of the form

$$v_c(t) = K_1 e^{-t/0.883} + K_2$$

The constant K_2 is evaluated by using the steady state value of v_c, which is 70.6 V. (Note that the circuit cannot foresee the change in circuit conditions and consequently will go toward the steady state voltage given by the present circuit conditions.) Therefore

$$K_2 = 70.6$$

Therefore,

$$v_c(t) = K_1 e^{-t/0.883} + 70.6$$

The constant K_1 is evaluated by using the initial condition in the expression for v_c.

$$\text{initial condition: } v_c(0+) = v_c(0-) = 0 \text{ (given)}$$

Therefore

$$K_1 = -70.6$$

so that

$$v_c(t) = 70.6(1 - e^{-1.13t}) \text{ V} \qquad (0 < t < 0.5) \qquad (6\text{-}40)$$

Interval $t > 0.5$ s: The open switch makes R_2 a part of the circuit. Again, replace the circuit by a Thevenin equivalent, as seen from the capacitor terminals:

$$R_{Th} = (R_1 + R_2) \| R_3 = 4.8 \text{ k}$$
$$v_{Th} = v_a[R_3/(R_1 + R_2 + R_3)] = 60 \text{ V}$$

The resulting circuit is shown in Fig. 6-14(c). For this circuit

$$\text{time constant } \tau = 1.2 \text{ s}$$

and the response is of the form

$$v_c(t) = K_3 e^{-t/1.2} + K_4$$

The constant K_4 is given by the steady state value of v_c, which is 60 V, so that

$$v_c(t) = K_3 e^{-t/1.2} + 60 \qquad (6\text{-}41)$$

For the initial condition for this interval, the voltage continuity condition requires the value of v_c at the instant *just after* opening the switch to equal the value of v_c at the instant *just before* opening the switch. That is, v_c at $t = 0.5$ s obtained from Eq. (6-40) should be the same as v_c at $t = 0.5$ s in Eq., (6-41). From Eq. (6-40)

$$v_c = 30.5 \text{ V} \quad \text{at} \quad t = 0.5 \text{ s}$$

and this is the initial condition for the interval $t > 0.5$. Putting $t = 0.5$ s and $v_c = 30.5$ V in Eq. (6-41)

$$30.5 = K_3 e^{-0.5/1.2} + 60$$

or $K_3 = -44.7$ V. Therefore

$$v_c(t) = -44.7 e^{-0.833t} + 60 \quad (t > 0.5) \tag{6-42}$$

The voltage across the capacitor in the given circuit is therefore given by

$$v_c(t) = \begin{cases} 70.6(1 - e^{-1.13t}) \text{ V} & 0 < t < 0.5 \text{ s} \\ -44.7 e^{-0.833t} + 60 \text{ V} & t > 0.5 \text{ s} \end{cases}$$

The graph of $v_c(t)$ is shown in Fig. 6-14(d). Charging is at a slower rate in the interval $t > 0.5$ s due to the larger time constant in that interval. ∎

Exercise 6-19 Use the expressions for v_c obtained in the example above and find (a) the current through R_3; (b) the current through the capacitor; (c) the current in the resistor R_2; and (d) the current in the resistor R_1.

Exercise 6-20 Suppose that, in the circuit of Example 6-7, the switch is open in the interval $0 < t < 0.5$ s and then closed. Obtain the expressions for $v_c(t)$ and the various currents mentioned in Exercise 6-19.

Example 6-8 (Basic Automatic Exposure Control Circuit) The circuit of Fig. 6-15 represents a basic automatic exposure control circuit for a camera. The resistor R_p is a photosensitive resistance that varies as a function of the incident light intensity. If P denotes the intensity of light (measured in lumens), then it is possible to write

$$R_p = A - B \log P$$

where A and B are constants. The switch is at position 1 until $t = 0$ and is assumed to be charged to V_B. At $t = 0$, the switch is moved to position 2 by the operator of the

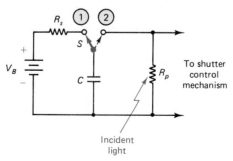

Fig. 6-15 Basic automatic exposure control circuit for Example 6-8.

camera. This action also activates an electromechanical system in the camera and opens the shutter. The capacitor discharges through R_p and when its voltage reaches $0.2\ V_B$, the electromechanical system closes the shutter.

Obtain an expression for the time of exposure as a function of the light intensity.

Solution   When the switch is moved to position 2 at $t = 0$, the capacitor discharges through R_p and

$$v_c(t) = V_B e^{-t/R_pC}$$

The shutter closes at $t = t_1$ when $v_c = 0.2\ V_B$. Therefore

$$t_1 = -(R_pC)\ \ln(v_c/V_B) = -(R_pC)\ \ln(0.2)$$
$$= 1.61 R_pC$$

The time of exposure is therefore

$$t_1 = 1.61 R_pC = 1.61 C(A - B\ \text{Log}\ P)$$

For example, suppose $A = 10^4$ and $B = 9 \times 10^3$ (when P is measured in lumens) for a certain resistor R_p, and let $C = 5\ \mu\text{F}$. Then the expression for the exposure time becomes

$$t = 8.05(10 - 9\ \text{Log}\ P)\ \text{ms}$$

If the light level is low with $P = 1$ lumen, $t_1 = 80.5$ ms (or 1/12 s). If the light level is bright with $P = 10$ lumens, $t_1 = 8.05$ ms (or 1/120 s). The exposure time decreases as the light level increases. Of course, the actual exposure control systems used in modern cameras are far more sophisticated than the circuit in this example, but the underlying principle is the same.

Exercise 6-21   Suppose the initial voltage on the capacitor in the exposure control circuit is only $0.8\ V_B$ due to weak batteries, and the shutter still closes when the capacitor voltage reaches $0.2\ V_B$. Obtain the new expression for the exposure time and find its numerical values for $P = 1$ and 10 lumens.

6-4   STEP RESPONSE OF A PARALLEL RC CIRCUIT

The principles developed for the step response of a series RC circuit are also applicable to simple parallel RC circuits. The time constant is $\tau = RC$; the voltage continuity condition is applied to the capacitor voltage to obtain the initial condition; and the steady state value of the response is determined by making the capacitor current equal to zero. The response is in the same form as before:

$$K_1 e^{-t/\tau} + K_2$$

where K_2 is the steady state value and K_1 is determined by using the initial condition on the complete expression.

Actually, regardless of whether a circuit containing a single capacitor is in the form of a series RC circuit, a parallel RC circuit, or any other configuration, a convenient approach would be to *replace the circuit by a Thevenin equivalent as seen from the capacitor terminals*, and use the procedures developed in the previous sections of this chapter.

Example 6-9 Determine the voltage $v_c(t)$ in the circuit of Fig. 6-16(a). Note that the battery V_B is active for all $t > -\infty$.

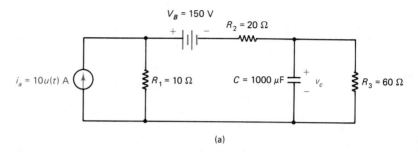

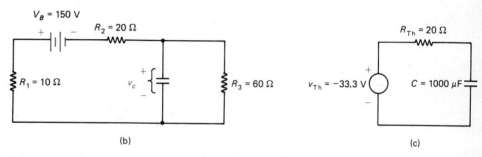

Fig. 6-16 (a) Circuit for Example 6-9. (b) Interval $t < 0$. (c) Thevenin equivalent for $t > 0$.

Solution *Interval* $t < 0$: The circuit is in a steady state, since the V_B source has been active for a long time. The current due to the current source is

$$i_a(t) = 10\,u(t) = 0 \quad \text{when} \quad t < 0$$

so that the circuit assumes the form shown in Fig. 6-16(b). The voltage across the capacitor is found to be

$$v_c = -100 \text{ V} \quad \text{at} \quad t = 0-$$

Interval $t > 0$: The Thevenin equivalent as seen from the capacitor terminals is

$$v_{Th} = [66.7u(t) - 100] = -33.3 \text{ V} \quad (t > 0)$$
$$R_{Th} = 20\,\Omega$$

The resulting circuit is as shown in Fig. 6-16(c).

- Time constant = $R_{Th}C = 0.02$ s
- Initial condition $v_c(0+) = v_c(0-) = -100$ V
- Steady state value of $v_c = -33.3$ V

The capacitor voltage is therefore given by

$$v_c(t) = (-66.7e^{-t/0.02} - 33.3) \text{ V}$$

6-4 Step Response of a Parallel RC Circuit

Exercise 6-22 A circuit consists of the parallel combination of a current source $i_s(t) = 50u(t)$ A, a capacitor $C = 5000$ pF, and a resistor $R = 20\ \Omega$. It is given that the voltage across the capacitor is 500 V at $t = 0-$. Determine (a) the voltage across the capacitor; (b) the current through the capacitor; and (c) the current through the resistor, all as functions of time for $t > 0$.

Exercise 6-23 Determine the voltage $v_c(t)$ in each of the circuits of Fig. 6-17.

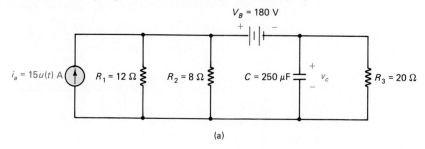

(a)

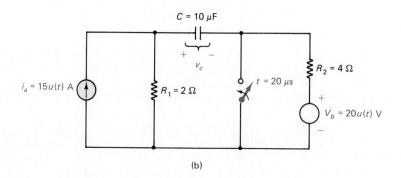

(b)

Fig. 6-17 Circuits for Exercise 6-23.

6-5 STEP RESPONSE OF RL CIRCUITS

Consider a series RL circuit driven by a step voltage function, as shown in Fig. 6-18(a). The differential equation for the circuit is, since $v_L(t) = L[di(t)/dt]$,

$$L\frac{di(t)}{dt} + Ri(t) = V_B \quad (t > 0)$$

Fig. 6-18 (a) Series RL Circuit under step input. (b) and (c): Step response of a series RL circuit.

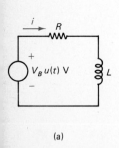

(a)

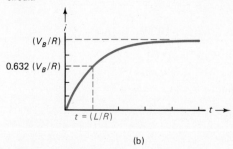

(b)

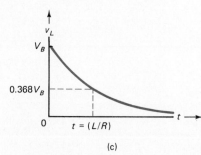

(c)

288 Circuits Containing Capacitors and Inductors

which is rewritten in the form

$$\frac{di(t)}{dt} + \frac{R}{L}(i(t)) = \frac{V_B}{L} \qquad (6\text{-}43)$$

Comparing the last equation with Eq. (6-15), the general solution is of the form

$$i(t) = K_1 e^{-(R/L)t} + K_2 \qquad (6\text{-}44)$$

Substitution of $i(t)$ and di/dt from Eq. (6-44) in Eq. (6-43) gives

$$K_2 = (V_B/R)$$

leading to

$$i(t) = K_1 e^{-(R/L)t} + (V_B/R) \qquad (6\text{-}45)$$

The constant K_1 is evaluated by using the initial condition $i(t)$ at $t = 0$ in Eq. (6-45). Assuming that $i = 0$ in the interval $t < 0$ and using the *current continuity condition*

$$i(0) = 0$$

which, when substituted along with $t = 0$ in Eq. (6-45), leads to

$$K_1 = -(V_B/R)$$

The response of the series *RL* circuit to a step input is therefore

$$i(t) = (V_B/R)[1 - e^{-(R/L)t}] u(t) \qquad (6\text{-}46)$$

The graph of the current in the circuit as a function of time is shown in Fig. 6-18(b). It is seen that the current starts at zero at $t = 0$ and builds up gradually to a final value of (V_B/R).

The voltage across the inductance is given by

$$v_L(t = L[di(t)/dt] \qquad (6\text{-}47)$$
$$= V_B e^{-(R/L)t}$$

and the graph of the voltage is shown in Fig. 6-18(c). The voltage across the inductor is seen to jump instantaneously from zero to V_B at $t = 0$ and then decay toward zero.

There is a strong similarity between the *current buildup in an inductor* and the *voltage buildup in a capacitor* discussed earlier. Similarly, there is a strong similarity between the behavior of the voltage in an inductance and the current in a capacitance. These similarities are to be expected because of the duality between an inductor and a capacitor.

6-5-1 Discussion of the Step Response of an RL Circuit

The current through the the circuit is zero just before $t = 0$. At $t = 0$, the voltage V_B is applied to the circuit. Due to the current continuity condition of an inductance, $i = 0$ at $t = 0+$. Since $i = 0$, the voltage across R, $v_R = 0$ at $t = 0+$ and, by KVL, the voltage across the inductance $v_L = V_B$. The situation at $t = 0+$ is shown in Fig. 6-19(a).

The voltage across the inductance causes a current to flow, and the current $i(t)$ increases from zero. As $i(t)$ increases, v_R increases and v_L decreases. The reduction in v_L causes

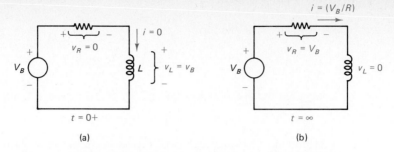

Fig. 6-19 Snapshots of the series RL circuit at (a) $t = 0+$ and (b) $t = \infty$.

a reduction of the rate at which the current builds up in the inductor, which explains the gradual reduction in the slope of the curve of $i(t)$ in Fig. 6-18(b).

Eventually, the current $i(t)$ increases to a point where $v_R = V_B$, as shown in Fig. 6-19(b). At this point

$$v_R = V_B \quad \text{and} \quad v_L = 0$$

and the current in the circuit is

$$i = (V_B/R) = \text{a constant}$$

Since the current through the inductance is a constant, the voltage across it is zero. This situation corresponds to the *steady state:* the current in the inductance is a constant, and the voltage across it is zero.

6-5-2 Time Constant of a Series RL Circuit

The exponential term $e^{-(R/L)t}$ appearing in $i(t)$ and $v_L(t)$ starts at 1 at $t = 0$ and decays toward zero as t increases. The rate at which the exponential term decreases is seen to depend upon the quantity (R/L). By using the basic dimensions (mass, length, time, and charge) for resistance R and inductance L, it is possible to show that the ratio (L/R) has the dimensions of time (seconds). The quantity (L/R) is called the *time constant* of the RL circuit. That is

$$\tau = (L/R) \tag{6-48}$$

The larger the value of the time constant, the longer it takes for the current to build up to the steady state value. Theoretically, it takes an infinite amount of time for the circuit to reach the steady state. As a practical matter, however, the time to reach the steady state is taken as 5 *time constants*.

Determination of the Step Response of an RL Circuit

The step response of a circuit containing a single inductance and any combination of resistances is obtained by using the following procedure (similar to that used in the case of RC circuits.)

1. If necessary, first use Thevenin's theorem to reduce the circuit to a series RL circuit.
2. Write the response in the form: (transient component + steady state component) $= K_1 e^{-t/\tau} + K_2$ where the time constant $\tau = (L/R)$.
3. Evaluate K_2 by using the steady state value of the response. For this, take the current in the inductor as a constant and the voltage across it as zero.

4. Evaluate the constant K_1 by using the initial condition. Initial condition is determined in terms of the *current through the inductance* and using the current continuity condition. K_1 is evaluated by substituting the initial condition in the *complete response*.

The following examples illustrate the procedure.

Example 6-10 The source V_B in the circuit in Fig. 6-20(a) has been active since $t = -\infty$. Obtain the expressions for (a) the current $i_L(t)$ in the inductance and (b) the voltage $v_L(t)$ across the inductance. (c) Determine the instant of time at which the current in the inductance is 75 percent of its steady state value.

Fig. 6-20 (a) Circuit for Example 6-10. (b) and (c): Circuits, interval $t < 0$. (d) and (e): Circuits, interval $t > 0$. (f) Graph, current in the inductor.

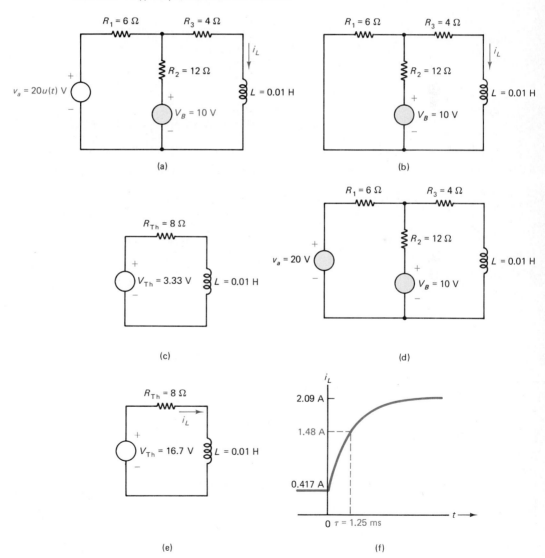

6-5 Step Response of RL Circuits

Solution *Interval $t < 0$:* In this interval, only the V_B source is active, and the circuit is as shown in Fig. 6-20(b). Replace the circuit as seen from the terminals of the inductor by a Thevenin equivalent, as shown in Fig. 6-20(c). The circuit is in the steady state at $t = 0-$: $v_L = 0$ and

$$i_L = 0.417 \text{ A} \quad \text{at} \quad t = 0- \quad (6\text{-}49)$$

Interval $t > 0$: The circuit in this interval is as shown in Fig. 6-20(d). Replace the circuit as seen from the terminals of the inductor by a Thevenin equivalent, as shown in Fig. 6-20(e). The current $i_L(t)$ can be written in the form

$$i_L(t) = K_1 e^{-t/\tau} + K_2 \quad (6\text{-}50)$$

Time constant: $\tau = (L/R_{Th}) = 1.25 \times 10^{-3}$ s

Steady state: $v_L = 0$, $v_R = v_{Th}$, and $i_L = (v_{Th}/R) = 2.09$ A.
The expression for $i_L(t)$ now reads

$$i_L(t) = K_1 e^{-800t} + 2.09 \text{ A} \quad (6\text{-}51)$$

Initial condition: $i_L(0+) = i_L(0-) = 0.417$ A
Putting $t = 0$ and $i_L = 0.417$ in Eq. (6-51), we find that

$$K_1 = -1.67$$

Therefore, the current i_L in the given circuit is described by

$$i_L(t) = \begin{cases} 0.417 \text{ A} & (t < 0) \\ -1.67 e^{-800t} + 2.09 \text{ A} & (t > 0) \end{cases} \quad (6\text{-}52)$$

(b) The voltage across the inductance is obtained from

$$v_L(t) = L[di(t)/dt]$$
$$= 13.3 e^{-800t} \text{ V}$$

(c) Seventy-five percent of the steady state current is $(0.75 \times 2.09) = 1.57$ A. If t' is the instant at which $i_L = 1.57$, then from Eq. (6-52)

$$1.57 = -1.67 e^{-800t'} + 2.09$$

which gives

$$t' = -(1/800) \ln(0.311) = 1.46 \times 10^{-3} \text{ s}$$

The graph of $i_L(t)$ is shown in Fig. 6-20(f). ∎

Exercise 6-24 Obtain the expression for $i_L(t)$ in the circuit of Example 6-10, shown in Fig. 6-20(a), if the locations of the two sources are interchanged.

Exercise 6-25 (Current Decay in an Inductor) The input to the RL circuit of Fig. 6-21(a) is as shown in Fig. 6-21(b). Determine and sketch $i_L(t)$.

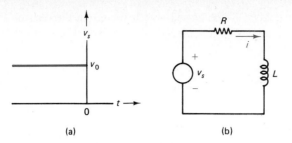

(a) (b)

Fig. 6-21 Circuit and input for Exercise 6-25.

Exercise 6-26 The current in an inductor is 10 A at $t = 0$ and made to decay through a resistance. If the current has decreased to 5 A in 12 ms, determine the time constant.

Example 6-11 Determine $i_L(t)$ and $i_2(t)$ in the circuit of Fig. 6-22(a), where the switch S remains closed until $t = 0.25$ s and then opened. Assume $i_L = 0$ at $t = 0-$.

Solution A solution for $i_L(t)$ is first obtained, which is then used for finding $i_2(t)$. Since there is a change of circuit conditions at $t = 0.25$ s, there are two intervals of interest: $0 < t < 0.25$ s and $t > 0.25$ s.

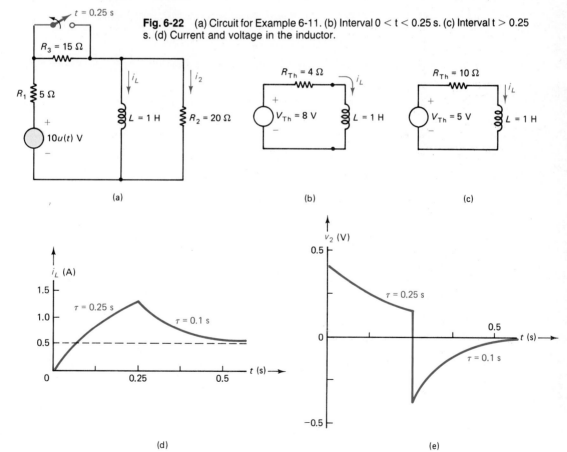

Fig. 6-22 (a) Circuit for Example 6-11. (b) Interval $0 < t < 0.25$ s. (c) Interval $t > 0.25$ s. (d) Current and voltage in the inductor.

6-5 Step Response of RL Circuits

Interval $t < 0.25$ s: The closed switch shorts out the resistor R_3. Replace the circuit by a Thevenin equivalent, as seen by the inductor shown in Fig. 6-22(b).

Time constant: $\tau = (L/R_{Th}) = 0.25$ s. The current $i_L(t)$ is of the form

$$i_L(t) = K_1 e^{-t/0.25} + K_2 \tag{6-53}$$

The constant K_2 is given by the steady state value. In the steady state, $v_L = 0$ and

$$i_L = (v_{Th}/R_{Th}) = 2.0 \text{ A}$$

Therefore, $K_2 = 2$ and Eq. (6-53) becomes

$$i_L(t) = K_1 e^{-4t} + 2 \text{ A} \tag{6-54}$$

K_1 is determined by using the initial condition

$$i_L(0+) = i_L(0-) = 0 \quad \text{(given)}$$

Using $i = 0$ at $t = 0$ in Eq. (6-54)

$$K_1 = -2$$

and Eq. (6-54) leads to

$$i_L(t) = -2e^{-4t} + 2 \text{ A} \quad (t < 0.25 \text{ s}) \tag{6-55}$$

Interval $t > 0.25$ s: The circuit now includes the resistor R_3. Replace the circuit by a Thevenin equivalent, as seen from the inductor terminals; see Fig. 6-22(c):

Time Constant: $\tau = (L/R'_{Th}) = 0.1$ s. The current $i_L(t)$ is of the form

$$i_L(t) = K_3 e^{-t/0.1} + K_4 \tag{6-56}$$

In the steady state for this interval, $v_L = 0$ and $i_L = (v'_{Th}/R'_{Th}) = 0.5$ A. Therefore

$$K_4 = 0.5$$

and Eq. (6-56) becomes

$$i_L(t) = K_3 e^{-10t} + 0.5 \text{ A} \tag{6-57}$$

The value of K_3 is found by using the initial condition for this interval, which occurs at $t = 0.25$ s. Due to the current continuity condition for an inductor, the value of i_L at $t = 0.25$ s given by Eq. (6-55) from the previous interval will be the same as i_L at $t = 0.25$ given by Eq. (6-57). From Eq. (6-55)

$$i_L = 2(1 - e^{-1}) = 1.26 \text{ A} \quad \text{at} \quad t = 0.25 \text{ s}$$

Substituting $i_L = 1.26$ and $t = 0.25$ in Eq. (6-57), we find that

$$K_3 = 9.26$$

Therefore, the expression for $i_L(t)$ becomes

$$i_L(t) = 9.26 e^{-10t} + 0.5 \text{ A} \quad (t > 0.25) \tag{6-58}$$

The current through the inductance in the given circuit is therefore given by

$$i_L(t) = \begin{cases} -2e^{-4t} + 2 \text{ A} & (t < 0.25 \text{ s}) \\ 9.26e^{-10t} + 0.5 & (t > 0.25 \text{ s}) \end{cases} \quad (6\text{-}59)$$

Returning to Fig. 6-22(a), we see that the current $i_2(t)$ is given by

$$i_2(t) = v_L(t)/R_2$$

Using $v_L(t) = L[di(t)/dt]$, we find that

$$i_2(t) = \begin{cases} 0.4e^{-4t} & (t < 0.25 \text{ s}) \\ -4.63e^{-10t} & (t > 0.25 \text{ s}) \end{cases} \quad (6\text{-}60)$$

The curves of i_L and i_2 are shown as functions of time in Figs. 6-22(d) and (e). The current through R_2 shows a reversal of direction after $t = 0.25$ s. At $t = 0.25$ s, the current in the inductor is 1.26 A, which is greater than its final steady state value of 0.5 A. The energy stored in the inductor is unloaded through the resistor R_2. The decrease in current accompanying the reduction of energy causes the voltage v_L to become negative. This results in the reversal of current in R_2. ∎

Exercise 6-27 Obtain expressions for the current in and the voltage across the inductor in the circuit of Fig. 6-23. Assume $i_L = 0$ at $t = 0-$.

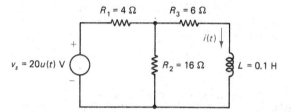

Fig. 6-23 Circuit for Exercise 6-27.

Exercise 6-28 A rectangular voltage pulse of amplitude V_m and duration t_p is applied to a series RL circuit. Assuming the current through the inductor is zero at $t = 0-$, determine the current $i_L(t)$. Sketch $i_L(t)$ for the following cases: (a) $t_p = 20(L/R)$; (b) $t_p = (L/R)$; (c) $t_p = 0.05(L/R)$.

6-6 ZERO INPUT AND ZERO STATE RESPONSE OF FIRST-ORDER CIRCUITS

The circuits discussed to this point in the chapter have a single inductor or a single capacitor, along with a combination of resistors. Such circuits are called *first-order circuits*, since their response involves the solution of a first-order differential equation. In the preceding sections, the response of a first-order circuit was separated into two components: a transient component and a steady state component. The complete response may also be viewed as consisting of two components called the *zero input response* and the *zero state response*.

The *zero input response* of a first-order circuit is defined as the response when there is no external forcing function acting on the circuit.

For a series *RL* circuit, the zero input current response is obtained by solving the differential equation

$$L\frac{di(t)}{dt} + Ri(t) = 0 \tag{6-61}$$

Thus the zero input response is the *complementary function* of the general solution of the differential equation. Therefore, the zero input response is given by

$$i(t) \quad \text{or} \quad v(t) = K_0 e^{-t/\tau}$$

where τ is the time constant of the circuit. The value of K_0 in the zero input response is *immediately* evaluated by using the initial condition. (This is different from the manner in which the coefficient of the transient term was evaluated in the preceding sections.) Therefore, K_0 is simply the value of the current or voltage at $t = 0+$.

The zero input current response of a series *RL* circuit (with an initial current I_0 in the inductance) is

$$i(t) = I_0 e^{-(R/L)t}$$

and the zero input voltage response of a series *RC* circuit (with an initial voltage V_0 across the capacitor) is

$$v(t) = V_0 e^{-t/RC}$$

The **zero state response** of a first-order circuit is defined as the response when the circuit has **zero initial energy stored:** that is, the initial current is zero for an inductor and initial voltage is zero for a capacitor.

The zero state current response of a series *RL* circuit driven by a source $v_s(t)$ is the solution of the differential equation

$$L\frac{di(t)}{dt} + Ri(t) = v_s(t)$$

under the stipulation that $i = 0$ at $t = 0$. When $v_s(t) = V_B u(t)$, a step voltage source, the zero state response is

$$i(t) = (V_B/R)[1 - e^{-(R/L)t}]$$

The zero state voltage response of a series *RC* circuit driven by a step voltage $V_B u(t)$ is

$$v(t) = V_B(1 - e^{-t/RC})$$

The zero state response of a first-order circuit is, in general, given by

$$i(t) \quad \text{or} \quad v(t) = K_{ss}(1 - e^{-t/\tau})$$

where K_{ss} is the steady state value of the response and τ is the time constant.

The complete response of a first-order circuit can be written as

$$\text{Complete response} = \text{zero input response} + \text{zero state response}$$

In the case of a first-order circuit subjected to a step voltage function, the complete response is given by

$$i(t) \quad \text{or} \quad v(t) = K_0 e^{-t/\tau} + K_{ss}(1 - e^{-t/\tau})$$
$$\underbrace{\phantom{K_0 e^{-t/\tau}}}_{\text{zero input response}} \underbrace{\phantom{K_{ss}(1 - e^{-t/\tau})}}_{\text{zero state response}}$$

Example 6-12 The initial voltage on the capacitor in the circuit of Fig. 6-24 is given as $v_c = -50$ V. Determine the voltage $v_c(t)$ by finding the zero input and zero state responses.

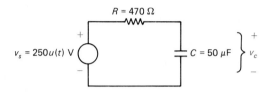

Fig. 6-24 Circuit for Example 6-12.

Solution Time constant: $\tau = 0.0235$ s.

The zero input response is given by $v_c(t) = -50 e^{-(t/0.0235)}$ V.

The zero state response is given by (since the steady state voltage on the capacitor is 250 V: $v_c(t) = 250[1 - e^{-(t/0.0235)}]$ V.

The complete response is

$$v_c(t) = -50 e^{-(t/0.0235)} + 250[1 - e^{-t(t/0.0235)}]$$

which reduces to

$$v_c(t) = 250 - 300 e^{-t/0.0235} \text{ V}$$

This is the same result obtained by using the procedure outlined in Section 6-3. ∎

Exercise 6-29 The voltage applied to a series combination of $R = 470 \, \Omega$ and $L = 15$ mH is given by $150 \, u(t)$ V. If the initial current in the inductor is given as 0.2 A, determine the current response of the circuit by finding the zero input and zero state responses.

The advantage of using the zero state and zero input responses for finding the response of a circuit is that the problem is split into two simpler problems.

6-7 STEP RESPONSE OF A SECOND-ORDER CIRCUIT

A circuit in which a resistance, capacitance, and inductance are in series (or in parallel) is described by a second-order differential equation; that is, the highest derivative is of second order. Such a circuit is called a *second-order circuit*. A capacitor stores energy in its electric field, and the energy is proportional to the square of the current; an inductor stores energy in its magnetic field, and the energy is proportional to the square of the voltage. The presence of the two different types of energy storage elements in a second-order circuit leads to some interesting and even unexpected results. For example, it is possible to have a response in the form of sinusoidal oscillations even though the circuit is driven by a constant forcing function.

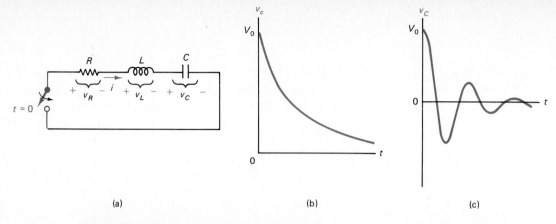

Fig. 6-25 (a) Zero input response of an *RLC* circuit. The response may be a steady decay, as in (b), or oscillatory, as in (c).

6-7-1 Zero Input Response of a Series RLC Circuit

Consider a series *RLC* circuit, as shown in Fig. 6-25(a), in which the capacitor has an initial voltage V_0 and the inductor has an initial current of zero. The switch is closed at $t = 0$. When the switch is closed, a current starts to flow in the circuit due to the initial stored energy from the capacitor. Eventually all the available energy is dissipated in the form of heat by the resistor. Therefore, in the steady state, the current in the circuit and the voltage across the capacitor must be zero. The manner in which the current in the circuit and the voltage across the capacitor voltage reach a steady state value of zero depends upon the relative values of *R*, *L*, and *C*. In some circuits, the capacitor voltage monotonically decreases to zero and the current in the circuit starts from zero, builds up to a maximum, and then decays monotonically to zero, as indicated in Fig. 6-25(b). In other circuits, the capacitor voltage and the current in the circuit undergo oscillations (of diminishing amplitudes) before becoming zero, as indicated in Fig. 6-25(c).

With the current direction chosen as shown in Fig. 6-25(a), the differential equation of the circuit is

$$L\frac{di(t)}{dt} + Ri(t) + \frac{1}{C}\int i(t)dt = 0$$

After differentiating once, the above equation becomes

$$\frac{d^2i(t)}{dt^2} + \frac{R}{L}\frac{di(t)}{dt} + \frac{1}{LC}i(t) = 0 \qquad (6\text{-}62)$$

which is a homogeneous differential equation whose solution is of the form

$$i(t) = Ke^{pt} \qquad (6\text{-}63)$$

where *K* and *p* are constants to be determined.

Substitution of Eq. (6-63) and derivatives obtained from Eq. (6-63) in (6-62) leads to the equation

$$p^2 + (R/L)p + (1/LC) = 0 \qquad (6\text{-}64)$$

which is known as the *characteristic equation* of the series *RLC* circuit. The quadratic equation (6-64) has two roots, p_1 and p_2, given by

$$p_1 = -\frac{R}{2L} + \sqrt{\frac{R^2}{4L^2} - \frac{1}{LC}} \qquad (6\text{-}65a)$$

and (6-65b)

$$p_2 = -\frac{R}{2L} - \sqrt{\frac{R^2}{4L^2} - \frac{1}{LC}}$$

Depending upon the relative values of R, L, and C, the quantity within the square root may be positive, zero, or negative. The three possible cases are:

Case 1: $(R^2/4L^2) = (1/LC)$. Then $p_1 = p_2 = -(R/2L)$.

Case 2: $(R^2/4L^2) > (1/LC)$. Then p_1 and p_2 are both real, but different from each other.

Case 3: $(R^2/4L^2) < (1/LC)$. Then p_1 and p_2 are complex conjugates of each other.

In Case (1), the solution in Eq. (6-63) has to be modified to the form

$$i(t) = (K_1 t + K_2) e^{-(R/2L)t} \qquad (6\text{-}66)$$

In Cases (2) and (3), the solution in Eq. (6-63) has to be modified to the form

$$i(t) = K_1 e^{p_1 t} + K_2 e^{p_2 t} \qquad (p_1 \neq p_2) \qquad (6\text{-}67)$$

In both Eqs. (6-66) and (6-67), K_1 and K_2 are coefficients to be determined from the initial conditions. Two initial conditions are required, since there are two undetermined coefficients. The two initial conditions available are:

$$i(0+) = i(0-) = 0 \qquad \text{(current continuity in inductor)} \qquad (6\text{-}68)$$
$$v_c(0+) = v_c(0-) = V_0 \qquad \text{(voltage continuity in capacitor)}$$

Instead of using the second initial condition (on v_c) directly, it is more convenient to find the initial value of di/dt by using that condition. At $t = 0+$, $i = 0$, and therefore $v_R(0+) = 0$ also. The KVL equation of the circuit in Fig. 6-25(a) is

$$v_R + v_L + v_c = 0$$

which leads to (at $t = 0+$):

$$v_L(0+) = -v_c(0+) = -V_0$$

and since $v_L(t) = L[di(t)/dt]$, the initial value of the derivative di/dt is

$$di/dt = (v_L/L) = -(V_0/L) \qquad \text{at} \qquad t = 0+ \qquad (6\text{-}69)$$

Eqs. (6-68) and (6-69) are the two initial conditions to be used in the determination of K_1 and K_2 in Eqs. (6-66) and (6-67).

Case 1: $(R^2/4L^2) = (1/LC)$ or $R = 2\sqrt{(L/C)}$. The zero input current response is, from Eq. (6-66):

$$i(t) = (K_1 t + K_2) e^{-(R/2L)t} \qquad (6\text{-}66)$$

Using the initial condition, Eq. (6-68), $i(0+) = 0$, in Eq. (6-66)

$$K_2 = 0$$

6-7 Step Response of a Second-Order Circuit

and Eq. (6-66) becomes

$$i(t) = K_1 t e^{-(R/2L)t} \tag{6-70}$$

In order to use the initial condition in Eq. (6-69), $di/dt = -(V_0/L)$ at $t = 0+$, differentiate the expression for $i(t)$ in Eq. (6-70):

$$di/dt = K_1 e^{-(R/2L)t} - (R/2L)K_1 t e^{-(R/2L)t}$$

Putting $t = 0$ and $di/dt = -(V_0/L)$, we find that

$$K_1 = -(V_0/L)$$

so that the solution for $i(t)$ becomes

$$i(t) = -(V_0/L)t e^{-(R/2L)t} \tag{6-71}$$

The negative sign in the expression for $i(t)$ arises as follows: The capacitor has an initial voltage V_0 with reference polarities, as indicated in Fig. 6-25(a). When the switch closes, the capacitor discharges through R and L and its voltage v_c decreases. Therefore, dv_c/dt is negative, and the current $i(t) = C(dv_c/dt)$ is negative also.

The voltage across the capacitor is obtained by using

$$v_c(t) = (1/C) \int_0^t i(u)\,du + V_0 \tag{6-72}$$

From Eqs. (6-71) and (6-72), we find that

$$v_c(t) = V_0[1 + (R/2L)t] e^{-(R/2L)t} \tag{6-73}$$

The graphs of $v_c(t)$ and the *magnitude* of the current $i(t)$ are shown in Fig. 6-26. The magnitude of the current at first increases, reaches a maximum, and then decays to zero. The capacitor voltage decays monotonically to zero.

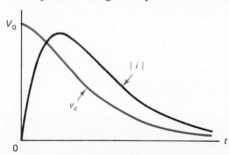

Fig. 6-26 Zero input response curves in a critically damped *RLC* circuit.

The zero input response of the circuit for the case $R = 2\sqrt{(L/C)}$ can be interpreted as follows. At $t = 0+$, the current in the circuit is zero due to the current continuity condition in the inductor, and the voltage across the capacitor is V_0 due to the voltage continuity condition in the capacitor. Since $i(0+) = 0$, the voltage across the resistor is also zero at $t = 0+$ and, from KVL, the voltage across the inductor $v_L(0+)$ has a magnitude of V_0. This voltage across the inductor causes a current to flow. As the current in the inductor builds up, energy is stored in its magnetic field. The current flows through the resistor as well, and some energy is dissipated in the form of heat. Thus part of the initial stored energy from the capacitor is transferred to the magnetic field of the inductor, and part of it is dissipated by the resistor. The presence of a voltage across the resistor due to current flow causes a reduction in the voltage, and hence di/dt across the inductor.

This slows down the rate of current buildup in the inductor, as seen from the gradual flattening of the $i(t)$ curve in Fig. 6-26. Eventually, the magnitude of the current $i(t)$ reaches a maximum and begins to decrease. The dissipation due to the resistor is happening at such a rate that the stored energy in the inductor now decreases steadily toward zero. This explains the decay of the current to a final value of zero.

An *RLC* circuit in which $R = 2\sqrt{(L/C)}$ is said to be *critically damped*. It is *damped* since there are no oscillations of the current or capacitor voltage; it is *critically* damped since the condition $R = 2\sqrt{(L/C)}$ forms the boundary between a nonoscillatory and an oscillatory response (as will be seen shortly).

Example 6-13 A series *RLC* circuit with $L = 10$ mH and $C = 400$ µF has an initial capacitor voltage of 180 V. Calculate the value of R needed for critical damping and obtain the zero input responses: $i(t)$ and $v_c(t)$.

Solution For critical damping, $R = 2\sqrt{(L/C)} = 10\ \Omega$. The root of the characteristic equation is therefore

$$p = -(R/2L) = -500$$

From Eqs. (6-71) and (6-73), the responses are given by

$$i(t) = -(18 \times 10^3)te^{-500t}\ \text{A}$$
$$v_c(t) = 180(1 + 500t)e^{-500t}\ \text{V}$$ ∎

Exercise 6-30 Determine the instant of time at which $i(t)$ reaches a maximum in the previous example and calculate the maximum value of the current.

Exercise 6-31 Given a series *RLC* circuit with $L = 20$ mH and $C = 50$ µF. Calculate the value of R needed to make the circuit critically damped. If the initial capacitor voltage is 80 V, obtain the zero input responses.

Case 2: $(R^2/4L^2) > (1/LC)$ or $R > 2\sqrt{(L/C)}$.

The zero input current response is given by Eq. (6-67), repeated below:

$$i(t) = K_1 e^{p_1 t} + K_2 e^{p_2 t} \tag{6-74}$$

where p_1 and p_2 are given by Eqs. (6-65), repeated below:

$$p_1 = -\frac{R}{2L} + \sqrt{\frac{R^2}{4L^2} - \frac{1}{LC}} \tag{6-75a}$$

$$p_2 = -\frac{R}{2L} - \sqrt{\frac{R^2}{4L^2} - \frac{1}{LC}} \tag{6-75b}$$

Use of the initial condition $i(0+) = 0$ in Eq. (6-74) leads to

$$K_1 + K_2 = 0 \tag{6-76}$$

The other initial condition is $di/dt = -(V_0/L)$ from Eq. (6-69). Differentiating Eq. (6-74), and putting $t = 0$ and $di/dt = -(V_0/L)$ leads to

$$p_1 K_1 + p_2 K_2 = -(V_0/L) \tag{6-77}$$

6-7 Step Response of a Second-Order Circuit

The solution of Eqs. (6-76) and (6-77) gives

$$K_1 = -K_2 = -\frac{V_0}{L(p_1 - p_2)} \quad (6\text{-}78)$$

The zero input current response is therefore

$$i(t) = \frac{V_0}{L(p_1 - p_2)}[e^{p_2 t} - e^{p_1 t}] \quad (6\text{-}79)$$

where p_1 and p_2 are given by Eqs. (6-75).

The voltage across the capacitor is given by

$$v_c(t) = (1/C)\int_0^t i(u)du + V_0$$

$$= \frac{V_0}{p_1 - p_2}(p_1 e^{p_2 t} - p_2 e^{p_1 t}) \quad (6\text{-}80)$$

The zero input responses are shown graphically in Fig. 6-27. The behavior of the circuit is seen to be quite similar to the critically damped case, and the explanation of the behavior follows the same general argument presented earlier. The damping of the responses due to the dissipation of energy in the resistor is faster here than in the critically damped case.

A series *RLC* circuit in which $R > 2\sqrt{L/C}$ is said to be *overdamped*, since it offers more damping than the critically damped circuit.

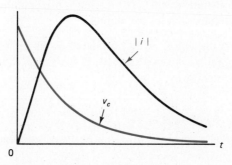

Fig. 6-27 Zero input response curves in an overdamped *RLC* circuit.

Example 6-14 A series *RLC* circuit with $R = 50\ \Omega$, $L = 5$ mH, and $C = 12.5\ \mu$F has an initial capacitor voltage of 10 V. Obtain the zero input responses.

Solution First check the damping in the circuit. For *critical* damping

$$R = 2\sqrt{(L/C)} = 40\ \Omega$$

and the given value of R is higher than 40 Ω. Therefore, the circuit is overdamped. The values of p_1 and p_2 are, from Eqs. (6-75), $p_1 = -2000$ and $p_2 = -8000$. The zero input responses are, from Eqs. (6-79) and (6-80)

$$i(t) = 0.333(e^{-8000t} - e^{-2000t})\ \text{A}$$

and

$$v_c(t) = 13.3e^{-8000t} - 3.33e^{-2000t}\ \text{V}\quad\blacksquare$$

Exercise 6-32 Repeat the work of the example above for the circuit with $R = 100\,\Omega$, $L = 20$ mH, $C = 50\,\mu\text{F}$. $v_c(0) = 10$ V.

Exercise 6-33 Determine the instant of time at which the current $i(t)$ of Example 6-14 reaches a maximum, and calculate the maximum value of the current.

Case 3: $(R^2/4L^2) < (1/LC)$, or $R < 2\sqrt{L/C}$. In this case, the quantities under the radical sign in the expressions for p_1 and p_2 in Eqs. (6-65) are *negative*. Therefore, p_1 and p_2 are *complex*. It is convenient to rewrite the expressions for p_1 and p_2 by using $j^2 = -1$.

$$p_1 = -\frac{R}{2L} + j\sqrt{\frac{1}{LC} - \frac{R^2}{4L^2}} \tag{6-81a}$$

and

$$p_2 = -\frac{R}{2L} - j\sqrt{\frac{1}{LC} - \frac{R^2}{4L^2}} \tag{6-81b}$$

If we let

$$\alpha = (R/2L) \tag{6-82}$$

and

$$\beta = \sqrt{(1/LC) - (R^2/4L^2)} \tag{6-83}$$

then

$$p_1 = -\alpha + j\beta \tag{6-84}$$

and

$$p_2 = -\alpha - j\beta \tag{6-85}$$

Then the zero input current response is, from Eq. (6-67),

$$i(t) = K_1 e^{-(\alpha - j\beta)t} + K_2 e^{-(\alpha + j\beta)t} \tag{6-86}$$

Use of the initial condition $i(0+) = 0$ in Eq. (6-85) gives

$$K_1 + K_2 = 0 \tag{6-87}$$

The other initial condition is $di/dt = -(V_0/L)$ from Eq. (6-69). Differentiating Eq. (6-86), putting $t = 0$ and $di/dt = -(V_0/L)$ in the derivative, leads to:

$$-(\alpha - j\beta)K_1 - (\alpha + j\beta)K_2 = -(V_0/L) \tag{6-88}$$

Solving Eqs. (6-87) and (6-88)

$$K_1 = -K_2 = -(V_0/2j\beta L) \tag{6-89}$$

The zero input current response of the circuit is therefore

$$i(t) = -(V_0/2j\beta L)\,[e^{-(\alpha - j\beta)t} - e^{-(\alpha + j\beta)t}]$$

$$= -\frac{V_0}{\beta L}e^{-\alpha t}\frac{(e^{j\beta t} - e^{-j\beta t})}{2j} \tag{6-90}$$

6-7 Step Response of a Second-Order Circuit

By using Euler's identity:

$$e^{j\theta} - e^{-j\theta} = 2j \sin \theta$$

Eq. (6-90) becomes

$$i(t) = -(V_0/\beta L)e^{-\alpha t} \sin \beta t \qquad (6\text{-}91)$$

where α and β are given by Eqs. (6-82) and (6-83).

The voltage across the capacitor is obtained from the relationship

$$v_c(t) = (1/C) \int_0^t i(u)du + V_0$$

Using a table of integrals, we find that (after a fair amount of manipulation)

$$v_c(t) = (V_0/\beta\sqrt{LC})e^{-\alpha t} \cos[\beta t - \arctan(\alpha/\beta)] \qquad (6\text{-}92)$$

The graphs of $v_c(t)$ and the magnitude of $i(t)$ are shown in Fig. 6-28. Both the current in the inductor and the voltage across the capacitor *oscillate*, instead of monotonically reaching a steady state value of zero. The amplitudes of the oscillations decrease (at the rate of $e^{-\alpha t}$), and eventually both v_c and i reach a value of zero.

A circuit in which $R < 2\sqrt{(L/C)}$ is said to be *underdamped* or *oscillatory*.

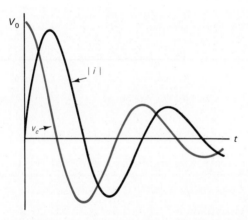

Fig. 6-28 Zero input response curves in an underdamped *RLC* circuit.

Example 6-15 A series *RLC* circuit with $R = 2\ \Omega$, $L = 10$ mH, and $C = 400\ \mu$F has an initial capacitor voltage of 200 V. Find the zero input responses.

Solution First, check the damping in the circuit. The value of R for critical damping is

$$R_{crit} = 2\sqrt{(L/C)} = 10\ \Omega$$

Since $R = 2\ \Omega$ in the given circuit, the circuit is underdamped.

The values of α and β are given by Eqs. (6-81) and (6-82) and found to be

$$\alpha = 100 \quad \text{and} \quad \beta = 490$$

Therefore, from Eq. (6-92)

$$i(t) = -40.8e^{-100t} \sin 490t \text{ A}$$

and from Eq. (6-93)

$$v_c(t) = 204e^{-100t} \cos(490t - 11.5°) \text{ V}$$

■

Exercise 6-34 Repeat Example 6-15 when $R = 0.5\ \Omega$ and the other elements and initial capacitor voltage are the same as before.

Exercise 6-35 An inductance of 16 mH and a capacitance of 400 pF are in series with a resistance R. Find the maximum value of R that will permit an oscillatory response.

Exercise 6-36 Consider a series LC circuit. Modify the results obtained in the discussion above to fit this circuit. Sketch $i(t)$ and $v_c(t)$. Note that (a) the oscillations remain constant in amplitude; (b) $i(t)$ is a maximum when $v_c(t)$ is a minimum; and (c) $v_c(t)$ is a maximum when $i(t)$ is a minimum.

The oscillatory behavior can be explained by noting that the resistance in the circuit is smaller than the value for critical damping. The rate at which the resistor is dissipating available energy is slower than for a critically damped circuit. The result is that there is an excess of energy available to permit a *continuous* exchange of energy between the magnetic field of the inductor and the electric field of the capacitor. As the energy in the inductor decreases (due to a decreasing current), the energy in the capacitor increases (causing the capacitor voltage to increase). Similarly, as the energy in the capacitor decreases (due to a decreasing voltage), the energy in the inductor increases (causing an increase in the current). This continuous exchange is responsible for the oscillations. The amplitudes of the oscillations diminish because the resistor is dissipating a certain amount of energy in each cycle, making a smaller amount of energy available for exchange between the fields of the inductor and capacitor in successive cycles. Eventually all the available energy is dissipated by the resistor, the oscillations die down, and the current and the capacitor voltage become zero.

The persistence of the oscillations depends upon $e^{-\alpha t}$: the larger the value of α, the faster is the decay of the oscillations.

If $\alpha = 0$; that is, when the circuit has zero resistance, the oscillations will be at a constant amplitude and last indefinitely. As the resistance increases (for a given L and C), the attenuation increases and the oscillations die down at a faster rate. When the resistance reaches the value given by $2\sqrt{(L/C)}$, the resistor dissipates available energy at such a rate that oscillations cannot get started. This is, of course, the case of critical damping.

6-7-2 Zero State Response of an RLC Circuit (Step Input)

A qualitative discussion of the response of a series RLC circuit (with no initial energy stored) to a step input voltage will be presented here. The detailed mathematical steps are quite similar to those in the discussion of the zero input response.

Consider the circuit shown in Fig. 6-29(a). The initial state of the circuit is zero: that is, $v_c(0) = 0$ and $i_L(0) = 0$. At $t = 0$, when the voltage V_B becomes active, the current in the circuit and the voltage across the capacitor both increase. In the steady state, the voltage across the capacitor becomes a constant and equal to V_B, and the current in the

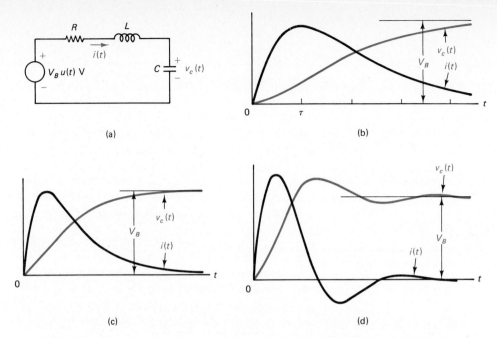

Fig. 6-29 (a) Step input to an *RLC* circuit. Zero state response curves: (b) Critical damping. (c) Overdamping. (d) Underdamping.

circuit becomes zero. The manner in which the voltage $v_c(t)$ and the current $i(t)$ approach the steady state depends upon the relative values of the elements R, L, and C.

When $R = 2\sqrt{(L/C)}$, the circuit is *critically damped*. The current increases from zero to a maximum value and then decays monotonically toward zero. The capacitor voltage increases from zero monotonically toward the final value of V_B. The response of the critically damped circuit is as shown in Fig. 6-29(b).

When $R > 2\sqrt{(L/C)}$, the circuit is *overdamped*. The response is similar to that of a critically damped circuit, except that the rate at which the capacitor voltage builds up is slower in this case. Figure 6-29(c) shows typical curves of $i(t)$ and $v_c(t)$ for an overdamped circuit.

When $R < 2\sqrt{(L/C)}$, the circuit is *underdamped* and exhibits an *oscillatory* response. As Fig. 6-29(d) shows, the current increases at first and then oscillates about the value of zero before settling down to the steady state value of zero. The capacitor voltage increases to V_B and shoots past that value. Then it oscillates about the value V_B before settling down to the steady state value of V_B. The amplitudes of the oscillations of $i(t)$ diminish between successive cycles, and the rate at which they decay depends upon $e^{-\alpha t}$, where $\alpha = (R/2L)$.

The zero state behavior of the *RLC* circuit is thus seen to bear a strong resemblance to the zero input response.

A question of practical interest is whether it is desirable to have an oscillatory or a nonoscillatory response in a second-order circuit. Oscillatory response has the disadvantage that it takes a certain length of time for the circuit to reach the constant values of the steady state. If the resistance is much smaller than that required for critical damping, the oscillations persist for a large number of cycles, and this may be undesirable in a given circuit.

Another disadvantage of the oscillatory response is the presence of the overshoot: The capacitor voltage becomes higher than V_B by an amount determined by the level of damping in the circuit. On the other hand, the overshoot of the oscillatory response has an advantage. The rate at which the capacitor voltage first increases from zero toward V_B is faster than in a critically damped or overdamped circuit. *That is, the initial speed of response of the circuit is faster when the circuit is underdamped, leading to a smaller rise time.* Some circuits are made faster by means of a small amount of underdamping. In such cases, the value of α is chosen so as to make the oscillations die down within one cycle.

6-8 INTEGRATING OP AMP CIRCUIT

The voltage-current relationship of a capacitor makes it possible to design an op amp circuit to perform *mathematical integration* of functions of time. Consider the circuit shown in Fig. 6-30(a) in which a capacitor is used as the feedback link between the input and the output of the op amp.

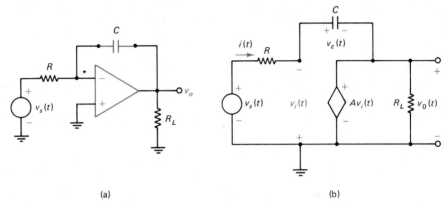

Fig. 6-30 (a) Integrating op amp circuit and (b) its equivalent circuit.

The equivalent circuit of the integrator is shown in Fig. 6-30(b), where the op amp has been replaced by the ideal model.

The equations are:

$$Ri(t) = v_s(t) + v_i(t) \quad \text{(using KVL)} \tag{6-93}$$

$$i(t) = C[dv_c(t)/dt] \tag{6-94}$$

$$v_c(t) = -v_i(t) - v_0(t) \tag{6-95}$$

$$v_0(t) = Av_i(t) \tag{6-96}$$

If we combine the four equations above, the following equation is obtained:

$$-\frac{RC(A+1)}{A}\frac{dv_0(t)}{dt} - \frac{1}{A}v_0(t) = v_s(t) \tag{6-97}$$

In practical op amps, the gain A is very much larger than 1, so that $(A + 1)/A \approx 1$ and $(1/A) \approx 0$. With these approximations, Eq. (6-97) becomes

$$-RC\frac{dv_0(t)}{dt} = v_s(t)$$

or
$$v_0(t) = -(1/RC) \int v_s(t)\, dt \quad (6\text{-}98)$$

The output of the circuit in Fig. 6-30(a) is therefore proportional to the integral of the input and the circuit performs integration.

From the discussion above and that in Chapter 3, we can see that op amp circuits perform the following operations on a signal: (a) multiplication by a scale factor using the inverting op amp circuit; (b) summing of two or more signals using the summing op amp circuit; and (c) mathematical integration of a signal using the integrating op amp circuit. It is therefore possible to set up a network of op amps (with capacitors and resistors) to solve a system of simultaneous differential equations. Such a network is called an *analog computer*. Therefore, the response of a linear system may be studied in *real time* by means of analog computers.

6-9 SUMMARY OF CHAPTER

The current through a capacitor is proportional to the rate of change of the voltage across it. When the voltage across the capacitor is a constant, the current through it is zero. Energy is stored in the electric field of a capacitor and is given by $(1/2)Cv^2$. Since energy in a finite physical system must be continuous, the energy stored in a capacitor is a continuous function of time, and consequently the voltage across a capacitor must be a continuous function of time unless an infinite amount of power is available.

An inductor is the dual of a capacitor. The voltage across it is proportional to the rate of change of the current through it. When the current in an inductor is a constant, the voltage across it is zero. Energy is stored in the magnetic field of an inductor and is given by $(1/2)Li^2$. Energy in an inductor is a continuous function of time, and consequently the current in an inductor is a continuous function of time unless an infinite amount of power is available.

The energy storage property and the voltage continuity condition of a capacitor dictate the manner in which the voltage across a capacitor varies. When a series *RC* circuit is excited by a step input voltage, the capacitor voltage is found to have two components: a transient component (or natural response term), which exponentially decays toward zero, and a steady state component that remains constant at all times. The rate of decay of the transient component depends upon the time constant *RC*. The larger the time constant, the slower the decay of the transient component. The time taken for the capacitor voltage to increase from 10% of its final value to 90% of its final value is called the rise time. The rise time equals 2.2 time constants.

The analysis of a simple *RC* series circuit can be done by examining the initial value and the steady state value of the capacitor voltage and calculating the time constant *RC*. The initial condition for the capacitor voltage is obtained from the voltage continuity condition. The steady state value of the capacitor voltage (with a step input) is obtained by noting that the capacitor voltage is constant, and the current through it is zero in the steady state. These principles are applied to a more complicated circuit (containing a single capacitor) by first reducing it to a series *RC* configuration through Thevenin's theorem.

When a rectangular pulse of duration t_p is applied to a series *RC* circuit, the output pulse measured across the capacitor is a rounded-off version of the input pulse when the pulse duration is very large compared with the time constant of the circuit. When the pulse duration is short compared with the time constant, it becomes difficult to distinguish the presence of a pulse in the output.

The behavior of the current in an *RL* circuit is quite similar to that of the voltage in an *RC* circuit. The time constant is given by (*L/R*). The initial condition for the current in the inductor is evaluated by using the current continuity condition. The steady state value of the inductor current (for a step input) is obtained by noting that the current in an inductor is a constant and the voltage across it is zero in the steady state. Thevenin's theorem is used to reduce complicated circuits containing a single inductor to a series RL configuration.

The terms zero input response and zero state response are used in the analysis of systems containing capacitors and inductors. For a first-order circuit, the zero input response (capacitor voltage or inductor current) is an exponential term whose exponent is $-(t/\tau)$, where τ is the time constant and the coefficient of the exponential term is the initial value of the capacitor voltage or inductor current. The zero state response (for a step input) consists of a constant term equal to the steady state value of the capacitor voltage or inductor current and an exponential term with an exponent $-(t/\tau)$ and coefficient equal to the steady state value of the capacitor voltage or inductor current.

The response of a second-order *RLC* circuit may be oscillatory or nonoscillatory, depending upon the relative values of *R*, *L*, and *C*. When $R < 2\sqrt{(L/C)}$, the circuit is underdamped and has an oscillatory response; that is, the response oscillates about the steady state value before settling down to the steady state. The amplitude of the oscillations decreases from one cycle to the next at a rate determined by $e^{-(R/2L)t}$. In such circuits, the rate at which the resistor dissipates the energy is small enough to permit a continuous exchange of energy between the capacitor and inductor, causing the oscillations. When $R > 2\sqrt{(L/C)}$, the circuit is overdamped and the response is nonoscillatory. In such circuits, the rate at which the energy is being dissipated by the resistor is so large that there is no exchange of energy between the inductor and the capacitor, but only a gradual decay of stored energy.

The use of a capacitor in the feedback path from the output to the inverting input of an op amp leads to an integrating circuit. Such a circuit is useful for mathematically integrating a given signal.

Answers to Exercises

6-1 $i(t) = CA$, a constant.

6-2 $i(t) = 3 \times 10^{-5}$ A $(0 < t < 0.01$ s$)$; 0 $(0.01 < t < 0.02$ s$)$; -3×10^{-5} A $(0.02 < t < 0.04)$.

6-3 $v_c(t) = (4 \times 10^7) \sin 1000t + 50$ V.

6-4 $w(t) = (75 \times 10^{-12})(10t - 5)^2$ J. 1.88×10^{-9} J at $t = 0$; 1.58×10^{-6} J at $t = 15$ s; 1.875×10^{-9} at $t = 1$ s. Energy change from $t = 0$ to $t = 1$ s is zero.

6-5 $v(t) = 2.5$ V $(0 < t < 0.03)$; -0.938 V $(0.03 < t < 0.11$ s$)$; -2.5 V $(0.11 < t < 0.14$ s$)$; 0.938 V $(0.14 < t < 0.22)$.

6-6 $i(t) = (V_0/Lp)(1 - e^{-pt}) = (d/dt)$.

6-7 $p(t) = i(L di/dt) = (d/dt)(Li^2/2)$.

6-8 $w(t) = 5e^{-4t}$ J. 5 J at $t = 0$, 0.677 J at $t = 0.5$ s, 0.0916 J at $t = 1$ s. Loses 4.91 J from $t = 0$ to $t = 1$ s.

6-9 $i(t) = -24e^{-2t} + 20$ A.

6-10 $v_c(t) = 120 - 120e^{-426t}$ V; $i(t) = 2.55 \times 10^{-4} e^{-426t}$ A.

6-11 $v_c(t) = 120 - 70e^{-426t}$ V; $i(t) = 1.49 \times 10^{-4} e^{-426t}$ A.

6-12 $v_c(t) = 25(1 - e^{-t/\tau})$. $(t_1/\tau) = 0.693$, where $t_1 = 100$ ms. $\tau = 144$ ms; $R = 4\,\Omega$; $C = 36 \times 10^{-3}$ F.

6-13 $t_1 = 0.1054\,\tau$; $t_2 = 2.303\,\tau$; $t_R = 2.198\,\tau$ or $2.2\,\tau$.

6-14 1.1×10^{-5} s.

6-15 $\tau = 5$ ms; $t_R = 11$ ms.

6-16 $i_c(t) = 0\,(t<0)$; $4.45 \times 10^{-2} e^{-1.78t}$ A $(t > 0)$. $v_1(t) = 112.5$ V $(t < 0)$; $(162.5 - 16.6e^{-1.78t})$ V $(t > 0)$. i_b (downward through R_b) $= 0.075$ A $(t < 0)$; $-(0.025 + 0.0111 e^{-1.78t})$ A $(t > 0)$. i_a (downward through R_a) $= -0.075$ A $(t < 0)$; $(0.025 - 0.0332 e^{-1.78t})$ A $(t > 0)$.

6-17 $v_c = V_B\,(t < 0)$; $V_B e^{-t/RC}\,(t > 0)$. $i(t) = 0\,(t < 0)$; $-(V_B/R)e^{-t/RC}\,(t > 0)$.

6-18 $v_R = V_m e^{-t/\tau}\,(0 < t < t_p)$; $-V_m (e^{t_p/\tau} - 1) e^{-t/\tau}\,(t > t_p)$.

6-19 $i_c(t) = 0.02 e^{-1.13t}$ A $(0 < t < 0.5$ s$)$; $0.00931 e^{-0.833t}$ A $(t > 0.5$ s$)$. $i_3(t) = (v_c/R_3) = 0.00588(1 - e^{-1.13t})$ A $(0 < t < 0.5$ s$)$; $(0.005 - 0.00372 e^{-0.833t})$ A $(t > 0.5$ s$)$. $i_2(t) = 0$ $(t < 0.5$ s$)$; $(0.005 + 0.00559 e^{-0.833t})$ $(t > 0.5$ s$)$. $i_1(t) = (0.00588 + 0.0141 e^{-1.13t})$ $(t < 0.5$ s$)$; $(0.005 + 0.00559 e^{-0.833t})$ A $(t > 0.5)$.

6-20 $v_c(t) = 60(1 - e^{-0.833t})$ V $(t < 0.5$ s$)$; $(70.6 - 88.4 e^{-1.13t})$ $(t > 0.5$ s$)$. $i_c(t) = 0.0125 e^{-0.833t}$ A $(t < 0.5$ s$)$; $0.025 e^{-1.13t}$ A $(t > 0.5$ s$)$. $i_3 = 0.005(1 - e^{0.0833t})$ A $(t < 0.5$ s$)$; $0\,(t > 0.5$ s$)$. $(0.00588 - 0.00737 e^{-1.13t})$ A $(t > 0.5$ s$)$. $i_2(t) = (0.005 + 0.0075 e^{-0.833t})$ A $(t < 0.5$ s$)$; $0\,(t > 0.5$ s$)$. $i_1(t) = (0.005 + 0.0075 e^{-0.833t})$ A $(t < 0.5$ s$)$; $(0.00588 + 0.0176 e^{-1.13t})$ A $(t > 0.5$ s$)$.

6-21 $t_1 = 1.39\,\tau$; 69.3 ms for $P = 1$; 6.93 ms for $P = 10$.

6-22 $v_c(t) = (1000 - 500 e^{-10^7 t})$ V; $i_c(t) = 25 e^{-10^7 t}$ A.

6-23 (a) $t < 0$ Thevenin equivalent: -145 V in series with $3.87\,\Omega$. $v_c(0) = 145$ V. $t > 0$: Thevenin equivalent: -87.1 V in series with $3.87\,\Omega$. $v_c(t) = -(87.1 + 57.9 e^{-1034t})$ V. (b) $t < 20\,\mu$s: Thevenin equivalent: 10 V in series with $6\,\Omega$. $v_c(t) = 10(1 - e^{-1.67 \times 10^4 t})$ V. v_c at $t = 20\,\mu$s is 2.83 V. $t > 20\,\mu$s: Thevenin equivalent: 30 V in series with $2\,\Omega$. $v_c = (30 - 73.8 e^{-5 \times 10^4 t})$ V.

6-24 $t < 0$: Thevenin equivalent: 6.67 V in series with $8\,\Omega$. $i_L(0-) = 0.833$ A. $t > 0$: Thevenin equivalent: 13.3 V in series with $8\,\Omega$. $i_L(t) = (1.66 - 0.827 e^{-800t})$ A.

6-25 $(V_0/R) e^{-(R/L)t}$.

6-26 17.3 ms.

6-27 $t > 0$: Thevenin equivalent: 16 V in series with $9.2\,\Omega$. $i_L(t) = 1.74 e^{-92t}$ A. $v_L(t) = 16 e^{-92t}$ V.

6-28 $0 < t < t_p$: $i_L(t) = (V_m/R)(1 - e^{-t/\tau})$. At $t = t_L$, $i_L = (V_m/R)(1 - e^{-t_p/\tau})$. $t > t_p$: $i_L(t) = (V_m/R)(1 - e^{-t_p/\tau}) e^{-(t-t_p)/\tau}$.

6-29 Zero input response: $i_L(t) = 0.2 e^{-3.13 \times 10^4 t}$ A.
Zero state response: $i_L(t) = 0.319(1 - 1e)^{-3.13 \times 10^4 t}$ A.
Total response: $i_L(t) = (0.319 - 0.119 e^{-3.13 \times 10^4 t})$ A.

6-30 $i_{max} = 13.2$ mA occurs at $t = 2$ ms.

6-31 $R_{crit} = 40\,\Omega$. $i(t) = -4000 t e^{-2000t}$ A. $v_c(t) = 80(1 + 1000t) e^{-1000t}$ V.

6-32 $R > R_{crit} = 40\,\Omega$. $p_1 = -209$; $p_2 = -4791$. $i(t) = 0.109(e^{-4791t} - e^{-209t})$ A. $v_c(t) = (-0.456 e^{-4791t} + 10.46 e^{-209t})$ V.

6-33 Maximum occurs at t $= (p_2 - p_1)/\ell n(p_1/p_2) = 2.31 \times 10^{-4}$ s. $|i_{max}| = 0.157$ A.

6-34 $\alpha = 25$; $\beta = 499.4$. $i(t) = -40 e^{-25t} \sin 499.4t$ A.
$v_c(t) = 200 e^{-25t} \cos (499.4t - 2.87°)$ V.

6-35 $R < 12.6 \text{ k}\Omega$.

6-36 $\alpha = 0$. $p_1 = j/\sqrt{LC}$. $p_2 = -j/\sqrt{LC}$. $i(t) = -V_0\sqrt{(C/L)} \sin t/\sqrt{LC}$. $v_c(t) = V_0 \cos t/\sqrt{LC}$.

PROBLEMS

Sec. 6-1 Properties and Relationships of a Capacitor

6-1 The waveform of the voltage across a capacitor is shown in Fig. 6-31. (a) Determine and sketch the current in the capacitor if $C = 150$ pF. (b) Determine and sketch the energy stored in the capacitor. (c) Determine and sketch the power associated with the capacitor.

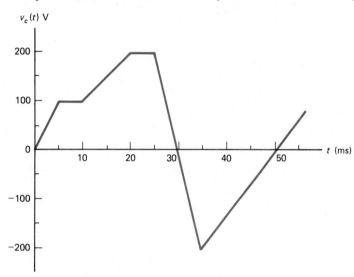

Fig. 6-31 Waveform for Problem 6-1.

6-2 The voltage across a capacitor is given by

$$v_c(t) = V_m (1 + m \cos At) \sin Bt$$

where V_m, m, A, and B are constants. Determine the current in the capacitor.

6-3 The current through a capacitor ($C = 1600$ μF) is as shown in Fig. 6-32. Assuming the voltage across the capacitor to be zero at $t = 0$, determine $v_c(t)$ and sketch it. Repeat if the voltage $v_c = -15$ V at $t = 0$.

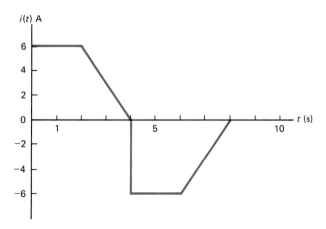

Fig. 6-32 Waveform for Problem 6-3.

Problems 311

6-4 The current through a capacitor is given by $i(t) = te^{-bt}$ ($t > 0$). (a) Assuming the initial voltage across the capacitor is zero, find $v_c(t)$. Sketch $i(t)$ and $v_c(t)$. (b) Find and sketch the energy and power associated with the capacitor.

6-5 The energy stored in a capacitor C is as shown in Fig. 6-33. Determine and sketch the voltage and current in the capacitor.

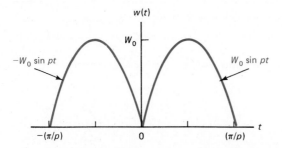

Fig. 6-33 Waveform for Problem 6-5.

Sec. 6-2 Properties and Relationships of an Inductor

6-6 The waveform of the current in an inductance is shown in Fig. 6-34. (a) Determine and sketch the voltage across the inductance if $L = 0.05$ H. (b) Determine and sketch the energy stored in the inductance. (c) Determine and sketch the power associated with the inductance.

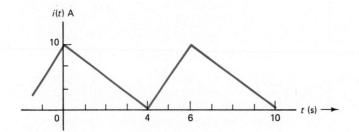

Fig. 6-34 Waveform for Problem 6-6.

6-7 The current in an inductance L is given by $i(t) = I_0 e^{-at} \cos bt$ A where I_0, a, and b are constants. Obtain an expression for the voltage $v_L(t)$ across the inductance. Sketch $i(t)$ and $v_L(t)$ in the interval $t = 0$ to $t = 50$ ms for the following numerical values: $I_0 = 20$ A; $a = 8$; $b = 377$, and $L = 5$ mH.

6-8 The voltage across an inductance ($L = 15$ H) is as shown in Fig. 6-35. Assuming that the current in the inductance is zero at $t = 0$, determine and sketch $i(t)$. Repeat if the initial current is 3 A.

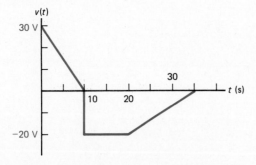

Fig. 6-35 Waveform for Problem 6-8.

6-9 The voltage across an inductance is given by $v(t) = e^{-|t|}$. That is, $v(t) = e^t$ when $t < 0$ and e^{-t} when $t > 0$. (a) Determine and sketch $i(t)$. (b) Determine the energy and power associated with the inductance.

Sec. 6-3 Step Response of RC Circuits

6-10 Find the solution of the differential equation $dx(t)/dt + 10x(t) = 250$ if it is given that $x(0) = 100$.

6-11 Solve the differential equation of the previous problem if it is given that $dx/dt = -10$ at $t = 0$.

6-12 (a) Solve the differential equation $di(t)/dt + Ai(t) = 0$ where A is a constant, given that $i(0) = I_0$. (b) Solve the differential equation $di(t)/dt + Ai(t) = B$ where A and B are constants, given that $I(0) = 0$. (c) Show that the *sum* of the two expressions obtained in parts (a) and (b) is a solution to the differential equation given in part (b) when the initial condition is $i(0) = I_0$.

6-13 Determine the responses $v_c(t)$ and $i(t)$ of a series RC circuit ($R = 100\ \Omega$; $C = 250\ \mu F$) driven by $v_s(t) = 120u(t)$ V if $v_c(0) = -120$ V.

6-14 A series RC circuit is driven by $v_s(t) = 25u(t)$ V. Given that $i(0) = 2$ mA, $v_c(0) = 4$ V, and $v_c = 10$ V at $t = 80$ ms, determine the values of R and C.

6-15 A step voltage $v_s(t) = 150u(t)$ V is applied to a series RC circuit with $C = 500\ \mu F$. Given that the energy stored in the capacitor is 20 mJ at $t = 0$ and 60 mJ at $t = 16$ s, determine the time constant of the circuit.

6-16 Determine the currents $i_1(t)$ and $i_2(t)$ in the circuit of Fig. 6-36.

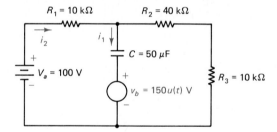

Fig. 6-36 Circuit for Problem 6-16.

6-17 Assume that the circuit in Fig. 6-37 is in the steady state before $t = 0$. The switch is closed at $t = 0$. (a) Obtain the expression for $v_c(t)$ for $t > 0$. (b) Obtain the expression for the current supplied by the V_b source.

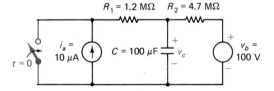

Fig. 6-37 Circuit for Problem 6-17.

6-18 Determine the voltage $v_c(t)$ and the current $i_1(t)$ in the circuit in Fig. 6-38.

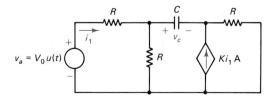

Fig. 6-38 Circuit for Problem 6-18.

6-19 Determine the voltage $v_c(t)$ in the circuit in Fig. 6-39.

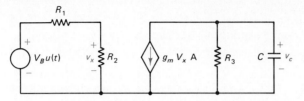

Fig. 6-39 Circuit for Problem 6-19.

6-20 The pulse train shown in Fig. 6-40 is applied to a series RC circuit. Draw neat and scaled sketches of $v_c(t)$ and $v_R(t)$ for each of the following cases: (a) $RC = 10$ ms; (b) $RC = 100$ ms. Discuss the results.

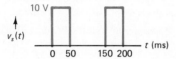

Fig. 6-40 Pulse train for Problem 6-20.

6-21 In the circuit of Fig. 6-41, the switch is off until $t = 0$, moved to position 1 at $t = 0$, and then moved to position 2 at $t = 10$ ms. Determine and sketch the voltage across the capacitor.

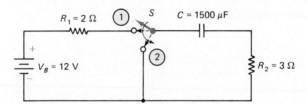

Fig. 6-41 Circuit for Problem 6-21.

6-22 The initial voltages on the capacitors in Fig. 6-42 are $v_{c1} = 80$ V and $v_{c2} = 120$ V. Obtain the expressions for v_{c1}, v_{c2}, and $i(t)$ for $t > 0$.

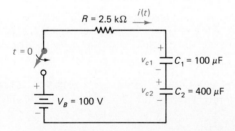

Fig. 6-42 Capacitors for Problem 6-22.

6-23 Consider a parallel circuit consisting of a current source $i_s(t) = I_0 u(t)$A, a conductance G, and a capacitance C. Assume that the initial voltage on the capacitor is zero. Set up the differential equation for the voltage $v(t)$ across the parallel combination by using KCL. Solve the differential equation using the steps outlined in Subsection 6-3-1.

Sec. 6-5 Step Response of RL Circuits

6-24 A series RL circuit is driven by a voltage source $v_s(t) = 100u(t)$ V. If it is given that the maximum current in the circuit should be 50 mA and that the current should reach 20 mA at $t = 10$ ms, find the values of R and L.

6-25 A series RL circuit (with $L = 10$ mH) is driven by $v_s(t) = 250u(t)$ V. The initial energy stored in the inductance is 2 J. The current is 25 A at $t = 2\tau$. Obtain an expression for $i(t)$.

6-26 Determine $i_L(t)$ in the circuit of Fig. 6-43.

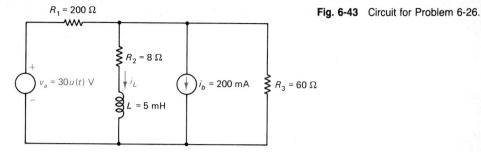

Fig. 6-43 Circuit for Problem 6-26.

6-27 Assume that the circuit of Fig. 6-44 is in the steady state of $t = 0-$. At $t = 0$, the switch is closed. (a) Obtain an expression for $i_L(t)$. (b) Obtain an expression for the current supplied by the V_B source.

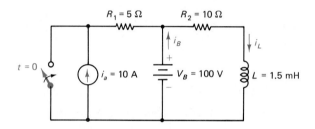

Fig. 6-44 Circuit for Problem 6-27.

6-28 Determine the currents $i_L(t)$ and $i_1(t)$ in the circuit of Fig. 6-45.

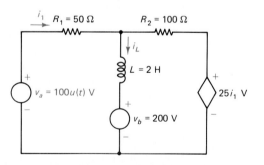

Fig. 6-45 Circuit for Problem 6-28.

6-29 Determine the currents $i_L(t)$ and $i_1(t)$ in the circuit of Fig. 6-46.

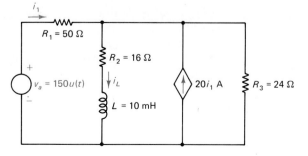

Fig. 6-46. Circuit for Problem 6-29.

6-30 A rectangular pulse of amplitude V_m and duration t_p is applied to a series RL circuit. Assume that $i(t) = 0$ at $t = 0$. Obtain expressions for $i_L(t)$ and $v_L(t)$. Sketch the two waveforms for each of the following two cases: (a) $t_p = (L/8R)$ and (b) $t_p = (8L/R)$.

6-31 Consider the expression $i_L(t) = I_0 e^{-(t/\tau)}$, which describes the decay of current in an inductance with an initial current I_0. (a) Show that a tangent to the curve drawn at $t = 0$ intersects the time axis at $t = \tau$. (b) If $i = I_1$ at $t = t_1$ and $i = I_2$ at $t = t_2$, obtain an expression for $(t_2 - t_1)$ in terms of I_1, I_2, and τ.

6-32 When the current in the coil in the circuit of Fig. 6-47 reaches a value of 12.5 mA, the magnetic field set up in the coil makes the contacts close. Assuming the initial current to be zero, the contacts are to close 15 ms after the switch S is moved to the ON position. The maximum current in the circuit should not exceed 30 mA. Determine the values of R and L.

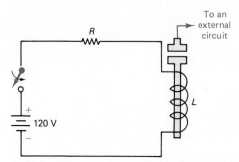

Fig. 6-47. Circuit for Problem 6-32.

Sec. 6-6 Zero Input and Zero State Response of First Order Circuits

6-33 Given a series RC circuit ($R = 1500\,\Omega$, $C = 500\,\mu F$) driven by a voltage source $v_s(t) = 20u(t)$ V. If the initial voltage on the capacitor is 25 V, determine: (a) the zero input responses $i(t)$ and $v_c(t)$; (b) the zero state responses; and (c) the total responses by using the results of (a) and (b).

6-34 Given a parallel RC circuit ($R = 1500\,\Omega$, $C = 500\,\mu F$) driven by a current source $i_s(t) = 30u(t)$ mA. If the initial voltage on the capacitor is 25 V, determine (a) the zero input responses $v_c(t)$ and $i_c(t)$; (b) the zero state responses; and (c) the total responses by using the results of (a) and (b).

6-35 The initial current in a series RL circuit ($R = 470\,\Omega$, $L = 15$ mH) is 30 mA. The circuit is driven by a voltage source $v_s = 100u(t)$ V. Determine (a) the zero input responses $i(t)$ and $v_L(t)$; (b) the zero state responses; and (c) the total responses by using the results of (a) and (b).

6-36 The initial current in the inductance in a parallel RL circuit ($R = 470\,\Omega$, $L = 15$ mH) is 30 mA. The circuit is driven by a current source $i_s(t) = 0.2u(t)$ A. Determine (a) the zero input responses $i_L(t)$ and $v_L(t)$; (b) the zero state responses; and (c) the total responses using the results of (a) and (b).

Sec. 6-7 Step Response of Second-Order Circuits

6-37 (a) The initial conditions of a circuit made up of $L = 0.5$ H and $C = 0.002$ F are: $i(t) = 5$ A and $v_c(t) = 80$ V. Determine the zero input responses $i(t)$ and $v_c(t)$. (b) Suppose a resistance R equal to one-half that needed for critical damping is placed in series with L and C. Determine the zero input responses $i(t)$ and $v_c(t)$.

6-38 The initial voltage on the capacitor in a series RLC circuit (with $L = 0.25$ H and $C = 100\,\mu F$) is 75 V, and the initial current is zero. For each of the following values of R, obtain the expressions for the zero input response. Draw neat sketches of $i(t)$ and $v_c(t)$ for the following cases: (a) $R = 10\,\Omega$; (b) $R = 100\,\Omega$; (c) $R = 1000\,\Omega$.

6-39 The data obtained for the zero input response of an underdamped series RLC circuit are as follows: $v_c(0) = 100$ V; the first positive peak of $v_c(t)$ is 122 V and occurs at $t = 150$ ms. The zero crossings of the curve occur at intervals of 1.05 s. Determine the values of R, L, and C. Write the expressions for $i(t)$ and $v_c(t)$.

6-40 The peak value of the zero input response $i(t)$ of a *critically damped* series RLC circuit is given as 50 A, occurring at $t = 25$ ms. If the initial voltage on the capacitor is 500 V and the initial current is zero, determine the values of R, L, and C. Write the expressions for the zero input responses.

6-41 The characteristic equations of several series RLC circuits are given below. In each case, assume that $v_c(0) = 100$ V, and $i(0) = 2$ A. Obtain the expressions for the zero input responses $i(t)$ and $v_c(t)$ in each case. (a) $p^2 + 25 = 0$; (b) $p^2 + 8p + 25 = 0$; (c) $p^2 + 8p + 16 = 0$; (d) $p^2 + 8p + 15 = 0$.

6-42 The zero input response of a circuit made up of a conductance G, an inductance L, and a capacitance C in *parallel* can be obtained by using the principle duality, along with the results obtained in the text for the series RLC circuit. (a) Write the characteristic equation. (b) Determine the relationship of G, L, and C for critical damping. (c) Write the expressions for the zero input responses $v(t)$, $i_c(t)$, and $i_L(t)$ for these three cases: overdamped, critically damped, and underdamped.

Sec. 6-8 Integration Using Op Amps

6-43 The signal $v_s(t)$ shown in Fig. 6-48 is fed to an integrating op amp circuit. Sketch the output waveform for (a) $R = 10$ kΩ, $C = 10$ μF; (b) $R = 1$ MΩ, $C = 1$ μF.

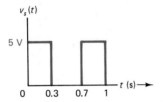

Fig. 6-48 Circuit for Problem 6-43.

6-44 An integrating op amp circuit can also be built by replacing the resistor R in Fig. 6-30(a) by an inductance L and the capacitor in Fig. 6-30(a) by a resistor R. Obtain the equation for the output $v_0(t)$ in terms of the input $v_s(t)$ and show that when the gain A is very large compared with 1, the circuit acts as an integrator. (This configuration is not used in practice, since an inductance has several disadvantages: weight, limited linear range, and inability to fabricate in integrated circuit, IC form.)

6-45 The network in Fig. 6-49 shows an analog computer set up to solve a differential equation involving the variable $x(t)$. The function $f(t)$ is a forcing function assumed to be known. Set up the equation for $f(t)$ in terms of $x(t)$ and its derivatives.

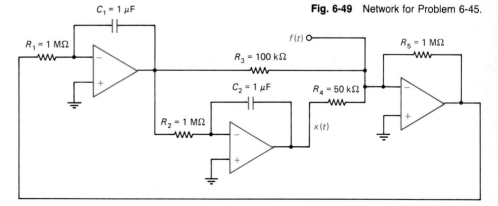

Fig. 6-49 Network for Problem 6-45.

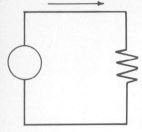

CHAPTER 7
SINUSOIDAL STEADY STATE–TIME DOMAIN ANALYSIS

The preceding chapters focused on the response of circuits to constant forcing functions. Signals of practical importance and interest, however, do not stay constant, but vary with time. Even though there is a wide variety of time-varying signals, the sinusoidal signal (that is, a signal described by a sine or cosine expression) occupies a special place in circuit analysis. Sinusoidal signals are usually referred to as *alternating currents*, abbreviated *ac*, and circuits driven by such signals are called *ac circuits*.

The analysis of ac circuits requires the solution of a set of differential equations. Solution of differential equations in the time domain is performed through classical procedures for solving ordinary linear differential equations, and this method is the topic of the present chapter. It will quickly become apparent, however, that the classical approach is time-consuming, cumbersome, and painful, except in the case of simple circuits. Fortunately, there is a more efficient method using complex algebra and the concept of phasors, which we develop in the next chapter.

7-1 IMPORTANCE OF SINUSOIDS IN CIRCUIT ANALYSIS

There are several reasons why the sinusoidal signal is of special significance in the analysis of circuits subjected to time-varying forcing functions.

1. The voltages generated by alternators in power systems and by most of the commercially available generators have an essentially sinusoidal waveform. Therefore, the use of sinusoidal signals in analysis covers most practical situations.

2. Certain mathematical properties of a sinusoid make it a convenient function to work with. (a) The addition of two or more sinusoidal functions of the *same frequency* results in a single sinusoidal function of that frequency. (b) The operations of differentiation and integration do not affect the sinusoidal nature of the function; they only lead to a shift in time and a change in amplitude. (c) The sinuoidal function is closely related to the complex exponential function, which leads to a simplification of the analysis procedures.
3. It was shown by Fourier that a periodic function may be expressed as the sum of a number of components, each of which is sinusoidal. It is therefore possible to determine the response of a circuit to a periodic nonsinusoidal forcing function by extending the techniques developed for sinusoidal forcing functions.

The practical relevance, mathematical convenience, and potential for extension to other time-varying signals are the reasons for choosing the sinusoidal signal as the ideal vehicle to study the response of circuits to time-varying signals. In the words of Guillemin, the sinusoid is "the one that shall forever be king and ruler within the realm of network theory."

7-2 BASIC DEFINITIONS AND RELATIONSHIPS OF SINUSOIDAL FUNCTIONS

A sinusoidal waveform as shown in Fig. 7-1(a), is described by the equation:

$$f(\theta) = A \cos \theta \tag{7-1}$$

where θ is measured in radians. In the case of time-varying signals, the angle θ varies in proportion to the time t. That is

$$\theta = \omega t \tag{7-2}$$

where ω is called the *angular velocity* or *angular frequency*, measured in *radians/second* (abbreviated r/s). A sinusoidal signal is therefore described by:

$$f(t) = A \cos \omega t \tag{7-3}$$

where A is the *amplitude* or *peak value* of the signal.

The waveform repeats itself after every T seconds, and T is called the *period*. The replica of the signal that is repeated every period is called a *cycle*. Since the angle θ described in one cycle is 2π radians, Eq. (7-2) leads to the following relationship between the period T and the angular frequency ω:

$$T = (2\pi/\omega) \tag{7-4}$$

The number of cycles completed each second is known as the *frequency*, f, of the signal. The unit of frequency is the *hertz* (abbreviated Hz). Since the time taken per cycle is T, the frequency f and period T are related by

$$f = (1/T) \tag{7-5a}$$

and, from Eqs. (7-4) and (7-5a)

$$\omega = 2\pi f \tag{7-5b}$$

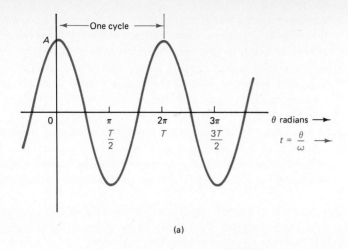

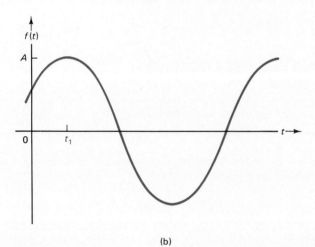

Fig. 7-1 Sinusoidal waveforms.

Equations (7-5 a and b) lead to the following alternative expressions for a sinusoidal signal:

$$f(t) = A \cos (2\pi t/T) \qquad (7\text{-}6a)$$
$$f(t) = A \cos (2\pi ft) \qquad (7\text{-}6b)$$

Normally, an expression of the form in Eq.(7-3) is commonly used for a sinusoidal signal.

7-2-1 Phase Angle

The peak value of the signal in Fig. 7-1(a) was chosen as occurring at $t = 0$. In the usual case, the peak value may occur at any arbitrary instant of time t_1, as indicated in Fig. 7-1(b), and the expression for such a signal is given by Eq. (7-3), with t replaced by $(t - t_1)$:

$$\begin{aligned} f(t) &= A \cos \omega(t - t_1) \\ &= A \cos (\omega t - \phi) \end{aligned} \qquad (7\text{-}7)$$

Sinusoidal Steady State–Time Domain Analysis

where

$$\phi = \omega t_1 \text{ radians} \quad (7\text{-}8)$$

is called the *initial phase angle* of the function. Since $\omega = (2\pi/T)$, an alternative expression for the initial phase angle is

$$\phi = (2\pi t_1/T) \text{ radians} \quad (7\text{-}9)$$

Note that, since ω is in radians per second, the quantity (ωt) in Eqs. (7-3) and (7-6) through (7-7) is always in radians. The phase angle ϕ given by Eq. (7-8) or (7-9) is also in radians. Common practice, however, condones the use of *degrees* for the phase angle, since we have a "physical feel" for angles measured in degrees. Therefore, the phase angle will usually be expressed in degrees. Be careful in numerical calculations, however; remember that (ωt) always yields radians.

Example 7-1 For each of the sinusoidal waveforms shown in Fig. 7-2, find the peak value, period, frequency, and angular frequency, and write an expression.

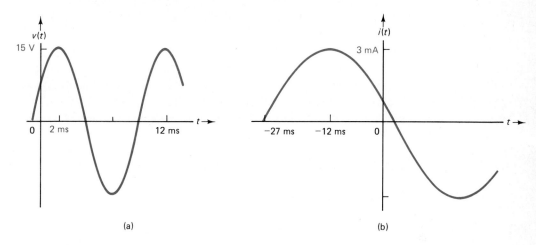

Fig. 7-2 Waveforms for Example 7-1.

Solution *Waveform (a):* Peak value = 15 V, $T = 10$ ms, $f = 100$ Hz, $\omega = 200\pi = 628$ r/s. Since the peak value occurs at $t = 2$ ms, the phase angle is, from Eq. (7-10)

$$\phi = 2\pi(2/10) = 0.4\pi \text{ radians}$$

which corresponds to 72°. The expression for $v_a(t)$ becomes

$$v_a(t) = 15 \cos(628t - 72°) \text{ V}$$

Waveform (b): Peak value = 3 mA; $T = 60$ ms, since it takes 15 ms for a quarter of a cycle; $f = 16.7$ Hz; $\omega = 377$ r/s. Since the peak value occurs at $t = -12$ ms, the phase angle is, from Eq. (7-10)

$$\phi = 2\pi(-12/60) = -0.4\pi \text{ radians} = -72°$$

The expression for the waveform is

$$i_b(t) = 3 \cos(377t + 72°) \text{ mA}$$

Exercise 7-1: For each of the waveforms shown in Fig. 7-3, find the peak value, period, frequency, and angular frequency, and write an expression.

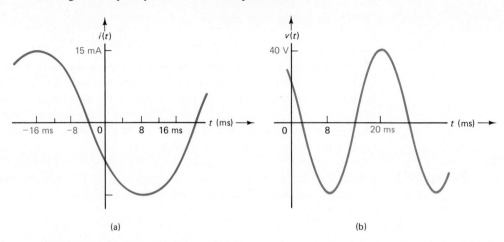

Fig. 7-3 Waveforms for Exercise 7-1.

The instant of time at which the positive peak occurs is used for determining the phase angle of a sinusoidal signal. Since a sinusoidal signal has an unlimited number of positive peaks, however, the phase angle can have more than one value. That is, since

$$A \cos (\omega t + \phi) = A \cos (\omega t + \phi + 2n\pi)$$

where n is any integer, the phase angle is given by $(\phi + 2n\pi)$ also. It is convenient, however, to choose the positive peak *closest* to the origin as the basis of the phase angle to keep the magnitude of ϕ within 180°.

7-2-2 Phase Difference between Two Signals

When two signals of the *same frequency* do not have their peak values occurring at the same instant of time, a *phase angle difference* or simply *phase difference* is said to exist between them. Consider two signals

$$a(t) = A \cos (\omega t + \phi_a)$$

and

$$b(t) = B \cos (\omega t + \phi_b)$$

The phase difference between the two signals is defined as $(\phi_a - \phi_b)$ or $(\phi_b - \phi_a)$:

If $\phi_a > \phi_b$, then $a(t)$ is said to *lead* $b(t)$ by $(\phi_a - \phi_b)$. If $\phi_a < \phi_b$, then $a(t)$ is said to *lag behind* $b(t)$ by $(\phi_b - \phi_a)$.

Example 7-2 (a) Given the two signals $v(t) = 100 \cos (1000t + 45°)$ V and $i(t) = 15 \cos (1000t + 13°)$ A, the phase difference between them is $(45 - 13) = 32°$. $v(t)$ leads $i(t)$ by 32°. $i(t)$ lags behind $v(t)$ by 32°.
(b) Given two signals $i_1(t) = 8 \cos (500t - 60°)$ A, and $i_2(t) = 19 \cos (500 t + 36°)$ A, the phase difference is $[36 - (-60)] = 96°$. $i_1(t)$ lags behind $i_2(t)$ or $i_2(t)$ leads $i_1(t)$ by 96°. ∎

Example 7-3 Calculate the phase difference between $i(t) = 20 \cos(700t + 10°)$ A, and $v(t) = 18 \sin(700t + 70°)$ V.

Solution Note that one expression is a cosine while the other is a sine. It is necessary first to put both expressions in the same form: both must be cosines or both must be sines before trying to calculate the phase difference. Convert $v(t)$ to a cosine expression.

The rule for converting a *sine* expression to a *cosine* expression is *subtract 90°*. That is,

$$\sin X = \cos(X - 90°)$$
$$v(t) = 18 \cos[700t + (70° - 90°)]$$
$$= 18 \cos(700t - 20°) \text{ V}$$

Since $i(t) = 20 \cos(700t + 10°)$, the phase difference between $v(t)$ and $i(t)$ is 30°, with $i(t)$ leading. ∎

Remember that phase difference is meaningful only when signals of the *same frequency* are being compared.

As we mentioned in connection with phase angle, the phase difference between two signals can have more than one value due to the multiplicity of instants of time at which positive peaks occur. Again, it is convenient to choose the instants of peaks closest to one another to prevent angles from becoming inconveniently large in numerical computations.

Exercise 7-2 Given $v(t) = 100 \cos(50t + 36°)$ V. Write an expression of a current $i(t)$ with an amplitude of 15 A and leading $v(t)$ by 70°.

Exercise 7-3 Find the phase difference between every possible pair of the following set of signals. In each case, state which signal is *leading*.

$$v_1(t) = 100 \cos(10t - 20°) \text{ V}; \quad v_2(t) = -100 \cos(10t + 160°) \text{ V};$$
$$v_3(t) = 10 \sin(10t + 45°) \text{ V}; \quad v_4(t) = -50 \sin 10t \text{ V}.$$

7-2-3 Decomposition of a General Sinusoidal Signal

A sinusoidal signal of the form

$$f(t) = A \cos(\omega t + \phi) \tag{7-10a}$$

can be decomposed as the sum of the two terms, a cosine and a sine term, to the form

$$f(t) = P_c \cos \omega t + P_s \sin \omega t \tag{7-10b}$$

where

$$P_c = A \cos \phi \quad \text{and} \quad P_s = -A \sin \phi \tag{7-11}$$

(Note the *minus* sign for P_s.) The validity of the relationships in Eq. (7-11) is established by expanding the expression in Eq. (7-10a) and comparing it with the expression in (7-10b).

Decomposition of the form Eq. (7-10b) is necessary when combining sinusoidal functions for the resultant signal.

Example 7-4 Given $v_1(t) = A \cos(\omega t + \phi_a)$ and $v_2(t) = B \cos(\omega t + \phi_b)$, write an expression for $v(t) = [v_1(t) + v_2(t)]$ in the form $(C \cos \omega t + D \sin \omega t)$.

Solution $v(t) = (A \cos \phi_a + B \cos \phi_b) \cos \omega t - (A \sin \phi_a + B \sin \phi_b) \sin \omega t$

A sinusoidal signal described by

$$A \sin(\omega t + \phi)$$

has the decomposition

$$Q_c \cos \omega t + Q_s \sin \omega t$$

where

$$Q_c = A \sin \phi \quad \text{and} \quad Q_s = A \cos \phi \qquad (7\text{-}12) \ \blacksquare$$

Example 7-5 In the arrangement shown in Fig. 7-4, $i_1(t) = 10 \cos(200t + 30°)$ A, $i_2(t) = 5 \cos(200t - 70°)$, and $i_3(t) = 15 \sin(200t + 25°)$ A. Express $i_4(t)$ in the form $[I_1 \cos 200t + I_2 \sin 200t]$.

Solution Using KCL, $i_4 = i_1 - i_2 + i_3$

$$\begin{aligned}
i_4(t) &= (8.66 \cos 200t - 5 \sin 200t) \\
&\quad + (-1.71 \cos 200t - 4.70 \sin 200t) \\
&\quad + (6.34 \cos 200t + 13.6 \sin 200t) \\
&= (13.3 \cos 200t + 3.90 \sin 200t) \text{ A}
\end{aligned}$$

Fig. 7-4 Circuit configuration for Example 7-5.

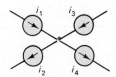

$\blacksquare$

Exercise 7-4 Given $f_1(t) = 10 \cos 100t$ and $f_2(t) = 12 \cos(100t - 40°)$, find $g(t) = [f_1(t) + f_2(t)]$ and express it in the form $(A \cos 100t + B \sin 100t)$.

Exercise 7-5 The source voltages in the series connection of Fig. 7-5 are given by $v_1(t) = 100 \cos 377t$ V, $v_2(t) = 100 \cos(377t + 120°)$ V, and $v_3(t) = 100 \cos(377t - 120°)$ V. Calculate the total voltage $v_s(t)$ and put it in the form $(V_1 \cos 377t + V_2 \sin 377t)$.

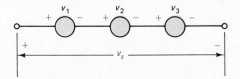

Fig. 7-5 Series connection for Exercise 7-5.

7-2-4 Combination of Sinusoidal Functions

Consider the addition of two sinusoidal components, $P \cos \omega t$ and $Q \sin \omega t$. Let

$$P \cos \omega t + Q \sin \omega t = A \cos(\omega t + \phi) \qquad (7\text{-}13)$$

The relationships between A, ϕ, P and Q are given by

$$A = \sqrt{(P^2 + Q^2)} \tag{7-14}$$

and

$$\phi = \arctan(-Q/P) \tag{7-15}$$

Note the minus sign in the expression for ϕ.

In numerical calculations, note that the value of ϕ obtained by using Eq. (7-15) may not always place ϕ in the correct quadrant, since the *ratio* of $(-Q/P)$ is involved. For example, the value of ϕ obtained from Eq. (7-15) will be the same whether $P = 3$ and $Q = -4$, or $P = -3$ and $Q = 4$.

The general rule for determining the correct angle ϕ is as follows:

If P is positive in Eq. (7-13), then Eq. (7-15) yields the correct value of ϕ directly.

If P is negative in Eq. (7-13), then the angle obtained from Eq. (7-13) has to be shifted by 180° by adding or subtracting 180°.

Note that addition of 180° and the subtraction of 180° are equivalent operations.

Exercise 7-6 Prove the rule for the determination of the correct angle ϕ. For this, study the effect of changing the angle ϕ in the expression $A \cos \omega t + \phi$ by 180° on the expansion $(P \cos \omega t + Q \sin \omega t)$.

Example 7-6 If $v_1(t) = 100 \cos 1500t$ V and $v_2(t) = 70 \sin 1500t$ V, find $v_3(t) = [v_1(t) + v_2(t)]$ and express it in the form $V_m \cos(1500t + \phi)$ V.

Solution Using Eq. (7-14) with $P = 100$ and $Q = 70$, $V_m = \sqrt{(100^2 + 70^2)} = 122$ V.

From Eq. (7-15), $\phi = \arctan(-70/100) = -35°$. Since $P = 100$ is positive, the angle obtained is the correct angle. Therefore

$$v_3(t) = 122 \cos(1500t - 35°) \text{ V}$$

∎

Example 7-7 Given $i_1(t) = -30 \cos(190t + 45°)$ A and $i_2(t) = -40 \sin 190t$ A, express their sum i_3 in the form $I_m \cos(190t + \phi)$.

Solution It is first necessary to decompose $i_1(t)$ into a cosine and a sine term before adding $i_2(t)$ to it.

$$i_1(t) = -(21.2 \cos 190t - 21.2 \sin 190t)$$

Therefore,

$$i_3(t) = i_1(t) + i_2(t)$$
$$= -21.2 \cos 190t - 18.8 \sin 190t$$

Using Eq. (7-14) with $P = -21.2$ and $Q = -18.8$, $I_m = \sqrt{(21.2^2 + 18.8^2)} = 28.3$.

The angle given by Eq. (7-15) is $\phi = \arctan(18.8/-21.2) = -41.6°$. But $P = -21.2$, which is negative. Therefore, the correct value of $\phi = (-41.6 + 180)° = 138.4°$, and

$$i_3(t) = 28.3 \cos(190t + 138.4°) \text{ A}$$

∎

Exercise 7-7 Given $v_1(t) = 10 \cos 150t$ V, and $v_2(t) = 15 \sin 150t$ V, determine (a) $v_a = (v_1 + v_2)$; (b) $v_b = (v_1 - v_2)$; (c) $v_c = (-v_1 + v_2)$ and (d) $v_d = (-v_1 - v_2)$. In each case, the final expression should be of the form $V_m \cos(150t + \phi)$.

Exercise 7-8 Given $i_1(t) = 10 \cos(50t + 60°)$ A, $i_2(t) = -10 \cos 50t$ A, and $i_3(t) = 10 \cos(50t - 60°)$ A, determine $i_s = i_1 + i_2 + i_3$.

7-3 VOLTAGE, CURRENT, AND POWER IN SINGLE COMPONENTS

We start with the voltage-current relationships of the basic passive electrical components and the relationship for power in the discussion of the response of electric circuits in the sinusoidal steady state.

Resistance

For a resistor,

$$v(t) = Ri(t)$$

If the voltage across a resistor is

$$v(t) = V_m \cos \omega t$$

then the current in the resistor is

$$i(t) = (V_m/R) \cos \omega t$$

Refer to Fig. 7-6(a).

The phase difference between the voltage and current in a resistor is zero; that is, $v(t)$ and $i(t)$ are *in phase* with each other.

The *instantaneous power* received by a resistor is

$$p(t) = v(t)i(t) = (V_m^2/R) \cos^2 \omega t$$

By using the formula

$$\cos^2 X = (1/2)(1 + \cos 2X)$$

the expression for $p(t)$ becomes

$$p(t) = (V_m^2/2R)(1 + \cos 2\omega t)$$

The power consumed by the resistor R varies from a maximum of (V_m^2/R) to a minimum of zero, and is non-negative at all times. The average value of power is $(V_m^2/2R)$.

Capacitance

For a capacitor

$$i(t) = C[dv(t)/dt]$$

If the voltage across a capacitor is:

$$v(t) = V_m \cos \omega t$$

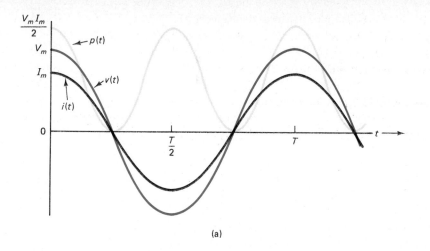

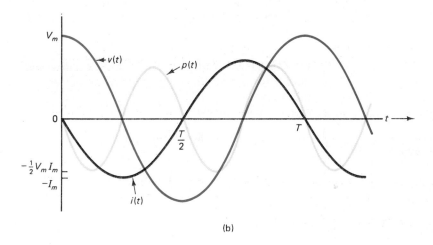

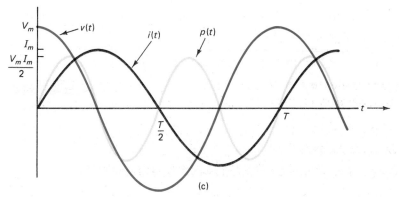

Fig. 7-6 Current, voltage, and power waveforms in (a) a resistance; (b) a capacitance; and (c) an inductance.

7-3   Voltage, Current, and Power in Single Components

then the current in the capacitor is

$$i(t) = -(\omega C V_m) \sin \omega t = (\omega C V_m) \cos (\omega t + 90°)$$

Refer to Fig. 7-6(b).

The *current in a capacitor leads the voltage by 90°*.

The instantaneous power received by a capacitor is

$$p(t) = v(t)i(t) = -(\omega C V_m^2) \cos \omega t \sin \omega t$$
$$= -(\omega C V_m^2/2) \sin 2\omega t$$

where the formula

$$\cos x \sin x = (1/2) \sin 2x$$

has been used.

The power $p(t)$ varies sinusoidally at twice the frequency of the voltage or current. *The power received by the capacitor over any one cycle is positive for half a cycle (when the voltage is increasing) and negative for half a cycle (when the voltage is decreasing). The average value of the power received by a capacitor is zero.*

Inductance

Since inductance is the dual of capacitance, the principle of duality is used to extend the results obtained for a capacitor to the case of the inductor. (The results may also be obtained directly.)

If the current in an inductor is

$$i(t) = I_m \cos \omega t$$

then

$$v(t) = (\omega L I_m) \cos (\omega t + 90°)$$

Refer to Fig. 7-6(c).

The *voltage in an inductor leads the current by 90°*.

The power received by an inductor is given by

$$p(t) = -(\omega L I_m^2/2) \sin 2\omega t$$

which varies sinusoidally at twice the frequency of the voltage or current. *The power received by the inductor is positive during one half of each cycle (when the current is increasing) and negative during the other half of each cycle (when the current is decreasing). The average value of the power received by an inductor is zero.*

Exercise 7-9 The current through a resistance $R = 15 \, \Omega$ is $i(t) = 14.1 \cos 5t$ A. Draw neat sketches of the voltage and power over the interval 0 to T.

Exercise 7-10 The voltage across a capacitance $C = 50 \, \mu F$ is $v(t) = 20 \cos 10t$ V. Draw neat sketches of the current and power over the interval 0 to T.

Exercise 7-11 The current in an inductance $L = 16$ mH is $i(t) = 25 \cos 400t$ A. Draw neat sketches of the voltage and power over the interval 0 to T.

7-4 THE SERIES RL CIRCUIT–TIME DOMAIN SOLUTION

As we saw in Chapter 6, the application of Kirchhoff's laws to a circuit containing inductors and capacitors leads to differential equations. We also saw that the general solution of the differential equation consisted of two components: a transient component (or natural response), and a steady state component (or forced response). The discussion in this and most of the following chapters will be restricted to the *steady state component* (or forced response); we will ignore the transient component. The reasons for the restriction are (1) The steady state response is of interest in a large number of problems of practical importance; and (2) the determination of the complete response is analytically cumbersome without the use of Laplace transform (which will be presented in the final chapter).

Consider a circuit made up of a resistance R and an inductance L in series driven by a sinusoidal voltage source (Fig. 7-7). The KVL equation for the circuit is

$$L\frac{di}{dt} + Ri = v_s \qquad (7\text{-}16)$$

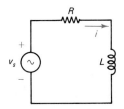

Fig. 7-7 Series *RL* circuit under sinusoidal excitation.

In some cases the current $i(t)$ may be given, and it is desired to find the voltage v_s; in other cases the voltage v_s is specified, and it is desired to find the current $i(t)$.

When $i(t)$ is given and $v_s(t)$ is to be determined, $i(t)$ is substituted in the left-hand side of Eq. (7-16) and some manipulations lead to a single expression for $v_s(t)$, as illustrated by Example 7-8.

Example 7-8 The current in a series *RL* circuit with $R = 9 \ \Omega$ and $L = 6$ mH is $i(t) = 8 \cos 2000t$ A. Determine the voltage v_T across the series combination.

Solution The KVL equation of the circuit is

$$(6 \times 10^{-3})(di/dt) + 9i = v_T$$

$$i(t) = 8 \cos 2000t \quad \text{and} \quad di/dt = -(16 \times 10^3) \sin 2000t$$

Therefore

$$v_T = -96 \sin 2000t + 72 \cos 2000t$$
$$= 120 \cos (2000t + 53.1°) \ \text{V} \qquad \blacksquare$$

Exercise 7-12 Determine the voltage across a series combination of $R = 15 \ \Omega$ and $L = 8$ mH when the current $i(t) = 4 \sin 500t$ A.

Classical Solution of the Differential Equation

When the voltage $v_s(t)$ applied to a series circuit is specified and we want to find $i(t)$, it becomes necessary to solve a differential equation for the steady state response.

Linear operations (such as addition, multiplication by a constant, differentiation, and integration) do not change the sinusoidal nature of a signal or its frequency, as we saw in the preceding sections of this chapter. Consequently, the steady state response of a *linear* circuit driven by a sinusoidal forcing function may safely be assumed to be sinusoidal at the same frequency as the forcing function. Therefore, the classical procedure for solving the differential equation of a linear circuit starts by assuming a sinusoidal response of the same frequency as the forcing function, with an amplitude and phase angle that are to be determined.

For the series RL circuit driven by

$$v_s(t) = V_m \cos \omega t \qquad (7\text{-}17)$$

Eq. (7-16) becomes

$$L\frac{di}{dt} + Ri = V_m \cos \omega t \qquad (7\text{-}18)$$

Assume a solution of the form

$$i(t) = I_m \cos(\omega t + \phi) \qquad (7\text{-}19)$$

where I_m and ϕ are unknown quantities to be determined, or of the form

$$i(t) = A \cos \omega t + B \sin \omega t \qquad (7\text{-}20)$$

where A and B are unknown coefficients to be determined.

The form in Eq. (7-20) is somewhat preferable for the present discussion. To evaluate the coefficients A and B, substitute the expression in Eq. (7-20) and its derivative in Eq. (7-18). This step leads to

$$-(\omega LA)\sin \omega t + (\omega LB)\cos \omega t + RA \cos \omega t + RB \sin \omega t = V_m \cos \omega t \qquad (7\text{-}21)$$

For Eq. (7-21) to be valid for all values of time t, it is necessary for the coefficients of $\cos \omega t$ on both sides of the equation to be equal and for the coefficients of $\sin \omega t$ to be equal to each other.

Equating the coefficients of $\cos \omega t$ on both sides of Eq. (7-21)

$$\omega LB + RA = V_m \qquad (7\text{-}22)$$

Equating the coefficients of $\sin \omega t$ on both sides of Eq. (7-21)

$$-\omega LA + RB = 0 \qquad (7\text{-}23)$$

Equations (7-22) and (7-23) are solved for A and B, and it is found that

$$A = RV_m/(R^2 + \omega^2 L^2) \qquad (7\text{-}24)$$

and

$$B = \omega LV_m/(R^2 + \omega^2 L^2) \qquad (7\text{-}25)$$

Therefore, the expression for $i(t)$ in Eq. (7-20) becomes

$$i(t) = [V_m/(R^2 + \omega^2 L^2)] (R \cos \omega t + \omega L \sin \omega t)$$

Using Eqs. (7-14) and (7-15), the expression above is rewritten in the form of a single sinusoid with a phase angle

$$i(t) = [V_m/\sqrt{(R^2 + \omega^2 L^2)}] \cos(\omega t - \arctan \omega L/R) \qquad (7\text{-}26)$$

which is the response of a series RL circuit to a sinusoidal forcing function.

The amplitude or peak value of $i(t)$ is given by $[V_m/\sqrt{(R^2 + \omega^2 L^2)}]$ and the current lags behind $v_s(t)$ by $\arctan \omega L/R$. Both the amplitude and phase angle are seen to depend upon the frequency as well as the element values. The waveforms of $v_s(t)$ and $i(t)$ are shown in Fig. 7-8.

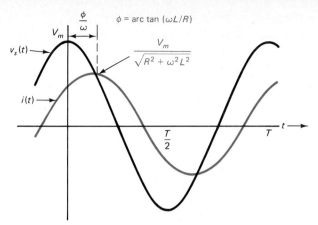

Fig. 7-8 Current and voltage waveforms in a series RL circuit.

Once the current $i(t)$ in the series circuit is determined, other quantities such as the voltage across the resistance or inductance can be found by using appropriate relationships. For example, the voltage across the inductance is given by

$$v_L(t) = L(di/dt) = [\omega L V_m/\sqrt{(R^2 + \omega^2 L^2)}] \cos(\omega t - \arctan \omega L/R - 90°)$$

Exercise 7-13 Derive the expression for $v_L(t)$ given above.

Exercise 7-14 For each of the following sets of values for a series RL circuit driven by a sinusoidal source of amplitude 100 V, determine the amplitude and phase angle of the current and write the expression for $i(t)$.

(a) $R = 30\ \Omega$, $C = 40\ \mu F$, $\omega = 1000$ r/s.
(b) $R = 30\ \Omega$, $C = 40\ \mu F$, $\omega = 250$ r/s.
(c) $R = 100\ \Omega$, $C = 40\ \mu F$, $\omega = 1000$ r/s.
(d) $R = 30\ \Omega$, $C = 200\ \mu F$, $\omega = 1000$ r/s.

Exercise 7-15 Find the values of R and L in a series RL circuit to make the amplitude of the current equal to twice the amplitude of the voltage and the current lag behind the voltage by 36° at a frequency of 159.2 Hz.

Exercise 7-16 Consider a parallel combination of a conductance G and a capacitance C driven by a sinusoidal current source $i_s(t) = I_m \cos \omega t$. This is the dual of the series RL circuit discussed earlier. Use the principles of duality and modify the expressions obtained for the series RL circuit to obtain the corresponding expressions for the parallel GC circuit.

7-5 THE SERIES RC CIRCUIT–TIME DOMAIN SOLUTION

The determination of the steady state response of a series RC circuit follows exactly the same lines as that of a series RL circuit. The KVL equation of a series RC circuit driven by a source $v_s = V_m \cos \omega t$ is

$$Ri + (1/C) \int i\, dt = V_m \cos \omega t \qquad (7\text{-}27)$$

Consider the integral $\int i\,dt$, which will normally give rise to an arbitrary constant. The arbitrary constant is ignored (or taken as zero) in sinusoidal *steady state* solutions, since the steady state response will be a pure sinusoid, and a constant term does not belong in such a solution. The solution of Eq. (7-27) is (and Exercise 7-17):

$$i(t) = [V_m/\sqrt{R^2 + 1/\omega^2 C^2}] \cos [\omega t + \arctan (1/\omega RC)] \qquad (7\text{-}28)$$

The current in the series RC circuit is seen to lead the applied voltage by arc tan $(1/\omega RC)$.

Exercise 7-17 Assume a solution of the form $i(t) = (A \cos \omega t + B \sin \omega t)$ for Eq. (7-27). Use steps exactly similar to those in the solution of the series RL circuit and derive Eq. (7-28).

Exercise 7-18 Obtain the expression for the voltage across the capacitor in the series RC circuit.

Exercise 7-19 For each of the following sets of values for a series RC circuit driven by a sinusoidal source of amplitude 100 V, determine the amplitude and phase angle of the current and write the expression for $i(t)$.

(a) $R = 30\ \Omega$, $C = 40\ \mu F$, $\omega = 1000$ r/s.
(b) $R = 30\ \Omega$, $C = 40\ \mu F$, $\omega = 250$ r/s.
(c) $R = 100\ \Omega$, $C = 40\ \mu F$, $\omega = 1000$ r/s.
(d) $R = 30\ \Omega$, $C = 200\ \mu F$, $\omega = 1000$ r/s.

Exercise 7-20 The current in a series RC circuit with $R = 12\ \Omega$ and $C = 20\ \mu F$ is $i(t) = 6 \cos 10^4 t$ A. Determine the voltage across the resistance, the capacitance, and the series combination.

Exercise 7-21 Find the values of R and C in a series RC circuit such that the phase angle between the current and the voltage is 29° and the amplitude of the current is one-third of the amplitude of the applied voltage. $\omega = 10^3$ r/s.

7-6 THE PARALLEL GLC CIRCUIT–TIME DOMAIN SOLUTION

The steady state response of first-order circuits (RL and RC) was discussed in the preceding section. Consider now a second-order circuit: a conductance G, an inductance L, and a capacitance C in parallel and driven by a current source, as indicated in Fig. 7-9. The KCL equation of the circuit is

$$C\frac{dv}{dt} + Gv + \frac{1}{L}\int v\,dt = I_m \cos \omega t \qquad (7\text{-}29)$$

Assume a solution of the form

$$v(t) = P \cos \omega t + Q \sin \omega t$$

Substitution of $v(t)$, its derivative, and its integral in Eq. (7-29) leads to

$$\left[GQ + P\left(\frac{1}{\omega L} - \omega C\right)\right] \sin \omega t + \left[GP + Q\left(\omega C - \frac{1}{\omega L}\right)\right] \cos \omega t = I_m \cos \omega t \qquad (7\text{-}30)$$

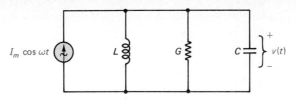

Fig. 7-9 Parallel GLC circuit under sinusoidal excitation.

Equating the coefficients of $\cos(\omega t)$ on both sides of Eq. (7-30) gives

$$GP + (\omega C - 1/\omega L)Q = I_m \qquad (7\text{-}31)$$

and equating the coefficients of $\sin \omega t$ on both sides of Eq. (7-30) gives

$$-(\omega C - 1/\omega L)P + GQ = 0 \qquad (7\text{-}32)$$

Solving Eqs. (7-31) and (7-32)

$$P = \frac{GI_m}{G^2 + (\omega C - 1/\omega L)^2}$$

and

$$Q = \frac{(\omega C - 1/\omega L)I_m}{G^2 + (\omega C - 1/\omega L)^2} \qquad (7\text{-}33)$$

Therefore, the solution of the differential equation becomes

$$v(t) = \frac{I_m}{\sqrt{G^2 + \left(\omega C - \dfrac{1}{\omega L}\right)^2}} \cos(\omega t + \phi) \qquad (7\text{-}34)$$

where

$$\phi = -\arctan \frac{\omega C - 1/\omega L}{G} \qquad (7\text{-}35)$$

The phase angle ϕ may be positive, zero, or negative (that is, the voltage may lead, be in phase with, or lag behind the current), depending on the relative values of ω, L, and C.

Exercise 7-22 A parallel *GLC* circuit with $G = 0.5$ S is driven by a current source $i_s = 12 \cos 16t$ A. For each of the following cases, determine the amplitude and phase angle of the voltage across the circuit. State which of the two quantities, voltage or current, leads in each case. (a) $L = 0.125$ H, $C = 0.0625$ F; (b) $L = 0.0625$ H, $C = 0.0625$ F; (c) $L = 0.0625$ H, $C = 0.03125$ F.

Exercise 7-23 Repeat the work of the previous exercise when i_s is changed to $12 \cos 8t$ A.

The principle of duality is used to modify the solution of a parallel *GLC* circuit driven by a sinusoidal current source and obtain the solution of the series *RLC* circuit driven by a sinusoidal voltage source.

Exercise 7-24 Modify the expressions obtained for the parallel *GLC* circuit to obtain the expression for the current in a series *RLC* circuit driven by $V_m \cos \omega t$.

7-6 The Parallel GLC Circuit—Time Domain Solution

Exercise 7-25 Repeat the work of Exercise 7-22 for a series *RLC* circuit with $R = 2\,\Omega$ driven by $v_s = 12 \cos 16t$ V.

For a general circuit with n independent voltage or current variables, it becomes necessary to solve a system of n simultaneous differential equations. The solution starts by assuming a set of n solutions (for the n variables), with each solution containing two unknown quantities: amplitude and phase. Substitution of the assumed solutions in the system of differential equations leads to a set of $2n$ simultaneous equations that must be solved to obtain the amplitudes and phases of the n independent variables. It is not difficult to imagine the amount of work involved, and the numbing tediousness of such an endeavor. Fortunately, however, there is an easier method of attack to solve circuits in the sinusoidal steady state, and such a method will be presented in the next chapter.

7-7 INSTANTANEOUS AND AVERAGE POWER IN AC CIRCUITS

Power is one of the most important quantities in electronic and communication systems, as well as electric power systems. The design of such systems often centers on the transmission of power from one point to another. An important question in ac circuit analysis is the definition and measurement of power.

We discussed instantaneous power $p(t)$ for individual components in Sec. 7-3. Consider now the general case of instantaneous power across an arbitrary combination of circuit components under sinusoidal excitation. Let the black box of Fig. 7-10(a) contain an arbitrary network of resistances, capacitances, and inductances, and let the voltage $v(t)$ and the current $i(t)$ associated with the black box be

$$v(t) = V_m \cos(\omega t + \theta)\ \text{V}$$

and

$$i(t) = I_m \cos(\omega t + \phi)\ \text{A}$$

where V_m and I_m are the amplitudes (or peak values), and θ and ϕ are the initial phase angles of the voltage and current, respectively.

The instantaneous power *received* by the circuit in the black box is then given by

Fig. 7-10 Power in sinusoidal steady state. $p(t) < 0$ implies that the circuit is receiving negative power (or delivering positive power). (a) Black box. (b) Waveform of $p(t)$.

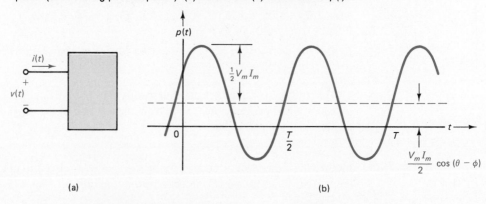

$$p(t) = v(t)i(t)$$
$$= V_m I_m \cos(\omega t + \theta) \cos(\omega t + \phi)$$

Using the trigonometric identity

$$(\cos X)(\cos Y) = (1/2)[\cos(X+Y) + \cos(X-Y)]$$

the expression for $p(t)$ is rewritten in the form

$$p(t) = (V_m I_m/2) \cos(2\omega t + \theta + \phi) + (V_m I_m/2) \cos(\theta - \phi) \qquad (7\text{-}36)$$

The instantaneous power $p(t)$ is seen to consist of two components. The component expressed by the first term in Eq. (7-36) is a time-varying sinusoidal function whose frequency is *twice* that of the voltage or current. The second term in $p(t)$ is a *constant* at all times. The constant value depends upon the peak values of the voltage and current, as well as the *phase difference* between the voltage and the current.

The waveform of $p(t)$ is shown in Fig. 7-10(b). One can see that $p(t)$ is positive for a part of each cycle and negative for the remainder of the cycle. *When $p(t)$ is positive, the power received by the circuit in the black box is positive, and there is a flow of power into the black box. When $p(t)$ is negative, the power received by the circuit in the black box is negative, and there is a flow of power out of the black box.* The outward flow of power is possible due to the presence of energy storage elements (inductors and capacitors) that deliver part or all of their stored energy during a portion of each cycle.

For the special case, $\theta = \phi$, that is, when the voltage and current are in phase (which implies a purely resistive circuit in the black box), the constant term in $p(t)$ becomes $(V_m I_m/2)$. In such a situation, $p(t)$ in Eq. (7-36) has a minimum value of zero, and $p(t)$ is always non-negative. Therefore, there is a flow of power into the black box at all times only when the circuit is purely resistive.

7-7-1 Average Power

The measurement of power in an ac circuit by an indicating instrument is difficult, since it varies as a function of time. It is more convenient to use the *average* value of power in practical measurements.

Average power is defined as the mean value of $p(t)$ over an internal of T seconds (where T is the period of the sinusoidal voltage or current).

One of the reasons for using average power is that a wattmeter (of the analog type used for measuring power) responds to average power. A *wattmeter* consists of a mechanism that responds to a torque proportional to the product of the currents in two coils, the potential coil and the current coil (see Fig. 7-11). The current in the potential coil is proportional to the voltage $v(t)$, while the current in the current coil equals $i(t)$. The torque is therefore proportional to the product $p(t) = v(t)i(t)$. The inertia of the meter mechanism, however, makes it difficult for the meter needle to track the time-varying torque, and it comes to rest at a position determined by the average value of the torque or the average value of $p(t)$. Thus, with proper connections, the wattmeter reading equals the average power received by a circuit.

The average power is defined by

$$P_{av} = (1/T) \int_0^T p(t)\, dt$$

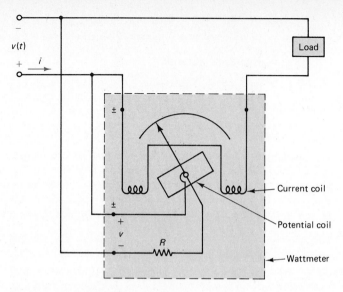

Fig. 7-11 Basic wattmeter mechanism.

Using Eq. (7-36) for $p(t)$ in the last equation, and noting that the integral of $\cos(2\omega t + \theta + \phi)$ over the interval 0 to T is zero, the average power is found to be

$$P_{av} = (V_m I_m/2)\cos(\theta - \phi) \tag{7-37}$$

That is, the average power received by an electric circuit is:

$P_{av} = (1/2)(\text{voltage amplitude}) \times (\text{current amplitude})$
$\quad\quad \times (\text{cosine of the phase difference between the voltage and current})$

Example 7-9 The voltage in Fig. 7-10(a) is $v(t) = 100 \cos 500t$ V, and the current is $i(t) = 24 \cos(500t + \theta_1)$ A. For each of the following values of θ_1, plot the instantaneous power $p(t)$ and calculate the average power: (a) $-120°$, (b) $60°$, (c) $0°$, (d) $90°$, (e) $120°$.

Solution The instantaneous power received by the circuit in the black box is

$$p(t) = 2400 \cos(500t) \cos(500t + \theta_1)$$
$$= 1200 \cos(1000t + \theta_1) + 1200 \cos \theta_1$$

and the waveforms are shown in Fig. 7-12.

The average power is given by

$$P_{av} = 1200 \cos \theta_1$$

(a) $\theta_1 = -120°$: $P_{av} = -600$ W. (b) $\theta_1 = 60°$: $P_{av} = 600$ W. (c) $\theta_1 = 0°$: $P_{av} = 1200$ W. (d) $\theta_1 = 90°$: $P_{av} = 0$. (e) $\theta_1 = 120°$: $P_{av} = -600$ W.

The negative values of average power in cases (a) and (d) mean that the given circuit *delivers* (on the average) 600 W to some components connected to the input of the black box. Case (c) corresponds to a purely resistive circuit, and the average power is seen to have the highest value for this case. Case (d) represents a purely capacitive circuit (since the current is leading the voltage by 90°), and the average power is zero. This means that the circuit is delivering as much energy as it returns in each cycle. ∎

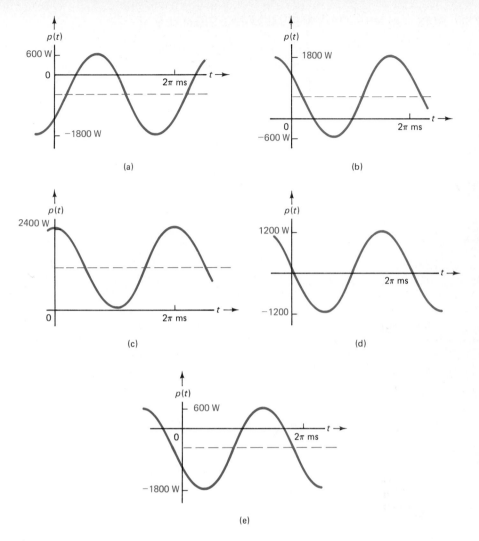

Fig. 7-12 Power waveforms for Example 7-9.

Exercise 7-26 Given that $v(t) = 800 \cos 377t$ V and $i(t) = 20 \cos(377t + \theta_2)$ in Fig. 7-10(a), find the instantaneous power $p(t)$ and calculate the average power when θ_2 equals (a) 30° and (b) 150°.

Exercise 7-27 If $v(t) = 150 \cos 200t$ V and $i(t) = 80 \cos(200t + \theta_3)$ in Fig. 7-10(a), determine the value of θ_3 for each of the following cases: (a) The circuit in the black box *receives* an average power of 1500 W. (b) The circuit in the black box *delivers* an average power of 4000 W.

Power in an ac circuit will be discussed again in Chapter 8 from a different perspective.

7-7 Instantaneous and Average Power in AC Circuits

7-8 RMS VALUES OF TIME-VARYING SIGNALS

In the case of power, its average value is convenient from the viewpoint of measurement. In the case of sinusoidal voltages and currents, the average value is zero (since the integral of a sinusoidal function over a period is zero) and is consequently not useful. A better measure is the *root mean square value*, abbreviated *rms value*. Voltmeters and ammeters (of the analog type) are calibrated to read the rms value of a sinusoidal signal. Even though the term *rms value* has been introduced here in connection with sinusoidal signals, it applies to any periodic signal (as well as to a wide variety of other types of signals).

The rms value of a voltage (or current) is defined on the basis of the average power received by a resistor. Consider a periodic voltage $v(t)$ applied to a resistor R. The average power dissipated in the resistor is

$$P_{av} = (1/T) \int_0^T (v^2/R) dt$$

$$= (1/R)[(1/T) \int_0^T v^2 dt] \qquad (7\text{-}37)$$

Suppose the same amount of average power is to be dissipated in R by applying a constant voltage V_0 to it. The power dissipated is a constant at (V_0^2/R), and the average power equals the constant value. Therefore,

$$(V_0^2/R) = (1/R)[(1/T) \int_0^T v^2 dt]$$

or

$$V_0 = \sqrt{(1/T) \int_0^T v^2 dt} \qquad (7\text{-}38)$$

Since V_0 produces the same average power as $v(t)$, it is called the *effective value* of $v(t)$. The quantity under the square root sign is seen to be the *mean square value* of the voltage $v(t)$. Consequently, V_0 is referred to as the *root mean square* or *rms* value of $v(t)$, and denoted by V_{rms}:

$$V_{rms} = \sqrt{(1/T) \int_0^T v^2 dt} \qquad (7\text{-}39)$$

Note that the integration may be taken over any interval spanning one period—such as, for example, from $-(T/2)$ to $(T/2)$, and need not always be (0 to T). If $i(t)$ is used in Eq. (7-39) instead of $v(t)$, the resulting expression yields the rms value of a periodic current.

Example 7-10 Determine the rms value of $v(t)$ shown in Fig. 7-13.

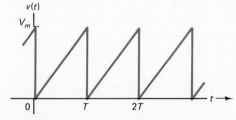

Fig. 7-13 Triangular waveform for Example 7-10.

Solution The first step is to describe the voltage over a period of T seconds by means of an expression.

$$v(t) = (V_m/T)t \quad (0 < t < T)$$

Using Eq. (7-39)

$$V_{rms}^2 = (1/T) \int_0^T v^2(t)dt$$

$$= (1/T) \int_0^T (V_m^2/T^2)t^2 dt$$

$$= (1/T)(V_m^2/T^2)(T^3/3)$$

$$= (1/3)V_m^2$$

Therefore,

$$V_{rms} = V_m/\sqrt{3} = 0.577 V_m \quad \blacksquare$$

Exercise 7-28 Determine the rms value of the voltage shown in Fig. 7-14. Note that two separate expressions are needed to describe $v(t)$ over one period: one expression valid from 0 to $T/2$, and the other valid from $T/2$ to T.

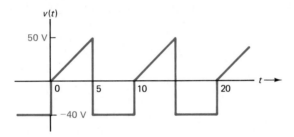

Fig. 7-14 Voltage for Exercise 7-28.

Exercise 7-29 Determine the rms value of the voltage of Exercise 7-28 if the positive portions are erased.

Exercise 7-30 Determine the rms value of the voltage of Exercise 7-28 if the negative portions are erased.

7-8-1 RMS Value of a Sinusoidal Voltage

The rms value of a sinusoidal signal

$$v(t) = V_m \cos(\omega t + \theta)$$

is given by

$$V_{rms} = \left[(1/T) \int_0^T V_m^2 \cos^2(\omega t + \theta) \, dt \right]^{1/2}$$

The integral is evaluated by using the trigonometric identity

$$\cos^2 X = (1/2)(1 + \cos 2X)$$

and the result is

$$V_{rms} = V_m/\sqrt{2} = 0.707 V_m$$

The rms value of a sinusoidal voltage (or current) is *0.707 times the peak value*.

The commonly available ac power in the United States has a frequency of 60 Hz ($\omega = 120\pi = 377$ r/s) and a nominal rms value of 110 V. Its peak value is therefore $(110/0.707) = 156$ V, and it is described by the expression $156 \cos 377t$.

Exercise 7-31 Perform the integration needed to obtain the result given in Eq. (7-40).

In ammeters and voltmeters constructed on what is known as the *electrodynamometer* principle, the deflection of the needle is due to the torque between two current-carrying coils. When both coils carry the same current, the torque is a function of i^2 and the meter deflection is proportional to the mean square value of the current. The scale is calibrated to read the square root of the mean square—that is, the rms value. In such meters, the reading will represent the rms value of any signal applied to it. There are other meters in which the electrodynamometer principle is not used, and the reading on the so-called rms scale in such meters may be valid only for sinusoidal signals.

7-9 SUMMARY OF CHAPTER

The sinusoidal signal is an important type of time-varying signal because of its availability and its mathematical properties. Among the mathematical properties is its relationship to complex exponential functions, which makes it possible to introduce the concept of phasors (to be discussed in Chapter 8). Phasors permit the solution of the differential equations of a circuit by converting them to algebraic equations.

The quantities of interest in a sinusoidal signal are the peak value (or amplitude), the period, the frequency, and the initial phase angle. Sinusoidal voltages (or currents) of the same frequency can be combined into a single sinusoidal function with an amplitude and phase angle. The time delay between two sinusoidal signals of the same frequency is expressed in the form of a phase difference between them.

The voltage and current in a resistor are in phase with each other. The ratio (voltage amplitude/current amplitude) equals the resistance. The instantaneous power received by a resistance is always non-negative.

The voltage and current in a capacitor are 90° out of phase, with the current leading. The ratio (voltage amplitude/current amplitude) equals $(1/\omega C)$. The instantaneous power received by a capacitance is positive for one-half of each cycle and negative for the other half, and the net power consumption over a complete cycle is zero. This is due to the energy storage property of a capacitor: the capacitor receives energy when the voltage increases and delivers energy when the voltage decreases, and the two cancel each other in a complete cycle.

Inductance is the dual of capacitance. The current in an inductor lags the voltage by 90°, and the ratio (voltage amplitude/current amplitude) is (ωL). The instantaneous power received by an inductance is positive when the current is increasing and negative when the current is decreasing. The net power consumption over a complete cycle is zero.

One method of determining the steady state sinusoidal response of a circuit is through the classical method of solution in the time domain. A sinusoidal solution with unknown amplitude and phase angle is assumed, and the two unknowns are evaluated by substituting the assumed solution in the differential equation. The process is time-consuming except for simple circuits, since each independent variable gives rise to a sinusoidal expression with two unknowns.

The instantaneous power received by a circuit varies with time, and consists in general of two components: a constant component, and a component that varies sinusoidally at twice the frequency of the voltage or current. The constant term depends upon the amplitudes of the voltage and current, as well as the cosine of the phase difference between the voltage and current.

Average power is commonly used in measurements and equals one-half of the product of voltage amplitude, current amplitude, and the cosine of the phase difference between voltage and current. For a given set of amplitudes of voltage and current, the average power is a maximum when the voltage and current are in phase, and zero when the voltage and current are 90° out of phase.

The rms (or root mean square) value is commonly used to specify currents and voltages in ac circuits. The rms value of a voltage is the constant value of voltage that would cause a resistor to dissipate the same amount of power as a sinusoidal voltage. It is evaluated by taking the average of the square of the signal and then taking the square root. For a sinusoidal voltage, the rms value is 0.707 times the amplitude.

Answers to Exercises

7-1 (a) $I_m = 15$ mA, $T = 48$ ms, $f = 20.8$ Hz, $\omega = 131$ r/s. $i(t) = 15 \cos(131t + 120°)$ mA. (b) $V_m = 40$ V, $T = 24$ ms, $f = 41.6$ Hz, $\omega = 262$ r/s. $v(t) = 40 \cos(262t - 300°)$ or $40 \cos(262t + 60°)$ V.

7-2 $i(t) = 15 \cos(50t + 106°)$ A.

7-3 v_1 and v_2 are in phase; v_1 leads v_3 by 25°; v_4 leads v_1 by 110°; v_4 leads v_3 by 135°.

7-4 $g(t) = 19.2 \cos 100t + 7.71 \sin 100t$.

7-5 $v_s(t) = 0$.

7-6 When the angle is changed by 180°, P and Q both change in sign. The ratio (P/Q) remains the same. When $P > 0$, the angle must be in the first or fourth quadrant. When $P < 0$, the angle must be in the second or third quadrant.

7-7 (a) $v_a = 18.0 \cos(150t - 56.3°)$ V; (b) $v_b = 18.0 \cos(150t + 56.3°)$ V; (c) $v_c = 18.0 \cos(150t - 124°)$ V; (d) $18.0 \cos(150t + 124°)$ V.

7-8 $i_s = (5 - 10 + 5) \cos 50t + (-8.66 + 8.66) \sin 50t = 0$.

7-9 $v(t) = 212 \cos 5t$ V; $p(t) = 1491 + 1491 \cos 10t$ W.

7-10 $i(t) = -0.01 \sin 10t$ A; $p(t) = -0.1 \sin 20t$ W.

7-11 $v(t) = -160 \sin 400t$ V; $p(t) = -2000 \sin 800t$ W.

7-12 $v(t) = 62.1 \sin(500t + 14.9°) = 62.1 \cos(500t - 75.1°)$ V.

7-13 $v_L(t) = -(\omega L V_m/\sqrt{R^2 + \omega^2 L^2}) \sin[\omega t - \arctan(\omega L/R)]$
$= (\omega L V_m/\sqrt{R^2 + \omega^2 L^2}) \cos[\omega t - \arctan(\omega L/R) - 90°]$.

7-14 (a) $i(t) = 2 \cos(1000t - 53.1°)$ A; (b) $i(t) = 3.16 \cos(250t - 18.4°)$ A; (c) $i(t) = 0.928 \cos(1000t - 21.8°)$ A; (d) $i(t) = 0.494 \cos(1000t - 81.5°)$ A.

7-15 $(R^2 + \omega^2 L^2)^{\frac{1}{2}} = 0.5$. $(\omega L/R) = \tan 36° = 0.726$. $R = 0.404$ Ω; $L = 0.294$ mH.

7-16 $v(t) = (I_m/\sqrt{G^2 + \omega^2 C^2}) \cos[\omega t - \arctan(\omega C/G)]$,
$i(t) = (\omega C I_m/\sqrt{G^2 + \omega^2 C^2}) \cos[\omega t - \arctan(\omega C/G) - 90°]$.

7-17 Equating coefficients of cos ωt: $RA - (B/\omega C) = V_m$. Equating coefficients of sin ωt: $RB + (A/\omega C) = 0$. $A = V_m \omega^2 RC^2/(\omega^2 R^2 C^2 + 1)$; $B = -V_m \omega C/(\omega^2 R^2 C^2 + 1)$. This leads to Eq. (7-28).

7-18 $v_c = V_m/[\omega C \sqrt{R^2 + 1/\omega^2 C^2}] \cos [\omega t + \arctan (1/\omega RC) - 90°]$

7-19 (a) $i = 2.56 \cos (1000t + 39.8°)$ A; (b) $i = 0.958 \cos (250t + 73.3°)$ A; (c) $i = 0.970 \cos (1000t + 14°)$ A; (d) $i = 3.29 \cos (1000t + 9.46°)$ A.

7-20 $v(t) = 78 \cos (10^4 t - 22.6°)$ V.

7-21 $\sqrt{R^2 + (1/\omega^2 C^2)} = 3$. $(1/\omega RC) = \tan 29° = 0.554$. $R = 2.62\Omega$. $C = 688$ μF.

7-22 $v(t) = 17.0 \cos (16t - 45°)$ V. $v(t) = 24 \cos (16t)$ V; $v(t) = 17.0 \cos (16t + 45°)$ V.

7-23 $v(t) = 17.0 \cos (8t + 45°)$ V; $v(t) = 7.59 \cos (8t + 71.6°)$ V; $v(t) = 6.59 \cos (8t + 74°)$ V.

7-24 $i(t) = \dfrac{V_m}{\sqrt{R^2 + \left(\omega L - \dfrac{1}{\omega}C\right)^2}} \cos \left[\omega t - \arctan \dfrac{\left(\omega L - \dfrac{1}{\omega C}\right)}{R}\right]$

7-25 (a) $i(t) = 5.37 \cos (16t - 26.6°)$ A; (b) $i(t) = 6 \cos 16t$ A; (c) $i(t) = 5.37 \cos (16t + 26.6°)$ A.

7-26 $p(t) = 16 \times 10^3 \cos (377t) \cos (377t + \theta_2) = 8000 \cos (754t + \theta_2) + 8000 \cos \theta_2$.
(a) $P_{av} = 6928$ W. (b) -6928 W.

7-27 (a) 75.5°. (b) 132°.

7-28 $V_{rms}^2 = (1/10)[\int_0^5 100t^2 \, dt + \int_5^{10} 1600 \, dt]$. $V_{rms} = 34.9$ V.

7-29 28.3 V.

7-30 20.4 V.

7-31 $V_{rms}^2 = (V_m^2/T) \int_0^T [(1/2) + (1/2) \cos (2\omega t + \theta)] \, dt = V_m^2/2$. $V_{rms} = 0.707 V_m$.

PROBLEMS

Sec. 7-2 Basic Definitions and Relationships for Sinusoidal Functions

7-1 Given: $i_1(t) = 10 \cos (5t + 30°)$ A and $i_2(t) = 50 \sin (5t + 30°)$ A. Draw neat and carefully scaled sketches of i_1 and i_2 to the same scale on a single graph sheet. Show exactly two cycles. Obtain the graph of $i_3 = i_1 + i_2$ from the plots (using point by point addition). Write an expression for $i_3(t)$ from the graph obtained.

7-2 Given $v(t) = 10 \cos (3000t - 30°)$ V. (a) Calculate the frequency of $v(t)$ in Hz. (b) Write an equation for a current of amplitude 1.5 A and lagging behind $v(t)$ by one-sixth of a cycle.

7-3 A sinusoidal function is known to have an amplitude of 20 V. At $t = 0$, $v(t) = -10$ V. At $t = 0.5$ ms, $v(t) = +10$ V. Write an expression for $v(t)$. Choose the smallest frequency that will fit the data. Is there more than one solution?

7-4 Two sinusoidal currents of the same frequency, i_1 and i_2, have amplitudes 20 A and 16 A, respectively. At $t = 0$, $i_1(t) = 20$ A and $i_2(t) = -8$ A. At $t = 2$ ms, $i_1(t) = 10$ A, and $i_2(t) = -16$ A. The frequency is known to be less than 100 Hz. Find the expressions for $i_1(t)$ and $i_2(t)$.

7-5 Determine the phase angle between each of the following pairs of functions. In each case, state which is leading. (a) $v(t) = 30 \cos(100t + 75°)$ V, $i(t) = 40 \cos(100t + 15°)$ A.
(b) $i_1(t) = -20 \cos(200t + 75°)$ A, $i_2(t) = 10 \cos(200t - 15°)$ A.
(c) $v_1(t) = 20 \sin(40t + 10°)$ V, $v_2(t) = 10 \cos(40t - 10°)$ V.

7-6 Express each of the following functions in the form $[P \cos \omega t + Q \sin \omega t]$, with P and Q reduced to single numbers. (a) $v_1(t) = 120 \cos(100t - 150°)$ V; (b) $i_2(t) = 30 \sin(450t + 60°)$ A; (c) $v_3(t) = [30 \cos(50t - 75°) + 25 \sin(50t - 36.9°)]$ V.

7-7 Determine the phase angle between $v_a(t)$ and $v_b(t)$ in each of the following cases:
(a) $v_a(t) = [-10 \cos 40t + 10 \sin 40t]$ V; $v_b(t) = 20 \sin(40t + 45°)$ V.
(b) $v_a(t) = [10 \cos(30t - 60°) + 15 \cos(30t + 15°)]$ V; $v_b(t) = 20 \sin 30t$ V.

7-8 In each of the following cases, determine $v_3 = (v_1 - v_2)$. Simplify the expression to the form $V_m \cos(\omega t + \phi)$ in each case.
(a) $v_1(t) = 20 \cos 100t$ V. $v_2(t) = 30 \sin 100t$ V.
(b) $v_1(t) = -10 \cos(200t - 45°)$ V. $v_2(t) = 20 \sin 200t$ V.
(c) $v_1(t) = 10 \cos(300t - 120°)$ V. $v_2(t) = -10 \cos 300t + 10 \sin(300t - 30°)$ V.

7-9 In the arrangement shown in Fig. 7-15, it is given that: $i_1(t) = 100 \cos 200t$ A; $i_2(t) = 50 \sin 200t$ A; $i_3(t) = 80 \cos(200t - 60°)$ A. Find $i_c(t)$ and express it in the form $I_m \cos(200t + \theta)$.

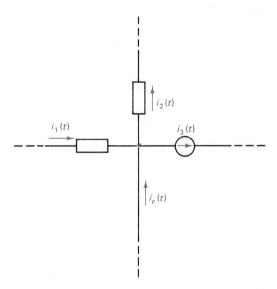

Fig. 7-15 System for Problem 7-9.

7-10 In the system shown in Fig. 7-16, a signal is passed through a phase-shifting network, and the output of the phase shifter is added to the original signal. If $v(t) = A \cos \omega t$: (a) obtain an expression (simplified as far as possible) for the amplitude of the output $v_0(t)$ of the system; (b) obtain an expression (simplified) for the phase angle ϕ of the output of the system; (c) determine the values of the phase shift ϕ for which the output amplitude is a maximum and a minimum.

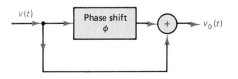

Fig. 7-16 System for Problem 7-10.

Sec. 7-3 Voltage, Current and Power in Single Components

7-11 The voltage across a resistor, $R = 10\ \Omega$, is given by $v(t) = 141.4 \cos(377t)$ V. Write the expression for the current and the instantaneous power $p(t)$. What are the minimum and maximum values of the voltage, current, and power?

7-12 The maximum value of the instantaneous power in a resistor, $R = 25\ \Omega$, is to be 500 W. The duration of each cycle of $p(t)$ is 30 ms. Obtain the expressions for the voltage and current. What is the average power?

7-13 The average power consumed by a resistance is 300 W. The voltage across it has a peak value of 156 V and the frequency is 60 Hz. Obtain the expressions for the current $i(t)$ and instantaneous power $p(t)$; $p(t)$ should be in the form of a constant term plus a sinusoidal term. Calculate the maximum power consumed by the appliance.

7-14 The voltage across a capacitor, $C = 1200$ pF, is $v(t) = 12 \cos(5 \times 10^6)$ V. Obtain the expressions for the current $i(t)$ and the instantaneous power $p(t)$ in the capacitor. What are the minimum and maximum values of the voltage, current, and power?

7-15 The current through a capacitor, $C = 50\ \mu\text{F}$, is $i(t) = 2.5 \cos(5000t)$ A. Obtain the expressions for the voltage $v(t)$ and the instantaneous power $p(t)$ in the capacitor. What are the minimum and maximum values of the voltage, current, and power?

7-16 The maximum value of power in a capacitor, $C = 25\ \mu\text{F}$, is 50 W. The duration of each cycle of power is 0.25 ms. Obtain the expressions for the voltage, current, and instantaneous power.

7-17 Obtain the expression for $w(t)$, the energy stored in a capacitor in the sinusoidal steady state. Draw a neat sketch of $w(t)$. If the maximum value of energy in a capacitor is 0.5 J and the peak value of the voltage is 50 V, calculate the capacitance and write the expressions for $v(t)$, $i(t)$, and $p(t)$. Assume $\omega = 1000$ r/s.

7-18 The current in an inductor, $L = 15$ mH, is $i(t) = 4 \cos(500t)$ A. Determine the voltage $v(t)$ and the instantaneous power $p(t)$ in the inductor. What are the maximum and minimum values of the voltage, current, and power?

7-19 The maximum value of power in an inductor, $L = 25$ mH, is 50 W. The duration of each cycle of power is 0.25 ms. Obtain the expressions for the voltage, current, and instantaneous power.

7-20 Repeat problem 7-17 for an inductor. On the same graph, show $w(t)$ for the inductor as well as the capacitor.

7-21 Consider a resistor, an inductor, and a capacitor. Let the current in each case be $I_m \cos(\omega t + \phi)$ A. Discuss the effect of each of the following changes on the instantaneous power (its peak value, frequency, and constant component) for the three elements. Treat each change as being independent of the others. (a) The amplitude of the current is increased by a factor of K_1. (b) The frequency of the current is increased by a factor of K_2. (c) The phase angle of the current is shifted by a value of ϕ_1. Present the results in a tabular form.

7-22 Repeat the previous problem when the voltage $v(t)$ is given as $V_m \cos \omega t$ and the changes pertain to $v(t)$.

Secs. 7-4, 7-5, 7-6 Time Domain Solution of Simple Series and Parallel Circuits

[Note: In all problems, the final result must be reduced to a single sinusoidal term of the form $A \cos(\omega t + \phi)$.]

7-23 The current $i(t)$ in a series circuit made up of $R = 25\ \Omega$, $L = 0.2$ H, and $C = 8\ \mu\text{F}$ is given as $i(t) = 20 \cos(500t)$ A. Determine the voltage across the series circuit.

7-24 The voltage $v(t)$ across a parallel circuit made up of $G = 0.1$ S, $L = 0.02$ H, and $C = 10\ \mu\text{F}$ is given as $v(t) = 100 \cos 2000t$ V. Determine the total current in the circuit.

7-25 In a series RLC circuit, it is given that the voltage across the capacitor is $v_c(t) = V_{cm} \cos \omega t$ V. Obtain an expression for the total voltage $v_s(t)$ across the circuit.

7-26 Consider a parallel combination of a conductance G and an inductance L driven by a current source $i_s = I_m \cos \omega t$. Set up the differential equation for the voltage $v(t)$ across the circuit. Assume a solution of the form $v(t) = [P \cos (\omega t) + Q \sin (\omega t)]$. Determine the values of P and Q. Write the final expression for $v(t)$. Verify your result by using duality on the series RC circuit discussed in Sec. 7-5.

7-27 A series combination of L and C is driven by a voltage source $v_s(t) = V_m \cos (\omega t)$. Set up the differential equation for the current $i(t)$ in the circuit. Assume a solution of the form $i(t) = P \cos \omega t + Q \sin \omega t$) and determine the values of P and Q. Write the final expression for $i(t)$. Obtain the expressions for $v_c(t)$, the voltage across the capacitor, and $v_L(t)$, the voltage across the inductor. Draw neat sketches of v_s, v_c, and v_L on the same set of axes.

7-28 In solving the differential equation of a series RL circuit, an alternative form of the assumed solution is $i(t) = I_m \cos (\omega t + \phi)$. Substitute this solution in the differential equation and evaluate I_m and ϕ by equating the coefficients of cosine terms on both sides and the sine terms on both sides. This form of assumed solution gives the amplitude and phase angle directly.

7-29 A resistance $R = 30 \, \Omega$ is in series with an inductor L. The applied voltage to the series combination is $v(t) = 100 \cos (628t)$ V, and the resulting current has an amplitude of 2 A. Determine the value of L. Write the expression for $i(t)$.

7-30 The voltage across a parallel combination of a conductance G and capacitance C is $v(t) = 32 \cos (15t)$ V and the total current to the circuit is $i(t) = 12.5 \cos (15t + 36.9°)$ A. Determine the values of G and C.

7-31 A series RLC circuit with $R = 12 \, \Omega$, $L = 25$ mH, and $C = 400 \, \mu F$ is driven by a voltage source $v_s = 100 \cos (\omega t)$ V. Determine the value of ω for each of the following cases: (a) v_s leads the current in the circuit by 60°; (b) v_s is in phase with the current; (c) v_s lags behind the current by 30°.

7-32 A parallel GLC circuit with $G = 0.25$ S, $L = 25$ mH, and $C = 400 \, \mu F$ is driven by a current source $i_s = 100 \cos \omega t$. Determine the value of ω for each of the following cases: (a) i_s leads the voltage across the circuit by 60°; (b) i_s is in phase with the voltage; (c) i_s lags behind the voltage by 30°.

7-33 Let $v_c(t)$ be the voltage across the capacitance in a series RC circuit and let $v_s(t)$ be the total voltage across the circuit. If the phase angle between v_c and v_s is 45° at $\omega = \omega_0$, obtain the relationship between ω_0, R, and C. Is v_c leading or lagging behind v_s at ω_0? Calculate the ratio of the amplitude of v_c to the amplitude of v_s at (a) $\omega = \omega_0$; (b) $\omega = 10 \, \omega_0$; (c) $\omega = 0.1 \, \omega_0$.

Sec. 7-7 Instantaneous and Average Power

7-34 The voltage applied to a series RL circuit with $R = 20 \, \Omega$ and $L = 15$ mH is $v(t) = 60 \cos (2500t)$ V. Obtain the expression for the instantaneous power $p(t)$ and write it as the sum of a constant term and a sinusoidal term. What is the value of average power in the circuit?

7-35 Repeat the work of the previous problem for a series RC circuit with $R = 20 \, \Omega$ and $C = 40 \, \mu F$.

7-36 The average power received by a circuit is 500 W. The applied voltage has an amplitude of 250 V. For each of the following cases, determine the amplitude of the current in the circuit and write the expression for $p(t)$.
(a) The voltage leads the current by 30°; (b) the current leads the voltage by 30°;
(c) the voltage is in phase with the current.

7-37 Consider a series RL circuit with an applied voltage $v_s(t) = V_m \cos \omega t$. Write the expression for the instantaneous power in each of the two individual elements in the form of a constant term plus a sinusoidal term. Show that the total average power received by the circuit equals the sum of the average power consumed by the resistance and the average power consumed by the inductance.

Sec. 7-8 RMS Value of Time Varying Signals

7-38 Determine the rms value of each of the signals shown in Fig. 7-17.

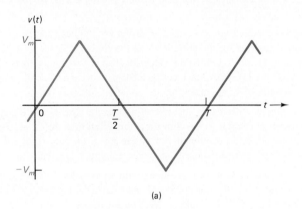

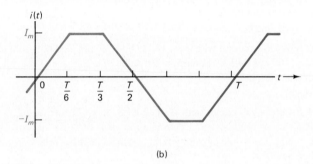

Fig. 7-17 Signals for Problem 7-38.

7-39 The waveform shown in Fig. 7-18 is the output of a half-wave rectifier. Determine its rms value. How does this value compare with that of a regular sinusoidal signal?

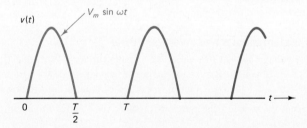

Fig. 7-18 Waveform for Problem 7-39.

7-40 Show that the average power received by a circuit with $v(t) = V_m \cos(\omega t)$ V and $i(t) = I_m \cos(\omega t + \phi)$ A can be written in the form $(V_{rms} I_{rms} \cos \phi)$.

7-41 An electronic voltmeter is designed so that it senses the positive peak value of the applied signal, and its scale is calibrated to read ($0.707 \times$ peak value) so that the scale reading will give the correct rms values for sinusoidal signals. Suppose the meter reading is 75 V. For each of the following cases, calculate the true rms value of the signal: (a) The signal is a pure sinusoid. (b) The signal is a half-wave rectified sinusoid as in Fig. 7-18. (c) The signal is a dc voltage.

Sinusoidal Steady State–Time Domain Analysis

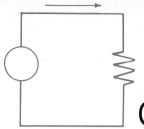

CHAPTER 8
PHASORS, IMPEDANCE, AND ADMITTANCE

The previous chapter showed that the classical solution of the differential equations of circuits under sinusoidal excitation becomes quite laborious, except in the case of simple circuits. Not only does the effort of solving a set of simultaneous differential equations become enormous, but the significance of the results gets buried in the details of analysis. In this chapter, we develop a more efficient and convenient approach using the concept of phasors.

Euler's theorem provides a relationship between sinusoidal functions and complex exponential functions and permits the replacement of sinusoidal functions by equivalent complex exponential functions. This replacement leads to a phasor representation of sinusoidal voltages and currents. The phasor concept permits the conversion of linear *differential* equations to simultaneous linear *algebraic* equations. The simultaneous equations involve complex numbers and are therefore more cumbersome than those involving only real numbers. Nevertheless, they are definitely easier to solve than a set of simultaneous differential equations!

The use of phasors in sinusoidal steady state analysis leads to the concepts of impedance and admittance, which are analogous to the resistance and conductance encountered in purely resistive circuits. The methods of analysis and the network theorems developed in earlier chapters for purely resistive networks can be readily modified for the sinusoidal steady state analysis of circuits containing capacitors and inductors as well as resistors.

8-1 COMPLEX EXPONENTIAL FUNCTIONS

The *Maclaurin's series* expansion of the exponential function e^x is given by

$$e^x = 1 + x + (x^2/2!) + \ldots + (x^n/n!) + \ldots \qquad (8\text{-}1)$$

If the variable x is replaced by the complex quantity* $(j\theta)$, then Eq. (8-1) becomes

$$e^{j\theta} = 1 + j\theta + (j\theta)^2/2! + (j\theta)^3/3! + \ldots + (j\theta)^n/n! + \ldots$$

Using the relationship $j^2 = -1$, the last series is rewritten in the form

$$e^{j\theta} = [1 - (\theta^2/2!) + (\theta^4/4!) - (\theta^6/6!) + \ldots]$$
$$+ j[\theta - (\theta^3/3!) + (\theta^5/5!) - (\theta^7/7!) + \ldots] \quad (8\text{-}2)$$

Reference to a mathematical handbook shows that the two parts of the series in Eq. (8-2) represent the Maclaurin's series for the trigonometric functions $(\cos \theta)$ and $(\sin \theta)$, respectively. Thus Eq. (8-2) leads to the identity:

$$e^{j\theta} = \cos \theta + j \sin \theta \quad (8\text{-}3a)$$

which is known as *Euler's theorem*. A similar procedure leads to the identity:

$$e^{-j\theta} = \cos \theta - j \sin \theta \quad (8\text{-}3b)$$

A combination of Eqs. (8-3a) and (8-3b) gives the following two relationships:

$$(e^{j\theta} + e^{-j\theta}) = 2 \cos \theta \quad (8\text{-}4a)$$

and

$$(e^{j\theta} - e^{-j\theta}) = 2j \sin \theta \quad (8\text{-}4b)$$

The relationships in Eqs. (8-4a) and (8-4b) permit the writing of a *real* sinusoidal function as the sum or difference of complex exponential functions.

If the notation **Re**[ ] is used to denote the operation "take the real part of the quantity within the brackets," then it follows from Eqs. (8-3a) and (8-3b) that

$$\mathbf{Re}[e^{j\theta}] = \mathbf{Re}[e^{-j\theta}] = \cos \theta \quad (8\text{-}5)$$

If the notation *Im*[ ] denotes the operation "take the imaginary part of the quantity within the brackets," then it follows from Eqs. (8-3a) and (8-3b) that

$$\mathbf{Im}[e^{j\theta}] = \sin \theta \quad (8\text{-}6)$$

and

$$\mathbf{Im}[e^{-j\theta}] = -\sin \theta \quad (8\text{-}7)$$

Eqs. (8-3) through (8-7) are also valid when θ is a function of time. If θ is replaced by (ωt), for example, the above equations become

$$e^{j\omega t} = \cos \omega t + j \sin \omega t \quad (8\text{-}8)$$
$$(e^{j\omega t} + e^{-j\omega t}) = 2 \cos \omega t \quad (8\text{-}9)$$
$$(e^{j\omega t} - e^{-j\omega t}) = 2j \sin \omega t \quad (8\text{-}10)$$
$$\mathbf{Re}[e^{\pm j\omega t}] = \cos \omega t \quad (8\text{-}11)$$
$$\mathbf{Im}[e^{\pm j\omega t}] = \pm \sin \omega t \quad (8\text{-}12)$$

Eq. (8-3) through (8-12) are constantly used in ac circuit analysis as well as the analysis of linear systems. One must be fully familiar with their use, as illustrated by the following examples.

---

*A brief summary of the algebra of complex numbers is presented in Appendix B.

Example 8-1 Evaluate $\mathbf{Re}[(a + jb)e^{-jct}]$ where a, b, and c are real quantities.

Solution First, expand the expression in brackets and reduce it to a single complex quantity.

$$(a + jb)e^{-jct} = (a + jb)(\cos ct - j \sin ct)$$
$$= (a \cos ct + b \sin ct) + j(-a \sin ct + b \cos ct)$$

Taking the real part

$$\mathbf{Re}[(a + jb)e^{-jct}] = a \cos ct + b \sin ct \qquad \blacksquare$$

Example 8-2 Determine $\mathbf{Im}[K_1 e^{jX} e^{-jBt}]$ where k_1, X, and B are real quantities.

Solution
$$K_1 e^{jX} e^{-jBt} = K_1 e^{-j(Bt - X)}$$
$$= K_1[\cos (Bt - X) - j \sin (Bt - X)]$$

Therefore

$$\mathbf{Im}[K_1 e^{jX} e^{-jBt}] = -K_1 \sin (Bt - X) \qquad \blacksquare$$

Example 8-3 Given $f(t) = K_2 \cos (K_3 t + \theta_1)$, express $f(t)$ in the form $\mathbf{Re}[\mathbf{A} e^{j\omega t}]$, where $\mathbf{A}$ is a *complex* quantity.

Solution From Eq. (8-11)

$$K_2 \cos (K_3 t + \theta_1) = \mathbf{Re}[K_2 e^{j(K_3 t + \theta_1)}]$$
$$= \mathbf{Re}[(K_2 e^{j\theta_1}) e^{jK_3 t}]$$
$$= \mathbf{Re}[\mathbf{A} e^{j\omega t}]$$

where $\mathbf{A} = K_2 e^{j\theta}$ and $\omega = K_3$. $\qquad \blacksquare$

Example 8-4 Express $f(t) = 3 \cos 10t + 4 \sin 10t$ in the form $\mathbf{Re}[\mathbf{A} e^{jBt}]$ where $\mathbf{A}$ is a complex number and B is a real number.

Solution First reduce $f(t)$ to the form of a single cosine term with a phase angle by using the procedure presented in the previous chapter:

$$f(t) = 5 \cos (10t - 53.1°)$$
$$= \mathbf{Re}[5 e^{j(10t - 53.1°)}] \text{ from Eq. (8-11)}$$
$$= \mathbf{Re}[(5 e^{-j53.1°}) e^{j10t}]$$

Therefore $\mathbf{A} = 5 e^{-j53.1°}$, and $B = 10$. $\qquad \blacksquare$

In complex exponentials, angles such as 53.1° in the previous example should be expressed in radians to be consistent with the quantity ($10t$) which is in turn radians. As we noted in Chapter 7, however, the use of degrees is an accepted practice in electrical circuits.

Exercise 8-1 Reduce each of the following quantities to a single real number: (a) $\mathbf{Re}[(10 + j10) e^{j50t}]$. (b) $\mathbf{Re}[(10 - j10) e^{-j50t}]$. (c) $\mathbf{Im}[(10 + j10) e^{j50t}]$. (d) $\mathbf{Im}[(10 - j10) e^{-j50t}]$.

Exercise 8-2 Evaluate the real and imaginary parts of each of the following expressions: (a) $(3 + j4)(je^{j2t})$. (b) $(-3 + j4)(-1 - j)e^{j(10t + 45)}$.

Exercise 8-3 Express each of the following functions in the form $\mathbf{Re}[\mathbf{A}e^{jBt}]$ where $\mathbf{A}$ is a complex number and B is a real number. (a) $f_1(t) = 100 \cos (60t - 30°)$; (b) $f_2(t) = (86.6 \cos 60t + 50 \sin 60t)$; (c) $f_3(t) = (50 \cos 60t + 86.6 \sin 60t)$

8-2 THE PHASOR CONCEPT

The three parameters that describe a sinusoidal voltage or current are amplitude, phase angle, and frequency. When a linear circuit is driven by one or more sinusoidal sources, all at the same frequency, the amplitudes and phase angles of the currents and voltages will differ from one branch to the next; but all the voltages and currents will be at the single frequency dictated by the forcing functions. The description of a circuit's response must therefore give the amplitude and phase angle of each current and voltage. The frequency need not be listed separately for each quantity, since it is the same throughout the circuit. It is therefore possible to set up a notation that focuses on the amplitudes and phase angles of currents and voltages while ignoring the frequency (which needs to be specified once in each problem). Such a notation leads to the concept of *phasors*.

To develop the concept of phasors, consider an arrow length of V_m, at an angle of θ at $t = 0$, rotating at an angular velocity of ω r/s, as shown in Fig. 8-1(a). The arrow may be viewed as the graphical representation of a complex number of magnitude V_m and angle θ: $V_m e^{j\theta}$. As the arrow rotates, it assumes new angles given by $(\omega t + \theta)$, as shown in Fig. 8-1(b), and it is described by $[V_m e^{(j\theta + j\omega t)}] = [V_m e^{j\theta}]e^{j\omega t}$.

Let

$$\mathbf{V} = V_m e^{j\theta} \qquad (8\text{-}13)$$

Then a complex function

$$\mathbf{v}(t) = \mathbf{V}e^{j\omega t} \qquad (8\text{-}14)$$

describes the arrow at any time t. The horizontal projection of the rotating arrow gives the real part of $\mathbf{v}(t)$:

$$\mathbf{Re}[\mathbf{v}(t)] = \mathbf{Re}[\mathbf{V}e^{j\omega t}] = \mathbf{Re}[V_m e^{j\theta}e^{j\omega t}]$$
$$= V_m \cos (\omega t + \theta) \qquad (8\text{-}15)$$

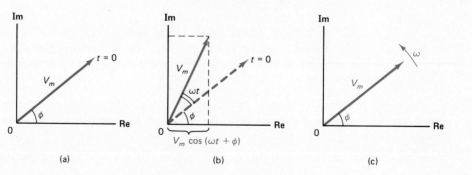

Fig. 8-1 A sinusoidal voltage can be treated as (a) the real or (b) imaginary part of a complex exponential function. (c) Phasor representation.

That is, *a physically realizable voltage function*, $V_m \cos(\omega t + \theta)$, *may be thought of as the real part of a complex voltage function* $\mathbf{v}(t)$ *described by a rotating arrow of magnitude* V_m *and an initial phase angle* θ.

Instead of showing the position of the rotating arrow at different instants of time, it is sufficient to show the arrow in its *initial* position and imply that it is rotating at the rate of ω radians per second, as in Fig. 8-1(c), which is a snapshot of the rotating arrow at $t = 0$.

The *complex coefficient* $\mathbf{V} = V_m e^{j\theta}$ of $\mathbf{v}(t)$ in Eq. (8-14) is called a *phasor*. The phasor $V_m e^{j\theta}$ contains explicit information about the amplitude and the initial phase angle of $\mathbf{v}(t)$.

Only the frequency is missing, since, as we noted at the beginning of this section, the frequency is a constant throughout the circuit, it needs to be listed only once as a single parameter for the entire circuit. A complex voltage function $\mathbf{v}(t)$ can therefore be represented by a phasor $\mathbf{V}$ with a magnitude V_m and an angle θ.

A similar argument leads to the current phasor $\mathbf{I} = I_m e^{j\theta}$, which is the representation of a complex current function $\mathbf{i}(t) = \mathbf{I} e^{j\omega t}$.

The diagrams of Fig. 8-2 show several examples of voltage and current phasors. Each phasor is described by a complex number and represents a complex time function $\mathbf{v}(t)$ or $\mathbf{i}(t)$.

Let us summarize the discussion: A complex (voltage or current) exponential function of time is represented by a complex number called a *phasor*, which is the value of the complex function at $t = 0$. The phasor describes the complex function completely except for the frequency, which must be specified separately.

Given a phasor, the corresponding complex exponential function is obtained by multiplying the phasor $e^{j\omega t}$.

Fig. 8-2 Examples of voltage and current phasors.

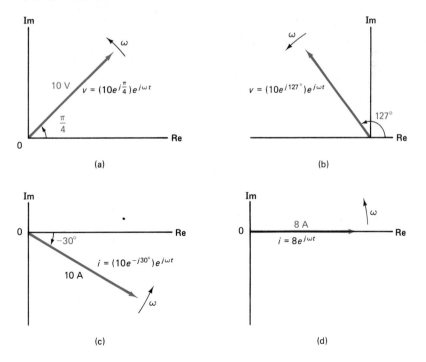

8-2 The Phasor Concept

8-2-1 Phasors and Sinusoidal Forcing Functions

Eq. (8-15) shows that the real part of a complex voltage or current exponential function gives a physically attainable sinusoidal driving function described by a *cosine* term with a phase angle. The imaginary part of a complex exponential function also gives a physically realizable sinusoidal function described by a *sine* term with a phase angle. That is

$$\mathbf{Re}[\mathbf{v}(t)] = V_m \cos(\omega t + \theta) \qquad (8\text{-}16)$$

and

$$\mathbf{Im}[\mathbf{v}(t)] = V_m \sin(\omega t + \theta) \qquad (8\text{-}17)$$

where $\mathbf{v}(t)$ is defined by Eqs. (8-13) and (8-14). Thus any given sinusoidal forcing function can be expressed as either the real part or the imaginary part of a complex exponential forcing function.

The *standard convention* used in circuit analysis is to express a sinusoidal forcing function as the *real part* of a complex exponential forcing function, as in Eq. (8-16).

Example 8-5 Find the complex exponential forcing function whose real part corresponds to each of the following sinusoidal forcing functions. (a) $v_1(t) = 10 \cos 50t$ V. (b) $i_2(t) = 10 \sin 50t$ A. (c) $v_3(t) = (10 \cos 50t + 10 \sin 50t)$ V. (d) $i_4(t) = 30 \sin(100t + 60°)$ A.

Solution (a) $v_1(t) = \mathbf{Re}[10e^{j50t}]$. Therefore, $\mathbf{v}_1(t) = 10e^{j50t}$ V. (b) $i_2(t)$ must be first expressed as a cosine term. Subtract 90° to go from sine form to cosine form. Then $i_2(t) = 10 \cos(50t - 90°) = \mathbf{Re}[10e^{j(50t - 90°)}]$. Therefore $\mathbf{i}_2(t) = 10e^{j(50t - 90°)}$ A. (c) $v_3(t) = (10 \cos 50t + 10 \sin 50t) = 14.14 \cos(50t - 45°) = \mathbf{Re}[14.14e^{j(50t - 45°)}]$. Therefore, $\mathbf{v}_3(t) = 14.14e^{j(50t - 45°)}$ (d) $i_4(t) = 30 \sin(100t + 60°) = 30 \cos(100t - 30°) = \mathbf{Re}[30e^{j(100t - 30°)}]$. Therefore, $\mathbf{i}_4(t) = 30e^{j(100t - 30°)}$ A. ∎

Exercise 8-4 Find the complex exponential forcing function whose real part is given by (a) $v_1(t) = -10 \cos 80t$ V; (b) $v_2(t) = 8 \sin(190t - 75°)$ V; (c) $i_3(t) = (3 \cos 80t - 4 \sin 80t)$ A; (d) $i_4(t) = (-3 \cos 80t - 4 \sin 80t)$ A.

Since a sinusoidal function is related to a complex exponential function, it may be represented by the phasor that corresponds to the complex exponential function. Using Eqs. (8-13) and (8-14) in Eq. (8-16), a sinusoidal forcing function is written as

$$V_m \cos(\omega t + \theta) = \mathbf{Re}[\mathbf{v}(t)] = \mathbf{Re}[(V_m e^{j\theta})e^{j\omega t}] \qquad (8\text{-}18)$$

The sinusoidal voltage function $[V_m \cos(\omega t + \theta)]$ is then represented by the phasor $(V_m e^{j\theta})$. A similar comment is valid when a sinusoidal current function is used.

That is, given a sinusoidal forcing function (voltage or current), its phasor representation is a complex quantity of magnitude equal to the amplitude and an angle equal to the phase angle of the given function.

Remember that the phasor is a *notation* for representing sinusoidal functions of time; it is not identical to the sinusoidal forcing function. A phasor is a *complex number*, whereas a sinusoidal forcing function is a *real function of time*. The link between a phasor and a sinusoidal function is a complex exponential function, as expressed by Eq. (8-18).

The *quick conversion* process between a sinusoidal function and its phasor representation is as follows:

Actual Function		Phasor Representation
$v(t) = V_m \cos(\omega t + \theta)$	$\leftrightarrow$	$\mathbf{V} = V_m e^{j\theta}$
$i(t) = I_m \cos(\omega t + \phi)$	$\leftrightarrow$	$\mathbf{I} = I_m e^{j\phi}$

Note that the actual function must be in the form of a *single cosine* term with a phase angle; that is, in the form $A \cos(\omega t + \alpha)$, for use with the conversion formulas.

Example 8-6 Obtain the phasor representing each of these forcing functions: (a) $i_1(t) = 10 \cos(377t - 50°)$ A. (b) $v_2(t) = 80 \sin 500t$ V. (c) $i_3(t) = (8 \cos 10t + 13 \sin 10t)$ A.

Solution (a) $i_1(t) = 10 \cos(377t - 50°)$ is represented by $\mathbf{I}_1 = 10 e^{-j50°}$ A. (b) $v_2(t) = 80 \sin 500t = 80 \cos(500t - 90°)$ is represented by $\mathbf{V}_2 = 80 e^{-j90°}$ V. (c) $i_3(t) = (8 \cos 10t + 13 \sin 10t) = 15.3 \cos(10t - 58.4°)$ is represented by $\mathbf{I}_3 = 15.3 e^{-j58.4°}$ A. ∎

Exercise 8-5 Obtain the phasors corresponding to the following forcing functions: (a) $v_a(t) = -100 \cos 40t$ V. (b) $v_b(t) = -100 \sin 40t$ V. (c) $i_c(t) = (-100 \cos 40t + 100 \sin 40t)$ A.

Example 8-7 Determine the sinusoidal functions represented by the following phasors: (a) $\mathbf{V}_1 = 100 e^{j45°}$ V. (b) $\mathbf{I}_2 = (-4 + j8)$ A. (c) $\mathbf{I}_3 = [(3 - j7)(10 e^{j30°})]$ A. (d) $\mathbf{V}_4 = [(4 + j5) - (3 + j8)]$ V.

Solution The formal derivation involves the use of Eq. 8-18. Multiply the phasor by $e^{j\omega t}$, simplify the product, and then take the real part. The quick conversion formulas listed earlier can also be used.

(a) $v_a(t) = \mathbf{Re}[100 e^{j45°} e^{j\omega t}] = \mathbf{Re}[100 e^{j(\omega t + 45°)}] = 100 \cos(\omega t + 45°)$ V.
(b) $i_2(t) = \mathbf{Re}[(-4 + j8) e^{j\omega t}] = \mathbf{Re}[8.94 e^{j(\omega t + 116°)}] = 8.94 \cos(\omega t + 116°)$ A.
(c) $i_3(t) = \mathbf{Re}[(3 - j7)(10 e^{j30°}) e^{j\omega t}] = \mathbf{Re}[(76.2 e^{-j36.8°}) e^{j\omega t}]$
$= 76.2 \cos(\omega t - 36.8°)$ A.
(d) $v_4(t) = \mathbf{Re}[(1 - j3) e^{j\omega t}] = \mathbf{Re}[3.16 e^{-j71.6°} e^{j\omega t}] = 3.16 \cos(\omega t - 71.6°)$. ∎

Exercise 8-6 Determine the time functions represented by the following phasors: (a) $\mathbf{V}_1 = 100 e^{-j60°}$ V. (b) $\mathbf{I}_2 = (-5 - j8.66)$ A. (c) $\mathbf{V}_3 = [(5 + j8.66)(100 e^{-j60°})]$ V. (d) $\mathbf{I}_4 = j e^{j45}$ A.

Phasors are helpful in the addition and subtraction of sinusoidal functions of the *same frequency*. The time functions are first converted to phasors; the phasors added or subtracted; and the resultant phasor is then converted back to a time function. Note that manipulations (addition, subtraction, multiplication, and division) using phasors are valid only when they represent currents and voltages of the *same frequency*. With currents and voltages of different frequencies, manipulations using phasors are meaningless; the complete complex exponential function must be used in such cases.

Example 8-8 If $i_1(t) = 10 \cos(220t + 36.9°)$ A and $i_2(t) = 20 \cos(220t - 90°)$, determine $i_3 = i_1 + i_2$.

Solution The phasor representations of the two currents are $\mathbf{I}_1 = 10e^{j36.9°} = (8 + j6)$ A, and $\mathbf{I}_2 = 20e^{-j90°} = -j20$. Therefore

$$\mathbf{I}_3 = \mathbf{I}_1 + \mathbf{I}_2 = (8 - j14) = 16.1e^{-60°} \text{ A}$$

and

$$i_3(t) = 16.1 \cos(200t - 60°) \text{ A}$$

■

Exercise 8-7 Given $v_1(t) = 100 \cos(377t + 75°)$ and $v_2(t) = (30 \cos 377t - 40 \sin 377t)$, find their sum, v_3.

Exercise 8-8 The voltages in a three-phase power system are given by $v_a = 311 \cos 377t$ V, $v_b = 311 \cos(377t + 120°)$ V, and $v_c = 311 \cos(377t - 120°)$ V. Show that the sum of the three voltages is zero.

Forms of Writing a Phasor

There are three forms of writing a phasor:

- Exponential form: $Ae^{j\theta}$
- Polar form: $A\underline{/\theta}$
- Rectangular form: $(A \cos \theta + jB \sin \theta)$

The polar form is almost the same as the exponential form: the angle is written as angle θ in the polar form and as an exponent ($j\theta$) in the exponential form. The polar form is the most widely used notation in ac circuit analysis. The rectangular form is necessary for addition and subtraction of phasors.

8-3 THE SERIES RL CIRCUIT UNDER COMPLEX EXPONENTIAL EXCITATION

The concept of phasors allows considerable simplification in the sinusoidal steady state analysis of circuits. The simplification is due to the use of complex exponential functions as forcing functions, instead of sinusoidal forcing functions. With complex exponential forcing functions, differentiation and integration are replaced by multiplication and division by ($j\omega$), so that the differential equation becomes a linear algebraic equation that can be readily solved. The use of complex exponential forcing functions in sinusoidal steady state analysis will be illustrated first by means of the series RL circuit.

Consider a series RL circuit driven by a complex exponential function

$$\mathbf{v}_s(t) = V_m e^{j(\omega t + \theta)} = [V_m e^{j\theta}]e^{j\omega t} \quad (8\text{-}19)$$

Let $\mathbf{v}_s(t)$ be written in the form

$$\mathbf{v}_s(t) = \mathbf{V}_s e^{j\omega t} \quad (8\text{-}20)$$

where

$$\mathbf{V}_s = V_m e^{j\omega t}$$

is the phasor coefficient of $\mathbf{v}_s(t)$.

The differential equation of the RL circuit becomes

$$Ri + L\frac{di}{dt} = V_s e^{j\omega t} \quad (8\text{-}21)$$

Since the circuit is driven by a complex exponential function, the steady state response will also be a complex exponential function at the same frequency as $v_s(t)$, but with some phasor coefficient $\mathbf{I}_s$. Assume a solution of the form

$$\mathbf{i}(t) = \mathbf{I}_s e^{j\omega t} \quad (8\text{-}22)$$

where the magnitude and angle of the phasor $\mathbf{I}_s$ are to be determined. Differentiating Eq. (8-22)

$$[d\mathbf{i}(t)/dt] = (j\omega)\mathbf{I}_s e^{j\omega t} = (j\omega)\mathbf{i}(t) \quad (8\text{-}23)$$

That is, the process of *differentiation is equivalent to multiplication by the factor* $(j\omega)$ for a complex *exponential function.*

When Eqs. (8-22) and (8-23) are used, the differential equation (8-21) transforms to

$$R\mathbf{I}_s e^{j\omega t} + (j\omega)L\mathbf{I}_s e^{j\omega t} = \mathbf{V}_s e^{j\omega t}$$

which is an algebraic equation. Solving the equation for $\mathbf{I}_s$

$$\mathbf{I}_s = \mathbf{V}_s/(R + j\omega L) \quad (8\text{-}24)$$

and the solution of the differential equation Eq. (8-22) becomes

$$\mathbf{i}(t) = \frac{\mathbf{V}_s}{R + j\omega L} e^{j\omega t} \quad (8\text{-}25)$$

or

$$\mathbf{i}(t) = \frac{V_m}{\sqrt{R^2 + \omega^2 L^2}} e^{j(\omega t + \theta + \phi)} \quad (8\text{-}26)$$

where $\phi = \arctan -\omega L/R$.

Eq. (8-26) gives the complex exponential current response of a series RL circuit when driven by the complex exponential voltage $v_s(t)$ in Eq. (8-19). Now, consider the same circuit driven by $\mathbf{v}_s^*(t)$, the complex conjugate of $v_s(t)$. That is

$$\mathbf{v}_s^*(t) = V_m e^{-j(\omega t + \theta)} \quad (8\text{-}27)$$

Using the same procedure as before, the complex current response of the circuit is found to be $\mathbf{i}^*(t)$, the complex conjugate of $\mathbf{i}(t)$:

$$\mathbf{i}^*(t) = \frac{V_m}{\sqrt{R^2 + \omega^2 L^2}} e^{-j(\omega t + \theta + \phi)} \quad (8\text{-}28)$$

where the angle ϕ is still given by $\arctan(-\omega L/R)$.

Now if the circuit is driven by the *sum* of the two functions $v_s(t)$ and $\mathbf{v}_s^*(t)$, superposition dictates that the response will be the sum of the two responses $\mathbf{i}(t)$ and $\mathbf{i}^*(t)$. That is,

$$\text{forcing function} = [v_s(t) + \mathbf{v}_s^*(t)]$$

then

$$\text{response} = [\mathbf{i}(t) + \mathbf{i}^*(t)] \quad (8\text{-}29)$$

But
$$[\mathbf{v}_s(t) + \mathbf{v}_s^*(t)] = 2\mathbf{Re}[\mathbf{v}_s(t)] \qquad (8\text{-}30)$$

and
$$[\mathbf{i}(t) + \mathbf{i}^*(t)] = 2\mathbf{Re}[\mathbf{i}(t)] \qquad (8\text{-}31)$$

Therefore, it follows from Eq. (8-29) that if
$$\text{forcing function} = \mathbf{Re}[\mathbf{v}_s(t)]$$

then
$$\text{response} = \mathbf{Re}[\mathbf{i}(t)] \qquad (8\text{-}32)$$

For the case of $\mathbf{v}_s(t) = V_m e^{j(\omega t + \theta)}$,
$$\mathbf{Re}[\mathbf{v}_s(t)] = V_m \cos(\omega t + \theta) \qquad (8\text{-}33)$$

The real part of $\mathbf{i}(t)$ in Eq. (8-26) is found to be (after a fair amount of manipulation):
$$\mathbf{Re}[\mathbf{i}(t)] = \frac{V_m}{\sqrt{R^2 + \omega^2 L^2}} \cos(\omega t + \theta + \phi) \qquad (8\text{-}34)$$

The expression in Eq. (8-34) is the same as that obtained in the previous chapter through the classical time domain solution of the series RL circuit.

The procedure shows that for a complex exponential forcing function, the differential equation of a series RL circuit becomes an algebraic equation. The response of the circuit is a complex exponential function obtained by solving the algebraic equation. When the circuit is driven by the *real part of the complex exponential forcing function*, the response is the *real part of the complex exponential response*. It is possible, then, to find the response of a series RL circuit to a sinusoidal forcing function by first obtaining the response to a complex exponential function and then taking the real part of the response.

Exercise 8-9 Consider a circuit containing a conductance $G = 2$ S in parallel with a capacitance $C = 0.005$ F driven by a current source $\mathbf{i}_s(t)$. Set up the differential equation for the voltage $\mathbf{v}(t)$ across the circuit. Let $\mathbf{i}_s(t) = 10 e^{j250t}$ A. Assume a solution of the form $\mathbf{v}(t) = \mathbf{V}_m e^{j250t}$. Solve the resulting algebraic equation for the phasor $\mathbf{V}_m$ and write the expression for $\mathbf{v}(t)$. Take the real part of $\mathbf{v}(t)$ and verify the result by using the classical time domain procedure presented in Chapter 7.

Exercise 8-10 Repeat the work of the previous exercise when an inductance of $L = 0.01$ H is added in parallel to G and C.

8-4 ANALYSIS OF CIRCUITS IN THE SINUSOIDAL STEADY STATE: PROCEDURE USING PHASORS

The discussion of the previous section can be extended to a general linear circuit in the following way. Let the circuit be driven by a sinusoidal forcing function
$$f_s(t) = A \cos(\omega t + \theta) \qquad (8\text{-}35)$$

and it is required to find the response $g(t)$. Let $\mathbf{f}_e(t)$ be a complex exponential function selected so that

$$f_s(t) = \text{Re}[\mathbf{f}_e(t)] \tag{8-36}$$

Then $\mathbf{f}_e(t)$ will be of the form

$$\mathbf{f}_e(t) = (\mathbf{F}_e)(e^{j\omega t}) \tag{8-37}$$

where $\mathbf{F}_e$ is the *phasor* associated with the given forcing function. That is

$$\mathbf{F}_e = Ae^{j\theta} \tag{8-38}$$

Let the response of the circuit, when driven by $\mathbf{f}_e(t)$, be given by

$$\mathbf{g}_e(t) = \mathbf{B}e^{j\omega t} \tag{8-39}$$

where $\mathbf{B}$ is a complex coefficient whose magnitude and angle are to be determined. $\mathbf{B}$ is the phasor associated with the response function $\mathbf{g}_e(t)$. When $\mathbf{g}_e(t)$ is substituted in the differential equation of the circuit

$$\text{each } [d\mathbf{g}(t)/dt] \text{ is replaced by } (j\omega)\mathbf{B}e^{j\omega t} \tag{8-40}$$

and

$$\text{each } \int \mathbf{g}(t)dt \text{ is replaced by } (1/j\omega)\mathbf{B}e^{j\omega t} \tag{8-41}$$

The differential equation becomes an algebraic equation to be solved for $\mathbf{B}$. $\mathbf{B}$ will be a function of the circuit elements and the angular frequency ω. Once $\mathbf{B}$ is determined, the response $\mathbf{g}_e(t)$ is known.

When the given circuit is excited by a complex exponential forcing function $\mathbf{f}_e^*(t)$, the complex conjugate of $\mathbf{f}_e(t)$,

$$\mathbf{f}_e^*(t) = Ae^{-j\theta}e^{-j\omega t}$$

the complex exponential response will be $\mathbf{g}_e^*(t)$:

$$\mathbf{g}_e^*(t) = \mathbf{B}^*e^{-j\omega t}$$

where $\mathbf{B}^*$ is the complex conjugate of $\mathbf{B}$ in Eq. (8-39).

These relationships and the principle of superposition lead to the following table:

Forcing Function	Response Function
$\mathbf{f}_e(t) = (Ae^{j\theta})e^{j\omega t}$ $\rightarrow$	$\mathbf{g}_e(t) = \mathbf{B}e^{j\omega t}$
$\mathbf{f}_e^*(t) = (Ae^{j\theta})e^{j\omega t}$ $\rightarrow$	$\mathbf{g}_e^*(t) = \mathbf{B}^*e^{-j\omega t}$
$[\mathbf{f}_e(t) + \mathbf{f}_e^*(t)]$ $\rightarrow$	
$= 2A\cos(\omega t + \theta)$	$[\mathbf{g}_e(t) + \mathbf{g}_e^*(t)] = 2\text{Re}[\mathbf{g}_e(t)] = 2\text{Re}[\mathbf{B}e^{j\omega t}]$
$A\cos(\omega t + \theta)$ $\rightarrow$	$\text{Re}[\mathbf{g}_e(t)] = \text{Re}[\mathbf{B}e^{j\omega t}]$

The discussion shows that the real part of the complex response to a complex forcing function gives the response of the circuit to the real part of the complex forcing function.

The response of a linear circuit to a sinusoidal forcing function can therefore be determined by first obtaining the response to a complex forcing function and then taking the real part of the complex response. This procedure is summarized in Fig. 8-3. The key to the procedure is that the phasor serves as a tool which *transforms* a differential equation into an algebraic equation. This transformation technique is used frequently in the analysis of linear circuits and systems. The phasor method is only one of several such transformation procedures; others such as Laplace and Fourier transformations will be introduced in later chapters.

As we will see shortly, not all the steps shown in Fig. 8-3 are required for the solution of every problem. In fact, the solution usually involves only the operations in the two boxes on the right side of the diagram. But the theoretical justification of the procedure involves the operations in all four boxes.

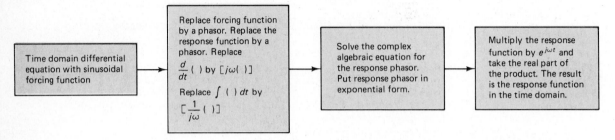

Fig. 8-3 Phasor method of finding the response of a circuit in the sinusoidal steady state.

Example 8-9 A series circuit consisting of $L_1 = 0.2$ H, $R_1 = 10\Omega$, $L_2 = 0.3$ H, $C = 0.0025$ F, and $R_2 = 15\Omega$ is driven by a voltage $v_s(t) = 100 \cos 25t$ V. Determine the current $i(t)$ in the circuit.

Solution The differential equation of the circuit is

$$L_1(di/dt) + R_1 i + L_2(di/dt) + (1/C)\int i\,dt + R_2 i = 100 \cos 25t \quad (8\text{-}42)$$

The phasor corresponding to the given $v_s(t)$ is $100e^{j0°} = 100$. The complex forcing function is therefore $100e^{j25t}$. Using the symbol $\mathbf{i}_e$ to denote the response to the complex forcing function, Eq. (8-42) is rewritten as

$$(L_1 + L_2)(d\mathbf{i}_e/dt) + (R_1 + R_2)\mathbf{i}_e + (1/C)\int \mathbf{i}_e\,dt = 100e^{j25t} \quad (8\text{-}43)$$

The response $\mathbf{i}_e(t)$ is one of the form

$$\mathbf{i}_e(t) = \mathbf{B}e^{j25t}$$

In Eq. (8-43), replace $(d\mathbf{i}_e/dt)$ by $(j25)\mathbf{B}e^{j25t}$ and $\int \mathbf{i}_e\,dt$ by $(1/j25)\mathbf{B}e^{j25t}$. Then Eq. (8-43) becomes

$$(j25)(L_1 + L_2)\mathbf{B}e^{j25t} + (R_1 + R_2)\mathbf{B}e^{j25t} + (1/j25C)\mathbf{B}e^{j25t} = 100e^{j25t}$$

which reduces to

$$[(R_1 + R_2) + j25(L_1 + L_2) + (1/j25C)]\mathbf{B} = 100$$

or

$$\mathbf{B} = \frac{100}{(R_1 + R_2) + j25(L_1 + L_2) + (1/j25C)}$$

358   Phasors, Impedance, and Admittance

Substituting the numerical values of the components, and using $(1/j) = -j$ (see Appendix B), the last equation becomes

$$B = 100/(25 + j12.5 - j16)$$
$$= 100/(25.2 \underline{/-7.97°}) = 3.97e^{j7.97°}$$

The response $i_e(t)$ is therefore

$$i_e(t) = Be^{j25t} = 3.97e^{j(25t + 7.97°)}$$

Taking the real part, the response $i(t)$ to the original sinusoidal forcing function is

$$i(t) = 0.0397 \cos(25t + 7.97°) \text{ A}$$

■

Exercise 8-11 Determine $v(t)$ in a circuit whose differential equation is given by

$$0.1(dv/dt) + 2\int v\,dt = 10 \cos 8t$$

by using the phasor approach.

Exercise 8-12 Determine $i(t)$ in a circuit whose differential equation is given by

$$150 \int i\,dt + 20i + 5(di/dt) = 80 \cos(4t + 45°)$$

by using the phasor approach.

The procedure using phasors leads to the response as a function of time in two steps:

1. Calculate the phasor representation of the response.
2. Convert the phasor response to the time function representation of the response.

Compare the two results:

Phasor Representation: $\quad B = |B|e^{j\phi}$ where $|B|$ is the magnitude of the phasor and ϕ is its angle.

Time Function Representation: $\quad \text{Re}[Be^{j\omega t}] = \text{Re}[|B|e^{j\phi}e^{j\omega t}] = |B|\cos(\omega t + \phi)$

For instance, the phasor representation in Example 8-9 was $3.97e^{j7.97°}$ and the time function was $3.97\cos(25t + 7.97°)$.

The phasor form of the response provides the same information as the time function, except for the frequency. As we noted in Sec. 8-2, the frequency is a single fixed quantity in the circuit and needs to be specified only once. Thus, *the phasor representation of the response is sufficient for all practical purposes, and it is not necessary to write the time function explicitly each time.* This is the general practice followed in ac circuit analysis: the forcing function and the response are left in the form of phasors, and the corresponding time functions are usually omitted (unless specifically asked for). Note, however, that in the case of circuits driven by sources of *different frequencies,* the combination of the responses due to the different sources should not be done with phasors, but *must be done with the complete time functions.*

8-5 IMPEDANCE, RESISTANCE, AND REACTANCE

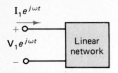

Fig. 8-4 Impedance is the ratio of the complex exponential voltage to the complex exponential current.

Since the use of phasors in sinusoidal steady state analysis leads to algebraic equations, the ratio (voltage/current) in a circuit leads to an expression involving complex quantities. This leads to the concept of impedance and a version of Ohm's law applicable to ac circuits. These enable us to analyze a circuit directly with phasor equations (using steps like those used for purely resistive circuits), thus bypassing the need to write the differential equations of a circuit.

Consider the circuit contained in a black box (see Fig. 8-4). Let the complex exponential voltage $v_1(t)$ be given by $V_1 e^{j\omega t}$ and the complex exponential current $i_1(t)$ by $I_1 e^{j\omega t}$.

Impedance is defined as the ratio of the complex exponential voltage function to the complex exponential current function. That is, if Z denotes impedance, then

$$Z = (V_1 e^{j\omega t} / I_1 e^{j\omega t})$$

or

$$Z = (V_1 / I_1)$$

where V_1 and I_1 are phasors. That is,

$$Z = \frac{\text{voltage expressed as a phasor}}{\text{current expressed as a phasor}}$$

Note that impedance is the *ratio of two complex exponential time functions*, or of *two phasors*. Impedance is *not* the ratio of two *sinusoidal* time functions.

The unit of impedance is *ohms*.

In the general case, voltage and current phasors are complex quantities. Their ratio, Z, is therefore also a complex quantity, and its magnitude and angle are given by the relations:

magnitude of impedance = (magnitude of the voltage phasor)/(magnitude of the current phasor)

angle of impedance = (angle of voltage phasor) − (angle of current phasor)

In Example 8-9, the voltage phasor was $100 e^{j0°}$ V and the current phasor was $3.97 e^{j7.97°}$ A. The impedance of the circuit is therefore

$$Z = (100 e^{j0°} / 3.97 e^{j7.97°})$$
$$= 25.2 e^{-j7.97°}$$

which is more commonly written $Z = 25.2 \underline{/-7.97°}\ \Omega$. The magnitude of Z is $(100/3.97) = 25.2\ \Omega$ and the angle of Z is $(0 - 7.97) = -7.97°$.

Exercise 8-13 Determine the impedances of the circuits in Exercises 8-9 through 8-12.

Based on the definition of impedance, the relationship between voltage and current in the sinusoidal steady state is

$$V = ZI \quad \text{(Ohm's law of ac circuits)} \tag{8-44}$$

If $V = |V|\underline{/\theta}$, $I = |I|\underline{/\phi}$, and $Z = |Z|\underline{/\alpha}$, then

$$|V| = |Z||I| \quad \text{and} \quad \theta = \phi + \alpha \tag{8-45}$$

Example 8-10 Determine the current **I** in the following cases: (a) $\mathbf{V} = 100\underline{/-30°}$ V, $\mathbf{Z} = 5\underline{/45°}\ \Omega$; (b) $v(t) = 80 \cos(100t - 75°)$ V, $\mathbf{Z} = 13\underline{/-60°}\ \Omega$.

Solution (a) $\mathbf{I} = (\mathbf{V}/\mathbf{Z}) = (100/5°)/(-30 - 45°) = 20\underline{/-75°}$ A. (b) The voltage V *must* first be expressed as a phasor. $\mathbf{V} = 80\underline{/75°}$ V. Therefore, $\mathbf{I} = (80/13)/(75 + 60)° = 6.15\underline{/135°}$ A.

Exercise 8-14 In each set of values given here, find the value of the missing item (voltage, current, or impedance). (a) $v(t) = 100 \cos 50t$ V; $i(t) = 25 \cos(50t - 90°)$ A. (b) $\mathbf{V} = 48\underline{/-36.9°}$ V; $\mathbf{Z} = 5\underline{/53.1°}\ \Omega$. (c) $i(t) = 3.25 \cos 500t$ A; $\mathbf{Z} = 8\underline{/-20°}\ \Omega$.

The rectangular form of an impedance is written as

$$\mathbf{Z} = R + jX \tag{8-46}$$

where

R, the real part of **Z**, is called the *resistance*

and

X, the imaginary part, is called the *reactance*

The unit of reactance is *ohms*, the same as for resistance. Note that the standard convention for the rectangular form of **Z** is always R plus jX. The reactance X may be *positive* or *negative*.

For example, if $\mathbf{Z} = (2.02\underline{/8.5°}) = (2 + j0.299)\ \Omega$, then the resistance $R = 2\Omega$ and the reactance $X = 0.299\ \Omega$. On the other hand, if $\mathbf{Z} = 8.3\underline{/-40°} = (6.36 - j5.34)\ \Omega$, then the resistance $R = 6.36\ \Omega$ and the reactance $X = -5.34\ \Omega$.

If **Z** is written in polar form,

$$\mathbf{Z} = |\mathbf{Z}|\underline{/\alpha}$$

then $|\mathbf{Z}|$ and α are related to R and X through the equations

$$|\mathbf{Z}| = \sqrt{R^2 + X^2} \quad \text{and} \quad \alpha = \arctan(X/R)$$
$$R = |\mathbf{Z}| \cos\alpha \quad \text{and} \quad X = |\mathbf{Z}| \sin\alpha$$

These relationships are conveniently displayed by using a triangle called the *impedance triangle*, which is shown in Fig. 8-5.

Fig. 8-5 Impedance triangles: (a) inductive circuit, (b) capacitive circuit.

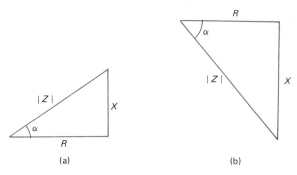

(a)   (b)

8-5 Impedance, Resistance, and Reactance

Exercise 8-15 Determine the resistance and reactance of each of the impedances calculated in Exercise 8-13.

8-5-1 Impedance of Single Elements

Resistor

If $v(t) = V_m \cos \omega t$ is applied to a resistor R, the equation $Ri(t) = V_m \cos \omega t$ becomes

$$RI = V$$

where $\mathbf{V} = V_m \underline{/0°}$. Then the impedance is

$$\mathbf{Z} = (\mathbf{V}/\mathbf{I}) = R\underline{/0°} = (R + j0)$$

The impedance of a resistor is therefore simply its resistance.

Inductor

If $v(t) = V_m \cos \omega t$ is applied to an inductance L, the equation $L(di/dt) = V_m \cos \omega t$ transforms to

$$(j\omega L)\mathbf{I} = \mathbf{V}$$

and the impedance is

$$\mathbf{Z} = (\mathbf{V}/\mathbf{I}) = j\omega L$$

or

$$\mathbf{Z} = 0 + j\omega L = \omega L\underline{/90°}$$

The impedance of an inductor contains only a *positive* reactance component:

$$X = \omega L$$

The reactance of an inductance increases linearly with frequency, being zero at $f = 0$ and infinity at $f = \infty$. The current in an inductance lags behind the voltage by 90° at all frequencies.

Capacitor

When $v(t) = V_m \cos \omega t$ is applied to a capacitance C, the equation $(1/C) \int i\,dt = v(t)$ transforms to

$$(1/j\omega C)\mathbf{I} = \mathbf{V}$$

and the impedance is

$$\mathbf{Z} = (\mathbf{V}/\mathbf{I}) = 1/j\omega C = -j(1/\omega C)$$

since $(1/j) = -j$. An alternative form of writing the impedance of a capacitance is

$$\mathbf{Z} = 0 - j(1/\omega C) = (1/\omega C)\underline{/-90°}$$

The impedance of a capacitor contains only a *negative* reactance component.

$$X = -(1/\omega C)$$

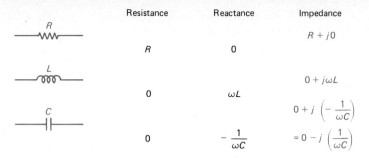

Fig. 8-6 Impedance, resistance, and reactance of single elements.

The reactance of a capacitance decreases as frequency increases: infinity at $f = 0$ and zero at $f = \infty$. The current in a capacitance leads the voltage by 90° at all frequencies.

Figure 8-6 shows the three elements, with information on impedance, resistance, and reactance.

Exercise 8-16 Calculate the impedance of an inductance $L = 5$ mH at (a) $f = 60$ Hz; (b) $f = 100$ kHz; (c) $f = 180$ MHz. Write your answers in polar and rectangular forms.

Exercise 8-17 Repeat the calculations of Exercise 8-16 for a capacitance $C = 50$ μF.

8-5-2 Series Circuits

Consider a series connection of components, as shown in Fig. 8-7, and let the voltages across the components be $v_1(t), v_2(t), \ldots, v_n(t)$, all of which are sinusoidal functions of the same frequency. Then the total voltage $v_s(t)$ will also be sinusoidal at the same frequency. The KVL equation of the circuit is

$$v_s(t) = v_1(t) + v_2(t) + \ldots + v_n(t) \qquad (8\text{-}47)$$

When complex exponential forms of the voltages are used, Eq. (8-47) becomes

$$\mathbf{V}_s e^{j\omega t} = \mathbf{V}_1 e^{j\omega t} + \mathbf{V}_2 e^{j\omega t} + \ldots + \mathbf{V}_n e^{j\omega t}$$

which reduces to

$$\mathbf{V}_s = \mathbf{V}_1 + \mathbf{V}_2 + \ldots + \mathbf{V}_n \qquad (8\text{-}48)$$

Fig. 8-7 A series connection of impedances.

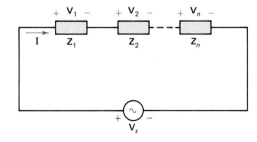

8-5 Impedance, Resistance, and Reactance

Eq. (8-48) is the phasor form of Eq. (8-47). If **I** is the (phasor) current in the circuit, then the voltages across the individual components are written as **ZI** products, and Eq. (8-48) becomes

$$\mathbf{V}_s = \mathbf{Z}_1\mathbf{I} + \mathbf{Z}_2\mathbf{I} + \ldots + \mathbf{Z}_n\mathbf{I}$$
$$= (\mathbf{Z}_1 + \mathbf{Z}_2 + \ldots + \mathbf{Z}_n)\mathbf{I} \qquad (8\text{-}49)$$

The current **I** is obtained in phasor form by solving Eq. (8-49). Since phasor forms of voltages and currents provide sufficient information (except for frequency), the solution is usually left in phasor form.

This discussion shows that a series circuit may be analyzed by writing the KVL equation directly in phasor form, as in Eq. (8-48), replacing the voltages across the individual components by **ZI** products, and then solving for the current as a phasor. But note that this procedure utilizes only the boxes on the right side of Fig. 8-3, and that the time domain is not explicitly used.

Example 8-11 If three impedances $\mathbf{Z}_1 = (10 + j10)\Omega$, $\mathbf{Z}_2 = (5 - j5)\Omega$, and $\mathbf{Z}_3 = (10 - j25)\Omega$ are in series and the voltage $v_s(t) = 100 \cos 377t$ is applied across them, calculate the current in the circuit and the voltage across each impedance.

Solution The phasor form of the applied voltage is $\mathbf{V}_s = 100\underline{/0°}$ V and the equation of the circuit is

$$(\mathbf{Z}_1 + \mathbf{Z}_2 + \mathbf{Z}_3)\mathbf{I} = \mathbf{V}_s$$

where **I** is the current (in phasor form). Substituting the numerical values, the last equation becomes

$$(25 - j20)\mathbf{I} = 100\underline{/0°}$$

or

$$\mathbf{I} = \frac{100}{(25 - j20)} = \frac{100}{(32.0\underline{/-38.7°})}$$
$$= 3.12\underline{/38.7°} \text{ A}$$

The current has an amplitude of 3.12 A and *leads* the voltage by 38.7°, which means that the circuit is effectively capacitive.

The voltages across the individual impedances are given by

- $\mathbf{V}_1 = \mathbf{Z}_1\mathbf{I} = (14.1\underline{/45°})(3.12\underline{/38.7°}) = 44.2\underline{/83.7°}$ V
- $\mathbf{V}_2 = \mathbf{Z}_2\mathbf{I} = 22.1\underline{/-6.3°}$ V
- $\mathbf{V}_3 = \mathbf{Z}_3\mathbf{I} = 84.0\underline{/-29.5°}$ V

The three voltages should add up to $100\underline{/0°}$ (within roundoff error).
The time function forms of the answers are:

- $i(t) = 3.12 \cos(377t + 38.7°)$ A
- $v_1(t) = 44.2 \cos(377t + 83.7°)$ V
- $v_2(t) = 22.1 \cos(377t - 6.3°)$ V
- $v_3(t) = 84.0 \cos(377t - 29.5°)$ V

Exercise 8-18 Three impedances, $\mathbf{Z}_1 = (10 + j0)\Omega$, $\mathbf{Z}_2 = (0 + j30)\Omega$, and $\mathbf{Z}_3 = (0 - j10)\Omega$, are in series, and the total voltage across them is $v(t) = 25 \cos 100t$ V. Calculate the current in the circuit and the voltages across the individual impedances. Write the answers in phasor and time function forms.

Exercise 8-19 Two impedances $\mathbf{Z}_1 = (10\underline{/30°})$ and $\mathbf{Z}_2 = (10\underline{/-60°})$ are in series. If the voltage across $\mathbf{Z}_1$ is given as $\mathbf{V}_1 = 25\underline{/-90°}$ V, find the current in the circuit, the voltage across $\mathbf{Z}_2$, and the total voltage across the circuit.

Discussion of the Series Circuit

Let us return to Eq. (8-49). The total impedance $\mathbf{Z}_T$ of the circuit defined by the equation

$$\mathbf{V}_s = \mathbf{Z}_T \mathbf{I}$$

is seen to be

$$\mathbf{Z}_T = (\mathbf{Z}_1 + \mathbf{Z}_2 + \ldots + \mathbf{Z}_n) \quad (8\text{-}50)$$

Therefore,

Total impedance of a series connection of impedances = sum of the individual impedances.

This result is like the series connection of resistances obtained in earlier chapters. If R_T is the total resistance and X_T the total reactance of the circuit; that is,

$$\mathbf{Z}_T = R_T + jX_T$$

then

$$R_T = \text{sum of the resistance components}$$

and

$$X_T = \text{sum of the reactance components}$$

of the individual impedances.

When the components connected in series are explicitly given as single elements (R, L, and C), the impedances of the individual elements are written using Fig. 8-6, and then the total impedance is calculated.

The voltage across any individual impedance $\mathbf{Z}_k$ in a series circuit is

$$\mathbf{V}_k = \mathbf{Z}_k \mathbf{I}$$

and

$$(\mathbf{V}_k/\mathbf{V}_s) = (\mathbf{Z}_k/\mathbf{Z}_T) \quad (8\text{-}51)$$

which is the *voltage division* relationship for series ac circuits.

Example 8-12 A resistor R, an inductance L, and a capacitance C are in series. Determine the total impedance, resistance, and reactance. Find the frequency at which the total reactance of the circuit is zero.

Solution The impedance is given by

$$Z = R + j\omega L - j(1/\omega C) = R + j(\omega L - 1/\omega C)$$

The resistance of the circuit is R and the reactance is

$$X = (\omega L - 1/\omega C)$$

which may be positive, zero, or negative depending upon whether ωL is greater than, equal to, or less than $(1/\omega C)$.

The reactance is zero when

$$\omega L = 1/\omega C$$

or

$$\omega = 1/\sqrt{LC}$$

■

Example 8-13 Determine the current and the voltages marked V_1, V_2, and V_3 in the circuit of Fig. 8-8.

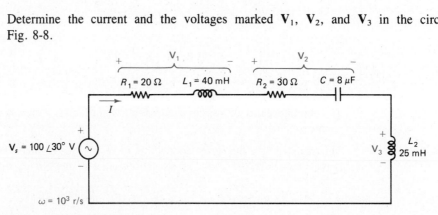

Fig. 8-8 Circuit for Example 8-13.

Solution The total impedance of the circuit is

$$Z_T = R_1 + j\omega L_1 + R_2 - j(1/\omega C) + j\omega L_2$$
$$= (R_1 + R_2) + j(\omega L_1 + \omega L_2 - 1/\omega C)$$
$$= (50 - j60) = 78.1\underline{/-50.1°}$$

The current in the circuit is

$$I = (V_s/Z_T) = (100\underline{/30°})/(78.1\underline{/-50.1°})$$
$$= 1.28\underline{/80.1°} \text{ A}$$

The voltages V_1, V_2, and V_3 are given by

$$V_1 = (R_1 + j\omega L_1)I = (20 + j40)(1.28\underline{/80.1°})$$
$$= 57.2\underline{/144°} \text{ V}$$

$$V_2 = [R_2 - j(1/\omega C)]I = (30 - j125)(1.28\underline{/80.1°})$$
$$= 164\underline{/3.59°} \text{ V}$$

$$V_3 = j\omega L_2 I = j25(1.28\underline{/80.1°}) = 32\underline{/170°} \text{ V}$$

Note that the voltage V_2 has a higher amplitude than the applied voltage. Such a situation is possible in ac circuits when both inductors and capacitors are present in the same circuit. The three voltages should add up to $100\underline{/30°}$ V. ∎

Exercise 8-20 A series *RLC* circuit contains $R = 5\Omega$, $L = 0.4$ H, and $C = 0.0125$ F, and the applied voltage is $100\underline{/0°}$ V. Determine the total impedance, reactance, resistance, and the current in the circuit for (a) $\omega = 5$ r/s; (b) $\omega = 20$ r/s; (c) $\omega = 14.14$ r/s. Discuss the variation of the calculated quantities as a function of frequency.

Exercise 8-21 Three impedances $\mathbf{Z}_1$, $\mathbf{Z}_2$, and $\mathbf{Z}_3$ are in series with a voltage $v_s(t) = 100 \cos 500t$ applied across the series connection. Determine the current in the circuit and the voltage across each impedance for each of the following cases:

(a) $\mathbf{Z}_1$ is a 10Ω resistor; $\mathbf{Z}_2$ is a 20 mH inductor; $\mathbf{Z}_3$ is a 100 μF capacitor.
(b) $\mathbf{Z}_1 = (30 + j40)\Omega$; $\mathbf{Z}_2 = (10 - j30)\Omega$; $\mathbf{Z}_3 = (10 - j10)\Omega$.
(c) $\mathbf{Z}_1 = 10\underline{/60°}\ \Omega$; $\mathbf{Z}_2 = 20\underline{/30°}\ \Omega$; $\mathbf{Z}_3 = 5\underline{/-30°}\ \Omega$.

Exercise 8-22 Two impedances are in series. The voltage across $\mathbf{Z}_1$ is $200\underline{/75°}$ V, and the voltage across the series connection is $\mathbf{V}_T = 100\underline{/15°}$ V. Find the voltage across $\mathbf{Z}_2$.

8-5-3 Impedance Triangles and Phasor Diagrams

Even though the analysis of a circuit is usually performed through algebraic equations and expressions, it is often convenient to use graphic representations as computational aids. Graphical representations of impedance, called *impedance triangles*, and diagrams showing the relative magnitudes and phase angles of voltages and currents, called *phasor diagrams*, are frequently used in connection with the analysis of series circuits.

As Fig. 8-5 shows, an *impedance triangle* is a right-angled triangle whose horizontal side represents the resistance of a circuit. The vertical side represents the reactance of the circuit: it is upward for an inductance (since the angle of the impedance of an inductance is $+90°$) and downward for a capacitance (since the angle of the impedance of a capacitance is $-90°$). The hypotenuse represents the impedance of the circuit: Its length is the magnitude of the impedance, and its angle with the horizontal is the angle of the impedance.

For a series circuit made up of a number of different components, the impedance triangle is obtained by drawing a horizontal line for each resistor and a vertical line for each inductance and capacitance. The lines are then added vectorially to obtain the total impedance of the circuit. Fig. 8-9 shows an example of an impedance diagram.

The impedance triangle is useful in studying the qualitative effect of varying one of the parameters in the circuit on the impedance. For example, the variation of the impedance of a series *RLC* circuit as a function of frequency is illustrated by the impedance triangles in Fig. 8-10. The diagrams show the general manner in which the reactance, the magnitude, and the angle of the impedance change with frequency. Such qualitative observations provide an insight into the behavior of certain circuits, as we will see in later chapters.

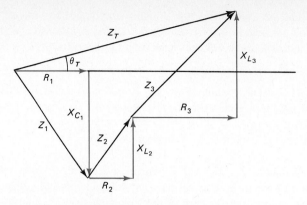

Fig. 8-9 Impedance triangle of a series circuit.

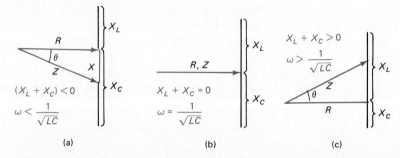

Fig. 8-10 Impedance triangles of a series *RLC* circuit at different frequencies: (a) reactance < 0 when $\omega < 1/\sqrt{LC}$; (b) reactance = 0 when $\omega = 1/\sqrt{LC}$; and (c) reactance > 0 when $\omega > 1/\sqrt{LC}$.

Exercise 8-23 Draw the impedance triangles of the circuits of Examples 8-12 and 8-13.

The relationship between the current and the voltage phasors in a circuit is graphically displayed by means of a diagram similar to the impedance triangle. Each phasor is drawn as a vector, with length proportional to the magnitude and angle equal to the angle of the phasor. Such a diagram is called the *phasor diagram*. The phasor diagram for Example 8-13 is shown in Fig. 8-11.

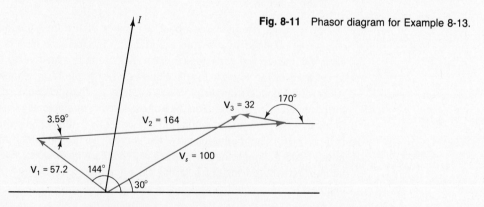

Fig. 8-11 Phasor diagram for Example 8-13.

Whenever possible, it is advisable to choose the current in a series circuit as the horizontal reference in a phasor diagram, since the current is the same in all the components. When the current in a series circuit is used as the horizontal reference, phasor and impedance diagrams can be obtained one from the other by merely relabeling the lines of the diagram.

Example 8-14 Draw the phasor diagram showing the current and the voltages in the circuit of Fig. 8-12(a).

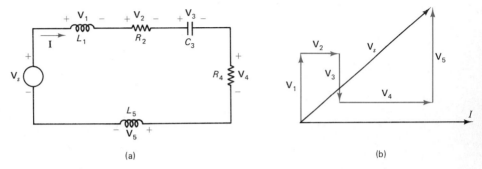

Fig. 8-12 (a) Circuit of Example 8-14 and (b) its phasor diagram.

Solution Using the current as the horizontal reference, the voltages across the individual components are drawn as vectors: horizontal for each resistance (since voltage is in phase with current), vertically upward for each inductance (since voltage leads current by 90°), and vertically downward for each capacitance (since voltage lags current by 90°). The lines are then added vectorially to obtain the total voltage in the circuit. The phasor diagram is shown in Fig. 8-12(b). ∎

Exercise 8-24 A series circuit consists of $R = 10\Omega$, $L = 0.05$ H, and $C = 0.05$ F, driven by a source $V_s = 100$ V. Draw the phasor diagram of the circuit for the following values of ω: (a) 10 r/s, (b) 20 r/s, and (c) 40 r/s.

8-6 ADMITTANCE, CONDUCTANCE, AND SUSCEPTANCE

The *admittance* of a circuit is defined as the ratio of the current phasor to the voltage phasor.

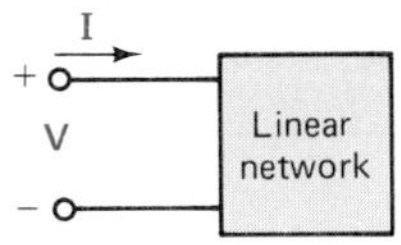

If **I** is the current in a component and **V** the voltage across it, as indicated in Fig. 8-13, then its admittance **Y** is given by

$$\mathbf{Y} = \frac{\mathbf{I}}{\mathbf{V}} \tag{8-52}$$

Fig. 8-13 Admittance is the ratio of the phasor current **I** to phasor voltage **V**.

The unit of admittance is the *siemens* (S).

A comparison of the defining equations of admittance and impedance shows that

$$\mathbf{Y} = 1/\mathbf{Z} \quad \text{and} \quad \mathbf{Z} = 1/\mathbf{Y}$$

The impedance and admittance of a circuit are reciprocals of each other.

Since admittance and impedance are complex quantities, their magnitudes and angles satisfy the following relationships:

magnitude of admittance = reciprocal of the magnitude of impedance

angle of admittance = negative of the angle of the impedance

Exercise 8-25 Calculate the admittances of the circuits whose impedances are given by: (a) $\mathbf{Z}_a = 5\underline{/-30°}$ Ω. (b) $\mathbf{Z}_b = (3 + j4)$ Ω. (c) $\mathbf{Z}_c = 10e^{j25}$ Ω.

Exercise 8-26 In each of the following sets of values, determine the admittance of the circuit. (a) $\mathbf{V}_a = 100\underline{/0°}$ V; $\mathbf{I}_a = 10\underline{/-45°}$ A. (b) $v_b(t) = 100 \cos(50t + 75°)$ V; $i_b(t) = 500 \cos(50t - 15°)$ A. (c) Amplitude of $\mathbf{V}_c = 60$ V; amplitude of $\mathbf{I}_c = 25$ A; current leads voltage by 63°.

The rectangular form of the admittance of a circuit is written as

$$\mathbf{Y} = G + jB$$

where

$$G = \text{Re}[\mathbf{Y}] \text{ is the } \textit{conductance}$$

and

$$B = \text{Im}[\mathbf{Y}] \text{ is the } \textit{susceptance}$$

The units of conductance and susceptance are siemens, just as for admittance.

The susceptance B may be positive or negative: $B > 0$ for a capacitive circuit, and $B < 0$ for an inductive circuit.

Note that the standard form of writing $\mathbf{Y}$ in rectangular form is always G plus jB, where B may be *positive* or *negative*.

For example, an admittance given by $\mathbf{Y} = 0.850\underline{/-50°}$ S $= (0.546 - j0.651)$ S has a conductance $= 0.546$ S, and susceptance $= -0.651$ S.

Exercise 8-27 Find the conductance and susceptance of each of the circuits in Exercises 8-25 and 8-26.

Example 8-15 The impedance of a circuit is given by

$$\mathbf{Z} = R + j(\omega L - 1/\omega C)$$

Determine its conductance and susceptance.

Solution The admittance of the circuit is

$$\mathbf{Y} = \frac{1}{\mathbf{Z}} = \frac{1}{R + j(\omega L - 1/\omega C)}$$

The real and imaginary parts of this function are found by rationalizing—that is, by multiplying both the numerator and the denominator by the *complex conjugate of the denominator*. This process makes the denominator a real expression without changing the original function. Then the admittance becomes

$$\mathbf{Y} = \frac{R - j(\omega L - 1/\omega C)]}{R^2 + (\omega L - 1/\omega C)^2}$$

Therefore,

$$G = \text{Re}[Y] = \frac{R}{R^2 + (\omega L - 1/\omega C)^2}$$

and

$$B = \text{Im}[Y] = \frac{-(\omega L - 1/\omega C)}{R^2 + (\omega L - 1/\omega C)^2}$$

Note that

1. The *conductance is a function of frequency* in this example, even though the resistance R of the given impedance is a constant.
2. G is not simply equal to $(1/R)$.

That is, the term *conductance is not simply the reciprocal of resistance*, as it was in purely resistive circuits. The possibility of confusing the term conductance with the reciprocal of a resistance instead of treating it as the real part of an admittance does exist, and it is important to watch out for it. ∎

8-6-1 Admittance of Single Elements

Resistor

For a resistor

$$Y = G = (1/R)$$

For a single pure resistance, the conductance is indeed the reciprocal of the resistance.

Inductor

Since $\mathbf{Z} = j\omega L$ for an inductance, the admittance is

$$Y = 1/j\omega L = -j(1/\omega L) = \omega L \underline{/-90°}$$

The admittance of an inductance is a pure (negative) susceptance given by

$$B = -(1/\omega L)$$

The susceptance of an inductor is infinity at $\omega = 0$ and decreases with frequency to a value of zero at $\omega = \infty$.

Capacitor

For a capacitance, $\mathbf{Z} = (1/j\omega C)$, and hence

$$Y = j\omega C = \omega C \underline{/90°}$$

The admittance of a capacitance is a pure (positive) susceptance given by

$$B = \omega C$$

The susceptance of a capacitor is zero at $\omega = 0$ and increases linearly with frequency to a value of ∞ at $\omega = \infty$.

Exercise 8-28 Determine the admittance (in both rectangular and polar forms) of an inductance of $L = 15$ mH at the following frequencies: (a) 50 Hz. (b) 10 kHz. (c) 16 MHz. What is the value of the susceptance in each case?

Exercise 8-29 Determine the admittance (in both polar and rectangular forms) of a capacitance $C = 250$ pF at the following frequencies: (a) 10 Hz. (b) 10 kHz. (c) 10 MHz. What is the value of the susceptance in each case?

Table 8-1 summarizes the expressions for the impedance and admittance of single elements.

TABLE 8-1

Element	Z	Y
R	$R + j0 = R\,\underline{/0°}$	$\frac{1}{R} + j0 = \frac{1}{R}\,\underline{/0°}$
L	$0 + j\omega L = \omega L\,\underline{/90°}$	$0 - j\frac{1}{\omega L} = \frac{1}{\omega L}\,\underline{/-90°}$
C	$0 - j\frac{1}{\omega C} = \frac{1}{\omega C}\,\underline{/-90°}$	$0 + j\omega C = \omega C\,\underline{/90°}$

8-6-2 Parallel Circuits

Use of Duality

The defining equations for impedance and admittance are

$$V = ZI$$
$$I = YV$$

Since each relationship can be obtained from the other by replacing voltage by current, current by voltage, impedance by admittance, and admittance by impedance, it follows that

Impedance and admittance are a dual pair.

Comparing the two relationships

$$Z = R + jX$$
$$Y = G + jB$$

it follows that

Resistance and conductance are a dual pair.

and

Reactance and susceptance are a dual pair.

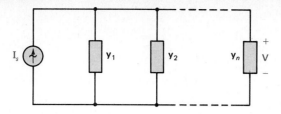

Fig. 8-14 A parallel connection of admittances.

The principle of duality will be used here to study a parallel circuit by modifying the results already obtained for a series circuit.

Consider a parallel combination of admittances $\mathbf{Y}_1, \mathbf{Y}_2, \ldots \mathbf{Y}_n$, as shown in Fig. 8-14, with a driving function $\mathbf{I}_s = I_m e^{j\phi}$. The total admittance seen by the source is

$$\mathbf{Y}_T = \mathbf{Y}_1 + \mathbf{Y}_2 + \ldots + \mathbf{Y}_n$$

Therefore,

total admittance of a parallel connection of admittances = sum of the individual admittances

The result is analogous to the parallel connection of conductances obtained in earlier chapters. If G_T is the total conductance and B_T the total susceptance of the circuit—that is,

$$\mathbf{Y}_T = G_T + jB_T$$

then

$$G_T = \text{sum of the conductance components}$$

and

$$B_T = \text{sum of the susceptance components}$$

of the individual admittances.

The magnitude and angle of the admittance are given in terms of G_T and B_T by the following relationships:

$$|\mathbf{Y}_T| = \sqrt{G_T^2 + B_T^2}$$

$$\text{Arg } \mathbf{Y}_T = \arctan(B_T/G_T)$$

Example 8-16 Determine the voltage across and the branch currents in the circuit of Fig. 8-15.

Fig. 8-15 Circuit for Example 8-16.

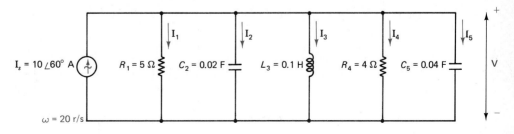

8-6 Admittance, Conductance, and Susceptance

Solution Total admittance

$$\mathbf{Y}_T = \mathbf{Y}_1 + \mathbf{Y}_2 + \mathbf{Y}_3 + \mathbf{Y}_4 + \mathbf{Y}_5$$
$$= G_1 + j\omega C_2 - j(1/\omega L_3) + G_4 + j\omega C_5$$
$$= (1/5) + j0.4 - j0.5 + (1/4) + j0.8$$
$$= (0.45 + j0.7) = 0.832\underline{/57.3°} \text{ S}$$

Voltage across the circuit:

$$\mathbf{V} = (\mathbf{I}_s/\mathbf{Y}_T)$$
$$= \frac{10\underline{/60°}}{0.832\underline{/57.3°}}$$
$$= 12.0\underline{/2.7°} \text{ V}$$

The branch currents are:

$$\mathbf{I}_1 = G_1\mathbf{V} = 2.4\underline{/2.7°} \text{ A}$$
$$\mathbf{I}_2 = j\omega C_2\mathbf{V} = 4.8\underline{/92.7°} \text{ A}$$
$$\mathbf{I}_3 = -j(1/\omega L_3)\mathbf{V} = 6.0\underline{/-87.3°} \text{ A}$$
$$\mathbf{I}_4 = G_4\mathbf{V} = 3.0\underline{/2.7°} \text{ A}$$
$$\mathbf{I}_5 = j\omega C_5\mathbf{V} = 9.6\underline{/92.7°} \text{ A}$$

■

Exercise 8-30 Repeat the analysis of the circuit in Fig. 8-15 by keeping the element values same as in Example 8-16 for the following values of ω: (a) 10 r/s. (b) 30 r/s.

Exercise 8-31 Analyze the circuit of Fig. 8-15 when the value of the inductance is changed to 0.01 H, with the values of the other elements and frequency the same as in Example 8-16.

Exercise 8-32 Analyze the circuit of Fig. 8-15 when the value of the capacitance C_5 is changed to 0.8 F, with the values of the other elements and frequency the same as in Example 8-16.

Example 8-17 Determine the voltage across the parallel combination and the current in each admittance of the circuit in Fig. 8-16.

Fig. 8-16 Circuit for Example 8-17.

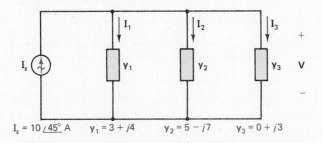

$I_s = 10\underline{/45°}$ A $y_1 = 3 + j4$ $y_2 = 5 - j7$ $y_3 = 0 + j3$

Solution

$$\mathbf{Y}_T = \mathbf{Y}_1 + \mathbf{Y}_2 + \mathbf{Y}_3 = (8 - j0) = 8\underline{/0°}\ \text{S}$$

$$\mathbf{V} = \frac{\mathbf{I}_s}{\mathbf{Y}_T} = \frac{10\underline{/45°}}{8\underline{/0°}} = 1.25\underline{/45°}\ \text{V}$$

$$\mathbf{I}_1 = \mathbf{Y}_1\mathbf{V} = 6.25\underline{/98°}\ \text{A}$$

$$\mathbf{I}_2 = \mathbf{Y}_2\mathbf{V} = 10.8\underline{/-9.7°}\ \text{A}$$

$$\mathbf{I}_3 = \mathbf{Y}_3\mathbf{V} = 3.75\underline{/135°}\ \text{A}$$

■

Exercise 8-33 Determine the current in each admittance and the voltage across the parallel combination of three admittances $\mathbf{Y}_1$, $\mathbf{Y}_2$, and $\mathbf{Y}_3$, driven by $\mathbf{I}_s = 10\underline{/45°}$ A for each of the following cases: (a) $\mathbf{Y}_1$ is a 10Ω resistor; $\mathbf{Y}_2$ is a 150 μF capacitor; $\mathbf{Y}_3$ is a 10 mH inductor. (b) $\mathbf{Y}_1 = (2 - j4)$ S; $\mathbf{Y}_2 = (4 - j2)$ S; $\mathbf{Y}_3 = (5 + j5)$ S. (c) $\mathbf{Y}_1 = 10\underline{/-30°}$ S; $\mathbf{Y}_2 = 10\underline{/-60°}$ S; $\mathbf{Y}_3 = 10\underline{/90°}$ S.

Exercise 8-34 If the current in $\mathbf{Y}_2$ in the parallel combination of two admittances $\mathbf{Y}_1$ and $\mathbf{Y}_2$ is to be $10\underline{/0°}$ A, calculate the current in $\mathbf{Y}_1$ and the total current in the parallel circuit for each of the following cases: (a) $\mathbf{Y}_1 = (0.05 + j0.1)$ S; $\mathbf{Y}_2 = (0.05 - j0.1)$ S. (b) $\mathbf{Y}_1 = (3 + j4)$ S; $\mathbf{Y}_2 = 4\underline{/-90°}$ S. (c) $\mathbf{Y}_1 = 2\underline{/53.1°}$ S; $\mathbf{Y}_2 = 8\underline{/36.9°}$ S.

8-6-3 Admittance Triangles and Phasor Diagrams

The admittance triangle is the dual of the impedance triangle: Its horizontal side represents the conductance, its vertical side the susceptance, and its hypotenuse the admittance. The vertical side is directed upward for a capacitance and downward for an inductance.

Examples of the admittance triangle are shown in Fig. 8-17.

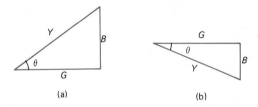

Fig. 8-17 Admittance triangles: (a) capacitive circuit (b) inductive circuit.

Exercise 8-35 Draw the admittance triangles of the circuits of Examples 8-16 and 8-17.

The phasor diagram for a parallel circuit shows the currents in the various branches and the voltage across the parallel combination. The phasor diagram of the circuit of

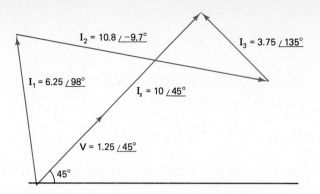

Fig. 8-18 Phasor diagram for Example 8-17.

Example 8-17 is shown in Fig. 8-18. Whenever possible, the voltage phasor should be chosen as the horizontal reference, since the voltage is the same for all the branches.

Example 8-18 Draw the phasor diagram showing the branch currents and the voltage across the circuit of Fig. 8-19(a).

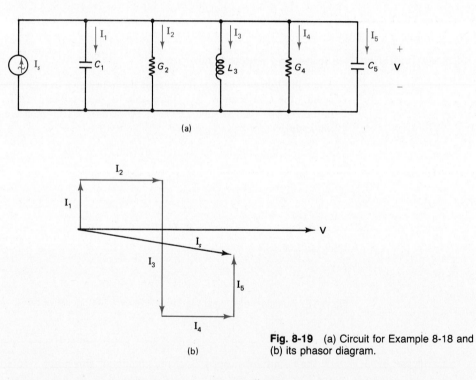

Fig. 8-19 (a) Circuit for Example 8-18 and (b) its phasor diagram.

Solution Vectors are drawn to represent the different branch currents, using the voltage as the horizontal reference. Note that the current phasor for an inductance is vertically *downward*, since the current lags the voltage by 90° in an inductance. The current phasor for a capacitance is vertically *upward*, since the current leads the voltage by 90° in a capacitance. The phasor diagram is shown in Fig. 8-19(b). ∎

Phasors, Impedance, and Admittance

Exercise 8-36 Draw the phasor diagrams of the circuits in Exercises 8-30 and 8-31.

Exercise 8-37 A resistance $R = 2\Omega$, an inductance $L = 0.05$ H, and a capacitance $C = 0.05$ F are in parallel and driven by a current source $\mathbf{I}_s = 10$ A. Draw the phasor diagram of the circuit for the following values of ω: (a) 10 r/s. (b) 20 r/s. (c) 40 r/s.

8-6-4 Relationships between Impedance and Admittance Components

If a circuit has an impedance $\mathbf{Z}$ and admittance $\mathbf{Y}$ given by

$$\mathbf{Z} = R + jX \quad \text{and} \quad \mathbf{Y} = G + jB$$

then using the formula

$$\mathbf{Y} = 1/\mathbf{Z}$$

the following relationships between (R, X) and (G, B) are obtained:

$$G = \frac{R}{R^2 + X^2} \qquad B = \frac{-X}{R^2 + X^2}$$
$$R = \frac{G}{G^2 + B^2} \qquad X = \frac{-B}{G^2 + B^2} \tag{8-52}$$

The important point to note in these relationships is that

the resistance and conductance of a circuit are, in general, functions of frequency.

That is, even though a physical resistor has a constant value of resistance, the *real part* of the impedance or admittance of a circuit in which the resistor is used may vary with frequency.

As an example, consider a series RL circuit. The impedance of the circuit is

$$\mathbf{Z} = R + j\omega L$$

and its admittance is

$$\mathbf{Y} = \frac{1}{R + j\omega L}$$

The four components—resistance, reactance, conductance, and susceptance—are given by

$$R(\omega) = R \qquad X(\omega) = \omega L$$
$$G(\omega) = \frac{R}{R^2 + \omega^2 L^2}$$
$$B(\omega) = \frac{-\omega L}{R^2 + \omega^2 L^2}$$

Note that the conductance is not simply the reciprocal of the resistance; nor is the susceptance the reciprocal of the reactance. Fig. 8-20 shows how the four components vary with frequency.

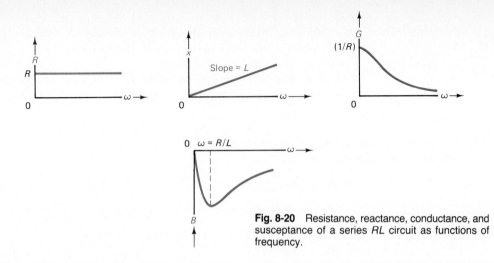

Fig. 8-20 Resistance, reactance, conductance, and susceptance of a series RL circuit as functions of frequency.

Exercise 8-38 Three circuit elements, $R = 10\,\Omega$, $L = 0.05$ H, and $C = 800\,\mu$F are available. Calculate the resistance, reactance, conductance, and susceptance for each of the following cases: (a) The three elements are in series at $\omega = 100$ r/s. (b) The three elements are in series at $\omega = 500$ r/s. (c) The three elements are in parallel at $\omega = 100$ r/s. (d) The three elements are in parallel at $\omega = 500$ r/s.

8-7 SIMPLE EQUIVALENT CIRCUITS

A linear time invariant circuit enclosed in a black box, as shown in Fig. 8-21(a), may be treated as a single impedance or admittance and replaced by a simple equivalent circuit: either a resistance and reactance in series, see Fig. 8-21(b), or a conductance and susceptance in parallel, see Fig. 8-21(c). For a given value of frequency, the element values of the equivalent circuit are uniquely determined. Such simple equivalent circuits are frequently used to replace portions of a larger network, thus simplifying the analysis of the larger network.

A *series equivalent* of a given circuit is obtained by first finding its *impedance*, separating the impedance into real and imaginary parts, and calculating the values of the resistance and reactance elements.

A *parallel equivalent* of a given circuit is obtained by first finding its *admittance*, separating the admittance into real and imaginary parts, and calculating the values of the conductance and susceptance elements.

Fig. 8-21 Series and parallel equivalents of a circuit.

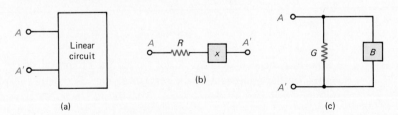

Phasors, Impedance, and Admittance

The two circuits obtained are equivalent to the given circuit and consequently to each other. Remember that the equivalence is valid *only at one frequency* (unless the element values are in the form of algebraic expressions that are functions of frequency).

Example 8-19 Given that the input voltage to a black box with two terminals is $V_s = 100\underline{/20°}$ V at $\omega = 500$ r/s and the resulting current is $I_s = 8\underline{/-33.1°}$ A, obtain (a) a simple series equivalent, and (b) a simple parallel equivalent of the circuit in the black box.

Solution (a) *Series Equivalent*

$$Z = (V_s/I_s) = 12.5\underline{/53.1°} = (7.5 + j10) \ \Omega$$

$$R = 7.5 \ \Omega \quad \text{and} \quad X = 10 \ \Omega.$$

Since $X > 0$, an inductance is needed with a value of $L = (X/\omega) = 0.02$ H.
(b) *Parallel Equivalent*

$$Y = (I_s/V_s) = 0.08\underline{/-53.1°} = (0.048 - j0.064) \ S$$

$$G = 0.048 \ S \quad \text{and} \quad B = -0.064 \ S.$$

Since $B < 0$, an inductance is needed with a value of $L = -(1/\omega B) = 0.031$ H. The two equivalent circuits are shown in Fig. 8-22. ∎

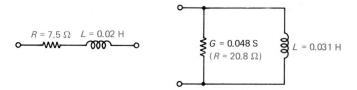

Fig. 8-22 Series and parallel equivalents of the circuit for Example 8-19.

Example 8-20 Obtain a series equivalent of the circuit shown in Fig. 8-23(a) at the following values of ω: (a) 100 r/s. (b) 400 r/s. (c) 1000 r/s.

Solution The admittance of the given circuit is

$$Y_T = G_p + j(\omega C_p - 1/\omega L_p)$$
$$= 2 + j(5 \times 10^{-3} \omega - 800/\omega)$$

Fig. 8-23 (a) Circuit for Example 8-20. (b), (c), (d): Series equivalents at different frequencies. Note that *both* resistance and reactance are different at different frequencies.

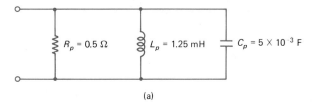

8-7 Simple Equivalent Circuits

(a) $\omega = 100$ r/s: Using $\omega = 100$ r/s, the value of $\mathbf{Y}_T$ becomes

$$\mathbf{Y}_T = (2 - j7.5) = 7.76\underline{/-75.1°}\text{ S}$$

Therefore, the impedance of the circuit is

$$\mathbf{Z}_T = (1/\mathbf{Y}_T) = 0.129\underline{/75.1°} = (0.0331 + j0.124)\,\Omega$$

The series equivalent consists of a resistance $R = 0.0331\,\Omega$ and an inductance $L = (0.124/100) = 1.24$ mH, as Fig. 8-23(b) shows.

(b) $\omega = 400$ r/s:

$$\mathbf{Y}_T = (2 + j0)\text{ S}$$
$$\mathbf{Z}_T = (0.5 + j0)\,\Omega$$

The series equivalent is simply a resistance $R = 0.5\,\Omega$, as shown in Fig. 8-23(c).

(c) $\omega = 1000$ r/s:

$$\mathbf{Y}_T = (2 + j4.2) = 4.65\underline{/64.5°}\text{ S}$$
$$\mathbf{Z}_T = 0.215\underline{/-64.5°} = (0.0926 - j0.194)\,\Omega$$

The series equivalent consists of a resistance $R = 0.0926\,\Omega$, and a capacitance $C = -(1/1000 \times 0.194) = 5.15 \times 10^{-3}$ F, as Fig. 8-23(d) shows. ∎

Exercise 8-39 The sets of values of $\mathbf{V}_s$ and $\mathbf{I}_s$ given below are from measurements made at the terminals of a black box known to contain a linear network. For each set of values, obtain a simple series equivalent and a simple parallel equivalent of the circuit in the black box. (a) $\mathbf{V}_s = 100\underline{/-10°}$ V; $\mathbf{I}_s = 4\underline{/-60°}$ A; $\omega = 500$ r/s. (b) Same $\mathbf{V}_s$ and $\mathbf{I}_s$ as in (a), but $\omega = 100$ r/s. (c) $\mathbf{V}_s = 100\underline{/-45°}$ V; $\mathbf{I}_s = 10\underline{/45°}$ A; $\omega = 377$ r/s. (d) $\mathbf{V}_s = 10\underline{/45°}$ V; $\mathbf{I}_s = 100\underline{/-45°}$ A; $\omega = 377$ r/s.

Exercise 8-40 A resistance $R = 100\,\Omega$ and a capacitance $C = 50\,\mu$F are in series. Obtain a parallel equivalent circuit at the following frequencies: (a) 100 Hz. (b) 500 Hz.

Exercise 8-41 Add an inductance $L = 4$ mH in series with the elements of the circuit in the previous exercise (8-40). Repeat the calculations.

8-8 AN IMPEDANCE BRIDGE

The notion of simple equivalent circuits discussed in Sec. 8-7 is useful in the measurement of impedance by means of a *bridge* circuit like that shown in Fig. 8-24, where $\mathbf{Z}_x$ is an unknown impedance. The elements R_1, R_2, C_v, and R_v are precision components. R_1 and R_2 are fixed in value, while R_v and C_v are adjustable and calibrated components. The components R_v and C_v are adjusted until the voltage $\mathbf{V}_d = 0$. When $\mathbf{V}_d = 0$, $\mathbf{I}_d = 0$. The null condition is detected by a pair of earphones or a galvanometer. The bridge is said to be balanced when the null condition is achieved.

$$\mathbf{V}_d = 0 \text{ implies that } \mathbf{V}_a = \mathbf{V}_x$$

Using voltage division, the relationships for $\mathbf{V}_a$ and $\mathbf{V}_x$ are

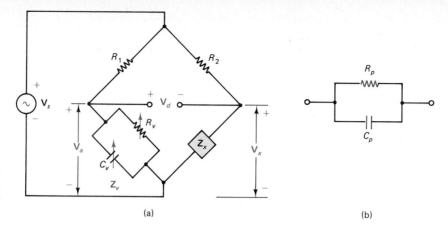

Fig. 8-24 (a) Impedance bridge. (b) Parallel equivalent of the unknown impedance Z_x.

$$\mathbf{V}_a = \frac{\mathbf{V}_s \mathbf{Z}_v}{\mathbf{Z}_v + R_1} \tag{8-53}$$

$$\mathbf{V}_x = \frac{\mathbf{V}_s \mathbf{Z}_x}{\mathbf{Z}_x + R_2} \tag{8-54}$$

When $\mathbf{V}_a = \mathbf{V}_x$ (null condition), Eqs. (8-53) and (8-54) lead to the equality

$$\mathbf{Z}_v R_2 = \mathbf{Z}_x R_1 \quad \text{(balanced bridge equation)}$$

Therefore, the unknown impedance $\mathbf{Z}_x$ is given by

$$\mathbf{Z}_x = (R_2/R_1)\mathbf{Z}_v$$

If a *parallel* equivalent is to be obtained for the unknown impedance, as Fig. 8-24(b) shows, then

$$R_p = (R_2/R_1)R_v \quad \text{and} \quad C_p = (R_2/R_1)C_v$$

Since (R_2/R_1) is a constant, it simply acts as a scale factor between R_p and R_v and between C_p and C_v. This is the reason for first finding the parallel equivalent circuit. Once the parallel equivalent is found, it also leads to the series equivalent of the unknown impedance.

Analysis will show that the bridge of Fig. 8-24(a) will not balance if the reactance of the unknown impedance is positive. A modification will be needed for such situations; see Problem (8-48).

Exercise 8-42 Given $R_1 = 100\Omega$, $R_2 = 250\Omega$ in the bridge circuit, the null condition occurs when $R_v = 50\Omega$ and $C_v = 10\ \mu\text{F}$ at a frequency of $f = 1$ kHz. Obtain the parallel and series equivalents of the unknown impedance.

8-9 POWER IN THE SINUSOIDAL STEADY STATE

The concepts of instantaneous power and average power were discussed in Chapter 7, where it was shown that if

$$v(t) = V_m \cos(\omega t + \theta) \quad \text{and} \quad i(t) = I_m \cos(\omega t + \phi)$$

then the instantaneous power is given by

$$p(t) = (1/2)V_m I_m [\cos(\theta - \phi) + \cos(2\omega t + \theta + \phi)]$$

while the average power, which is the constant term in the expression for $p(t)$, is given by

$$P_{av} = (1/2) V_m I_m \cos(\theta - \phi) \tag{8-55}$$

That is, given a circuit with voltage $\mathbf{V} = V_m \underline{/\theta}$ and current $I_m \underline{/\phi}$, the average power is given by Eq. (8-55).

It is sometimes convenient to use the *rms values* of voltage and current instead of the peak values. Using the relationships $V_{rms} = (V_m/\sqrt{2})$ and $I_{rms} = (I_m/\sqrt{2})$ obtained in the previous chapter, Eq. (8-55) becomes

$$P_{av} = V_{rms} I_{rms} \cos(\theta - \phi) \tag{8-56}$$

When V_m and V_{rms} are in volts, and I_m and I_{rms} are in amperes, the average power given by Eqs. (8-55) and (8-56) is in watts.

In the solution of problems, pay careful attention to whether the given values of voltages and currents are peak values—in which case Eq. (8-55) should be used—or rms values—in which case Eq. (8-56) should be used. Unless otherwise stated, the given values of voltages and currents in this and the following chapters should be treated as *peak values*.

8-9-1 Power Factor

For a given V_m and I_m, the highest value of average power is $(V_m I_m/2)$ obtained when $(\theta - \phi) = 0$ and $\cos(\theta - \phi) = 1$ in Eq. (8-55). This situation pertains to a circuit containing only resistors, since the voltage and current are in phase in such circuits. For any other circuit, the average power is smaller by a factor of $[\cos(\theta - \phi)]$. The quantity $[\cos(\theta - \phi)]$ is therefore called the *power factor* of a circuit. That is,

power factor (or pf) = cosine of the angle between the voltage and current.

The formula for average power becomes

$$\text{average power} = (1/2)(V_m)(I_m)(\text{power factor}) \tag{8-57}$$

The minimum value of power factor is -1 (when the phase angle between voltage and current is 180°), and its maximum value is $+1$ (when the voltage and current are in phase).

For a *single* impedance or admittance,

power factor = cosine of the angle of the impedance or admittance.

This statement follows from the fact that the angle between the voltage and current phasors for an impedance or admittance equals the angle of the impedance or admittance.

Since a cosine is an *even function* of its argument; that is, $\cos X = \cos(-X)$, the *power factor* will be the same whether the current is lagging behind the voltage or leading the voltage by a certain angle. The terms *current leading* and *current lagging* should be used to distinguish between the two cases.

Example 8-21 A voltage $\mathbf{V} = 100\underline{/30°}$ V is applied to an impedance $\mathbf{Z} = 5\underline{/-53.1°}$ Ω. Calculate the average power received by the impedance.

Solution

$$I = (V/Z) = 20\underline{/83.1°}\ A$$

Using Eq. (8-55),

$$P_{av} = (1/2)(100)(20)(\cos 83.1°) = 120\ W$$

Example 8-22 A circuit has a power factor of 0.75 (current leading) and consumes an average power of 1000 W when the voltage across it is 90 V (*rms*) at a frequency of 60 Hz. Obtain a simple series and a simple parallel equivalent for the given circuit.

Solution Using Eq. (8-56),

$$P_{av} = V_{rms} I_{rms}\ \text{(power factor)}$$

$$I_{rms} = 1000/(90 \times 0.75) = 14.8\ A$$

The *magnitude* of the impedance is therefore

$$|Z| = (90/14.8) = 6.08\ \Omega$$

The angle of the impedance is the angle of the power factor. Since it is specified that the current is leading, the circuit must be *capacitive*, and the angle of the impedance must be *negative*.

$$\text{impedance angle} = -\arctan 0.75 = -41.4°$$

and

$$Z = 6.08\underline{/-41.4°} = (4.56 - j4.02)\ \Omega$$

and the admittance is

$$Y = (1/Z) = 0.164\underline{/41.4°} = (0.123 + j0.108)\ S$$

The two equivalent circuits are shown in Fig. 8-25.

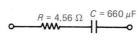
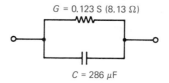

Fig. 8-25 Equivalent circuits for Example 8-22.

Exercise 8-43 A voltage $100\underline{/0°}$ V is applied to an impedance Z. Calculate the power factor and the average power consumed by the impedance when (a) $Z = (8 + j8)\Omega$; (b) $Z = (9 + j81)\Omega$; (c) $Z = (8 - j8)\Omega$; and (d) $Z = (81 - j9)\Omega$.

Exercise 8-44 A current $10\underline{/0°}$ A is fed to an admittance Y. Calculate the power factor and the average power consumed by the circuit when (a) $Y_1 = (0.3 + j0.4)$ S and (b) $Y_2 = (0.4 - j0.3)$ S.

Exercise 8-45 The average power consumed by a circuit is 1500 W when the applied voltage is 100 V at $\omega = 1000$ r/s. For each of the following values of power factor, obtain a simple series and a simple parallel equivalent of the circuit. (a) $pf = 0.5$ (current lagging). (b) $pf = 0.8$ (current leading).

8-9 Power in the Sinusoidal Steady State

8-9-2 Apparent Power and Reactive Power

The average power is of practical importance, since it is the power measured by wattmeters and forms the basis of the charges paid by a consumer to a power company, for example. From the point of view of the power-generating system, however, it is not enough to consider only the average power, since the power factor of the load affects the current-carrying capacity of the system for a given transmission voltage. For a given voltage and average power demand, a load with a smaller power factor requires a larger current. The equipment used in the generation, transmission, and distribution of power must be designed on the basis of the current demand, I_m or I_{rms}, of the load for a given voltage V_m or V_{rms}. The product $(V_{rms}I_{rms})$, called *apparent* power, thus forms an important specification of all such equipment. If P_{app} denotes the apparent power, then

$$P_{app} = (V_{rms}I_{rms}) = (1/2)V_m I_m \tag{8-58}$$

The name *volt-ampere* (abbreviated VA) is used for the units for apparent power to distinguish it from average power. (Generically, the product of volts and amperes leads to watts, but different names are used for different power quantities in ac circuits for purposes of distinction.)

The average power and apparent power are related by the equation

$$P_{av} = P_{app} \cos \alpha \tag{8-59}$$

where $\cos \alpha$ is the power factor. The last relationship is shown graphically by means of a *power triangle* with P_{app} as the hypotenuse and P_{av} as the horizontal side; see Figs. 8-26(a), (b).

The vertical side of the power triangle represents another possible component of power called *reactive power* or P_r:

$$P_r = P_{app} \sin \alpha \tag{8-60}$$

The name *volt-ampere-reactive* (abbreviated VAR) is used for reactive power to distinguish it from the other two power quantities, apparent and average power. Reactive power may be positive or negative, depending upon whether α is positive or negative. *If* the angle α in Eq. (8-60) is the angle by which the *voltage leads the current*, then a *positive* reactive power represents an *inductive* circuit and a *negative* reactive power a *capacitive* circuit. It is generally safer, however, to specify the magnitude of reactive power and state whether the current is leading or lagging.

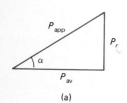

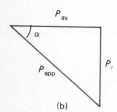

Fig. 8-26 Power triangles: horizontal side = average power, vertical side = reactive power, hypotenuse = apparent power, angle = power factor angle.

Example 8-23 A circuit with a power factor of 0.8 (current leading) consumes a reactive power of 1000 VAR when a voltage of 120 V at 60 Hz is applied. Calculate the element values of (a) a series equivalent circuit and (b) a parallel equivalent circuit.

Solution The angle for a power factor of 0.8 has a *magnitude* of [arc tan 0.8] = 36.9°. Set up a power triangle with 1000 VAR as the vertical side and an angle of 36.9°, as shown in Fig. 8-27(a). (It does not matter whether the vertical side is drawn upward or downward.) Then

$$P_{app} = \frac{P_r}{\sin \alpha} = \frac{1000}{0.6} = 1667 \text{ VA}$$

Using Eq. (8-58), $P_{app} = (1/2)V_m I_m$,

Fig. 8-27 Power triangles for Example 8-23

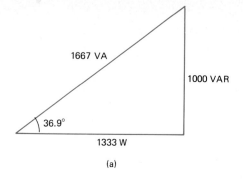

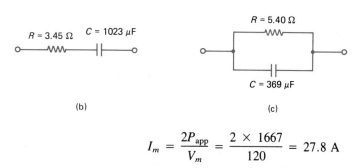

$$I_m = \frac{2P_{app}}{V_m} = \frac{2 \times 1667}{120} = 27.8 \text{ A}$$

(a) The angle of the impedance is $-36.9°$ (since current is leading) and

$$|Z| = (V_m/I_m) = (120/27.8) = 4.32 \text{ }\Omega$$

Therefore,

$$\mathbf{Z} = 4.32\underline{/-36.9} = (3.45 - j2.59) \text{ }\Omega$$

The series equivalent consists of a resistance of 3.45 Ω and a capacitance whose reactance is -2.59 Ω. Since $\omega = 377$ r/s,

$$C = 1/(377 \times 2.59) = 1023 \text{ }\mu\text{F}$$

Thus $R = 3.45$ Ω and $C = 1023$ μF in the series equivalent, shown in Fig. 8-27(b).
(b) $\mathbf{Y} = (1/\mathbf{Z}) = 0.2332\underline{/36.9°} = (0.185 + j0.139)$ S. The parallel equivalent consists of a conductance of 0.185 S and a capacitance whose susceptance is 0.139 S. Since $\omega = 377$ r/s,

$$C = (0.139/377) = 369 \text{ }\mu\text{F}$$

The elements of the parallel equivalent circuit are $R = (1/0.185) = 5.40$ Ω and $C = 369$ μF, as Fig. 8-27(c) shows. ∎

Exercise 8-46 Determine the apparent power and reactive power in the circuits of Exercises 8-43 and 8-44. Draw the power triangle in each case.

Exercise 8-47 Calculate the elements of (a) a simple series equivalent, and (b) a simple parallel equivalent of the circuit for each of the following sets of data. $f = 60$ Hz.

8-9 Power in the Sinusoidal Steady State

(i) $V_m = 90$ V. $P_{app} = 1800$ VA. $P_{av} = 400$ W. Current lagging.
(ii) $I_m = 16$ A. $P_{av} = 4$ kW. $P_r = 1.5$ kVAR. Current leading.
(iii) $P_{app} = 5$ kVA. $pf = 0.9$ (current lagging). $I_m = 125$ A.

8-9-3 Modification of Power Factor

For a given value of apparent power, the amount of average power consumed decreases as the power factor decreases. It is desirable, especially in power systems, to have a power factor as close to unity as possible to increase the average power delivered to the load for a given apparent power.

The power factor of a circuit is increased by moving the angle of the impedance (or admittance) of the circuit closer to zero; that is, reducing the *magnitude of the angle* of the impedance (or admittance).

Such a reduction is accomplished by decreasing the magnitude of the reactance of an impedance or susceptance of an admittance. The customary procedure is to place an inductance or capacitance *in parallel* with the existing load so as to *reduce the magnitude of the susceptance* of the circuit to a desired level. (The power factor can also be modified by placing an inductance or capacitance in series with a circuit. But the parallel connection is preferable, since adding a component in parallel to a given circuit is more convenient in a power distribution system.)

Consider a circuit with the admittance

$$Y_1 = G_1 + jB_1$$

The power factor of Y_1 is given by

$$pf_1 = \text{arc tan}(B_1/G_1)$$

Choose a capacitance or inductance with a susceptance B_2 *opposite* in sign to B_1 and let

$$Y_2 = jB_2$$

When Y_2 is connected in parallel with Y_1, the total admittance is

$$Y_T = G_1 + j(B_1 + B_2)$$

and the new power factor is

$$pf_T = \text{arc tan}[(B_1 + B_2)/G_1] \qquad (8\text{-}61)$$

Since B_1 and B_2 are opposite in sign,

$$|B_1 + B_2| < |B_1| \qquad \text{and} \qquad pf_T > pf_1$$

Equation (8-61) is used to calculate the value of B_2 to meet the new power factor specified.

Example 8-24 A reactive element is to be added in parallel to the circuit of Example 8-23 to increase the power factor to 0.95. Determine the value of the reactive element.

Solution The admittance of the circuit in Example 8-23 is

$$Y_1 = (0.185 + j0.139) \text{ S}$$

which has a positive susceptance (capacitive). An inductance L must be added in parallel to increase the power factor. Let the new admittance be

$$\mathbf{Y}_T = 0.185 + jB_T$$

where

$$B_T = (0.139 - 1/\omega L)$$

The power factor of $\mathbf{Y}_T$ is to be 0.95; therefore, the angle α of $\mathbf{Y}_T$ must satisfy the equation

$$\cos \alpha = 0.95$$

There are two possible values of α: $+18.2°$ and $-18.2°$, as indicated in Figs. 8-28(a) and (b).

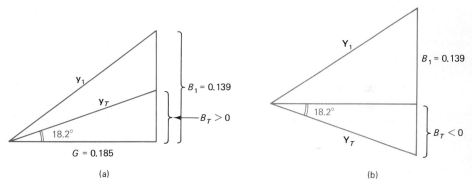

Fig. 8-28 Power factor modification for Example 8-24.

Case 1: $\alpha = +18.2°$. $B_T > 0$. $(B_T/G_1) = \tan 18.2° = 0.329$. $B_T = 0.0608$ S. Therefore,

$$(1/\omega L) = 0.0782 \quad \text{or} \quad L = 1/(377 \times 0.0782) = 33.9 \text{ mH}$$

The circuit is still capacitive in this case, and the power factor is 0.95, with current leading.

Case 2: $\alpha = -18.2°$. $B_T < 0$. $(B_T/G_1) = \tan(-18.2°) = -0.329$. $B_T = -0.0608$ S. Therefore, from Eq. (8-62),

$$(1/\omega L) = -0.200 \quad \text{or} \quad L = 1/(377 \times 0.2) = 13.3 \text{ mH}$$

The circuit is inductive in this case and the power factor is 0.95 with current lagging. Even though both answers are correct, the 13.3 mH might be preferable because of its smaller value. ∎

Exercise 8-48 For each circuit of Exercise 8-47, calculate both possible values of a *shunt* element in order to increase the power factor to 0.95.

Industrial loads in a power distribution system use a large number of induction motors, resulting in a lower power factor with current lagging. The improvement of the power factor using actual shunt capacitors is usually impractical, since very large capacitors are needed. A more practical solution is to use a special type of motor called a *synchronous motor* that operates as either an inductive load or a capacitive load, depending upon the field excitation levels. Synchronous motors are useful not only for increasing the power factor, but also for driving loads as motors. The term *synchronous condenser* is used for synchronous motors used to improve the power factor of a system.

Example 8-25 A manufacturing plant is rated at 4000 kW at a power factor of 0.5 (current lagging). Determine the VAR specifications of a synchronous motor to be added in parallel to the system to change the power factor to 0.75 (current lagging). Assume that the apparent power rating of the synchronous motor is given as 3000 kVA.

Solution Set up a power triangle for the original system: 4000 W as the horizontal side, and an angle of (arc cos 0.5) = 60°; see Fig. 8-29(a). The reactive power (vertical side) is found to be 4000 tan 60° = 6928 kVAR.

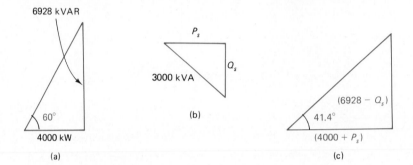

Fig. 8-29 Power triangles for Example 8-25. (a) Original system. (b) Synchronous motor. (c) Modified system.

Let P_s (in kW) and Q_s (in kVAR) denote the average and reactive power components of the synchronous motor. The power triangle of the synchronous motor is shown in Fig. 8-29(b).

$$P_s^2 + Q_s^2 = 3000^2 \tag{8-62}$$

The power triangle of the modified system will have an angle of (arc cos 0.75) = 41.4°. Its horizontal side equals $(4000 + P_s)$ and its vertical side equals $(6928 - Q_s)$, as indicated in Fig. 8-29(c). Therefore,

$$\frac{6928 - Q_s}{4000 + P_s} = \tan 41.4° = 0.882 \tag{8-63}$$

Equations (8-62) and (8-63) are to be solved for P_s and Q_s. A combination of the two equations leads to the following quadratic equation:

$$1.78 P_s^2 - 5998 P_s + 2.56 \times 10^6 = 0$$

There are two possible values:

$$P_s = 501 \text{ kW} \quad \text{and} \quad P_s = 2872 \text{ kW}$$

Therefore, there are two solutions: (a) $P_s = 501$ kW, $Q_s = 2958$ kVAR; and (b) $P_s = 2872$ kW and $Q_s = 867$ kVAR. The solution $P_s = 501$ kW, $Q_s = 2958$ kVAR is preferable, since the total apparent power of the system is considerably smaller than in the other case. ∎

Exercise 8-49 A manufacturing plant is load rated at 10^4 kW with a power factor of 0.8 (current lagging). A synchronous motor designed to operate at a power factor of 0.5 (current leading) is added in parallel to the plant to increase the overall power factor to 0.95 (current lagging). Determine the total average power, reactive power, and apparent power of the modified system.

8-9-4 Complex Power

The power triangle used in the preceding discussion suggests the possibility that the average power may be thought of as the *real part* and the reactive power as the *imaginary part* of a complex quantity represented by the hypotenuse. This viewpoint leads to the definition of *complex power*:

$$\text{magnitude of complex power} = \text{apparent power}$$
$$\text{angle of complex power} = \text{power factor angle}$$

More formally, given $\mathbf{V} = V_m\underline{/\theta}$ and $\mathbf{I} = I_m\underline{/\phi}$, and using $(\theta - \phi)$ as the power factor angle, complex power is defined by

$$\mathbf{P}_{\text{comp}} = (1/2)V_m I_m \underline{/\theta - \phi} \qquad (8\text{-}64)$$

Since $\mathbf{I}^*$, the complex conjugate of $\mathbf{I}$, is given by $\mathbf{I}^* = I_m\underline{/-\phi}$, it follows that

$$\mathbf{P}_{\text{comp}} = (1/2)\mathbf{V}\mathbf{I}^* \qquad (8\text{-}65)$$

The relationships among average power, reactive power, and complex power are given by

$$P_{\text{av}} = \text{Re}[\mathbf{P}_{\text{comp}}] = (1/2)\text{Re}[\mathbf{V}\mathbf{I}^*] \qquad (8\text{-}66)$$

$$P_r = \text{Im}[\mathbf{P}_{\text{comp}}] = (1/2)\text{Im}[\mathbf{V}\mathbf{I}^*] \qquad (8\text{-}67)$$

$$\mathbf{P}_{\text{comp}} = P_{\text{av}} + jP_r \qquad (8\text{-}68)$$

The concept of complex power is particularly useful in theoretical analysis, since it permits the use of all three power components (apparent, average, and reactive) in a combined analytical form. For example, it is possible to show that

The total complex power in a network = sum of the complex powers of the individual components.

Therefore,

The total average power = sum of the average powers. The total reactive power = sum of the reactive powers of the individual components.

But

The total apparent power ≠ sum of the individual apparent powers.

(The proof of these statements is the topic of Problem 8-57.)

Example 8-26 Determine the complex power, the average power, and the reactive power in each of the following cases: (a) $\mathbf{V}_1 = 100\underline{/-36.9°}$ V, $\mathbf{I}_1 = 16\underline{/16.2°}$ A. (b) $v_2(t) = 156 \cos 377t$, $i_2(t) = 12 \cos(377t - 45°)$ A. (c) $\mathbf{V}_3 = 12\underline{/-60°}$ V, $\mathbf{I}_3 = 5\underline{/-60°}$ A.

Solution (a) $V_1 = 100\underline{/-36.9°}$ V. $I_1^* = 16\underline{/-16.2°}$ A.

$$P_{comp} = (1/2)(100\underline{/-36.9°})(16\underline{/-16.2°}) = 800\underline{/-53.1°}$$
$$= (480 - j640) \text{ VA}.$$

$P_{av} = \text{Re}[P_{comp}] = 480$ W. $P_r = \text{Im}[P_{comp}] = -640$ VAR. (In complex power notation, a *minus* sign for reactive power indicates *current leading*.)

(b) Put V_2 and I_2 in phasor form first. $V_2 = 156\underline{/0°}$ V. $I_2 = 12\underline{/-45°}$ A. $I_2^* = 12\underline{/45°}$ A. Therefore,

$$P_{comp} = 936\underline{/45°} = (662 + j662) \text{ VA}.$$
$$P_{av} = 662 \text{ W}. \ P_r = 662 \text{ VAR}.$$

(In complex power notation, a *plus* sign for reactive power indicates *current lagging*.)

(c) $P_{comp} = 30\underline{/0°} = (30 + j0)$ VA. $P_{av} = 30$ W. $P_r = 0$. ∎

Example 8-27 Two loads are connected in parallel across a 220 V, 60 Hz source. Load 1 consumes an average power of 4000 W and a reactive power of 3000 VAR (current leading). Load 2 consumes a complex power of $9000\underline{/60°}$ VA. (a) Determine the total average power, the total reactive power, and the total apparent power in the circuit. (b) Obtain a simple series equivalent of the circuit.

Solution (a) The complex power in load 1 is

$$P_{comp1} = (4000 - j3000) \text{ VA}$$

The total complex power in the circuit is therefore

$$P_{comp} = (4000 - j3000) + 9000\underline{/60°}$$
$$= (8500 + j\,4794) - 9758\underline{/29.4°} \text{ VA}$$

- Total average power = 8500 W
- Total reactive power = 4794 VAR (current lagging)
- Total apparent power = 9758 VA
- Power factor = 0.871 (current lagging)

(b) Since $P_{comp} = 9758\underline{/29.4°}$ VA and $V = 220\underline{/0°}$, and $P_{comp} = (VI^*\,2$, the current in the circuit is

$$I = 88.7\underline{/-29.4°} \text{ A}$$

Therefore,

$$Z = 2.48\underline{/29.4°} = (2.16 + j1.22)$$

The series equivalent consists of $R = 2.16\Omega$, and $L = 3.23$ mH. ∎

Exercise 8-50 Repeat the calculations of the previous example (for the same applied voltage and frequency) in the case of the following sets of loads connected in parallel. (a) Load 1: $P_{comp} = 1900\underline{/-30°}$ VA; Load 2: $P_{comp} = 2400\underline{/30°}$ VA. (b) Load 1: 400 W at 0.9 pf (current lagging); Load 2: 800 W at unity pf.

Exercise 8-51 A voltage of 150 V is applied to each of the following impedances. In each case, calculate the complex power, average power, reactive power, apparent power, and power factor. (a) $Z_1 = 50\underline{/-36.9°}\ \Omega$. (b) $Z_2 = 30\underline{/-90°}\ \Omega$. (c) $Z_3 = 48\underline{/63.5°}\ \Omega$. (d) $Z_4 = 1.5\underline{/0°}\ \Omega$.

Exercise 8-52 A current of 15 A is sent through each of the following admittances. In each case, calculate the complex power, average power, reactive power, apparent power, and power factor. (a) $Y_1 = 0.02\underline{/36.9°}\ S$. (b) $Y_2 = 30\underline{/90°}\ S$. (c) $Y_3 = 0.195\underline{/-45°}\ S$. (d) $Y_4 = 100\underline{/-90°}\ S$.

8-10 SUMMARY OF CHAPTER

A real sinusoidal forcing function can be expressed in terms of complex exponential forcing functions by means of Euler's theorem. A phasor is a complex quantity with the magnitude and angle of the complex exponential forcing function and represents the corresponding sinusoidal function. The differential equation of a circuit excited by a complex exponential forcing function transforms into an algebraic equation, since differentiation and integration are replaced by multiplication and division by the factor $(j\omega)$. The solution of the algebraic equation gives the complex exponential response of the circuit. When the circuit is excited only by the real part of a complex exponential function, the response is only the real part of the complex exponential response obtained by solving the algebraic equation. The solution of circuits in the sinusoidal steady state is expedited by using phasors for voltages and currents and replacing derivatives and integrals of voltages and currents by $(j\omega)$ and $(1/j\omega)$, respectively. The resulting phasor responses contain all the essential information except the frequency.

The impedance of a circuit is given by the ratio of a voltage phasor to a current phasor. The version of Ohm's law for ac circuits is: The product of a current phasor and complex impedance equals the voltage phasor. The real and imaginary parts of an impedance are, respectively, its resistance and its reactance. The reactance may be positive (an inductive circuit) or negative (a capacitive circuit). The impedance of a resistor is simply its resistance; of a capacitor, it is $(1/j\omega C)$ or $-j(1/\omega C)$; of an inductor, it is $(j\omega L)$.

The total impedance of a series connection of impedances is the sum of the individual impedances. The voltage across each individual impedance is proportional to that impedance. Graphic representations of the components of an impedance (the impedance triangle) and the voltages in a series circuit (phasor diagrams) are useful visual aids in the study of a circuit.

Admittance is the dual of impedance, conductance the dual of resistance, and susceptance the dual of reactance. Susceptance is positive for a capacitive circuit and negative for an inductive circuit. Admittance and impedance are mutual reciprocals, and the product of a voltage phasor and complex admittance gives the current phasor in a circuit. In general, conductance, resistance, susceptance, and reactance of a given circuit are all functions of frequency.

Given a circuit, it is possible to find a simple series equivalent circuit (a resistance in series with a reactance element) or a simple parallel equivalent circuit (a conductance in parallel with a susceptance element). It is important to remember that the total admittance and total impedance of the two equivalent circuits are mutual reciprocals at a specific frequency.

The average power in the sinusoidal steady state is given by the product of the rms value of the voltage, the rms value of the current, and the cosine of the angle between them. The cosine of the angle between voltage and current in a circuit is the power factor. The average power consumed by a circuit is at a maximum when the power factor is 1 and zero when the power factor is zero. For single impedances and admittances, the power factor angle is the same as the angle of impedance or admittance.

Apparent power is the product of the rms voltage and rms current in a circuit, and is a measure of the capacity of the power source needed to feed a circuit of an arbitrary power factor. Reactive power is the vertical component of a power triangle.

The power factor of a circuit can be changed by adding a suitable reactance element in series or parallel with the circuit; the shunt connection is preferred. The aim of the parallel addition is to reduce the magnitude of the effective susceptance of the circuit. In industrial systems, a synchronous motor is used as a capacitive load to achieve power factor correction.

Complex notation for power uses the average power as the real component, and the reactive power as the imaginary components (with a negative sign for current leading the voltage). The magnitude of the complex power equals the apparent power, and its angle is the power factor angle. In a circuit, the average power, the reactive power, and the complex power in the individual components are added to obtain the totals of those quantities. The total apparent power, however, is not equal to the sum of the individual components of apparent power (except when the circuit is purely resistive or purely reactive).

Answers to Exercises

8-1 Write each complex coefficient in the form $Ae^{j\theta}$ first. (a) $14.14 \cos(50t + 45°)$. (b) $14.14 \cos(50t + 45°)$. (c) $14.14 \sin(50t + 45°)$. (d) $-14.14 \sin(50t + 45°)$.

8-2 (a) $5e^{j(2t+143.1°)}$. Real part: $5 \cos(2t + 143.1°)$. Imaginary part: $5 \sin(2t + 143.1°)$.
(b) $7.07e^{j(\omega t+45+126.9°-135°)}$. Real part: $7.07 \cos(10t + 36.9°)$.
Imaginary part: $7.07 \sin(10t + 36.9°)$.

8-3 (a) $\mathbf{A} = 100e^{-j30°}$. $B = 60$. In (b) and (c), reduce the expression first to a single cosine term. (b) $\mathbf{A} = 100e^{-j30°}$. $B = 60$. (c) $\mathbf{A} = 100e^{-j60°}$. $B = 60$.

8-4 See comment in answer to Exercise (8-3). (a) $10e^{j(80t-180°)}$ V. (b) $8e^{j(190t-165°)}$ V. (c) $5e^{j(80t+53.1°)}$ A. (d) $5e^{j(80t+126.9°)}$ A.

8-5 (a) $100\underline{/180°}$ V. (b) $100\underline{/90°}$ V. (c) $141.4\underline{/-135°}$ A.

8-6 (a) $100 \cos(\omega t - 60°)$ V. (b) $10 \cos(\omega t - 120°)$ A. (c) $1000 \cos \omega t$ V. (d) $\cos(\omega t + 135°)$ A.

8-7 $\mathbf{V}_3 = [(25.9 + j96.6) + (30 + j40)]$. $v_3(t) = 147.6 \cos(377t + 67.7°)$ V.

8-8 Verification.

8-9 $\mathbf{V}_m = 10/(2 + j1.25)$ V. $v(t) = 4.24 \cos(250t - 32°)$ V.

8-10 $\mathbf{V}_m = 10/(2 + j1.25 - j0.4)$. $v(t) = 4.61e^{j(250t-23)}$ V.

8-11 $\mathbf{V} = 10/j0.55$. $v(t) = 18.2 \cos(8t - 90°)$ V.

8-12 $\mathbf{I} = 80\underline{/45°}/(20 - j17.5)$. $i(t) = 3 \cos(4t + 86.2°)$ A.

8-13 $0.424\underline{/-32°}$ Ω. $0.461\underline{/-23}$ Ω. $1.82\underline{/-90°}$ Ω. $26.6\underline{/-41.2°}$ Ω.

8-14 (a) $(100\underline{/0°}/25\underline{/-90°}) = 4\underline{/-90°}$ Ω. (b) $9.6\underline{/90°}$ A. (c) $26\underline{/-20°}$ V.

8-15 (R, X): $(0.36, -0.225)$; $(0.424, -0.180)$; $(0, -1.82)$, $(20, -17.5)$.

8-16 (a) $j1.88 = 1.88\underline{/90°}$ Ω. (b) $j3142 = 3142\underline{/90°}$ Ω. (c) $j5.66 \times 10^6 = 5.66 \times 10^6\underline{/90°}$ Ω.

8-17 (a) $-j53 = 53\underline{/-90°}$ Ω. (b) $-j0.0318 = 0.0318\underline{/-90°}$ Ω. (c) $-j1.77 \times 10^{-5} = 1.77 \times 10^{-5}\underline{/-90°}$ Ω.

8-18 $\mathbf{I} = 1.12\underline{/-63.4°}$ A. Voltages: $11.2\underline{/-63.4°}$, $33.6\underline{/26.6°}$, $11.2\underline{/-153.4°}$ V.

8-19 $\mathbf{I} = 25\underline{/-120°}$ A. $\mathbf{V}_2 = 25\underline{/-180°}$ V. $\mathbf{V}_T = 35.4\underline{/-135°}$ V.

8-20 (a) $\mathbf{Z} = 14.9\underline{/-70.3°}$ Ω. $R = 5$Ω. $X = -14$Ω. $\mathbf{I} = 6.71\underline{/70.3°}$ A. (b) $\mathbf{Z} = 6.40\underline{/38.7°}$ Ω. $R = 5$Ω. $X = 4$Ω. $\mathbf{I} = 15.6\underline{/-38.7°}$ A. (c) $\mathbf{Z} = 5\underline{/0°}$ Ω. $R = 5$Ω. $X = 0$. $\mathbf{I} = 20\underline{/0°}$ A.

8-21 (a) $7.07\underline{/45°}$ A, $70.7\underline{/45°}$ V, $70.7\underline{/135°}$ V, $141.4\underline{/-45°}$ V. (b) $2\underline{/0°}$ A, $100\underline{/53.1°}$, V, $63.2\underline{/-71.6°}$ V, $28.2\underline{/-45°}$ V. (c) $3.20\underline{/-31.3°}$ A, $32.0\underline{/28.7°}$ V, $64\underline{/-1.3°}$ V, $16\underline{/-61.3°}$ V.

8-22 $173\underline{/-75°}$ V.

8-23 (a) Horizontal side = R. Vertical side = $\left(\omega L - \dfrac{1}{\omega C}\right)$. (b) Horizontal side = 50Ω. Vertical side = 60Ω (downward).

8-24 See Fig. 8-30(a).

8-25 (a) $0.2\underline{/30°}$ S. (b) $0.2\underline{/-53.1°}$ S. (c) $0.1\underline{/-25°}$ S.

8-26 (a) $0.1\underline{/-45°}$ S. (b) $5\underline{/-90°}$ S. (c) $0.397\underline{/63°}$ S.

8-27 (G, B): $(0.173, 0.1)$, $(0.12, -0.16)$, $(0.0906, 0.0423)$, $(0.0707, -0.0707)$, $(0, -5)$, $(0.180, 0.354)$.

8-28 (a) $-j0.212 = 0.212\underline{/-90°}$ S. (b) $-j1.06 \times 10^{-3} = 1.06 \times 10^{-3}\underline{/-90°}$ S. (c) $-j6.63 \times 10^{-7} = 6.67 \times 10^{-7}\underline{/-90°}$ S.

8-29 (a) $j1.57 \times 10^{-8} = 1.57 \times 10^{-8}\underline{/90°}$ S. (b) $j1.571 \times 10^{-5} = 1.57 \times 10^{-5}\underline{/90°}$ S. (c) $j0.0157 = 0.0157\underline{/90°}$.

8-30 (a) $0.602\underline{/-41.6°}$ S; $16.6\underline{/101.6°}$ V, $3.32\underline{/101.6°}$ A, $3.32\underline{/191.6°}$ A, $16.6\underline{/11.6°}$ A, $4.15\underline{/101.6°}$ A, $6.64\underline{/191.6°}$ A. (b) $1.54\underline{/73°}$ S, $6.49\underline{/-13°}$ V, $1.30\underline{/-13°}$ A, $3.89\underline{/77°}$ A, $2.17\underline{/-103°}$ A, $1.62\underline{/-13°}$ A, $7.79\underline{/77°}$ A.

8-31 $3.82\underline{/-84°}$ S, $2.62\underline{/144°}$ V, $0.524\underline{/144°}$ A, $1.05\underline{/234°}$ A, $13.1\underline{/54°}$ A, $0.655\underline{/144°}$ A, $2.1\underline{/234°}$ A.

8-32 $15.9\underline{/88.4°}$ S, $0.629\underline{/-28.4°}$ V, $0.126\underline{/-28.4°}$ A, $0.252\underline{/61.6°}$ A, $0.315\underline{/-118.4°}$ A, $0.157\underline{/-28.4°}$ A, $10.1\underline{/61.6°}$ A.

8-33 (a) $0.289\underline{/-69.7°}$ S, $34.6\underline{/114.7°}$ V, $3.46\underline{/114.7°}$ A, $1.63\underline{/-155.3°}$ A, $11.0\underline{/24.7°}$ A. (b) $11.0\underline{/-5.2°}$ S, $0.906\underline{/50.2°}$ V, $4.05\underline{/-13.2°}$ A, $4.05\underline{/23.6°}$ A, $6.40\underline{/90.2°}$ A. (c) $14.14\underline{/-15°}$ S, $0.707\underline{/60°}$ V, $7.07\underline{/30°}$ A, $7.07\underline{/0°}$ A, $7.07\underline{/150°}$ A.

8-34 $(\mathbf{I}_1/\mathbf{I}_2) = (\mathbf{Y}_1/\mathbf{Y}_2)$. (a) $\mathbf{I}_1 = 10\underline{/126.8°}$ A. $\mathbf{I}_T = 8.96\underline{/63.4°}$ A. (b) $\mathbf{I}_1 = 12.5\underline{/143.1°}$ A. $\mathbf{I}_T = 7.5\underline{/90°}$ A. (c) $\mathbf{I}_1 = 2.5\underline{/16.2°}$ A. $\mathbf{I}_T = 12.4\underline{/3.2°}$ A.

8-35 (a) Horizontal side = 0.45 S. Vertical side = 0.7 S (upward). (b) Horizontal side = 8 S.

8-36 See Fig. 8-30(b).

8-37 See Fig. 8-31.

8-38 Listing of answers is (R, X), (G, B) in all four cases: (a) $(10, -7.5)$, $(6.4 \times 10^{-2}, 4.8 \times 10^{-2})$. (b) $(10, 22.5)$, $(1.65 \times 10^{-2}, -3.71 \times 10^{-2})$ (c) $(4.1, 4.92)$, $(0.1, -0.12)$ (d) $(0.716, -2.58)$, $(0.1, 0.36)$.

8-39 (a) 16.1Ω in series with 38.4 mH; 38.9Ω in parallel with 65.4 mH. (b) 16.1Ω in series with 0.192 H; 38.9 Ω in parallel with 0.327 H. (c) 265 μF. (d) 26.5 mH.

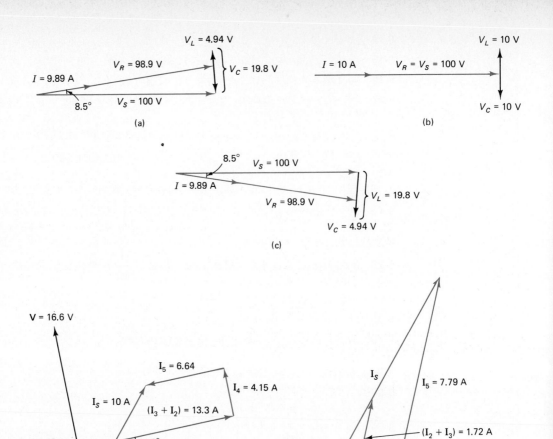

Fig. 8-30 Answers for Exercises 8-24 and 8-36.

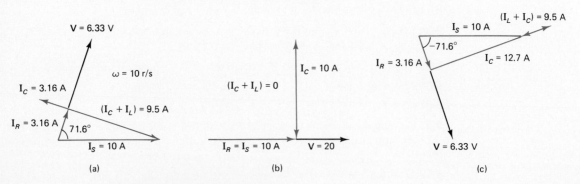

Fig. 8-31 Answer for Exercise 8-37.

Phasors, Impedance, and Admittance

8-40 (a) 110Ω, 4.6 μF. (b) 100Ω, 0.202 μF.

8-41 (a) 109Ω, 4.3 μF. (b) 100Ω, 0.515 H.

8-42 $Z_x = (11.5 - j36.1)Ω$. Series: 11.5Ω, 4.41 μF. Parallel: 125Ω, 3.99 μF.

8-43 (a) 312 W, 0.707. (b) 6.75 W, 0.110. (c) 312 W, 0.707. (d) 61.1 W, 0.994.

8-44 (a) 60 W, 0.6. (b) 80 W, 0.8.

8-45 (a) $\mathbf{I} = 60\underline{/-60°}$ A. Series: 0.834Ω, 1.44 mH. Parallel: 3.33Ω, 1.92 mH.
(b) $\mathbf{I} = 37.5\underline{/36.9°}$ A. Series: 2.13Ω, 625 μF. Parallel: 3.33Ω, 225μF.

8-46 441.3 VA, 312 VAR; 61.4 VA, 61.0 VAR; 441.3 VA, 312 VAR; 61.5 VA, 6.72 VAR; 100 VA, 80 VAR; 100 VA, 60 VAR.

8-47 (i) $\mathbf{I} = 40\underline{/-77.2°}$ A. Series: 0.498Ω, 5.81 mH. Parallel: 10.2Ω, 6.13 mH.
(ii) $\mathbf{V} = 312.5\underline{/-36.9°}$ V. Series: 15.6Ω, 226 μF. Parallel: 24.4Ω, 81.7 μF. (iii) $\mathbf{V} = 80\underline{/25.8°}$ V. Series: 0.576Ω, 0.740 mH. Parallel: 0.709Ω, 3.9 mH.

8-48 (i) Add C. $B_c = 0.401$ or 0.465 S. $C = 1063$ μF or 1234 μF. (ii) Add L. $B_L = 1.732 \times 10^{-2}$ or 4.428×10^{-2} S. $L = 153$ mH or 59.9 mH. (iii) Add C. $B_c = 1.144$ or 0.216. $C = 3034$ μF or 573 μF.

8-49 Plant: 10^4 kW; 0.75 kVAR. Motor: $Q_s = -1.732 P_s$. Total: $(-1.732 P_s + 0.75 \times 10^4)/(P_s + 10^4) = +0.329$. $P_s = 0.771 \times 10^4$ kW or 0.204×10^4 kW. Total: 1.771×10^4 kW, 0.585×10^4 kVAR (current leading), 1.865×10^4 kVA. Or, 1.204×10^4 kW, 0.396 kVAR (current lagging), 1.267×10^4 kVA.

8-50 (a) 3715 W, 250 VAR, 3723 VA. 6.49Ω, 1.15 mH. (b) 1200 W, 194 VAR, 1216 VA. 19.7Ω, 8.39 mH.

8-51 (a) $225\underline{/-36.9°}$ VA, 180 W, 135 VAR, 225 VA, 0.8. (b) $375\underline{/-90°}$ VA, 0 W, 375 VAR, 375 VA, 0. (c) $234\underline{/63.5°}$ VA, 105 W, 210 VAR, 234 VA, 0.446. (d) $7500\underline{/0°}$ VA, 7500 W, 0 VAR, 7500 VA, 1.

8-52 (a) $5625\underline{/-36.9°}$ VA, 4498 W, -3377 VAR, 5625 VA, 0.8. (b) $3.75\underline{/-90°}$ VA, 0 W, 3.75 VAR, 3.75 VA, 0. (c) $76.9\underline{/45°}$ VA, 54.4 W, 54.4 VAR, 76.9 VA, 0.707.
(d) $1.125\underline{/-90°}$ VA, 0 W, 1.125 VAR, 1.125 VA, 0.

PROBLEMS

Sec. 8-1: Complex Exponential Functions

8-1 Evaluate each of the following and simplify the result as far as possible.
(a) $|(3 - j4)(10\underline{/20°})(e^{j30°})|$.
(b) $\mathrm{Re}[A^2 \div |A|^2]$ where $\mathbf{A} = a + jb$.
(c) Imaginary part of the expression in (b).

8-2 If $\mathbf{V}_1 = (-7.07 - j7.07)$, evaluate and sketch
(a) $v_1(t) = \mathrm{Re}[\mathbf{V}_1 e^{j200t}]$.
(b) $v_2(t) = \mathrm{Im}[\mathbf{V}_1 e^{j(200t + \pi/3)}]$.
(c) $v_3(t) = \mathrm{Re}[\mathbf{V}_1 e^{j200(t - 0.004)}]$.

8-3 Given $\mathbf{X} = (-4 - j9)$ and $\mathbf{Y} = 11.2\underline{/105°}$, evaluate and simplify the following.
(a) $(\mathbf{X} - j\mathbf{Y})$. (b) $\mathrm{Im}[\mathbf{X}e^{-j(100t + 26°)}]$. (c) $\mathrm{Re}[\mathbf{X}e^{j100t}] + \mathrm{Im}[\mathbf{Y}e^{-j100t}]$.

8-4 Evaluate each of the following functions, where $\mathbf{A} = (p + jq)$; p, q, r, and b are real numbers. Simplify the result as far as possible.
(a) $f_1(t) = \mathbf{A}e^{jbt} + \mathbf{A}^* e^{-jbt}$.
(b) $f_2(t) = |\mathbf{A}|e^{jbt} + |\mathbf{A}^*|e^{-jbt}$.

(c) $f_3(t) = Ae^{j(bt+r)} + A^*e^{-j(bt+r)}$.
(d) $f_4(t) = |Ae^{jr}|e^{jbt} + |A^*e^{-jr}|e^{-jbt}$.

8-5 Write the complex exponential function whose *real* part is:
(a) $g_1(t) = 10 \cos 100t - 8 \sin 100t$. (b) $g_2(t) = 10 \cos 100t - 20 \cos(100t + 90°)$.
(c) $g_3(t) = 30 \cos(100t - 36.9°) + 40 \sin(100t - 53.1°)$.

8-6 Write the complex exponential function whose imaginary part is given by: (a) $h_1(t) = 10 \sin 500t + 20 \cos 500t$. (b) $h_2(t) = 10 \cos(500t + 53.1°) + 10 \sin(500t + 53.1°)$.
(c) $h_3(t) = -10 \cos 500t - 15 \sin 500t$.

8-7 If $X = |X|\underline{/\theta}$ and $Y = |Y|\underline{/\phi}$, evaluate and simplify: (a) $(Xe^{j\omega t})(Y^* e^{-j\omega t})$. (b) $(Xe^{j\omega t})/(Ye^{j\omega t})$.
(c) $[Re(XY^*)](e^{j\omega t} + e^{-j\omega t})$. (d) $Re[(X^*Y)(e^{j\omega t} - e^{-j\omega t})]$

Sec. 8-2: Phasors

8-8 Write the phasor representations of: (a) $v(t) = 10 \cos(100t + 30°) + 10 \sin 100t$ V.
(b) $i(t) = 10 \cos(100t + 30°) + Re[5e^{j100t}] + Im[15e^{j(100t+45°)}]$.

8-9 Write the sinusoidal functions represented by each of the following phasors. Assume the period $T = 10$ ms for each signal. (The final answer must be a single cosine function.)
(a) $V_1 = 10e^{j60°} + 20e^{j30°}$. (b) $V_2 = (10\underline{/50°})(30 - j40) + (-12 + j5)$.
(c) $V_3 = [(30 + j40)(100e^{j90°})] \div (50 - j50)$.

8-10 If $i_1(t) = 2 \cos 100t$ A, $i_2(t) = 2.828 \cos(100t - 45°)$ A, and $i_3(t) = 5 \cos(100t + 90°)$ A, evaluate (the final answer must be a single cosine function of time):
(a) $i_a = (i_1 + i_2 + i_3)$ (b) $i_b = (i_1 + i_2 - i_3)$. (c) $i_c = (i_1 - i_2 - i_3)$.

8-11 Determine the complex quantity X in each of the following equations. Write the final answer in the form $|X|e^{j\theta}$. (a) $(10\underline{/-30°})(30e^{j100t}) + X = (130\underline{/90°})$. (b) $[(10\underline{/25°}) + (3 + j4)]e^{j(35t+50°)} = X[e^{j35t}]$. (c) $[(10\underline{/36.9°}) + (10\underline{/-36.9°})] = X^*$.

Secs. 8-3, 8-4: Complex Exponential Excitation and Use of Phasors:

8-12 In a certain network, it is found that when the input is $v_s(t) = 4e^{j(3t+\pi/6)}$ V, the resulting current is $i(t) = 0.5e^{j(3t-0.1\pi)}$. (a) Find the current when the input is $v_a(t) = 16e^{j(3t+\pi/4)}$ V. (b) Find the voltage needed to produce the current $i_b(t) = 2 \sin(3t - 60°)$ A.

8-13 When the input current to a network is $i_s(t) = 10e^{j40t}$ A, the output voltage is found to be $v_0(t) = 30e^{j(40t-60°)}$ V. Determine the output voltage in the following cases: (a) Input $i_a(t) = 10e^{-j40t}$ A. (b) Input $i_b(t) = 10 \cos 40t$ A. (c) Input $i_c(t) = 10 \sin 40t$ A. (d) Input $i_d(t) = 10 \cos(40t + 60°) + 10e^{j40t}$.

8-14 A resistor $R = 47\Omega$, a capacitor $C = 120$ μF, and an inductor $L = 14$ mH are in series with a voltage source $v_s(t) = 160 \cos 377t$ V. (a) Write the differential equation of the circuit. (b) Assume the forcing function to be $v_s(t) = 160e^{j377t}$, and a suitable complex exponential current. Substitute the assumed solution and forcing function in the differential equation. (c) Evaluate the complex exponential current. (d) Repeat the steps (b) and (c) for a forcing function $160e^{-j377t}$. Write all the steps. (e) Combine the results of parts (c) and (d) to find the current in the original circuit.

8-15 Repeat the work of the previous problem if the three elements are in parallel across a current source $i_s(t) = 20 \cos 1500t$ A. In parts (b), (c), and (d), use forcing functions that correspond to the given source current.

8-16 Given the differential equation $(d^2 i/dt^2) + 2(di/dt) + i = 5 \cos(2t + 30°)$, transform the equation to algebraic form by using phasors. Solve for the phasor I. Write the solution $i(t)$ of the given equation.

8-17 Given the differential equation $v(t) + 0.05(dv/dt) + 400 \int v(t)dt = 10 \cos 100t$, transform the equation to algebraic form by using phasors. Solve for the phasor V. Write the solution $v(t)$ of the given equation.

8-18 A resistance R and a capacitance C are in parallel across a current source $i_s(t)$. (a) Write the differential equation relating the voltage $v(t)$ across the circuit to $i_s(t)$. (b) Transform the equation by using phasors when $i_s(t) = I_0 \cos \omega t$. (c) Solve the equation for the voltage $\mathbf{V}$. (d) If it is given that $v(t) = 2 \cos(10t - 30°)$ V when $i_s(t) = 5 \cos 10t$ A, calculate the values of R and C. (e) Find the voltage $v(t)$ if $i_s(t) = 5 \cos 50t$ A for R and C found in (d).

8-19 An inductor L and a capacitor C are in series across a voltage source $v_s(t) = V_m \cos \omega t$. Set up *one* differential equation that relates the voltage v_L across the inductance to the source voltage v_s. Transform the differential equation to an algebraic equation by using phasors. Solve the equation and determine $v_L(t)$.

8-20 An inductor L and a capacitor C are in parallel across a current source $i_s(t) = I_m \cos(t/2\sqrt{LC})$. Set up one differential equation that relates the current i_L in the inductance to the source current. Transform the differential equation to an algebraic equation by using phasors. Solve the equation and determine $i_L(t)$.

[In the following problems, answers pertaining to voltages and currents should be left in phasor form unless otherwise specified.]

Sec. 8-5 Impedance and Series Circuits

[Impedance triangles and phasor diagrams will often be helpful in solving these problems.]

8-21 In each of the sets of values given below, one item (voltage, current, or impedance) is missing. Evaluate the unknown quantity. (a) $v(t) = 100 \cos(500t + 20°)$ V. $i(t) = 12.5 \cos(500t - 60°)$ A. (b) $v(t) = 100 \cos(500t - 60°)$ V. $\mathbf{Z} = (30 - j40)\Omega$. (c) $i(t) = 12.5 \cos(500t + 36.9°)$ A. $\mathbf{Z} = (5 + j12)\Omega$.

8-22 A resistor $R = 47\Omega$, an inductance $L = 25$ mH, and a capacitance $C = 625$ μF are in series. Determine the impedance, resistance, and reactance of the circuit at the following values of ω: (a) 10 r/s. (b) 100 r/s. (c) 10^3 r/s. (d) 10^4 r/s.

8-23 A resistor $R = 25\Omega$ is in series with an inductance $L = 50$ mH and a capacitance C. $\omega = 1000$ r/s. Determine the value of C for each of the following cases: (a) reactance X of the circuit $= 20\Omega$; (b) $X = -20\Omega$; (c) angle of the impedance is $+45°$; (d) angle of the impedance is $-45°$.

8-24 A series circuit consists of $R_1 = 15\Omega$, $C_1 = 10$ μF, $L_1 = 0.2$ H, $R_2 = 10\Omega$, $C_2 = 40$ μF. (a) If the *current* $\mathbf{i}(t)$ through the circuit is given as $20 \cos 500t$ A, determine the voltage across each element and the total voltage. (b) Find the value of ω at which the total voltage is in phase with the current. (c) Find the value of ω at which the total voltage leads the current by $45°$. (d) Find the value of ω at which the total voltage lags behind the current by $45°$.

8-25 The voltage $\mathbf{V}_2$ in the circuit of Fig. 8-32 is found to be $100/\underline{60°}$ V at $\omega = 2$ r/s. Calculate the current $\mathbf{I}$ and the voltages $\mathbf{V}_1$ and $\mathbf{V}_s$.

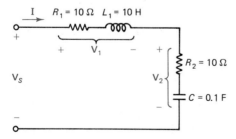

Fig. 8-32 See Problem 8-25.

8-26 A resistance R and capacitance C are in series with a voltage $\mathbf{V}_s$ across them. (a) Determine the frequency ω at which the phase angle between the current $\mathbf{I}$ in the circuit and $\mathbf{V}_s$ is $45°$ and evaluate the current **I** at that frequency in terms of V_s and R. (b) Determine the frequency at which the amplitude of the current is one-half of that in part (a) and evaluate the current at that frequency.

8-27 A resistor $R = 470\,\Omega$ is in series with an inductance L and a capacitance C. The applied voltage is $\mathbf{V}_s = 12\underline{/0°}$ V at $\omega = 1000$ r/s. It is found that the amplitudes of the voltages across L and C are equal to each other and 25 times as high as the applied voltage. Calculate the values of L and C.

8-28 An impedance $\mathbf{Z}_x$ is in series with a capacitance C whose reactance is $-10\,\Omega$. The voltage across the capacitance is found to be $v_c(t) = 5\cos(1000t - 30°)$ V and the voltage across the series combination is $v_s(t) = 10\cos(1000t + 60°)$ V. (a) Determine the value of C. (b) Determine the value of $\mathbf{Z}_x$ and obtain a simple series equivalent.

8-29 An inductance of $L = 0.02$ H is in series with an impedance $\mathbf{Z}_1$. The current in the circuit $i(t) = 20\cos(200t + \theta)$ A, the voltage across $\mathbf{Z}_1$ is $v_1(t) = 100\cos(200t + \phi)$ V, and the voltage across the series combination is $v_s(t) = 100\cos 200t$ V. Determine the values of the angles θ and ϕ.

8-30 The impedance of a circuit is found to have a *magnitude* of $|\mathbf{Z}_1| = 100\,\Omega$ at $\omega = 100$ r/s. A capacitance of $C = 100\,\mu\text{F}$ is connected in series with $\mathbf{Z}_1$ and it is found that the *magnitude* of the new impedance is $|\mathbf{Z}_2| = 50\,\Omega$. Determine $\mathbf{Z}_1$ and obtain a simple series equivalent.

8-31 Measurements were made in a laboratory (which did not have an oscilloscope) on a circuit made up of a resistance $R_1 = 10\,\Omega$ in series with a coil that has a resistance R_L and an inductance L. Only the amplitudes of the voltages were measured (at $\omega = 1000$ r/s) and found to be: total voltage = 125 V, voltage across R_1 = 50 V, and voltage across the coil = 103 V. Phase angles were not measured. Determine the values of R_L and L.

Sec. 8-6: Admittance and Parallel Circuits

[*Admittance triangles and phasor diagrams will often be helpful in solving these problems.*]

8-32 In each of the sets of values given below, one item (voltage, current, or admittance) is missing. Evaluate the unknown quantity. (a) $v(t) = 100\cos(500t + 20°)$ V. $i(t) = 12.5\cos(500t - 60°)$ A. (b) $v(t) = 100\cos(500t - 60°)$ V. $\mathbf{Y} = (30 - j40)$ S. (c) $i(t) = 12.5\cos(500t + 36.9°)$ A. $\mathbf{Y} = (5 + j12)$ S.

8-33 A resistor $R = 47\,\Omega$, an inductance $L = 25$ mH, and a capacitance $C = 625\,\mu\text{F}$ are in parallel. Determine the admittance, conductance, and susceptance of the circuit at the following values of ω: (a) 10 r/s. (b) 100 r/s. (c) 10^3 r/s. (d) 10^4 r/s.

8-34 A resistor $R = 25\,\Omega$ is in parallel with an inductance $L = 10$ mH and a capacitance C. $\omega = 1000$ r/s. Determine the value of C for each of the following cases: (a) susceptance B of the circuit is 0.01 S; (b) $B = -0.01$ S; (c) angle of the admittance is $+45°$; (d) angle of the admittance is $-45°$.

8-35 A parallel circuit consists of $R_1 = 15\,\Omega$, $C_1 = 10\,\mu\text{F}$, $L_1 = 0.2$ H, $R_2 = 10\,\Omega$, $C_2 = 40\,\mu\text{F}$. (a) If the voltage $v(t)$ across the circuit is given as $20\cos 500t$ V, determine the current in each element and the total current. (b) Find the value of ω at which the voltage is in phase with the total current. (c) Find the value of ω at which the voltage leads the total current by $45°$. (d) Find the value of ω at which the voltage lags behind the total current by $30°$.

8-36 Determine the resistance, reactance, conductance, and susceptance of the circuits described by the following data: (a) A series combination of $10\,\Omega$, 25 mH, and $500\,\mu\text{F}$ at $\omega = 377$ r/s. (b) A parallel combination of the elements in part (a). (c) Input voltage $\mathbf{V} = 100\underline{/-30°}$ V produces an input current of $\mathbf{I} = 50\underline{/-100°}$ A. (d) The current has an amplitude 5 times as high as the input voltage and leads the input voltage by $53.1°$.

8-37 The current $\mathbf{I}_2$ in the circuit of Fig. 8-33 is found to be $10\underline{/-30°}$ A at $\omega = 12$ r/s. Calculate the currents $\mathbf{I}_1$ and $\mathbf{I}_s$ and the voltage $\mathbf{V}$.

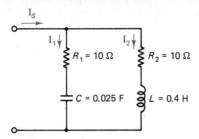

Fig. 8-33 See Problem 8-37.

8-38 A resistance R and the capacitance C are in parallel with a current I_s through the combination. (a) Determine the frequency ω at which the phase angle between the current I_s in the circuit and V is $45°$ and evaluate the voltage V at that frequency. (b) Determine the frequency at which the amplitude of the voltage is one-half of that in part (a) and evaluate the voltage at that frequency.

8-39 A resistor $R = 0.470\Omega$ is in parallel with an inductance L and a capacitance C. The applied current is $I_s = 12\underline{/0°}$ A at $\omega = 1000$ r/s. It is found that the amplitudes of the currents in L and C are equal to each other and three times as high as the applied current. Calculate the values of L and C.

8-40 An admittance Y_x is in parallel with a capacitance C whose susceptance is 0.25 S. The voltage across the capacitance is found to be $v_c(t) = 5 \cos(1000t - 30°)$ V and the total current in the parallel combination is $i_s(t) = 10 \cos(1000t + 60°)$ V. (a) Determine the value of C. (b) Determine the value of Y_x and obtain a simple parallel equivalent.

8-41 The admittance of a circuit is found to have a *magnitude* of $|Y_1| = 100$ S at $\omega = 100$ r/s. A capacitance of $C = 0.8$ F is connected in parallel with Y_1, and it is found that the *magnitude* of the new admittance is $|Y_2| = 50$ S. Determine Y_1.

8-42 A resistance $R = 600\Omega$ is placed in parallel with a capacitance C. Determine the value of C so that at $f = 1$ kHZ, the input *impedance* is given by $Z = (400 + jX)$, where X is not specified.

Sec. 8-7: Simple Equivalent Circuits

8-43 Measurements are made at the terminals of a black box, and the following data are obtained. For each case, obtain (a) a simple series equivalent circuit, and (b) a simple parallel equivalent circuit.
(i) $V_s = 50\underline{/-45°}$ V. $I_s = 4\underline{/30°}$ A. $\omega = 1000$ r/s.
(ii) Same V_s and I_s as in part (a), but $\omega = 200$ r/s.
(iii) $V_s = 150\underline{/45°}$ V. $I_s = 300\underline{/-30°}$ A. $\omega = 1000$ r/s.
(iv) Same V_s and I_s as in part (iii), but $\omega = 200$ r/s.

8-44 A resistance R, a capacitance C, and an inductance L are in series. Obtain expressions for the conductance and susceptance of the circuit.

8-45 A conductance G, a capacitance C, and an inductance L are in parallel. Obtain expressions for the resistance and reactance of the circuit.

8-46 A resistance $R = 100\Omega$ is in series with a capacitance $C = 1500$ μF. (a) Determine the frequency at which the susceptance of the circuit is 0.005 S. (b) Determine the frequency at which the conductance = one-half of the susceptance. (c) Determine the conductance and susceptance of the circuit at the frequency calculated in (b).

8-47 A resistance $R = 100\Omega$ is in parallel with a capacitance $C = 1500$ μF. (a) Determine the frequency at which the magnitude of the reactance of the circuit is 50Ω. (b) Determine the frequency at which the resistance equals the magnitude of the reactance. (c) Determine the conductance and susceptance at the frequency calculated in (b).

Sec. 8-8: Impedance Bridge

8-48 Show that the bridge in the circuit of Fig. 8-24 cannot balance if the reactance of the unknown impedance is positive. Make the necessary modification to permit the use of the bridge circuit for measuring an unknown impedance with a positive reactance.

Sec. 8-9: Power

8-49 Determine the average power and power factor for the circuits for which the following data are available. (a) $\mathbf{V}_s = 100\underline{/-30°}$ V. $\mathbf{I}_s = 12.5\underline{/45°}$ A. (b) $\mathbf{V}_s = 100\underline{/-60°}$ V. $\mathbf{Z} = (10 + j16)$ Ω. (c) $\mathbf{I}_s = 100\underline{/-60°}$ A. $\mathbf{Y} = (3 + j4)$ S.

8-50 When a voltage of 100 V at $\omega = 200$ r/s is applied to a series RL circuit, the voltage across the inductance L is found to have an amplitude of 50 V, and the average power consumed by the circuit is 200 W. Calculate the values of R and L.

8-51 When a current of 10 A at $\omega = 50$ r/s is fed to a parallel RC circuit, the average power consumed by the circuit is 400 W at a power factor of 0.8. Calculate the values of R and C.

8-52 A voltage $\mathbf{V}_s = 100\underline{/0°}$ V at $\omega = 100$ r/s is applied to a series connection of an impedance $\mathbf{Z}_1$ and an inductance L. The angle of the impedance $\mathbf{Z}_1$ has a *magnitude* of 45°. The average power delivered by the source is 1000 W at a power factor of unity. Determine the values of L and the impedance $\mathbf{Z}_1$.

8-53 For each of the circuits specified in Problem 8-49, determine the apparent power and reactive power.

8-54 For each of the circuits specified in Problem 8-49, determine the value of the reactive element to be connected in parallel with it so as to change the power factor to 0.90. Assume $\omega = 377$ r/s. Calculate both possible values in each case.

8-55 An industrial load is rated at 500 kVA and consumes an average power of 300 kW. Assume the load to have a net positive reactance. (a) If the applied voltage is $220\underline{/0°}$ V, obtain a simple series equivalent circuit for the load. (b) If a pure reactive load with a reactive power of 100 kVAR (current leading) is connected in parallel with the load, determine the values of the apparent power, average power, and reactive power of the combination. (c) Obtain a simple series equivalent of the combination.

8-56 A factory has a total load of 500 kW at a power factor of 0.7 (current lagging). The applied voltage is 220 V. Determine the average power and reactive power components of a synchronous motor to be added in parallel to the factory load so as to change the power factor to 0.85. Assume that the synchronous motor is operating at a power factor of 0.75 (current leading).

8-57 Consider a parallel connection of admittances $\mathbf{Y}_1, \mathbf{Y}_2, \ldots, \mathbf{Y}_n$ with a voltage $\mathbf{V}_s$ applied across it. Let $\mathbf{I}_s$ be the current supplied to the circuit. Write an expression for the complex power in each branch in terms of its admittance and $\mathbf{V}_s$. Write an expression for the total complex power in terms of $\mathbf{I}_s$ and $\mathbf{V}_s$. Show that the total complex power equals the sum of the complex powers in the individual branches and a similar statement is true for the total average power and for the total reactive power. Why is a similar statement not valid for the total apparent power?

8-58 Two impedances $\mathbf{Z}_1$ and $\mathbf{Z}_2$ are in series. $\mathbf{Z}_1$ consumes 1 kW with a power factor of 0.5 (current leading), while $\mathbf{Z}_2$ consumes 2 kW with a power factor of 0.75 (current lagging). The voltage across $\mathbf{Z}_2$ has an amplitude of 100 V. (a) Write the complex power in $\mathbf{Z}_1$, $\mathbf{Z}_2$, and the series combination. (b) Determine the total average power, power factor, reactive power, and apparent power in the circuit. (c) Obtain a simple series equivalent of the circuit at $\omega = 377$ r/s.

8-59 Two impedances, $\mathbf{Z}_1$ and $\mathbf{Z}_2$, are in parallel. $\mathbf{Z}_1$ consumes 1 kW with a power factor of 0.5 (current leading). The apparent power in $\mathbf{Z}_2$ is 3 kVA and the voltage across the circuit has an amplitude of 250 V. The power factor of the parallel combination is given as 0.75 (current leading). Determine the complex power for each of the impedances and the parallel combination.

8-60 Redo Problem 8-54, using the complex power notation.

8-61 Redo Problem 8-55, using the complex power notation.

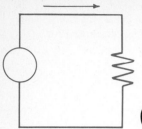

CHAPTER 9
NETWORKS IN THE SINUSOIDAL STEADY STATE

We saw in the last chapter that, in sinusoidal steady state analysis using phasors, impedance and admittance performed a function exactly analogous to resistance and conductance in resistive circuits with constant forcing functions. This analogy is applicable not only to the KVL and KCL equations of simple circuits (as was done in the last chapter), but also to the more systematic equations of a network using nodal, loop, and mesh analysis. It is simply a matter of using $\mathbf{Z}$ instead of R and $\mathbf{Y}$ instead of G in the development shown in Chapters 3 and 4. The equations in sinusoidal steady state are somewhat more cumbersome due to the complex coefficients, and the results are more complicated (and more interesting) than in purely resistive circuits driven by a constant forcing function. But the principles and procedures of analysis are exactly the same in both types of circuits.

This chapter discusses the applications of circuit analysis methods to the sinusoidal steady state. Instead of repeating all the material developed for circuit analysis in Chapters 3, 4, and 5, a brief review of the relevant principles and procedures will be presented here at appropriate points, along with a number of illustrative examples.

9-1 NETWORKS WITH SINGLE SOURCES

When a network consists of passive elements (resistors, capacitors, and inductors) driven by a single source, it is analyzed in most cases by using either of the following methods.

1. *Series-parallel Combinations:* **Series and parallel combinations of impedance and admittances are used to reduce the network to a single impedance. Voltage and current divisions can then be used to solve for the currents and voltages in individual branches.**

2. *Ladder Network Method:* **An assumption is made about the voltage or current in a suitably selected branch of the circuit. Voltages and currents in the remaining branches are found by successive steps involving KCL, KVL, and voltage-current equations. Linearity is then invoked to find the currents and voltages in the various branches of the circuit for the given forcing function.**

The following examples illustrate the application of these methods to circuits driven by sinusoidal sources.

Example 9-1   **Series-parallel Reduction**
Determine the currents in all the branches of the network shown in Fig. 9-1(a).

Solution   The steps in the reduction of the network are shown in Fig. 9-1(b), (c). Note that **Y** and **Z** are used interchangeably, depending upon whichever is convenient.
The total admittance is seen to be

$$\mathbf{Y}_T = \mathbf{Y}_1 + \cfrac{1}{\mathbf{Z}_2 + \cfrac{1}{\mathbf{Y}_3 + \mathbf{Y}_4}}$$

$$= \frac{\mathbf{Y}_1 \mathbf{Z}_2 (\mathbf{Y}_3 + \mathbf{Y}_4) + \mathbf{Y}_1 + \mathbf{Y}_3 + \mathbf{Y}_4}{\mathbf{Z}_2 (\mathbf{Y}_3 + \mathbf{Y}_4) + 1}$$

The voltages and currents are found by retracing the steps in Fig. 9-1, and using voltage and current divisions as appropriate.

Fig. 9-1 Series-parallel reduction (Example 9-1).

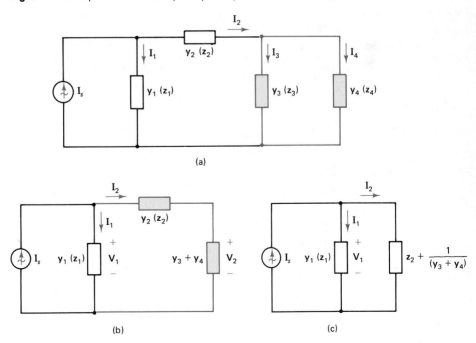

9-1   **Networks with Single Sources**

$$V_1 = (I_s/Y_T) \qquad I_1 = V_1 Y_1 = I_s (Y_1/Y_T)$$

$$I_2 = \cfrac{V_1}{Z_2 + \cfrac{1}{Y_3 + Y_4}} = \cfrac{(Y_3 + Y_4)}{Y_T[Z_2(Y_3 + Y_4) + 1]} I_s$$

$$V_2 = \cfrac{[1/(Y_3 + Y_4)]}{Z_2 + \cfrac{1}{Y_3 + Y_4}} V_1 = \cfrac{1}{Y_T[Z_1(Y_3 + Y_4) + 1]} I_s$$

$$I_3 = V_2 Y_3 = \cfrac{Y_3}{Y_T[Z_2(Y_3 + Y_4) + 1]} I_s$$

$$I_4 = Y_4 V_2 = \cfrac{Y_4}{Y_T[Z_2(Y_3 + Y_4) + 1]} I_s$$

Example 9-2 Series-parallel Reduction

Determine the branch currents in the circuit of Fig. 9-2(a).

Solution The steps in the reduction are shown in Figs. 9-2(b), (c), and (d).

$$Z_1 = R_1 - j(1/\omega C_1) = 5 - j5 \; \Omega$$
$$Y_2 = (1/R_2) + j(\omega C_2) = 0.5 + j\,0.25 \; S$$
$$Z_3 = j\omega L_2 + (1/Y_2) = 1.6 + j\,1.2 \; \Omega$$
$$Y_4 = (1/Z_1) + (1/Z_3) = 0.5 - j\,0.2 \; S$$
$$Z_T = R_3 + (1/Y_4) = 2.224 + j\,0.6896 \; \Omega$$

Fig. 9-2 (a) Circuit for example 9-2. (b)-(d) Reduction of network.

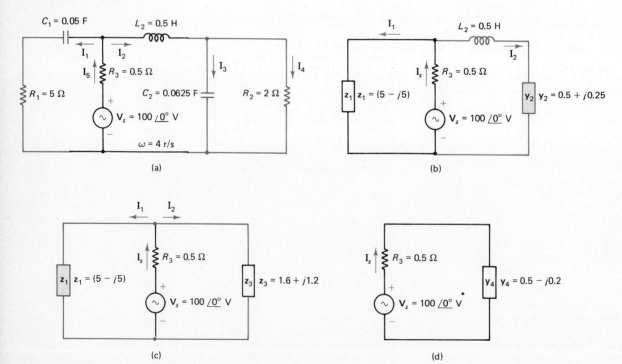

The current supplied by the voltage source V_s is

$$I_s = (V_s/Z_T) = 42.96 \underline{/-17.2°} \text{ A}$$

The branch currents are found by using current division and KCL and going backward through the diagrams of Fig. 9-2.

$$I_1 = \frac{Z_3}{Z_3 + Z_1} I_s = 11.28 \underline{/49.6°} \text{ A}$$

$$I_2 = I_s - I_1 = 33.72 - j\,21.29 = 39.88 \underline{/-32.3°} \text{ A}$$

$$I_3 = \frac{R_2}{R_2 - j(1/\omega C_2)} I_2 = 17.83 \underline{/31.1°} \text{ A}$$

$$I_4 = I_2 - I_3 = 18.45 - j\,30.5 = 35.65 \underline{/-58.8°} \text{ A}$$ ∎

Exercise 9-1 Determine the total impedance seen by the source and the average power delivered by it in each of the circuits of Fig. 9-3.

Fig. 9-3 Circuits for Exercise 9-1.

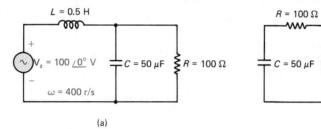

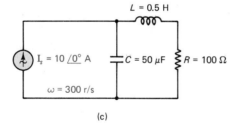

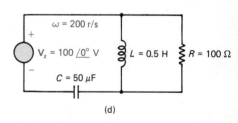

Exercise 9-2 Determine all the branch currents in each of the circuits of Fig. 9-4.

Fig. 9-4 Circuits for Exercise 9-2.

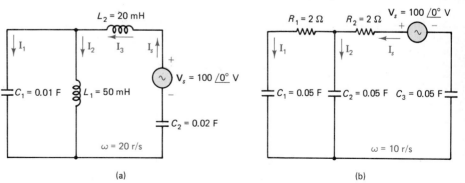

9-1 Networks with Single Sources

Example 9-3

Ladder Network Method

Determine the currents and the voltages in the circuit of Fig. 9-5.

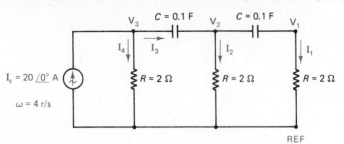

Fig. 9-5 Ladder network for Example 9-3.

Solution Assume $I_1 = 1$ A. Then $V_1 = 2$ V.

$$V_2 = V_1 + I_1[-j(1/\omega C)] = 2 - j\,2.5 \text{ V}$$
$$I_2 = (V_2/R) = 1 - j\,1.25 \text{ A}$$
$$I_3 = I_1 + I_2 = 2 - j\,2.5 \text{ A}$$
$$V_3 = V_2 + I_3(-j/\omega C) = -1.125 - j\,7.5 \text{ V}$$
$$I_4 = (V_3/R) = -0.5625 - j\,3.75 \text{ A}$$
$$I_s = I_4 + I_3 = 1.438 - j\,5 = 5.203\,\underline{/-74°} \text{ A}$$

Since the *given* current source is $20\underline{/0°}$ A, linearity requires that all the currents and voltages obtained must be multiplied by the ratio:

$$\frac{\text{given value of } I_s}{\text{value of } I_s \text{ obtained when } I_1 = 1 \text{ A}} = 3.844\underline{/74°}$$

The values of the branch currents and voltages in the given circuit are therefore:

$$I_1 = 3.84\,\underline{/74°} \text{ A}; \quad V_1 = 7.69\,\underline{/74°} \text{ V}; \quad V_2 = 12.3\,\underline{/22.7°} \text{ V};$$
$$I_2 = 6.15\,\underline{/22.7°} \text{ A}; \quad V_3 = 29.2\,\underline{/-24.5°} \text{ V}; \quad I_4 = 14.6\,\underline{/24.5°} \text{ A}$$

■

Example 9-4

Ladder Network Method

Redo the circuit of Example 9-2 by using the ladder network procedure.

Solution Refer to Fig. 9-6 for the currents and voltages assigned.

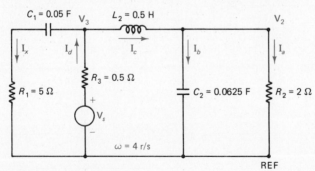

Fig. 9-6 Circuit for Example 9-4 (same circuit as in Example 9-2).

Assume $I_a = 1$ A. Then $V_2 = 2$ V. $I_b = V_2(j\omega C_2) = j\,0.5$ A

$$I_c = I_a + I_b = 1 + j\,0.5 \text{ A}$$
$$V_3 = j\omega L_2 I_c + V_2 = 1 + j\,2 \text{ V} \qquad (9\text{-}1)$$

Starting from the left, assume the current through R_1 to be I_x. Then

$$V_3 = I_x\,[R_1 - j(1/\omega C_1)] = (5 - j\,5)\,I_x \qquad (9\text{-}2)$$

Equating the two values of V_3 from Eqs. (9-1) and (9-2):

$$I_x = -0.1 + j\,0.3 \text{ A when } I_a = 1 \text{ A}$$
$$I_d = I_x + I_c = 0.9 + j\,0.8 \text{ A}$$
$$V_s = R_3 I_d + V_3 = 1.45 + j\,2.4 = 2.804\,\underline{/58.9°}\text{ V}$$

Since the given voltage $V_s = 100\,\underline{/0°}$ V, linearity requires that all the currents and voltages obtained above must be multiplied by $(100\,\underline{/0°} \div 2.804\,\underline{/58.9°}) = 35.66\,\underline{/-58.9°}$.

Therefore, the branch currents in the given circuit are:

$$I_a = 35.7\,\underline{/-58.9°}\text{ A; } I_b = 17.8\,\underline{/31.1°}\text{ A; } I_c = 39.9\,\underline{/32.3°}\text{ A}$$
$$I_d = 42.9\,\underline{/-17.3°}\text{ A; } I_x = 11.3\,\underline{/49.5°}\text{ A.} \qquad \blacksquare$$

Exercise 9-3 Use the ladder network method to determine the branch currents in each of the circuits shown in Fig. 9-7.

Fig. 9-7 Circuits for Exercise 9-3.

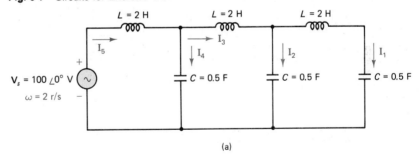

(a)

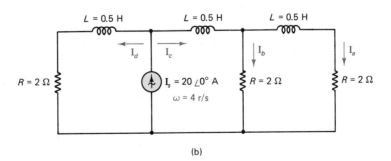

(b)

9-2 NETWORKS WITH MULTIPLE SOURCES

When a network contains two or more sources, all at the *same* frequency, all the voltages and currents in it are represented by phasors of the same frequency, and voltages and currents in such a circuit can be manipulated completely in terms of phasors. The circuit is then analyzed by using impedances and admittances in nodal, loop, or mesh analysis.

If a network contains sources of *different* frequencies, the principle of superposition is employed for its analysis. The sources are separated into groups, each of which operates at the same frequency. The response of the network to each group of resources is found by using phasors of a single frequency, impedance, and admittance. Then the individual responses are transformed to real sinusoidal functions of time and combined to obtain the total response.

The number of independent node voltages and independent loop currents in a circuit are determined by using the topological principles discussed in Chapters 3 and 4.

Exercise 9-4 For each of the circuits shown in Fig. 9-8, draw the graph, set up three trees for each graph, and for each tree, set up the independent loops.

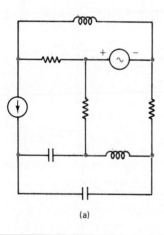
(a)

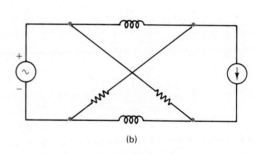
(b)

Fig. 9-8 Circuits for Exercise 9-4.

9-3 NODAL ANALYSIS

As we discussed in Chapter 3, there are n independent node voltages in a network with $(n + 1)$ nodes. One of the nodes is chosen as the reference node, and nodal equations are written for the remaining n nodes.

Consider an arbitrary node as shown in Fig. 9-9, where the voltages $V_1, V_2 \ldots$ are measured with respect to the reference node REF. Writing KCL at node 1 in the same form as presented in Chapter 3:

$\sum$ currents *leaving* node 1 through the *admittances*

$= \sum$ source currents *entering* node 1

The equation for node 1 in Fig. 9-9 is

$$Y_1(V_1 - V_2) + Y_2(V_1 - V_2) + Y_3(V_1 - V_3) + Y_4V_1 + Y_5V_1 = I_a - I_b$$

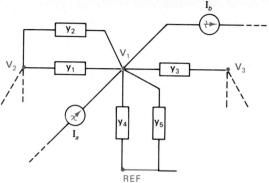

Fig. 9-9 Portion of a circuit for nodal equations.

When the coefficients of the different voltages are collected together, the equation above becomes

$$(Y_1 + Y_2 + Y_3 + Y_4 + Y_5)V_1 - (Y_2 + Y_3)V_2 - Y_3 V_3 = I_a - I_b$$

The coefficient of V_1 is the sum of all the admittances connected to node 1, the coefficient of V_1 is the negative sum of the admittances linking node 2 with node 1, and the coefficient of V_3 is the negative of the admittance linking node with node 1. These properties of the coefficients in a nodal equation can be generalized:

In the equation for node j in a network, the coefficient of V_j will be the sum of all the admittances connected to node j and the coefficient of V_k, $k \neq j$, will be the *negative* sum of the admittances linking node k with node j.

Example 9-5 Write the nodal equations of the circuit shown in Fig. 9-10.

Fig. 9-10 Circuit for Example 9-5. Note that the elements in the shaded box are in *series* and that their *impedances* must be added.

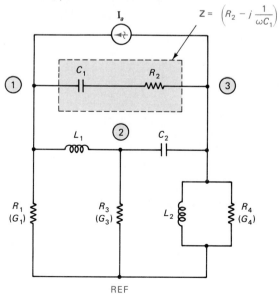

9-3 Nodal Analysis

Solution Using the rules for writing nodal equations discussed above, the equations of the various nodes are:

Node 1:

$$\left(G_1 - j\frac{1}{\omega L_1} + \frac{1}{R_2 - j\frac{1}{\omega C_1}}\right)\mathbf{V}_1 - \left(-j\frac{1}{\omega L_1}\right)\mathbf{V}_2 - \frac{1}{R_2 - j\frac{1}{\omega C_1}}\mathbf{V}_3 = \mathbf{I}_a$$

Node 2:

$$-\left(-j\frac{1}{\omega L_1}\right)\mathbf{V}_1 + \left(G_3 + j\omega C_2 - j\frac{1}{\omega L_1}\right)\mathbf{V}_2 - j\omega C_2\mathbf{V}_3 = 0$$

Node 3:

$$-\frac{1}{R_2 - j\frac{1}{\omega C_1}}\mathbf{V}_1 - j\omega C_2\mathbf{V}_2 + \left(G_4 + j\omega C_2 - j\frac{1}{\omega L_2} + \frac{1}{R_2 - j\frac{1}{\omega C_1}}\right)\mathbf{V}_3 = -\mathbf{I}_a$$

■

Example 9-6 Set up the nodal equations of the circuit shown in Fig. 9-11. Note the presence of a dependent source.

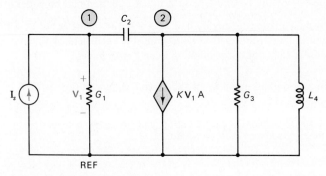

Fig. 9-11 Circuit for Example 9-6.

Solution The nodal equations are:

$$(G_1 + j\omega C_2)\mathbf{V}_1 - j\omega C_2\mathbf{V}_2 = \mathbf{I}_s$$

$$-j\omega C_2\mathbf{V}_1 + \left(G_3 + j\omega C_2 - j\frac{1}{\omega L_4}\right)\mathbf{V}_2 = -K\mathbf{V}_1 \qquad (9\text{-}3)$$

In the last equation, the voltage $\mathbf{V}_1$ appears on the right side because of the dependent source. Move it to the left side, where it belongs.

$$(K - j\omega C_2)\mathbf{V}_1 + \left(G_3 + j\omega C_2 - j\frac{1}{\omega L_4}\right)\mathbf{V}_2 = 0 \qquad (9\text{-}4)$$

Equations (2) and (4) are the nodal equations of the given circuit.

■

Exercise 9-5 Set up the nodal equations of each of the circuits shown in Fig. 9-12.

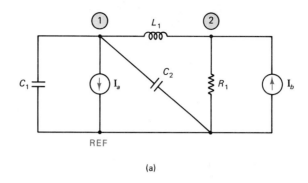

(a)

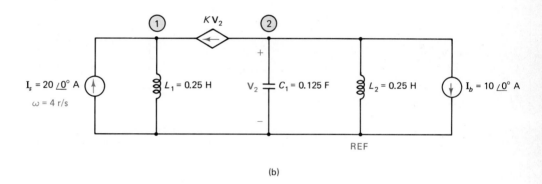

(b)

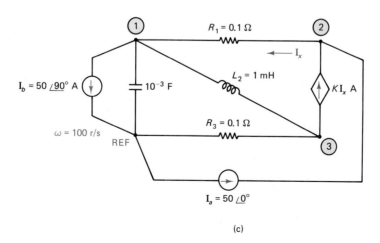

(c)

Fig. 9-12 Circuits for Exercise 9-5.

Exercise 9-6 Determine the voltage at node 2 of each circuit in Fig. 9-13.

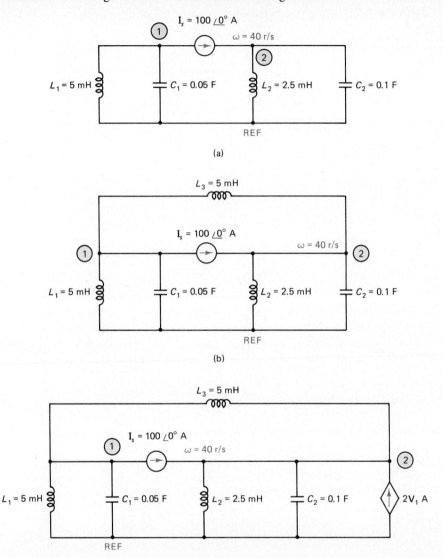

Fig. 9-13 Circuits for Exercise 9-6.

9-3-1 Nodal Analysis in the Presence of Voltage Sources

The principles of nodal analysis of a circuit containing one or more voltage sources are the same as those discussed above, but some modification is necessary in the procedure because of the voltage sources. The following methods can be used for such circuits:

1. If the voltage source is in series with an impedance, then it can be converted to a current source in parallel with an admittance. The principle of *source conversion* is analogous to that used for purely resistive circuits and is shown in Fig. 9-14. Nodal equations are written for the modified circuit.

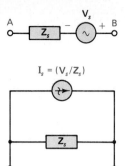

Fig. 9-14 Source conversion. Z_s is the same in both models and $V_s = Z_s I_s$.

2. A judicious choice of the reference node can lead to the reduction in the number of independent unknown nodal voltages and eliminate the need to write the equation for a node at either terminal of a voltage source.

The following examples illustrate the procedures.

Example 9-7 Use nodal analysis and determine the voltage V_2 in the circuit of Fig. 9-15(a).

Fig. 9-15 (a) Circuit for Example 9-7. (b) Circuit after conversion of the voltage source to a current source.

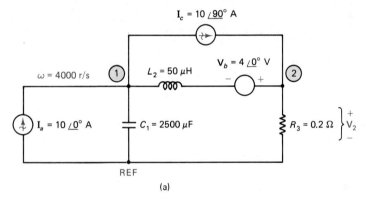

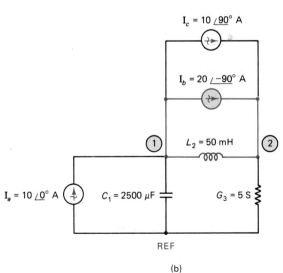

9-3 Nodal Analysis

413

Solution The $V_b - L_2$ combination is replaced by a current source in shunt with an admittance:

$$I_b = \frac{V_b}{j\omega L_2} = -j20 \text{ A}$$

The modified circuit is shown in Fig. 9-15(b) with the admittance values of the different components. The nodal equations are as follows:

Node 1: $\quad j 5 V_1 + j 5 V_2 = (10 + j 10)$
Node 2: $\quad j 5 V_1 + (5 - j 5) V_1 = -j 10$

The solution of the two equations above leads to

$$V_2 = 2 \underline{/-53.1°} \text{ V}$$

∎

Exercise 9-7 Determine the voltage at node 2 for the circuit shown in Fig. 9-16.

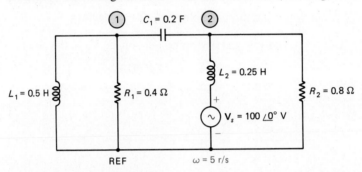

Fig. 9-16 Circuit for Exercise 9-7.

Exercise 9-8 Use nodal analysis and determine the average power supplied by the 10A source in Fig. 9-17.

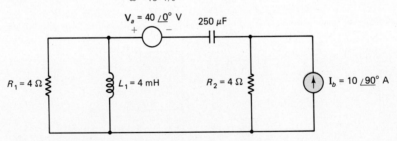

Fig. 9-17 Circuit for Exercise 9-8.

Example 9-8 Determine the average power supplied by the voltage source V_a in the circuit of Fig. 9-18.

Solution In order to calculate the average power supplied by the voltage source, it is necessary to find the current I_1 supplied by it, which is also the current through Z_4. Therefore,

$$I_1 = (V_1 - V_2)Y_4 \tag{9-5}$$

Networks in the Sinusoidal Steady State

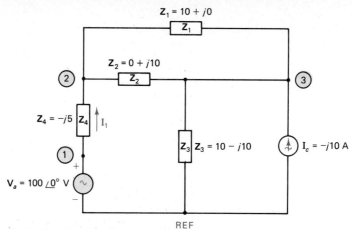

Fig. 9-18 Circuit for Example 9-8.

With the reference node as given, the voltage at node 1 becomes a known quantity.

$$V_1 = V_a = 100\underline{/0°}$$

Equation for Node 2:

$$-Y_4V_1 + (Y_4 + Y_2 + Y_1)V_2 - (Y_2 + Y_1)V_3 = 0$$
$$(0.1 + j0.1)V_2 - (0.1 - j0.1)V_3 = j20 \quad (9\text{-}6)$$

Equation for Node 3:

$$-(Y_1 + Y_2)V_2 + (Y_1 + Y_2 + Y_3)V_3 = I_c$$
$$-(0.1 - j0.1)V_2 + (0.15 - j0.05)V_3 = -j10 \quad (9\text{-}7)$$

Solving Eqs. (9-6) and (9-7) for V_2

$$V_2 = 55.4\underline{/33.7°} = (46.1 + j30.7) \text{ V}$$

Using Eq. (9-5), the current supplied by the source V_a is

$$I_1 = 12.4\underline{/60.3°} \text{ A}$$

Therefore, the average power supplied by the source V_a is

$$P_a = (1/2)(100 \times 12.4 \times \cos 60.3°) = 307 \text{ W} \quad \blacksquare$$

Exercise 9-9 Make a judicious choice of the reference node in each of the circuits of Fig. 9-19(a),(b) and determine the power delivered by each of the sources in the circuit.

Exercise 9-10 Calculate the average power supplied by each of the sources in the circuit of Fig. 9-19(c).

Exercise 9-11 Determine the average power delivered by each of the sources in the circuit of Fig. 9-19(d).

Fig. 9-19 Circuits for Exercises 9-9, 9-10, and 9-11.

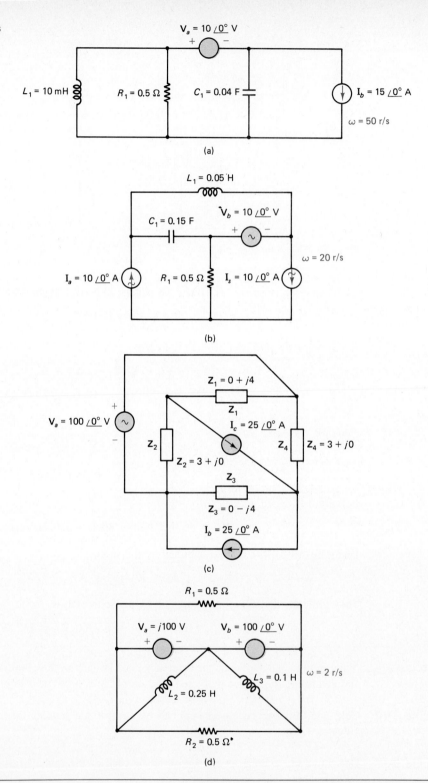

The general discussion of nodal analysis through a nodal conductance matrix and the concepts of driving point and transfer resistances of Chapter 3 is directly translated for sinusoidal steady state by using admittance in place of conductance.

9-4 LOOP AND MESH ANALYSIS

Loop and mesh analysis of a network in the sinusoidal steady state consists of the selection of an independent set of loops or meshes, which are defined by a set of loop currents or mesh currents, and the setting up of an equation for each loop or mesh to satisfy Kirchhoff's voltage law. Each term in a loop equation or mesh equation will be in the form of a voltage—either a voltage due to an independent source, or the product of an impedance and a current. The procedure in loop and mesh analysis in the sinusoidal steady state is again exactly similar to that used in purely resistive circuits driven by constant voltage sources and discussed in Chapter 4.

Consider a loop in an arbitrary network as shown in Fig. 9-20. The KVL equation is written in the form:

algebraic sum of the voltages in the impedances in the loop
$$= \text{algebraic sum of the voltages due to the sources in the loop}$$

using the same rules as discussed in Chapter 4.

For loop 1 shown in Fig. 9-20, the KVL equation is

$$\mathbf{Z}_1(\mathbf{I}_1 + \mathbf{I}_2) + \mathbf{Z}_2(\mathbf{I}_1 + \mathbf{I}_2) + \mathbf{Z}_3(\mathbf{I}_1 + \mathbf{I}_2) + \mathbf{Z}_4\mathbf{I}_1 + \mathbf{Z}_5(\mathbf{I}_1 + \mathbf{I}_2 - \mathbf{I}_3) = \mathbf{V}_a - \mathbf{V}_b$$

When the coefficients of the different currents are collected together, the equation becomes

$$(\mathbf{Z}_1 + \mathbf{Z}_2 + \mathbf{Z}_3 + \mathbf{Z}_4 + \mathbf{Z}_5)\mathbf{I}_1 + (\mathbf{Z}_1 + \mathbf{Z}_2 + \mathbf{Z}_5 + \mathbf{Z}_3)\mathbf{I}_2 - \mathbf{Z}_5\mathbf{I}_3 = \mathbf{V}_a - \mathbf{V}_b$$

The coefficient of $\mathbf{I}_1$ in the equation for loop 1 is the sum of all the impedances connected in that loop, the coefficient of $\mathbf{I}_2$ is the *positive* sum of the impedances common to loops 1 and 2, and the coefficient of $\mathbf{I}_3$ is the *negative* of the impedance common to loops 1 and 3.

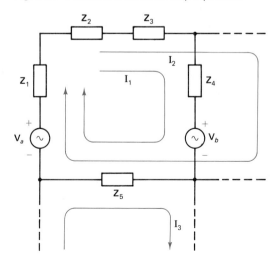

Fig. 9-20 Portion of a circuit for loop equations.

These properties of the coefficients of the loop currents in a loop equation can be made more general:

> In the equation for loop j in a network, the coefficient of I_j will be the sum of all the impedances connected in loop j; and the coefficient of I_k, $k \neq j$, will be either the positive or negative sum of the impedances common to loops j and k, where a positive sum is used when I_k and I_j flow in the same direction in the common branch, and a negative sum is used when the currents I_k and I_j flow in opposite directions through the common branch.

Example 9-9 Write the loop equations of the network shown in Fig. 9-21.

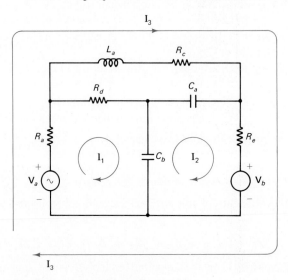

Fig. 9-21 Network for Example 9-9.

Solution: *Loop 1:*
$$\left(R_a + R_d - j\frac{1}{\omega C_b}\right)I_1 - \left(-j\frac{1}{\omega C_b}\right)I_2 + R_a I_3 = V_a$$

Loop 2:
$$-\left(-j\frac{1}{\omega C_b}\right)I_1 + \left(R_e - j\frac{1}{\omega C_a} - j\frac{1}{\omega C_b}\right)I_2 + R_e I_3 = -V_b$$

Loop 3:
$$R_a I_1 + R_e I_2 + (R_a + R_c + R_e + j\omega L_a)I_3 = V_a - V_b$$

Pay close attention to the coefficient of I_2 in the equation for loop 1, and the coefficient of I_1 in the equation for loop 2. In both cases, it is $-Z_c$ where Z_c is the impedance of C_b. ∎

Example 9-10 For the network shown in Fig. 9-22(a): (a) Draw the graph; (b) set up a tree using the branches with the capacitors in them; (c) set up the resulting loop currents; and (d) write the loop equations.

Solution: The graph, the required tree, and the set of loop currents are all shown in Figs. 9-22(b),(c),(d).

The loop equations are:

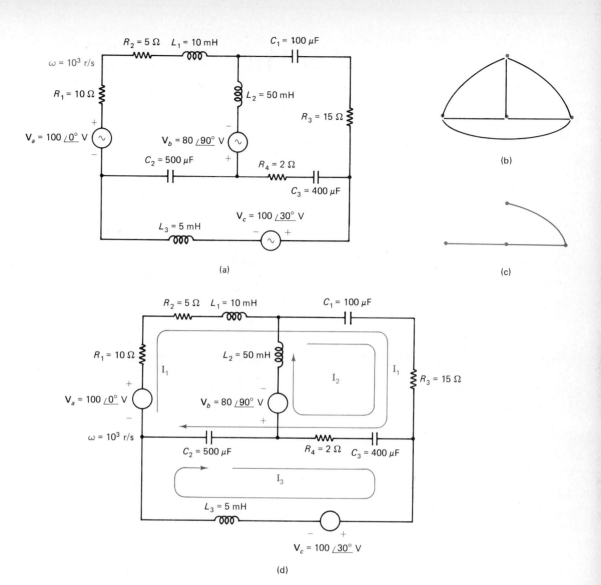

Fig. 9-22 (a) Circuit for Example 9-10. (b) Graph of the circuit. (c) Assigned tree. (d) The loop currents conforming to the tree.

Loop 1:

$$\left(R_1 + R_2 + j\omega L_1 - j\frac{1}{\omega C_1} + R_3 - j\frac{1}{\omega C_3} + R_4 - j\frac{1}{\omega C_2}\right)\mathbf{I}_1$$
$$+ \left(-j\frac{1}{\omega C_1} + R_3 - j\frac{1}{\omega C_3} + R_4\right)\mathbf{I}_2 - \left(-j\frac{1}{\omega C_3} + R_4 - j\frac{1}{\omega C_2}\right)\mathbf{I}_3 = \mathbf{V}_a$$

which becomes

$$(32 - j4.5)\mathbf{I}_1 + (17 - j12.5)\mathbf{I}_2 - (2 - j4.5)\mathbf{I}_3 = 100\,\underline{/0°} \qquad (9\text{-}8)$$

9-4 Loop and Mesh Analysis

Loop 2:

$$\left(-j\frac{1}{\omega C_1} + R_3 - j\frac{1}{\omega C_3} + R_4\right)\mathbf{I}_1 + \left(j\omega L_2 - j\frac{1}{\omega C_1} + R_3 - j\frac{1}{\omega C_3} + R_4\right)\mathbf{I}_2$$
$$- \left(-j\frac{1}{\omega C_3} + R_4\right)\mathbf{I}_3 = -\mathbf{V}_b$$

which becomes

$$(17 - j12.5)\mathbf{I}_1 + (17 + j37.5)\mathbf{I}_2 - (2 - j2.5)\mathbf{I}_3 = -j80 \qquad (9\text{-}9)$$

Loop 3:

$$-\left(-j\frac{1}{\omega C_2} + R_4 - j\frac{1}{\omega C_3}\right)\mathbf{I}_1 - \left(R_4 - j\frac{1}{\omega C_3}\right)\mathbf{I}_2$$
$$+ \left(-j\frac{1}{\omega C_2} + R_4 - j\frac{1}{\omega C_3} + j\omega L_3\right)\mathbf{I}_3 = -\mathbf{V}_c$$

which becomes

$$-(2 - j4.5)\mathbf{I}_1 - (2 - j2.5)\mathbf{I}_2 + (2 + j0.5)\mathbf{I}_3 = -100\underline{/30°} \qquad (9\text{-}10)$$

Eqs. (9-8), (9-9), and (9-10) are the required loop equations. ∎

Exercise 9-12 For the network in the example above, set up a tree so as to make the branches with the capacitors the *links*. Write the resulting loop equations.

Exercise 9-13 For the network in the example above, set up the mesh currents and write the resulting mesh equations.

Example 9-11 For the circuit shown in Fig. 9-23(a), determine the transfer function ($\mathbf{V}_0 / \mathbf{V}_s$) by loop analysis.

Solution Since the voltage of the dependent source is controlled by the current $\mathbf{I}_x$, it is desirable to choose the R_2 branch as a *link*. The choice of loop currents is shown in Fig. 9-23(b), and the resulting loop equations are as follows:

Loop 1: $\qquad (R_1 + R_2)\mathbf{I}_1 + R_1\mathbf{I}_2 = \mathbf{V}_s$

Loop 2: $\qquad (R_1 - K)\mathbf{I}_1 + \left(R_1 + R_3 - j\frac{1}{\omega C_1}\right)\mathbf{I}_2 - R_3\mathbf{I}_3 = \mathbf{V}_s$

Loop 3: $\qquad K\mathbf{I}_1 - R_3\mathbf{I}_2 + (R_3 + j\omega L_1)\mathbf{I}_3 = 0$

Using the numerical values given, the equations become:

Loop 1: $\qquad 20\mathbf{I}_1 + 10\mathbf{I}_2 = \mathbf{V}_s$

Loop 2: $\qquad (10 - K)\mathbf{I}_1 + (20 - j10)\mathbf{I}_2 - 10\mathbf{I}_3 = \mathbf{V}_s$

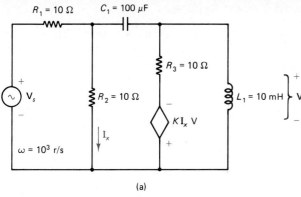
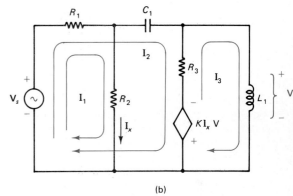

Fig. 9-23 (a) Circuit for Example 9-11.
(b) Set up for loop analysis.

Loop 3: $\qquad K\mathbf{I}_1 - 10\mathbf{I}_2 + (10 + j10)\mathbf{I}_3 = 0$

The last three equations are solved for $\mathbf{I}_3$. It is found that

$$\mathbf{I}_3 = \frac{100 + j10K}{3000 + j(1000 + 100K)}$$

and the output voltage is

$$\mathbf{V}_O = j\omega L_3 \mathbf{I}_3 = j10\mathbf{I}_3$$

Therefore, the transfer function is

$$\frac{\mathbf{V}_O}{\mathbf{V}_s} = -\left[\frac{(100K - j1000)}{3000 + j(1000 + 100K)}\right]$$

∎

Exercise 9-14: Determine the ratio $\mathbf{V}_O / \mathbf{V}_s$ of the circuit shown in Fig. 9-24.

Fig. 9-24 Circuit for Exercise 9-14.

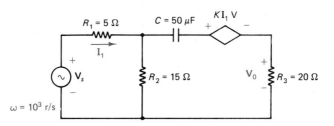

9-4 **Loop and Mesh Analysis**

9-4-1 Loop and Mesh Analysis in the Presence of Current Sources

The principles of loop and mesh analysis of a circuit containing current sources are the same as those discussed above, but some modification is necessary in the procedure because of the current sources. The following methods can be used for such circuits.

1. **If the current source is in parallel with an admittance, it can be converted to a voltage source in series with an impedance.** The principle of source conversion is like that used for purely resistive circuits and is shown in Fig. 9-14. Loop equations are written for the modified circuit.
2. **A judicious choice of loops can lead to the reduction in the number of independent unknown loop currents** and eliminate the need to write the equation for a loop involving the current source.

The following examples illustrate these procedures.

Example 9-12 Set up the two mesh equations needed to analyze the circuit of Fig. 9-25(a) after performing a source conversion.

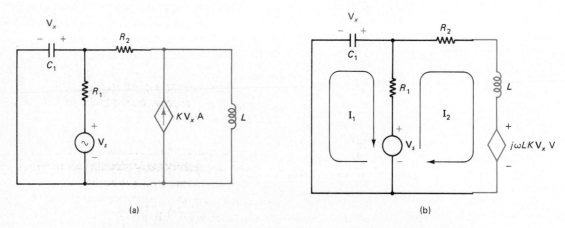

Fig. 9-25 (a) Circuit for Example 9-12. (b) Mesh currents after source conversion.

Solution A conversion of the dependent current source to a dependent voltage source leads to the circuit of Fig. 9-25(b). The voltage V_x that controls the dependent sources is given by:

$$V_x = -\left(-j\frac{1}{\omega C}\right)I_1 \qquad (9\text{-}11)$$

The mesh equations are:

Mesh 1: $\qquad \left(R_1 - j\dfrac{1}{\omega C_1}\right)I_1 - R_1 I_2 = -V_s \qquad (9\text{-}12)$

Mesh 2: $\qquad -R_1 I_1 + (R_1 + R_2 + j\omega L)I_2 = V_s - j\omega L\, K\, V_x$

which becomes, on using Eq. (9-11) for V_x,

$$-\left(R_1 + \frac{LK}{C}\right)I_1 + (R_1 + R_2 + j\omega L)I_2 = V_s \qquad (9\text{-}13)$$

Networks in the Sinusoidal Steady State

Equations (9-11) and (9-13) are the two mesh equations needed to analyze the given circuit. ∎

Exercise 9-15 After a source conversion in each of the circuits, as shown in Fig. 9-26, write the mesh equations for the circuit.

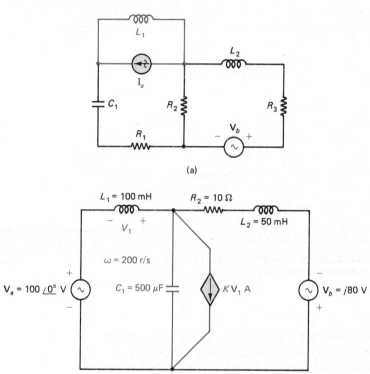

Fig. 9-26 Circuits for Exercise 9-15.

Example 9-13 Use loop analysis to find the average power supplied by the current source I_b in the circuit of Fig. 9-27(a).

Solution It is desirable to choose the I_b branch as a link. The tree and the choice of loop currents are shown in Figs. 9-27(b),(c). The loop current $I_2 = 10\underline{/0°}$ A. There is no need to write an equation for loop 2.

Loop 1: $\left(R_1 - j\dfrac{1}{\omega C_1} + j\omega L_1 + R_2\right)I_1 + (j\omega L_1 + R_2)I_2 - R_2 I_3 = V_a$

which becomes on using the numerical values and $I_2 = 10$,

$$(15 - j35)I_1 - 5I_3 = (50 - j50) \qquad (9\text{-}14)$$

Loop 3: $\qquad -R_2 I_1 - R_2 I_2 + \left(R_2 - j\dfrac{1}{\omega C_2}\right)I_3 = j90$

9-4 Loop and Mesh Analysis

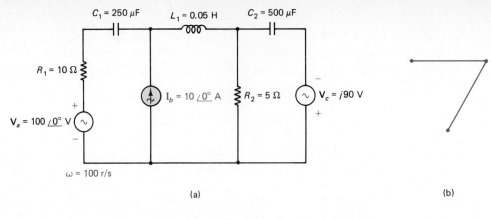

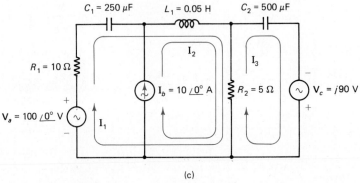

Fig. 9-27 (a) Circuit for Example 9-13. (b) Tree chosen to make I_b a link; (c) resulting loop currents.

which becomes on using the numerical values and $I_2 = 10$,

$$-5I_1 + (5 - j20)I_3 = (50 + j90) \qquad (9\text{-}15)$$

The solution of Eqs. (9-14) and (9-15) leads to

$$I_1 = 1.172 \underline{/22°} \text{ A}$$

- Voltage across I_b source $= V_a - \left(R_1 - j\dfrac{1}{\omega C_1}\right)I_1$

$$= (71.6 + j39.1) = 81.6 \underline{/28.6°} \text{ V}$$

- Average power supplied by $I_b = \dfrac{1}{2} \times 81.6 \times 10 \times \cos 28.6$

$$= 358.2 \text{ W}$$

Suppose the circuit in the example above is tackled through mesh analysis (see Fig. 9-28). An additional unknown quantity, the voltage V_x across the current source I_b, is introduced in order to write the KVL equations for meshes 1 and 2. Also, the two mesh currents I_1 and I_2 are related through the current I_b:

$$I_2 - I_1 = I_b = 10$$

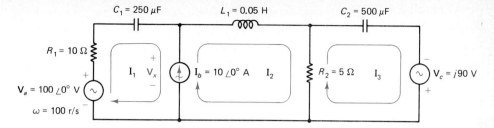

Fig. 9-28 Mesh analysis of the circuit for Example 9-13. The voltage V_x across the current source is an additional unknown appearing in the equations.

The three mesh equations are:

Mesh 1: $\left(R_1 - j\dfrac{1}{\omega C_1}\right)I_1 = V_a - V_x$ or $(10 - j40)I_1 + V_x = 100$

Mesh 2: $(R_2 + j\omega L_1)I_2 - R_2 I_3 = V_x$ or $(5 + j5)I_2 - 5I_3 - V_x = 0$

Mesh 3: $-R_2 I_2 + \left(R_2 - j\dfrac{1}{\omega C_2}\right)I_3 = V_c$ or $-5I_2 + (5 - j20)I_3 = j90$

There are four equations in four unknowns to be solved. Clearly, a great deal of work is saved by a suitable choice of loop currents, as we saw in the earlier solution.

Exercise 9-16 Determine the voltage across the current source in each of the circuits shown in Fig. 9-29.

Fig. 9-29 Circuits for Exercise 9-16.

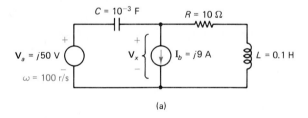

(a)

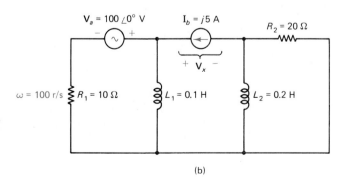

(b)

9-4 Loop and Mesh Analysis

Exercise 9-17 Determine the input impedance Z_{in} of each of the circuits shown in Fig. 9-30.

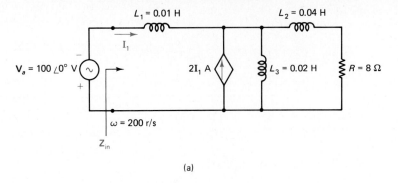

(a)

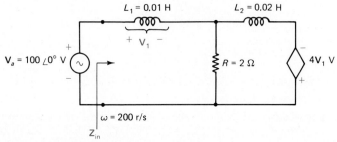

Fig. 9-30 Circuits for Exercise 9-17.

The general discussion of mesh analysis through the mesh impedance matrix and the concepts of driving point and transfer conductances of Chapter 4 is directly translated for sinusoidal steady state by using impedance in place of resistance. An application of the mesh impedance matrix in sinusoidal steady state will be found in Sec. 9-6.

9-5 TRANSFER FUNCTIONS OF A NETWORK

The ratio of the voltage or current in some branch (the response branch) to the voltage or current of a source in another branch (the source branch) is called the transfer function of a network.

For networks with sinusoidal signals, such ratios are *functions of frequency,* and the variation of the magnitude and phase angle of a transfer function with frequency has important practical applications. Transfer function analysis is used to predict the range of frequencies that will be transmitted with little or no attenuation and the range that will be attenuated significantly by a given network such as an amplifier. Conversely, the frequency characteristics to be met by a network may be specified, and a network designed to meet those specifications. The determination of the transfer functions and some of their properties will be discussed in this section. A further treatment of transfer functions in terms of the complex frequency will be presented in Chapter 12.

If V_s and I_s denote the voltage and current in the source branch, and V_L and I_L those in the response branch, there are four possible transfer functions: V_L/V_s, I_L/V_s, V_L/I_s, and I_L/I_s. The voltage transfer functions V_L/V_s and I_L/I_s are dimensionless; I_L/V_s is a transfer *admittance*, and V_L/I_s is a transfer *impedance*.

The general symbol for a transfer function is $H(j\omega)$. Recall that frequency appears in sinusoidal analysis in the form of $(j\omega)$ or $(1/j\omega)$ in impedances and admittances. When such terms in an expression are manipulated, powers of $(j\omega)$ appear in the resulting expressions; hence the parameter $(j\omega)$ is used in the notation $H(j\omega)$.

Example 9-14 Determine the voltage transfer function V_L/V_s of the network shown in Fig. 9-31.

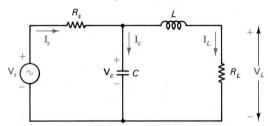

Fig. 9-31 Circuit for Example 9-14.

Solution Using the ladder network approach, start by assuming $V_L = 1$. Then the following expressions are obtained.

$$V_L = 1$$

$$I_L = 1/R_L$$

$$V_c = \frac{(R_L + j\omega L)}{R_L}$$

$$I_c = \frac{-\omega^2 LC + j\omega CR_L}{R_L}$$

$$I_s = \frac{(1 - \omega^2 LC) + j\omega CR_L}{R_L}$$

$$V_s = \frac{R_L + R_s(1 - \omega^2 LC) + j\omega(R_s R_L C + L)}{R_L}$$

Therefore, the voltage transfer function is (since V_L is assumed to be 1):

$$\frac{V_L}{V_s} = \frac{R_L}{[R_L + R_s(1 - \omega^2 LC)] + j[\omega(R_s R_L C + L)]}$$ ∎

Exercise 9-18 Obtain the expression for the current transfer function I_L/I_s of the network in the example above.

Exercise 9-19 Assume that, in the network of the example above, $R_L = R_s = 5\Omega$, $L = 0.1$ H, and $C = 4 \times 10^{-3}$ F. Suppose the input source voltage and frequency are: (a) 10 V at $\omega = 5$ r/s; (b) 10 V at $\omega = 50$ r/s; (c) 10 V at $\omega = 500$ r/s. For each of these cases, determine the output voltage phasor.

9-5 Transfer Functions of a Network

Exercise 9-20 Determine the voltage transfer function V_L/V_s and the current transfer function I_L/I_s of the network shown in Fig. 9-32.

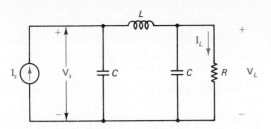

Fig. 9-32 Network for Exercise 9-20.

9-5-1 Frequency Response Characteristics of a Transfer Function

Since a transfer function is a complex quantity, it has a magnitude $|\mathbf{H}(j\omega)|$ and a phase $\theta(j\omega)$. That is,

$$\mathbf{H}(j\omega) = |\mathbf{H}(j\omega)|e^{j\theta(j\omega)}$$

Since $\mathbf{H}(j\omega)$ is in the form of a ratio of polynomials, the most convenient method of finding the magnitude and phase functions is through the following formulas.

$|\mathbf{H}(j\omega)|$ = (magnitude of the numerator polynomial)

÷ (magnitude of the denominator polynomial)

$\theta(j\omega)$ = [(angle of the numerator polynomial) − (angle of denominator polynomial)]

Plots showing the variation of the magnitude and phase of a transfer function with frequency are called *frequency response* characteristics. There are two components of a frequency response:

(1) Variation of $|\mathbf{H}(j\omega)|$ as a function of frequency, called the *amplitude response*.
(2) Variation of $\theta(j\omega)$ as a function of frequency, called the *phase response*.

A typical set of frequency response curves is shown in Fig. 9-33. The frequency response of a network (when the appropriate transfer function is considered) provides a powerful tool for the study of the behavior of the network when the input signal contains a band of frequencies. Conversely, the first step in the design of a network for the processing of a signal is the determination of the desired frequency response.

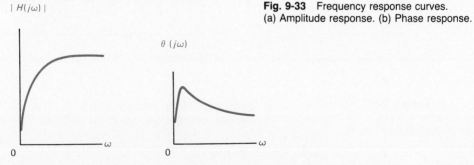

Fig. 9-33 Frequency response curves.
(a) Amplitude response. (b) Phase response.

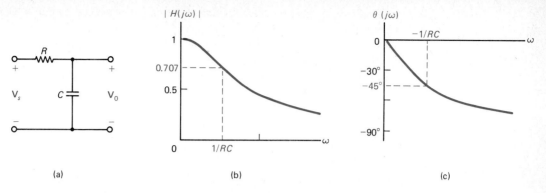

Fig. 9-34 A low-pass filter and its frequency response characteristics.

Example 9-15 Determine and sketch the frequency response of the network shown in Fig. 9-34(a) when the transfer function of interest is $V_o/\mathbf{V}_s$.

Solution The transfer function is given by

$$\mathbf{H}(j\omega) = \frac{-j(1/\omega C)}{R - j(1/\omega C)}$$

$$= \frac{1}{(1 + j\omega RC)}$$

The amplitude response is given by

$$|\mathbf{H}(j\omega)| = \frac{1}{\sqrt{(1 + \omega^2 R^2 C^2)}}$$

and the phase response is given by

$$\theta(j\omega) = -\arctan \omega RC$$

The frequency response plots are shown in Figs. 9-34(b) and (c). The physical message in the two plots is important and should be understood. Suppose, for the sake of convenience, the input $\mathbf{V}_s$ has an amplitude of 1 V and a phase angle of 0° at all frequencies. As the frequency is increased (starting at *dc*), the amplitude response shows that the amplitude of the output voltage decreases monotonically. At low frequencies, the amplitude of the output is roughly equal to the amplitude of the input. At high frequencies, the amplitude of the output decreases and approaches zero. Thus the network transmits low-frequency signals with only a small degree of attenuation, while it suppresses high-frequency signals and acts as a *low-pass filter*. The phase response shows that the output voltage is in phase with the input at *dc*. As frequency increases, the output voltage lags behind the input voltage, with the angle of lag increasing as the frequency increases. In the limit of infinite frequency, the phase angle becomes 90°.

The frequency response sketches, especially the amplitude response, serve as a visual guide to the behavior of the network as the frequency of the input signal is varied over all possible values. ∎

Exercise 9-21 In the circuit of the previous example, define a transfer function $\mathbf{H}(j\omega) = V_R/\mathbf{V}_s$, where V_R is the voltage across the resistance R. Determine and sketch the corresponding fre-

9-5 Transfer Functions of a Network

quency response. Discuss the behavior of the circuit when the output is taken across R rather than C.

Exercise 9-22 Suppose a series combination of R and L is to be used to obtain a low-pass filter. Set up the proper configuration, determine the relevant transfer function, and verify by means of frequency response sketches that the circuit is indeed performing as a low-pass filter.

9-5-2 Complex Exponential Driving Functions and the Concept of Negative Frequency

When the forcing and response functions are to be studied in the real time domain, it becomes necessary to introduce the complex conjugate $\mathbf{H}^*(j\omega)$ of the transfer function. This leads to the concept of negative frequency, as we will show in the following discussion. Consider the input to a linear network given by the complex exponential function

$$\mathbf{v}_s(t) = \mathbf{A}e^{j\omega t} \tag{9-16}$$

As noted in Chapter 8, the differential equations of the network become algebraic equations. When the resulting algebraic equations are solved for $\mathbf{v}_0(t)$, the result will be of the form

$$\mathbf{v}_0(t) = \mathbf{H}(j\omega)[\mathbf{A}e^{j\omega t}] \tag{9-17}$$

Remember that the form of Eq. (9-17), in which a function of time is given directly as the product of a transfer function and another function of time, is valid if and only if the functions are *complex exponential functions* of time, and not real sinusoidal functions.

To find the response of the network to a real sinusoidal voltage, it is necessary to consider an input $\mathbf{v}_s^*(t)$, the complex conjugate of $\mathbf{v}_s(t)$, as well as $\mathbf{v}_s(t)$ itself. $\mathbf{v}_s^*(t)$ is given by

$$\mathbf{v}_s^*(t) = \mathbf{A}^* e^{-j\omega t} \tag{9-18}$$

and the corresponding output is the complex conjugate of $\mathbf{v}_0(t)$ in Eq. (9-17):

$$\mathbf{v}_0^*(t) = \mathbf{H}(-j\omega)\mathbf{A}^* e^{-j\omega t} \tag{9-19}$$

since

$$\mathbf{H}^*(j\omega) = \mathbf{H}(-j\omega) \tag{9-20}$$

Then, since a real sinusoidal input is given by the sum of the two inputs $\mathbf{v}_s$ and $\mathbf{v}_s^*$, the output will be the sum of the two responses $\mathbf{v}_0$ and $\mathbf{v}_0^*$ given by Eqs. (9-17) and (9-19).

Exercise 9-23 Let $\mathbf{A} = |A|e^{j\theta}$ and $\mathbf{H}(j\omega) = |H|e^{j\phi}$ in the analysis above. Show that the sum of the expressions in Eqs. (9-17) and (9-19) is a real sinusoidal function of time and write it in the form of a single cosine term with a phase angle.

That is,

if real sinusoidal functions are to be considered in a network, then the time domain solution requires the inclusion of two terms: one involving the transfer function $H(j\omega)$, and the other its complex conjugate $H^*(j\omega) = H(-j\omega)$.

Therefore, the transfer function $H(j\omega)$ itself, in a sense, tells us only half the story, and it is necessary to include its counterpart $H(-j\omega)$ to find the response of networks in the time domain.

Since $H(-j\omega)$ is obtained from $H(j\omega)$ by replacing ω by $-\omega$, the frequency ω assumes both *positive* and *negative* values! Clearly, it is not possible to generate a sinusoidal voltage of frequency -1 kHz, for example, in the laboratory. So what does this negative frequency mean? It is simply a mathematical necessity: the complex conjugate response functions for a pair of complex conjugate input functions need to be added to produce a real sinusoidal function. The two response functions are related to the two input functions by a pair of transfer functions, one evaluated for positive values of ω and the other for negative values of ω.

Example 9-16 Consider the transfer function of a network

$$H(j\omega) = \frac{V_0}{V_s} = \frac{12j\omega + 6}{-4\omega^2 + 14j\omega + 8}$$

Obtain the output of the network when the input is (a) $v_s(t) = 20e^{j5t}$ V; (b) $v_s^*(t) = 20e^{-j5t}$ V; and (c) $v_s(t) = 40 \cos 5t$ V.

Solution

(a)
$$v_{o1}(t) = [(j60 + 6)/(-100 + j70 + 8)]20e^{j5t}$$
$$= (60.3e^{j84.3}/116e^{j143})20e^{j5t}$$
$$= 10.4e^{j(5t - 58.4°)} \text{V}$$

(b)
$$v_{o1}^*(t) = [-j60 + 6)/(-100 - j70 + 8)]20e^{-j5t}$$
$$= (60.3e^{-j84.3}/116e^{-j143})20e^{-j5t}$$
$$= 10.4e^{-j(5t - 58.4°)} \text{V}$$

(c) Since

$$v_s(t) = 40 \cos 5t = 20(e^{j5t} + e^{-j5t})$$

the output is

$$v_o(t) = v_{o1}(t) + v_{o1}^*(t)$$
$$= 20.8 \cos(5t - 58.4°) \text{ V} \qquad \blacksquare$$

Exercise 9-24 Determine the output of the network of the previous example when the input is $v_s = 10 \cos 2t$ V using complex exponential components of the given input.

Exercise 9-25 Repeat the previous exercise when the input is $10 \sin 2t$ V.

When negative frequencies are used, the frequency response curves should be extended to negative frequencies also. Since

$$H(-j\omega) = H^*(j\omega) \qquad (9\text{-}21)$$

it follows that

$$|\mathbf{H}(-j\omega)| = |\mathbf{H}(j\omega)| \quad (9\text{-}22)$$

and

$$\text{angle of } \mathbf{H}(-j\omega) = -[\text{angle of } \mathbf{H}(j\omega)] \quad (9\text{-}23)$$

The last two equations are used in plotting frequency response curves that extend into negative as well as positive frequencies. The frequency response curves corresponding to those in Figs. 9-33 and 9-34 are shown redrawn in Fig. 9-35, where negative frequencies are included.

Fig. 9-35 Frequency response characteristics of Figs. 9-33 and 9-34 with negative frequencies included. Note the even symmetry in the amplitude response and the odd symmetry in the phase response.

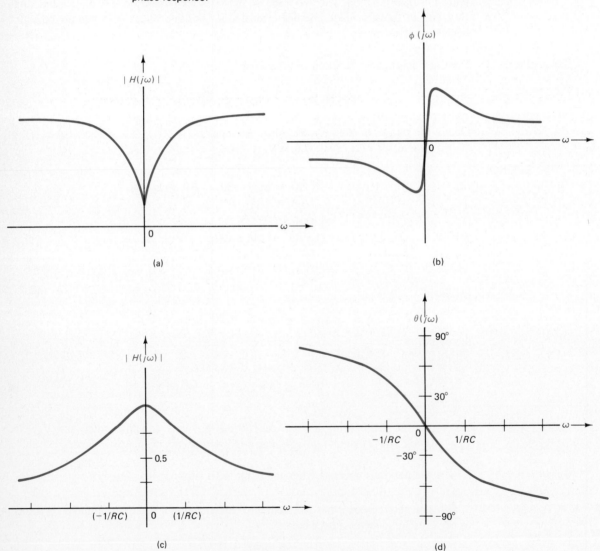

Networks in the Sinusoidal Steady State

Even and Odd Symmetries of the Frequency Response Curves

The frequency response curves in Fig. 9-35 are seen to have certain features:

> The *amplitude response* is an even function of ω; that is, the portion of the curve on the left side of the origin is a mirror image of that on the right side. This is a general property that arises from Eq. (9-22), which states that $|\mathbf{H}(-j\omega)| = |\mathbf{H}(j\omega)|$.
>
> The *phase response* is an odd function of ω; that is, if we rotate the left portion by 180° about the origin as pivot, it coincides with the right portion. This is also a general property due to Eq. (9-23), which states that [angle of $\mathbf{H}(-j\omega)$] = $-$[angle of $\mathbf{H}(j\omega)$].

Strictly speaking, negative frequencies should be included in the discussion of transfer functions and frequency response diagrams. As a practical matter, however, the frequency response of networks (such as filters) can be studied by using only the positive frequency axis, as we did in the previous section. This is an accepted practice. We will see in Chapter 11 that Bode diagrams (a graphic tool for studying the frequency response of certain circuit configurations) are customarily drawn *only* for positive frequencies.

9-6 TRANSFER FUNCTIONS FROM MESH IMPEDANCE MATRICES

For a relaxed linear network (that is, a network without any independent sources), a knowledge of the mesh or loop impedance matrix is sufficient for finding the transfer function between a response branch and a source branch.

Suppose it is required to find the transfer function

$$\mathbf{H}(j\omega) = (\mathbf{V}_m/\mathbf{V}_k)$$

where $\mathbf{V}_m$ is the voltage across a load impedance $\mathbf{Z}_L$ in mesh m and $\mathbf{V}_k$ is a voltage source placed in mesh k of a network (see Fig. 9-36). The matrix form of the mesh equations of the network is:

$$\begin{bmatrix} \mathbf{Z}_{11} & \mathbf{Z}_{12} & \cdots & \mathbf{Z}_{1m} & \cdots & \mathbf{Z}_{1n} \\ \mathbf{Z}_{21} & \mathbf{Z}_{22} & \cdots & \mathbf{Z}_{2m} & \cdots & \mathbf{Z}_{2n} \\ \cdot & \cdot & & \cdot & & \cdot \\ \mathbf{Z}_{k1} & \mathbf{Z}_{2k} & \cdots & \mathbf{Z}_{km} & \cdots & \mathbf{Z}_{kn} \\ \cdot & \cdot & & \cdot & & \cdot \\ \mathbf{Z}_{1n} & \mathbf{Z}_{2n} & \cdots & \mathbf{Z}_{nm} & \cdots & \mathbf{Z}_{nn} \end{bmatrix} \begin{bmatrix} \mathbf{I}_1 \\ \mathbf{I}_2 \\ \cdot \\ \mathbf{I}_k \\ \cdot \\ \mathbf{I}_n \end{bmatrix} = \begin{bmatrix} 0 \\ 0 \\ \cdot \\ \mathbf{V}_k \\ \cdot \\ 0 \end{bmatrix}$$

where the only nonzero element in the voltage matrix on the right-hand side is $\mathbf{V}_k$.

The current $\mathbf{I}_m$ in mesh m is given by Cramer's rule:

$$\mathbf{I}_m = \frac{\mathbf{A}_{km}}{\|\mathbf{z}\|} \mathbf{V}_k \tag{9-24}$$

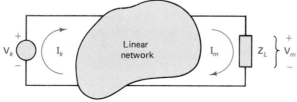

Fig. 9-36 Relaxed linear network driven by a voltage source.

where

$$A_{km} = \text{cofactor of the element } z_{km} \text{ of the mesh impedance matrix}$$

and

$$\|z\| = \text{the determinant of the mesh impedance matrix}$$

The voltage V_m across the load is given by

$$V_m = Z_L I_m$$

and

$$H(j\omega) = V_m/V_k = \frac{A_{km}Z_L}{\|z\|} \qquad (9\text{-}25)$$

A similar procedure yields any other transfer function when the mesh impedance matrix is known.

Example 9-17   A network is shown in Fig. 9-37 with an assignment of mesh currents. Determine the voltage transfer function, V_c/V_s, when a voltage source V_s is inserted in series with L_2.

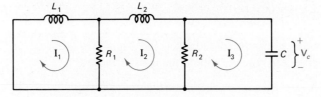

Fig. 9-37   Circuit for Example 9-17.

Solution   The mesh impedance matrix is written directly by inspection, since no dependent sources are present.

$$[z] = \begin{bmatrix} R_1 + j\omega L_1 & -R_1 & 0 \\ -R_1 & R_1 + R_2 + j\omega L_2 & -R_2 \\ 0 & -R_2 & R_2 - j/\omega C \end{bmatrix}$$

The determinant of the above matrix is found to be

$$\|z\| = (1/j\omega C)[-j\omega^3 L_1 L_2 R_2 C \\ - \omega^2(L_1 L_2 + R_1 R_2 L_2 C + R_1 R_2 L_1 C) \\ + j\omega(R_1 L_2 + R_1 L_1 + R_2 L_1) + R_1 R_2]$$

The voltage source is in mesh 2, and the output voltage is due to the current in mesh 3.

$$I_3 = \frac{A_{23}}{\|z\|}V_s$$

$$= \frac{R_2(R_1 + j\omega L_1)}{\|z\|}V_s$$

The voltage transfer function is

$$H(j\omega) = V_c/V_s$$
$$= -j(1/\omega C)I_3/V_s$$
$$= \frac{R_1R_2 + j\omega L_1R_2}{-j\omega^3 L_1L_2R_2C - \omega^2(L_1L_2 + R_1R_2L_2C + R_1R_2L_1C) + j\omega(R_1L_2 + R_1L_1 + R_2L_1) + R_1R_2}$$

∎

Exercise 9-26 Determine the transfer function V_0/V_s of the network shown in Fig. 9-38 for each of the following cases: (a) A voltage source V_s is interested in series with the resistor R_1 and the output voltage V_0 is measured across the capacitor C. (b) A voltage source V_s is inserted in series with the resistor R_2 and the output voltage is measured across the resistor R_1.

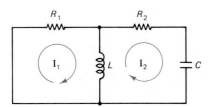

Fig. 9-38 Circuit for Exercise 9-25.

Once the relevant transfer function is determined for a given network, the output phasor for any given input phasor at a specified frequency can be calculated. Therefore,

the transfer function contains all the essential information about a network as far as its input/output terminal behavior is concerned.

In fact it is possible to think of the network as a sort of a black box with a pair of input terminals and a pair of output terminals, as in Fig. 9-39, with a specified transfer function. This *block diagram* approach is commonly used in the analysis of linear systems.

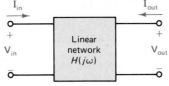

Fig. 9-39 Block diagram of a network.

Example 9-18 Determine the transfer function V_L/V_s of the network shown in Fig. 9-40. Evaluate the output voltage for the following cases: V_s is a sinusoidal voltage source of 100 V amplitude and angular frequency: (a) 0.2 r/s. (b) 2 r/s. (c) 20 r/s.

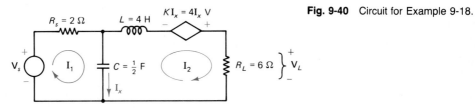

Fig. 9-40 Circuit for Example 9-18.

9-6 Transfer Functions from Mesh Impedance Matrices

Solution The mesh impedance matrix of the network is obtained by writing the mesh equations and moving the voltage of the dependent source to the left side in the equation of mesh 2. It is found to be

$$[\mathbf{z}] = \begin{bmatrix} R_s - j(1/\omega C) & +j(1/\omega C) \\ -K + j(1/\omega C) & K + R_L + j(\omega L - 1/\omega C) \end{bmatrix}$$

which becomes, after using the given numerical values

$$[\mathbf{z}] = \begin{bmatrix} 2 - j(2/\omega) & j(2/\omega) \\ -4 + j(2/\omega) & 10 + j(4\omega - 2/\omega) \end{bmatrix}$$

The output voltage $\mathbf{V}_L$ is given by

$$\mathbf{V}_L = R_L \mathbf{I}_2$$

$$= \frac{R_L A_{12}}{\|\mathbf{z}\|} \mathbf{V}_s$$

and

$$\mathbf{H}(j\omega) = \frac{12(j\omega) + 6}{4(j\omega)^2 + 14(j\omega) + 8}$$

$\omega = 0.2\ r/s$: $\mathbf{H}(j0.2) = (6 + j2.4)/(7.84 + j2.8) = 0.776\ \underline{/2.2°}$

$\mathbf{V}_0 = 77.6\ \underline{/2.2°}\ V$

$\omega = 2\ r/s$: $\mathbf{H}(j2) = (6 + j24)/(-8 + j28) = 0.847\ \underline{/-30°}$

$\mathbf{V}_0 = 8.49\ \underline{/-30°}\ V$

$\omega = 20\ r/s$: $\mathbf{H}(j20) = (6 + j240)/(-1592 + j280) = 0.148\ \underline{/-81.4°}$

$\mathbf{V}_0 = 14.8\ \underline{/-81.4°}\ V$ ∎

Exercise 9-27 Find the transfer function $\mathbf{V}_0/\mathbf{I}_s$ in the circuit shown in Fig. 9-41. Evaluate the output voltage when the input current is a sinusoid of amplitude 10 A and angular frequency: (a) 10 r/s. (b) 100 r/s. (c) 1000 r/s.

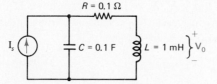

Fig. 9-41 Circuit for Exercise 9-27.

Transfer functions will be discussed again in Chapter 12, where the ideas developed in this chapter will be extended by using complex frequency.

9-7 NETWORK THEOREMS

The theorems presented in Chapter 5 for resistive networks can be translated to the sinusoidal steady state by using impedances and admittances instead of resistances and conductances. There are no significant new aspects to Thevenin's and Norton's theorems in the sinusoidal steady state. The maximum power transfer theorem, on the other hand, becomes more elaborate, since it is possible to vary one or more of the following components of the load impedance: real part, imaginary part, magnitude, or angle. The principle of superposition becomes more interesting and important than in the case of constant voltage and current sources and is an indispensable tool for analyzing circuits driven by sources of different frequencies. The proofs of the theorems are exactly like those presented in Chapter 5 and will not be repeated here.

9-7-1 Thevenin's Theorem

Thevenin's theorem for networks in the sinusoidal steady state can be stated as follows:
Given a network made up of linear components and sources, the voltages and currents in any general load connected to a terminal pair A-B of the network can be determined by replacing the network by an equivalent circuit consisting of *a voltage source in series with an impedance*. The voltage of the source in the equivalent circuit is equal to the voltage that would appear across the terminal pair A-B if those terminals were open-circuited. The impedance in the equivalent circuit is the impedance of the network, as seen from the terminal pair A-B after the network is relaxed.

Finding the Thevenin Equivalent

1. Open-circuit the terminals A-B (unless they are already given in the form of an open circuit) and determine the voltage across A-B. This requires setting up a suitable set of equations and solving them.
2. Make the network relaxed by deactivating all the *independent* sources (but leave the dependent sources as they are). Determine the total impedance as seen from the terminal pair A-B. In networks *without* dependent sources, this can usually be done by series and parallel combinations. In networks *with* dependent sources, it is necessary to use a test source and determine the impedance seen by that source by using proper analysis procedures. The Thevenin impedance of networks with dependent sources *may* contain a *negative resistance* component.

Example 9-19 Determine the Thevenin equivalent of the circuit in Fig. 9-42(a) as seen from the terminals A-B.

Solution *Determination of V_{Th}*: Choosing a current **I** as shown, the *KVL* equation is

$$(10 + j15 + 15 - j25)\mathbf{I} = 100 \underline{/0°}$$

which gives

$$\mathbf{I} = 3.72 \underline{/21.8°} \text{ A}$$

The Thevenin voltage $\mathbf{V}_{Th}$ is the voltage across A-B: $\mathbf{V}_{AB}$ in Fig. 9-42(a). The *KVL* equation for $\mathbf{V}_{AB}$ is

$$\mathbf{V}_{AB} - 50 \underline{/90°} + 10\mathbf{I} - 100 \underline{/0°} = 0$$

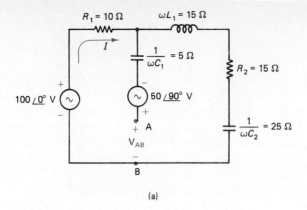

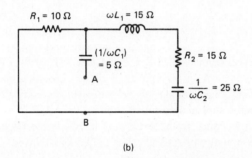

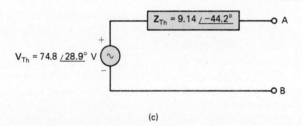

Fig. 9-42 (a) Circuit for Example 9-19. (b) Calculation of Thevenin voltage. (c) Calculation of Thevenin impedance.

which leads to

$$\mathbf{V}_{Th} = \mathbf{V}_{AB} = 74.8 \underline{/28.9°} \text{ V}$$

Determination of Z_{Th}: The relaxed network is shown in Fig. 9-42(b). The Thevenin impedance can be determined by series-parallel combinations.

$$Z_{Th} = -j5 + [10 \parallel (15 - j10)]$$
$$= 9.14 \underline{/-44.2°} \text{ } \Omega$$

The Thevenin equivalent circuit is shown in Fig. 9-42(c).

Exercise 9-28 Determine the Thevenin equivalent circuit of the network shown in Fig. 9-43 as seen from (a) terminal pair A-B; (b) terminal pair C-D; (c) terminal pair E-F.

Fig. 9-43 Circuit for Exercise 9-28.

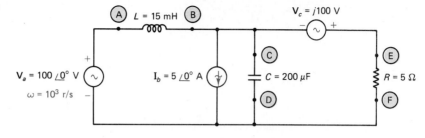

Example 9-20 Determine the Thevenin equivalent of the network in Fig. 9-44(a) as seen from terminal pair A-B.

Fig. 9-44 (a) Circuit for Example 9-20. (b) Calculation of Thevenin voltage. (c) Calculation of Thevenin impedance.

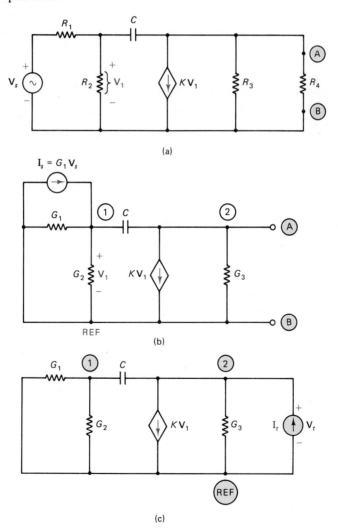

9-7 Network Theorems

Solution First remove the resistance R_4 in order to open-circuit the terminals A-B. The resulting circuit is shown in Fig. 9-44(b). The voltage source model has been replaced by a current source model and the elements relabeled to facilitate nodal analysis.

Determination of V_{Th}:

Node 1:
$$(G_1 + G_2 + j\omega C)V_1 - j\omega C V_2 = I_s \quad (9\text{-}26)$$

Node 2:
$$-j\omega C V_1 + (G_3 + j\omega C)V_2 = -KV_1$$

which is rewritten as

$$(K - j\omega C)V_1 + (G_3 + j\omega C)V_2 = 0 \quad (9\text{-}27)$$

The determinant of this system of equations is found to be

$$\Delta = (G_1 + G_2)G_3 + j\omega C(G_1 + G_2 + G_3 + K)$$

The voltage V_2, which is V_{Th} in the present case, is then found to be

$$V_{Th} = V_2 = \frac{(-K + j\omega C)I_s}{(G_1 + G_2)G_3 + j\omega C(G_1 + G_2 + G_3 + K)} \quad (9\text{-}28)$$

Determination of Z_{Th}: The relaxed network is shown in Fig. 9-44(c). Since there is a dependent source, a test source I_t has been added. Then

$$Z_{Th} = V_2/I_t$$

Node 1: $\quad (G_1 + G_2 + j\omega C)V_1 - j\omega C V_2 = 0 \quad (9\text{-}29)$

Node 2: $\quad (K - j\omega C)V_1 + (G_3 + j\omega C)V_2 = I_t \quad (9\text{-}30)$

Solving for V_2

$$Z_{Th} = V_2/I_t = \frac{(G_1 + G_2 + j\omega C)}{(G_1 + G_2)G_3 + j\omega C(G_1 + G_2 + G_3 + K)} \quad \blacksquare$$

Exercise 9-29 Determine the Thevenin equivalent of the circuit shown in Fig. 9-45 as seen from (a) the terminal pair A-B; (b) the terminal pair C-D; and (c) the terminal pair E-F.

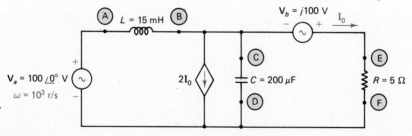

Fig. 9-45 Circuit for Exercise 9-29.

9-7-2 Norton's Theorem

Norton's theorem is the dual of Thevenin's theorem. Instead of a voltage source in series with an impedance, the Norton equivalent contains a current source in parallel with an admittance. Given a fixed linear network with an arbitrary load, as indicated in Fig. 9-

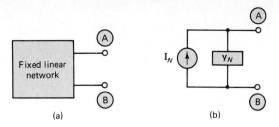

Fig. 9-46 Norton equivalent.

46(a), the Norton equivalent as seen from the terminal pair A-B consists of a current source $\mathbf{I}_N$ in parallel with an admittance $\mathbf{Y}_N$, as indicated in Fig. 9-46(b). The current $\mathbf{I}_N$ is equal to the current that would flow from A to B through a short circuit placed across A-B. The admittance $\mathbf{Y}_N$ is equal to the admittance seen from the terminal pair A-B.

The Norton equivalent circuit can be obtained from the Thevenin equivalent circuit for a given terminal pair by using the relationships

$$\mathbf{I}_N = \frac{\mathbf{V}_{Th}}{\mathbf{Z}_{Th}} \tag{9-31}$$

and

$$\mathbf{Y}_N = \frac{1}{\mathbf{Z}_{Th}} \tag{9-32}$$

Example 9-21 Determine the Norton equivalent circuit of the network in Fig. 9-47(a) as seen from the terminal pair A-B.

Solution A short circuit is placed across the terminals A, B, as indicated in Fig. 9-47(b). The short circuit current is given by

$$\mathbf{I}_N = \mathbf{I}_1 - \mathbf{I}_2 \tag{9-33}$$

$$\mathbf{I}_1 = \frac{\mathbf{V}_s}{R_1 + j\omega L} = \frac{100\ \underline{/0°}}{3 + j4} = (12 - j\,16)\ \text{A}$$

$$\mathbf{I}_2 = \frac{8\mathbf{I}_1}{R_3 - j1/\omega C} = \frac{(96 - j\,128)}{2 - j2} = (56 - j\,8)\ \text{A}$$

Therefore, the short circuit current $\mathbf{I}_N$ is, from Eq. (9-33),

$$\mathbf{I}_N = (\mathbf{I}_1 - \mathbf{I}_2) = (-44 - j\,8) = 44.7\ \underline{/-170°}\ \text{A}$$

which is the current for the Norton source. ∎

To determine the admittance $\mathbf{Y}_N$, the network is made relaxed and a test source used as indicated in Fig. 9-47(c). We need to calculate the ratio $\mathbf{V}_t/\mathbf{I}_t$.

$$\mathbf{I}_t = \mathbf{I}_2 - \mathbf{I}_1 \tag{9-34}$$

$$\mathbf{I}_1 = -\frac{\mathbf{V}_t}{3 + j4} = (-0.12 + j0.16)\mathbf{V}_t \tag{9-35}$$

The *KVL* equation for the other loop is

$$(2 - j2)\mathbf{I}_2 = \mathbf{V}_t + 8\mathbf{I}_1$$

9-7 Network Theorems

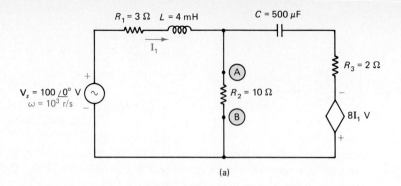

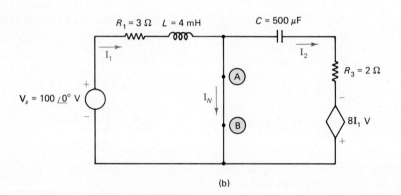

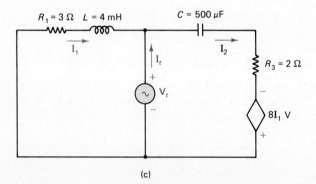

Fig. 9-47 (a) Circuit for Example 9-21. (b) Calculation of Norton current. (c) Calculation of Norton admittance.

which becomes, on using Eq. (9-35),

$$(2 - j2)\mathbf{I}_2 = (0.04 + j1.28)V_t$$

Therefore

$$\mathbf{I}_2 = (-0.31 + j0.33)V_t \tag{9-36}$$

From Eqs. (9-34), (9-35), and (9-36)

$$\mathbf{I}_t = [(-0.31 + j0.33) - (-0.12 + j0.16)]V_t$$

so that

$$\mathbf{Y}_N = \mathbf{I}_t/\mathbf{V}_t = (-0.19 + j0.17) \text{ S}$$

Note that the real part of the admittance is *negative*, which is possible due to the presence of the dependent source. The Norton equivalent circuit is shown in Fig. 9-47(c).

Exercise 9-30 Obtain the Norton equivalent of the network of the example above, as shown in Fig. 9-47(a), as seen from the terminals of (a) the capacitor C; (b) the inductor L.

Exercise 9-31 Determine the Norton source currents *directly* in the circuits of Fig. 9-45. Show that in each case, the relationship $\mathbf{I}_N = \mathbf{V}_N/\mathbf{Z}_{\text{Th}}$ is satisfied.

Exercise 9-32 A linear circuit is inside a black box with two terminals A-B available for making measurements. When a short circuit is placed across the terminals, the short circuit current is found to be $20/\!-30°$ A. If, after removing the short circuit, an impedance of $4/53.1$ Ω is connected between A and B, the voltage across that impedance is found to be $20/0°$ V. Determine the Thevenin and Norton equivalents of the circuit in the black box.

9-7-3 Maximum Power Transfer Theorem

Consider a *fixed linear* network connected to a *variable* load impedance $\mathbf{Z}_L$ as indicated in Fig. 9-48(a). It is necessary to determine the value of $\mathbf{Z}_L$ in terms of the parameters of the fixed network such that the average power consumed by $\mathbf{Z}_L$ is a maximum.

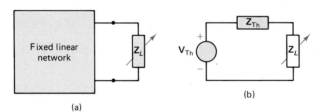

Fig. 9-48 Maximum power transfer from a fixed linear network to a variable load.

The fixed linear network is replaced by its Thevenin equivalent, as seen from the load terminals, as in Fig. 9-48(b). If

$$\mathbf{Z}_{\text{Th}} = R_{\text{Th}} + jX_{\text{Th}} \text{ and } \mathbf{Z}_L = R_L + jX_L$$

then the average power consumed by the load can be written as

$$P_L = I^2_{L(\text{rms})} R_L \qquad (9\text{-}37)$$

where $I_{L(\text{rms})}$ is the rms value of the current in $\mathbf{Z}_L$.

$$\mathbf{I}_L = \frac{\mathbf{V}_{\text{Th}}}{(R_{\text{Th}} + R_L) + j(X_{\text{Th}} + X_L)}$$

The magnitude of $\mathbf{I}_L$ is given by

$$|\mathbf{I}_L| = \frac{|\mathbf{V}_{\text{Th}}|}{\sqrt{(R_{\text{Th}} + R_L)^2 + (X_{\text{Th}} + X_L)^2}}$$

so that Eq. (9-37) leads to

$$P_L = \frac{1}{2} \frac{|V_{Th}|^2 R_L}{(R_{Th} + R_L)^2 + (X_{Th} + X_L)^2} \quad (9\text{-}38)$$

When we want to maximize the value of P_L, the question is: what are the variables? Since the impedance has two parts, R_L and X_L, either of which can be varied with certain constraints imposed, the maximization of P_L has to be considered for different cases. Even though a number of cases can be visualized, we will confine ourselves to the following two.

Case 1: Both R_L and X_L can be varied over any range of values, and each independently of the other. This is the most flexible situation of a variable load.

Case 2: Only the *magnitude* of the impedance; that is, the quantity $\sqrt{R_L^2 + X_L^2}$, can be varied, but not the angle. This means that the ratio X_L/R_L remains fixed. This is an important practical case. In a number of situations, the load may be a pure resistance, which means that the angle of the load impedance is fixed at $0°$, whereas the magnitude R_L can be varied. And in the case of loads which are coils, the ratio of the inductance to resistance is usually a constant, again corresponding to the case of a fixed angle of the impedance.

Case 1: R_L and X_L are both variable, each independently of the other.

In order to maximize P_L in Eq. (9-38) under this condition, it is necessary to have

$$\frac{\partial P_L}{\partial R_L} = 0 \quad \text{and} \quad \frac{\partial P_L}{\partial X_L} = 0$$

When $(\partial P_L/\partial R_L)$ is zero, the resulting condition is found to be

$$R_{Th}^2 - R_L^2 + (X_{Th} + X_L)^2 = 0 \quad (9\text{-}39)$$

When $(\partial P_L/\partial X_L)$ is zero, the resulting condition is found to be

$$(X_{Th} + X_L) = 0$$

Therefore, it is necessary to make

$$X_{Th} = -X_L$$

which makes Eq. (9-39) give the condition

$$R_{Th} = R_L \quad (9\text{-}40)$$

Therefore, *in order to transfer maximum power from a fixed linear network to a variable load impedance with both its real part and imaginary part variable and independent of each other, the load impedance must be the complex conjugate of the Thevenin impedance* seen from the terminals of the load impedance.

$$\mathbf{Z}_L = \mathbf{Z}_{Th}^* \quad (9\text{-}41)$$

Examining the above result, we can see that when X_L is made to cancel X_{Th}, the total reactance of the network with the load is made zero. This makes the power factor seen by the Thevenin source equal to 1, which is the highest power factor that can be obtained. The power factor has been made as high as possible in order to increase the average power supplied by the source. Once the reactance of the circuit is reduced to zero, the situation reverts to the case of purely resistive networks. For such networks, it is necessary to make the load resistance equal to the Thevenin resistance for maximum power transfer.

When the condition in Eq. (9-41) is satisfied, the maximum average power delivered to the load becomes

$$P_{L(max)} = \frac{|V_{Th}|^2}{8R_{Th}} \qquad (9\text{-}42)$$

where $|V_{Th}|$ represents the amplitude of the voltage of the Thevenin source. An alternative form of Eq. (9-42) is obtained by using $V_{Th(rms)} = |V_{Th}|/\sqrt{2}$:

$$P_{L(max)} = \frac{V_{Th(rms)}^2}{4R_{Th}}$$

which is identical to the expression obtained for purely resistive networks in Chapter 4.

Example 9-22 Suppose a load impedance with variables R_L and X_L is connected to the terminals of the network in Fig. 9-42 (Example 9-19). Determine the load impedance for maximum power transfer and the maximum power consumed by the load.

Solution In Example 9-19, the Thevenin equivalent of the network was given by

$$\mathbf{V}_{Th} = 74.8\underline{/28.9°} \text{ V}$$

$$\mathbf{Z}_{Th} = 9.14\underline{/-44.2°} \text{ } \Omega$$

Therefore, the desired value of the load impedance is

$$\mathbf{Z}_L = \mathbf{Z}_{Th}^* = 9.14\underline{/+44.2°} = (6.55 + j6.37) \text{ } \Omega$$

The maximum average power delivered to the load is, from Eq. (9-42),

$$P_{L(max)} = \frac{74.8^2}{8 \times 6.55} = 107 \text{ W} \qquad \blacksquare$$

Case 2: Only the *magnitude* of the load impedance is variable, and its angle is held constant.

Since the angle of Z_L is a constant, the ratio

$$X_L/R_L = \text{a constant}$$

Therefore, we can write

$$X_L = KR_L$$

where K is a constant.

Then the expression for the average power, Eq. (9-38), becomes

$$P_L = \frac{1}{2} \frac{|V_{Th}|^2 R_L}{(R_{Th} + R_L)^2 + (X_{Th} + KR_L)^2} \qquad (9\text{-}43)$$

and it is necessary to make $dP_L/dR_L = 0$ for maximum P_L.

Differentiating Eq. (9-43) with respect to R_L and equating the derivative to zero leads to

$$(R_{Th} + R_L)^2 + (X_{Th} + KR_L)^2 - 2R_L(R_{Th} + R_L) - 2KR_L(X_{Th} + KR_L) = 0$$

which leads to

$$R_L^2 + K^2 R_L^2 = R_{Th}^2 + X_{Th}^2 \qquad (9\text{-}44)$$

Since $X_L = KR_L$, the left side of Eq. (9-44) can be recognized as $|Z_L|^2$ while the right hand side is $|Z_{Th}|^2$.

Therefore, the *condition for maximum power transfer when only the magnitude of the load impedance can be varied, but not its angle,* is

$$|Z_L| = |Z^{Th}| \qquad (9\text{-}45)$$

In the special case when the *load is a pure resistance, the condition for maximum power transfer* becomes

$$R_L = |Z_{Th}|$$

since $R_L = |Z_L|$ here. The maximum power transferred to the load in the present case will not be as large as it was when both the magnitude and angle of the load impedance were allowed to vary, since the overall power factor is not unity when the angle of the load impedance cannot be varied.

Example 9-23 Suppose a load impedance with a fixed angle of θ but variable magnitude is used in the network of Fig. 9-42 (Examples 9-19 and 9-22). Determine the value of the load impedance for maximum power transfer. Discuss the effect of the angle θ on the maximum power transferred.

Solution Since

$$\mathbf{Z}_{Th} = 9.14\underline{/-44.2°}\ \Omega$$

for the network under consideration, the condition to be satisfied is

$$|Z_L| = 9.14\ \Omega$$

so that

$$\mathbf{Z}_L = 9.14\underline{/\theta}\ \Omega$$

for maximum power transfer.

As the angle θ is given different (fixed) values, the maximum average power transferred to the load is zero when the angle θ is $-90°$. It increases as θ increases toward zero. At $\theta = 0$, which corresponds to a purely resistive load, the maximum average power transferred is 89.2W. When θ is increased further, the highest maximum average power is reached when it equals $+44.2°$ (as is to be expected), and then it decreases again, reaching a value of zero at $+90°$. ∎

Exercise 9-33 The Thevenin equivalent of a circuit as seen from the load terminals consists of $\mathbf{V}_{Th} = 35\underline{/160°}$ V and $\mathbf{Z}_{Th} = 50\underline{/80°}\ \Omega$. Determine the load impedance for maximum power transfer and the power consumed by the load when (a) the real and imaginary parts of the load are independently adjustable, and (b) when the load is a resistance.

Exercise 9-34 Consider a load resistance R_L connected to the terminal pairs given in the circuit of Exercise 9-28. Determine the value of R_L for maximum power transfer.

9-7-4 The Superposition Theorem

The superposition theorem states that the response of a linear circuit to an input which is a sum of component functions is equal to the sum of the responses of the circuit to the individual components of the input. Superposition has already been used in this chapter in the section on transfer functions to find the response of a network to an input which was the sum of complex exponential functions.

In using superposition in the sinusoidal steady state, the response of the network to the individual components of the forcing function are determined by using phasors, impedance, and admittance. The response to each input component is then written as a time function, either as a complex exponential function (if the component is a complex exponential function), or as a real sinusoidal function (if the component is a real sinusoidal function). The responses are added in the *time domain* to obtain the total response.

Example 9-24 The current source in the circuit of Fig. 9-49 produces a current $i_s(t)$ given by

$$i_s(t) = 10 + 20 \cos 2t + 30 \cos (4t - 30°) \text{ A}$$

Determine the voltage $v_0(t)$ across the capacitor.

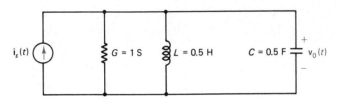

Fig. 9-49 Circuit for Example 9-24 (use of superposition).

Solution The admittance of the circuit is given by

$$\mathbf{Y} = G + j\left(\frac{-1}{\omega L} + \omega C\right)$$

and the voltage $\mathbf{V}_0$ at any frequency is given by

$$\mathbf{V}_0 = \mathbf{I}_s/\mathbf{Y}$$

where $\mathbf{I}_s$ is the phasor representation of the component of the current at the frequency under consideration. Consider each of the three frequencies present in the current $i_s(t)$.

1. DC: $\mathbf{Y} = -j\infty$
  $\mathbf{V}_0 = 0$ (dc component)

2. $\omega = 2$ r/s: $\mathbf{Y} = 1 + j0$ S
  $\mathbf{V}_0 = 20\underline{/0°}$ V
  $v_0(t) = 20 \cos 2t$ V

3. $\omega = 4$ r/s: $\mathbf{Y} = 1 + j1.5 = 1.80\underline{/56.4°}$ S
  $\mathbf{V}_0 = 16.7\underline{/-86.4°}$ V
  $v_0(t) = 16.7 \cos (4t - 86.4°)$ V

Therefore, the total output voltage is given by
$$v_0(t) = 20 \cos 2t + 16.7 \cos(4t - 86.4°) \text{ V}$$

Example 9-25 The input to the circuit shown in Fig. 9-50 is specified as
$$v_s(t) = 50 + 50 \cos 1000t + 50 \cos 3000t \text{ V}$$
Determine the output voltage $v_0(t)$.

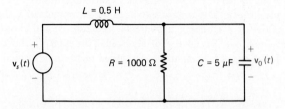

Fig. 9-50 Circuit for Example 9-25 and Exercise 9-35.

Solution The relevant transfer function is
$$\mathbf{H}(j\omega) = \frac{\mathbf{V}_0}{\mathbf{V}_s} = \frac{\mathbf{Z}_2}{\mathbf{Z}_1 + \mathbf{Z}_2}$$
where
$$\mathbf{Z}_1 = j\omega L$$
and
$$\mathbf{Z}_2 = [R \,||\, -j(1/\omega C)] = \frac{R}{1 + j\omega RC}$$

Therefore
$$\mathbf{H}(j\omega) = \frac{R}{R(1 - \omega^2 LC) + j\omega L}$$
$$= \frac{1000}{(1000 - 2.5 \times 10^{-3} \omega^2) + j0.5\omega}$$

Now consider each of the three components of the input voltage.

DC: The transfer function is
$$\mathbf{H}(j\omega) = 1$$
so that the output component becomes
$$\mathbf{V}_0 = 50 \text{ V}$$

$\omega = 1000$: The transfer function becomes
$$\mathbf{H}(j\omega) = \frac{1000}{-1500 + j500} = 0.632\underline{/-162°}$$

so that the output component becomes

$$v_o(t) = 31.6 \cos(1000t - 162°) \text{ V}$$

$\omega = 3000$: The transfer function becomes

$$\mathbf{H}(j\omega) = \frac{1000}{-21500 + j1500} = 0.0464\underline{/-176°}$$

so that the output component becomes

$$v_o(t) = 2.32 \cos(3000t - 176°) \text{ V}$$

The total output voltage is given by

$$v_o(t) = 50 + 31.6 \cos(1000t - 162°) + 2.32 \cos(3000t - 176°) \text{ V}$$

The situation in the last example is encountered in practical electronic circuits. Transistors require dc power in order to function in the desired manner in amplifiers. The usual practice is to convert available ac power into dc by *rectification* (that is, the removal of the negative half cycles of the sinusoidal waveform, or by flipping them over so that they become positive half cycles also). But a rectified sinusoid has a dc component as well as sinusoidal components at different frequencies. In order to eliminate, or at least suppress as much as possible, the sinusoidal components, a circuit of the type in Fig. 9-50 is used. The amplitudes of the different components of the input are equal to one another, but in the output voltage, the components at 1000 r/s and 3000 r/s have been significantly attenuated. As we will see in the following exercise, the two sinusoidal components can be suppressed even further by a suitable selection of the values of the circuit elements.

Exercise 9-35 In the circuit of Fig. 9-50, change L to 5 H and C to 50 μF. If the input is

$$v_s(t) = 31.8 + 50 \cos 377t + 21.2 \cos 754t \text{ V}$$

determine the output voltage. Compare the relative amplitudes of the 60 Hz and 120 Hz components in the output and input using the dc component as the basis of comparison.

Exercise 9-36 Determine the current supplied by the source in the circuit of Fig. 9-51 if

$$v_s(t) = 100 + 80 \cos 500t + 40 \cos 1000t \text{ V}$$

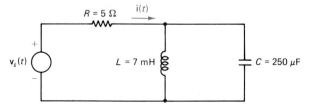

Fig. 9-51 Circuit for Exercise 9-36.

We will see in Chapter 14 that a periodic signal can be written as the sum of sinusoidal components, and we will use superposition again in circuit analysis in that chapter. When signals are not periodic, the principles of superposition is restated in the form of an integral called the *convolution integral*.

9-7-5 The Reciprocity Theorem

The reciprocity theorem states that, in a passive linear network, the ratio of the current response in one branch to a forcing voltage function in another remains unaffected when the source branch and response branch are interchanged.

That is, in the situation shown in Figs. 9-52 (a) and (b), the ratio I_k/V_m equals I_m/V_k. The theorem is also valid for the ratio of the voltage response in one branch to a forcing current function in another. In Figs. 9-52 (c) and (d) the ratios V_m/I_k and V_k/I_m are equal to each other.

As we noted in Chapter 5, the theorem does not apply to the ratio of response current in one branch to the forcing current in another, or response voltage in one branch to the forcing voltage in another.

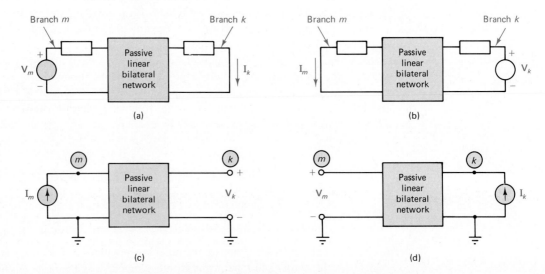

Fig. 9-52 Reciprocity theorem. (a)-(b): $(I_k/V_m) = (I_m/V_k)$. (c)-(d): $(V_k/I_m)(V_m/I_k)$.

Example 9-26 In the circuit of Fig. 9-53(a): (a) Connect a current source of current I_1 between node 1 and the reference node and calculate the ratio V_2/I_1; (b) connect a current source of current I_2 between node 2 and the reference node and calculate the ratio V_1/I_2 Note that in each part, only one current source is active. The results should verify the reciprocity theorem.

Solution (a) Referring to Fig. 9-53(b) we see that

$$V_2 = \frac{R_2}{R_2 + j\omega L} V_1$$

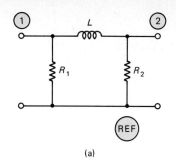

(a)

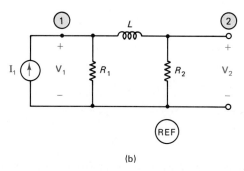

(b)

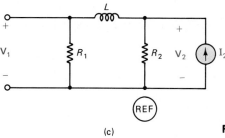

(c)

Fig. 9-53 Reciprocity theorem (Example 9-26).

and

$$\mathbf{V}_1 = \frac{R_1(R_2 + j\omega L)}{R_1 + R_2 + j\omega L} \mathbf{I}_1$$

$$\frac{\mathbf{V}_2}{\mathbf{I}_1} = \frac{R_1 R_2}{R_1 + R_2 + j\omega L}$$

(b) Referring to Fig. 9-53(c), we have

$$\mathbf{V}_1 = \frac{R_1}{R_1 + j\omega L} \mathbf{V}_2$$

and

$$\mathbf{V}_2 = \frac{R_1(R_2 + j\omega L)}{R_1 + R_2 + j\omega L} \mathbf{I}_2$$

9-7 Network Theorems

so that

$$\frac{V_1}{I_2} = \frac{R_1 R_2}{R_1 + R_2 + j\omega L}$$

We see that V_2/I_1 is equal to V_1/I_2, thus verifying the reciprocity theorem. ∎

Exercise 9-37 Verify the reciprocity theorem in the circuit of Fig. 9-54 by: (a) cutting into branch A, inserting a voltage source V_a into the opening, and calculating the ratio I_b/V_a; and (b) cutting into branch B, inserting a voltage source V_b into the opening, and calculating the ratio I_a/V_b. In each situation, only one voltage source is active.

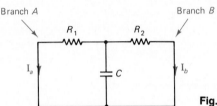

Fig. 9-54 Circuit for Exercise 9-37.

The combination of the reciprocity theorem and the principle of superposition is useful in the analysis of linear networks, as illustrated by the following example.

Example 9-27 The box in Fig. 9-55 contains a passive, linear, bilateral network. When a current source $I_1 = 1\underline{/0°}$ A is connected between node 1 and the reference node, the nodal voltages V_2 and V_3 are given by

$$V_2 = 16\underline{/-30°} \text{ V} \quad \text{and} \quad V_3 = 28\underline{/75°} \text{ V}$$

Determine the voltage V_1 (at node 1) when two current sources are connected to the network, as follows: A source of current $12.5\underline{/-90°}$ A is connected between node 2 and the reference node and a source of current $15\underline{/45°}$ A is connected between node 3 and the reference node. (No source is connected between node 1 and the reference node.)

Solution Consider first the case when only a single current source of current $I_2 = 12.5\underline{/-90°}$ A is connected between node 2 and the reference node.

Using the reciprocity theorem and the principle of superposition, the contribution to the voltage V_1 from the I_2 source acting alone becomes

$$V_1/I_2 = V_2/I_1 = 16\underline{/-30°}$$

Fig. 9-55 Use of reciprocity and superposition (Example 9-27).

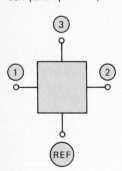

so that

$$V_1 = (16\underline{/-30°})(12.5\underline{/-90°})$$
$$= 200\underline{/-120°} \text{ V}$$

due to the source I_2.

Consider next the case when only a single current source of current $I_3 = 15\underline{/45°}$ A is connected between node 3 and the reference node.

Then

$$V_1/I_3 = V_3/I_1 = 28\underline{/75°}$$

so that
$$V_1 = (28\underline{/75°})(15\underline{/45°}) = 420\underline{/120°} \text{ V}$$

Therefore, the total voltage at node 1 (when both current sources are connected simultaneously) is, by the principle of superposition,
$$V_1 = 200\underline{/-120°} + 420\underline{/120°}$$
$$= 363\underline{/148°} \text{ V} \qquad \blacksquare$$

Exercise 9-38 Suppose in the network of the previous example that the following additional piece of information is provided. With a single current source $I_1 = 1\underline{/0°}$ A, the voltage at node 1 is found to be $V_1 = 100\underline{/90°}$ V. Suppose there are three current sources, I_1, I_2, I_3, connected, respectively, between nodes 1, 2, 3 and the reference node. If $I_1 = 20\underline{/-120°}$ A and $I_2 = 40\underline{/-90°}$ A, determine the value of I_3 such that the total voltage at node 1 is $100\underline{/0°}$ V.

9-8 SUMMARY OF CHAPTER

The analysis of circuits in sinusoidal steady state uses the same procedures as those for the resistive networks discussed in Chapters 2, 3, and 4. The main difference is that impedances are used instead of resistances, and admittances instead of conductances. Voltages and currents are used in their phasor forms. Series parallel combination and reduction of a network to a single impedance or admittance is possible for circuits driven by a single forcing function. The ladder network approach is especially effective when the currents and voltages in several or all of the branches of a circuit are to be determined.

Networks with multiple sources are analyzed by means of nodal, loop, or mesh analysis. Again the steps are exactly like those used in resistive networks. Even though the steps are not new, the results obtained in sinusoidal steady state are interesting because of their variation with frequency.

Transfer functions of a network are ratios of a response function (voltage or current in some branch) to a source function (voltage or current due to a source placed in another branch). Once the appropriate transfer function is known, the response to any complex exponential or sinusoidal signal can be readily determined. The frequency dependency of transfer functions is an important factor in the analysis and design of filters and other communication circuits. The curves showing the variation of the magnitude and phase of a transfer function as a function of frequency constitute the frequency response characteristics of a network. Frequency response curves are convenient and useful visual aids in studying the behavior of a network for signals of different frequencies.

When the forcing function is in the form of a complex exponential forcing function, the response of a network may be written as the product of the appropriate transfer function and the forcing function. For real sinusoidal inputs, it is necessary to add two complex conjugate response functions. This leads to the mathematical concept of negative frequencies.

Transfer functions of a network can be obtained by direct analysis or from the mesh or loop impedance matrix or the nodal admittance matrix. The transfer function is related to the ratio of an appropriate cofactor and the determinant of the matrix. Matrix representations of a system of equations of a network are therefore powerful tools in circuit analysis.

Thevenin's and Norton's theorems permit the reduction of a complicated network to a simple series or a parallel equivalent circuit made up of a source and an impedance. The equivalents and the procedures for finding them are exactly like those presented in Chapter 5. The maximum power transfer theorem leads to different choices of load impedance, depending on the constraints placed upon the variation of the load impedance. When there is complete freedom in choosing the load impedance, maximum power is transferred when the load impedance is the complex conjugate of the Thevenin impedance seen looking in from the load impedance. When only the magnitude of the load impedance can be varied (but not its angle), the load impedance magnitude must equal the magnitude of the Thevenin impedance.

Superposition is a more interesting and useful principle for time-varying inputs than for constant inputs. It is essential in finding the response of a circuit when it is simultaneously subjected to sources of different frequencies. The individual responses can be obtained by using phasors, impedances, and admittances; the sum of the response functions in the time domain gives the total response. The reciprocity theorem is exactly like the case of resistive circuits.

Answers to Exercises

9-1 (a) $Z_T = 160.7 \underline{/82.7°}\ \Omega$. $P_{av} = 3.95$ W. (b) $Y_T = 1.61 \times 10^{-2} \underline{/-82.9°}$ S. $Z_T = 62.1 \underline{/82.9°}\Omega$. $P_{av} = 384$ W. (c) $Y_T = 1.08 \times 10^{-2} \underline{/73.5°}$ S. $Z_T = 92.2 \underline{/-73.5°}\Omega$. $P_{av} = 1309$ W. (d) $Z_T = 70.7 \underline{/-45°}\Omega$. $P_{av} = 50$ W.

9-2 (a) $Z_T = -j0.850\Omega$. Branch currents are: $j118, -j29.4, j147$ A.
(b) $Z_T = 4\underline{/-53.1°}\Omega$. Branch currents are: $25\underline{/53.1°}, 11.2\underline{/26.5°}, 15.9\underline{/71.5°}$ A.

9-3 (a) With 1 A assumed in the capacitor at right, $V_s = j7$ V. Ratio to multiply all values $= -j14.3$. Branch currents are (from right to left): $-j14.3, j42.9, j28.6, -j71.5, -j42.8$ A.
(b) With 1 A assumed in the resistor at right and x in the resistor at left, $x = 2.12\underline{/45°}$A. $I_s = 4.3\underline{/35.5°}$A. Ratio: $4.65\underline{/-35.5°}$. Branch currents are: $4.65\underline{/-35.5°}, 6.58\underline{/9.5°}, 10.4\underline{/-8.9°}, 9.86\underline{/9.5°}$ A.

9-5 (a) Node 1: $[j\omega(C_1 + C_2) - j(1/\omega L_1)]V_1 + (j/\omega L_1)V_2 = -I_a$. Node 2: $(j/\omega L_1)V_1 + [G_1 - j(1/\omega L_1)]V_2 = I_b$. (b) Node 1: $-jV_1 - KV_2 = 20$. Node 2: $(K - j0.5)V_2 = -10$. (c) Node 1: $(10 - j9.9)V_1 - 10V_2 = -j50$. Node 2: $-10(1 - K)V_1 + 10(1 - K)V_2 = 50$. Node 3: $10(K - j1)V_1 + 10KV_2 + (10 - j10)V_3 = 0$.

9-6 (a) Equations: $-j3V_1 = -100$. $-j6V_2 = 100$. $V_2 = j16.7$ V. (b) Equations: $-j8V_1 + j5V_2 = -100$. $j5V_1 - j11V_2 = 100$. $V_2 = j4.76$ V. (c) Equations: $(-j8)V_1 + j5V_2 = -100$. $(-2 + j5)V_1 - j11V_2 = 100$. $V_2 = 5.65\underline{/65°}$V.

9-7 After source conversion, equations are: $(2.5 + j0.6)V_1 - jV_2 = 0$. $-jV_1 + (1.25 + j0.2)V_2 = -j80$. $V_2 = 49.1\underline{/-93.9°}$ V.

9-8 After source conversion, equations are: $(0.25V_1 - j0.25V_2) = j10$. $-j0.25V_1 + (0.25 + j0.25)V_2 = 0$. $V_2 = 17.9\underline{/153°}$ V. $P_{av} = 40.6$ W.

9-9 (a) $V_1 = 10$ V. $V_2 = (17.5 - j10)$ V. 132 W. $I_a = 35.4\underline{/81.9°}$ A. 24.9 W.
(b) $V_1 = 10$ V. $V_2 = (15 - j5)$ V. $V_3 = 10$ V. $I_b = 21.2\underline{/-135°}$ A. 25 W. 50 W. -75 W.

9-10 Node voltages (with left node chosen as reference) are: 100 V, $85.1\underline{/-98.1°}$ V, $80\underline{/-36.9°}$ V. Source V_a: current $= 35.2\underline{/-19.9°}$ A; power delivered $= 1655$ W. I_b receives 800 W and I_c delivers 950 W.

9-11 Source V_a: current $= 721\underline{/33.7°}$ A. Delivers 20 kW. Source V_b: current $= 412\underline{/-14°}$ A. Delivers 20 kW.

9-12 Loop 1: $(15 + j58)I_1 + (15 + j60)I_2 + (15 + j10)I_3 = (100 + j80)$.
Loop 2: $(15 + j60)I1 + (17 + j62.5)I_2 + (15 + j15)I_3 = (13.4 + j30)$.
Loop 3: $(15 + j10)I_1 + (15 + j15)I_2 + (30 + j5)I_3 = (13.4 - j50)$.

9-13 *Mesh 1:* $(15 + j58)I_1 - j50I_2 + j_2I_3 = (100 + j80)$. *Mesh 2:* $-j50I_1 + (17 + j37.5)I_2 - (2 - j2.5)I_3 = -j80$. *Mesh 3:* $j2I_1 - (2 - j2.5)I_2 + (2 + j0.5)I_3 = -100\underline{/30°}$.

9-14 Equations: $20I_1 - 15I_2 = V_s$. $(K - 15)I_1 + (35 - j20)I_2 = 0$. $V_o/V_s = (300 - 20K)/[(15K + 475) - j400]$.

9-15 (a) *Mesh 1:* $[R_1 + R_2 + j(\omega L_1 - 1/\omega C_1)]I_1 - R_2I_2 = -j\omega L_1 I_a$. *Mesh 2:* $-R_2I_1 + (R_2 + R_3 + j\omega L_2)I_2 = -V_b$. (b) *Mesh 1:* $(200K + j10)I_1 + j10I_2 = 100$. *Mesh 2:* $(-200K + j10)I_1 + 10I_2 = j80$.

9-16 (a) $V_x = 146\underline{/-164°}$ V. (b) Currents in the inductors are: $(7.5 - j2.5)$ A and $(2.5 + j2.5)$ A. $V_x = 127\underline{/101°}$ V.

9-17 (a) Equations: $j14I_1 - j4I_2 = 100$. $-j12I_1 + (8 + j12)I_2 = 0$. $Z_{in} = 11.4\underline{/80.7°}\Omega$. (b) Equations: $(2 + j2)I_1 - 2I_2 = V_a$. $-(2 + j8)I_1 + (2 + j4)I_2 = 0$. $Z_{in} = 2\underline{/143°}\Omega$. Note the negative resistance, which is due to the dependent source.

9-18 $(I_L/I_s) = 1/[(1 - \omega^2 LC) + j\omega CR_L]$.

9-19 (a) $5\underline{/-5.74°}$ V. (b) $4.47\underline{/-63.4°}$ V. (c) $0.1\underline{/-168°}$ V.

9-20 $(V_L/V_s) = R/(-\omega^2 LC + j\omega L + R)$. $(I_L/I_s) = 1/[-j\omega^3(RLC^2) - \omega^2(LC) + j\omega(2RC) + 1]$.

9-21 $H(j\omega) = j\omega RC/(1 + j\omega RC)$. $|H|$ is 0 at $\omega = 0$ and 1 as $\omega \to \infty$. It acts as a high pass filter.

9-22 $H(j\omega) = R/(R + j\omega L)$.

9-23 Sum $= 2|A||H|\cos(\omega t + \theta + \phi)$.

9-24 Response components: $8.49e^{j(2t - 29.9°)}$ and $8.49e^{-j(2t - 29.9°)}$. Total response $= 17.0 \cos(2t - 29.9°)$ V.

9-25 Using the components from the previous exercise, response $= 17.0 \sin(2t - 29.9°)$ V.

9-26 Elements of the mesh impedance matrix are: $(R_1 + j\omega L)$, $-j\omega L$, $-j\omega L$, and $(R_2 + j\omega L - j1/\omega C)$. (a) $H(j\omega) = j\omega L/[-\omega^2 LC(R_2 + R_1) + j\omega(CR_1R_2 + L) + R_1]$. $H(j\omega) = -\omega^2 LCR_1/[\text{same denominator as in part (a)}]$.

9-27 $H(j\omega) = j10^{-3}\omega/[-\omega^2(10^{-4}) + j\omega(0.01) + 1]$. (a) $0.101\underline{/-84.2°}$ V. (b) $1\underline{/0°}$ V. (c) $0.101\underline{/-84°}$ V.

9-28 (a) $167\underline{/13°}$ V, $3.54\underline{/-45°}\Omega$; (b) $128.2\underline{/-82.1°}$ V, $4.74\underline{/18.4°}\Omega$; (c) $146\underline{/110°}$ V, $-j7.5\Omega$.

9-29 (a) $I_o = 6.329\underline{/162°}$ A, $V_{Th} = 158\underline{/34.7°}$ V, $Z_{Th} = 1.581\underline{/-18.4°}\Omega$. (b) $I_o = 3.123\underline{/-38.7°}$ A, $V_{Th} = 110.7\underline{/-83.6°}$ V, $Z_{Th} = 1.66\underline{/6.34°}\Omega$. (c) Note that the dependent source is dead for the calculation of V_{Th}, but *not* for the calculation of Z_{Th}! $V_{Th} = 112\underline{/117°}$ V. $Z_{Th} = -j22.5\Omega$.

9-30 (a) $I_N = 33.5\underline{/-116°}$ A; $Y_N = 0.253\underline{/-99.5°}$ S. (b) $I_N = 34.5\underline{/123°}$ A, $Y_N = 0.345\underline{/124°}$ S.

9-31 (a) $I_N = 100\underline{/53.1°}$ A. (b) $I_N = -j66.6$ A. (c) $I_N = 4.96\underline{/-153°}$ A.

9-32 Start with a Norton model. $I_N = 20\underline{/-30°}$ A (given). $Y_N = (I_N - I_L)/V_L = 0.775\underline{/-22.7°}$ S. $V_{Th} = 25.8\underline{/-7.3°}$ V. $Z_{Th} = 1.29\underline{/22.7°}\Omega$.

9-33 (a) $Z_L = 50\underline{/-80°}\Omega$. $P_L = 35.3$ W. (b) $R_L = 50\ \Omega$. $P_L = 5.22$ W.

9-34 (a) 3.54Ω. (b) 4.74Ω. (c) 7.5Ω.

9-35 At 377 r/s: $V_o/V_s = (52.9\underline{/-87°})/(1832\underline{/89.9°})$. At 754 r/s: $V_o/V_s = (26.5\underline{/-88.5°})/(3743\underline{/90°})$. $v_o(t) = [31.8 + 1.44\cos(377t - 176.9°) + 0.15\cos(754t - 178.5°)]$ V.

9-36 At 500 r/s, $Z_{in} = 7.98\underline{/51.2°}\Omega$. At 1000 r/s, $Z_{in} = 10.6\underline{/-61.8°}\Omega$. $i(t) = [20 + 10\cos(500t - 51.2°) + 3.78\cos(1000t + 61.8°)]$ A.

9-37 $1/(R_1 + R_2 + j\omega CR_1R_2)$.

9-38 $\mathbf{V}_1 = (j100)\mathbf{I}_1 + (16\underline{/-30°})\mathbf{I}_2 + (28\underline{/75°})\mathbf{I}_3$. $\mathbf{I}_3 = (2034\underline{/130°})/(28\underline{/75°}) = 72.6\underline{/55°}$ A.

PROBLEMS

Sec. 9-1 Networks with Single Sources

9-1 Determine the current supplied by the source and the voltage across the inductance in the circuit shown in Fig. 9-56.

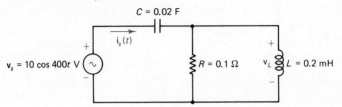

Fig. 9-56 Circuit for Problem 9-1.

9-2 Determine the total impedance $\mathbf{Z}_{in}$ of each of the circuits shown in Fig. 9-57. Use $\omega = 10$ r/s.

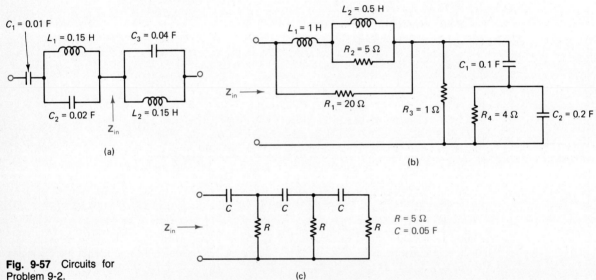

Fig. 9-57 Circuits for Problem 9-2.

9-3 (a) Write the expressions for the input impedances $\mathbf{Z}_1$ and $\mathbf{Z}_2$ in the circuits shown in Fig. 9-58, (b) Obtain the expressions for the real part and the imaginary part of the two impedances. (c) Calculate $\mathbf{Z}_1 + \mathbf{Z}_2$.

Fig. 9-58 Circuits for Problem 9-3.

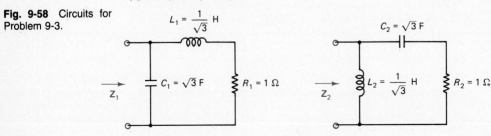

9-4 (a) Determine the impedance $\mathbf{Z}_{in}$ of the circuit in Fig. 9-59 at $\omega = 1000$ r/s. (b) If $\mathbf{Z}_{in}$ is to be increased by a factor of 25 (with frequency remaining at 1000 r/s), calculate the new values of the individual elements in the circuit. (c) Returning to the element values in part (a), calculate the values of the elements in the circuit so as to have the same impedance as in part (a) at a frequency of $\omega = 5000$ r/s.

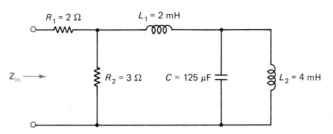

Fig. 9-59 Circuit for Problem 9-4.

9-5 Assume $\mathbf{V}_2 = 1\underline{/0°}$ V in the circuit of Fig. 9-60. (a) Determine the resulting voltages $\mathbf{V}_1$ and $\mathbf{V}_s$ (as a function of the element values and ω). (b) Write the expressions for the magnitude and phase of the voltage $\mathbf{V}_1$. (c) Calculate the phase angle between $\mathbf{V}_1$ and $\mathbf{V}_2$ when $R_1C_1 = R_2C_2$.

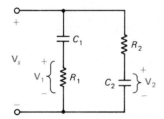

Fig. 9-60 Circuit for Problem 9-5.

9-6 In the circuit of Fig. 9-61, it is given that (a) $\omega RC = 2$ and (b) $\mathbf{V}_2$ leads $\mathbf{V}_1$ by 45°. Use a ladder network approach and determine the reactance (ωL) in terms of R.

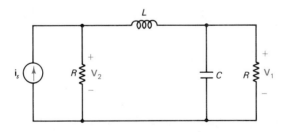

Fig. 9-61 Circuit for Problem 9-6.

9-7 Determine all the branch currents in the circuit of Fig. 9-62. The values given in the diagram are impedances.

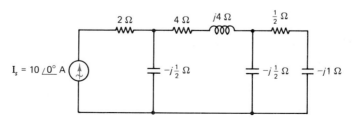

Fig. 9-62 Circuit for Problem 9-7.

Problems 457

9-8 In the circuit of Fig. 9-63, Load 1 consumes 30 kW of average power at a power factor of 0.45 (current lagging) and load 2 consumes 30 kW of average power at a power factor of 1. The amplitude of the voltage $\mathbf{V}_L$ is given as 1000 V. Determine the value of: (a) $\mathbf{V}_s$; (b) the total impedance seen by $\mathbf{V}_s$; (c) the total average power supplied by $\mathbf{V}_s$ and the corresponding power factor.

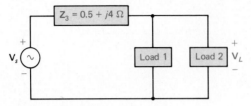

Fig. 9-63 Circuit for Problem 9-8.

9-9 Given that $\mathbf{I}_3 = 10\underline{/-60°}$ A, determine $\mathbf{V}_s$ in the circuit of Fig. 9-64.

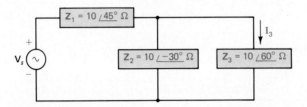

Fig. 9-64 Circuit for Problem 9-9.

9-10 A resistance $R = 10\Omega$ is in series with an unknown impedance $\mathbf{Z}_x$. A voltage $\mathbf{V}_s$ of amplitude 150 V is applied to the series combination. In making measurements, the phase angle information was unfortunately overlooked, and only the amplitudes of the voltages were recorded: voltage across $R = 100$ V and voltage across $\mathbf{Z}_x = 80$ V. Determine the value of the unknown impedance $\mathbf{Z}_x$ (both magnitude and angle).

9-11 The box marked X in the circuit of Fig. 9-65 contains a pure reactance (L or C). The input impedance at $\omega = 1000$ r/s is to be $(150 + j0)\Omega$. Determine the value of R, the nature of the element in X, and its value (in henries or farads).

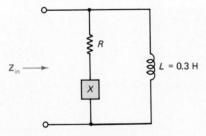

Fig. 9-65 Circuit for Problem 9-11.

Sec. 9-3 Nodal Analysis

9-12 In the circuit of Fig. 9-66: (a) redraw the circuit showing single boxes between the various nodes, with each box having a single admittance value in it. (b) Write the equations for nodes 1

Fig. 9-66 Circuit for Problem 9-12.

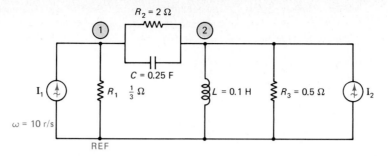

and 2. Set up the coefficients of the equations in a matrix form (this is the nodal admittance matrix). The idea of redrawing a given circuit with single admittance boxes between various nodes is an advisable first step in solving nodal analysis problems. It helps minimize the probability of errors.

9-13 Repeat the work of the previous problem for the circuit shown in Fig. 9-67.

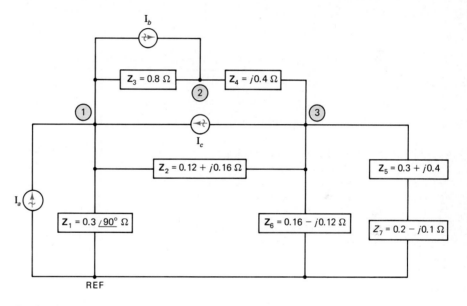

Fig. 9-67 Circuit for Problem 9-13.

9-14 Determine the node voltages V_1 and V_2 in the circuit of Fig. 9-68.

Fig. 9-68 Circuit for Problem 9-14.

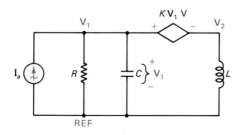

Problems

9-15 Convert the $V_a - L_1$ combination in Fig. 9-69 to a current source in parallel with an admittance. Write the nodal equations and set up the nodal admittance matrix. Determine the voltages V_1 and V_2.

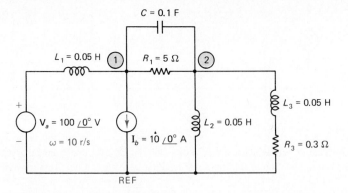

Fig. 9-69 Circuit for Problem 9-15.

9-16 Determine the average power delivered by each of the sources in the circuit of Fig. 9-70.

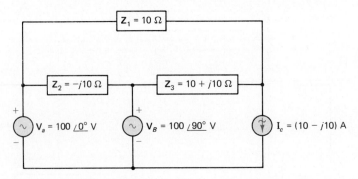

Fig. 9-70 Circuit for Problem 9-16.

9-17 Obtain an expression for the voltage V_1 in the circuit of Fig. 9-71.

Fig. 9-71 Circuit for Problem 9-17.

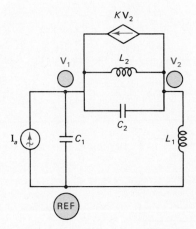

9-18 Determine the input admittance in the circuit of Fig. 9-72.

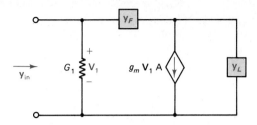

Fig. 9-72 Circuit for Problem 9-18.

9-19 Calculate the value of I_a in the circuit of Fig. 9-73 so that the node marked A is at the same potential as the reference node.

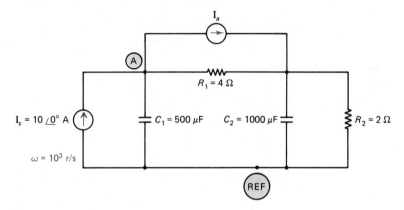

Fig. 9-73 Circuit for Problem 9-19.

Sec. 9-4 Loop and Mesh Analysis

9-20 In the circuit of Fig. 9-74: (a) Redraw the circuit showing single boxes in the various meshes or loops, with each box having a single impedance value in it. (b) Write the equations for meshes 1 and 2. Set up the coefficients of the equations in a matrix form (this is the mesh impedance matrix). The idea of redrawing a given circuit with single impedance boxes in the various meshes or loops is an advisable first step in solving mesh or loop analysis problems. It helps minimize the probability of errors.

Fig. 9-74 Circuit for Problem 9-20.

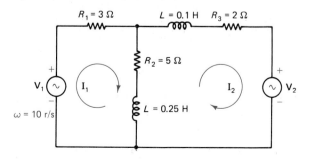

Problems 461

9-21 Repeat the work of the previous problem for the loop currents in the circuit shown of Fig. 9-75.

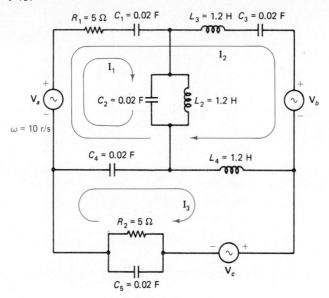

Fig. 9-75 Circuit for Problem 9-21.

9-22 Determine the currents I_1 and I_2 in the circuit of Fig. 9-76.

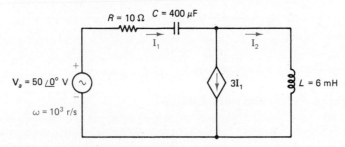

Fig. 9-76 Circuit for Problem 9-22.

9-23 Convert the $I_a - L_1$ combination in Fig. 9-77 to a voltage source in series with an impedance. Write the mesh equations and set up the mesh impedance matrix. Determine the current I_1 and I_2.

Fig. 9-77 Circuit for Problem 9-23.

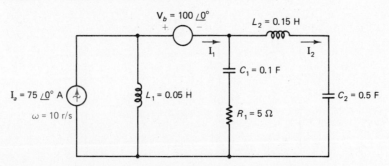

462   Networks in the Sinusoidal Steady State

9-24 Determine the average power delivered by each of the sources in the circuits of Fig. 9-78.

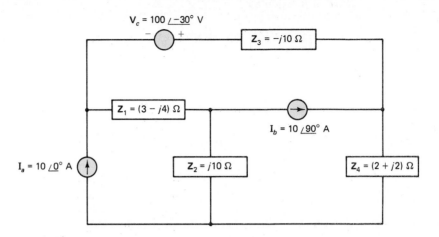

Fig. 9-78 Circuit for Problem 9-24.

9-25 Obtain an expression for the current I_1 in the circuit of Fig. 9-79.

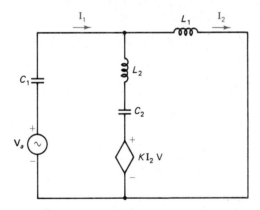

Fig. 9-79 Circuit for Problem 9-25.

9-26 Determine the input impedance the circuit of Fig. 9-80.

Fig. 9-80 Circuit for Problem 9-26.

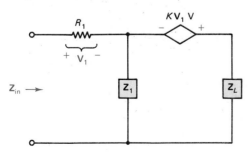

Problems

9-27 Calculate the value of V_a in the circuit of Fig. 9-81 so that the currents in loops 2 and 3 are equal to each other.

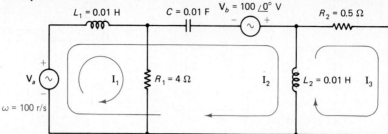

Fig. 9-81 Circuit for Problem 9-27.

9-28 Determine the voltage V_c in the circuit of Fig. 9-82: (a) by using nodal analysis, and (b) by using loop analysis.

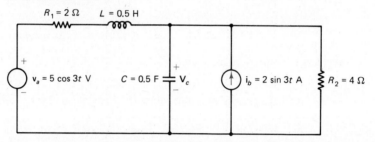

Fig. 9-82 Circuit for Problem 9-28.

9-29 The current source I_b in the circuit of Fig. 9-83 is known to deliver an average power of 90 W with a power factor of 0.5 (current leading). Determine the voltage V_a.

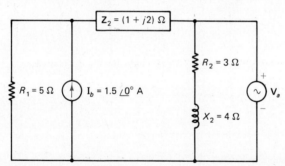

Fig. 9-83 Circuit for Problem 9-29.

9-30 Interchange the positions of the two sources I_b and V_a in the circuit of the previous problem and determine the new value of V_a.

Sec. 9-5, 9-6 Transfer Functions

9-31 Determine the transfer function V_o/V_i for each of the circuits of Fig. 9-84.

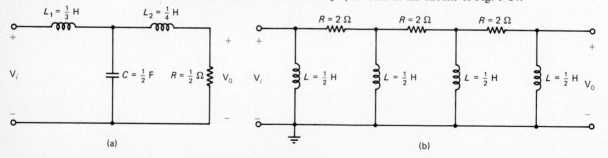

Fig. 9-84 Circuits for Problem 9-31.

464   Networks in the Sinusoidal Steady State

9-32 Determine the transfer function I_o/V_s in the circuit of Fig. 9-85.

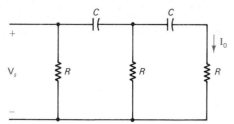

Fig. 9-85 Circuit for Problems 9-32 and 9-34.

9-33 The transfer function V_o/V_s of a network is given by

$$\frac{-\omega^2 + j4\omega + 2}{-\omega^2 + j9\omega + 1}$$

Determine the output function $v_o(t)$ for each of the following inputs: (a) $v_i = 30$ V (a constant).
(b) $v_i = 30e^{-j4t}$ V. (c) $v_i = 30e^{j(4t + 45°)}$ V. (d) $v_i = 30\cos(4t + 45°)$ V.
(e) $v_i = 30\sin 4t$ V.

9-34 Consider the circuit of Problem 9-32, with $R = 2\,\Omega$, $C = 0.8$F.
(a) Determine i_o when the input is

$$v_1 = 10 + 10e^{j2t} + 10e^{j4t} \text{ V}$$

(b) Write i_o when the input is

$$v_2 = 10 + 10e^{-j2t} + 10e^{-j4t} \text{ V}$$

(c) Write i_o when the input is

$$v_2 = 10 + 10\cos 2t + 10\cos 4t \text{ V}$$

9-35 When the input to a network is $v_i(t) = 10e^{-j(40t + 75°)}$ V, the output is found to be $v_o(t) = 2.5e^{-j(40t - 36.9°)}$ V. (a) Calculate the value of the transfer function $\mathbf{H}(j40)$, its magnitude and phase. (b) Find the input $v_i(t)$ needed to produce an output of $100\cos 40t$ V.

9-36 Determine the amplitude function $|\mathbf{H}(j\omega)|$ and phase function $\theta(j\omega)$ for each of the following functions.

(a) $\mathbf{H}(j\omega) = \dfrac{-\omega^2 + 9j\omega + 10}{-\omega^2 + 3j\omega + 2}$

(b) $\mathbf{H}(j\omega) = \dfrac{1000(j\omega + 1)(1000\omega + j1)}{\omega + j1000}$

(c) $\mathbf{H}(j\omega) = \dfrac{j30\omega + 30}{-10\omega^2 + j30\omega}$

9-37 In the circuit of Fig. 9-86: (a) Determine the transfer function V_2/V_1. (b) Determine the

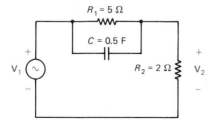

Fig. 9-86 Circuit for Problem 9-37.

frequency at which the *phase* of H(jω) is a maximum. The answer must be a single number.
(c) For the frequency calculated in (b), calculate |H(jω)|.

9-38 (a) Determine the transfer function V_2/V_1 of the circuit in Fig. 9-87. (b) Plot the amplitude response and phase response of the circuit.

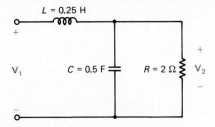

Fig. 9-87 Circuit for Problem 9-38.

9-39 Repeat the work of the previous problem for the circuit of Fig. 9-88.

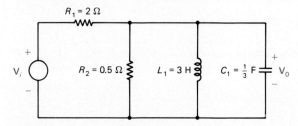

Fig. 9-88 Circuit for Problem 9-39.

9-40 For each of the transfer functions given below, sketch the frequency response (both amplitude and phase) characteristics.

(a) $\mathbf{H}(j\omega) = \dfrac{100(j\omega + 1)}{-\omega^2 + j2\omega + 4}$

(b) $\mathbf{H}(j\omega) = \dfrac{5}{(j\omega)(1 + j\omega)(2 + j\omega)}$

9-41 (a) Determine the transfer function V_o/I_s in the circuit of Fig. 9-89. (b) Find the output when the input is $i_s(t) = \mathbf{I}_o e^{-jt/\sqrt{LC}}$. (c) Sketch the amplitude response of the circuit.

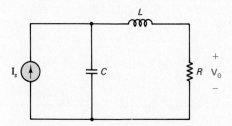

Fig. 9-89 Circuit for Problem 9-41.

9-42 (a) Determine the transfer function V_2/V_1 of the circuit shown in Fig. 9-90. (b) Plot the amplitude response. (c) Evaluate the output phasor for an input of $1\underline{/0°}$ V at the following values of ω: (i) (5/RC); (ii) (2/RC); (iii) (1/RC); (iv) (1/2RC); (v) (1/5RC).

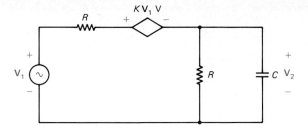

Fig. 9-90 Circuit for Problem 9-42.

9-43 The mesh impedance matrix of a network is given as

$$[Z] = \begin{bmatrix} (10 + j10) & -j4 & 0 \\ 0 & (8 - j8) & 5 \\ 0 & 5 & j6 \end{bmatrix}$$

For each of the following situations, determine the transfer function I_L/V_s, where I_L is the current in the mesh designated as the load mesh and V_s is the voltage of the source: (a) V_s in mesh 1, and mesh 3 taken as the load mesh; (b) V_s in mesh 3, and mesh 1 taken as the load mesh; (c) V_s in mesh 2, and mesh 3 taken as the load mesh.

9-44 Given the nodal admittance matrix [y] of a linear network with no independent sources, obtain an expression in terms of the appropriate cofactor for the transfer function V_m/I_j where V_m is the voltage at node m and I_j is a current source connected to node j. Apply this expression to the circuit of Fig. 9-37 (used in Example 9-17), choosing the bottom node as reference, and the node at left of L_1 as node 1 and that at right of L_2 as node 2. Find the following transfer functions: V_2/I_1 and V_1/I_2.

9-45 In the circuit of Fig. 9-37, insert a dependent current source in parallel with R_2. The current in the source is given by KV_1, where V_1 is the voltage at node 1. Recalculate the transfer functions V_2/I_1 and V_1/I_2 for this case.

Sec. 9-7 Network Theorems

9-46 Obtain the Thevenin equivalent of the circuit shown in Fig. 9-91 as seen from (a) the terminals of R_L and (b) the terminals of C.

Fig. 9-91 Circuit for Problems 9-46 and 9-51.

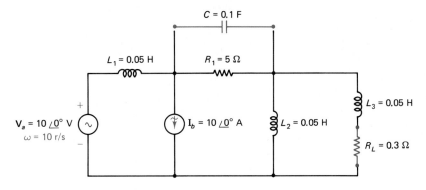

9-47 Determine the Thevenin equivalent of the circuit shown in Fig. 9-92 as seen from the terminals A-B.

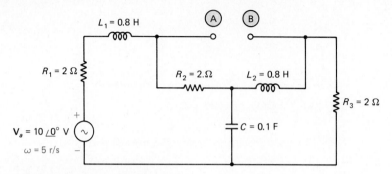

Fig. 9-92 Circuit for Problem 9-47.

9-48 Determine the Thevenin equivalent of each of the circuits shown in Fig. 9-93 as seen from (a) the terminal pair A-A' and (b) the terminal pair B-B'.

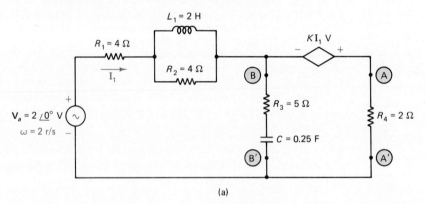

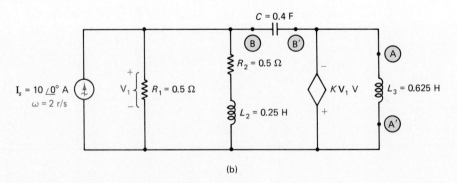

Fig. 9-93 Circuit for Problems 9-48 and 9-52.

9-49 Determine the Thevenin impedance of the circuit shown in Fig. 9-94 as seen from (a) the terminal pair A-A' and (b) the terminal pair B-B'.

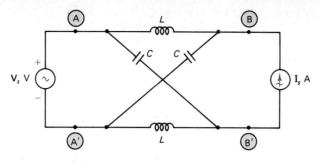

Fig. 9-94 Circuit for Problem 9-49.

9-50 Determine the Thevenin impedance of the circuit shown in Fig., 9-95 as seen from the terminal pair A-B.

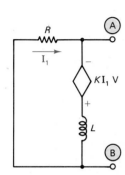

Fig. 9-95 Circuit for Problem 9-50.

9-51 Determine the Norton source current directly in the circuit of Problem 9-46 for each terminal pair specified.

9-52 Determine the Norton source current directly in the circuits of Problem 9-48.

9-53 Measurements are made at the terminals of a black box containing a network. The open circuit voltage is measured to be $120/45°$ V. When a load impedance $Z_L = (30 + j40)\Omega$ is connected to the terminals, the voltage across the load is measured to be $80/-30°$ V. Obtain the Thevenin and Norton equivalent circuits of the network in the black box.

9-54 Measurements are made at the terminals of a black box. When a load impedance of $(10 + j10)\Omega$ is connected to the terminals, the current in it is found to be $1.56/-81.3°$ A. When a load impedance of $(10 - j10)\Omega$ is connected to the terminals, the current in it is found to be $2/-66.9°$ A. Obtain the Thevenin and Norton equivalents of the network in the black box.

9-55 Determine the value of the load impedance for maximum power transfer to the load in the circuits of Problem 9-48 when the load is fully adjustable.

9-56 (a) Determine the values of R and X in the circuit shown in Fig. 9-96 to maximize the power dissipated in R. Calculate the power dissipated in R. (b) Suppose the angle of the load impedance is fixed so as to make $(X/R) = 0.5$. Redo part (a) for this case.

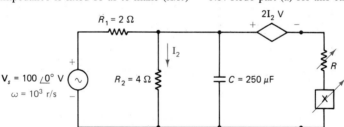

Fig. 9-96 Circuit for Problem 9-56.

9-57 Consider a load impedance whose magnitude is fixed, while its angle α can assume any value between $-90°$ and $+90°$. Obtain the relationship between the load impedance and the Thevenin impedance for maximum power transfer.

9-58 A series LC circuit is driven by a voltage $v_s(t) = V_0 + V_1 \cos(t/2\sqrt{LC}) + V_2\cos(2t/\sqrt{LC})$ V. Obtain an expression for the current $i(t)$.

9-59 A series RC circuit is driven by a voltage source $v_s(t) = V_0 + V_1 \cos(t/2RC) + V_2 \cos(t/RC) + V_3 \cos(2t/RC)$ V. Obtain expressions for the voltage $v_c(t)$ across the capacitor and $v_R(t)$ across the resistor.

Problems

9-60 A parallel RC circuit is driven by a current source $i_s(t) = I_0 + I_1 \cos(t/2RC) + I_2 \cos(t/RC) + I_3 \cos(2t/RC)$ A. Obtain expressions for the current $i_c(t)$ in the capacitor and $i_R(t)$ in the resistor.

9-61 The voltage v_s from the source in the circuit of Fig. 9-97 is given by $v_s(t) = 30 + 50 \cos 377t + 25 \cos 754t$ V. Determine the output voltage $v_o(t)$.

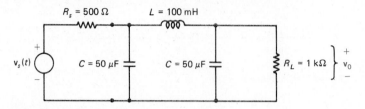

Fig. 9-97 Circuit for Problems 9-61 and 9-62.

9-62 Replace the voltage source in the circuit of the previous problem by a current source $i_s(t) = 10 + 10 \cos 500t + 10 \cos 1000t$ A. Determine the output voltage $v_o(t)$.

9-63 Determine the currents in all the branches of the circuit shown in Fig. 9-98.

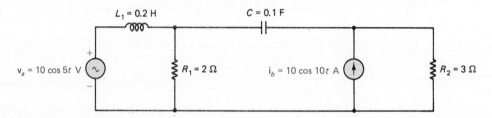

Fig. 9-98 Circuit for Problem 9-63.

9-64 Determine the voltage $V_c(t)$ across the capacitor in the circuit in Fig. 9-99.

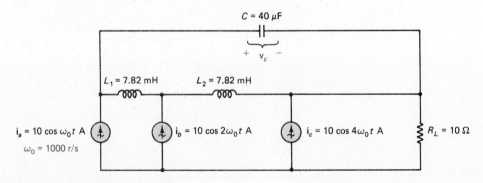

Fig. 9-99 Circuit for Problem 9-64.

9-65 Determine the output voltage of the operational amplifier circuit shown in Fig. 9-100 when $v_a(t) = 10 \cos 1000t$ V and $v_b = 20 \cos 2000t$ V. Use the approximate model of the op amp discussed in Chapter 3.

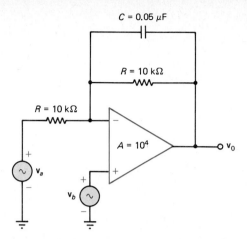

Fig. 9-100 Circuit for Problem 9-65.

9-66 The following data are available for the network in the box of Fig. 9-54 (Example 9-27 of the text). When a source $i_1(t) = 10e^{j50t}$ A is connected between node 1 and ground, the resulting node voltages are: $v_1(t) = 25e^{j(50t - 30°)}$ V, $v_2(t) = 4e^{j(50t + 90°)}$ V, and $v_3(t) = 10e^{j(50t + 60°)}$ V. When a current source $i_2(t)$ is connected to node 2, the voltage at node 2 is found to have twice the amplitude of i_2 and lags 40° behind $i_2(t)$, while the voltage at node 3 is found to have the same amplitude as $i_2(t)$ and is in phase with it. When a current source $i_3(t)$ is connected to node 3, the input impedance seen by it is measured to be $(30 + j40)\Omega$. Assume that only one source is active in each of the above sets of measurements and that all of them operate at $\omega = 50$ r/s. Determine the node voltages when the following three sources are simultaneously connected to the network: $i_1(t) = 20 \cos(50t + 45°)$ A, $i_2(t) = 30e^{j(50t - 75°)}$ A, and $i_3(t) = 25e^{-j50t}$ A.

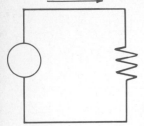

CHAPTER 10
Magnetic Circuits, Coupled Coils, and Three-Phase Circuits

The magnetic field set up by a current in a structure made up of magnetic materials plays an important part in many practical applications. For example, the magnetic recording of a signal (as in tape recorders) depends upon the conversion of a varying electrical signal to varying magnetization levels of the recording medium (the magnetic tape). Conversely, the reproduction of a signal from a magnetic tape depends upon the conversion of the varying levels of magnetization on the tape into varying voltage levels. The analysis of a structure of magnetic materials used in such systems is possible (to a certain degree of approximation) by using principles similar to those used in electric circuits, and such structures are therefore called *magnetic circuits*. In this chapter, we develop the principles of analysis of magnetic circuits.

When the magnetic field set up in a coil due to a time-varying current links the wires of another coil, there is a voltage induced in the latter coil. This is the phenomenon of *mutual induction*. The two coils are said to be *magnetically coupled*. Mutual induction is the underlying mechanism in transformer action. We discuss analysis of circuits with coupled coils and the principles of linear transformers in this chapter.

Three-phase networks are commonly used in electric machines, power systems, and networks. The balanced three-phase system has some special relationships because of the particular amplitude and phase angle distribution between the component voltages. We also treat balanced three-phase systems in this chapter.

10-1 BASIC RELATIONSHIPS OF MAGNETIC FIELDS

A detailed treatment of the basic relationships of magnetic fields can be found in any standard physics text. Here, we present just a brief overview.

Flux and Flux Density

The space surrounding a magnet or current-carrying conductor is the site of a magnetic field, and the magnetic field is represented by lines of induction, or magnetic flux lines. The unit of flux is the *weber* (abbreviated *Wb*). In the case of a conductor in which a magnetic field is set up by a uniformly wound coil, as indicated in Fig. 10-1(a), the direction of the flux lines is determined by using the *right-hand rule:*

> if the fingers of one's right hand are curled around the conductor in the direction of the current flow, then the flux is in the direction of the thumb.

For the sake of convenience, the flux is assumed to be uniformly distributed through the core of the coil. At the two extremities of the coil, however, there is a spreading of the flux lines, known as the *fringing effect,* which should be included in calculations involving the neighborhoods of the extremities. In a core in the form of a toroid (or doughnut), the flux lines are assumed to be uniformly distributed along concentric circles, as indicated in Fig. 10-1(b).

Fig. 10-1 Magnetic flux due to electric current. (a)-(b) Flux direction is determined by the right-hand rule. (c)-(d) Relationship between flux and flux density.

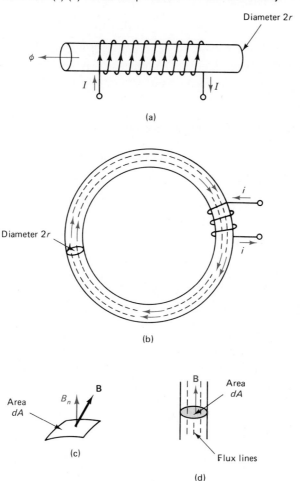

10-1   Basic Relationships of Magnetic Fields   473

The magnetic field is characterized by a vector **B** called the *flux density*. The unit of flux density is the *tesla* (abbreviated T). One tesla corresponds to one weber per square meter. The magnetic flux $d\phi$, shown in Fig. 10-1(c), passing through an elementary area dA of the magnetic field, is related to the flux density **B** by the equation

$$d\phi = B_n dA$$

where B_n is the component of the vector **B** perpendicular to the element dA. If the element dA is perpendicular to the flux density vector **B** as shown in Fig. 10-1(d), then

$$d\phi = B dA \qquad (10\text{-}1)$$

For example, the flux passing through any section in the middle of the core in Fig. 10-1(a) or the toroid in Fig. 10-1(b) is $B(\pi r^2)$, where B is the flux density in the core. One important problem is how to determine the flux when the current I is known, and vice versa.

Magnetic Field Intensity

For an isotropic material, a vector **H**, called the *magnetic field intensity*, is defined by means of the relationship

$$\mathbf{B} = \mu \mathbf{H} \qquad (10\text{-}2)$$

where μ is called the *magnetic permeability* of the material. The unit of magnetic field intensity is the *ampere per meter* (A/m), as we will see from Ampere's law. For a vacuum, the permeability equals the *permeability constant* μ_0:

$$\mu_0 = 4\pi \times 10^{-7} \text{ tesla-meter/ampere.}$$

The *relative permeability* of any material μ_r is defined by

$$\mu_r = (\mu/\mu_0) \qquad (10\text{-}3)$$

An alternative form of Eq. (10-2) is

$$\mathbf{B} = \mu_0 \mu_r \mathbf{H} \qquad (10\text{-}4)$$

The relative permeability of a ferromagnetic material (such as iron) is not a constant, but depends upon the flux density itself. In such cases, the graph showing B as a function of H (called the *magnetization curve*) is not a straight line, but is of the form shown in Fig. 10-1(e). We can see that, for a given magnetic intensity, the material with the higher permeability has the higher flux density. The relative permeability of materials other than ferromagnetic ones can be taken as 1.

Ampere's Law

Consider a magnetic field set up by a current I in a conductor. Let dl be an element of a closed path in the magnetic field, as shown in Fig. 10-2(a). Then $\mathbf{H}.d\mathbf{l}$ is the product of the component of the magnetic field intensity in the direction of the path and the length of the element dl. The line integral around the closed path is $\oint \mathbf{H}.d\mathbf{l}$.

Ampere's law states:

$$\oint \mathbf{H}.d\mathbf{l} = I$$

When the conductor carrying the current I is in the form of a coil of N turns, as in Fig. 10-2(b), then each turn of the coil contributes a current I to the equation for Ampere's

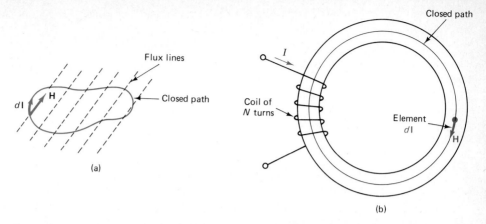

Fig. 10-2 Ampere's law. For coils with N turns as in (b), the product NI is used in Ampere's law.

law, and the product *NI* represents the number of *ampere turns* associated with the path of integration. Then Ampere's law becomes

$$\oint \mathbf{H} \cdot d\mathbf{l} = NI \tag{10-5}$$

The product (*NI*) is called the *magnetomotive force*, abbreviated *mmf* and measured in ampere turns (abbreviated *AT*). Equations (10-1) through (10-6) provide the tools for the solution of problems in magnetic circuits.

10-2 A BASIC MAGNETIC CIRCUIT

Consider a toroid with a ferromagnetic core, as shown in Fig. 10-3. A coil of *N* turns carrying a current *I* is wound closely and uniformly over the toroid. The flux lines are assumed to be uniformly distributed in the core in the form of concentric circles. Choosing a circular path of radius *R* in the middle of the core, the magnetic field intensity vector **H** is tangential to the path at all points, and the integral in Ampere's law becomes

$$\oint \mathbf{H} \cdot d\mathbf{l} = Hl$$

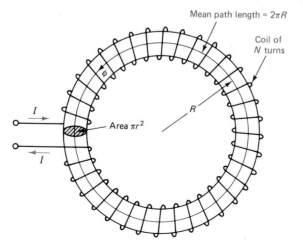

Fig. 10-3 A basic magnetic circuit—the toroidal core.

10-2   A Basic Magnetic Circuit

where l is the length of the circular path and equals $(2\pi R)$, and Ampere's law becomes

$$H(2\pi R) = NI$$

or

$$H = NI/2\pi R \tag{10-6}$$

If μ is the permeability of the core, then the flux density B in the core is, from Eq. (10-2),

$$B = \mu H = \frac{\mu NI}{2\pi R}$$

The flux ϕ in the core is, from Eq. (10-1),

$$\phi = B(\pi r^2)$$

which leads to

$$\phi = (\mu NI/2\pi R)\pi r^2$$
$$= \frac{\mu r^2}{2R}(NI) \tag{10-7}$$

Equation (10-7) permits the calculation of the magnetic flux due to a given *mmf NI* if the dimensions and composition of the toroid are known. Conversely, the mmf required to set up a specified magnetic flux in the toroid is also determined by using Eq. (10-7).

Example 10-1 Determine the magnetic flux set up in a toroidal core of mean radius 0.05 m and area of cross section 2×10^{-4} m² due to a current of 2.5 A in a coil of 2000 turns wound closely on the toroid. Assume that the core material has a relative permeability of 1.

Solution
$$mmf = NI = 5000 \text{ AT}$$

Choosing a circular path of radius $R = 0.05$ m,

$$\text{length of path } l = 2\pi R = 0.314 \text{ m}$$
$$\text{magnetic field intensity } H = (NI/l) = 15.9 \times 10^3 \text{ AT/m}$$
$$\text{flux density } B = \mu_0 H = (4\pi \times 10^{-7})(15.9 \times 10^3) = 0.02 \text{ T}$$
$$\text{flux } \phi = B \times \text{area of cross section} = 0.02 \times 2 \times 10^{-4}$$
$$= 4 \times 10^{-6} \text{ Wb}$$

Exercise 10-1 Calculate the flux in the toroid of the example above if the core material has a relative permeability of $\mu_r = 2400$.

Exercise 10-2 Calculate the current needed to establish a flux of 20×10^{-6} Wb in the core of Example 10-1 (with the relative permeability of the core equal to 1).

Exercise 10-3 For a given *mmf* in the core, how will each of these changes affect the flux established in the core: (a) The radius of the toroid is doubled; (b) the area of cross section of the toroid is doubled; (c) the relative permeability is doubled.

10-2-1 Reluctance of a Magnetic Circuit

In the case of a magnetic circuit with a core permeability μ through the entire core, a magnetic path of length l, area of cross section A, and an *mmf* of NI AT,

$$\text{intensity of the magnetic field } H = (NI/l)$$

$$\text{flux density } B = \mu H = (\mu/l)(NI)$$

and

$$\text{flux } \phi = BA = \frac{\mu A}{l}(NI)$$

or

$$(NI) = \left(\frac{l}{\mu A}\right)\phi \qquad (10\text{-}8)$$

It is possible to create an analogy between a magnetic circuit and an electric circuit by thinking of the magnetic flux being analogous to electric current and the magnetomotive force (NI) as analogous to electromotive force (or voltage). The relationship between the *mmf* and flux, Eq. (10-8), is seen to be similar to Ohm's law for a resistance. Consequently, Eq. (10-8) is called the Ohm's law of *magnetic circuits*.

The quantity $(l/\mu A)$ in Eq. (10-8) plays the same part as resistance in an electric circuit and is called the *reluctance* of the magnetic circuit.

$$\text{reluctance} = (l/\mu A) \qquad (10\text{-}9)$$

Then Eq. (10-8) states that

$$mmf = (\text{reluctance})(\text{flux})$$

Example 10-2 Determine the reluctance of the magnetic circuit shown in Fig. 10-4.

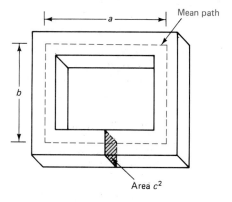

Fig. 10-4 Magnetic circuit for Example 10-2.

Solution Assume that the flux in the core is uniformly distributed so that the mean magnetic path can be taken as that shown by the dashed lines. Then the length of the path is

$$l = 2(a + b)$$

and the area of cross section of the core is

$$A = c^2$$

Therefore

$$\text{reluctance} = \frac{l}{\mu A} = \frac{2(a + b)}{\mu c^2}$$

∎

Exercise 10-4 Calculate the reluctance of the toroid in Example 10-1 and the toroid in Exercise 10-1.

Exercise 10-5 If the dimensions of the core in Example 10-2 are: $a = 0.10$ m, $b = 0.08$ m, $c = 0.02$ m, and the relative permeability of the core is $\mu_r = 2500$, calculate (a) the flux established by a coil of 500 turns carrying a current of 250 mA; and (b) the current in a 500-turn coil needed to establish a flux of 0.05 Wb.

10-3 PRINCIPLES OF ANALYSIS OF MAGNETIC CIRCUITS

The magnetic circuits considered in the previous examples were simple circuits with a single path of the same material. Magnetic circuits of practical interest usually consist of more than one path or have cores made up of two or more materials, or a combination of both. The analysis procedure is similar to that used for electric circuits, except that instead of Kirchhoff's voltage law, Ampere's law is used.

Consider the magnetic circuit shown in Fig. 10-5. Even though the flux is the same through the circuit, the flux densities are different in different sections due to the different cross-sectional areas. The values of the magnetic intensity H are therefore also different in the different sections. In such a case, the integral in Ampere's law, $\oint \mathbf{H} \cdot d\mathbf{l}$, is replaced by a summation of the products of H and path lengths of the different sections. The

Fig. 10-5 Magnetic circuit in which the integral of Ampere's law is replaced by a summation of (Hl) products.

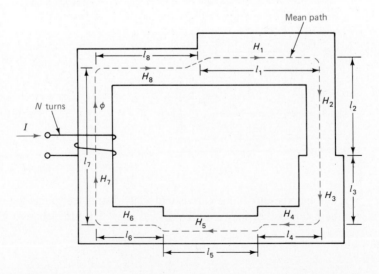

Ampere's law equation for the circuit in Fig. 10-5 is

$$H_1 l_1 + H_2 l_2 + \ldots + H_n l_n = NI \tag{10-10}$$

which is analogous to the *KVL* equation of a series resistive circuit.

For a general magnetic circuit, equations similar to Eq. (10-10) are set up in a manner similar to the *KVL* equations of an electric circuit. These equations are solved in conjunction with equations of the form $B = \mu H$ in order to determine the flux in the core due to a given *mmf* or the *mmf* needed to establish a specified flux. There are some serious difficulties in carrying out this task, however. The relationship

$$B = \mu H$$

might lead one to think that the permeability μ of a material is a constant and the graph of B versus H is a straight line. Unfortunately, the B versus H curve of a ferromagnetic material is not a straight line, as is evident from the typical curve shown in Fig. 10-6.

In problems where the flux in the circuit is specified and one needs to find the *mmf* needed to establish the flux, the magnetization curve is used to find the value of the magnetic intensity H corresponding to the known value of flux density B. The solution of equations like Eq. (10-10) is fairly straightforward in such cases.

The reverse of this type of problem—that is, where the *mmf* is known and the flux is to be determined—is more complicated. In some cases, it is possible to draw a load line (similar to that used in electric circuits with a nonlinear element) and solve the problem graphically. In other cases, it is necessary to use trial and error.

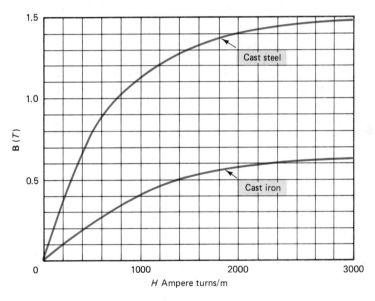

Fig. 10-6 *B-H* curves of cast steel and cast iron.

10-3-1 Determination of *mmf* for Specified Flux

The method of determination of the *mmf* needed to establish a specified flux in a magnetic circuit is illustrated in the following example.

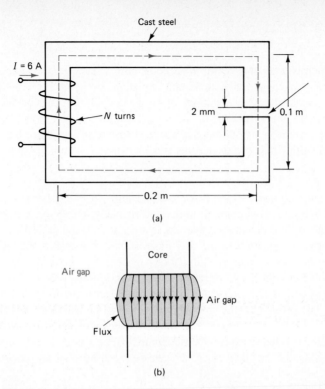

Fig. 10-7 (a) Magnetic circuit with an air gap (Example 10-3). (b) Fringing of flux at the air gap.

Example 10-3 Determine the number of turns of coil needed in the magnetic circuit of Fig. 10-7(a) in order to establish a flux of $\phi = 5$ mWb in the air gap. The core material is cast steel. The area of the cross section equals $49 \times 10^{-4} \text{m}^2$.

Solution When magnetic flux passes through an air gap, there is a tendency for the lines to diverge, as shown in Fig. 10-7(b). This fringing increases the effective cross-sectional area of the air gap. The fringing effect is ignored when the width of the air gap is very small compared with the length of the path in the ferromagnetic material.

The flux has the same value in the ferromagnetic core and in the air gap. Ignoring the fringing effect, the flux density also has the same value in the ferromagnetic core and the air gap.

$$B = (\phi/A) = (5 \times 10^{-3}/49 \times 10^{-4}) = 1.02 \text{ T}$$

If the intensity of the magnetic field in the core is denoted by H_c and that in the air gap by H_a, then Ampere's law leads to

$$H_c l_c + H_a l_a = NI \qquad (10\text{-}11)$$

where

$$l_c = \text{length of the path in the core} = 0.6 \text{ m}$$

and

$$l_a = \text{length of the air gap} = 0.002 \text{ m}$$

so that Eq. (10-11) becomes

$$0.6H_c + 0.002H_a = NI \qquad (10\text{-}12)$$

Using the magnetization curve of cast steel from Fig. 10-6, the intensity in the core H_c, when $B = 1.02$ T, is found to be

$$H_c = 800 \text{ AT/m}$$

Since the permeability of air is μ_0, the intensity in the air gap, H_a, is found from the equation (using $\mu_0 = 4\pi \times 10^{-7}$):

$$H_a = (B/\mu_0) = 8.12 \times 10^5 \text{ AT/m}$$

Therefore, Eq. (10-12) leads to

$$NI = 2104 \text{ AT}$$

and, since $I = 6$ A, $N = 351$ turns.

Note that the portion of the *mmf* in the ferromagnetic core is $H_c l_c = (0.6 \times 800) = 480$ AT, while that in the air gap is $H_a l_a = (0.002 \times 8.12 \times 10^5) = 1624$ AT. The *mmf* drop in the air gap is considerably larger than that in the ferromagnetic core, even though the length of the magnetic path in the air gap is much smaller than that in the ferromagnetic material. This is due to the fact that the reluctance of an air gap is much higher than that of a ferromagnetic core. ∎

Exercise 10-6 Suppose the air gap width is 2.5 mm in the core of Exercise 10-5. Recalculate the results of that example.

Exercise 10-7 Introduce an air gap of width 3 mm in the core of Exercise 10-1. Calculate the mmf needed for a flux of 9.6 mWb.

Exercise 10-8 Repeat Exercise 10-7 with the air gap width reduced to half its previous value.

10-3-2 Determination of Flux for Specified *mmf*

The examples and exercises in Sec. 10-3-1 dealt with the determination of the *mmf* needed to establish a specified flux in a magnetic circuit with different values of H in different sections. Now consider the reverse problem: Given a magnetic circuit in which the *mmf* is known and the flux has to be determined, Ampere's law equations similar to Eq. (10-10) are set up. They need to be solved for the different intensities H in those equations. Suppose the structure is such that the flux is ϕ throughout the circuit. The flux densities in different sections should first be expressed in terms of ϕ and the areas of the sections. But since B is not linearly related to H, it is not easy to introduce the flux densities into the Ampere's law equations.

Graphic Analysis Using Load Line

Consider the magnetic circuit set up by a ferromagnetic toroid with an air gap of width l_a. Let the length of the path in the ferromagnetic core be l_c. If the fringing in the air gap is ignored, then the flux ϕ in the circuit and the flux density B are the same throughout the circuit. If H_c is the intensity in the ferromagnetic core with a path length l_c, and H_a is the intensity of the air gap, then the Ampere's law equation is

$$H_c l_c + H_a l_a = NI \tag{10-13}$$

where NI is the *mmf* due to a current I flowing through N turns of a coil. Using $H_a = (B/\mu_0)$, Eq. (10-13) becomes

$$(B/\mu_0)l_a + H_c l_c = NI$$

or

$$B = \frac{NI\mu_0}{l_a} - \frac{\mu_0 l_c}{l_a} H_c \tag{10-14}$$

Equation (10-14) represents a straight line on a two-dimensional plane with B as ordinate and H_c as abscissa. The straight line may be called the *load line* in analogy with the load line in electric circuits.

The load line intersects the horizontal H_c axis at (NI/l_c) and the vertical B axis at $(NI\mu_0/l_a)$, and its slope is $-(\mu_0 l_c/l_a)$.

The load line is shown in Fig. 10-8. The intersection of the magnetization curve and the load line gives the value of B and H_c that will satisfy the load line equation and the magnetization curve of the material of the core.

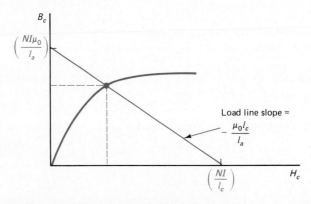

Fig. 10-8 Graphic analysis of a magnetic circuit.

Example 10-4 A magnetic circuit of the form shown in Fig. 10-7 has the following data: $l_c = 1.2$ m; $l_a = 0.001$ m; area of cross section $= 0.01$ m². $NI = 1000$ AT. Determine the flux in the core. The core material is cast steel.

Solution The equation of the load line is

$$B = (\mu_0 NI/l_a) - (\mu_0 l_c/l_a) H_c$$
$$= 1.26 - 0.0015 H_c$$

The load line intersects the H axis at $(1.26/0.0015) = 840$ AT/meter and the B axis at 1.26 T, as shown in Fig. 10-9. The intersection of the magnetization curve and the load line is found to be: $H = 420$ AT/m and $B = 0.65$ T. The flux density in the core is therefore 0.65 T, and the flux is 6.5 mWb. ∎

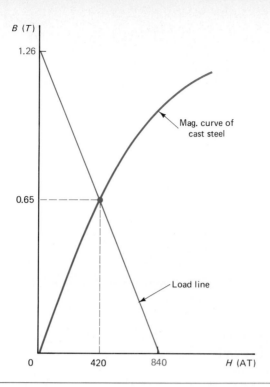

Fig. 10-9 Solution for Example 10-4.

Exercise 10-9 Determine the flux established in the magnetic circuit of the previous example for an *mmf* of 1200 AT.

Exercise 10-10 Return to Example 10-4, and change the air gap width to 0.002 m. Recalculate the flux in the circuit.

Trial and Error Procedure for the Determination of Flux

When the magnetic circuit contains two or more nonlinear components (due to two or more different ferromagnetic materials), a load line cannot be drawn, and the problem must be solved by trial and error. A trial value of the flux is selected first, the flux density is calculated, the values of H in the different sections are found from the magnetization curve, and then the Ampere's law equation is checked. If the equation is not satisfied, a different trial value of flux is selected and the process repeated until the Ampere's law equation is satisfied within a specified percentage of error (e.g., 5 percent). Even though there is no prescribed rule for the initial trial value of the flux, it is a good idea to assume initially that the total applied *mmf* is across the portion with the highest reluctance (or lowest permeability). In a magnetic circuit with an air gap, for example, the *mmf* is initially assumed to be across the air gap. The procedure is illustrated in the following example (10-5).

Example 10-5 Determine the flux (within an error of 5 percent in the Ampere's law equation) set up by 1000 AT in the core of the magnetic circuit shown in Fig. 10-10. Use the magnetization curves given in Fig. 10-6 for the two materials.

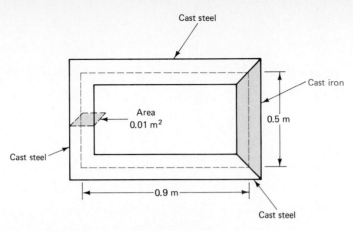

Fig. 10-10 Magnetic circuit for Example 10-5.

Solution Since there are two nonlinear components in the circuit due to the two different ferromagnetic materials, a load line cannot be used, and it is necessary to use a trial and error procedure.

path length in cast steel: $l_1 = 2.3$ m
path length in cast iron: $l_2 = 0.5$ m
Ampere's law equation: $H_1 l_1 + H_2 l_2 = NI$

or

$$2.3H_1 + 0.5H_2 = 1000 \tag{10-15}$$

Trial 1: Assume that the *mmf* of 1000 AT is across the portion made up of cast iron, since it has the lower permeability. Then $H_2 l_2 = 1000$, or $H_2 = 2000$ AT/m. From the magnetization curve of material II, $B_2 = 0.58$ T. Since the flux and flux density are the same in both sections of the circuit, $B_1 = 0.58$ T and, from the magnetization curve for cast steel, $H_1 = 350$ AT/m. Then the left side of Eq. (10-15) becomes

$$2.3H_1 + 0.5H_2 = 2.3(350) + 0.5(2000) = 1805$$

which is quite a bit higher than the value of 1000 needed to satisfy Eq. (10-15).

Trial 2: Try a value of H_2 smaller than the previous value of 2000. Let $H_2 = 1400$ AT/m for cast iron. Then $B_2 = 0.51$ T from the curve for cast iron. $B_1 = B_2 = 0.51$ T. Therefore, $H_1 = 300$ AT/m from the curve for cast steel. The left side of Eq. (10-15) becomes

$$2.3(300) + 0.5(1400) = 1390$$

and Eq. (10-15) is not satisfied.

Trial 3: Try a value of $H_2 = 950$ AT/m. Then $B_2 = B_1 = 0.38$ T, and $H_1 = 220$ AT/m. The left side of Eq. (10-15) adds up to 981, which is only 1.9 percent less than the desired value of 1000. Therefore, the final approximate value of flux density is $B = 0.38$ T, and the flux is 3.8 mWb. ∎

Exercise 10-11 Switch the two materials in the magnetic circuit of the previous example and recalculate the flux.

Exercise 10-12 Make the length of each section in the magnetic circuit of the previous example 1.4 m and calculate the new value of flux.

10-4 MAGNETICALLY COUPLED COILS

In Chapter 6, we discussed the phenomenon of self-induction and introduced the coefficient of self-induction (or inductance). In self-induction, we are interested in the voltage induced in a coil due to a time-varying current in that coil. There is also an induced voltage in one coil when there is a time-varying current in another coil.

Consider two coils (coil 1 and coil 2) placed near each other. If there is a current in coil 1, then the magnetic flux surrounding it also links the turns of coil 2. If the current in coil 1 varies with time, then the time-varying magnetic flux induces a voltage in coil 2. Similarly, if there is a time-varying current in coil 2, it induces a voltage in coil 1. This phenomenon is called *mutual induction:*

> A time varying current in one coil induces a voltage in another coil when there is a linking of magnetic flux between the coils. Just as in the case of self-induced voltage, mutually induced voltage is proportional to the *rate of change of magnetic flux* causing it.

That is, if $i_1(t)$ is the current in coil 1, as Fig. 10-11(a) shows, then the voltage $v_2(t)$

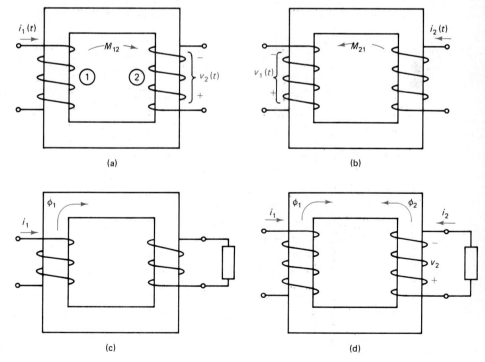

Fig. 10-11 Mutually induced voltages in coupled coils. (a) Current in coil 1 induces a voltage in coil 2. (b) Current in coil 2 induces a voltage in coil 1. (c)-(d) Determination of the polarities of the induced voltage v_2.

induced in coil 2 due to the current in coil 1 is proportional to $di_1(t)/dt$, and the relationship for v_2 is given by

$$v_2(t) = M_{21}\frac{di_1(t)}{dt} \tag{10-16}$$

where M_{21} is called the coefficient of mutual induction (or simply mutual inductance) of coil 2 with respect to coil 1.

Similarly, the voltage v_1 induced in coil 1 due to a current in coil 2, as Fig. 10-11(b) shows, is given by

$$v_1(t) = M_{12}\frac{di_2(t)}{dt} \tag{10-17}$$

where M_{12} is the mutual inductance of coil 1 with respect to coil 2. We will see in Sec. 10-6, on energy in coupled coils, that

$$M_{12} = M_{21}$$

Therefore, a single symbol M is used for both M_{12} and M_{21} so that Eqs. (10-16) and (10-17) become, respectively,

$$v_2(t) = M\frac{di_1(t)}{dt} \tag{10-18}$$

and

$$v_1(t) = M\frac{di_2(t)}{dt} \tag{10-19}$$

The unit of mutual inductance is the *henry*, the same as for self-inductance L.

10-4-1 Polarity of Mutually Induced Voltages

The polarity of the mutually induced voltages in a pair of coupled coils is determined by using the principle that the induced voltage must be in such a direction as to *oppose* the current that causes it. (This is similar to Lenz's law for self-induction).

Consider a current i_1 in coil 1 inducing a voltage v_2 in coil 2. Suppose there is a closed path on the side of coil 2, as indicated in Fig. 10-11(c). The induced voltage v_2 should generate a current i_2 in coil 2 in such a direction as to set up a flux ϕ_2 *opposing* the flux ϕ_1 set up by i_1. Therefore, we assign a direction to i_2 by taking a flux direction opposing the flux due to i_1 and using the right-hand rule. This gives the direction of i_2 shown in Fig. 10-11(d). Since coil 2 acts as a *voltage source*, the current i_2 must be directed from the negative reference terminal to the positive reference terminal of v_2 in coil 2; this determines the direction of v_2.

Therefore, the polarity of v_2 is chosen as follows:

> Determine the direction of a current i_2 in coil 2 needed to set up a flux that will oppose the flux due to i_1. Place the positive polarity mark of v_2 on the side of the coil through which i_2 *leaves* coil 2.

A similar set of steps leads to the determination of the polarity of v_1 induced in coil 1 due to a current in coil 2.

The procedure requires a knowledge of the winding directions of the coils on a core, which is not always convenient to show in circuit diagrams. A more convenient notation, called the *dot polarity notation*, is commonly used and will be presented in Sec. 10-4-4.

Equations in the Sinusoidal Steady State

In the *sinusoidal steady state*, the derivative d/dt is replaced by $j\omega$, so that Eqs. (10-18) and (10-19) can be rewritten in the form

$$\mathbf{V}_2 = j\omega M \mathbf{I}_1$$

and

$$\mathbf{V}_1 = j\omega M \mathbf{I}_2$$

where $\mathbf{V}_1$, $\mathbf{V}_2$, $\mathbf{I}_1$, and $\mathbf{I}_2$ are phasors.

10-4-2 Coefficient of Coupling

A parameter called the *coefficient of coupling*, denoted by k, is defined for a pair of coupled coils:

$$k = \frac{M}{\sqrt{L_1 L_2}} \qquad (10\text{-}20)$$

The value of k lies in the range $0 \leq k \leq 1$. (The proof of this statement will be presented in Sec. 10-6.) The coupling coefficient is a measure of the tightness of coupling: The closer k is to 1, the tighter the coupling. Note that, since $k \leq 1$, Eq. (10-22) leads to the inequality

$$M^2 \leq L_1 L_2 \qquad (10\text{-}21)$$

Exercise 10-13 Calculate the mutual inductance between two coils whose inductances are 10 mH and 25 mH when the coefficient of coupling is (a) 0.1, (b) 0.5, and (c) 0.9. What is the maximum value of M that can be attained for the two coils?

10-4-3 Voltage Components in Coupled Coils

Consider now the situation in Fig. 10-12(a) where both coils are carrying current. There are two components of voltage in coil 1: a self-induced voltage $j\omega L_1 \mathbf{I}_1$ due to the current in itself, and a mutually induced voltage $j\omega M \mathbf{I}_2$ due to the current in the other coil. Depending upon the relative directions of the windings of the two coils and the magnetic flux, the two components may be additive or subtractive. That is, the total voltage across coil 1 is of the form

$$\mathbf{V}_1 = j\omega L_1 \mathbf{I}_1 \pm j\omega M \mathbf{I}_2 \qquad (10\text{-}22)$$

where a choice has to be made for the sign $\pm$. Similarly, the total voltage across coil 2 is

$$\mathbf{V}_2 = j\omega L_2 \mathbf{I}_2 \pm j\omega M \mathbf{I}_1 \qquad (10\text{-}23)$$

Suppose the relative winding directions of the two coils are as shown in Fig. 10-12(b). Consider the flux set up in the core by each of the currents $\mathbf{I}_1$ and $\mathbf{I}_2$ acting separately: ϕ_1 due to $\mathbf{I}_1$ and ϕ_2 due to $\mathbf{I}_2$. The two fluxes are in the *same direction* through the core

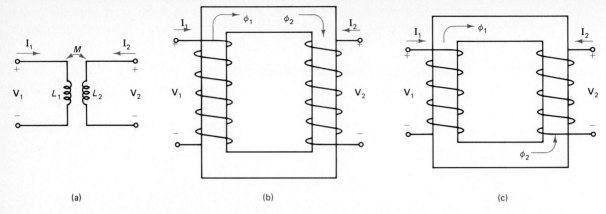

Fig. 10-12 (a)-(c): Determination of the sign of the mutually induced voltage term in *KVL* equations of coupled coils.

and are *additive*. In such a case, the *mutually induced voltage and self-induced voltage are also additive*. The $\pm$ signs in Eqs. (10-22) and (10-23) are therefore replaced by plus signs.

On the other hand, in the situation shown in Fig. 10-12(c), the fluxes due to the two currents are in *opposite directions* and are *subtractive*. In this case, the *mutually induced voltage and self-induced voltage are also subtractive*. The $\pm$ signs in Eqs. (10-22) and (10-23) are therefore replaced by minus signs.

If the winding directions are known, it is therefore possible to determine whether the self-induced and mutually induced voltage components should be added (if the fluxes aid each other) or subtracted (if the fluxes oppose each other).

Example 10-6 Let the currents in the coils shown in Fig. 10-12(b) be given by $i_1(t) = K_1 e^{-pt}$ and $i_2(t) = K_2(1 - e^{-pt})$ where K_1, K_2, and p are constants. Obtain the expressions for $v_1(t)$ and $v_2(t)$.

Solution Since the given currents are not sinusoidal, phasors cannot be used. The fluxes due to $i_1(t)$ and $i_2(t)$ in Fig. 10-12(b) *aid* each other; therefore, the self-inductance term and mutual inductance term in each coil are *added*.

$$v_1(t) = L_1 \frac{di_1(t)}{dt} + M\frac{di_2(t)}{dt}$$
$$= -pL_1 K_1 e^{-pt} + pMK_2 e^{-pt}$$
$$= -p(L_1 K_1 - MK_2)e^{-pt}$$

Similarly

$$v_2(t) = -p(L_2 K_2 - MK_1)e^{-pt} \qquad \blacksquare$$

Example 10-7 Let the currents in the coils shown in Fig. 10-12(c) be given by $i_1(t) = 100 \cos 377t$ A and $i_2(t) = 50 \cos (377t + 45°)$ A. If $L_1 = 20$ mH, $L_2 = 50$ mH, and $M = 30$ mH, determine the voltages $v_1(t)$ and $v_2(t)$ across the two coils.

Solution The problem may be solved either directly in the time domain or by using phasors.

Using phasors:

$$\mathbf{I}_1 = 100\underline{/0°} \text{ A} \qquad \mathbf{I}_2 = 50\underline{/45°} \text{ A} \qquad \text{and} \qquad \omega = 377 \text{ r/s}$$

Since the fluxes due to the two currents in Fig. 10-12(c) oppose each other, the mutual inductance term is subtracted from the self-inductance term. Therefore

$$V_1 = j\omega L_1 I_1 - j\omega M I_2 = j754 - j566\underline{/45°}$$
$$= (400 + j354) = 534\underline{/41.5°} \text{ V}$$

and

$$V_2 = j\omega l_2 I_2 - j\omega M I_1 = j942\underline{/45°} - j1131$$
$$= -666 - j465 = 812\underline{/-145°} \text{ V}$$

The voltages in the two coils are therefore

$$v_1(t) = 534 \cos(377t + 41.5°) \text{ V}$$

and

$$v_2(t) = 812 \cos(377t - 145°) \text{ V}$$

■

Exercise 10-14 If the currents in the two coils in Fig. 10-12(b) are $i_1(t) = (K_1 t + K_2)$ A and $i_2(t) = -(1/t)$ A, obtain the expressions for the voltages across the coils.

Exercise 10-15 Verify the results of Example 10-7 by working directly in the time domain.

Exercise 10-16 Two coupled coils have these data: $L_1 = 0.1$ H, $L_2 = 0.4$ H, $M = 0.15$ H (coupling additive), $\omega = 100$ r/s. When $V_1 = 1000\underline{/0°}$ V, $I_1 = 50\underline{/36.9°}$ A. Calculate V_2.

10-4-4 Dot Polarity Notation

In order to determine whether the self- and mutual inductance terms should be added or subtracted, the relative winding directions must be known. Showing the winding directions in circuit diagrams, however, is not always convenient, so a special notation called the *dot polarity convention* is used instead.

Consider the two coils shown in Fig. 10-13(a). A dot mark is placed on each coil, as follows:

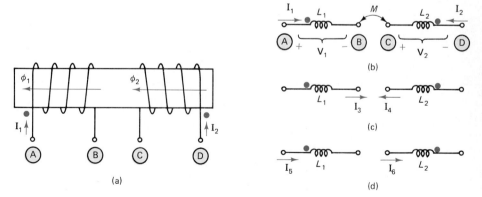

Fig. 10-13 Dot polarity notation. (a) Currents entering the dotted terminals set up an additive flux in the core. (b)-(c) Mutual inductance term adds to the self-inductance term if currents enter similarly marked terminals. (d) Mutual inductance term subtracts from the self-inductance term if currents enter dissimilarly marked terminals.

Suppose currents are chosen through the two coils such that the fluxes due to the two currents *add*. A dot mark is placed at the points where the currents *enter* the two coils. Once the dots have been assigned, the winding directions need not be shown, and the coils are shown instead with the dot assignments as indicated in Fig. 10-13(b). Even though currents were chosen in a particular direction so as to assign dots, it does not mean that, in analyzing a given circuit, current directions must always be chosen in the same manner. Currents are chosen in any arbitrary manner, and the following rules are used in writing the equations.

If the currents in two coupled coils are chosen (or assigned) so that they *both enter the dotted terminals*, as in Fig. 10-13(b), then the dot convention states that the fluxes due to them will aid each other. The *mutual inductance term is therefore added to the self-inductance term* in this case.

The same rule applies if the currents are chosen so that they *both leave* the dotted terminals, as in Fig. 10-13(c). The voltages across the two coils in Fig. 10-13(b) or (c) are given by:

$$V_1 = j\omega L_1 I_1 + j\omega M I_2$$
$$V_2 = j\omega L_2 I_2 + j\omega M I_1$$

If the currents in two coupled coils are chosen (or assigned) so that *one enters the dotted terminal* of a coil and the *other enters the undotted terminal* of the other coil, the fluxes will oppose each other. The *mutual inductance term is subtracted from the self-inductance term* in this case.

The voltages across the two coils in Fig. 10-13(d) are given by:

$$V_1 = j\omega L_1 I_1 - j\omega M I_2$$
$$V_2 = j\omega L_2 I_2 - j\omega M I_1$$

Exercise 10-17 Place a dot at the appropriate terminal of coil 2 in each arrangement of Fig. 10-14.

Fig. 10-14 (a)-(b) Coils for Exercise 10-17

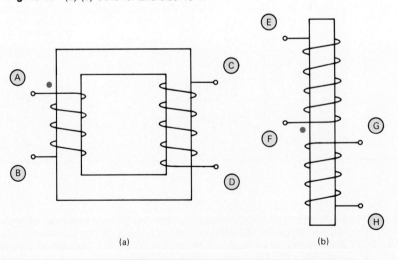

490 Magnetic Circuits, Coupled Coils, and Three-Phase Circuits

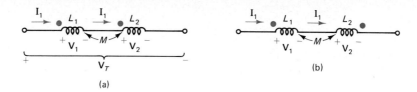

Fig. 10-15 Coupled coils in series. The two connections lead to a determination of M.

10-4-5 Coupled Coils in Series

Consider the series connection of two coupled coils as shown in Fig. 10-15(a). Assuming sinusoidal steady state, the voltage $\mathbf{V}_T$ across the series combination is given by

$$\mathbf{V}_T = \mathbf{V}_1 + \mathbf{V}_2$$
$$= (j\omega L_1 \mathbf{I}_1 + j\omega M \mathbf{I}_1) + (j\omega L_2 \mathbf{I}_1 + j\omega M \mathbf{I}_1)$$
$$= j\omega(L_1 + L_2 + 2M)\mathbf{I}_1 \tag{10-24}$$

If the series combination of the two coupled coils in Fig. 10-15(a) is replaced by a single equivalent inductance L_{eq}, so that

$$\mathbf{V}_T = j\omega L_{eq} \mathbf{I}_1 \tag{10-25}$$

then, by comparing Eqs. (10-24) and (10-25),

$$L_{eq} = (L_1 + L_2 + 2M) \tag{10-26}$$

Now suppose one of the coils is reversed, leading to the situation shown in Fig. 10-15(b). Then the voltage $\mathbf{V}_T'$ is given by

$$\mathbf{V}_T' = j\omega(L_1 + L_2 - 2M)\mathbf{I}_1$$

and the equivalent inductance for this connection is

$$L_{eq}' = (L_1 + L_2 - 2M) \tag{10-27}$$

Subtracting Eq. (10-27) from (10-26)

$$L_{eq} - L_{eq}' = 4M \quad \text{or} \quad M = (L_{eq} - L_{eq}')/4 \tag{10-28}$$

Thus the mutual inductance between two coils can be determined by measuring the equivalent inductances L_{eq} and L_{eq}'.

Exercise 10-18 The total inductance of two coils is measured to be 13 mH. If one of the coils is reversed, the total inductance is found to be 18 mH. If the inductance L_1 of one coil is known to be 5 mH, calculate the inductance of the other coil, the mutual inductance, and the coefficient of coupling between the two coils.

10-4-6 Coupled Coils in Parallel

Consider two coils connected in parallel, as shown in Fig. 10-16. Choose two *branch currents* $\mathbf{I}_1$ and $\mathbf{I}_2$ and two closed paths, as shown. Since $\mathbf{V}_T$ is the voltage across L_1,

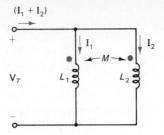

Fig. 10-16 Coupled coils in parallel.

$$\mathbf{V}_T = j\omega L_1 \mathbf{I}_1 + j\omega M \mathbf{I}_2 \qquad (10\text{-}29)$$

Since $\mathbf{V}_T$ is also the voltage across L_2,

$$\mathbf{V}_T = j\omega M \mathbf{I}_1 + j\omega L_2 \mathbf{I}_2 \qquad (10\text{-}30)$$

Solving Eqs. (10-29) and (10-30)

$$\mathbf{I}_1 = \frac{j(L_2 - M)\mathbf{V}_T}{\omega(M^2 - L_1 L_2)} \qquad (10\text{-}31)$$

and

$$\mathbf{I}_2 = \frac{j(L_1 - M)\mathbf{V}_T}{\omega(M^2 - L_1 L_2)} \qquad (10\text{-}32)$$

The total current supplied by $\mathbf{V}_T$ is

$$\mathbf{I}_T = \mathbf{I}_1 + \mathbf{I}_2 = \frac{j\mathbf{V}_T(L_1 + L_2 - 2M)}{\omega(M^2 - L_1 L_2)} \qquad (10\text{-}33)$$

and the equivalent inductance is

$$L_{eq} = \frac{\mathbf{V}_T}{j\omega \mathbf{I}_T} = \frac{L_1 L_2 - M^2}{L_1 + L_2 - 2M} \qquad (10\text{-}34)$$

Exercise 10-19 Reverse the coil L_2 in Fig. 10-16. Obtain the expression for the new equivalent inductance.

10-5 ANALYSIS OF CIRCUITS WITH COUPLED COILS

Loop or mesh analysis is applied to circuits with coupled coils using the same principles as any other circuit, with the only difference being that there are *two voltage components* to be included for each coil and that the addition or subtraction of the mutual inductance term is decided by means of the dot polarity notation.

Consider, for example, the circuit shown in Fig. 10-17. Mesh currents are assigned as shown. Note that the current $\mathbf{I}_1$ enters the dotted terminal of coil L_1 while $\mathbf{I}_2$ leaves the dotted terminal of coil L_2. The mesh equations are

Mesh 1: $\qquad R_1 \mathbf{I}_1 + (j\omega L_1 \mathbf{I}_1 - j\omega M \mathbf{I}_2) = \mathbf{V}_s$

or

$$(R_1 + j\omega L_1)\mathbf{I}_1 - j\omega M \mathbf{I}_2 = \mathbf{V}_s \qquad (10\text{-}35)$$

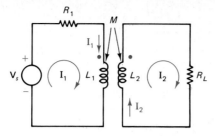

Fig. 10-17 Mesh analysis of circuit with coupled coils.

Mesh 2: $R_L \mathbf{I}_2 + (j\omega L_2 \mathbf{I}_2 - j\omega M \mathbf{I}_1) = 0$

or

$$-j\omega M \mathbf{I}_1 + (R_L + j\omega L_2)\mathbf{I}_2 = 0 \tag{10-36}$$

Eqs. (10-35) and (10-36) are the mesh equations of the given circuit and can be solved for $\mathbf{I}_1$ and $\mathbf{I}_2$.

Note that, in the example above, the two voltage components across each coil were written together at first and an additional step was taken to collect the coefficients. This approach is recommended to avoid possible errors in sign for mutual inductance terms in KVL equations of circuits with coupled coils.

Example 10-8 Write the mesh equations for the circuit of Fig. 10-18(a).

Solution It is convenient to identify the *branch current* through each inductance (in terms of the mesh currents). This makes it easier to determine the sign of the mutual inductance terms for the coils.

Redraw the diagram as shown in Fig. 10-18(b), where the current through L_1 is taken as $(\mathbf{I}_1 - \mathbf{I}_2)$ entering the dotted terminal and the current in L_2 is $\mathbf{I}_2$ entering the undotted terminal. Then the voltage across coil L_1 is

$$\begin{aligned}\mathbf{V}_1 &= j\omega L_1(\mathbf{I}_1 - \mathbf{I}_2) - j\omega M \mathbf{I}_2 \\ &= j\omega L_1 \mathbf{I}_1 - j\omega(L_1 + M)\mathbf{I}_2\end{aligned} \tag{10-37}$$

The voltage across coil L_2 is

$$\begin{aligned}\mathbf{V}_2 &= j\omega L_2 \mathbf{I}_2 - j\omega M(\mathbf{I}_1 - \mathbf{I}_2) \\ &= -j\omega M \mathbf{I}_1 + j\omega(L_2 + M)\mathbf{I}_2\end{aligned} \tag{10-38}$$

Fig. 10-18 (a) Circuit for Example 10-8. (b) Using branch currents in setting up the mesh equations.

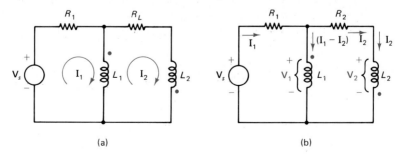

10-5 Analysis of Circuits with Coupled Coils

Equation for Mesh 1: $\qquad R_1\mathbf{I}_1 + \mathbf{V}_1 = \mathbf{V}_T$

Substituting for $\mathbf{V}_1$ from Eq. (10-37) and collecting coefficients, the equation for mesh 1 becomes

$$(R_1 + j\omega L_1)\mathbf{I}_1 - j\omega(L_1 + M)\mathbf{I}_2 = \mathbf{V}_T \qquad (10\text{-}39)$$

Equation for Mesh 2: $\qquad -\mathbf{V}_1 + R_2\mathbf{I}_2 + \mathbf{V}_2 = 0$

Substituting for $\mathbf{V}_1$ and $\mathbf{V}_2$ from Eqs. (10-37) and (10-38), and collecting coefficients, the equation for mesh 2 becomes

$$-j\omega(L_1 + M)\mathbf{I}_1 + [R_2 + j\omega(L_1 + L_2 + 2M)]\mathbf{I}_2 = 0 \qquad (10\text{-}40)$$

Equations (10-39) and (10-40) are the mesh equations for the given circuit. ∎

Exercise 10-20 Write the mesh equations of each of the circuits shown in Fig. 10-19.

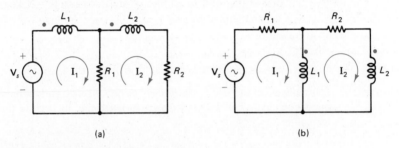

Fig. 10-19 Circuits for Exercise 10-20.

Exercise 10-21 Determine the voltage ratio (V_2/V_s) in the circuit shown in Fig. 10-20.

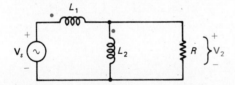

Fig. 10-20 Circuit for Exercise 10-21.

10-6 ENERGY STORED IN COUPLED COILS

The energy stored in the magnetic field of a pair of coupled coils has components due to the self-inductance of each coil and the mutual inductance between them. The energy stored in each coil is given by

$$w(t) = \int v(t)i(t)dt$$

If L_1 and L_2 are the inductances of two coils with a mutual inductance of M, and the currents in the coils are $i_1(t)$ in L_1 and $i_2(t)$ in L_2, then the total energy in the two coils at any time t is given by

$$w(t) = \int_0^t v_1(u)i_1(u)du + \int_0^t v_2(u)i_2(u)du \qquad (10\text{-}41)$$

where u is a dummy variable. It has been assumed for the sake of convenience that the initial stored energy $w(0) = 0$.

Let M_{21} denote the coefficient of mutual induction from coil 1 to coil 2 and M_{12} the coefficient of mutual induction from coil 2 to coil 1. (The aim of the present development is to prove that $M_{12} = M_{21}$, to justify the use of a single symbol M for M_{12} and M_{21}, a pair of coupled coils.)

Assuming that the *mutual inductance term adds to the self-inductance term*, the voltage in coil L_1 is

$$v_1(t) = L_1 \frac{di_1(t)}{dt} + M_{12} \frac{di_2(t)}{dt} \qquad (10\text{-}42)$$

and the voltage in coil L_2 is

$$v_2(t) = L_2 \frac{di_2(t)}{dt} + M_{21} \frac{di_1(t)}{dt} \qquad (10\text{-}43)$$

Consider an instant of time t_0 at which the current in coil L_1 is I_1 and the current in coil L_2 is I_2, and it is desired to determine the energy at $t = t_0$.

Suppose the following experiment is performed. Both coils start with zero current at $t = 0$. At first i_2 is maintained at zero while i_1 increases from zero to I_1 over the interval $0 < t < t_1$. Then i_1 is maintained constant at I_1 while i_2 increases from zero to I_2 over the interval $t_1 < t < t_0$.

Interval $0 < t < t_1$: Since $i_2(t)$ and $[di_2(t)/dt]$ are both zero, Eqs. (10-42) and (10-43) become

$$v_1(t) = L_1 \frac{di_1(t)}{dt}$$

$$v_2(t) = M_{21} \frac{di_1(t)}{dt}$$

Using the above equations for v_1 and v_2 and the fact $i_2 = 0$, the energy stored in the system during this interval is, from Eq. (10-41),

$$w(0, t_1) = \int_0^{t_1} L_1 \frac{di_1(t)}{dt} i_1(t) dt$$

$$= (1/2)L_1 i_1^2 \Big|_0^{t_1}$$

$$= (1/2)L_1 I_1^2 \qquad (10\text{-}44)$$

since $i_1(t_1) = I_1$.

Interval $t_1 < t < t_0$: Since $i_1(t) = I_1$, a constant, $[di_1(t)/dt] = 0$ in this interval, and Eqs. (10-42) and (10-43) become

$$v_1(t) = M_{12} \frac{di_2(t)}{dt}$$

$$v_2(t) = L_2 \frac{di_2(t)}{dt}$$

Using the equations above, the energy stored in the system during this interval is, from Eq. (10-41),

$$w(t_1, t_0) = \int_{t_1}^{t_0} M_{12}\frac{di_2(t)}{dt}I_1 dt + \int_{t_1}^{t_0} L_2\frac{di_2(t)}{dt}i_2(t)dt$$

Note that in the first integral, I_1 is a constant and is therefore taken outside the integral, leading to

$$w(t_1, t_0) = I_1 M_{12} i_2(t)\Big|_{t_1}^{t_0} + (1/2)L_2 i_2^2 \Big|_{t_1}^{t_0}$$
$$= M_{12}I_1 I_2 + (1/2)L_2 I_2^2 \tag{10-45}$$

since $i_2(t_0) = I_2$ and $i_2(t_1) = 0$.

The total energy in the system at $t = t_0$ is given by

$$w(t_0) = w(0, t_1) + w(t_1, t_0)$$

and, from Eqs. (10-44) and (10-45),

$$w(t_0) = (1/2)L_1 I_1^2 + (1/2)L_2 I_2^2 + M_{12}I_1 I_2 \tag{10-46}$$

The total energy has components due to self-induction L_1 and L_2, which are of the form $(1/2)LI^2$, and a component due to mutual induction of the form $M_{12}I_1 I_2$.

Now suppose the building up of the currents from zero to I_1 and I_2 was done in the following manner: The current i_1 is maintained at zero while i_2 increases from zero to I_2 in the interval $0 < t < t_1$; then the current i_2 is maintained at I_2 while i_1 increases from zero to I_1 in the interval $t_1 < t < t_0$. Then the total energy of the system will have the expression

$$w(t_0) = (1/2)L_1 I_1^2 + (1/2)L_2 I_2^2 + M_{21}I_1 I_2 \tag{10-47}$$

Since the total energy in the system at $t = t_0$ as given by the two equations (10-46) and (10-47) must be the same, it follows that

$$M_{21} = M_{12}$$

and a single symbol is therefore used to denote the mutual inductance between a pair of coils.

In general, the total energy at any time t in a pair of coupled coils where the mutual inductance term *adds* to the self inductance terms is

$$w(t) = (1/2)L_1[i_1(t)]^2 + (1/2)L_2[i_2(t)]^2 + Mi_1(t)i_2(t) \quad \text{(additive } M \text{ term)} \tag{10-48}$$

Exercise 10-22 Derive Eq. (10-47).

Now consider the total energy in a pair of coils in which the mutual inductance term *subtracts* from the self-inductance term. The energy stored in the system at any time t can be obtained by replacing M in Eq. (10-48) by $-M$.

$$w(t) = (1/2)L_1[i_1(t)]^2 + (1/2)L_2[i_2(t)]^2$$
$$- Mi_1(t)i_2(t) \quad \text{(subtractive } M \text{ term)} \tag{10-49}$$

The total stored energy in the system must be non-negative for all t and regardless of whether i_1 and i_2 are individually positive or negative. The expressions for $w(t)$ in Eqs. (10-48) and (10-49) are said to be in *quadratic form*. The necessary and sufficient conditions for the quadratic forms in Eqs. (10-48) and (10-49) to be positive are:

$$L_1 \geq 0$$
$$L_2 \geq 0$$
$$L_1 L_2 \geq M^2$$

Using the coefficient of coupling defined earlier,

$$k = M/\sqrt{(L_1 L_2)}$$

it can be seen that if $L_1 L_2 \geq M^2$, then

$$k \leq 1$$

that is, the maximum value of the coefficient of coupling is unity.

Exercise 10-23 The currents in a pair of coupled coils ($L_1 = 0.2$ H, $L_2 = 0.3$ H, and $M = 0.15$ H) are given by $i_1(t) = e^{-0.5t}$ A and $i_2(t) = 0.2t$ A. Determine the energy stored in the system at $t = 2$ s if (a) the mutual inductance term adds to the self-inductance term, and (b) the mutual inductance term subtracts from the self-inductance term.

Exercise 10-24 Repeat the previous exercise if the coefficient of coupling is changed to 0.3 with L_1 and L_2 having the same values as before.

10-7 THE T EQUIVALENT CIRCUIT FOR COUPLED COILS

It is possible to replace a pair of coupled coils in a circuit by an equivalent circuit made up of three *uncoupled* coils in the form of a T. The coils in the equivalent circuit are *not magnetically coupled* to one another, and this offers an advantage in writing its equations.

Fig. 10-21 shows pairs of coupled coils and their T equivalent circuits. The equivalences

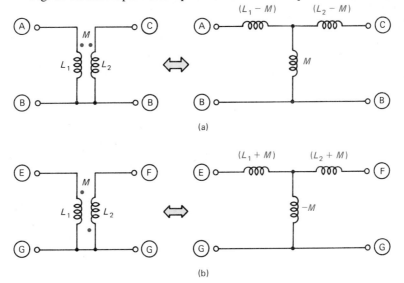

Fig. 10-21 Coupled coils and their T equivalent circuits.

between the circuits of Fig. 10-21(a) and (b) and between those of Fig. 10-21(c) and (d) are independent of the current directions and voltage polarities chosen.

Exercise 10-25 For the two circuits in Fig. 10-21(a), choose currents I_1, I_2 in the coils and assign voltages V_1 and V_2 across the coils. Write KVL equations for the circuits and establish their equivalence.

Exercise 10-26 Repeat the work of the previous exercise for the two circuits in Fig. 10-21(b).

The use of the T equivalent circuit is illustrated by the following example.

Example 10-9 Determine the input current I_1 and the output voltage V_2 of the circuit shown in Fig. 10-22(a). The coefficient of coupling between the coils is 0.6, and the angular frequency is 500 r/s.

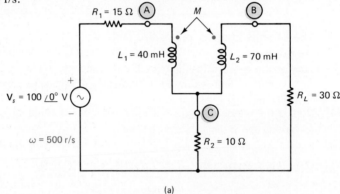

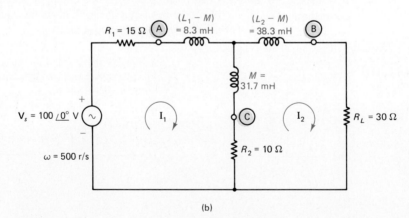

Fig. 10-22 (a) Circuit for Example 10-9. (b) Modified circuit using T equivalent for the coupled coils.

Solution Since $L_1 = 40$ mH, $L_2 = 70$ mH, and $k = 0.6$

$$M = 0.6 \sqrt{(40)(70)} = 31.7 \text{ mH}$$

Replace the coupled coils by their T equivalent, as shown in Fig. 10-22(b). Choose mesh currents as shown, and write the mesh equations.

Mesh 1: $\quad [R_1 + j\omega(L_1 - M) + j\omega M + R_2]I_1 - (j\omega M + R_2)\mathbf{I}_2 = \mathbf{V}_s$

or

$$(25 + j20)\mathbf{I}_1 - (10 + j15.9)\mathbf{I}_2 = 100 \tag{10-50}$$

Mesh 2: $\quad -(R_2 + j\omega M)\mathbf{I}_1 + [R_2 + j\omega M + j\omega(L_2 - M) + R_L]\mathbf{I}_2 = 0$

or

$$-(10 + j15.9)\mathbf{I}_1 + (40 + j35)\mathbf{I}_2 = 0 \tag{10-51}$$

Solving Eqs. (10-50) and (10-51),

$$\mathbf{I}_1 = 3.72\underline{/-30.3°} \text{ A}$$

$$\mathbf{I}_2 = 1.31\underline{/-13.7°} \text{ A}$$

and

$$\mathbf{V}_2 = R_L\mathbf{I}_2 = 39.3\underline{/-13.7°} \text{ V}$$

■

Exercise 10-27 Verify the results of the example above by solving it without using a T equivalent circuit.

Exercise 10-28 Two coils L_1 and L_2, with a coefficient of coupling k, are in parallel (with dots on the top terminals of the two coils) and driven by a current source $\mathbf{I}_s$. Set up the T equivalent circuit for the coupled coils. Use nodal analysis and find the total inductance seen by the current source.

Exercise 10-29 Reverse one of the coils in the previous exercise and recalculate the total inductance seen by the current source.

10-8 TRANSFORMERS

One of the most important uses of a pair of coupled coils is for *transforming an available voltage (or current) level to a different voltage (or current) level*. Therefore, a pair of coupled coils is referred to as a *transformer*. The analysis of the preceding sections dealt with coupled coils having finite values of inductance and coupling coefficients less than 1. Such circuits are called *linear transformers*, since their analysis uses the procedures of linear circuits.

When the two inductances are made to approach infinity and the coefficient of coupling becomes equal to 1, the voltages across the two coils and the currents in the two coils are found to be proportional to the number of turns in the coils. Such a transformer is known as an *ideal transformer*. Even though no practical transformer is completely ideal, it is possible to develop models for a practical transformer using an ideal transformer and other components such as inductances and resistances.

10-8-1 The Ideal Transformer

Consider two coupled coils wound on a ferromagnetic core. The following assumptions are made for an ideal transformer.

Assumption 1: The permeability of the core is infinite. The flux set up in the core depends upon the permeability of the ferromagnetic material. As the permeability increases, the current needed in either coil to set up a given flux in the core decreases. In the limit where the *permeability becomes infinite, the current needed to set up a given flux becomes zero.* Since the inductance of a coil is given by the ratio (flux linkage/current), an infinite permeability means that the inductance of the coil is also infinite. That is

$$L_1 = L_2 = \infty$$

It will be assumed, however, that the *ratio* L_1/L_2 is *finite*.

Assumption 2: The coupling between the two coils is so complete that the coefficient of coupling equals 1. This means that

$$M = \sqrt{L_1 L_2}$$

and M is also infinite, since L_1 and L_2 are infinite.

Consider the pair of coupled coils in Fig. 10-23. The equations are

$$j\omega L_1 \mathbf{I}_1 - j\omega M \mathbf{I}_2 = \mathbf{V}_1 \qquad (10\text{-}52)$$

$$-j\omega M \mathbf{I}_1 + j\omega L_2 \mathbf{I}_2 = -\mathbf{V}_2 \qquad (10\text{-}53)$$

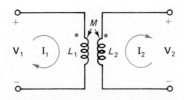

Fig. 10-23 Development of the ideal transformer.

Solving Eqs. (10-52) and (10-53)

$$\frac{\mathbf{V}_2}{\mathbf{V}_1} = \frac{-L_2 \mathbf{I}_2 + M \mathbf{I}_1}{L_1 \mathbf{I}_1 - M \mathbf{I}_2}$$

Using the condition $M = \sqrt{L_1 L_2}$ in the equation above, the expression for the magnitude of the voltage ratio reduces to

$$\left| \frac{\mathbf{V}_2}{\mathbf{V}_1} \right| = \sqrt{\frac{L_2}{L_1}} \qquad (10\text{-}54)$$

Solving Eqs. (10-52) and (10-53) for $\mathbf{I}_1$ and $\mathbf{I}_2$, the ratio $\mathbf{I}_2/\mathbf{I}_1$ becomes

$$\frac{\mathbf{I}_2}{\mathbf{I}_1} = \frac{\mathbf{V}_2 L_1 - \mathbf{V}_1 M}{-\mathbf{V}_1 L_2 + \mathbf{V}_2 M} \qquad (10\text{-}55)$$

Using the condition $M = \sqrt{L_1 L_2}$, the magnitude of the current ratio is found to be

$$\left| \frac{\mathbf{I}_2}{\mathbf{I}_1} \right| = \sqrt{\frac{L_1}{L_2}} \qquad (10\text{-}56)$$

which is the inverse of the ratio for $\mathbf{V}_2/\mathbf{V}_1$.

Note that the magnitude of the voltage ratio given by Eq. (10-54) and the magnitude of the current ratio given by Eq. (10-56) are not affected if one of the coils is reversed.

Now, consider the relationship of the inductance of a coil to the number of turns in its winding. The inductance L of a coil with N turns is given by

$$L = N\frac{d\psi}{dt}$$

where ψ, the flux linkage, is related to the flux ϕ by $\psi = N\phi$. Therefore,

$$L = N^2\frac{d\phi}{dt}$$

When two coils have a coupling coefficient of 1, then the flux linking each coil also links the other coil completely. In such a case,

$$L_1 = N_1^2\frac{d\phi}{dt} \tag{10-57}$$

and

$$L_2 = N_2^2\frac{d\phi}{dt} \tag{10-58}$$

where N_1 and N_2 are the values of the number of turns in L_1 and L_2, respectively.

Using Eqs. (10-57) and (10-58) in Eqs. (10-54) and (10-55),

$$\left|\frac{\mathbf{V}_2}{\mathbf{V}_1}\right| = \frac{N_2}{N_1} \tag{10-59}$$

and

$$\left|\frac{\mathbf{I}_2}{\mathbf{I}_1}\right| = \frac{N_1}{N_2} \tag{10-60}$$

Therefore, the magnitude of the voltage ratio $|\mathbf{V}_2/\mathbf{V}_1|$ equals the turns ratio (N_2/N_1) in an ideal transformer. Remember that the higher voltage is across the coil with the larger number of turns.

The magnitude of the current ratio $|\mathbf{I}_2/\mathbf{I}_1|$ equals the turns ratio (N_1/N_2). The higher current is across the coil with the smaller number of turns.

Equations (10-59) and (10-60) also lead to the condition

$$|\mathbf{V}_1\mathbf{I}_1| = |\mathbf{V}_2\mathbf{I}_2| \tag{10-61}$$

That is, the instantaneous power in the two coils is equal, and there is no loss of power in the ideal transformer.

One of the coils (usually the one on the input side) of the ideal transformer is called the *primary coil*, and the other is called the *secondary coil*.

Voltage Polarities and Current Directions

Equation (10-55) shows that when $\mathbf{V}_1$ and $\mathbf{V}_2$ are positive, $\mathbf{I}_1$ and $\mathbf{I}_2$ are also positive in the coils of Fig. 10-23. This leads to the following statement about the current directions and voltage polarities of an ideal transformer:

Voltage Polarities: If the voltages across the coils of an ideal transformer are chosen with their *positive* references at the *dotted terminals* of the two coils, the two voltages will be *in phase* with each other.

Current Directions: If the currents in the coils of an ideal transformer are chosen so that *one of them is entering the dotted terminal* of a coil and the *other is leaving the dotted terminal* of the other coil, the two currents will be *in phase* with each other.

Figure 10-24 shows several different cases, in all of which V_2 and V_1 are in phase with each other and I_2 and I_1 are in phase with each other.

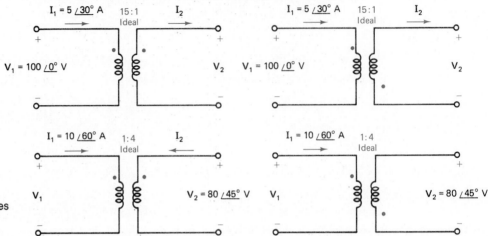

Fig. 10-24 Examples of the ideal transformer.

Example 10-10 Determine the current supplied by V_s and the voltage across the resistor R_L in the circuit of Fig. 10-25(a). The quantity a denotes the turns ratio (N_1/N_2).

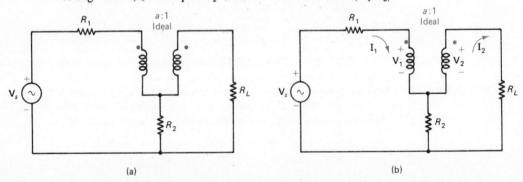

Fig. 10-25 (a) Circuit for Example 10-10. (b) Assignment of voltages and currents for analysis.

Solution Assign voltages and currents in the coils of the ideal transformer as shown in Fig. 10-25(b). Note that V_1 and V_2 are chosen with their positive references at the dotted terminals, and the currents are chosen with a current entering the dotted terminal of one coil and a current leaving the dotted terminal of the other coil. Then

$$\frac{V_2}{V_1} = \frac{1}{a} \quad \text{or} \quad V_2 = \frac{V_1}{a} \tag{10-62}$$

and

$$\frac{I_2}{I_1} = a \quad \text{or} \quad I_2 = aI_1 \tag{10-63}$$

The mesh equations of the circuit are

Mesh 1: $\quad (R_1 + R_2)\mathbf{I}_1 - R_2\mathbf{I}_2 = (\mathbf{V}_s - \mathbf{V}_1)$

which becomes, after using Eq. (10-63),

$$[R_1 + R_2(1 - a)]\mathbf{I}_1 + \mathbf{V}_1 = \mathbf{V}_s \qquad (10\text{-}64)$$

Mesh 2: $\quad -R_2\mathbf{I}_1 + (R_2 + R_L)\mathbf{I}_2 = \mathbf{V}_2$

which becomes, after using Eqs. (10-62) and (10-63),

$$[R_2(a - 1) + aR_L]\mathbf{I}_1 - (1/a)\mathbf{V}_1 = 0 \qquad (10\text{-}65)$$

Solving Eqs. (10-64) and (10-65),

$$\mathbf{I}_1 = \frac{\mathbf{V}_s}{R_1 + R_2(1 - a)^2 + a^2 R_L}$$

which is the current supplied by $\mathbf{V}_s$.

The current through R_L is $\mathbf{I}_2$, given by

$$\mathbf{I}_2 = a\mathbf{I}_1 = \frac{a\mathbf{V}_s}{R_1 + R_2(1 - a)^2 + a^2 R_L}$$

and the voltage across R_L is

$$\mathbf{V}_o = R_L \mathbf{I}_2$$

$$= \frac{aR_L \mathbf{V}_s}{R_1 + R_2(1 - a)^2 + a^2 R_L} \qquad \blacksquare$$

Exercise 10-30 If, in the circuit of Example 10-10, the dot in one of the coils is moved to the other terminal of that coil, what will be the new expressions for $\mathbf{I}_1$ and $\mathbf{V}_o$? Try to answer this without actually going through the analysis. Would it make a difference which coil has the dot shifted?

Exercise 10-31 Determine the voltage $\mathbf{V}_2$ in the circuit of Fig. 10-26(a) when (a) the terminals 2-2′ are shorted, and (b) a resistance of $R_L = 25\Omega$ is connected to the terminals 2-2′.

Exercise 10-32 Find the current $\mathbf{I}_s$ supplied by the source in the circuit of Fig. 10-26(b).

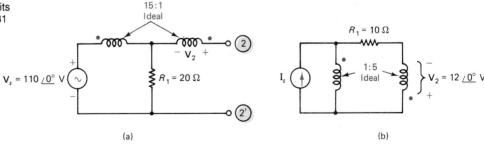

Fig. 10-26 Circuits for Exercises 10-31 and 10-32.

10-8 Transformers

10-8-2 Impedance Transformation Using Ideal Transformers

We saw in the discussion of maximum power transfer theorems (Chapters 5 and 9) that it is desirable to match the load impedance to the Thevenin impedance of the network. In a number of practical situations, however, the available range of load impedance may not fit the matching condition. For example, the output resistance of an amplifier may be in the order of kilohms, while the resistance of a loudspeaker is around 8 ohms. *The ideal transformer has the property of changing the impedance level of a network and is useful in providing matching between a network and a load.*

Consider the circuit shown in Fig. 10-27(a), where a load resistance Z_L is connected to the secondary coil of an ideal transformer with turns ratio $a:1$. It is desired to determine the impedance Z_{in} seen from the primary side.

$$V_2 = Z_L I_2 \tag{10-66}$$

But $V_2 = (V_1/a)$ and $I_2 = aI_1$ and Eq. (10-66) becomes $(V_1/a) = Z_L(aI_1)$. Or

$$Z_{in} = (V_1/I_1) = a^2 Z_L$$

That is, the effective impedance seen from the primary side is a^2 times the impedance connected as the load on the secondary side. Thus the ideal transformer is useful for transforming a given impedance to another value.

The quantity $Z_{in} = a^2 Z_L$ is called the *reflected impedance* seen from the primary side.

A procedure similar to the one above will show that when a load impedance Z_L is connected to the primary side, as in Fig. 10-27, the reflected impedance seen from the secondary side is $(1/a^2)Z_L$.

In general, when a load impedance Z_L is connected to either coil of an ideal transformer, the reflected impedance seen from the other coil equals Z_L multiplied by the square of the turns ratio. It is convenient to remember that the reflected impedance is larger than the actual Z_L when seen from the coil with the larger number of turns.

Fig. 10-27 (a) Impedance transformation using an ideal transformer. (b)-(c) Reflected impedances.

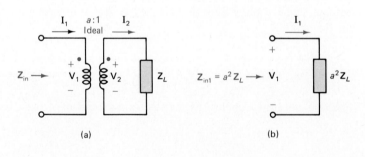

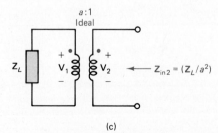

Exercise 10-33 Show that the impedance labeled $\mathbf{Z}_{in2}$ in Fig. 10-27 (c) equals $(1/a^2)\mathbf{Z}_L$.

Exercise 10-34 An ideal transformer has a primary to secondary turns ratio of 24:1. If an impedance $\mathbf{Z}_1 = (3 + j4)$ is connected to the primary side, determine the reflected impedance seen from the secondary side. If the impedance is connected to the secondary side (instead of the primary), determine the reflected impedance seen from the primary side.

Exercise 10-35 A resistance of 16 Ω is available. If it is necessary to transform it to a 5 kΩ resistance, what is the turns ratio of the transformer needed? Should the 16 Ω be connected to the side with the larger coil or the smaller coil?

The impedance transformation property of an ideal transformer is useful in matching a load impedance to a given network, as is evident from the preceding exercises. The use of a transformer for matching is not without disadvantages, however. A practical transformer has losses and nonlinearities. Therefore, it may not fully meet the idealizing assumptions made earlier.

10-9 THREE-PHASE SYSTEMS

In the generation, transmission, and distribution of power, it is common practice to use three sinusoidal voltages of equal amplitude, but differing in phase by 120° from each other. Such a system is known as a *three-phase system*. There are polyphase systems with other than three phases, but the three-phase system is the most commonly used. Three-phase systems have certain advantages over single-phase systems. For a given size frame, a three-phase motor or generator has a greater power capacity than a single-phase unit. A polyphase motor has a uniform torque, whereas most single-phase motors have pulsating torques. Rectification (that is, conversion of ac to dc) of three-phase ac voltages yields an output with considerably less ripple than the rectification of single-phase ac voltages.

10-9-1 Double Subscript Notation

It is convenient to use a special notation, called the *double subscript* notation, using two subscript symbols in the analysis of three-phase systems.

In the case of a voltage, the *first subscript* identifies the *positive reference terminal*. That is, if a and b are two terminals of an element, as in Fig. 10-28(a), the symbol $\mathbf{V}_{ab}$ denotes the voltage across the element with terminal a chosen as the positive reference.

Note that a reversal of the subscripts reverses the voltage; that is,

$$\mathbf{V}_{ba} = -\mathbf{V}_{ab}$$

Consider now a situation with several elements connected, as shown in Fig. 10-28(b). The equation for the voltage $\mathbf{V}_{ad}$ is given by

$$\mathbf{V}_{ad} = \mathbf{V}_{ab} + \mathbf{V}_{bc} + \mathbf{V}_{cd}$$

The pattern of the subscripts in the last equation should be noted carefully: the path is from a to b, b to c, and c to d, and the subscripts appear in the same order. Because of this pattern, it is not always necessary to mark the polarities of voltages in a circuit.

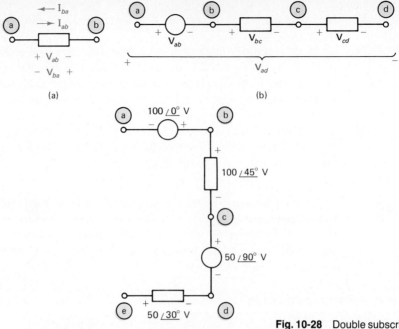

Fig. 10-28 Double subscript notation for voltage and current.

Example 10-11 Determine the voltages V_{ac}, V_{da}, and V_{ae} in the arrangement of Fig. 10-28(c).

Solution To go from a to c (for finding V_{ac}), the path is from a to b and b to c. Note that $V_{ab} = -100\underline{/0°}$ in the diagram. Therefore,

$$V_{ac} = V_{ab} + V_{bc} = (-100\underline{/0°}) + (100\underline{/45°})$$
$$= -29.3 + j70.7 = 76.5\underline{/112°} \text{ V}$$

To find V_{da}, it is necessary to start at d and finish at a. Therefore, paying careful attention to the order of subscripts and the voltages in the diagram,

$$V_{da} = V_{dc} + V_{cb} + V_{ba} = (-50\underline{/90°}) + (-100\underline{/45°}) + (100\underline{/0°})$$
$$= 29.3 - j120.7 = 124.2\underline{/-76.4°} \text{ V}$$

$$V_{ae} = V_{ab} + V_{bc} + V_{cd} + V_{de}$$
$$= -100\underline{/0°} + 100\underline{/45°} + 50\underline{/90°} - 50\underline{/30°}$$
$$= -72.6 + j95.7 = 120.1\underline{/127.2°} \text{ V}$$

V_{ae} could also have been found by using V_{ac} found earlier and V_{cd} and V_{de}. ∎

It is important to realize that the basic principle is still Kirchhoff's voltage law; the double subscript serves only to make the application of the law more systematic.

Exercise 10-36 Determine the voltages V_{bd} and V_{ec} in the arrangement of Fig. 10-28.

Exercise 10-37 Given that $V_{ab} = 100\underline{/0°}$ V and $V_{bc} = 100\underline{/-120°}$ V in a circuit, determine the voltage V_{ac}.

Magnetic Circuits, Coupled Coils, and Three-Phase Circuits

A double subscript notation is also used for currents. If an element has terminals a and b, then a current $\mathbf{I}_{ab}$ denotes the current through the element in the direction a to b. That is, the *first subscript identifies the assumed starting point of the current*. Note that

$$\mathbf{I}_{ab} = -\mathbf{I}_{ba}$$

10-9-2 Balanced Three-Phase Sources

A *balanced* three-phase source consists of three voltages of equal amplitude but separated in phase by 120°. There are two connections of a three-phase source, the Y connected source, as in Fig. 10-29(a), and the delta connected source, as in Fig. 10-29(b).

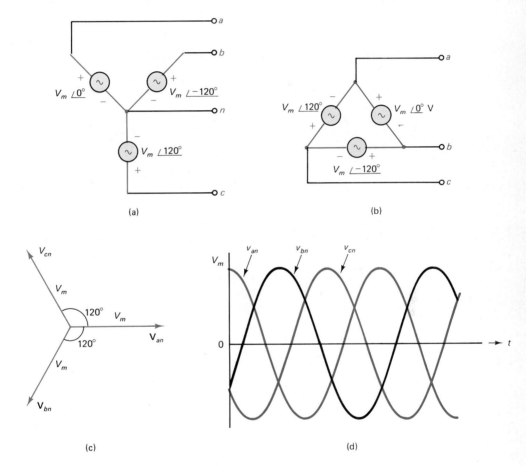

Fig. 10-29 Balanced three-phase sources: (a) Y connected. (b) Delta connected. (c) Phasor diagram of Y connected source. (d) Voltage waveforms.

In the Y connected source, the terminal labeled n is called the *neutral point*. The *line to neutral* voltages in Fig. 10-29(a) are $\mathbf{V}_{an} = V_m \underline{/0°}$; $\mathbf{V}_{bn} = V_m \underline{/-120°}$; $\mathbf{V}_{cn} = V_m \underline{/120°}$, where V_m is the amplitude of each of the three voltages. Use of the above expressions of the three line to neutral voltages will show that

$$\mathbf{V}_{an} + \mathbf{V}_{bn} + \mathbf{V}_{cn} = 0$$

The term *phase voltage* is also used to refer to the line to neutral voltages. The voltages between any two of the three lines a, b, c is called *line to line voltages* or simply *line voltages*. The three line to line voltages of a Y connected source also form a balanced set, as the following exercise shows.

Exercise 10-38 Obtain the values of the following line to line voltages in the Y connected source of Fig. 10-29(a). (a) $\mathbf{V}_{ab}$, (b) $\mathbf{V}_{bc}$, and (c) $\mathbf{V}_{ca}$. Reduce your result to a single phasor in each case and show that the three voltages form a balanced set. Verify that the three voltages add to zero.

In the delta connected source, shown in Fig. 10-29(b), the line to line voltages are $\mathbf{V}_{ab} = V'_m \underline{/0°}$; $\mathbf{V}_{bc} = V'_m \underline{/-120°}$; $\mathbf{V}_{ca} = V'_m \underline{/120°}$, where V'_m is the amplitude of each of the three voltages. Use of the above three expressions for the line to line voltages will show that

$$\mathbf{V}_{ab} + \mathbf{V}_{bc} + \mathbf{V}_{ca} = 0$$

The phasor diagram of the Y connected source is shown in Fig. 10-29(c). If the three voltages in the Y connection are expressed as sinusoidal functions of time, then

$$v_{an}(t) = V_m \cos(\omega t)$$
$$v_{bn}(t) = V_m \cos(\omega t - 120°)$$
$$v_{cn}(t) = V_m \cos(\omega t + 120°)$$

and these are shown in Fig. 10-29(d). It is seen that $v_{an}(t)$ attains its peak value first, v_{bn} second, and v_{cn} third. This sequence repeats itself. The sequence in which the three voltages attain their peaks is called their *phase* sequence and denoted by *a-b-c*.

Phase sequence can be determined directly from the phasor diagram also: *If the phasors in Fig. 10-29(c) are imagined to be rotating counterclockwise,* **then the order in which the phasors will be seen is:** *a-b-c-a-b-c. . .*; **hence the phase sequence is** *a-b-c*.

When the three voltages in a three-phase source do not meet the conditions of equal amplitude and 120° phase difference, the source is an *unbalanced* three-phase source. Circuits with unbalanced three-phase sources are analyzed by routine techniques of ac circuit analysis (see Chapters 8, 9, and 10). Circuits driven by balanced sources have special formulas and relationships that simplify their analysis considerably.

The relationship between the line to neutral voltages and line to line voltages is most conveniently remembered by means of an *equilateral triangle* with lines drawn from the vertices to the center, as shown in Fig. 10-30. The vertices are marked a, b, c and the

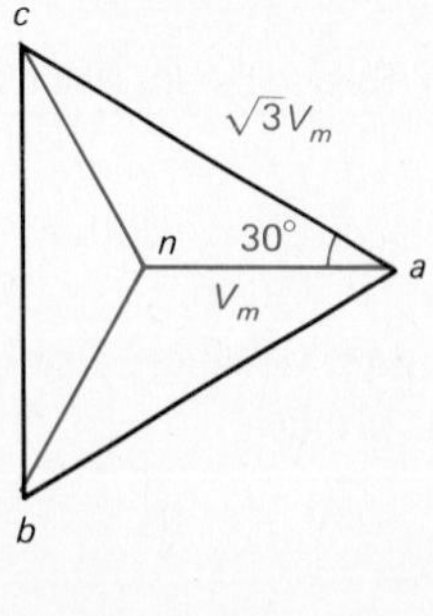

Fig. 10-30 Voltage triangle for a balanced three-phase system.

center represents the neutral point n. The sides of the triangle represent the line to line voltages and the lines from the vertices to the center represent the line to neutral voltages. Note that

$$|\text{line to line voltage}| = \sqrt{3}\,|(\text{line to neutral voltage})|$$

The load connected to a three-phase source may be balanced or unbalanced and connected in the form of a Y or a delta. A *balanced load* consists of three *identical* impedances connected in the form of a Y or a delta. Balanced loads lead to special formulas and relationships, as will be seen in the following discussion.

10-9-3 Y Connected Load

A Y connected load fed by a balanced three-phase source is indicated in Fig. 10-31(a). The neutral point of the source is connected to the center of the load, thus leading to a *three-phase, four-wire system*. The voltage across each load is directly equal to a line to neutral voltage, and the load currents are given by

$$\mathbf{I}_{an} = (\mathbf{V}_{an}/\mathbf{Z}_a) \tag{10-67a}$$

$$\mathbf{I}_{bn} = (\mathbf{V}_{bn}/\mathbf{Z}_b) \tag{10-67b}$$

$$\mathbf{I}_{cn} = (\mathbf{V}_{cn}/\mathbf{Z}_c) \tag{10-67c}$$

Equations (10-67) are valid whether the load in the three-phase, four-wire system is balanced or not (assuming that an unbalanced load does not upset a balanced source).

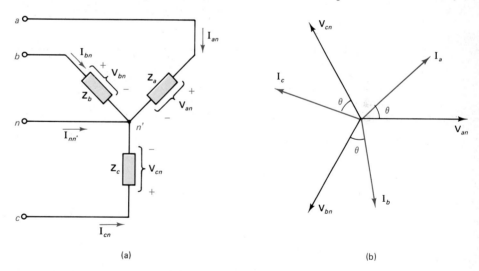

Fig. 10-31   (a) Y connected load. (b) Phasor diagram.

Balanced Load

When $\mathbf{Z}_a = \mathbf{Z}_b = \mathbf{Z}_c = \mathbf{Z}_L$ (for a balanced Y connected load), the three currents have equal amplitude $|\mathbf{V}_{an}/\mathbf{Z}_L|$ and a phase difference of 120° between them.

Figure 10-31(b) shows the phasor diagram of voltages and currents in the balanced Y connected load.

The average power consumed by each load is the same, since the amplitudes of the voltages and currents are equal in the three impedances. The power factor angle being equal to the impedance angle is also the same for all three. The average power consumed by *each load* is given by

$$P_{av} = (1/2)|V||I|\cos\theta \tag{10-68}$$

where $|V|$ is the amplitude of the line to neutral voltage, $|I|$ the amplitude of the line current, and θ is the angle of each impedance.

The total average power consumed by the load is three times that of each load.

Example 10-12 A balanced Y connected source (phase sequence a-b-c) with $\mathbf{V}_{an} = 100\underline{/0°}$ V is connected to a balanced load $\mathbf{Z}_L = 50\underline{/-36.9°}$, as indicated in Fig. 10-32(a). (a) Determine the currents in each of the load impedances. (b) Draw the phasor diagram showing the voltages and currents. (c) Calculate the average power consumed by each load and the average power consumed by the total load.

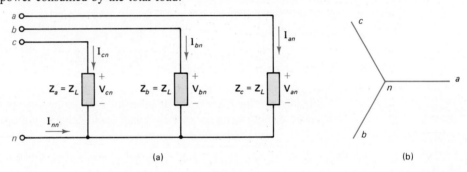

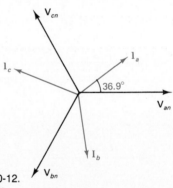

Fig. 10-32 (a) Circuit for Example 10-12. (b) Source voltages. (c) Phasor diagram.

Solution (a) The three line to neutral voltages are shown in the partial phasor diagram of Fig. 10-32(b), where $\mathbf{V}_{an}$ is the reference voltage (as given) and the other two phasors are drawn by using the given phase sequence a-b-c.

The currents are:

$$\mathbf{I}_{an} = (\mathbf{V}_{an}/\mathbf{Z}_L) = (100\underline{/0°}) \div (50\underline{/-36.9°}) = 2\underline{/36.9°} \text{ A}$$

$$\mathbf{I}_{bn} = (\mathbf{V}_{bn}/\mathbf{Z}_L) = 2\underline{/-83.1°} \text{ A}$$

$$\mathbf{I}_{cn} = (\mathbf{V}_{cn}/\mathbf{Z}_L) = 2\underline{/156.9°} \text{ A}$$

(b) The phasor diagram is shown in Fig. 10-32(c).
(c) Using Eq. (10-68), the average power consumed by each load is

$$P_{av} = (1/2)|V||I|(\cos \theta)$$
$$= (1/2)(100)(2)(\cos 36.9°) = 80 \text{ W}$$

Total average power consumed by the load is 240 W. ∎

Exercise 10-39 Repeat the work of the example above if the phase sequence of the supply voltages is reversed to *a-c-b*.

Exercise 10-40 Calculate the currents and the average power in a balanced, Y connected load with $Z_L = 10/45° \Omega$ if $V_{an} = 110/90°$ V and the phase sequence is *a-b-c*.

Exercise 10-41 A balanced three-phase source delivers an apparent power of 9 kVA and an average power of 6 kW to a balanced, Y connected load. If the line to neutral voltage of the source is 500 V, determine the magnitude and angle of each load impedance.

Current in the Neutral Wire

In the system where a balanced, Y connected load is fed by a balanced, Y connected source, the three line currents I_{an}, I_{bn}, and I_{cn} are, from Eq. 10-67,

$$I_{an} = V_{an}/Z_L, \quad I_{bn} = V_{bn}/Z_L, \quad I_{cn} = V_{cn}/Z_L$$

and since $(V_{an} + V_{bn} + V_{cn}) = 0$ for a balanced source, it follows that

$$I_{an} + I_{bn} + I_{cn} = 0$$

This means that the current I_{nn}' in the fourth wire (neutral wire connecting the neutral of the source to the center of the load) in the circuit of Fig. 10-31 or Fig. 10-32(a) is *identically equal to zero for a balanced load*. The calculations and results for a balanced load will be the same whether the neutral wire is present or not.

This conclusion does not apply to an unbalanced load, as will be seen from the following example.

Example 10-13 The impedances in the circuit of Fig. 10-32 (used in Example 10-12) are: $Z_a = (5 + j15)\Omega$, $Z_b = (15 - j25)\Omega$, and $Z_c = j60 \Omega$. Determine the current I_{nn}' through the neutral wire.

Solution The three load currents are:

$$I_{an} = (V_{an}/Z_a) = (100/\underline{0°})/(15.8/\underline{71.6°}) = 6.33/\underline{-71.6°} \text{ A}$$
$$I_{bn} = (V_{bn}/Z_b) = (100/\underline{-120°})/(29.2/\underline{-59.0°}) = 3.42/\underline{-61°} \text{ A}$$
$$I_{cn} = (V_{cn}/Z_c) = (100/\underline{120°})/(60/\underline{90°}) = 1.67/\underline{30°} \text{ A}$$

Using KCL at the central node of the load

$$I_{nn}' = -(I_{an} + I_{bn} + I_{cn}) = -(5.10 - j8.16) \text{ A} = 9.62/\underline{122°} \text{ A} \quad ∎$$

The results obtained above will change if the neutral wire is removed. The analysis of an unbalanced three-wire load (that is, without the neutral wire) is done by using routine techniques such as mesh analysis.

Exercise 10-42 Determine the average power consumed by each of the loads in Example 10-13 and the total average power in the load.

10-9-4 Delta Connected Load

Figure 10-33(a) shows three impedances connected in the form of a delta fed by a balanced three-phase source. The currents in the three impedances depend upon the *line to line voltages* rather than the line to neutral voltages (which was the case in a four-wire, Y connected load). For the directions shown, the branch currents are

$$\mathbf{I}_{ab} = (\mathbf{V}_{ab}/\mathbf{Z}_a) \qquad (10\text{-}69a)$$

$$\mathbf{I}_{bc} = (\mathbf{V}_{bc}/\mathbf{Z}_b) \qquad (10\text{-}69b)$$

$$\mathbf{I}_{ca} = (\mathbf{V}_{ca}/\mathbf{Z}_c) \qquad (10\text{-}69c)$$

Equations (10-69) are valid whether the load is balanced or not.

> For a balanced delta-connected load, the three branch currents have equal amplitudes $= |V_{ab}/Z|$ and differ in phase by 120° from one another.

The phasor diagram of voltages and currents is shown in Fig. 10-33(b) with $\mathbf{V}_{ab}$ chosen as reference. The average power consumed by *each load* is the same and is given by

$$P_{av} = (1/2)|V_L||I|\cos\theta \qquad (10\text{-}70)$$

where $|V_L|$ is the amplitude of the *line to line* voltage, $|I|$ is the amplitude of the branch current and θ is the angle of the impedance.

> *The total average power consumed by the load is three times that consumed by each impedance.*

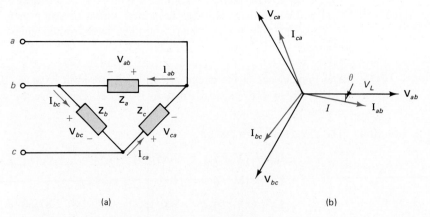

Fig. 10-33 (a) Delta connected load. (b) Phasor diagram.

Example 10-14 Let each of the impedances in Fig. 10-33(a) be $\mathbf{Z}_L = (30 - j40)\Omega$ and let $\mathbf{V}_{ab} = 173\underline{/0°}$ V (phase sequence a-b-c). Determine the branch currents, the average power consumed by each branch, and the total average power consumed by the load.

Solution The three line to line voltages are shown in the partial phasor diagram of Fig. 10-33(b). The branch currents are

$$\mathbf{I}_{ab} = (\mathbf{V}_{ab}/\mathbf{Z}_L) = (173\underline{/0°})/(50\underline{/-53.1°}) = 3.46\underline{/53.1°} \text{ A}$$

$$\mathbf{I}_{bc} = (\mathbf{V}_{bc}/\mathbf{Z}_L) = (173\underline{/-120°})/(50\underline{/-53.1°}) = 3.46\underline{/-66.9°} \text{ A}$$

$$\mathbf{I}_{ca} = 3.46\underline{/173.1°} \text{ A}$$

The average power consumed by each branch is

$$P_{ave} = (1/2)(173)(3.46) \cos 53.1° = 179.7 \text{ W}$$

and the total power is 539.2 W. ∎

Exercise 10-43 Determine the branch currents and the average power consumed by the load in the network of the previous example if $\mathbf{Z}_L = (5 + j15)\ \Omega$.

Exercise 10-44 A three-phase source with line to line voltage of 300 V delivers a total apparent power of 10 kVA and an average power of 0.95 kW to a balanced, delta-connected load. Calculate the magnitude and angle of each load impedance.

When the line to neutral voltages of a three-phase source are given instead of the line to line voltages needed for using Eqs. (10-69), some initial computations are necessary to obtain the line to line voltages. Such computations are made systematic by the equilateral triangle introduced in Fig. 10-30.

> Orient the triangle so that the lines from the vertices to the center conform to the given line to neutral voltages. Then the sides of the triangle provide both the amplitude and the phase of the line to line voltages.

Example 10-15 If the voltage from line a to the neutral in a three-phase source is $\mathbf{V}_{an} = 100\underline{/0°}$ V (phase sequence a-b-c), calculate the line to line voltages of the source.

Solution The vertex a of the equilateral triangle is situated so that the line from the center n to the vertex a is at 0°. Then $\mathbf{V}_{an}$ has an angle of 0°. Choose the other vertices on the basis of the given phase sequence. Remember that the phasors are rotating *counterclockwise* about n as the pivot. The triangle is shown in Fig. 10-34. Each side of the triangle is related to the length of a line from the vertex to the center by a factor of $\sqrt{3}$ or 1.732.

Fig. 10-34 Voltage triangle for Example 10-15.

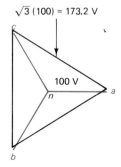

From the diagram, the line to line voltages are

$$V_{ab} = 173.2\underline{/30°} \text{ V}$$
$$V_{bc} = 173.2\underline{/-90°} \text{ V}$$
$$V_{ca} = 173.2\underline{/150°} \text{ V.}$$ ∎

Exercise 10-45 Suppose the phase sequence of the example is changed to *a-c-b*. Set up the triangle, and calculate the line to line voltages.

Exercise 10-46 The line to line voltage in a three-phase source is given as $V_{ab} = 500\underline{/0°}$ V. Calculate the three line to neutral voltage phasors if the phase sequence is (a) *a-b-c*, and (b) *a-c-b*.

Example 10-16 A source is connected to a balanced delta as shown in Fig. 10-35 with $Z_L = (5 + j12)$ Ω. Use the voltage triangle of the previous example (Fig. 10-34). Determine the branch currents and line currents marked on the diagram.

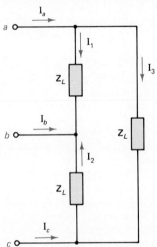

Fig. 10-35 Circuit for Example 10-16.

Solution The branch currents are calculated by using the appropriate voltage from the triangle. Careful attention must be paid to the given directions of the current and the corresponding angles of the voltages.

$$I_1 = I_{ab} = (V_{ab}/Z_L) = (173.2\underline{/30°}) \div (13\underline{/67.4°}) = 13.3\underline{/-37.4°} \text{ A}$$
$$I_2 = I_{cb} = (V_{cb}/Z_L) = (173.2\underline{/90°}) \div (13\underline{/67.4°}) = 13.3\underline{/22.6°} \text{ A}$$
$$I_3 = I_{ac} = (V_{ac}/Z_L) = (173.2\underline{/-30°}) \div (13\underline{/67.4°}) = 13.3\underline{/-97.4°} \text{ A}$$

The line currents are obtained by using *KCL* at the different nodes:

$$I_a = I_1 + I_3 = 23.0\underline{/-67.4°} \text{ A}$$
$$I_b = -I_1 - I_2 = 23.0\underline{/173°} \text{ A}$$
$$I_c = I_2 - I_3 = 23.0\underline{/52.6°} \text{ A}$$ ∎

Exercise 10-47 Change the phase sequence to *a-c-b* in Example 10-16 and recalculate the currents.

10-10 POWER IN THREE-PHASE CIRCUITS

As we have already seen in several examples and exercises, the total power in a three-phase load can be calculated by calculating the power in each load impedance and then adding them. This comment applies to the *average power* and to the *reactive power* in *all* loads, but it does not apply to the *total apparent power for unbalanced loads*.

For balanced loads, the total apparent power, the total average power, and the total reactive power are all three times the corresponding power component in each load.

In the case of *balanced loads,* it is possible to relate the total average power in a three-phase load directly to the line to line voltage and the line current. Such an expression is useful, since the line to line voltage and line currents are often directly measured or specified in a system.

Let $|V_{LN}|$ be the amplitude of the line to neutral voltage, and $|I_L|$ be the amplitude of the line current. Let the angle of each impedance in a balanced, Y connected load be θ. Since the load is Y connected, the average power consumed by each impedance is

$$P_{av} = (1/2)|V_{LN}||I_L|(\cos \theta)$$

and the total average power consumed by the load is

$$P_T = (3/2)|V_{LN}||I_L|(\cos \theta)$$

If $|V_{LL}|$ is the line to line voltage, then

$$|V_{LL}| = \sqrt{3}\,|V_{LN}|$$

from the voltage triangle of Fig. 10-30. Therefore,

$$P_T = (\sqrt{3}/2)|V_{LL}||I_L|(\cos \theta) \qquad (10\text{-}71)$$

The same expression can be derived for a delta connected load. Therefore,

Total average power in a balanced load = $(\sqrt{3}/2)$
$\times$ (amplitude of the line to line voltage)
$\times$ (amplitude of the line current) $\times$ (angle of each impedance)

An alternative version of the formula is in terms of the *rms values* of voltages and currents. Since the rms value is $(1/\sqrt{2})$ times the amplitude (or peak value), the total average power is given by

$$P_T = \sqrt{3}|V_{LLRMS}||I_{LRMS}|(\cos \theta) \qquad (10\text{-}72)$$

As is the practice in the rest of the text, *voltages and currents given are to be treated as peak values* unless otherwise stated.

Note very carefully that the angle θ is the angle of each impedance and *not* the angle between the line to line voltage and the line current.

Example 10-17 Calculate the total average power delivered by a three-phase source with line to line voltage of 500 V to each of the following balanced, Y connected loads with Z_L equal to: (a) $(30 + j0)\Omega$; (b) $(30 + j72)\Omega$; (c) $(30 - j12.5)\Omega$.

Solution Since only the amplitudes of the line currents need be determined, it is not necessary to worry about actual phase angles. For the Y connected load,

$$\text{voltage across each load} = (\text{line to line voltage}) \div \sqrt{3}$$
$$= (500/1.732) = 288.7 \text{ V}$$

Therefore, the current in each load, which is also the line current in the Y connected load, is

$$I_L = (288.8 \div |Z_L|)$$

(a) $\mathbf{Z}_L = (30 + j0) = 30\underline{/0°}\ \Omega$. $I_L = 9.62$ A.
Average power $P_T = (\sqrt{3}/2)(500)(9.62)(\cos 0°) = 4167$ W.

(b) $\mathbf{Z}_L = (30 + j72) = 78.0\underline{/67.4°}\ \Omega$. $I_L = 3.70$ A.
$P_T = (\sqrt{3}/2)(500)(3.70)(\cos 67.4°) = 616$ W.

(c) $\mathbf{Z}_L = (30 - j12.5) = 32.5\underline{/-22.6°}\ \Omega$. $I_L = 8.88$ A. $P_T = 3551$ W. ∎

Example 10-18 Repeat the calculations of the example above if the loads are delta connected.

Solution Since the voltage across each branch is the line to line voltage, the amplitude of the branch currents are $|V_{LL}/Z_L|$. The average power in each load can be calculated and multiplied by 3 to get the total average power. For a delta connected balanced load, this procedure is simpler than using Eq. (10-71).

(a) $\mathbf{Z}_L = 30\underline{/0°}\ \Omega$. Branch current $= (500/30) = 16.7$ A.
$P_T = 3 \times (1/2)(500)(16.7) \cos 0 = 12.5$ kW.

(b) $\mathbf{Z}_L = 78\underline{/67.4°}$. Branch current $= (500/78) = 6.41$ A.
$P_T = 3 \times (1/2)(500)(6.41) \cos 67.4° = 1848$ W.

(c) $\mathbf{Z}_L = 32.5\underline{/-22.6°}$. Branch current $= (500/32.5) = 15.4$ A.
$P_T = 3 \times (1/2)(500)(15.4) \cos 22.6° = 10.6$ kW.

Comparing the results of this example with those of the previous example, we see that the same impedance in a delta connection consumes three times the average power of the Y connection. ∎

Exercise 10-48 Calculate the total average power consumed by a balanced, Y connected load when the line to line voltage is 2300 V if $\mathbf{Z}_L$ is equal to: (a) $(20 + j20)\Omega$. (b) $(50 - j80)\Omega$. (c) $(80 + j0)\Omega$.

Exercise 10-49 By what factor will the answers change if the given voltage in the previous exercise is the value of the line to neutral voltage (instead of line to line)?

Exercise 10-50 Calculate the total average power consumed by a delta connected load when the line to line voltage is 1000 V if $\mathbf{Z}_L$ equals: (a) $(60 + j0)\Omega$. (b) $(10 + j90)\Omega$. (c) $(90 - j10)\Omega$.

Exercise 10-51 By what factor will the answers change if the given voltage in the previous exercise is the value of the line to neutral voltage (instead of line to line)? How does this change compare with that in Exercise 10-49?

Apparent and Reactive Power in Balanced Loads

For balanced loads, since the total apparent power equals three times the apparent power of each impedance, the following expression is valid:

$$\text{total apparent power} = (\sqrt{3}/2)|V_{LL}||I_L|$$

Similarly, the total reactive power can be written as

$$\text{total reactive power} = (\sqrt{3}/2)|V_{LL}||I_L|(\sin\theta)$$

where θ is once again the impedance angle.

The expressions above are not valid for unbalanced loads.

Exercise 10-52 Calculate the total apparent and total reactive power in Exercises 10-48 and (10-50).

10-10-1 Use of Wattmeters in Measuring Power

The wattmeter, which uses the principle of the electrodynamometer (based on the torque between two coils carrying current), was discussed in Chapter 7 in connection with the definition of average power. There are ± marks on the two coils in a wattmeter. When the wattmeter is connected to some arbitrary load, as in Fig. 10-36(a), let

$\mathbf{V}_w$ = voltage across the potential coil with the + as the positive reference (10-73a)

$\mathbf{I}_w$ = current in the current coil from the + terminal to the other terminal (10-73b)

and

$(\mathbf{V}_w, \mathbf{I}_w)$ = angle between the voltage $\mathbf{V}_w$ and the current $\mathbf{I}_w$ defined above (10-73c)

Fig. 10-36 (a) Connection of the coils of a wattmeter for measuring power. (b) Use of three wattmeters for measuring power in a three-phase Y connected load.

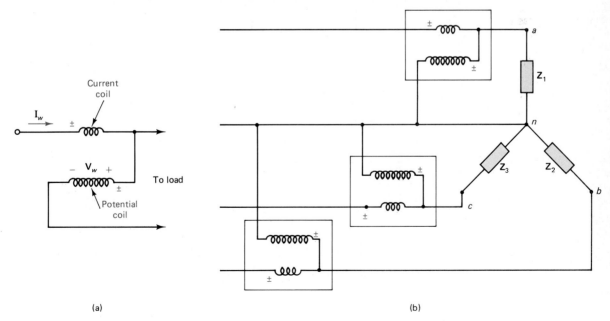

10-10 Power in Three-Phase Circuits

Then

$$\text{wattmeter reading} = (1/2)|V_w||I_w|[\cos{(\mathbf{V}_w, \mathbf{I}_w)}] \qquad (10\text{-}74)$$

Pay careful attention to the connections of the ± polarity marks of the instrument and the definitions of the voltage $\mathbf{V}_w$ and the current $\mathbf{I}_w$ in Eqs. (10-73). When using a *two wattmeter scheme* for measuring three-phase power (to be discussed below), the reading of a wattmeter can be *legitimately negative* and must be calculated or recorded as a negative value. To make sure that the reading is legitimately negative and not the result of errors in calculation or connections, it is critical to pay careful attention to the polarity of $\mathbf{V}_w$ and the direction of the current $\mathbf{I}_w$ as defined in Eqs. (10-73).

The use of three wattmeters for measuring the power in a three-phase Y connected load is indicated in Fig. 10-36(b) where the neutral point of the load serves as the common reference for all three potential coils. Equation (10-74) shows that each wattmeter measures the average power of the particular load branch to which it is connected, and the sum of the three readings will give the total average power. The connection is valid for both balanced and unbalanced Y connected loads.

Since the three wattmeter arrangement requires the presence of a common point in the load, it is not convenient for delta connected loads.

Two Wattmeter Method

The method of using two wattmeters for measuring average power in a three-phase load is useful for almost *all* types of three-phase loads: balanced or unbalanced, Y or delta connected. (The only exception is in a four-wire unbalanced load). In this method, two wattmeters are connected in a three-phase circuit by choosing *any one of the lines as the common reference for the potential coils*. The *current coils are connected in series with the other two lines* as well as the ± terminals of the potential coils. A typical arrangement is shown in Fig. 10-37. The following discussion will show that

total average power in the load = algebraic sum of the readings of the two wattmeters

Fig. 10-37 Use of two wattmeters for measuring power in a three-phase load.

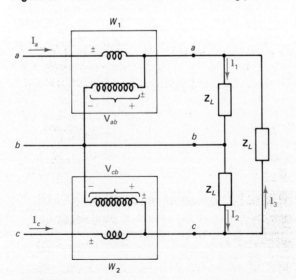

If we look at the three load impedances, the total average power is, in terms of complex power notation,

$$P_T = (1/2)\text{Re}[\mathbf{V}_{ab}\mathbf{I}_1^* + \mathbf{V}_{bc}\mathbf{I}_2^* + \mathbf{V}_{ca}\mathbf{I}_3^*] \qquad (10\text{-}75)$$

Now, since $(\mathbf{V}_{ab} + \mathbf{V}_{bc} + \mathbf{V}_{ca}) = 0$ (by *KVL*),

$$\mathbf{V}_{ca} = -(\mathbf{V}_{ab} + \mathbf{V}_{bc}) \qquad (10\text{-}76)$$

Also, the line currents $\mathbf{I}_a$ and $\mathbf{I}_c$ are related to the branch currents:

$$\mathbf{I}_a = (\mathbf{I}_1 - \mathbf{I}_3) \quad \text{and} \quad \mathbf{I}_c = (\mathbf{I}_3 - \mathbf{I}_2) \qquad (10\text{-}77)$$

Using Eqs. (10-76) and (10-77) in Eq. (10-75), it can be shown that

$$\mathbf{P}_T = (1/2)\text{Re}[\mathbf{V}_{ab}\mathbf{I}_a^* + \mathbf{V}_{cb}\mathbf{I}_c^*]$$

which is rewritten as

$$P_T = (1/2)[|\mathbf{V}_{ab}||\mathbf{I}_a| \cos{(\mathbf{V}_{ab},\mathbf{I}_a)} + |\mathbf{V}_{cb}||\mathbf{I}_c| \cos{(\mathbf{V}_{cb},\mathbf{I}_c)}] \qquad (10\text{-}78)$$

If we look at the wattmeter connections, their readings are

$$P_{w1} = (1/2)|\mathbf{V}_{ab}||\mathbf{I}_a| \cos{(\mathbf{V}_{ab},\mathbf{I}_a)} \qquad (10\text{-}79)$$

and

$$P_{w2} = (1/2)|\mathbf{V}_{cb}||\mathbf{I}_c| \cos{(\mathbf{V}_{cb},\mathbf{I}_c)} \qquad (10\text{-}80)$$

A comparison of Eqs. (10-78), (10-79), and (10-80) shows that

the *algebraic sum of the two wattmeter readings equals the total average power consumed by the load.*

Note that the only assumption made in the derivation above is that the sum of the three branch currents must be zero. No assumption is made about the load being balanced.

It is quite possible (and not at all uncommon) to have the angle in either Eq. (10-79) or Eq. (10-80), but not in both, greater than 90°, in which case *that particular wattmeter reading will be negative,* and it is important to use the *algebraic sum* of the two meter readings to find the total average power.

Example 10-19 Calculate the wattmeter readings in Fig. 10-37 if the line to line voltage $\mathbf{V}_{ab} = 500\underline{/0°}$ V, the phase sequence is *a-b-c*, and each impedance is $\mathbf{Z}_L = (10 + j60)\Omega$ (Delta). Verify by direct calculation that the total average power consumed by the load equals the algebraic sum of the two wattmeter readings.

Solution The voltage triangle is set up as shown in Fig. 10-38. The branch currents are:

$$\mathbf{I}_1 = \mathbf{V}_{ab}/\mathbf{Z}_L = (500\underline{/0°}) \div (60.83\underline{/80.5°})$$
$$= 8.22\underline{/-80.5°} \text{ A} = (1.351 - j8.108) \text{ A}$$

$$\mathbf{I}_2 = \mathbf{V}_{bc}/\mathbf{Z}_L = (500\underline{/-120°}) \div (60.83\underline{/80.5°})$$
$$= 8.22\underline{/-200.5°} \text{ A} = (-7.699 + j2.879) \text{ A}$$

$$\mathbf{I}_3 = \mathbf{V}_{ca}/\mathbf{Z}_L = (500\underline{/120°}) \div (60.83\underline{/80.5°})$$
$$= 8.22\underline{/39.5°} \text{ A} = (6.346 + j5.224) \text{ A}$$

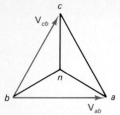

Fig. 10-38 Voltage triangle for Example 10-19.

The line currents through the currents coils of the wattmeter are:

$$\mathbf{I}_a = (\mathbf{I}_1 - \mathbf{I}_3) = (-4.995 - j13.33) = 14.24\underline{/-110°} \text{ A.}$$
$$\mathbf{I}_c = (\mathbf{I}_3 - \mathbf{I}_2) = (14.04 + j2.345) = 14.24\underline{/9.48°} \text{ A.}$$

The voltage across the potential coil of wattmeter 1 is $\mathbf{V}_{ab} = 500\underline{/0°}$ V and the current through its current coil is $\mathbf{I}_a = 14.24\underline{/-110°}$ A. Therefore,

$$\text{reading of wattmeter 1} = (1/2)(500)(14.24)(\cos 110°)$$
$$= -1247 \text{ W}$$

The voltage across the potential coil of wattmeter 2 is $\mathbf{V}_{cb} = 500\underline{/60°}$ V (from Fig. 10-38), and the current in its current coil is $\mathbf{I}_c = 14.24\underline{/9.48°}$ A. Therefore,

$$\text{reading of wattmeter 2} = (1/2)(500)(14.24)[\cos(60 - 9.48)] = 2263 \text{ W}$$

The total average power in the load is

$$P_T = (2264 - 1247) = 1017 \text{ W}$$

The average power in each branch impedance is:

$$(1/2)(500)(8.22) \cos 80.5° = 339.1 \text{ W}$$

and the total average power in the load is $(3 \times 339.1) = 1018$ W, which checks with the value obtained from the wattmeters. ∎

Exercise 10-53 Calculate the wattmeter readings of the previous example if the load is Y connected and find the total average power. Verify the result by adding the values of the individual average power.

Exercise 10-54 Calculate the wattmeter readings of Example 10-19 if the phase sequence is reversed to *a-c-b*.

10-11 SUMMARY OF CHAPTER

Magnetic circuits are structures of magnetic materials serving as paths for magnetic flux in devices based on the interaction of electric and magnetic fields. By using linear approximations, we can analyze such circuits in a manner similar to electric circuits. When the flux is known, the flux density is first calculated. The intensity of the magnetic field H is then found by using the permeability of the material. Since the permeability of ferromagnetic materials is not a constant, it is necessary to use a magnetization curve (B versus H) to find H. The product of H and the length of a section gives the *mmf* for the section. Ampere's law is used in magnetic circuits leads to: sum of the *mmf* components around a closed path $= 0$. The equation is used to find the *mmf* to be provided by a coil (or coils) placed in the magnetic circuit. The solution of the reverse problem presents difficulties because of the nonlinearity of ferromagnetic circuits; in this case, either a graphic procedure or a trial and error approach is used.

When two current coils are placed so that the magnetic flux of one links the other, there is a mutually induced voltage due to electromagnetic induction. The coefficient of coupling is a measure of the tightness of the coupling (maximum being 1 when

$M = \sqrt{L_1 L_2}$ and minimum being 0 when $M = 0$). In analyzing a circuit with coupled coils, the voltage across each coil is the algebraic sum of the self-induced voltage and the voltage(s) induced in it due to the current(s) in other coil(s). A dot polarity notation is used to guide the determination of the relative signs of the voltage components in a coil. If the current causing the self-induced voltage in a coil and the current causing the mutually induced voltage in that coil enter similarly marked terminals in the two coils, the mutual term and the self-induced term add to each other.

Coupled coils form the basis of transformer action. Ideal transformers have the property that their voltage ratio equals the turns ratio, and the current ratio is the reciprocal of the turns ratio. There are no losses in the ideal transformer. One of the applications of an ideal transformer is in the transformation of impedance levels for matching a load to a given network in order to attain maximum power transfer.

Three-phase systems are important in the generation, transmission, and distribution of electric power. The three voltages in a balanced three-phase source have equal amplitudes and a phase difference of 120° between each pair of voltages. The line to line voltages are also balanced, and the amplitude of the line to line voltage is $\sqrt{3}$ times the amplitude of the line to neutral voltage. The source may be connected as a Y or a delta. The load may be balanced (all three load impedances identically equal to one another) or unbalanced. The load may be connected in the form of a Y or a delta. In the case of balanced loads, the currents in the different branches and lines have special relationships which help simplify the analysis. For a balanced, delta connected load, the amplitude of each line current is $\sqrt{3}$ times the amplitude of each branch current.

The total average power consumed by a three-phase load is the sum of the average powers in each of the three impedances. For a balanced load, the three impedances consume the same amount of power, and the total is three times the power in the individual branches. The total average power is also given by the formula: ($\sqrt{3}/2$)(amplitude of the line to line voltage)(amplitude of the line current)(angle of each impedance). Average power is measured in the laboratory by using a two wattmeter method. The algebraic sum of the readings of the wattmeter gives the total average power.

Answers to Exercises

10-1 9.6 mWb.

10-2 $NI = 2.5 \times 10^4$ AT.

10-3 (a) H and ϕ decrease by a factor of 2. (b) B not affected. ϕ doubles. (c) B and ϕ double.

10-4 (a) 1.25×10^9. (b) 5.21×10^5.

10-5 (a) 4.36×10^{-4} Wb. (b) 28.6 A.

10-6 $NI = (347 \times 0.3575) + (8.67 \times 10^5 \times 0.0025) = 2292$ AT.

10-7 $NI = 1.2 \times 10^5$ AT.

10-8 $NI = 6.23 \times 10^4$ AT.

10-9 $B = 1.51 - 1.51 \times 10^{-3} H_c$. 7.77 mWb.

10-10 $B = 0.628 - 7.54 \times 10^{-4} H_c$. 4.3 mWb.

10-11 $H_2 = 400$, $H_1 = 110$. $B = 0.18$ T. 1.8 mWb.

10-12 $H_2 = 530$. $H_1 = 150$. $B = 0.25$ T. 2.5 mWb.

10-13 (a) 1.58 mH. (b) 7.90 mH. (c) 14.2 mH. Max $M = 15.8$ mH.

10-14 $v_1 = L_1 K_1 + M/t^2$. $v_2 = -L_2/t^2 + K_1 M$.

10-15 Verification.

10-16 $j10\mathbf{I}_1 + j15\mathbf{I}_2 = 1000$. $\mathbf{I}_2 = 90.7\underline{/-107°}$ A. $\mathbf{V}_2 = 3053\underline{/-8.67°}$ V.

10-17 (a) dot on terminal D. (b) dot on terminal H.

10-18 $M = 1.25$ mH. $L_2 = 10.5$ mH.

10-19 $L_{eq} = (L_1L_2 - M^2)/(L_1 + L_2 + 2M)$.

10-20 (a) $(R_1 + j\omega L_1)\mathbf{I}_1 - (R_1 - j\omega M)\mathbf{I}_2 = \mathbf{V}_s$.
$-(R_1 - j\omega M)\mathbf{I}_1 + (R_1 + R_2 + j\omega L_2)\mathbf{I}_2 = 0$. (b) $(R_1 + j\omega L_1)\mathbf{I}_1 - j\omega(L_1 - M)\mathbf{I}_2 = \mathbf{V}_s$. $-j\omega(L_1 - M)\mathbf{I}_1 + [R_2 + j\omega(L_1 + L_2 - 2M)]\mathbf{I}_2 = 0$.

10-21 Equations: $j\omega[(L_1 + L_2 + 2M)\mathbf{I}_1 - (L_2 + M)\mathbf{I}_2]$
$= \mathbf{V}_s$. $-j\omega(L_2 + M)\mathbf{I}_1 + (R + j\omega L_2)\mathbf{I}_2 = 0$. $(\mathbf{V}_2/\mathbf{V}_s) = [j\omega(L_2 + M)R]/[j\omega R(L_1 + L_2 + 2M) - \omega^2(L_1L_2 - M^2)]$.

10-22 Steps exactly similar to the discussion preceding the exercise.

10-23 (a) $w = 0.1e^{-t} + 0.006t^2 + 0.03te^{0.15t} = 0.0596$ J. (b) 0.0155 J.

10-24 0.0429 J. 0.0267 J.

10-25 Verification.

10-26 Verification.

10-27 $(25 + j20)\mathbf{I}_1 - (10 + j15.8)\mathbf{I}_2 = 100$. $-(10 + j15.8)\mathbf{I}_1 + (40 + j35)\mathbf{I}_2 = 0$. Answers same as in Example 10-9.

10-28 $(L_1 - M)$ in parallel with $(L_2 - M)$ and the combination in series with M.
$L_{eq} = (L_1L_2 - M^2)/(L_1 + L_2 - 2M)$.

10-29 $(L_1L_2 - M^2)/(L_1 + L_2 + 2M)$.

10-30 Replace a by $-a$.

10-31 (a) $\mathbf{V}_s - 15\mathbf{V}_2 + \mathbf{V}_2 = 0$. $\mathbf{V}_2 = 7.86$ V. (b) $-14R_1\mathbf{I}_1 + 15\mathbf{V}_2 = \mathbf{V}_s$. $(14R_1 + 15R_L)\mathbf{I}_1 = \mathbf{V}_2$. $\mathbf{V}_{2b} = 7.55$ V.

10-32 $\mathbf{I}_s = -6\mathbf{I}_2$. Mesh 2: $\mathbf{V}_2 + R_1\mathbf{I}_2 + 0.2\mathbf{V}_2 = 0$. $\mathbf{I}_s = 8.64$ A.

10-33 $\mathbf{Z}_L = (a\mathbf{V}_2)/(\mathbf{I}_2/a)$.

10-34 (a) $(5.21 + j6.94) \times 10^{-3}\,\Omega$ (b) $(1728 + j2304)\,\Omega$.

10-35 Ratio 17.7:1. Side with the fewer turns.

10-36 $\mathbf{V}_{bd} = \mathbf{V}_{bc} + \mathbf{V}_{cd} = 140\underline{/59.6°}$ V. $\mathbf{V}_{ec} = \mathbf{V}_{ed} + \mathbf{V}_{dc} = 50\underline{/-30°}$ V.

10-37 $100\underline{/-60°}$ V.

10-38 $\mathbf{V}_{ab} = \mathbf{V}_{an} + \mathbf{V}_{nb} = 1.732V_m\underline{/30°}$ V. $\mathbf{V}_{bc} = 1.732V_m\underline{/-90°}$ V.
$\mathbf{V}_{ca} = 1.732V_m\underline{/150°}$ V.

10-39 Interchange the values of $\mathbf{V}_{bn}$ and $\mathbf{V}_{cn}$ and those of $\mathbf{I}_{bn}$ and $\mathbf{I}_{cn}$ in the steps of Example 10-12. Power values are the same as before.

10-40 $\mathbf{I}_{an} = 11\underline{/45°}$ A. $\mathbf{I}_{bn} = 11\underline{/-75°}$ A. $\mathbf{I}_{cn} = 11\underline{/165°}$ A. 1283 W.

10-41 3 kVA and 2 kW per impedance. Current in each impedance = 12 A. pf = (2/3).
$\mathbf{Z} = 41.7\underline{/+48.2°}\,\Omega$. Insufficient data to choose the sign.

10-42 $\mathbf{Z}_a$: 99.9 W. $\mathbf{Z}_b$: 88.1 W. $\mathbf{Z}_c$: 0 W. Total: 188 W.

10-43 $\mathbf{I}_{ab} = 10.9\underline{/-71.6°}$ A. $\mathbf{I}_{bc} = 10.9\underline{/168.4°}$ A.
$\mathbf{I}_{ca} = 10.9\underline{/48.4°}$ A. $P_{av} = 892.8$ W.

10-44 Apparent power per branch = 3.333 kVA. Branch current = 12.2 A. pf = 0.095. $Z = 13.5/+84.5°\ \Omega$.

10-45 Interchange vertices b and c in Fig. 10-34. $\mathbf{V}_{ab} = 173.2/-30°$ V. $\mathbf{V}_{bc} = 173.2/90°$ V. $\mathbf{V}_{ca} = 173.2/-150°$.

10-46 (a) $\mathbf{V}_{an} = 288.7/-30°$ V. $\mathbf{V}_{bn} = 288.7/-150°$ V. $\mathbf{V}_{cn} = 288./90°$ V. (b) Change the signs of the angles in part (a).

10-47 $\mathbf{I}_1 = 13.3/-97.4°$ A. $\mathbf{I}_2 = 13.3/-157.4°$ A. $\mathbf{I}_3 = 13.3/-37.4°$ A. $\mathbf{I}_a = 23.0/-67.3°$ A. $\mathbf{I}_b = 23.0/52.6°$ A. $\mathbf{I}_c = 23.0/173°$ A.

10-48 Line to neutral voltage = 1328 V. (a) $|I| = 47.0$ A. $P = 66.1$ kW. (b) $|I| = 14.1$ A. $P = 14.9$ kW. (c) $|I| = 16.6$ A. $P = 33.1$ kW.

10-49 Each value of average power will increase by a factor of 1.732.

10-50 (a) $|I| = 16.7$ A. $P = 25.0$ kW. (b) $|I| = 11.0$ A. $P = 1.82$ kW. (c) $|I| = 11.0$ A. $P = 16.4$ kW.

10-51 Values increase by a factor of 1.732.

10-52 93.5 kVA, 66.1 kVAR; 28.0 kVA, 23.8 kVAR; 33.1 kVA, 0 kVAR; 25 kVA, 0 kVAR: 16.5 kVA, 16.2 kVAR; 16.5 kVA, 1.82 kVAR.

10-53 Wattmeter 1: $\mathbf{I}_a = 4.75/-110.5°$ A. $\mathbf{V}_{ab} = 500/0°$ V. Reading = -415.9 W. Wattmeter 2: $\mathbf{I}_c = 4.75/9.5°$ A, $\mathbf{V}_{cb} = 500/60°$ V. Reading = 755.3 W. Total power = 339.4 W.

10-54 Wattmeter 1: $\mathbf{V}_{ab} = 500/0°$ V. $\mathbf{I}_a = 14.24/-50.5°$ A. Reading = 2264 W. Wattmeter 2: $\mathbf{V}_{cb} = 500/-60°$ V. $\mathbf{I}_c = 14.24/-170.4°$ A. Reading = -1241 W. Total power = 1023 W.

PROBLEMS

Secs. 10-1, 10-2, 10-3 Magnetic Circuits

10-1 A toroidal core made of cast steel has a mean radius of 0.06 m and an area of cross section 3.2×10^{-3} m². (a) Calculate the *mmf* needed to establish a flux of 4 mWb in the core. (b) Calculate the reluctance of the circuit.

10-2 For the toroidal core of the previous problem: (a) Calculate the flux resulting from a coil of 400 turns carrying a current of 3 A. (b) Calculate the reluctance of the circuit. Is the reluctance different from the previous problem? Should it be different? Discuss.

10-3 Repeat Problem 10-1 if the toroidal core is made up of a nonmagnetic material.

10-4 Repeat Problem 10-2 if the toroidal core is made up of a nonmagnetic material.

10-5 A toroidal core made of cast steel has a mean circumference of 1.2 m and a cross sectional area of 4×10^{-4} m². A coil of 600 turns is wound on the core, and the current in the coil is 6 A. (a) Determine the flux in the core. (b) Calculate the width of an air gap to be cut radially in the core if it is desired to limit the flux in the core to 0.48 mWb.

10-6 A rectangular core of the form shown in Fig. 10-7(a) is made of cast iron. The width of the air gap is 1.5 mm. The length of the path in the ferromagnetic material is 0.8 m. The area of cross section is 50×10^{-4} m². Calculate the current in a coil of 1200 turns to establish a flux of 2.5 mWb in the core. Ignore the fringing effect.

10-7 For the magnetic circuit of the previous problem, calculate the flux established if the current in the coil is 1 A.

10-8 Repeat the calculations of the previous problem if the material of the core is cast steel instead of cast iron.

Secs. 10-4 to 10-8 Coupled Coils and Ideal Transformers

10-9 If $v_s(t) = 100 \cos 200t$ V in the circuit of Fig. 10-39, determine the current $i_1(t)$ if (a) the switch S is open and (b) the switch S is closed.

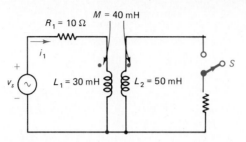

Fig. 10-39 Circuit for Problem 10-9.

10-10 For each of the circuits shown in Fig. 10-40, write the mesh or loop equations using the currents already assigned. Use phasors and impedances.

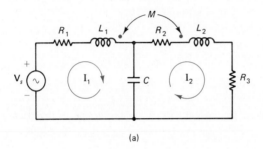

(a)

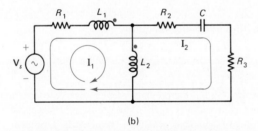

(b)

Fig. 10-40 Circuits for Problem 10-10.

10-11 Determine the input impedance of each of the circuits shown in Fig. 10-41. Do not use T equivalent circuits for the coupled coils.

Fig. 10-41 Circuits for Problem 10-11.

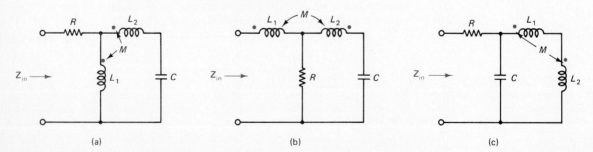

(a)   (b)   (c)

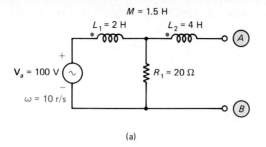

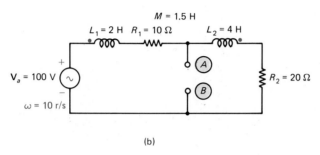

Fig. 10-42 Circuits for Problem 10-12.

10-12 Determine the Thevenin equivalent circuit as seen from the terminals A-B of each of the circuits shown in Fig. 10-42. Do not use T equivalent circuits for the coupled coils.

10-13 In Fig. 10-23, let $L_1 = 0.1$ H, $L_2 = 0.25$ H, and $M = 0.15$ H. Connect a source $V_s = 5\underline{/0°}$ V at $\omega = 100$ r/s across L_1 and a load $R_L = 25\Omega$ across L_2. (a) Determine the energy stored in the circuit at $t = 10$ ms. (b) Reverse the polarity of one of the coils and recalculate the energy of part (a).

10-14 Repeat the work of the previous problem if the coefficient of coupling is changed to 0.3 (with L_1 and L_2 having the same values as before).

10-15 Repeat the work of Problem 10-11 after replacing the coupled coils by their T equivalent circuits.

10-16 Repeat the work of Problem 10-12 after replacing the coupled coils by their T equivalent circuits.

10-17 An ideal transformer is designed to convert 110 V to 12 V. (a) If a resistance of 20 Ω is connected as a load to the low voltage side with 110 V applied to the high voltage side, calculate the average power supplied by the 110 V source. (b) If the 20 Ω resistor is connected to the high voltage side with 12 V applied to the low voltage winding, calculate the average power supplied by the 12 V source.

10-18 A voltage source $V_s = 100\underline{/0°}$ with a series impedance of $Z_s = (30 + j40)\Omega$ is available. The $V_s - Z_s$ combination is connected in series with one winding of an ideal transformer, while a load resistance of 20 Ω is connected to the other winding. For each of the following cases, calculate the current supplied by V_s and the average power delivered by it: (a) Turns ratio = 1:2.5 with the load on the side with the larger number of turns. (b) Turns ratio = 2.5:1 with the load on the side with the fewer turns.

10-19 A voltage source $V_s = 12\underline{/0°}$ V in series with an impedance $Z_s = (10 - j90)\Omega$ is connected to the side of an ideal transformer with N_1 turns. A load impedance Z_L is connected to the side with N_2 turns. For each of the following cases, determine Z_L for maximum average power transfer to the load and the maximum average power received by the load.

(a) $(N_1/N_2) = 5$. Both the resistance and reactance of Z_L can be varied independently.
(b) $(N_1/N_2) = 5$. Z_L is a resistance.
(c) $(N_1/N_2) = 0.1$. Both the resistance and reactance of Z_L can be varied independently.
(d) $(N_1/N_2) = 0.1$. Z_L is a pure reactance.

10-20 Determine the input impedance Z_{in} of the circuit shown in Fig. 10-43. The transformer is ideal.

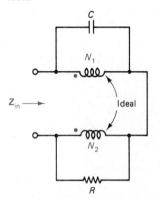

Fig. 10-43 Circuit for Problem 10-20.

10-21 Determine the current supplied by the source V_s in each of the circuits of Fig. 10-44.

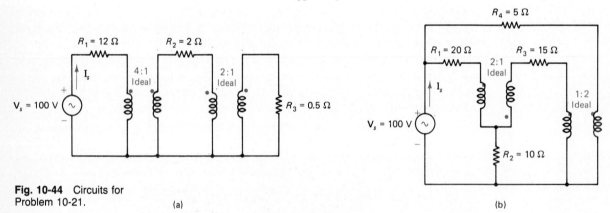

Fig. 10-44 Circuits for Problem 10-21.

Secs. 10-9, 10-10 Three Phase Systems

10-22 The voltage V_{an} in a balanced three-phase power supply is given by $V_{an} = 230\underline{/0°}$ V and the phase sequence is a-b-c. Calculate the following voltages (a) V_{ab}. (b) V_{ac}. (c) V_{bn}. (d) V_{ca}. (e) V_{cb}.

10-23 Repeat the work of the previous problem if the phase sequence is reversed to a-c-b.

10-24 A balanced Y connected load with $Z_L = (20 - j60)\Omega$ is connected to a three-phase source with $V_{ab} = 550\underline{/0°}$ V and phase sequence a-b-c. (a) Calculate the three line currents I_a, I_b, and I_c. (b) Calculate the average power received by each impedance. (c) Calculate the total average power delivered by the source.

10-25 Repeat the work of the previous problem if the impedances are connected in the form of a delta.

10-26 Three identical impedances Z_1 are connected in the form of a Y to a three-phase balanced source. If the load is to be replaced by a balanced delta connected load (each of impedance Z_2), how is Z_2 related to Z_1 if the line currents remain the same in both cases? Derive your answer.

10-27 A three-phase balanced source with $V_{an} = 250/\underline{0°}$ V feeds a Y connected load with the center of the load tied to the neutral point of the source. If $Z_a = (40 - j60)\Omega$, $Z_b = (60 - j40)\Omega$, and $Z_c = (100 + j100)\Omega$, determine: (a) The three line currents. (b) The average power received by each impedance. (c) The total average power delivered by the source.

10-28 A three-phase balanced source with $V_{an} = 250/\underline{0°}$ V feeds a delta connected load with $Z_{ab} = (40 - j60)\Omega$, $Z_{bc} = (60 - j40)\Omega$, and $Z_{ca} = (100 + j100)\Omega$. Determine (a) the three line currents; (b) the average power received by each impedance; (c) the total average power delivered by the source.

10-29 A balanced three-phase source with $V_{ab} = 220/\underline{90°}$ V feeds an unbalanced load consisting of $Z_1 = (5 + j15)\Omega$, $Z_2 = (15 + j15)\Omega$, $Z_3 = (15 + j5)\Omega$. The impedances are connected in the form of a Y with Z_1 connected to line a, Z_2 to line b, and Z_3 to line c. Determine (a) the current in each impedance; (b) the three line currents; (c) the average power received by each impedance; and (d) the total average power for each of the following cases: (i) phase sequence a-b-c, (ii) phase sequence a-c-b.

10-30 Repeat the work of the previous problem if the three impedances are connected in the form of a delta: Z_1 between lines a and b, Z_2 between b and c, and Z_3 between c and a.

10-31 A balanced three-phase system with line to line voltage of 325 V (frequency 60 Hz) delivers a total average power of 24 kW to a balanced load. The power factor of each impedance is 0.75 (current lagging). (a) Determine a simple series equivalent circuit for each impedance, assuming the load to be Y connected. (b) Calculate the total apparent power supplied by the source. (c) Calculate the total reactive power supplied by the source.

10-32 A balanced three-phase system with line to neutral voltage of 400 V (frequency 50 Hz) delivers a total apparent power of 30 kVA and a total average power of 15 kW to a balanced inductive load. (a) Obtain a simple parallel equivalent circuit for each branch of the load, assuming the load to be delta connected. (b) Determine the value of a reactance element that should be placed in parallel with each load to increase the power factor to 0.75 (current lagging). (c) Calculate the total apparent and total average power supplied by the system to the modified load.

10-33 Repeat the work of the previous problem, if the load is Y connected.

10-34 Two wattmeters are connected as follows in a balanced three-phase system. *Wattmeter 1:* current coil in series with line a, + terminal of potential coil to line a and the other terminal of the potential coil to line c. *Wattmeter 2:* current coil in series with line b, + terminal of potential coil to line b, and the other terminal of the potential coil to line c. Assume a balanced, delta connected load with each impedance $|Z_L|/\underline{\theta°}\,\Omega$ and $V_{an} = V_m/\underline{0°}$ V. (a) Obtain the expressions for the readings of the individual wattmeters. (b) Show that the algebraic sum of the two meter readings is the sum of the total average power consumed by the circuit. (c) Calculate the range of values of the angle θ for which one of the wattmeters will show negative values. (d) When will the readings of the two wattmeters be equal?

10-35 A balanced three-phase source with $V_{ab} = 600/\underline{0°}$ V feeds a balanced delta connected load with $Z_L = 200/\underline{-75°}\,\Omega$. Two wattmeters are connected with current coils in series on lines b and c and the common connection of the potential coils on line a. Calculate the individual wattmeter readings when the phase sequence is (a) a-b-c, and (b) a-c-b.

10-36 Three unequal loads are connected to form a delta and fed by a balanced three-phase source with line to line voltage of 320 V. The data are: load Z_{ab} consumes 10 kW at unity power factor; load Z_{bc} consumes 20 kW at a powerfactor of 0.7 (current lagging); and Z_{ca} consumes 15 kW at a power factor of 0.7 (current leading). Two wattmeters are connected as stated in Problem 10-34 (with phase sequence a-b-c). Calculate (a) the individual wattmeter readings. (b) The total average power consumed by the load. (c) The total reactive power consumed by the load. (d) The total apparent power consumed by the load.

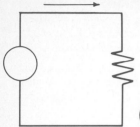

CHAPTER 11
RESONANT CIRCUITS

Networks containing both inductors and capacitors exhibit some interesting frequency response characteristics. For example, they present a negligibly small attenuation to signals over a band of frequencies and essentially suppress signals outside that band. The frequency response characteristics of such networks lead to a variety of practical applications.

Since the energy stored in the magnetic field of an inductance depends upon the current in the inductance, and the energy stored in the electric field of a capacitance depends upon the voltage across the capacitor, a mutual exchange of energy is possible in a circuit containing both inductors and capacitors. When the resistance of the circuit is zero or close to zero, the exchange of energy between the magnetic field of the inductance and the electric field of the capacitance continues for long periods of time in the form of oscillations at a particular frequency. The phenomenon whereby the exchange of energy between the two fields reaches an optimum condition is known as *resonance*. The properties of a circuit at resonance form the basis of frequency selective networks.

We develop resonance and the properties of resonant circuits in this chapter using phasors and sinusoidal steady state analysis. We discuss resonance again in terms of complex frequency in Chapter 12.

11-1 RESONANCE IN A PARALLEL GLC CIRCUIT

Consider a parallel combination of a conductance G, inductance L, and capacitance C driven by a current source of variable frequency—see Fig. 11-1(a).

The admittance of the circuit is

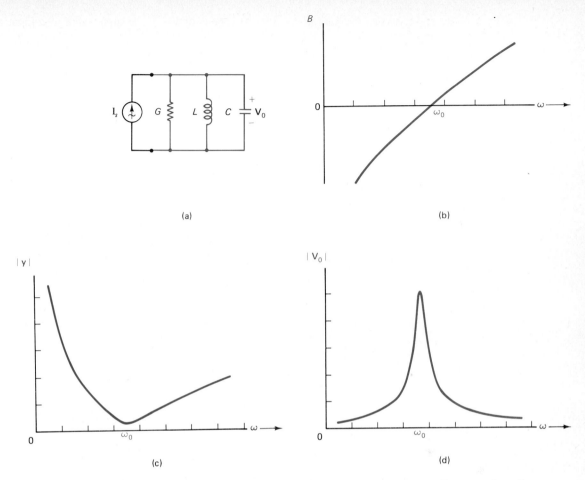

Fig. 11-1 (a) Parallel GLC circuit driven by a variable frequency source. Frequency dependence of (b) susceptance, (c) admittance, and (d) voltage across the circuit.

$$\mathbf{Y} = G + jB \qquad (11\text{-}1)$$

where G is a constant and B is a function of frequency given by

$$B = (\omega C - 1/\omega L) \qquad (11\text{-}2)$$

Consider the variation of B with frequency.

At $\omega = 0$, $\omega C = 0$, $(1/\omega L) = \infty$ and $B = -\infty$. As ω increases, ωC increases, $(1/\omega L)$ decreases, and B increases. At some frequency, ω_0, called the *resonant frequency*, B becomes 0. That is, at $\omega = \omega_0$,

$$B = (\omega_0 C - 1/\omega_0 L) = 0$$

which gives the expression for the resonant frequency:

$$\omega_0 = \frac{1}{\sqrt{LC}} \qquad (11\text{-}3)$$

As frequency increases above ω_0, ωC continues to increase and $(1/\omega L)$ continues to decrease, so that B continues to increase. The variation of the susceptance as a function of frequency is shown in Fig. 11-1(b).

Now consider the magnitude of the admittance $|Y|$. From Eq. (11-1),

$$|Y| = \sqrt{G^2 + B^2}$$

Therefore, $|Y|$ is always $\geq G$. It starts at ∞ at $\omega = 0$ and decreases as frequency increases. At resonance, $B = 0$ and

$$|Y| = G \text{ at } \omega = \omega_0 \qquad (11\text{-}4)$$

that is, $|Y|$ has its *minimum* value of G at resonance. As frequency increases above ω_0, $|Y|$ increases and $\rightarrow \infty$ as $\omega \rightarrow \infty$.

Figure 11-1(c) shows the variation of $|Y|$ as a function of frequency.

At resonance, $B = 0$ and

$$\mathbf{Y} = G + j0 = G\underline{/0°} \qquad (11\text{-}5)$$

Now consider the voltage $\mathbf{V}_0$ across the circuit.

$$\mathbf{V}_0 = \mathbf{I}_s/\mathbf{Y}$$

Assuming $\mathbf{I}_s$ to have the same amplitude at all frequencies, $|\mathbf{V}_0|$ is inversely proportional to $|\mathbf{Y}|$. As shown in Fig. 11-1(d), it starts at zero at $\omega = 0$, increases, reaches a maximum value of $|\mathbf{I}_s/G|$ at ω_0, and then decreases to zero. Note that since the angle of $\mathbf{Y}$ is zero at resonance, $\mathbf{V}_0$ and $\mathbf{I}_s$ are in phase at resonance.

If we collect the pieces of information in the preceding discussion, we have:

$$\text{Resonant frequency: } \omega_0 = \frac{1}{\sqrt{LC}} \text{ r/s}$$

or $\qquad (11\text{-}6)$

$$f_0 = \frac{1}{2\pi\sqrt{LC}} \text{ Hz}$$

At resonance,

$$\text{susceptance } B = 0$$
$$\text{admittance } \mathbf{Y} = G \text{ (minimum)}$$
$$\text{voltage } \mathbf{V}_0 = (\mathbf{I}_s/G) \text{ (maximum)}$$

$\mathbf{V}_0$ is in phase with $\mathbf{I}_s$.

A direct application of the property of maximum output voltage at resonance is illustrated in the following example.

Example 11-1 The source in the circuit of Fig. 11-1 produces a current containing several frequencies of equal amplitude:

$$i_s(t) = 0.10 + 0.10 \cos 1000t + 0.10 \cos 2000t + 0.10 \cos 3000t$$

(a) Design a parallel GLC circuit to pass the $\omega = 2000$ r/s component more freely than the other components, if it is given that $G = 0.001$ S and $L = 50$ mH.
(b) Repeat the calculations if G is changed to 2×10^{-4} S.

Solution (a) Since the voltage is a maximum at resonance, the resonant frequency of the circuit must be set at 2000 r/s. The required value of C is, from Eq. (11-3),

$$C = \frac{1}{\omega_0^2 L}$$
$$= 5 \times 10^{-6} \text{ F}$$

or 5 μF. The admittance of the circuit at any frequency is

$$\mathbf{Y} = 0.001 + j(5 \times 10^{-6}\omega - 20/\omega) \text{ S} \tag{11-7}$$

and

$$\mathbf{V}_0 = (\mathbf{I}/\mathbf{Y}) = (0.1/\mathbf{Y}) \tag{11-8}$$

since the amplitude of the given current is 0.1 A for all the components.

Consider the different frequency components of the input current:

$\omega = 0$ (dc): $\mathbf{Y} = -j\infty$ and $\mathbf{V}_0 = 0$

$\omega = 1000$ r/s: From Eqs. (11-7) and (11-8),

$$\mathbf{Y} = (0.001 - j0.015) = 0.015\underline{/-86.2°} \text{ S}$$
$$\mathbf{V}_0 = (0.1/\mathbf{Y}) = 6.67\underline{/86.2°} \text{ V}$$

$\omega = 2000$ r/s: From Eqs. (11-7) and (11-8),

$$\mathbf{Y} = 0.001 \text{ S}$$
$$\mathbf{V}_0 = (0.1/\mathbf{Y}) = 100\underline{/0°} \text{ V}$$

$\omega = 3000$ r/s: From Eqs. (11-7) and (11-8),

$$\mathbf{Y} = (0.001 + j0.00833) = 0.00839\underline{/83.2°} \text{ S}$$
$$\mathbf{V}_0 = (0.1/\mathbf{Y}) = 11.9\underline{/-83.2°} \text{ F}$$

The total output voltage is therefore,

$$v_0(t) = 6.67 \cos(1000t + 86.2°) + 100 \cos 2000t + 11.9 \cos(3000t - 83.2°)$$

The 2000 r/s component in the output is 15 times as high as the 1000 r/s component and 8.4 times as high as the 3000 r/s component. The resonant circuit has thus significantly attenuated the components at other than the resonant frequency.

(b) G changed to 2×10^{-4} S. From Eq. (11-3), G has no effect on the value of the resonant frequency. $C = 5$ μF as before, and the admittance of the circuit at resonance is given by

$$\mathbf{Y} = 2 \times 10^{-4} + j(5 \times 10^{-6}\omega - 20/\omega) \text{ S} \tag{11-9}$$

and the output voltage is again given by Eq. (11-8).

Consider the different frequencies of the input current.

$\omega = 0$: $\mathbf{Y} = -j\infty$ and $\mathbf{V}_0 = 0$ as before.

$\omega = 1000$ r/s: $\mathbf{Y} = (2 \times 10^{-4} - j0.015) = 0.015\underline{/-89.2°} \text{ S}$

$$\mathbf{V}_0 = 6.67\underline{/89.2°} \text{ V}$$

$\omega = 2000$ r/s: $\quad \mathbf{Y} = 2 \times 10^{-4}$ S
$\qquad\qquad\qquad \mathbf{V}_0 = 500\underline{/0°}$ V

$\omega = 3000$ r/s: $\quad \mathbf{Y} = (2 \times 10^{-4} + j83.3 \times 10^{-4}) = 83.3 \times 10^{-4}\underline{/88.6°}$ S
$\qquad\qquad\qquad \mathbf{V}_0 = 12\underline{/-88.6°}$ V

The total output voltage is

$$v_0(t) = 6.67 \cos(1000t + 89.2°) + 500 \cos 2000t + 12 \cos(3000t - 88.6°)$$

The amplitude of the component at 2000 r/s is seen now to be 75 times as high as that at 1000 r/s and 42 times as high as that at 3000 r/s.

The smaller value of G has resulted in a more efficient suppression of the components at the unwanted frequencies. It will become evident in the discussion that this is to be expected for the given set of values. ∎

Exercise 11-1 A parallel GLC resonant circuit is known to have a resonant frequency of 10^4 r/s, $C = 50$ nF, and its admittance at resonance is 10^{-4} S. Determine the values of L and G.

Exercise 11-2 Suppose the current source in a parallel GLC circuit produces the current

$$i_s(t) = 0.01 \cos(5 \times 10^3 t) + 0.01 \cos 10^4 t$$

Given that $C = 100$ nF, choose the values of L and G so that (a) resonance is at 5×10^3 r/s and the amplitude of the component in the output voltage at 5×10^3 r/s is 50 times as high as the other component; (b) resonance is at 10^4 r/s and the amplitude of the component in the output voltage at 10^4 r/s is 50 times as high as the other component.

11-1-1 Properties of the Parallel Resonant Circuit

The minimum admittance of the parallel resonant circuit depends only upon G, and not upon L and C.

When $G = 0$ in the circuit, $\mathbf{Y}_{\min} = 0$, and the circuit presents an *open circuit* to the source. As we will see shortly, this does not mean that the current in each individual branch is zero. It only means that the source supplies no current to the circuit at resonance when $G = 0$. The explanation for this rather startling situation is based on the exchange of energy between the electric field of the capacitance and the magnetic field of the inductance.

Consider a parallel GLC circuit driven by a voltage source $\mathbf{V}_s$ as shown in Fig. 11-2(a). Let the current supplied by the source be $\mathbf{I}_s$ and the currents in the three branches be $\mathbf{I}_G$, $\mathbf{I}_L$, and $\mathbf{I}_C$. Then

$$\mathbf{I}_s = \mathbf{V}_s \mathbf{Y} \text{ or } \mathbf{V}_s = \frac{\mathbf{I}_s}{\mathbf{Y}} \qquad (11\text{-}10)$$

$$\mathbf{I}_G = G\mathbf{V}_s = G\left(\frac{\mathbf{I}_s}{\mathbf{Y}}\right) \qquad (11\text{-}11)$$

$$\mathbf{I}_L = -j\frac{1}{\omega L}\mathbf{V}_s = -j\frac{1}{\omega L}\frac{\mathbf{I}_s}{\mathbf{Y}} \qquad (11\text{-}12)$$

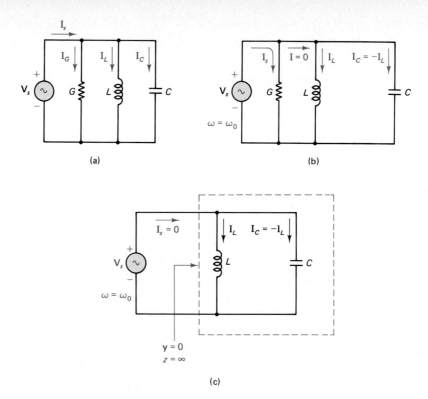

Fig. 11-2 Current distribution in a parallel resonant circuit. I_L and I_C are equal in magnitude but 180° out of phase.

$$\mathbf{I}_C = (j\omega C)\mathbf{V}_s = j\omega C \frac{\mathbf{I}_s}{\mathbf{Y}} \tag{11-13}$$

where $\mathbf{Y}$ is given by Eqs. (11-1) and (11-2).

At resonance, $\mathbf{Y} = G$ and, from Eq. (11-11)

$$\mathbf{I}_G = \mathbf{I}_s \tag{11-14}$$

Since

$$\mathbf{I}_G + \mathbf{I}_L + \mathbf{I}_C = \mathbf{I}_s$$

Eq. (11-14) leads to the following constraint between $\mathbf{I}_L$ and $\mathbf{I}_C$:

$$\mathbf{I}_L + \mathbf{I}_C = 0 \tag{11-15}$$

Eqs. (11-12), (11-13), and (11-15) show that $\mathbf{I}_L$ and $\mathbf{I}_C$ are not individually equal to zero, but their sum is zero at resonance.

Therefore,

at resonance, all the current supplied by the source goes through G, while the currents in L and C exactly cancel each other. If $G = 0$ (that is, a parallel LC circuit), there would be no path through which the source current could flow, and consequently there would be no current supplied by V_s.

These situations are shown in Figs. 11-2(b) and (c).

11-1   Resonance in a Parallel GLC Circuit

Example 11-2 In Fig. 11-2(a), assume that $V_s = 100$ V and is operating at the resonant frequency of the circuit. $L = 40$ mH, $C = 900$ pF. Evaluate the branch currents and I_s for the following values of G: (a) 10 S, and (b) 0.1 S.

Solution The resonant angular frequency is given by

$$\omega_0 = 1/\sqrt{LC} = 1.67 \times 10^5 \text{ r/s}$$

At resonance,

$$Y_L = -j(1/\omega_0 L) = -j1.5 \times 10^{-4} \text{ S}$$
$$Y_c = j(\omega_0 C) = j1.5 \times 10^{-4} \text{ S}$$

The currents in the inductance and capacitance are not affected by the value of G. They are:

$$I_L = Y_L V_s = -j1.5 \times 10^{-2} \text{ A}$$
$$I_C = Y_c V_s = j1.5 \times 10^{-2} \text{ A}$$

Note that I_L and I_c are equal in magnitude but 180° out of phase so that they cancel each other.

The current in G and the total current I_s are given by

$$I_s = I_G = GV_s$$

since

$$I_L + I_C = 0.$$

(a) $G = 10$ S: $I_s = I_G = 1000\underline{/0°}$ A
(b) $G = 0.1$ S: $I_s = I_G = 10\underline{/0°}$ A

Note that the current in the inductance and capacitance remain the same in both cases, while the current supplied by the voltage source decreases in proportion to the value of G. ∎

Exercise 11-3 Calculate the branch currents and the current supplied by the voltage source in a parallel resonant circuit, as in Fig. 11-2(a), with $G = 0.01$ S, $L = 50$ mH, and $C = 250$ pF, if $V_s = 100$ V operating at the resonant frequency. Repeat the calculations if $G = 0$.

11-1-2 Energy Considerations

The energy stored by the capacitor in its electric field and by the inductor in its magnetic field play an important part in the behavior of a resonant circuit. The resistive element G does not store energy, but dissipates it in the form of heat. At resonance, there is an exchange of energy between the inductor and the capacitor: As the energy in one of them increases, the energy in the other decreases. The source is needed to provide energy only to the dissipative element G. If $G = 0$, then the source need not supply any energy, which explains why it supplies no current. The exchange of energy between L and C will continue for an indefinitely long time even if the source is removed in such a case (assuming that the inductor and capacitor are ideal elements with no losses).

Let the voltage $v_s(t)$ of the source in the circuit of Fig. 11-2(a) be at the resonant frequency:

$$v_s(t) = V_m \cos \omega_0 t \qquad (11\text{-}16)$$

The energy stored in the capacitor is given by

$$w_c(t) = (1/2)Cv_c^2$$

or

$$w_c(t) = (1/2)CV_m^2 \cos^2 \omega_0 t \qquad (11\text{-}17)$$

since $v_c = v_s$. Therefore, the energy in the capacitor is

$$w_c(t) = \frac{CV_m^2}{4}(1 + \cos 2\omega_0 t) \qquad (11\text{-}18)$$

The energy stored in the inductor is given by

$$w_L(t) = (1/2)Li_L^2 \qquad (11\text{-}19)$$

The *phasor* $\mathbf{I}_L$ is given by

$$\mathbf{I}_L = \frac{\mathbf{V}_s}{j\omega_0 L} = \frac{\mathbf{V}_s}{\omega_0 L}\underline{/-90°}$$

so that the *time domain* form of i_L becomes

$$i_L(t) = \frac{V_m}{\omega_0 L} \cos(\omega_0 t - 90°)$$

$$= \frac{V_m}{\omega_0 L} \sin \omega_0 t \qquad (11\text{-}20)$$

Substitution of Eq. (11-20) in Eq. (11-19) leads to

$$w_L(t) = \frac{V_m^2}{2\omega_0^2 L} \sin^2 \omega_0 t$$

$$= \frac{V_m^2}{4\omega_0^2 L}(1 - \cos 2\omega_0 t) \qquad (11\text{-}21)$$

But from Eq. (11-3),

$$\omega_0 L = (1/\omega_0 C)$$

so that Eq. (11-21) becomes

$$w_L(t) = \frac{CV_m^2}{4}(1 - \cos 2\omega_0 t) \qquad (11\text{-}22)$$

The *total energy stored* in the inductor and capacitor becomes, from Eqs. (11-18) and (11-23)

$$w_L + w_C = (1/2)CV_m^2 \qquad (11\text{-}23)$$

That is,

the *total energy stored in the two elements L and C remains a constant at all times*.

The two energy components and the total stored energy are plotted in Fig. 11-3.

11-1 Resonance in a Parallel GLC Circuit

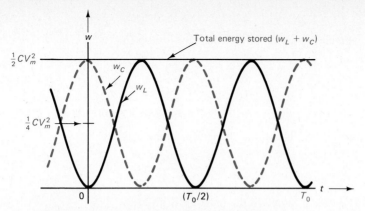

Fig. 11-3 Energy stored in a parallel resonant circuit.

Example 11-3 If $G = 0.5$ S, $L = 10$ mH, $C = 4$ pF in a parallel GLC circuit connected to a voltage $V_s = 16$ V at the resonant frequency, obtain the expressions for the energy stored in the inductor and the capacitor.

Solution The resonant angular frequency is

$$\omega_0 = 1/\sqrt{LC} = 5 \times 10^6 \text{ r/s}$$

The voltage across the capacitor is

$$v_C = 16 \cos (5 \times 10^6 \, t) \text{ V}$$

so that the energy stored in C is

$$\begin{aligned} w_c &= (1/2)Cv_c^2 \\ &= (1/2)(4 \times 10^{-12})[256 \cos^2 (5 \times 10^6 t)] \\ &= 2.56 \times 10^{-10}[1 + \cos (10^7 t)] \text{ J} \end{aligned}$$

The current in the inductance is given by

$$i_L = 3.2 \times 10^{-4} \sin (5 \times 10^6 t) \text{ A}$$

and the energy stored in L becomes

$$\begin{aligned} w_L &= (1/2)Li_L^2 \\ &= (1/2)(10 \times 10^{-3})[10.24 \times 10^{-8} \sin^2 (5 \times 10^6 t)] \\ &= 2.56 \times 10^{-10}[1 - \cos (10^7 t)] \text{ J} \end{aligned}$$

The energy stored in C varies from a maximum of (5.12×10^{-10}) J to a minimum of zero. The same is true for the energy stored in L. The total energy stored in L and C is equal to (5.12×10^{-10}) J for all t. ∎

Exercise 11-4 Repeat the calculations of the previous example (a) if G were doubled to its previous value; (b) if L were doubled to its previous value; (c) if C were doubled to its previous value. In each case, assume that only that particular change has been made, and the other element values are the same as in the previous example.

The curves of Fig. 11-3 show that as the energy in the electric field of the capacitance increases, that in the magnetic field of the inductance decreases, and vice versa. When one is a maximum, the other is a minimum. The two fields feed each other alternately, and since neither of them loses any energy, the total remains a constant. If there were no resistive element in the circuit, there would be no dissipation of energy at all and no need for any external source of energy. The exchange of energy between the inductor and capacitor could go on forever (at least in theory) and become self-sustaining.

When a resistive element is present in the circuit, it dissipates energy that must be provided from the voltage source. The *average* power P dissipated by the conductance G is given by

$$P = (1/2)GV_m^2 \tag{11-24}$$

and as the conductance increases, the power dissipated increases also.

Exercise 11-5 Calculate the average power dissipated by the circuits of Example 11-4 and Exercise 11-4.

Quality Factor Q

The ratio of the total stored energy to the average power dissipated by the circuit is a measure of the energy-storing quality of the circuit. A *quality factor Q* is defined for a resonant circuit by the equation

$$Q = \frac{(2\pi)(\text{total energy stored})}{(\text{energy dissipated per cycle})} \tag{11-25}$$

The energy dissipated per cycle is given by

$$PT_0$$

where P is the power dissipated in G, as in Eq. (11-24), and $T_0 = (2\pi/\omega_0)$ is the period of the voltage and current in the circuit at resonance.

The value of Q at resonance, called the *resonant Q*, denoted by Q_0, for the *parallel resonant GLC circuit*, is obtained from Eqs. (11-23), (11-24), and (11-25):

$$Q_0 = \frac{\omega_0 C}{G} \tag{11-26}$$

The higher the value of Q_0, the smaller the amount of energy dissipated per cycle for a given total of stored energy..

Q_0 is an important factor in the study of resonant circuits.

Example 11-4 A parallel GLC circuit is resonant at $f_0 = 1.59 \times 10^3$ Hz with $Q_0 = 20$. The input current is

$$i_s(t) = 40 \cos 10^4 t \text{ A}.$$

(a) If $L = 20$ mH, calculate the values of G and C. (b) Calculate the total energy stored in L and C. (c) Calculate the average power and the energy dissipated per cycle. (d) Calculate the ratio of the amplitudes of the currents in the different branches to the source current.

Solution (a) Using Eq. (11-6) for f_0, the value of C is given by

$$C = \frac{1}{(2\pi f_0)^2 L} = 0.5 \; \mu F$$

Using Eq. (11-26) for Q_0,

$$G = \frac{\omega_0 C}{Q_0} = 2.5 \times 10^{-4} \; S$$

which corresponds to a resistance of 4 kΩ.

(b) At resonance,

$$\mathbf{V}_0 = (\mathbf{I}_s/G) = 1.6 \times 10^5 \; V$$

Therefore, the maximum value of the voltage is 1.6×10^5 V. From Eq. (11-23), the total energy stored in the circuit is

$$(w_c + w_L) = (1/2)CV_m^2 = 6400 \; J \qquad (11\text{-}27)$$

This is the total energy stored in the circuit at any instant of time.

(c) The average power in the circuit is given by

$$P = (1/2)V_0^2 G = 3.2 \times 10^6 \; W$$

The energy dissipated per cycle is

$$w_G = PT_0$$

where T_0 is the period at resonance and is given by

$$T_0 = (2\pi/\omega_0) = (2\pi \times 10^{-4}) \; s$$

Therefore,

$$w_G = 2011 \; J \qquad (11\text{-}28)$$

Equation (11-25) could also have been used to find w_G.

(d) The currents in the three branches are

$$\mathbf{I}_G = \mathbf{I}_s = 40 \; A$$

$$\mathbf{I}_L = \frac{\mathbf{V}_0}{j\omega_0 L} = \frac{1.6 \times 10^5}{j200} = -j800 \; A$$

$$\mathbf{I}_C = -\mathbf{I}_L = j800 \; A$$

The ratio of the branch currents to the source current is

$$|I_G/I_s| = 1$$

$$|I_L/I_s| = |I_C/I_s| = 20$$

The last ratio has the same value as the Q of the circuit; this is not a coincidence, as will be shown by Exercise 11-7. ■

Exercise 11-6 The Q_0 of a parallel GLC circuit has a value of 20 at its resonant frequency of 12.5 kHz. If $G = 0.01$ S, calculate the values of L and C.

Exercise 11-7 Show that the ratio $|I_c|/|I_s|$ in a parallel GLC circuit at resonance equals Q_0.

11-2 BANDWIDTH OF THE PARALLEL GLC CIRCUIT

Example 11-1 showed that the effect of a resonant circuit is different on the different frequency components of an input current. If a parallel *GLC* circuit is fed by a current source containing different frequency components (all of which have the same amplitude), the component of the output voltage at the resonant frequency has the largest amplitude. Look at Fig. 11-1(d): there is a successive diminution of amplitude of the output voltage components on either side of the resonant frequency. The amplitudes of the voltage components at frequencies sufficiently far from resonance become negligibly small in comparison with that at the resonant frequency. The range of frequencies over which the amplitude of the output component remains significant compared with the resonant component represents the useful *bandwidth* of the circuit. The rate at which the amplitudes of the components roll off on either side of resonance and hence the bandwidth of the circuit depends directly upon the value of Q_0 of the circuit.

11-2-1 Definition of Bandwidth

The useful bandwidth of a resonant circuit is defined in terms of the average power consumed by the circuit. Consider a parallel *GLC* circuit fed by a current source of variable frequency. Assume that the amplitude of the current is kept the same at all frequencies. The average power consumed by the circuit is due to the conductance G and is given by

$$P = (1/2)|V_0|^2 G \qquad (11\text{-}29)$$

where V_0 is the voltage across the circuit, whose magnitude is given by

$$|V_0| = \frac{|I_s|}{|Y|} \qquad (11\text{-}30)$$

$$|Y| = \sqrt{G^2 + (\omega C - 1/\omega L)^2} \qquad (11\text{-}31)$$

Combining Eqs. (11-29), (11-30), and (11-31), we obtain an expression for P. The variation of P as a function of frequency is shown in Fig. 11-4. The maximum average power P_{max} occurs at resonance and is given by

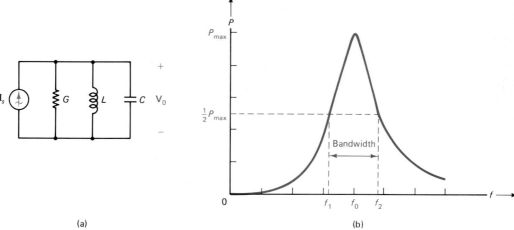

Fig. 11-4 Variation of average power with frequency in a parallel resonant circuit and the definition of bandwidth.

$$P_{max} = (1/2)|V_{0(res)}|^2 G$$
$$= \frac{|I_s|^2}{2G} \qquad (11\text{-}32)$$

where $V_{0(res)} = |I_s|/G$ is the voltage at resonance.

There are two frequencies, f_1 and f_2, called the *half-power frequencies*, at which the average power is one-half of that at resonance.

> **The rule for the definition of useful bandwidth is that the power output of a circuit may be treated as significant provided it is *greater than or equal to one-half the maximum power*. The bandwidth of a resonant circuit is therefore the range between the two half-power frequencies f_1 and f_2.** That is,

$$\text{bandwidth} = (f_2 - f_1) \text{ Hz} \qquad (11\text{-}33)$$

The bandwidth of a resonant circuit is often defined in terms of *decibels*. The ratio of two power quantities in a network is customarily expressed in *decibels (dB)*:

$$\text{power ratio in dB} = 10 \text{ Log (power ratio)} \qquad (11\text{-}34)$$

When the power ratio has a value of one-half, the corresponding dB value is:

$$10 \text{ Log}(0.5) = -3.01 \text{ dB}$$

which is usually taken as -3 dB. That is, *the power at either half-power frequency is 3 dB less than that at resonance*. The two half-power frequencies are therefore referred to as the -3 *dB frequencies*, and the bandwidth is called the -3 *dB bandwidth*.

Determining Half-power Frequencies

Equations (11-29) through (11-32) are combined to give the relationship

$$\frac{P}{P_{max}} = \frac{G^2}{|Y|^2} \qquad (11\text{-}35)$$

At the two half-power frequencies, f_1 and f_2, $P = 0.5 P_{max}$, and Eq. (11-35) leads to the equation

$$|Y|^2 = 2G^2 \qquad (11\text{-}36)$$

Since $Y = (G + jB)$, $|Y|^2 = (G^2 + B^2)$, and Eq. (11-36) leads to the condition

$$B^2 = G^2 \quad \text{or} \quad B = \pm G \qquad (11\text{-}37)$$

From Eq. (11-2), the susceptance B is

$$B = (\omega C - 1/\omega L)$$

and from Eq. (11-37),

$$(\omega C - 1/\omega L) = \pm G \qquad (11\text{-}38)$$

at the two half-power frequencies f_1 and f_2.

Since $B > 0$ above resonance, using $+G$ in Eq. (11-38) leads to the determination of the *upper half-power frequency* f_2, and since $B < 0$ below resonance, using $-G$ in Eq. (11-38) leads to the determination of the *lower half-power frequency* f_1.

It is convenient to remember that the *admittance* triangle has an angle of $-45°$ at f_1 and $+45°$ at f_2, as indicated in Fig. 11-5.

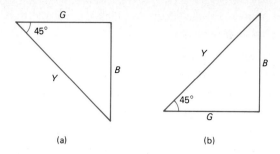

Fig. 11-5 Admittance triangles at the half-power frequencies: (a) $\omega < \omega_0$ (b) $\omega > \omega_0$.

The solution of Eq. (11-38) leads to this expression:

$$\omega_1, \omega_2 = \pm \frac{G}{2C} + \sqrt{\frac{G^2}{4C^2} + \frac{1}{LC}} \tag{11-39}$$

with the *minus* sign chosen for ω_1 and the *plus* sign for ω_2.

The bandwidth of the resonant circuit in *radians per second* is, from Eq. (11-39),

$$\text{bandwidth} = (\omega_2 - \omega_1) = G/C \text{ r/s} \tag{11-40}$$

Since the resonant Q of the circuit is given by

$$Q_0 = \omega_0 C/G$$

Eq. (11-40) also leads to

$$\text{bandwidth} = \frac{\omega_0}{Q_0} \text{ r/s} \tag{11-41a}$$

$$= \frac{f_0}{Q_0} \text{ Hz} \tag{11-41b}$$

The bandwidth of a resonant circuit is therefore inversely proportional to its resonant Q.

The product of the two half-power frequencies is equal to f_0^2; see Exercise 11-9. That is,

$$(f_1 f_2) = f_0^2 \tag{11-42}$$

The resonant frequency is the geometric mean of the two half-power frequencies.

For computational purposes, it is convenient to remember the two formulas:

$$\text{Difference: } (f_2 - f_1) = (f_0/Q_0) \tag{11-42a}$$

$$\text{Product: } (f_1 f_2) = f_0^2 \tag{11-42b}$$

These two simple relationships form the basis of calculating the half-power frequencies and the bandwidth and make it unnecessary to remember a number of complicated formulas.

Example 11-5 A parallel *GLC* circuit with $G = 0.05$ S, $L = 20$ mH, and $C = 45$ μF is driven by a current source of amplitude 25 A and variable frequency. Determine the resonant frequency, the admittance at resonance, the resonant Q, the average power at resonance, the half-power frequencies, the bandwidth, the admittance, the voltage across the circuit, the currents in the three branches, and the average power at each half-power frequency.

Solution

- Resonant frequency: $\omega_0 = 1/\sqrt{LC} = 1054$ r/s, or $f_0 = 168$ Hz.
- Admittance at resonance: $\mathbf{Y} = G = 0.05$ S.
- Resonant Q: $Q_0 = \omega_0 C/G = 0.949$.
- Average power at resonance: $P_{max} = |\mathbf{I}_s|^2/2G = 6250$ W.
- Half-power frequencies: Using $f_1 f_2 = f_0^2$,

$$f_1 f_2 = 2.822 \times 10^4 \tag{11-43}$$

and $(f_2 - f_1) = (f_0/Q_0)$ gives

$$\text{bandwidth} = (f_2 - f_1) = 177 \text{ Hz} \tag{11-44}$$

Equations (11-43) and (11-44) are combined into a single equation:

$$[f_2 - (2.822 \times 10^4/f_2)] = 177$$

which leads to the quadratic equation

$$f_2^2 - 177 f_2 - 2.822 \times 10^4 = 0$$

If we solve the quadratic equation above, we find that

$$f_2 = 278 \text{ Hz}$$

and, using Eq. (11-44),

$$f_1 = 101 \text{ Hz}$$

At the lower half-power frequency ($f_1 = 101$ Hz):

$$\mathbf{Y} = G(1 - j1) = 0.0707\underline{/45°} \text{ S}$$
$$\mathbf{V}_0 = \mathbf{I}_s/\mathbf{Y} = 354\underline{/45°} \text{ V}$$

The branch currents are

$$\mathbf{I}_G = \mathbf{V}_0/G = 17.7\underline{/45°} \text{ A}$$
$$\mathbf{I}_L = \mathbf{V}_0/j\omega L = 27.8\underline{/-45°} \text{ A}$$
$$\mathbf{I}_C = \mathbf{V}_0(j\omega C) = 10.1\underline{/135°} \text{ A}$$

The average power consumed by the circuit is

$$P = (1/2)|\mathbf{V}_0|^2 G = 3125 \text{ W}$$

which is one half of P_{max}, as expected.

At the upper half-power frequency: The values obtained for the lower half-power frequency are readily modified for the upper half-power frequency: $\mathbf{Y} = 0.0707\underline{/45°}$ S.

$$\mathbf{V}_0 = 354\underline{/-45°} \text{ V}$$
$$\mathbf{I}_G = 17.7\underline{/-45°} \text{ A}$$
$$\mathbf{I}_L = 10.1\underline{/-135°} \text{ A}$$
$$\mathbf{I}_C = 27.8\underline{/45°} \text{ A}$$
$$P = 3125 \text{ W}$$

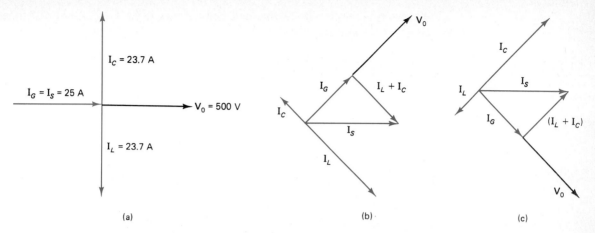

Fig. 11-6 Phasor diagrams for Example 11-5: (a) at resonance; (b) at the lower half-power frequency; and (c) at the upper half-power frequency.

The phasor diagrams at the resonant frequency and the two half-power frequencies are shown in Fig. 11-6. ∎

Exercise 11-8 For a circuit with $G = 0.01$ S, $L = 10$ mH and $C = 100$ pF, repeat the calculations of the previous example. Use $\mathbf{I}_s = 25/0°$ A.

Exercise 11-9 Show that $f_1 f_2 = f_0^2$ for a parallel GLC circuit.

Exercise 11-10 Design a parallel GLC circuit with $C = 1200$ pF, to have resonance at 30 kHz and a half-power frequency at (a) 20 kHz; (b) 29.5 kHz.

Exercise 11-11 Given a parallel GLC circuit with $G = 10^{-3}$ S and bandwidth of 2400 Hz, determine L and C when (a) the upper half-power frequency is at 3200 Hz, and (b) the lower half-power frequency is at 3200 Hz.

Summary of Relationships of a Parallel GLC Circuit Driven by a Current Source I_s

- *Resonant frequency:* $\omega_0 = \dfrac{1}{\sqrt{LC}}$ r/s   or $f_0 = \dfrac{1}{2\pi\sqrt{LC}}$ Hz.
- *Resonant Q:* $Q_0 = \omega_0 C/G$.
- *Admittance at resonance:* $\mathbf{Y}_{\min} = G + j0$.
- *Branch currents:* $\mathbf{I}_G = \mathbf{I}_s$ and $\mathbf{I}_L + \mathbf{I}_C = 0$.
- *Average power:* $P = |\mathbf{I}_s|^2/2G$.
- *Half-power frequencies* f_1 and f_2 satisfy the following equations: $(f_2 - f_1) = (f_0/Q_0)$ and $(f_1 f_2) = f_0^2$.

At the half-power frequencies, $G = |B|$, where B is the susceptance of the circuit. The admittance triangle has an angle of $-45°$ below resonance and $+45°$ above resonance.

11-2 Bandwidth of the Parallel GLC Circuit

Voltage across the circuit at either half-power frequency is $(1/\sqrt{2})$, or 0.707 times the voltage at resonance.

11-3 THE TWO-BRANCH PARALLEL RLC CIRCUIT

The parallel circuit with three parallel branches G, L, and C assumes the availability of a pure resistance, a pure inductance, and a pure capacitance. Even though such an assumption is somewhat justifiable for a resistance and capacitance (except at extremely high frequencies), practical coils do not have zero resistance. Coils are modeled by an inductance L in series with a resistance R_L.

Parallel resonant circuits in practical applications (such as radio receivers) use a coil and a capacitor in parallel, as indicated in Fig. 11-7, where R_L represents the resistance of the coil and L its inductance. Circuits of the form in Fig. 11-7 are called *two-branch parallel* resonant circuits and also *tuned circuits* or *tank circuits*. The general analysis of tuned circuits in and around their resonant frequency becomes rather messy unless certain approximations can be made. The aim of the approximations is to convert the two-branch parallel circuit to an *equivalent* three-branch GLC circuit to make use of the relationships already derived for *GLC* circuits.

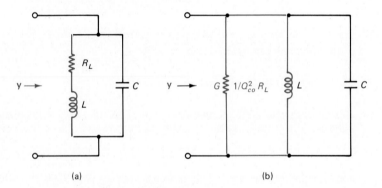

Fig. 11-7 (a) Two-branch parallel resonant circuit and (b) its approximate parallel GLC equivalent circuit.

The following assumptions are made:

1. The analysis is restricted to a range of frequencies in the neighborhood of the resonant frequency.
2. In the range of *frequencies in the neighborhood of the resonant frequency*, the reactance of a coil ωL is assumed to be *much greater* than its resistance. That is,

$$(\omega L/R_L) \gg 1 \qquad (11\text{-}45)$$

The section on series resonance in this chapter shows that the *resonant Q of a coil* (when a coil is taken by itself) is defined by

$$Q_{co} = \frac{\omega_0 L}{R_L} \qquad (11\text{-}46)$$

so that the condition in Eq. (11-45) can also be expressed as

$$Q_{co} \gg 1 \qquad (11\text{-}47)$$

in the neighborhood of the resonant frequency.

Resonant Circuits

The admittance of the circuit of Fig. 11-7 is given by

$$Y = \frac{1}{R_L + j\omega L} + j\omega C$$

After rationalizing the denominator of the first term, the last equation becomes

$$Y = \frac{R_L - j\omega L}{R_L^2 + \omega^2 L^2} + j\omega C$$

$$= \frac{R_L - j\omega L}{\omega^2 L^2 (R_L^2/\omega^2 L^2 + 1)} + j\omega C \qquad (11\text{-}48)$$

The assumption in Eq. (11-45) permits the approximation

$$(R_L^2/\omega^2 L^2) + 1 \approx 1 \qquad (11\text{-}49)$$

in the neighborhood of the resonant frequency. When this approximation is used in Eq. (11-50), it becomes

$$Y \approx = \frac{R_L - j\omega L}{\omega^2 L^2} + j\omega C$$

or

$$Y = (R_L/\omega^2 L^2) + j(\omega C - 1/\omega L) \qquad (11\text{-}50)$$

which is of the same form as the admittance of a *parallel GLC circuit,* as shown in Fig. 11-7(b), provided that

$$G = \frac{R_L}{\omega^2 L^2} \qquad (11\text{-}51)$$

Equation (11-51) is combined with Eq. (11-46) to provide the alternative expression for G:

$$G = \frac{1}{Q_{co}^2 R_L} \qquad (11\text{-}52)$$

Note that the *resistance* of the branch labeled G in the equivalent circuit has a value of

$$R_{eq} = (1/G) = Q_{co}^2 R_L \qquad (11\text{-}53)$$

That is, the resistance in the equivalent circuit may be very high due to the factor Q_{co}^2, even though the resistance R_L of the coil itself is small.

> **Based on the discussion above, a two-branch tuned circuit (a coil with inductance L and resistance R_L in parallel with a capacitance) is replaced by an equivalent three-branch parallel *GLC* circuit with $G = 1/Q_{co}^2 R_L$ and L and C having the same values as in the given circuit. The equivalent circuit is then analyzed by means of the formulas derived earlier for the parallel *GLC* circuit. The results obtained are acceptably accurate, provided (a) the analysis is confined to the neighborhood of the resonant frequency (that is, within the useful bandwidth of the circuit), and (b) the resonant Q of the *coil* is much greater than 1.**

In order to determine how high Q of the coil should be, rewrite the approximation in Eq. (11-49) in the form

$$(1/Q_{co}^2) + 1 \approx 1 \tag{11-54}$$

For an error of 10%, Eq. (11-54) requires that

$$Q_{co}^2 > 10 \text{ or } Q_{co} > 3.2$$

For an error of 1%, condition (11-55) leads to

$$Q_{co} > 10$$

Therefore, the minimum value of Q_{co} needed to justify the use of the equivalent GLC circuit will depend upon the tolerable error. *A minimum value of 5 for the resonant Q of the coil is a convenient number to use.*

Example 11-6 A coil with an inductance of 10 mH and resistance of 25Ω is connected in parallel with a capacitance of 50 pF, as Fig. 11-8(a) shows. (a) Obtain the equivalent parallel GLC circuit assuming a sufficiently high resonant Q for the coil. (b) Determine the resonant frequency and the admittance at resonance. Verify the assumption made in part (a).

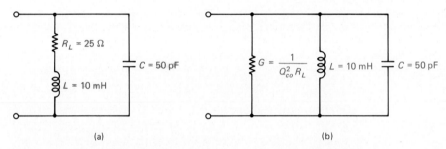

Fig. 11-8 (a) Circuit for Example 11-6 and (b) its parallel GLC equivalent.

Solution Under the assumption of high Q for the coil, the equivalent parallel circuit is of the form shown in Fig. 11-8(b). Note that G cannot yet be determined, since Q_{co} is not known until the resonant frequency is calculated.

Resonance occurs at

$$\omega_0 = 1/\sqrt{LC}$$
$$= 1.41 \times 10^6 \text{ r/s}$$

or 0.225 MHz. The resonant Q of the coil is

$$Q_{co} = \omega_0 L/R_L$$
$$= 566$$

which validates (in retrospect) the high Q assumption. The value of G in Fig. 11-8(b) becomes

$$G = 1/Q_{co}^2 R_L$$
$$= 1.25 \times 10^{-7} \text{ S}$$

(which corresponds to a resistance of 8 MΩ). At resonance,

$$\mathbf{Y} = G = 1.25 \times 10^{-7} \text{ S}$$

∎

Exercise 11-12 Suppose a current source of 25 µA at the resonant frequency feeds the circuit of the last example. Determine the currents in the three branches of the *equivalent* circuit. Also determine the current in the coil of the *original* circuit.

Exercise 11-13 Find a value of the capacitance to produce at resonance at 2.25 kHz when connected in parallel with the coil of Example 11-6. Repeat the calculations of Example 11-6 and Exercise 11-12 for this situation.

Bandwidth Calculations

Once the equivalent *GLC* circuit has been obtained for a given (high Q_{co}) tuned circuit, the formulas (11-41a) and (11-42) lead to the bandwidth and the half-power frequencies.

Exercise 11-14 Show that for a tuned circuit and its equivalent *GLC* circuit the resonant Q of the coil ($Q_{co} = \omega_0 L/R_L$) is equal to the resonant Q of the *GLC* circuit ($Q_0 = \omega_0 C/G$).

The result of the exercise above shows that either Q_0 defined by Eq. (11-26) or Q_{co} defined by Eq. (11-46) may be used in bandwidth calculations. Be careful, however, since additional branches may be placed in parallel with a tuned circuit in certain applications.

Example 11-7 A two-branch tuned circuit is made up of a coil with $L = 50$ mH and $R_L = 470\Omega$ in parallel with a capacitance $C = 400$ pF. Determine (a) the resonant frequency, assuming that the resonant Q of the coil is sufficiently high; (b) the half-power frequencies and the bandwidth of the tuned circuit; (c) the admittance of the circuit at each half-power frequency.

Solution (a) Assuming high Q for the coil, the resonant frequency is

$$\omega_0 = 1/\sqrt{LC} = 2.236 \times 10^5 \text{ r/s (35.59 kHz)}$$

The resonant Q of the coil is therefore

$$Q_{co} = \omega_0 L/R_L = 23.8$$

which justifies the high Q assumption. The G in the equivalent *GLC* circuit is given by

$$G = 1/Q_{co}^2 R_L = 3.76 \times 10^{-6} \text{ S}$$

(b) The half-power frequencies are obtained by using

$$\text{bandwidth} = (f_2 - f_1) = (f_0/Q_0) = 1495 \text{ Hz}$$

and

$$f_1 f_2 = f_0^2 = 1.267 \times 10^9$$

The last two relationships lead to the quadratic equation:

$$f_1^2 + 1495 f_1 - 1.267 \times 10^9 = 0$$

which gives

$$f_1 = 3.485 \times 10^4 \text{ Hz} \quad \text{or} \quad 34.85 \text{ kHz}$$

Since the bandwidth is 1495 Hz, the upper half-power frequency is

$$f_2 = 36.35 \text{ kHz}$$

For high Q circuits, the *resonant frequency is at the center of the useful bandwidth*, as is the case in the calculations above.

(c) At the half-power frequency, $|B| = G$. That is

$$|B| = 3.76 \times 10^{-6}$$

Therefore, the admittance is

$$\mathbf{Y} = 3.76 \times 10^{-6} (1 \pm j1)$$
$$= 5.32 \times 10^{-6} \underline{/\pm 45°} \text{ S}$$

where the plus sign is used at f_2 and the minus sign at f_1. ∎

Exercise 11-15 Determine the admittance of the tuned circuit of the previous example *directly* (that is, without using the equivalent *GLC* circuit) at the two half-power frequencies obtained in that example.

Exercise 11-16 Repeat the calculations of Example 11-7 if the capacitance is changed to 4000 pF.

Example 11-8 A parallel circuit consists of a resistance $R = 10 \text{ k}\Omega$, $C = 1500 \text{ pF}$, and a coil whose inductance $L = 40 \text{ mH}$ and resistance $R_L = 80\Omega$, see Fig. 11-9(a). Determine (a) the resonant frequency assuming that the resonant Q of the coil is sufficiently high; (b) the half-power frequencies and the bandwidth of the given circuit.

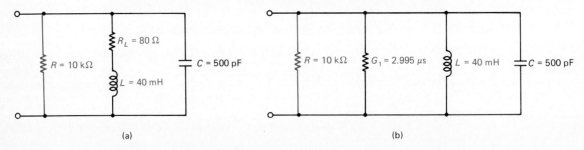

Fig. 11-9 (a) Circuit for Example 11-8 and (b) its parallel *GLC* equivalent.

Solution (a) Assuming the Q of the coil to be sufficiently high, the resonant frequency is

$$f_0 = 1/(2\pi\sqrt{LC}) = 20.55 \text{ kHz}$$

Resonant Q of the coil $= Q_{co} = \omega_0 L/R_L = 64.6$, which justifies the initial assumption.

The coil is replaced by an equivalent parallel combination of an inductance $L = 40$ mH and a conductance

$$G_1 = 1/Q_{co}^2 R_L = 2.995 \times 10^{-6} \text{ S}$$

and the equivalent circuit is shown in Fig. 11-9(b).

(b) The bandwidth is controlled by the resonant Q of the *whole circuit* in Fig. 11-9(b).

Because of the additional resistance $R = 10\text{ k}\Omega$, the resonant Q of the circuit is *not* the same as that of the coil.

The total conductance G_T of the circuit is

$$G_T = G + G_1 = 1.03 \times 10^{-4}\text{ S}$$

and the resonant Q of the *circuit* is

$$Q_0 = \omega_0 C/G_T = 1.88$$

Even though the Q of the circuit is small, the replacement of the given coil by an equivalent parallel combination of L and G_1 is still valid, since such an equivalence depends *only* upon the Q of the coil.

The equations for the determination of the half-power frequencies are:

$$f_2 - f_1 = (f_0/Q_0) = 1.093 \times 10^4\text{ Hz}$$

and

$$f_2 f_1 = f_0^2 = 4.223 \times 10^8$$

The resulting quadratic equation is

$$f_2^2 - 1.093 \times 10^4 f_2 - 4.223 \times 10^8 = 0$$

Therefore,

$$f_2 = 26.72\text{ kHz}$$

and

$$f_1 = 15.80\text{ kHz}$$

The bandwidth is 10.92 kHz. ∎

Exercise 11-17 Change the resistance R in the previous Example to 100 kΩ and repeat the calculations.

Exercise 11-18 Find the new value of R in the circuit of Fig. 11-9(a) for which the bandwidth of the circuit would be 100 kHz.

11-3-1 Application of Tuned Circuits

In communication circuits, the tuned circuit is used as a filter that transmits the resonant frequency or a range of frequencies centered at its resonant frequency while suppressing the other frequencies from an input signal. As a specific example, certain amplifiers (called class C amplifiers) produce a series of pulses of the form shown in Fig. 11-10(a), and a tuned circuit is used to produce an output voltage that is nearly sinusoidal.

Fourier analysis (discussed in Chapter 14) shows that the periodic pulse train in Fig. 11-10(a) contains components at dc, f_0, $2f_0$, . . . , kf_0, . . . , where $f_0 = (1/T_0)$ is called the *fundamental frequency*. A sinusoidal voltage at one of the frequencies can be produced by using a tuned circuit resonant at that frequency, as indicated in Fig. 11-10(b).

The tuned circuit presents a high impedance to the input component at the resonant frequency, and a much lower impedance to the other components. Consequently, the output voltage of the circuit will consist essentially of the resonant frequency component.

Fig. 11-10 The tuned circuit in (b) is used as a filter to pass a single frequency component of the input current in (a).

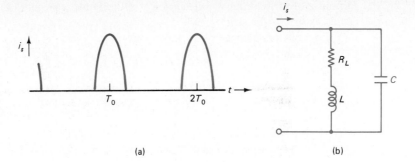

Example 11-9 Let $i_s(t)$ of Fig. 11-10(a) be given by

$$i_s(t) = 0.225 + 0.409 \cos(2\pi \times 10^3)t + 0.300 \cos(4\pi \times 10^3 t) + 0.159 \cos(6\pi \times 10^3 t) \text{ mA}$$

The current i_s is the input to a tuned circuit, as Fig. 11-10(b) shows.
(a) If $L = 400$ mH and $R_L = 250\Omega$, determine the value of C so that the circuit resonates at 1 kHz.
(b) Obtain the expression for the output voltage $v_0(t)$ of the circuit. Compare the amplitudes of the components at dc, 2 kHz, and 3 kHz with respect to that at 1 kHz in the output.

Solution
- Resonant frequency $f_0 = 1000$ Hz.
- Resonant Q of the coil: $Q_{co} = (2\pi f_0 L/R_L)$
$$= (2 \times 10^3 \times 400 \times 10^{-3}/250)$$
$$= 10.0$$

High Q approximation is therefore applicable, and the given circuit is replaced by the equivalent GLC circuit of Fig. 11-11, where

$$G = (1/Q_{co}^2 R_L) = 3.96 \times 10^{-5} \text{ S}$$

Fig. 11-11 Equivalent GLC circuit for Example 11-9.

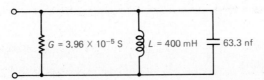

(a) Using the formula for resonant frequency, the value of C is

$$C = 1/(2\pi f_0)^2 L = 63.3 \text{ nF}$$

(b) The principle of superposition is used for finding the output voltage. The output voltage at any frequency is given by

$$\mathbf{V}_0 = (I_s/\mathbf{Y})$$

where I_s is the amplitude of the current. The admittance $\mathbf{Y}$ is given by

$$\mathbf{Y} = G + j(\omega C - 1/\omega L)$$
$$= 3.96 \times 10^{-5} + j\left(63.3 \times 10^{-9}\omega - \frac{1}{400 \times 10^{-3}\omega}\right)$$

Resonant Circuits

- DC: $Y = -j\infty$

  $V_0 = 0$

- 1 kHz: $Y = 3.96 \times 10^{-5}$ S

  $V_0 = (0.409 \times 10^{-3}/Y) = 10.3$ V

- 2 kHz: $Y = (3.96 \times 10^{-5} + j5.96 \times 10^{-4}) = 5.98 \times 10^{-4}\underline{/86°}$ S

  $V_0 = (0.3 \times 10^{-3}/Y) = 0.502\underline{/-86°}$ V

- 3 kHz: $Y = 3.96 \times 10^{-5} + j1.06 \times 10^{-3} = 1.06 \times 10^{-3}\underline{/88°}$ S

  $V_0 = (0.159 \times 10^{-3}/Y) = 0.150\underline{/-88°}$ V

Therefore, the output voltage is given by

$$v_0(t) = 10.3 \cos(2 \times 10^3 t) + 0.502 \cos(4 \times 10^3 t - 86°) + 0.150 \cos(6 \times 10^3 t - 88°)$$

In the output, the amplitude of the 1 kHz component is 10.3 V, while the 2 kHz and 3 kHz components are, respectively, 0.502 V and 0.150 V. In the input current, the amplitudes of the different components are fairly close in value, while the output has a much stronger component at 1 kHz than at the other frequencies. The tuned circuit essentially filters out all but the 1 kHz component of the input.

The filtering can be made even more pronounced by selecting a coil with a much higher Q_{co} than the one used here, but it will cause a corresponding reduction in the bandwidth of the circuit. In radio communication circuits, the ability of the circuit to filter out unwanted frequencies (its selectivity) as well as the bandwidth are both important factors. Therefore, a very narrow bandwidth may not always be desirable. ∎

Exercise 11-19 It is possible to design a tuned circuit to act as a frequency multiplier (called a *harmonic generator*). For instance, the circuit of the last example can be tuned to a frequency of 3 kHz instead of 1 kHz to make the output have a strong component at 3 kHz and relatively small components at the other frequencies. Redesign the circuit to do this and determine the output voltage $v_0(t)$. Compare the amplitude of the 3 kHz component with the others.

11-4 RESONANCE IN A SERIES RLC CIRCUIT

The phenomenon of resonance also occurs when a resistance, inductance, and capacitance are connected in series, as shown in Fig. 11-12(a).

Since a series *RLC* circuit driven by a voltage source is the *dual* of a parallel *GLC* circuit driven by a current source, the discussion of resonance in the series *RLC* circuit (properties as well as relationships) is based on the principle of duality. When using duality,

- Replace G by R.
- Replace L by C and C by L.
- Replace *current* by *voltage* and *voltage* by *current* (that is, i by v and v by i).
- Replace *impedance* by *admittance* and *admittance* by *impedance* (that is, Y by Z and Z by Y).

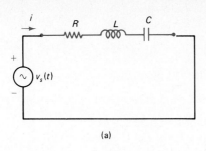
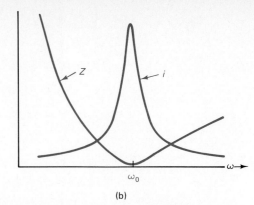

Fig. 11-12 (a) Series resonant circuit. Frequency dependence of (b) impedance and (c) current.

The following relationships and properties are obtained for the series RLC *circuit*. Resonant frequency:

$$\omega_0 = \frac{1}{\sqrt{LC}} \quad \text{or} \quad f_0 = \frac{1}{2\pi\sqrt{LC}} \quad \text{[dual of Eq. (11-6)]} \quad (11\text{-}55)$$

which turns out to be the same as for the parallel *GLC* circuit. The *impedance* of the circuit is a *minimum* at resonance:

$$\mathbf{Z}_{min} = R \quad \text{[dual of Eq. (11-4)]} \quad (11\text{-}56)$$

The *current in the circuit is a maximum* at resonance:

$$\mathbf{I}_{max} = \mathbf{V}_s/R \quad (11\text{-}57)$$

The variation of $|Z|$ and $|I|$ with frequency are shown in Fig. 11-12(b).

When the resistance $R = 0$ in a series *RLC* circuit, the circuit has an impedance of zero at resonance; that is, the circuit appears as a *short circuit* to the source. The current supplied by a voltage source at the resonant frequency is infinitely large in such a case.

Consider a series *RLC* circuit driven by a current source as shown in Fig. 11-13, and let $\mathbf{V}_s$ be the total voltage across the circuit. Then the voltages across the components are:

$$\mathbf{V}_R = \frac{R\mathbf{V}_s}{\mathbf{Z}} \quad \text{[dual of Eq. (11-11)]} \quad (11\text{-}58)$$

$$\mathbf{V}_C = -\frac{j\mathbf{V}_s}{\omega C \mathbf{Z}} \quad \text{[dual of Eq. (11-12)]} \quad (11\text{-}59)$$

$$\mathbf{V}_L = \frac{j\omega L \mathbf{V}_s}{\mathbf{Z}} \quad \text{[dual of Eq. (11-13)]} \quad (11\text{-}60)$$

where

$$\mathbf{Z} = R + j(\omega L - 1/\omega C)$$

is the total impedance of the circuit. The total voltage across the circuit is

$$\mathbf{V}_s = \mathbf{V}_R + \mathbf{V}_L + \mathbf{V}_c \quad (11\text{-}61)$$

At resonance,

$$\mathbf{Z} = R + j0$$

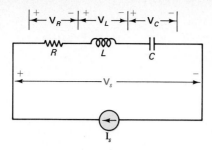

Fig. 11-13 Voltage components in a series resonant circuit.

and

$$\mathbf{V}_R = \mathbf{V}_s \qquad (11\text{-}62)$$

Therefore, Eq. (11-61) leads to

$$\mathbf{V}_C + \mathbf{V}_L = 0 \qquad [\text{dual of Eq. (11-15)}] \qquad (11\text{-}63)$$

Therefore, the voltages across C and L exactly cancel each other at resonance. If the resistance R were made equal to zero (series LC circuit), the voltage across the current source would become zero, even though there is a nonzero voltage across each individual component.

Example 11-10 A resistance $R = 25\,\Omega$ is in series with an inductance $L = 50$ mH and a capacitance C.
(a) Choose C to give a resonant frequency of $\omega_0 = 10^4$ r/s.
(b) Determine the impedance of the circuit at resonance.
(c) If a current source at the resonant frequency and amplitude 16 A is applied to the series circuit, determine the voltage across each of the three components. Verify that the voltages across the inductance and capacitance cancel each other.
(d) If the circuit is driven by a voltage source

$$v_s(t) = 10 + 10 \cos(0.5 \times 10^4 t) + 10 \cos 10^4 t + 10 \cos(2 \times 10^4 t) \text{ V}$$

determine the current $i_s(t)$ supplied by the source.

Solution (a) Using Eq. (11-55),

$$C = 1/(\omega_0^2 L) = 200 \text{ nF}$$

(b) Impedance at resonance: $\mathbf{Z}_{\min} = R = 25\,\Omega$
(c) Using Eqs. (11-58), (11-59), and (11-60), the voltages across the individual elements are

$$\mathbf{V}_R = R\mathbf{I} = 400\underline{/0°} \text{ V}$$
$$\mathbf{V}_C = -[j(1/\omega C)]\mathbf{I} = 8000\underline{/-90°} \text{ V}$$
$$\mathbf{V}_L = j\omega L\mathbf{I} = 8000\underline{/90°} \text{ V}$$

$\mathbf{V}_L$ and $\mathbf{V}_C$ are seen to have the same amplitude, but are 180° out of phase. Thus they add up to zero as expected. (d) The current at any frequency is given by

$$\mathbf{I} = \mathbf{V}_s/\mathbf{Z}$$

where

$$Z = R + j(\omega L - 1/\omega C)$$
$$= 25 + j(50 \times 10^{-3}\omega - 5 \times 10^6/\omega)$$

- DC: $Z = -j\infty$ and $I = 0$
- $\omega = 0.5 \times 10^4$: $Z = 25 - j750 = 750\underline{/-88°}\ \Omega$
$$I = (10/Z) = 0.0133\underline{/88°}\ A$$
- $\omega = 10^4$ r/s: $Z = 25 + j0 = 25\underline{/0°}\ \Omega$
$$I = (10/Z) = 0.4\underline{/0°}\ A$$
- $\omega = 2 \times 10^4$ r/s: $Z = 25 + j750 = 750\underline{/88°}\ \Omega$
$$I = (10/Z) = 0.0133\underline{/-88°}\ A$$

The total current supplied by the source is

$$i_s(t) = 0.0133 \cos(0.5 \times 10^4 t + 88°) + 0.4 \cos 10^4 t$$
$$+ 0.0133 \cos(2 \times 10^4 t - 88°)\ A$$

Even though the amplitudes of the different frequency components of the input voltage are equal, the component at the resonant frequency is much larger in the current than the components at the other frequencies. The circuit can therefore be used as a filter by taking the output across the resistor. ∎

Exercise 11-20 A series *RLC* circuit with $C = 5000$ pF resonates at 2.5×10^4 r/s and its impedance at resonance is 470Ω. Determine the values of R and L.

Exercise 11-21 The *RLC* circuit of Exercise 11-20 is driven by a voltage source

$$v_s(t) = 50 + 50 \cos 5000t + 50 \cos(2.5 \times 10^4 t) + 50 \cos(5 \times 10^4 t)\ V$$

Obtain the expression for the current $i_s(t)$ produced by the source.

Exercise 11-22 A current source of 10 A operating at resonant frequency is connected to a series *RLC* circuit with $R = 80\Omega$, $L = 250$ mH, and $C = 500\ \mu F$. Determine the voltage across each branch and the current source. Repeat the calculations if $R = 0$.

11-4-1 Energy Considerations in a Series RLC Circuit

Consider a series *RLC* circuit connected to a current source $i_s(t)$ given by

$$i_s(t) = I_m \cos \omega_0 t$$

Using duality on the results of the parallel *GLC* circuit, the expressions for the energy stored in L and C are given by

$$w_L(t) = (LI_m^2/4)(1 + \cos 2\omega_0 t) \quad \text{[dual of Eq. (11-18)]} \quad (11\text{-}64)$$
$$w_c(t) = (LI_m^2/4)(1 - \cos 2\omega_0 t) \quad \text{[dual of Eq. (11-21)]} \quad (11\text{-}65)$$

$$w_L + w_c = (1/2)LI_m^2 \qquad \text{[dual of Eq. (11-23)]} \qquad (11\text{-}66)$$

The total energy stored in a constant at all times, as in the parallel *GLC* circuit. The waveforms of energy stored in *L* and *C*, and the total energy, are exactly similar to those for the parallel *GLC* circuit.

Example 11-21   A series *RLC* circuit with $R = 12\Omega$, $L = 500$ mH, and $C = 250$ μF has a current $I_s = 0.4$ A at the resonant frequency. Write the expressions for the energy stored in *L*, *C*, and the circuit as a function of time.

Solution   Resonant frequency: $\omega_0 = 1/\sqrt{LC} = 89.4$ r/s. The current in the circuit is given as 0.4 A. Therefore,

$$i_L(t) = 0.40 \cos 89.4t \text{ A}$$

and the energy stored in the inductance is

$$\begin{aligned} w_L(t) &= (1/2)L[i_L(t)]^2 \\ &= (1/2)(500 \times 10^{-3})(0.4 \cos 89.4t)^2 \\ &= 0.02(1 + \cos 179t) \text{ J} \end{aligned}$$

The voltage across the capacitor is

$$\mathbf{V}_C = \mathbf{I}(-j1/\omega C) = 17.9 \underline{/-90°} \text{ V}$$

which leads to

$$v_c(t) = 17.9 \sin 89.4t \text{ V}$$

and the energy stored in the capacitor is

$$\begin{aligned} w_c(t) &= (1/2)C[v_c(t)]^2 \\ &= (1/2)(250 \times 10^{-6})(17.9 \sin 89.4t)^2 \\ &= 0.02(1 - \cos 179t) \text{ J} \end{aligned}$$

Total energy stored = 0.04 J.   ■

Exercise 11-23   Repeat the calculations of the previous example when (a) $R = 50\Omega$, $L = 500$ mH, $C = 250$ μF. (b) $R = 25\Omega$, $L = 250$ mH, $C = 250$ μF. (c) $R = 25\Omega$, $L = 500$ mH, $C = 125$ μF. Also, calculate the average power dissipated in the circuit.

Exercise 11-24   The total energy stored in a series resonant circuit is found to be 0.1 J when the current in the circuit is at the resonant frequency of 377 r/s and has an amplitude of 0.25 A. Determine the values of *L* and *C*.

11-4-2 Resonant Q of a Series RLC Circuit

The resistance in a series *RLC* circuit dissipates a certain amount of energy every cycle, and this energy must be supplied by the source connected to the circuit. The *average power* dissipated by the resistance is given by

$$P_R = (1/2)RI_m^2 \qquad (11\text{-}67)$$

where I_m is the amplitude of the current i_s and energy dissipated per cycle is

$$P_R T_0 = (2\pi/\omega_0) P_R \qquad (11\text{-}68)$$

since the period

$$T_0 = (2\pi/\omega_0)$$

The Q_0 of a series resonant circuit is defined by the same relationship as in the parallel *GLC* circuit:

$$Q_0 = (2\pi)(\text{energy stored})/(\text{energy dissipated per cycle})$$

From Eqs. (11-66), (11-67), and (11-68), the expression for the resonant Q is given by

$$Q_0 = \frac{\omega_0 L}{R} \qquad [\text{dual of (11-26)}] \qquad (11\text{-}69)$$

Recall that the ratio $(\omega_0 L/R)$ was used as the Q_0 of a coil in the discussion of the two-branch tuned circuit. In the case of a series *RLC* circuit, the Q_0 of the coil is also the Q_0 of the circuit.

Example 11-22 A series *RLC* circuit has a resonant frequency of 10^4 r/s, $L = 20$ mH, and $Q_0 = 8$. It is driven by a voltage source

$$v_s(t) = 80 \cos 10^4 t \text{ V}$$

(a) Determine the values of R and C. (b) Calculate the total energy stored in the circuit. (c) Calculate the average power and the energy dissipated per cycle. (d) Calculate the ratio of the amplitudes of the voltages across the individual branches to the voltage across the circuit.

Solution (a) Using Eq. (11-69)

$$R = (\omega_0 L/Q_0) = 25\,\Omega$$

Using Eq. (11-55),

$$C = 1/(\omega_0^2 L) = 0.5 \text{ μF}$$

(b) At resonance, the current in the circuit is

$$I_s = V_s/R = (80/25) = 3.2 \text{ A}$$

Therefore, the *maximum value* of the energy stored in the inductance is

$$w_{L(\max)} = (1/2)LI_s^2 = 0.102 \text{ J}$$

which is also the total energy stored in the circuit.

(c) The average power dissipated in the resistor is

$$P_R = (1/2)I_s^2 R = 128 \text{ W}$$

and the energy dissipated per cycle is, from Eq. (11-68),

$$w_R = (2\pi/\omega_0) P_R = 0.08 \text{ J}$$

(The value above could also have been obtained by using the definition of Q_0.)

(d) The voltages across the individual branches in the circuit are:

$$\mathbf{V}_R = R\mathbf{I}_s = \mathbf{V}_s = 80\underline{/0°}\text{ V}$$
$$\mathbf{V}_L = j(\omega_0 L)\mathbf{I}_s = 640\underline{/90°}\text{ V}$$
$$\mathbf{V}_c = -j(1/\omega_0 C)\mathbf{I}_s = 640\underline{/-90°}\text{ V}$$

Therefore

$$|V_R/V_s| = 1 \quad \text{and} \quad |V_L/V_s| = |V_c/V_s| = 8$$

The voltage across the inductance (or capacitance) is larger than the applied voltage by a factor equal to Q_0, which is a general result; see Exercise 11-26. ∎

Exercise 11-25 The Q_0 of a series RLC circuit has a value of 12, and its resonant frequency is 25 kHz. If $R = 50\,\Omega$, determine the values of L and C.

Exercise 11-26 Show that the ratio $|V_L/V_s|$ in a series RLC circuit is equal to Q_0 at resonance.

11-4-3 Bandwidth of a Series Resonant Circuit

The bandwidth of a series resonant circuit is defined as the difference between the two half-power frequencies f_1 and f_2:

$$\text{bandwidth} = (f_2 - f_1)\text{ Hz}$$

Consider a series RLC circuit driven by a voltage source $v_s(t)$ and let I_0 be the amplitude of the current at resonance. Then the amplitude of the current at f_1 or f_2 is $(I_0/\sqrt{2}) = 0.707 I_0$; see Fig. 11-14(a). At the two half-power frequencies, f_1 and f_2, the reactance

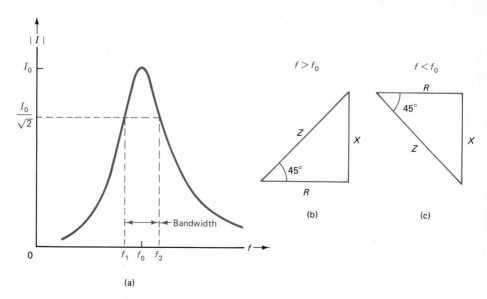

Fig. 11-14 Definition of bandwidth in a series resonant circuit and the impedance triangles at the half-power frequencies.

11-4 Resonance in a Series RLC Circuit

of the circuit is related to the resistance by

$$X = \pm R \quad \text{[dual of Eq. (11-37)]} \quad (11\text{-}70)$$

Above resonance, $(\omega L) > (1/\omega C)$ and $X > 0$. Therefore, the plus sign is used in Eq. (11-70) to find the upper half-power frequency. Below resonance, $(\omega L) < (1/\omega C)$ and $X < 0$. Therefore, the minus sign is used in Eq. (11-70) to find the lower half-power frequency f_1.

Since resistance and reactance are equal at the half-power frequencies, the angle of the impedance is $\pm 45°$ at the two half-power frequencies, as indicated in the impedance triangles of Fig. 11-14(b) and (c).

Using

$$X = (\omega L - 1/\omega C)$$

and Eq. (11-70), the relationships for ω_1 and ω_2 are

$$\omega_1, \omega_2 = \pm \frac{R}{2L} + \sqrt{\frac{R^2}{4L^2} + \frac{1}{LC}} \quad \text{[dual of Eq. (11-39)]} \quad (11\text{-}71)$$

with the *plus* sign yielding the value of f_2 and the *minus* sign the value of f_1.

The resonant frequency is the geometric mean of the two half-power frequencies: that is,

$$f_1 f_2 = f_0^2 \quad (11\text{-}72)$$

Also, from Eq. (11-71),

$$\text{bandwidth} = \frac{R}{2\pi L} \text{ Hz}$$

and, since $Q_0 = (2\pi f_0 L_0/R)$

$$\text{bandwidth} = (f_2 - f_1) = \frac{f_0}{Q_0} \text{ Hz} \quad (11\text{-}73)$$

which is exactly the same equation as for the parallel *GLC* circuit. The *bandwidth of the series RLC circuit is inversely proportional to the Q_0 of the circuit*.

Summary of Relationships of a Series RLC Circuit

Driven by a voltage source V_s

- *Resonant frequency:* $f_0 = \dfrac{1}{2\pi\sqrt{LC}}$ Hz
- *Resonant Q:* $Q_0 = \dfrac{\omega_0 L}{R}$
- *Impedance at resonance:* $\mathbf{Z}_{min} = R + j0$
- *Branch voltages at resonance:* $\mathbf{V}_R = \mathbf{V}_s$ and $\mathbf{V}_L + \mathbf{V}_c = 0$
- *Average power:* $P = \dfrac{|V_s|^2}{2R}$

The two *half-power frequencies* satisfy the equations:

$$f_1 f_2 = f_0^2 \quad \text{and} \quad (f_2 - f_1) = \frac{f_0}{Q_0}$$

At the half-power frequencies,

$$X = \pm R$$

where X is the reactance of the circuit. The circuit is capacitive below resonance and inductive above resonance:

$$\mathbf{Z} = R(1 \pm j1)$$

Example 11-23 A series RLC circuit with $R = 47$ kΩ, $L = 300$ mH, and $C = 20$ pF is driven by a source of voltage $V_s = 12$ V and variable frequency. Determine the resonant frequency, the impedance at resonance, the resonant Q, the average power at resonance, the half-power frequencies, and the bandwidth. At each half-power frequency, calculate the impedance of the circuit, the current in the circuit, the voltages across the three branches, and the average power consumed by the circuit.

Solution
- Resonant frequency: $f_0 = 1/2\pi\sqrt{LC} = 65.0$ kHz
- Impedance at resonance: $\mathbf{Z}_{min} = R\underline{/0°} = 47 \times 10^3$ Ω
- Resonant Q: $Q_0 = (2\pi f_0 L/R) = 2.60$

At resonance, the current in the circuit is

$$I_0 = (V_s/R) = 0.255 \times 10^{-3} \text{ A}$$

and the average power consumed is

$$P = (1/2)I_s^2 R = 1.53 \times 10^{-3} \text{ W}$$

Bandwidth:

$$(f_2 - f_1) = (f_0/Q_0) = 25.0 \text{ kHz} \qquad (11\text{-}74)$$

and the half-power frequencies are also related to f_0 through the equation

$$f_1 f_2 = f_0^2 = 4225 \times 10^6 \qquad (11\text{-}75)$$

Combining Eqs. (11-75) and (11-76), the following quadratic equation is obtained:

$$f_2^2 - (25 \times 10^3)f_2 - 4.225 \times 10^9 = 0$$

Solving the quadratic equation

$$f_2 = 78.7 \text{ kHz}$$

and since the bandwidth is 25 kHz

$$f_1 = (78.7 - 25) = 53.7 \text{ kHz}$$

Lower half-power frequency:

The circuit has a negative reactance.

$$X = -R = -47 \times 10^3$$

and

$$\mathbf{Z} = 47 \times 10^3 (1 - j)$$
$$= 66.4 \times 10^3 \underline{/-45°} \text{ }\Omega$$

The current is

$$\mathbf{I} = (V_s/\mathbf{Z}) = 0.181 \times 10^{-3}\underline{/45°} \text{ A}$$

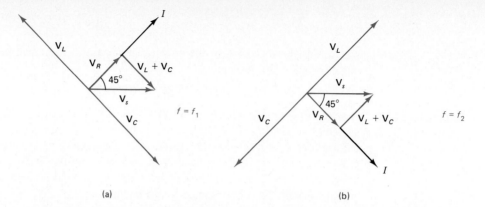

Fig. 11-15 Phasor diagrams for Example 11-23.

The amplitude of the current is seen to be 0.707 times the current at resonance. The voltages across the three branches are

$$\mathbf{V}_R = R\mathbf{I} = 8.50\underline{/45°} \text{ V}$$
$$\mathbf{V}_L = (j\omega_1 L)\mathbf{I} = 18.3\underline{/135°} \text{ V}$$
$$\mathbf{V}_c = -j(1/\omega_1 C)\mathbf{I} = 26.8\underline{/-45°} \text{ V}$$

and the corresponding phasor diagram is shown in Fig. 11-15(a).

Upper half-power frequency:

- $\mathbf{Z} = 47 \times 10^3(1 + j) = 66.4 \times 10^3\underline{/45°} \ \Omega$
- $\mathbf{I} = 0.181 \times 10^{-3}\underline{/-45°}$ A
- $\mathbf{V}_R = 8.50\underline{/-45°}$ V
- $\mathbf{V}_L = 26.8\underline{/45°}$ V
- $\mathbf{V}_c = 18.3\underline{/-135°}$ V

and the phasor diagram is shown in Fig. 11-15(b).

The average power consumed by the circuit at either half power frequency is

$$P = (1/2)|I|^2 R = 0.770 \times 10^{-3} \text{ W}$$

which is one-half of that at resonance. ∎

Exercise 11-27 Change the value of R to 470Ω in the previous example and repeat all the calculations.

Exercise 11-28 In Example 11-23, keep R at 47 kΩ, but change L and C so that the bandwidth is one-tenth of the value obtained in that example while the resonant frequency is the same as before. Recalculate the half-power frequencies.

Exercise 11-29 In the circuit of Example 11-23, determine the total energy stored and the energy dissipated per cycle at resonance. Use the value of Q_0 to verify your answer.

In the discussion of *parallel resonance*, we discussed the two-branch parallel (tuned) circuit at length. The dual of such a circuit is a pure inductance in series with a parallel

RC branch. Such a circuit is of no practical significance, however, since it calls for a coil with zero resistance.

11-5 REACTANCE/SUSCEPTANCE CURVES OF LOSSLESS NETWORKS

The analysis of circuits other than simple parallel or series connections of *R*, *L*, and *C* is quite cumbersome, and it is difficult to get a sense of the behavior of such circuits at resonance. The difficulty is alleviated, however, in the case of circuits containing only inductances and capacitances. Since none of the stored energy is lost in such circuits (due to zero resistance), they are called *lossless circuits*. Lossless circuits offer certain advantages in analysis and design, and consequently are useful in the initial phase in the design of passive filters. Adjustments and refinements are then made to account for the inevitable losses in practical coils.

The chief advantage of a lossless circuit is that all its impedances and admittances have only an imaginary component. The algebraic manipulation of purely imaginary quantities is considerably easier than those with both real and imaginary parts. Diagrams showing the variation of the reactance or susceptance as a function of frequency, called *reactance/susceptance curves,* are useful for studying the response of lossless circuits.

Critical Frequencies of a Lossless Network

At resonance, a parallel *LC* circuit has an admittance of zero since $G = 0$, while a series *LC* circuit has an impedance of zero since $R = 0$. The behavior of a lossless network is thus characterized by an *infinite susceptance* (zero reactance) or *zero susceptance* (infinite reactance) at certain values of ω determined by the structure of the circuit and element values.

> **Frequencies where the impedance or admittance of a network passes through zero or becomes infinitely large are known as *critical frequencies*. More specifically, frequencies at which an impedance (or admittance) is zero are called the *zeros* of the impedance (or admittance), and those at which an impedance (or admittance) becomes infinite are called the *poles* of the impedance (or admittance).**

A more formal discussion of critical frequencies, poles, and zeros of impedance and admittance functions will be presented in the next chapter. The informal definition given here for critical frequencies is sufficient for studying reactance/susceptance curves of lossless networks. The next chapter shows that critical frequencies provide the key to the behavior of a network. Reactance/susceptance curves are useful in the qualitative observation of critical frequencies of lossless networks.

Properties of Reactance/Susceptance Curves

The reactance of a single inductance is

$$X = \omega L$$

and its susceptance is

$$B = -(1/\omega L)$$

Figures 11-16(a) and (b) show the reactance and susceptance curves of a single inductance.

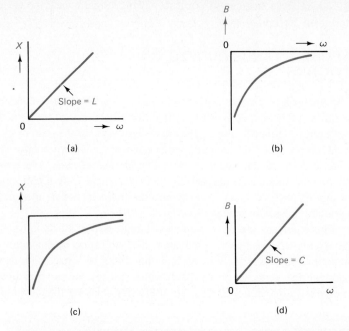

Fig. 11-16 Reactance and susceptance curves of single elements: (a)-(b) inductance; (c)-(d) capacitance.

For a single capacitance,

$$X = -(1/\omega C) \quad \text{and} \quad B = \omega C$$

and its reactance and susceptance curves are shown in Figs. 11-16(c) and (d).

The reactance/susceptance curves of the single elements are seen to have a *positive slope;* that is,

$$\frac{dX}{d\omega} > 0 \quad \text{and} \quad \frac{dB}{d\omega} > 0 \qquad (11\text{-}76)$$

which is a general property of the reactance/susceptance curves of lossless networks. Reactance/susceptance curves in the discussion here will all be seen to have a positive slope.

At the critical frequencies of a lossless network, reactance/susceptance curves pass through zero or become infinitely large ($-\infty$ or $+\infty$ in the diagrams). Poles where a reactance/susceptance curve becomes infinite are indicated by placing small crosses on the ω axis and zeros where a curve passes through zero are indicated by placing small circles on the ω axis.

For a lossless network, the impedance and admittance are of the form

$$\mathbf{Z} = jX \quad \text{and} \quad \mathbf{Y} = jB$$

Using $\mathbf{Z} = (1/\mathbf{Y})$, the susceptance and reactance of a lossless network are related to each other through the equations:

$$B = -(1/X) \quad \text{or} \quad X = -(1/B) \tag{11-77}$$

When $B = 0$, $X \to -\infty$, and when $X = 0$, $B \to -\infty$.

Thus *the poles of a reactance curve are the zeros of a susceptance curve, and the zeros of a reactance curve are the poles of a susceptance curve.* This property is seen to be valid in the curves of Fig. 11-16.

Reactance/Susceptance Curves of an LC Circuit

Now consider a series *LC* circuit, as in Fig. 11-17(a). The total reactance is the sum of the reactances of the individual components. Therefore, the *reactance curves* of the components are *added* to obtain the reactance curve of the circuit, as indicated in Fig. 11-17(b). The following critical frequencies are present: a pole at $\omega = 0$, a pole at $\omega = \infty$, and a zero at $\omega = \omega_0$. (ω_0 is, of course, the resonant frequency of a series *LC* circuit and can be calculated. But the primary aim in using reactance/susceptance curves is not the specific locations of the critical frequencies, but their locations with respect to one another.)

The susceptance curve of the series *LC* circuit is obtained by using Eq. (11-77): $B = -(1/X)$ and is shown in Fig. 11-17(c). The critical frequencies are: a zero at $\omega = 0$, a pole at ω_0, and a zero at $\omega = \infty$. Note that the poles and zeros reverse their roles between reactance and susceptance.

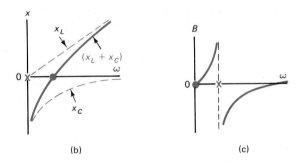

Fig. 11-17 A series *LC* circuit and its reactance/susceptance curves.

Exercise 11-30 Draw the reactance/susceptance curves of a parallel *LC* circuit. Examine the effects of duality.

The following rules lead to sketching the reactance/susceptance curves of more complicated networks:

(a) All the curves have a positive slope.
(b) Both $\omega = 0$ and $\omega = \infty$ are critical frequencies, since capacitances and inductances have zero or infinite reactance at those frequencies.
(c) Reactances add when components or branches are in series, while susceptances add when they are in parallel.
(d) The conversion of reactance to susceptance and vice versa involves a *negative reciprocal* relationship.
(e) The poles and zeros of a reactance curve reverse their roles in a susceptance curve, and vice versa.

Remember too that since the object here is a qualitative study, the exact locations of the poles and zeros are not important; only their relative distribution is.

The reactance/susceptance curves of two lossless networks are shown in Figs. 11-18 and 11-19. For the network in Fig. 11-18(a), the susceptance curve of the L_1C_1 branch is drawn first, using Fig. 11-17(c). The susceptance curve of the L_2 branch is also drawn, from Fig. 11-17(b). The two susceptances are then added to obtain the total susceptance B_a of the given network. The B_a curve is converted to a reactance curve X_a by using $X = -(1/B)$.

The series of steps for obtaining the reactance/susceptance curves of the network of Fig. 11-19 are similar to those used for Fig. 11-18.

Fig. 11-18 A lossless circuit and its reactance/susceptance curves.

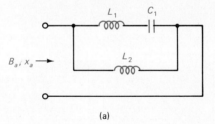

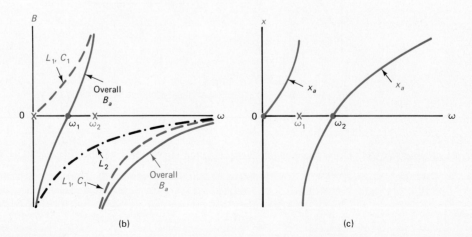

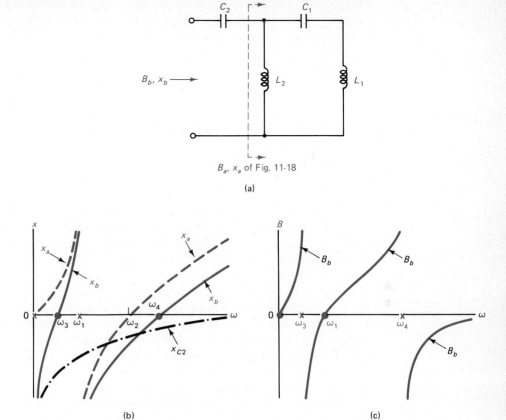

Fig. 11-19 A lossless circuit and its reactance/susceptance curves.

Exercise 11-31 Draw the reactance/susceptance curves of the circuits shown in Fig. 11-20.

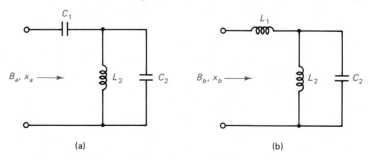

Fig. 11-20 Circuits for Exercise 11-31.

The reactance/susceptance curves obtained above show that at a pole, the curve to the left of a pole goes up to $+\infty$, while the curve to the right of the pole starts at $-\infty$ and goes upward. This is a result of the positive slope constraint mentioned in Eq. (11-76). A consequence of the positive slope constraint of the reactance/susceptance curve of a lossless network is that *the poles and zeros of the impedance or admittance of a lossless*

11-5   Reactance/Susceptance Curves of Lossless Networks

network alternate with each other on the ω axis. The converse is also true: if all the poles and zeros of an impedance or admittance function alternate on the ω axis, then the function can be realized by means of a lossless network.

11-6 SUMMARY OF CHAPTER

A circuit containing both inductance and capacitance exhibits resonance at a frequency determined by the values of the elements in the circuit. At resonance, the energy storage in the inductance and capacitance reach an optimal level.

In the case of a parallel *GLC* circuit, resonance occurs when $f_0 = 1/(2\pi\sqrt{LC})$. The susceptance of the circuit is zero; the admittance is a minimum and equals G. The total current in the circuit is in phase with the voltage across the circuit. The currents in L and C are equal in magnitude but 180° out of phase with each other, thus adding up to zero. The source does not notice the presence of the inductance and capacitance. The energy stored in L or C varies sinusoidally with a frequency equal to twice the input frequency. When the energy in L increases, that in C decreases; and when the energy in L decreases, that in C increases. The total energy stored in L and C remains constant at all times. The source is needed only to providethe energy being dissipated in the resistor. The energy storage quality of a circuit is measured by the factor Q. A higher value of Q at resonance implies that a smaller energy is dissipated per cycle in the circuit.

The bandwidth of a resonant circuit is defined as the difference between its two half-power frequencies (or -3 dB frequencies). The bandwidth is inversely proportional to Q_0. The resonant frequency is the geometric mean of the two half-power frequencies. At either half-power frequency, the magnitude of the susceptance equals the conductance, and the admittance triangle has a 45° angle.

A more practical circuit than the parallel *GLC* is the two-branch parallel circuit, also called the tuned circuit. The two-branch parallel circuit is conveniently analyzed by means of an equivalent parallel *GLC* circuit when the resonant Q of the coil is sufficiently high (≥ 5).

The series *RLC* circuit is the dual of the parallel *GLC* circuit, and the principles of duality permit the modification of the statements for the parallel *GLC* to the series *RLC* circuit. The series *RLC* circuit has a minimum impedance and zero reactance at resonance.

Reactance/susceptance curves of lossless circuits are useful in the study of resonant circuits more complicated than simple series and parallel structures. The basic principles involved in sketching the curves are these: The reactance of an inductance and the susceptance of a capacitance vary linearly with frequency; reactance and susceptance of a given element or lossless circuit are negative reciprocals of each other; the slopes of reactance and susceptance curves are all positive. The frequencies at which reactance/susceptance curves pass through zero are called the zeros of the reactance or susceptance function, and the frequencies at which they go to infinity are called the poles of the function. Poles and zeros together represent the critical frequencies of the reactance or susceptance function. For lossless networks, poles and zeros alternate on the ω axis.

Answers to Exercises

11-1 0.2 H, 10^{-4} S.

11-2 $L = 0.4$ H. $(0.01/G) = 50[0.01/\sqrt{G^2 + B^2}]$, where $B = 7.5 \times 10^{-4}$ S. $G = 1.5 \times 10^{-5}$ S.

11-3   (a) $I_s = 1$ A. $\mathbf{I}_L = 0.707 \times 10^{-4}\underline{/-90°}$ A. $\mathbf{I}_c = -\mathbf{I}_L$. (b) $I_s = 0$. Other two currents, same as before.

11-4   (a) Same resonant frequency and same energy stored. (b) $\mathbf{I}_L = 2.26 \times 10^{-4}\underline{/-90°}$ A. $w_L(t) = 2.56 \times 10^{-10}(1 - \cos 7.07 \times 10^6 t)$ J. $w_c(t) = 2.56 \times 10^{-10}(1 + \cos 7.07 \times 10^6 t)$ J. (c) $\mathbf{I}_L = 4.52 \times 10^{-4}\underline{/-90°}$ A. $w_L(t) = 5.12 \times 10^{-10}(1 - \cos 7.07 \times 10^6 t)$ J. $w_c(t) = 5.12 \times 10^{-10}(1 + \cos 7.07 \times 10^6 t)$ J.

11-5   64 W, 128 W, 64 W, 64 W.

11-6   63.7 µH, 2.55 µF.

11-7 $\mathbf{I}_c/\mathbf{I}_s = (j\omega_0 C \mathbf{V}_s)/(G\mathbf{V}_s)$. Hence the result.

11-8 $f_0 = 159$ kHz. $Y_{min} = 0.01$ S. $Q_0 = 0.01$. $V_0 = 2500$ V. $P_{max} = 3.125 \times 10^4$ W. Bandwidth = 15.9 MHz. Half-power frequencies: 1.59 kHz, 15.9 MHz. At the lower half-power frequency: $\mathbf{Y} = 1.414 \times 10^{-2}\underline{/-45°}$ S, $\mathbf{V}_0 = 1768\underline{/45°}$ V, $\mathbf{I}_G = 17.68\underline{/45°}$ A, $\mathbf{I}_L = 17.68\underline{/-45°}$ A, $\mathbf{I}_c = 1.768 \times 10^{-3}\underline{/135°}$ A, $P = 1.563 \times 10^4$ W. At the upper half-power frequency, $\mathbf{Y} = 1.414\underline{/45°}$ S, $\mathbf{V}_0 = 1768\underline{/-45°}$ V, $\mathbf{I}_G = 17.68\underline{/-45°}$ A, $\mathbf{I}_L = 1.768 \times 10^{-3}\underline{/-135°}$ A, $\mathbf{I}_c = 17.68\underline{/45°}$ A, $P = 1.563 \times 10^4$ W.

11-9   Use Eq. (11-39).

11-10 $L = 23.4$ mH. (a) $f_2 = 45$ kHz. $(\omega_0 C/G) = 1.2$. $G = 1.89 \times 10^{-4}$ S. (b) $f_2 = 30.5$ kHz. $(\omega_0 C/G) = 29.8$. $G = 7.59 \times 10^{-6}$ S.

11-11   (a) $f_1 = 800$ Hz. $Q_0 = 0.667$. $C = 66.3$ nF. $L = 0.149$ H. (b) $f_2 = 5600$ Hz. $Q_0 = 1.76$. $C = 66.3$ nF. $L = 21.3$ mH.

11-12 $\mathbf{I}_G = 25 \times 10^{-6}$ A. $\mathbf{V}_0 = 200$ V. $\mathbf{I}_L = 1.42 \times 10^{-2}\underline{/-90°}$ A. $\mathbf{I}_c = -\mathbf{I}_L$. Since Q_0 is extremely high, the value of $\mathbf{I}_L$ in the original circuit is the same as in the equivalent circuit.

11-13 $C = 0.5$ µF. $G = 1.25 \times 10^{-3}$ S. Currents: $25 \times 10^{-6}\underline{/0°}$ A, $1.41 \times 10^{-4}\underline{/-45°}$ A, $1.41 \times 10^{-4}\underline{/45°}$ A.

11-15   (a) $\mathbf{Y}_T = (3.98 \times 10^{-6} - j9.12 \times 10^{-5}) + j8.76 \times 10^{-5}$ S $= (3.98 \times 10^{-6}) - j3.62 \times 10^{-6})$ S.   (b) $\mathbf{Y}_T = (3.66 \times 10^{-6} + j3.91 \times 10^{-6})$ S.

11-16 $f_0 = 11.3$ kHz. $Q_0 = 7.53$. Bandwidth = 1495 Hz. $f_2 = 12.03$ kHz. $f_1 = 10.54$ kHz. $\mathbf{Y} = (3.76 + j3.78) \times 10^{-5}$ S.

11-17 $Q_0 = 14.9$. $f_1 = 19.87$ kHz. $f_2 = 21.25$ kHz.

11-18 $Q_0 = 0.206$. Add $G = 3.10 \times 10^{-4}$ S, or 3.22 kΩ in parallel.

11-19 $C = 7.036$ nF. $Q_{co} = 30.2$. $G = 4.4 \times 10^{-6}$ S. $v_0(t) = 1.16 \cos(2\pi \times 10^3 t + 89.3°) + 2.72 \cos(4\pi \times 10^3 t + 87.7°) + 36.1 \cos(6\pi \times 10^3 t)$.

11-20 $L = 0.32$ H. $R = 470$Ω.

11-21 $i_s(t) = 1.30 \times 10^{-3} \cos(5000t + 89.3°) + 0.106 \cos(2.5 \times 10^4 t) + 4.17 \times 10^{-3} \cos(5 \times 10^4 t - 88°)$ A.

11-22   89.4 r/s. Voltages: $800\underline{/0°}$ V, $224\underline{/90°}$ V, $224\underline{/-90°}$ V. When $R = 0$, total voltage is zero. $\mathbf{V}_L$ and $\mathbf{V}_c$ same as before.

11-23   (a) w_L and w_c are the same as in Example 11-21. $P_R = 4$ W.
(b) $w_L = 0.01(1 + \cos 253t)$ J. $w_c = 0.01(1 - \cos 253t)$ J. $P_R = 2$ W.
(c) $w_L = 0.02(1 + \cos 253t)$ J. $w_c = 0.02(1 - \cos 253t)$ J. $P_R = 2$ W.

11-24   Use Eq. (11-66). $L = 2$ H. $C = 2$ µF.

11-25   3.82 mH. 10.6 nF.

11-26 $\mathbf{V}_L = j\omega_0 L(\mathbf{V}_s/R)$. Hence the result.

11-27 $f_0 = 65$ kHz. $Z_{max} = 470/0°$ Ω. $Q_0 = 261$. $P = 0.153$ W. Bandwidth $= 0.249$ kHz. $f_2 = 65.1$ kHz. $f_1 = 64.86$ kHz. At the lower half-power frequency: $Z = 664.7/-45°$. $I = 1.8 \times 10^{-2}/45°$ A. Voltages: $8.484/45°$, $2210/135°$, $2210/-45°$ V. $P = 0.0766$ W. At the upper half-power frequency: $Z = 664.7/45°$ Ω. $I = 1.8 \times 10^{-2}/-45°$ A. Voltages: $8.484/-45°$, $2201/45°$, $2208-135°$ V.

11-28 $Q_0 = 26$. $f_2 = 66.27$ kHz. $f_1 = 63.77$ kHz.

11-29 $P_R = 1.53 \times 10^{-3}$ W. $w_R = 2.36 \times 10^{-8}$ J. Total energy $= 9.76 \times 10^{-9}$ J.

11-30 See Fig. 11-21(a).

11-31 See Figs. 11-21(b) and 11-21(c).

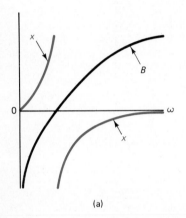
(a)
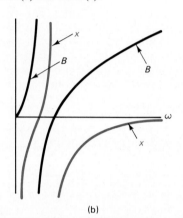
(b)
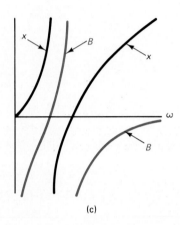
(c)

Fig. 11-21 Answers for Exercises 11-30 and 11-31.

PROBLEMS

Sec. 11-1 Resonance in a Parallel GLC Circuit

11-1 A conductance $G = 0.01$ S, an inductance $L = 50$ mH, and a capacitance $C = 250$ μF are in parallel. Calculate (a) the resonant frequency, (b) the admittance at resonance, (c) the admittance at one-half the resonant frequency, and (d) the admittance at twice the resonant frequency.

11-2 A conductance $G = 0.5$ S is in parallel with a capacitance $C = 5000$ nF. (a) Calculate the value of L needed to make the GLC circuit resonate at 100 kHz. (b) Calculate the current in each branch if the circuit is fed by a current source $\mathbf{I}_s = 10/0°$ A at resonant frequency.

11-3 A conductance $G = 10^{-3}$ S, a capacitance $C = 500$ nF, and an inductance $L = 500$ mH are in parallel and fed by a current source:

$$i_s(t) = 0.20 + 0.20 \cos 2000t + 0.20 \cos 4000t \text{ A}$$

Determine the voltage across the circuit and the current in each branch.

11-4 The circuit of the previous problem is fed by a voltage source:

$$v_s(t) = 25 \cos 1000t + 20 \cos 2000t + 15 \cos 3000t \text{ V}.$$

Determine the current in each branch and the total current.

11-5 Design a parallel GLC resonant circuit to meet the following specifications: resonant frequency $f_0 = 2800$ Hz; minimum admittance $= 0.470$ S; angle of the admittance $= 60°$ at 3600 Hz.

11-6 (a) Design a parallel *GLC* circuit resonant at 10^4 r/s so as to have a minimum admittance of 0.01 S. At resonance, the current through the inductance should be 50 times as large as the current supplied by the source. (b) Let $i_s = 0.02 \cos 10^4 t$ A. Write the expressions for the energy stored in the inductance and the capacitance at resonance. (c) Calculate the maximum stored energy.

11-7 Design a parallel *GLC* circuit to have a resonant frequency of 50 kHz and a minimum admittance of 0.040 S. The admittance has a magnitude of 0.1 S at 100 kHz.

11-8 Consider the current $\mathbf{I}_c$ through the capacitance in a parallel *GLC* circuit fed by a current source. (a) Determine and sketch the variation of $|I_c|$ and Arg I_c as a function of frequency. (b) Obtain an expression for the frequency at which $|I_c|$ reaches a maximum value.

11-9 Repeat the work of the previous problem for the current I_L through the inductance in a parallel *GLC* circuit.

11-10 In radio receivers, tuning is done by *varying the capacitance*. Assume that the input current source to a parallel *GLC* circuit has a given frequency of f_0 while the capacitance is varied. (a) Obtain an expression for C_0 (in terms of the circuit elements and the frequency f_0) when the circuit is at resonance. (b) Determine the value of the capacitance at which the voltage across the circuit is 50 percent of the voltage at resonance for a given input current amplitude I_0. The final expression for C should be in terms of G, L, and ω_0.

11-11 Given: a parallel *GLC* circuit with $G = 0.25$ S, $L = 50$ mH, and $C = 1200$ μF, driven by a current source $\mathbf{I}_s = 2.5$ A operating at the resonant frequency. Calculate (a) the admittance of the circuit; (b) the voltage across the circuit; (c) the current in each of the three branches; (d) the energy stored in L; (e) the energy stored in C; (f) the energy dissipated per cycle in the circuit; and (g) the resonant Q.

11-12 The following data are available for a parallel *GLC* circuit at resonance: maximum energy stored $= 10^{-3}$ J; resonant frequency $= 250$ kHz; resonant $Q = 6.25$; voltage across the circuit $= 150 \cos \omega_0 t$ V. (a) Determine the values of G, L, and C. (b) Calculate the current in each branch. (c) Write the expressions for the energy stored in L and in C. (d) Calculate the energy dissipated per cycle by the circuit.

11-13 Design a parallel *GLC* circuit to meet the following specifications: resonant frequency $= 625$ kHz, maximum stored energy $= 10^{-5}$ J at resonance, minimum admittance of the circuit $= 0.025$ S, current through G at resonance $= 150$ mA.

11-14 The concepts of *normalized frequency* and *normalized* admittance are useful in comparing parallel resonant circuits with different values of Q_0. The normalized admittance $\mathbf{Y}_n$ is defined by: $\mathbf{Y}_n = (\mathbf{Y}/G)$, where $\mathbf{Y}$ is the actual admittance of a parallel *GLC* circuit. The normalized frequency ω_n is defined by: $\omega_n = (\omega/\omega_0)$ where ω is any arbitrary frequency. Write the expression for $\mathbf{Y}_n$ and show that it can be reduced to the form

$$\mathbf{Y}_n = 1 + jQ_0(\omega_n - 1/\omega_n)$$

Determine and sketch the variation of $|\mathbf{Y}_n|$ and Arg $(\mathbf{Y}_n)$ as a function of normalized frequency for the following values of Q_0: (i) 0.1. (ii) 1.00 (iii) 10.

Sec. 11-2 Bandwidth of a Parallel GLC Circuit

11-15 It was shown that the resonant frequency is the geometric mean of the two half-power frequencies. Show that for a sufficiently high value of Q_0, the resonant frequency may be taken as the arithmetic mean of the two half-power frequencies. Then the two half-power frequencies will be symmetrical with respect to the resonant frequency (on a linear scale). Find the minimum value of Q_0 for which the approximation above is valid if the tolerable error is 1 percent. *Note:* This approximation (with 1 percent error) may be used wherever applicable in the following problems.

11-16 (a) Calculate the bandwidth and the half-power frequencies of a parallel GLC circuit with $G = 0.01$ S, $L = 500$ mH, and $C = 50$ μF. (b) Calculate the admittance at resonance. (c) Calculate the admittance at the two half-power frequencies.

11-17 Given: a parallel GLC circuit resonant at 250 kHz with $G = 10^{-4}$ S and $L = 20$ mH. Calculate (a) the value of C; (b) the bandwidth; (c) the half-power frequencies; (d) the frequencies at which the admittance has an angle of $+30°$.

11-18 (a) Design a parallel GLC circuit to have a resonant frequency of 250 kHz, a bandwidth of 25 kHz, and a maximum impedance of 5 kΩ. (b) Calculate the half-power frequencies. (c) Calculate the total energy stored at resonance if $i_s(t) = 0.5 \cos \omega_0 t$ mA. (d) Calculate the energy dissipated per cycle at resonance.

11-19 Obtain expressions for the energy stored in L and in C at the two half-power frequencies in a parallel GLC circuit. Calculate the total energy stored in the circuit at the two half-power frequences. Calculate the maximum stored energy at the two half-power frequencies and find its relationship to the maximum stored energy at resonance.

11-20 Show that the property that the resonant frequency is the geometric mean of the two half-power frequencies means that on a *logarithmic* frequency scale, the two half-power frequencies will be symmetrical with respect to the resonant frequency.

11-21 A parallel GLC circuit has a bandwidth of 10 kHz and upper half-power frequency at 955 kHz. The magnitude of the impedance at the upper half-power frequency is 500Ω. Determine the values of G, L, and C.

11-22 A parallel GLC circuit consists of $G = 5 \times 10^{-6}$ S, $L = 0.625$ mH, and $C = 40$ pF. (a) Calculate the bandwidth and the half-power frequencies. (b) Calculate the resistance that must be placed in parallel with the given circuit in order to increase the bandwidth to 120 percent of the value obtained in part (a). The resonant frequency is the same as before.

11-23 (a) The two half-power frequencies of a parallel GLC circuit are 20 kHz and 320 kHz. If the maximum *impedance* of the circuit is 100Ω, determine the values of G, L, and C. (b) It is desired to shift *both* half-power frequencies *upward* by 5 kHz. What modifications should be made to accomplish this? Keep the modifications as simple as possible, rather than redesigning the whole circuit.

Sec. 11-3 The Two-Branch Parallel RLC Circuit

11-24 A coil with resistance $R_L = 20Ω$ and inductance $L = 50$ mH is in parallel with a capacitance $C = 500$ nF. (a) Calculate the resonant frequency assuming the coil to have a high Q_0. (Check your assumption.) (b) Draw the equivalent parallel GLC circuit. (c) If the circuit is driven by a current source $I_s = 50\underline{/0°}$ mA, calculate the current through the capacitance and through the coil of the *original* circuit. (d) Calculate the bandwidth and the half-power frequencies.

11-25 Repeat the work of the previous problem if the resistance of the coil is reduced to 10Ω.

11-26 Design a tuned circuit to have a resonant frequency of 1280 kHz, a bandwidth of 10 kHz, and a maximum impedance of $5 \times 10^3 Ω$. Be sure to provide the values of the resistance and inductance of the coil. Calculate the half-power frequencies.

11-27 A coil with a resistance of 200Ω and inductance 100 mH is used in a tuned circuit. (a) Calculate the value of C needed to have resonance at 25 kHz. (b) Calculate the impedance of the circuit at resonance. (c) Calculate the current in the coil and in the capacitance if an input current of 5 mA at resonant frequency is applied to the circuit.

11-28 The coil in a tuned circuit has a resistance of 50 and an inductance of 50 mH. Capacitance $C = 1.2$ μF. (a) Calculate the resonant frequency assuming high Q_0. (b) Calculate the error introduced in using Eq. (11-49) by the Q of the coil at resonance. (c) Calculate the lower half-power frequency (by using the equivalent parallel GLC circuit). (d) Calculate the value of Q of the coil at the resonant frequency and the error introduced by it in using Eq. (11-49). (e) If the error introduced in part (d) is greater than 10 percent, find the lowest frequency at which the equivalent parallel GLC circuit can be used in the analysis with an error of no more than 10 percent.

11-29 Design a tuned circuit to act as a filter with the following specifications: it is to have a maximum output voltage when the input current has a frequency of 100 kHz with an impedance of 20 kΩ at that frequency. At 50 kHz and at 200 kHz, its impedance must be no more than 2 kΩ. Calculate the output voltage $v_0(t)$ when the input to the filter is

$$i_s(t) = 1.2 + 1.2 \cos(\pi \times 10^5 t) + 1.2 \cos(2\pi \times 10^5 t) + 1.2 \cos(4\pi \times 10^5 t) \text{ mA}$$

11-30 Consider a coil with resistance R_L and inductance L in parallel with a capacitance C, with no assumptions made about the value of Q of the coil. (a) Write the expression for the admittance $\mathbf{Y}$ of the circuit. (b) Determine the frequency at which $|\mathbf{Y}|$ is a minimum. (c) Determine the frequency at which the angle of $\mathbf{Y}$ is zero (that is, $B = 0$). (Note that these two frequencies coincide in a parallel GLC circuit.)

Sec. 11-4 Resonance in a Series RLC Circuit

11-31 A resistance $R = 100\Omega$, an inductance $L = 50$ mH, and a capacitance $C = 250$ μF are in series. Calculate (a) the resonant frequency; (b) the impedance at resonance; (c) the impedance at one-half the resonant frequency; and (d) the impedance at twice the resonant frequency.

11-32 A resistance $R = 2\Omega$ is in series with a capacitance $C = 5000$ nF. (a) Calculate the value of L needed to make the circuit resonate at 100 kHz. (b) Calculate the voltage across each branch if the circuit is fed by a voltage source $\mathbf{V}_s = 10\underline{/0°}$ V at resonant frequency.

11-33 Design a series RLC resonant circuit to meet the following specifications: resonant frequency $f_0 = 2800$ Hz; minimum impedance $= 470\Omega$; angle of the impedance $= 60°$ at 3600 Hz.

11-34 (a) Design a series RLC circuit resonant at 10^4 r/s so as to have a minimum impedance of 100Ω. At resonance, it is found that the voltage across the inductance is 50 times as large as the voltage of the source. (b) Write the expressions for the energy stored in the inductance and the capacitance when $v_s(t) = 10 \cos 10^4 t$ V. (c) Calculate the maximum stored energy.

11-35 Design a series RLC circuit to have a resonant frequency of 50 kHz and a minimum impedance of 400Ω. The impedance has a magnitude of 1000Ω at 100 kHz.

11-36 Consider the voltage $\mathbf{V}_c$ across the capacitance in a series RLC circuit. (a) Determine and sketch the variation of $|\mathbf{V}_c|$ and Arg $\mathbf{V}_c$ as a function of frequency. (b) Obtain an expression for the frequency at which $|\mathbf{V}_c|$ reaches a maximum value.

11-37 Given: a series RLC circuit with $R = 0.25\Omega$, $L = 50$ mH, and $C = 1200$ μF driven by a current source $\mathbf{V}_s = 250$ V operating at the resonant frequency. Calculate (a) the impedance of the circuit, (b) the voltage across each component, (c) the energy stored in L, (e) the energy stored in C, (f) the energy dissipated per cycle in the circuit, and (g) the resonant Q.

11-38 The following data are available for a series RLC circuit at resonance: maximum energy stored $= 10^{-3}$ J; resonant frequency $= 250$ kHz; resonant $Q = 6.25$; voltage across the circuit $= 150$ V. (a) Determine the values of R, L, and C. (b) Calculate the voltage across each component. (c) Write the expressions for the energy stored in L and in C. (d) Calculate the energy dissipated per cycle by the circuit.

11-39 Design a series RLC circuit to meet the following specifications: resonant frequency $= 625$ kHz, maximum stored energy $= 10^{-5}$ J at resonance, minimum impedance of the circuit $= 50\Omega$, voltage across R at resonance $= 150$ V.

11-40 (a) Calculate the bandwidth and the half-power frequencies of a series RLC circuit with $R = 50\Omega$, $L = 500$ mH, and $C = 500$ nF. (b) Calculate the impedance at resonance. (c) Calculate the impedance at the two half-power frequencies.

11-41 Given a series RLC circuit resonant at 250 kHz with $R = 10^4 \Omega$ and $L = 20$ mH. Calculate (a) the value of C; (b) the bandwidth; (c) the half-power frequencies; (d) the frequencies at which the impedance has an angle of $+30°$.

11-42 (a) Design a series RLC circuit to have a resonant frequency of 250 kHz, a bandwidth of 25 kHz, and a minimum impedance of 5 kΩ. (b) Calculate the half-power frequencies. (c) Calculate the total energy stored at resonance if $v_s(t) = 10 \cos \omega_0 t$ V. (d) Calculate the energy dissipated per cycle at resonance.

11-43 A series RLC circuit has a bandwidth of 10 kHz and upper half-power frequency at 955 kHz. The magnitude of the impedance at the upper half-power frequency is 500Ω. Determine the values of R, L, and C.

Sec. 11-5 Reactance and Susceptance Curves of Lossless Networks

11-44 Draw the reactance and susceptance curves of the lossless networks shown in Fig. 11-22.

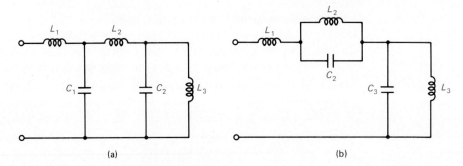

Fig. 11-22 Lossless networks for Problem 11-44.

11-45 An inductance L_1 and a capacitance C_1 are in series. Sketch the reactance/susceptance curves of the L_1C_1 branch and label the pole frequency as ω_1. If a susceptance B_2 is added in parallel with the L_1C_1 branch, a zero will be introduced. If the zero is to be at a frequency $< \omega_1$, should B_2 be > 0 or < 0? If the zero is to at a frequency $> \omega_1$, then should B_2 be > 0 or < 0? Calculate the values of L_1, C_1, and B_2 for the following two cases: (a) Susceptance function is to have a zero at 100 r/s and a pole at 200 r/s. (b) Susceptance function is to have a pole at 100 r/s and a zero at 200 r/s. In each case, are there other zeros and poles? Where are they?

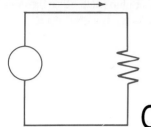

CHAPTER 12
COMPLEX FREQUENCY AND NETWORK FUNCTIONS

In the last several chapters, we discussed the steady state analysis of networks driven by sinusoidal signals. It is necessary to consider a more general type of signal in order to extend the discussion of network response. We do this through the concept of complex frequency, which is introduced on an informal basis in this chapter (the formal treatment appears in Chapter 15). Complex frequency signals cover sinusoidal and exponential signals as well as constant voltages and currents.

Network functions are ratios of the response in a branch of the network to the forcing function applied in that branch or some other branch of the network. An example of such functions is the transfer function (where the excitation is applied to one branch and the response is observed in another), which was discussed in Chapter 9. Driving point functions are obtained when the response of interest is in the same branch as the one in which the forcing function is applied.

The concept of complex frequency introduces the concept of network functions. This chapter will present the importance of critical frequencies in the study of network behavior. One of the important applications of critical frequencies is in the development of Bode diagrams of the frequency response of certain circuits. Bode diagrams are useful graphic aids in the analysis and design of amplifiers.

12-1 THE CONCEPT OF COMPLEX FREQUENCY

In Chapter 6 we noted that the complete response of a network to a forcing function consists of two parts: the *natural response* and the *forced response*. The form of the natural response is governed by the passive components in the network: for example, the response may decay exponentially, or oscillate with decaying amplitude. The form of the forced response, on the other hand, is dictated by the forcing function itself: a sinusoidal forcing function, for example, produces a sinusoidal response.

Consider the zero input response of a linear circuit; that is, the circuit has an initial energy and a switch is closed at $t = 0$, causing a current to flow in the circuit. The equation governing the circuit's response is a homogeneous linear differential equation with constant coefficients. The solution of such an equation is known as the *complementary function* and represents the natural response of the network. The complementary function is assumed to be of the form

$$f(t) = Ke^{st} \qquad (12\text{-}1)$$

where K and s are constants. When the assumed solution is substituted in the differential equation, the derivatives are replaced by a multiplication by s and integrals by $(1/s)$. The differential equation becomes an algebraic equation in s, called the *characteristic equation* of the circuit. The roots of the characteristic equation describe the form of the natural response of the circuit.

For some circuits, such as *RL* and *RC* circuits, the roots of the characteristic equation are found to be negative real numbers. For other circuits, such as underdamped *RLC* circuits, the roots turn out to be complex numbers of the form

$$s = \sigma + j\omega \qquad (12\text{-}2)$$

where

$$\sigma = \text{real part of } s \qquad (12\text{-}3)$$
$$\omega = \text{imaginary part of } s \qquad (12\text{-}4)$$

Depending upon the nature of the roots of the characteristic equation, the zero input response of a passive network may be a decaying exponential or an oscillatory function with a decaying amplitude. For example, when the characteristic equation has a single pair of complex conjugate roots s and s^*, given by

$$s = \sigma + j\omega$$

and

$$s^* = \sigma - j\omega$$

the response $f(t)$ in Eq. (12-1) assumes the form

$$f(t) = \mathbf{K}_1 e^{st} + \mathbf{K}_1^* e^{s^*t} \qquad (12\text{-}5)$$

As we will see in the following exercise, this expression can be written in the form of a sinusoid with decaying amplitude:

$$f(t) = Ae^{\sigma t} \cos(\omega t + \theta) \qquad (12\text{-}6)$$

where A and θ depend upon $\mathbf{K}_1$ and $\mathbf{K}_1^*$.

Exercise 12-1 Use Eq. (12-2) for s in Eq. (12-5) and let $\mathbf{K}_1 = (a + jb)$. Obtain expressions for A and θ in Eq. (12-6) in terms of a and b.

The expression in Eq. (12-6) represents a general response function. Specific functions are obtained from it by assigning suitable values to σ and ω in the expression for $s = \sigma \pm j\omega$.

Response is a constant: This corresponds to $\sigma = \omega = 0$ and $f(t) = A \cos \theta$, a constant for all t.

Response is an exponential function: This corresponds to $\omega = 0$ (but $\sigma \neq 0$) and $f(t) = (A \cos \theta)e^{\sigma t}$.

Response is a sinusoidal function of constant amplitude: This corresponds to $\sigma = 0$ (but $\omega \neq 0$) and $f(t) = A \cos(\omega t + \theta)$.

It is possible to cover the different response functions in a linear network by using the single symbol s. The cases here are valid for forced response functions also, when the forcing functions are constants, exponential functions, or sinusoids, respectively.

If we compare Eq. (12-6) and the definition of s in Eq. (12-2), the angular frequency ω of the sinusoid is seen to be the imaginary part of s. Since one component of s is a frequency, it is convenient to think of s itself as a *generalized frequency*, which is called the *complex frequency*. (The real part σ of s does not represent an actual frequency, since it leads to an exponential function of time of the form $Ae^{\sigma t}$ which does not oscillate. But, since σ is part of a complex frequency, it may also be thought of as a frequency. It is usually referred to as the *neper frequency*.)

12-2 IMPEDANCE AND ADMITTANCE IN THE COMPLEX FREQUENCY DOMAIN

Consider a series RL circuit driven by a *complex* exponential voltage

$$\mathbf{v}_1(t) = \mathbf{K}e^{st}$$

where $\mathbf{K}$ and s are, in general, complex quantities, and the complex frequency s is used instead of $(j\omega)$, as was done in Chapter 8.

The differential equation of the circuit is

$$L(d\mathbf{i}/dt) + R\mathbf{i} = \mathbf{v}_1(t) = \mathbf{K}e^{st} \tag{12-7}$$

where $\mathbf{i}(t)$ represents the *complex* response function. The steady state solution of Eq. (12-7) is of the form

$$\mathbf{i}(t) = \mathbf{I}_0 e^{st} \tag{12-8}$$

where $\mathbf{I}_0$ is the complex coefficient to be determined.

Then

$$\frac{d\mathbf{i}}{dt} = s\mathbf{i}(t) \tag{12-9}$$

so that Eq. (12-7) becomes

$$(Ls + R)\mathbf{i}(t) = \mathbf{v}_1(t) \tag{12-10}$$

The quantity $(Ls + R)$ acts like an impedance function that relates the applied voltage to the resulting current. It is therefore possible to define

$$Z(s) = Ls + R \tag{12-11}$$

as the impedance of a series RL circuit for complex frequency.

To extend the discussion of the simple RL circuit to the general case consider a circuit driven by a complex exponential function

$$\mathbf{f}(t) = \mathbf{F}_0 e^{st}$$

where s is the complex frequency and $\mathbf{F}_0$ is, in general, a complex coefficient. Then the steady state response is also a complex exponential function of the form

$$\mathbf{g}(t) = \mathbf{G}_0 e^{st}$$

where $\mathbf{G}_0$ is a complex coefficient to be determined. Differentiation and integration of the response function lead to:

$$\frac{d\mathbf{g}(t)}{dt} = s[\mathbf{g}(t)]$$

and

$$\int \mathbf{g}(t)dt = \frac{1}{s}[\mathbf{g}(t)]$$

It follows that:

Differentiation in the time domain transforms to multiplication by s in the complex frequency domain, and integration in the time domain transforms to division by s in the complex frequency domain.

These transformations are applied to the voltage-current relationships of single elements excited by forcing functions of complex frequency s, leading to the following expressions for the impedance and admittance of single elements:

Resistors:	$Z(s) = R$		$Y(s) = (1/R)$	(12-12)
Inductors:	$Z(s) = Ls$		$Y(s) = (1/Ls)$	(12-13)
Capacitors:	$Z(S) = (1/Cs)$		$Y(s) = Cs$	(12-14)

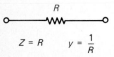
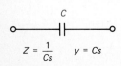

Fig. 12-1 Impedance and admittance of single components as functions of s.

These relationships are also shown in Fig. 12-1 for convenience. Note that the relationships in Eqs. (12-12) through (12-14) and in Fig. 12-1 assume *zero initial stored energy* in inductors and capacitors, since they were obtained from a consideration of the steady state response.

The impedance and admittance relationships in the complex frequency domain are exactly similar to those obtained by using phasors in sinusoidal steady state analysis. The difference is that when using phasors, the forcing functions involve terms of the form $e^{j\omega t}$, whereas in complex frequency analysis, the forcing functions involve terms of the form e^{st}. Thus s appears in the impedance and admittance expressions instead of $(j\omega)$.

12-2-1 Analysis of Circuits in the Complex Frequency Domain

The methods of analysis developed in Chapters 8 and 9 (series and parallel combinations, nodal analysis, mesh and loop analysis, and network theorems) are applicable to the analysis of circuits in the complex frequency domain by using the impedance and admittance relationships in the complex frequency domain. The importance of using complex frequencies, however, is not just the availability of another technique for network analysis, but the introduction of the concept of critical frequencies of network functions.

Example 12-1 For the network shown in Fig. 12-2, determine (a) $Z_{in}(s) = (V_i/I_i)$ and (b) $H(s) = V_0/V_i$.

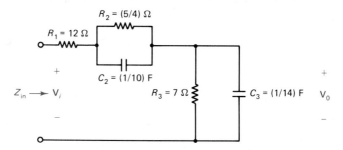

Fig. 12-2 Circuit for Example 12-1.

Solution Let Z_2 denote the parallel combination of C_2 and R_2 and Z_3 the parallel combination of R_3 and C_3.
Then

$$Z_{in}(s) = R_1 + Z_1(s) + Z_2(s)$$

and

$$H(s) = Z_3/Z_{in}$$

$$Z_2(s) = [1/(C_2 s) \| R_2] = \frac{R_2}{R_2 C_2 s + 1} = \frac{50}{5s + 40}$$

$$Z_3(s) = \frac{R_3}{R_3 C_3 s + 1} = \frac{14}{s + 2}$$

(a) $Z_{in}(s) = R_1 + Z_2 + Z_3 = \dfrac{60s^2 + 720s + 1620}{5s^2 + 50s + 80}$

(b) $H(s) = Z_3/Z_{in} = \dfrac{70s^2 + 700s + 1120}{60s^3 + 840s^2 + 3060s + 3240}$ ∎

Exercise 12-2 For each of the networks shown in Fig. 12-3, determine (a) the input impedance $Z_{in}(s)$; (b) the input admittance $Y_{in}(s)$, and (c) the transfer function $H(s) = V_0/V_i$.

Fig. 12-3 Networks for Exercise 12-2.

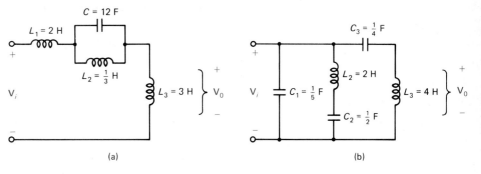

(a)   (b)

12-2   Impedence and Admittance in the Complex Frequency Domain

12-3 CRITICAL FREQUENCIES OF A NETWORK FUNCTION

Driving point and transfer impedances and admittances as well as ratios of voltages and currents in different branches of a network are called *network functions*. When analysis is done in the complex frequency domain, the network functions are functions of s. The standard form of writing a network function is a *rational function;* that is, as a ratio of polynomials with real coefficients and the powers of s being integers. It is customary to normalize the numerator and denominator polynomials by making the coefficient of the highest power of s equal to unity.

That is, if $F(s)$ is a network function, then it is written in the form:

$$F(s) = K \frac{s^m + a_{m-1}s^{m-1} + \ldots + a_1 s + a_0}{s^n + b_{n-1}s^{n-1} + \ldots + b_1 s + b_0} \quad (12\text{-}15)$$

where K is a constant.

For instance, the standard form of the function $H(s)$ of Example 12-1 is

$$H(s) = \frac{7}{6} \frac{s^2 + 10s + 16}{s^3 + 14s^2 + 51s + 54}$$

Since a network function $F(s)$ is complex, it has a magnitude $|F(s)|$ and phase $\theta(s)$. Their plots as a function of s are useful in studying the complex frequency response of the network. But s is a complex quantity in which both σ and ω are independently variable. Consequently, it will take *three-dimensional* plots to show the variation of $|F(s)|$ and $\theta(s)$. Such plots are usually done by means of suitable computer software, and we are not interested in them here.

The purpose of this discussion is to focus on those values of s that highlight the behavior of the network: those values of s for which the function $F(s)$ is zero, and those values of s for which the function $F(s)$ becomes infinite. For example, the impedance function $Z(s)$ of a network may become zero at some values of s, causing the network to act like a short circuit at those values. Similarly, the network acts like an open circuit for those values of s that make $Z(s)$ infinite. Examples of these situations were seen in the reactance/susceptance curves of Chapter 11.

The values of s at which the network function reaches the two extreme values (0 and ∞) are called its *critical frequencies*. A knowledge of the *critical frequencies of a network function* is not only helpful in predicting the behavior of the network, but is, in fact, *sufficient to describe the network function* (except for a constant multiplying factor).

Consider the function $F(s)$ given by Eq. (12-15). Suppose the roots of the numerator and denominator polynomials are found by equating each polynomial to zero and using appropriate formulas or computer routines. Let the roots of the *numerator polynomial* be denoted by $s_1, s_2, \ldots$ and the roots of the denominator polynomial by $s_a, s_b, \ldots$ Then $F(s)$ in Eq. (12-15) can be written in the form:

$$\frac{(s - s_1)(s - s_2)(s - s_3) \ldots}{(s - s_a)(s - s_b)(s - s_c) \ldots} \quad (12\text{-}16)$$

Not all the roots of either polynomial need to be distinct; that is, one or more of its roots may be repeated. Also, if any of the roots are complex, they must occur in complex conjugate pairs, since all the coefficients of the polynomials are real.

Example 12-2 Write the network functions of Example 12-1 in the form of Eq. (12-16).

Solution (a) To find the roots of the numerator polynomial of $Z_{in}(s)$, solve the equation:

$$s^2 + 12s + 27 = 0$$

Roots of numerator polynomial: -3 and -9.
Similarly

$$s^2 + 10s + 16 = 0$$

gives the roots of the denominator polynomial: -2 and -8.
Therefore

$$Z_{in}(s) = \frac{12(s + 3)(s + 9)}{(s + 2)(s + 8)} \quad (12\text{-}17)$$

(b) The roots of the numerator polynomial of $H(s)$ are: $-2, -8$.
To find the roots of the denominator polynomial, the equation to be solved is:

$$s^3 + 14s^2 + 51s + 54 = 0$$

Recall that in Example 12-1, the polynomial above was obtained by multiplying $(60s^2 + 720s + 1620)$ by $(s + 2)$. Therefore, $s = -2$ is a root of this polynomial, and the other two roots are found by solving the quadratic equation

$$60s^2 + 720s + 1620 = 0$$

or

$$s^2 + 12s + 27 = 0$$

and the roots of this equation are -3 and -9. Thus the roots of the denominator polynomial of $H(s)$ are $-2, -3,$ and -9. Therefore

$$H(s) = \frac{7}{6} \frac{(s + 2)(s + 8)}{(s + 2)(s + 3)(s + 9)}$$

$$= \frac{7}{6} \frac{(s + 8)}{(s + 3)(s + 9)}$$ ∎

Exercise 12-3 Put the functions obtained in Exercise 12-2 in the form of Eq. (12-16).

12-3-1 Poles and Zeros of Network Functions

The *zeros* of a network function $F(s)$ are defined as those values of s for which $F(s) = 0$. Since $F(s) = 0$ whenever the numerator is zero, the zeros of a network function are given by the roots of the numerator polynomial.

For instance, the zeros of $Z_{in}(s)$ in the previous example are $s = -3$ and $s = -9$, while $H(s)$ has a zero at $s = -8$.

The *poles* of a network function $F(s)$ are defined as those values of s for which $F(s) = \infty$. Since $F(s) = \infty$ whenever its denominator becomes zero, the poles of $F(s)$ are given by the roots of its denominator polynomial.

In the previous example, the poles of $Z_{in}(s)$ are given by $s = -2$ and -8, while those of $H(s)$ are at $s = -3$ and -9.

The picturesque name "pole" arises from the following analogy. Suppose a rubber sheet is stretched over the complex frequency plane to illustrate a network function. When the function becomes infinite, it would be necessary to insert a pole at that point to make the altitude of the rubber sheet very great; hence the name pole.

The poles and zeros of a network function represent the two extreme behaviors of a network: becoming infinitely large or becoming nothing. At first it may appear that, being extreme conditions, they should be of little interest, but they are actually the lifeblood of a network function.

Once the poles and zeros of a network function are known, the function is specified completely except for a constant multiplicative factor.

For instance, suppose the critical frequencies of an impedance function are: zeros at $s = -3$ and -9, and poles at $s = -2, -8$. Then it is possible immediately to write the ratio of two polynomials by using factors $(s + 3)$ and $(s + 9)$ in the numerator and the factors $(s + 2)$ and $(s + 8)$ in the denominator. This gives the function

$$F(s) = \frac{(s + 3)(s + 9)}{(s + 2)(s + 8)}$$

which is seen to be the same function as Z_{in} of Example 12-2 except for the multiplicative factor of 12. Thus Z_{in} is specified completely by its poles and zeros (except for a multiplicative factor). A constant factor attached to a network function simply represents a *scale factor* or *gain* which is easily accommodated by multiplying the elements of a network by appropriate scale factors.

Exercise 12-4 Write the network function for each of the following sets of critical frequencies. Note that a pole or zero at infinity may not be explicitly specified. In each case, write the final expression in the form of Eq. (12-15).
(a) Zero at $s = -1$. No finite poles.
(b) Pole at $s = -4$. No finite zeros.
(c) Zeros at $s = -1, -5$; poles at $s = 0, -3, -7$.
(d) Zeros at $s = 0$, $s = \pm j3$. Poles at $s = \pm j1, \pm j5$.
(e) Zeros at $s = (2 \pm j3)$; poles at $s = 0, (3 \pm j4)$.

Note too that a network function owes its importance and usefulness to the presence of poles and zeros. Otherwise the network function will be a plain, uninteresting constant that does not change under any conditions. To quote Van Valkenburg:* "Without poles and zeros, the three dimensional representation of the network function becomes a tedious expanse of mathematical desert—absolutely flat. But add a few poles and a few zeros and we have a land of spectacular peaks (elevation: ∞) and beautiful springs (elevation: 0)."

Since poles and zeros completely describe a network function (except for a gain constant), the reason for calling them "critical frequencies" is apparent.

---

*M. E. Van Valkenburg, *Network Analysis*. Englewood Cliffs, NJ: Prentice Hall, 1955.

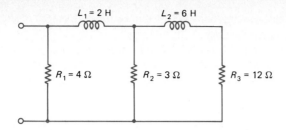

Fig. 12-4 Circuit for Example 12-3.

Example 12-3 Determine the critical frequencies of the input impedance of the network shown in Fig. 12-4.

Solution The expression for $Z_i(s)$ is obtained first, and then the critical frequencies are determined by factoring the numerator and denominator.

Starting from the right end of the network and proceeding toward the input, we obtain the following impedance components:

$$Z_a = L_2 s + R_3 = 6s + 12$$

$$Y_b = \frac{1}{R_2} + \frac{1}{Z_a} = \frac{1}{3} + \frac{1}{6s + 12}$$

$$= \frac{2s + 5}{6s + 12}$$

$$Z_c = L_1 s + \frac{1}{Y_b} = 2s + \frac{6s + 12}{2s + 5}$$

$$= \frac{4s^2 + 16s + 12}{2s + 5}$$

$$Y_i = \frac{1}{R_1} + \frac{1}{Z_c}$$

$$= \frac{1}{4} + \frac{2s + 5}{4s^2 + 16s + 12} = \frac{s^2 + 6s + 8}{4(s^2 + 4s + 3)}$$

Therefore, the input impedance is given by

$$Z_{in} = \frac{4(s^2 + 4s + 3)}{s^2 + 6s + 8}$$

The roots of the numerator polynomial are: $-1, -3$. The roots of the denominator polynomial are $-2, -4$.

- Zeros of $Z_i(s)$: $s = -1, -3$
- Poles of $Z_i(s)$: $s = -2, -4$ ∎

Checking Your Work:

Evaluate the final expression for the network function obtained in a problem at (a) $s = 0$ (dc) and (b) $s = \infty$ (infinite frequency) and see whether the values given by $F(s)$ at these two extreme frequencies match those obtained by inspecting the circuit. Treat inductances as short circuits and capacitances as open circuits for dc, and use the opposite situation for $s = \infty$.

Even though checking does not guarantee the correctness of the expression for $F(s)$, it at least decreases the probability of error.

In the circuit of the previous example, the expression gives $Z(0) = (12/8) = 1.5$, and an inspection of the circuit at dc shows three resistors (4Ω, 3Ω, and 12Ω) in parallel, which also leads to 1.5Ω. At $s = \infty$, the expression gives $Z(\infty) = 4$, and an inspection of the circuit at infinite frequency shows only the resistor $R_1 = 4\Omega$, since the inductances are open circuits.

Exercise 12-5 Determine the critical frequencies of the input impedance function of each of the networks shown in Fig. 12-5.

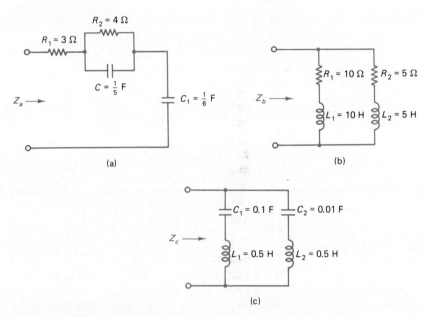

Fig. 12-5 Networks for Exercise 12-4.

The concept of critical frequencies opens up a wide field of topics dealing with the analysis and synthesis of networks. In this chapter we discuss transfer functions and frequency responses of certain useful networks, as well as the study of a resonant circuit from the viewpoint of complex frequency.

12-4 FREQUENCY RESPONSE AND BODE DIAGRAMS

Since critical frequencies of a network function provide sufficient information for writing the function, they are useful in the study of the frequency characteristics of transfer functions of networks in the sinusoidal steady state. For some simple circuits, it is possible to sketch the frequency response characteristics quickly from a knowledge of the critical frequencies by means of *Bode diagrams*.

Bode diagrams show the variation of a magnitude of a transfer function in *decibels* and the phase of a transfer function as functions of the *logarithm* of frequency.

12-4-1 Amplitude Response Diagrams

Consider a transfer function $H(s)$ with a single pole at $s = -s_0$, where s_0 is a constant. Then

$$H(s) = \frac{1}{s + s_0} \qquad (12\text{-}18)$$

where the multiplicative constant is taken as unity for the time being.

To study the behavior of the network in the *sinusoidal steady state*, the complex frequency variable s is replaced by $(j\omega)$, and Eq. (12-18) becomes

$$H(j\omega) = \frac{1}{j\omega + s_0} \qquad (12\text{-}19)$$

whose magnitude is given by

$$|H(j\omega)| = \frac{1}{\sqrt{\omega^2 + s_0^2}} \qquad (12\text{-}20)$$

In networks with wide bandwidths (on the order of kHz or more), it is difficult to make a meaningful plot of the frequency response on linear graph paper, since there will be a congestion of data points at low frequencies. If, on the other hand, a logarithmic scale is used for frequency, it is possible to accommodate a wide bandwidth, and the spacing between frequencies is more manageable.

Computations involving gains of an amplifier are also easier when the gains are expressed in the *decibel* (or dB) unit. The use of dB is particularly advantageous in the design of multistage amplifiers. Voltage gains in dB of multiple stages are *added*, while voltage gains as ratios have to be multiplied.

The conversion of a *voltage* (or *current*) gain A to its dB value is given by the formula

$$A \text{ in dB} = 20 \text{ Log } A \qquad (12\text{-}21)$$

where the logarithm is to base 10. For example, a voltage gain of 500 corresponds to $(20 \text{ Log } 500) = 53.98$ dB.

The frequency response of the transfer function of a network is plotted on *semilog graph paper*, where the frequency is plotted on the logarithmic (horizontal) axis and the dB values on the linear (vertical) axis.

Applying the formula of Eq. (12-21) to $|H(j\omega)|$ in Eq. (12-20)

$$|H(j\omega)|_{dB} = -10 \text{ Log } (\omega^2 + s_0^2) \qquad (12\text{-}22)$$

High-Frequency Asymptote

First, consider the range of high frequencies defined by $\omega \gg s_0$. (The term "much greater than" is generally interpreted as "greater by a factor of 10 or more" in numerical computations.) Then the term (s_0^2) may be ignored in comparison with ω^2 in Eq. (12-22), and

$$|H(j\omega)|_{dB} \approx -10 \text{ Log } \omega^2 = -20 \text{ Log } \omega \qquad (12\text{-}23)$$

Consider a semilog graph where

$$x = \text{Log } \omega \qquad (12\text{-}24a)$$

is the abscissa and

$$y = |H(j\omega)|_{dB} \tag{12-24b}$$

is the ordinate. Then Eq. (12-23) assumes the form

$$y = -20x \tag{12-25}$$

Equation (12-25) describes a straight line. That is, a plot of $|H(j\omega)|$ in dB as a function of Log ω, as given by Eq. (12-23), is a straight line. The straight line is called the *high-frequency asymptote*; see Fig. 12-6(a).

In Eq. (12-24), $y = 0$ when $x = 0$. Using Eq. (12-23), this corresponds to $|H(j\omega)|$ being 0 dB when $\omega = 1$ r/s.

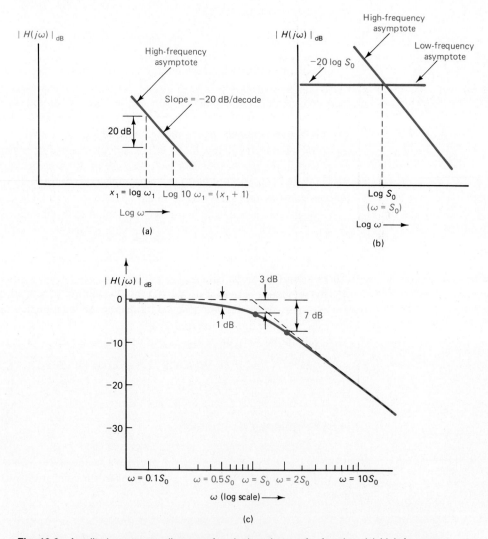

Fig. 12-6 Amplitude response diagram of a single pole transfer function: (a) high-frequency asymptote; (b) low- and high-frequency asymptotes; (c) actual response curve obtained by using three data points near the corner frequency.

The slope of the straight line is, from Eq. (12-25), -20. That is, if x increases by a value of 1, y changes by -20. Consider two frequencies, ω_1 and $10\omega_1$, as in Fig. 12-6(a). Then the values of x corresponding to these are

$$x_1 = \text{Log } \omega_1$$
$$x_2 = \text{Log } 10\omega_1 = 1 + \text{Log } \omega_1 = 1 + x_1$$

Thus an increase in the value of x by 1 corresponds to an increase in the value of ω by a factor of 10. Therefore, the slope of -20 in Eq. (12-24) means that $y = |H(j\omega)|_{dB}$ decreases by 20 dB for every tenfold increase in frequency; that is, the slope of the straight line in Eq. (12-23) is therefore stated as -20 *dB per decade*.

At *high frequencies*—that is, frequencies higher than the pole frequency by a factor of 10 or more—the plot of $|H(j\omega)|_{dB}$ versus Log ω is a straight line (called the high-frequency asymptote) of slope -20 dB per decade.

Exercise 12-6 Consider two frequencies ω_1 and $2\omega_1$ and calculate the loss in dB of the high-frequency asymptote between those two. This result is also used to specify the slope of the high-frequency asymptote: -6 dB per octave, where the term "octave" stands for a doubling of the frequency.

Low-Frequency Asymptote

Next consider the range of low frequencies defined by $\omega \ll s_0$. (The "much less than" is generally interpreted as "less than 10 percent" in numerical computations.) Then the term ω^2 may be ignored compared with (s_0^2) in Eq. (12-20), so that Eq. (12-22) becomes

$$|(H(j\omega)|_{dB} \approx -10 \text{ Log } s_0^2$$
$$= -20 \text{ Log } s_0 \qquad (12\text{-}26)$$

which is a constant; see Fig. 12-6(b).

Therefore, at *low frequencies*,—that is, frequencies less than the pole frequency by at least a factor of 10—the amplitude response is a horizontal straight line (called the *low-frequency asymptote*) at $-20 \text{ Log } s_0$ dB.

The point of intersection of the high-frequency and low-frequency asymptotes obtained by equating Eqs. (12-25) and (12-23) is given by

$$\omega = s_0$$

That is, the *two asymptotes intersect at the pole frequency*.

The term *corner frequency* or *break frequency* is commonly used for the pole frequency, since a straight line seems to turn a corner or break at that frequency.

The graph made up of the low- and high-frequency asymptotes is, of course, not a true representation of the amplitude response of the given transfer function at all frequencies. But at frequencies less than 10 percent of the break frequency and at frequencies higher than ten times the break frequency, the asymptotic plot coincides with the actual amplitude response. The only region where the actual plot significantly varies from the asymptotic plot is in the range between one-tenth of the break frequency and ten times the break frequency. This portion of the amplitude response is sketched by using a few data points obtained from Eq. (12-22), as follows:

First, consider the break frequency: $\omega = s_0$ in Eq. (12-22).

$$|H(j\omega)|_{dB} = -10 \text{ Log } (2s_0^2) = -10 \text{ Log } 2 - 20 \text{ Log } s_0 = -3.01 - 20 \text{ Log } s_0$$

Taking the -3.01 dB as -3 dB for convenience, the actual amplitude response is 3 *dB below the low-frequency asymptote at* $\omega = s_0$.

Two more data points are usually sufficient to draw the actual curve in the region between $(0.1s_0)$ and $(10s_0)$.

At twice the break frequency, $\omega = (2s_0)$:

$$\begin{aligned}|H(j\omega)|_{dB} &= -10 \text{ Log } (5s_0^2) \\ &= -10 \text{ Log } (5) - 20 \text{ Log } s_0 \\ &= -7 - 20 \text{ Log } s_0\end{aligned}$$

which is *7 dB below the low-frequency asymptote.*

At one-half the break frequency, $\omega = 0.5s_0$:

$$\begin{aligned}|H(j\omega)|_{dB} &= -10 \text{ Log } (1.25 s_0^2) \\ &= -10 \text{ Log } (1.25) - 20 \text{ Log } s_0 \\ &= -1 - 20 \text{ Log } s_0\end{aligned}$$

which is *1 dB below the low-frequency asymptote.*

The procedure for drawing the amplitude response of a transfer with a *single pole* is summarized below.

Given

$$H(s) = \frac{1}{s + s_0} \quad \text{or} \quad H(j\omega) = \frac{1}{j\omega + s_0} \qquad (12\text{-}27)$$

mark the vertical axis of a semilog paper in dB. Mark the horizontal (logarithmic) axis in ω in r/s. The logarithmic scale takes care of the Log in Log ω.

1. Starting at the low-frequency end of the graph, draw a horizontal straight line at $-20 \text{ Log } s_0$ dB. This is the low-frequency asymptote.
2. The break frequency is at $\omega = s_0$ on the low-frequency asymptote. Draw a downward straight line with a slope of -20 dB/decade or -6 dB per octave, whichever is convenient. This is the high-frequency asymptote.
3. Mark the following data points:
   1 dB below the low-frequency asymptote at $(s_0/2)$.
   3 db below the low-frequency asymptote at s_0.
   7 dB below the low-frequency asymptote at $(2s_0)$.
4. Draw a smooth curve through the data points and make the curve merge with the low-frequency asymptote at $(s_0/10)$ and with the high-frequency asymptote at $(10s_0)$. This curve gives the amplitude response, as shown in Fig. 12-6(c), of the single-pole transfer function.

Note on Units: Theoretical expressions use ω, the angular frequency in r/s, whereas laboratory measurements and practical specifications are usually in terms of frequency f in Hz. *The procedure for drawing the amplitude response* is exactly the same whether r/s or Hz is used. The only difference is in labeling the horizontal axis of the graph paper: Use r/s or Hz, depending upon the statement of the problem.

Example 12-4 Sketch the amplitude response of the transfer function:

$$H(s) = \frac{1}{s + 40}$$

Solution *Low-frequency asymptote:* Starting at some low frequency, draw a horizontal straight line at $-20 \log 40 = -32$ dB.

The break frequency is at $\omega = 40$ r/s (given by the pole at $s = -40$).

High-frequency asymptote: Starting at the break frequency of 40 r/s, draw a straight line downward with a slope of -20 dB/decade. That is, at $\omega = 400$ r/s, which is ten times the break frequency, the straight line will pass through $(-32-20) = -52$ dB, which is 20 dB lower than the value of -32 at the break frequency.

Mark the following three data points: 1 dB below the low-frequency asymptote at $(40/2)$ or 20 r/s; 3 dB below the low-frequency asymptote at 40 r/s; and 7 dB below the low-frequency asymptote at (2×40) or 80 r/s.

Join the three data points by a smooth curve and extend the curve in both directions to merge with the low-frequency asymptote at $(40/10) = 4$ r/s and with the high-frequency asymptote at $(40 \times 10) = 400$ r/s. The resulting curve (Fig. 12-7) is the amplitude response of the given transfer function.

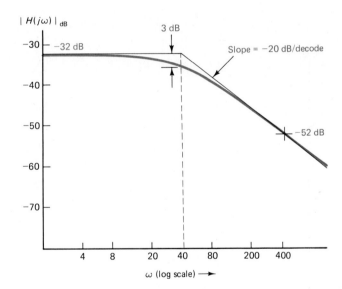

Fig. 12-7 Amplitude response curve for Example 12-4.

Exercise 12-7 Sketch the amplitude response of the following transfer functions: (a) $H(s) = 1/(s + 100)$. (b) $H(s) = 1/(s + 0.1)$.

Effect of a Constant Factor on the Amplitude Response Diagram

Consider a transfer function:

$$H(s) = \frac{K}{s + s_0} \quad \text{or} \quad H(j\omega) = \frac{K}{j\omega + s_0} \qquad (12\text{-}28)$$

12-4 Frequency Response and Bode Diagrams

which is different from that in Eq. (12-27) by the constant factor K. When converted to dB units, Eq. (12-28) becomes

$$|H(j\omega)|_{dB} = 20 \text{ Log } K + 20 \text{ Log } [1/(j\omega + s_0)] \qquad (12\text{-}29)$$

The second term is exactly the same form as in Eq. (12-27). The first term has a constant value of 20 Log K dB for all values of ω and has the *effect of shifting the amplitude response* due to the second term in Eq. (12-27) by 20 Log K dB.

Example 12-5 Determine the transfer function $H(s) = (V_o/V_s)$ of the circuit shown in Fig. 12-8(a) and sketch the amplitude response.

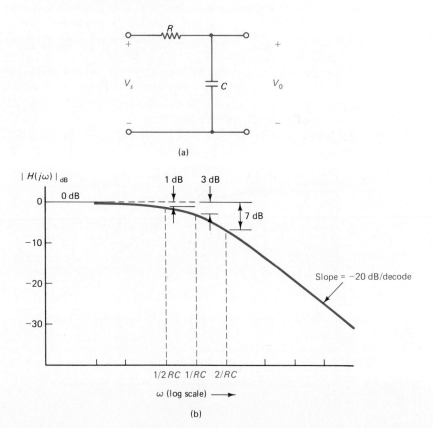

Fig. 12-8 (a) Circuit for Example 12-5; (b) the amplitude response of its transfer function.

Solution Transfer function $H(s) = \dfrac{1/Cs}{R + (1/Cs)} = \dfrac{1}{RC} \dfrac{1}{s + (1/RC)}$

There are two terms: a constant factor $(1/RC)$ and a single pole term $1/(s + 1/RC)$. The amplitude response of the single pole term is obtained by using the same steps as in the previous example: break frequency at $-(1/RC)$, low-frequency asymptote of value -20 Log $(1/RC)$ and the three data points: 1 dB below the low-frequency asymptote at $\omega = (1/2RC)$, 3 dB below the low-frequency asymptote at $\omega = (1/RC)$, and 7 dB below the low-frequency asymptote at $\omega = (2/RC)$.

The constant factor of (1/RC) introduces a uniform shift of 20 Log (1/RC) on the amplitude response obtained above. The final curve is shown in Fig. 12-8(b). Note that the low-frequency asymptote has now shifted to 0 dB. ∎

Since the RC circuit in Example 12-5 presents no attenuation (0 dB) at low frequencies but large attenuations at high frequencies, it serves as a *low pass filter* (*LPF*). At the break frequency of the transfer function, the loss is −3 dB. (Recall the use of −3 dB in the definition of bandwidth of a resonant circuit discussed in Chapter 11.) The frequency at which the response is down by −3 dB is defined as the *cutoff frequency of the filter:*

$$\text{cutoff frequency} = (1/RC) \text{ r/s}$$

The amplitude response of any low pass filter made up of a single resistance in series with a single capacitance will have the same shape and form as in Fig. 12-8(b). The only difference will be the location of the cutoff frequency as determined by the specific values of R and C.

> **Exercise 12-8** A low pass filter is to have a cutoff frequency of 10 kHz (note the units!). Sketch its amplitude response in the range 0.1 kHz to 1000 kHz. If $R = 10 \text{ k}\Omega$, determine the value of C.

12-4-2 Phase Response Diagrams

The phase response of the transfer function of Eq. (12-19) shows the variation of the angle $\theta(j\omega)$ of $H(j\omega)$ as a function of Log ω. From Eq. (12-19)

$$\theta(j\omega) = -\arctan(\omega/s_0) \qquad (12\text{-}30)$$

When $\omega = s_0$, that is at the break frequency,

$$\theta = -\arctan 1 = -45°$$

At low frequencies, $\omega \ll s_0$, the value of θ is approximately zero. At high frequencies, $\omega \gg s_0$, the value of θ is approximately −90°.

> **Therefore, an approximate phase response curve is drawn by setting up the following straight-line segments: a straight line at 0° at frequencies less than one-tenth of the break frequency, a straight line at −90° at frequencies higher than ten times the break frequency, and a straight line going from 0° at ($0.1s_0$) to −90° at ($10s_0$).**

The asymptotic plot is shown in Fig. 12-9(a). Note that the horizontal coordinate is Log ω, as in the amplitude response graph.

The actual phase response departs from the asymptotic plot and is drawn by selecting a few data points obtained from Eq. (12-29), as follows

At $\omega = 0.1s_0$, $\theta = -\arctan 0.1 = -6°$. Similarly, the angle is −26.5° at ($0.5s_0$); −63.4° at ($2s_0$); and −84° at ($10s_0$).

The actual phase response is shown in Fig. 12-9(b).

The set of two curves, $|H(j\omega)|_{\text{dB}}$ and $\theta(j\omega)$, as a function of Log ω are known as *Bode diagrams*. The asymptotic plots used as a first step in obtaining the actual plots are called the *idealized Bode plots*.

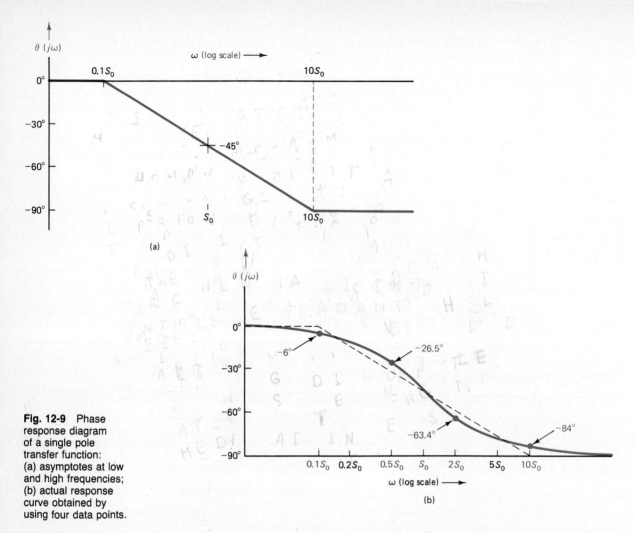

Fig. 12-9 Phase response diagram of a single pole transfer function: (a) asymptotes at low and high frequencies; (b) actual response curve obtained by using four data points.

Exercise 12-9 Draw the phase response curves of the functions in Exercise 12-6 and the low pass filter of Example 12-6.

12-4-3 Bode Diagrams for $H(s)$ with a Single Zero

The Bode diagrams for the transfer function with a zero at $s = -s_0$:

$$H(s) = (s + s_0) \quad \text{or} \quad H(j\omega) = (j\omega + s_0) \qquad (12\text{-}31)$$

are drawn by using steps similar to those of the transfer function with a pole at $s = -s_0$. The amplitude response of Eq. (12-31) is given by

$$|H(j\omega)|_{dB} = 10 \, \text{Log} \, (\omega^2 + s_0^2) \qquad (12\text{-}32)$$

which differs from Eq. (12-22) only by a sign. Therefore, the low-frequency asymptote is $+20 \, \text{Log} \, s_0$, and the high-frequency asymptote has a *positive* slope of 20 dB/decade starting at the break frequency $\omega = s_0$. The curve between $0.1s_0$ and $10s_0$ is *above the*

590   **Complex Frequency and Network Functions**

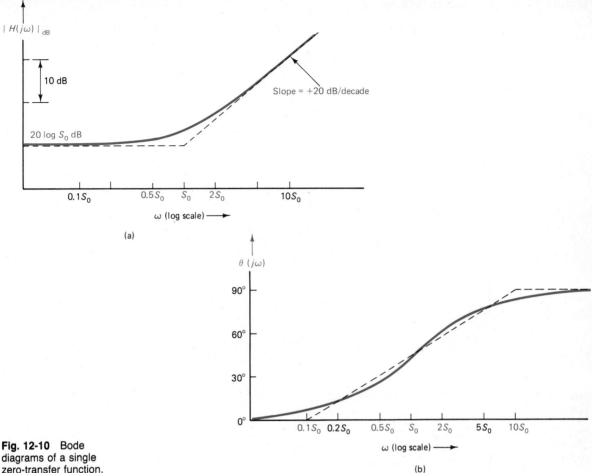

Fig. 12-10 Bode diagrams of a single zero-transfer function.

low-frequency asymptote by the values listed in connection with Eq. (12-27). The amplitude response is as shown in Fig. 12-10(a).

The phase response is given by

$$\theta(j\omega) = +\arctan(\omega/s_0)$$

which also differs from the phase function of Eq. (12-30) by a sign. Therefore, the phase response is made up of a low-frequency asymptote at $0°$ ($\omega < 0.1\ s_0$), a high-frequency asymptote at $+90°$ ($\omega > 10\ s_0$), and a straight line joining the two asymptotes from $0.1s_0$ to $10s_0$. The curve between $0.1s_0$ and $10s_0$ is obtained by using the positive counterparts of the values listed for Eq. (12-30). The phase response diagram is as shown in Fig. 12-10(b).

Exercise 12-10 Draw the Bode diagrams of each of the following transfer functions: (a) $H(s) = s$. (b) $H(s) = (1/s)$. Do these by using the appropriate values of s_0 in the transfer functions already discussed.

12-4 Frequency Response and Bode Diagrams

12-4-4 Bode Diagrams of a Network

When a transfer function has more than one critical frequency, the Bode diagram is drawn by taking each critical frequency separately, setting up the corresponding Bode diagram, and then adding them.

When the critical frequencies are sufficiently separated from one another on the ω axis (that is, two successive critical frequencies are different by a factor of 10 or more), the curve in the region of a critical frequency does not affect the curve in the region of the next critical frequency. In such cases, the asymptotes are added directly, and the curves near the critical frequencies are sketched in after the addition. An example of such a transfer function is presented in Exercise 12-10.

When the critical frequencies are not sufficiently far apart, then point by point addition of the individual diagrams is needed to obtain the overall diagram.

Example 12-6 The circuit of Fig. 12-11(a) shows a high pass filter (HPF). Determine the transfer function $H(s) = (V_0/V_s)$ and sketch the Bode diagrams.

Fig. 12-11 (a)–(f) A high pass filter and its Bode diagrams (Example 12-6).

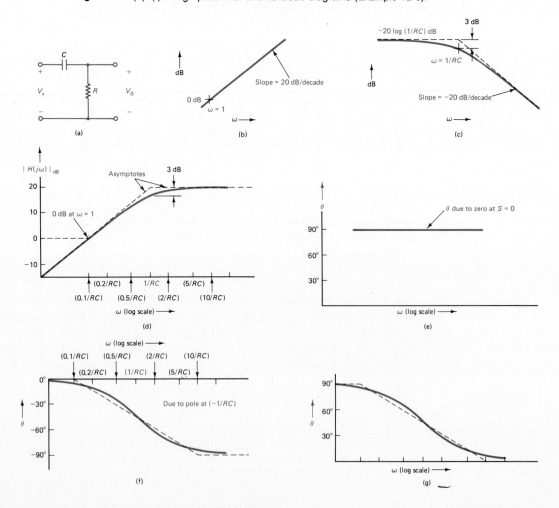

Complex Frequency and Network Functions

Solution Transfer function:

$$H(s) = \frac{R}{R + (1/Cs)} = \frac{s}{s + (1/RC)}$$

which has a zero at $s = 0$ and a pole at $s = -(1/RC)$.

The amplitude response for the zero at $s = 0$ was obtained in Exercise 12-9(a) and repeated in Fig. 12-11(b). The amplitude response for the pole at $-1/RC$ is shown in Fig. 12-11(c). The two component diagrams are added to obtain the total amplitude response shown in Fig. 12-11(d). A similar procedure is used for the phase diagram shown in Figs. 12-11(d), (e), (f). ∎

Exercise 12-11 Draw the Bode diagrams of the transfer function

$$H(s) = \frac{1}{(s + 10)(s + 200)}$$

Example 12-7 For the network shown in Fig. 12-12(a), determine $H(s)$, and plot the Bode diagrams.

Fig. 12-12 (a)–(k) Network for Bode diagrams for Example 12-7.

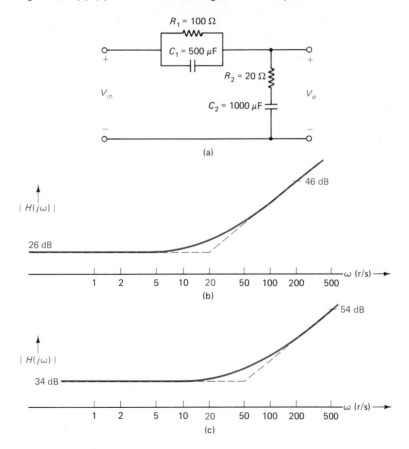

12-4 Frequency Response and Bode Diagrams

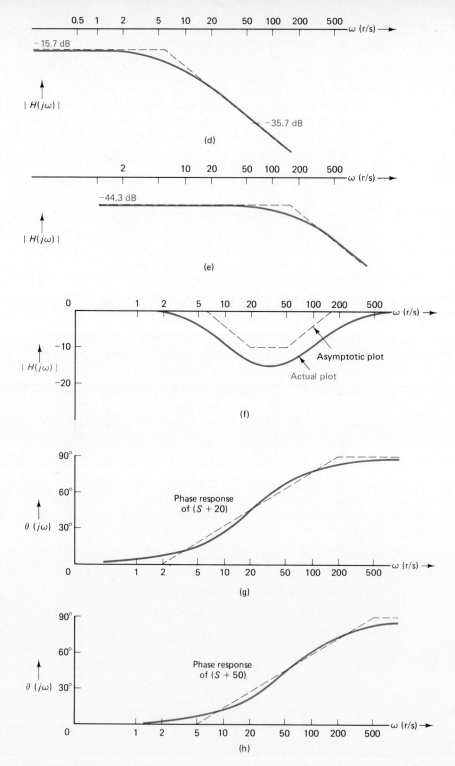

(Fig 12-12 cont.)

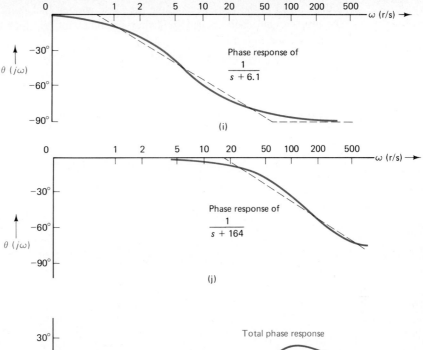

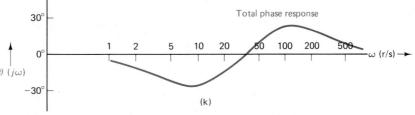

Solution

$$H(s) = \frac{Z_2}{Z_1 + Z_2}$$

where

$$Z_1 = \frac{R_1}{R_1 C_1 s + 1} = \frac{2000}{s + 20}$$

$$Z_2 = \frac{R_2 C_2 s + 1}{C_2 s} = \frac{20(s + 50)}{s}$$

Therefore

$$H(s) = \frac{s^2 + 70s + 1000}{s^2 + 170s + 1000}$$

The critical frequencies of $H(s)$ are:

- Zeros at: $-20, -50$
- Poles at: $-6.1, -164$

The Bode diagrams of the amplitude responses for the two zeros are shown in Figs. 12-12(b) and (c), and those for the two poles are shown in Figs. 12-12(d) and (e). Unlike the previous exercise, where the critical frequencies were sufficiently isolated from each

12-4 Frequency Response and Bode Diagrams

other, those in the present example are close enough that point by point addition must be used around each critical frequency in combining the four plots to obtain the total response, see Fig. 12-12(f).

The Bode diagrams for the phase response are shown in the remaining diagrams of Fig. 12-12:(g)–(k). ∎

Exercise 12-12 Determine $H(s) = V_o/V_s$ for the circuits shown in Fig. 12-13. Plot the Bode diagrams.

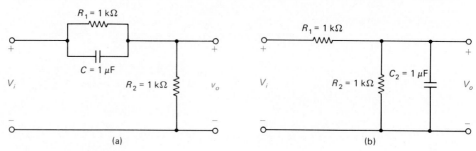

Fig. 12-13 Circuits for Exercise 12-11.

Exercise 12-13 The gain of an amplifier can be written in the form

$$A(s) = -\frac{10^4}{(s + 20)(s + 20 \times 10^4)}$$

Sketch its amplitude response.

The discussion and examples show that when the poles and zeros of a network are of the form $s = -s_0$, where s_0 is a real number, the frequency response is drawn by a careful use of low-frequency and high-frequency asymptotes. The frequency response is thus drawn quite quickly, without a great deal of computation. Bode diagrams are useful tools in the design of amplifiers and control systems because of the simplicity with which they can be sketched. This is particularly so when the critical frequencies are sufficiently separated from each other so that it is not necessary to draw the actual curve in the region of each critical frequency. The following example illustrates the application of Bode diagrams in a multistage amplifier.

Example 12-8 The gain of a three-stage amplifier is given by

$$A(s) = -\frac{10^{21}}{(s + 10^5)(s + 2 \times 10^6)(s + 2 \times 10^7)}$$

This amplifier is modified by the addition of a network whose transfer function is given by

$$H_1(s) = \frac{0.1(s + 10^5)}{(s + 0.1 \times 10^5)}$$

so that the overall transfer function of the modified network becomes $H(s) = A(s)H_1(s)$.

Sketch the asymptotic (amplitude response) Bode plots of the gain of the amplifier, the added network, and the combination.

Solution: The Bode diagrams of the functions $A(s)$ and $H_1(s)$ are shown in the diagrams of Fig. 12-14(a) and (b).

Fig. 12-14 Bode diagrams for Example 12-8.

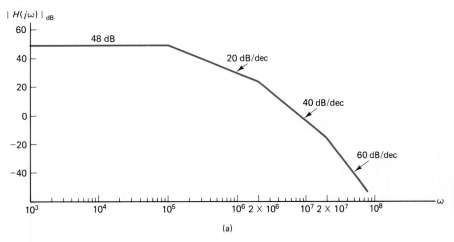

(a)

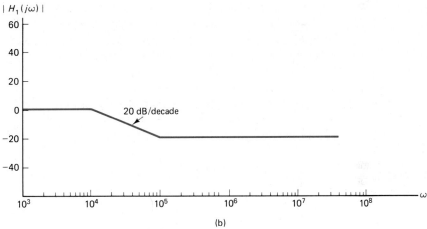

(b)

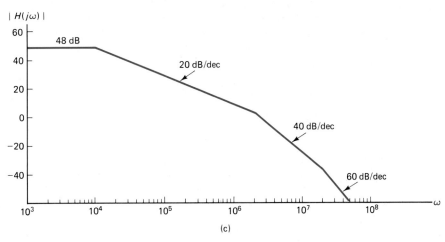

(c)

The transfer function of the modified network is

$$H(s) = A(s)H_1(s)$$

$$= -\frac{10^{20}}{(s + 0.1 \times 10^5)(s + 2 \times 10^6)(s + 2 \times 10^7)}$$

and the resulting Bode diagrams are shown in Fig. 12-14(c).

The purpose of the network added to the amplifier is to stabilize the amplifier against oscillating at a fixed frequency. Unfortunately, further discussion of this topic is beyond the scope of this text. ∎

12-5 COMPLEX FREQUENCY VIEWPOINT OF RESONANCE

The concept of complex frequency is applicable to driving point network functions also, and as an exmaple of such an application, resonance (which was discussed in Chapter 11) will be treated here from the viewpoint of complex frequency. The complex frequency approach will provide certain insights into the behavior of resonant circuits and a link between their natural responses and a forced response in the sinusoidal steady state.

Chapter 6 showed that the *natural response* of an *RLC* circuit is oscillatory when the value of the resistance in the circuit is less than the critical value: $R = 2\sqrt{L/C}$. Under sinusoidal excitation, the *forced (or steady state) response* of the circuit is always oscillatory, of course; but the properties of the resonant circuit are significantly affected by the relative values of R, L, C, and the frequency, as we saw in Chapter 11. A relationship must therefore exist between the forced response of an *RLC* circuit under sinusoidal excitation and its natural response. The use of complex frequency permits the study of the complete response of the circuit without restricting it to a specific type of forcing function, and consequently provides a more complete picture of the behavior of resonant circuits than the sinusoidal steady state approach.

12-5-1 Magnitude and Angle of a Driving Point Function

The frequency dependence of the amplitude and phase of a certain type of network function was examined by means of Bode diagrams in Sectiron 12-4. An alternative way of examining the magnitude and phase of a network function is by sketching *vectors in the complex frequency plane*.

In the complex freuqency or s plane, the horizontal axis represents σ, the real part of s, while the vertical axis represents ω, the imaginary part of s; see Fig. 12-15(a).

Fig. 12-15 (a) Complex frequency plane. (b) Pole-zero diagram.

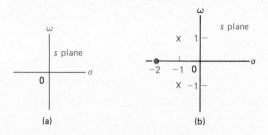

The locations of the poles and zeros of a function are shown in the s plane by means of small crosses and small circles, respectively. For example, the function

$$Z(s) = \frac{s + 2}{s^2 + 2s + 2}$$

$$= \frac{s + 2}{(s + 1 + j)(s + 1 - j)}$$

has a zero at $s = -2$ and poles at $s = -(1 - j)$ and $-(1 + j)$. Its pole-zero diagram is shown in Fig. 12-15(b).

Consider a network function

$$F(s) = \frac{(s - s_1)(s - s_2)}{(s - s_a)(s - s_b)}$$

At any point $s = s_p$ in the complex plane, the value of $F(s)$ is given by

$$F(s_p) = \frac{(s_p - s_1)(s_p - s_2)}{(s_p - s_a)(s_p - s_b)}$$

If P is the location of $s = s_p$ in the s plane, then each term in $F(s_p)$ is represented by a vector drawn from the corresponding critical frequency to the point P, as shown in Fig. 12-16. Then the magnitude of $F(s_p)$ is given by multiplying and dividing the vector lengths:

$$|F(s_p)| = \frac{|P_1P|\,|P_2P|}{|P_aP|\,|P_bP|}$$

while the angle of $F(s_p)$, denoted by Arg $F(s_p)$, is given by the sum and difference of the vector angles (measured from the real axis):

$$\text{Arg } F(s_p) = (\theta_1 + \theta_2) - (\theta_a + \theta_b)$$

Fig. 12-16 Using the pole-zero diagram to estimate the magnitude and phase of a transfer function.

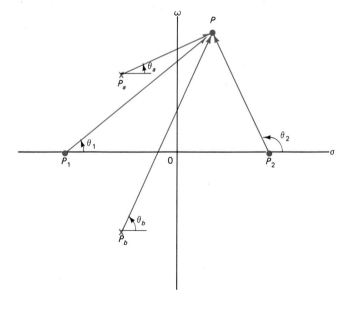

12-5 Complex Frequency Viewpoint of Resonance

The magnitude and angle of a network function at any specified value of the complex frequency s can therefore be evaluated graphically by drawing vectors from the poles and zeros in the s plane and then using the following relationships:

magnitude of a function = (product of the lengths of the vectors from the zeros to the given point) *divided by* (product of the lengths of the vectors from the poles to the given point) (12-33a)

angle of a function = (sum of the angles of the vectors from the zeros to the given point) *minus* (sum of the angles of the vectors from the poles to the given point) (12-33b)

The importance of the graphic approach does not lie in the actual evaluation of $F(s)$ at specific values of s; in fact, such an evaluation might be cumbersome in practice. The important advantage of the graphic approach is its ability to show the manner in which a given function varies without the need for detailed calculations and plots.

12-5-2 Response of a Tuned Circuit

The tuned circuit made up of a coil (with inductance L and resistance R) in parallel with a capacitance C will be used as the vehicle for discussing the use of pole-zero diagrams in the complex frequency plane to study the response of a resonant circuit.

The admittance of the tuned circuit is given by

$$Y(s) = \frac{1}{Ls + R} + Cs$$

$$= \frac{C[s^2 + (R/L)s + (1/LC)]}{[s + (R/L)]}$$

Using Q_{co} to denote the resonant Q of the coil

$$Q_{co} = (\omega_0 L/R)$$

and

$$\omega_0^2 = 1/LC$$

the expression for $Y(s)$ becomes

$$Y(s) = \frac{C[s^2 + (\omega_0/Q_{co})s + \omega_0^2)]}{[s + (\omega_0/Q_{co})]} \quad (12\text{-}34)$$

$Y(s)$ has a single pole at $s = -(\omega_0/Q_{co})$.

The zeros of $Y(s)$ are the solution of the quadratic equation

$$s^2 + (\omega_0/Q_{co})s + \omega_0^2 = 0$$

given by

$$s_1, s_2 = -\frac{\omega_0}{2Q_{co}} \pm \omega_0 \sqrt{\frac{1}{4Q_{co}^2} - 1} \quad (12\text{-}35)$$

Depending upon whether $(2Q_{co})$ is < 1, $= 1$, or > 1, there are three possible cases: (1) If $(2Q_{co}) < 1$; that is, $Q_{co} < (1/2)$, the zeros are real and distinct. (2) If $(2Q_{co}) = 1$;

that is, $Q_{co} = (1/2)$, the zeros are real and equal to each other. (3) If $(2Q_{co}) > 1$; that is, $Q_{co} > (1/2)$, the zeros are a complex conjugate pair.

12-5-3 Movement of Zeros of Y(s) as a Function of Q_{co}

Consider the locus of the zeros and poles of $Y(s)$ in the s plane as Q_{co} is varied while the resonant frequency ω_0 is kept constant.

Starting with $Q_{co} = \infty$, the zeros are

$$s_1, s_1 = j\omega_0$$

That is, they are both purely imaginary and lie on the vertical axis in the s plane: points A and B in Fig. 12-17.

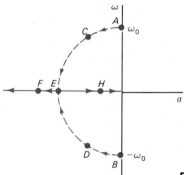

Fig. 12-17 Locus of the critical frequencies of a tuned circuit.

Now if Q_{co} is decreased but kept greater than $(1/2)$ at first, the zeros become complex conjugates given by

$$s_1, s_2 = -(\omega_0/2Q_{co}) \pm j\omega_0[\sqrt{1 - (1/4Q_{co}^2)}]$$

The zeros move away from the vertical axis along a semicircular path—points C and D in the sketch. The semicircular shape of the path is established by noting that the distance OC (or OD) is given by

$$|OC|^2 = \frac{\omega_0^2}{4Q_{co}^2} + \omega_0^2\left(1 - \frac{1}{4Q_{co}^2}\right)$$
$$= \omega_0^2$$

which is a constant.

Eventually, Q_{co} decreases to a value of $(1/2)$, making

$$s_1 = s_2 = -\frac{\omega_0}{2Q_{co}}$$

and the two zeros now coincide at the point E.

If Q_{co} is decreased further, the zeros remain on the negative real axis, points F and H move away from each other as Q_{co} decreases. When $Q_{co} = 0$, $s_1 = 0$, and $s_2 = \infty$.

The movement of the pole is confined to the negative real axis for all values of Q_{co}: it starts at the origin when $Q_{co} = \infty$ and moves away from the origin as Q_{co} decreases.

In terms of the terminology introduced in Chapter 6, when the zeros are on the vertical axis or on the semicircular path in Fig. 12-17, the circuit is *underdamped*. The natural response of the circuit is oscillatory due to the imaginary part ω_0 of the zeros. When the zeros coincide at point E, the circuit is *critically damped*. The circuit is *overdamped* when the zeros occur on the negative real axis.

Exercise 12-14 Given: a tuned circuit with $L = 50$ mH and $C = 20$ µF. Sketch the locus of the zeros and pole of the admittance function by considering values of R starting at 0 and increasing in steps of 20 Ω up to 200Ω.

12-5-4 Variation of Y(s) as a Function of s

The *magnitude and angle* of the admittance function of a tuned circuit can be studied by using vectors and the relationships in Eqs. (12-32) and (12-33). Consider the variation of $|Y(s)|$ and Arg $Y(s)$ for a fixed set of element values in the circuit (which makes the pole and zeros fixed also). For any given value of $s = s_p$ (point P in Fig. 12-18)

$$|Y(s_p)| = \frac{|P_1P| \, |P_2P|}{|P_aP|} \qquad (12\text{-}36)$$

and

$$\text{Arg } Y(s_p) = (\theta_1 + \theta_2) - \theta_a \qquad (12\text{-}37)$$

As s assumes different values, the variation of the magnitude and angle of $Y(s)$ are studied by examining the vectors and using Eqs. (12-36) and (12-37). As a case of special interest, the variation of $Y(s)$ for *sinusoidal* driving functions is studied by moving the point P along the $(j\omega)$ axis, since a sinusoidal driving function is obtained from e^{st} by making $s = j\omega$.

The point P may be chosen on the $j\omega$ axis *either above or below the origin*. Consequently, the plots showing the variation of the magnitude and angle of $Y(s)$ will include both positive and negative values of ω. That is, *negative* frequencies should be included in the study of the variation of $Y(s)$.

Variation of |Y(s)|

For small values of Q_{co}; that is, $(4Q_{co}^2) \ll 1$, the approximate locations of the zeros of $Y(s)$ are at 0 and $-(\omega_0/Q_{co})$, while the pole is at $-(\omega_0/Q_{co})$. Thus one of the zeros and the pole coincide, making the vectors P_2P and P_aP essentially equal to each other; see Fig. 12-18(b). Then

$$|Y(s_p)| \approx |P_1P|$$

As P starts from 0 and moves along the $j\omega$ axis in either direction, P_1P begins at a small value ($= 1/R_L$) and increases monotonically. Therefore, $|Y(s)|$ has a minimum value at $s = 0$ (that is, dc excitation) and shows a continuous increase as ω increases. Resonance as such is not observed when Q_{co} of the coil is *very low*.

As Q_{co} is increased (but still low), the pole no longer coincides with one of the zeros, but it stays sufficiently close to a zero so that the monotonic variation of $|Y(s)|$ persists. The response of the circuit does not show a clearly identifiable resonance; see Fig. 12-18(c).

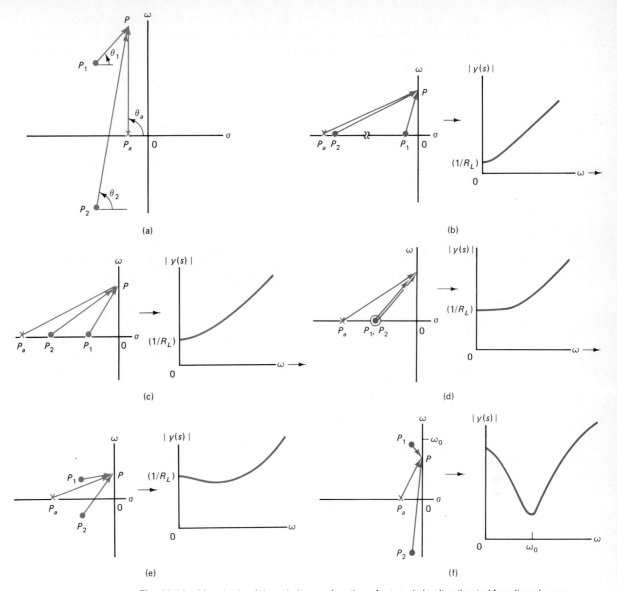

Fig. 12-18 Magnitude of the admittance function of a tuned circuit estimated from its pole-zero diagram. The Q of the coil starts at a very low value in (b) and increases successively to a very high value in (f).

When Q_{co} increases to a value of 0.5, the two zeros of $Y(s)$ occur at $-\omega_0$ while the pole is at $-2\omega_0$; see Fig. 12-18(d). Then

$$|Y| = \frac{|P_1 P|^2}{|P_a P|}$$

which starts at $(1/R_L)$ when $s = 0$ and increases monotonically as we move along the $j\omega$ axis. Again resonance is not observed, but the curve of $|Y|$ remains rather flat for a range of frequencies near $s = 0$.

When Q_{co} increases beyond 0.5, the zeros become complex conjugates; see Fig. 12-

12-5 Complex Frequency Viewpoint of Resonance

18(e). The value of $|Y|$ starts at $(1/R_L)$ at $s = 0$ (dc excitation) and at first decreases to a minimum and then increases again. When Q_{co} is small (but > 0.5), the frequency at which $|Y|$ is a minimum cannot be clearly pinpointed. That is, resonance is not *sharp* for low Q circuits.

When Q_{co} is much greater than 0.5, the zeros of $Y(s)$ are at approximately

$$-\frac{\omega_0}{2Q_{co}} \pm j\omega_0$$

and they are quite close to the $j\omega$ axis, as Fig. 12-18(f) shows. The behavior of $|Y(s)|$ when s is in the neighborhood of ω_0 in this case is of interest. When the point P is in the neighborhood of ω_0, the two vectors P_2P and P_aP do not show significant variation; that is, they remain substantially at the same values as when P is at ω_0. But the vector P_1P shows a dramatic variation, becoming quite small when P is at ω_0 and increasing significantly when P is moved away from ω_0 in either direcition. The admittance thus reaches a minimum value when P is at ω_0 and the resonant behavior is limited to a very small range of frequencies centered at ω_0; that is, the resonance is quite sharp. As Q_{co} is increased, the resonance becomes sharper and the variation of the admittance is rapid around the frequency of resonance. In the limit $Q_{co} \to \infty$, the zeros of $Y(s)$ are on the $j\omega$ axis at $j\omega_0$ while the pole is at the origin. The minimum value of $|Y|$ is zero in this case.

Now consider the sinusoidal steady state behavior as seen in the complex frequency plane in comparison with the transient analysis of Chapter 6. When the circuit is *overdamped*, $Q_{co} < 0.5$, the natural response of the circuit shows only an exponential decay, and it is now clear that the sinusoidal steady state response does not exhibit any resonant behavior when $Q_{co} < 0.5$. When the circuit is *underdamped*, $Q_{co} > 0.5$, the natural response is oscillatory, with decaying oscillations. The rate of decay of oscillations depends upon the value of (R/L). When the Q_{co} of the circuit is small, (R/L) is large and the oscillations die down more rapidly than when Q_{co} is high. The sinusoidal steady state response displays resonance in such cases, but the resonance is not sharp for small values of Q_{co}.

The sharpness of resonance is thus related to the rate at which the natural oscillations decay in the circuit: the sharper the resonance, the slower the rate of decay of the natural oscillations.

Exercise 12-15 Examine the behavior of the impedance $Z(s)$ of a series *RLC* circuit using the complex frequency plane.

Variation of the Angle of Y(s)

Return to Eq. (12-37). The angle of $Y(s)$ is given by

$$\text{Arg } Y(s) = \theta_1 + \theta_2 - \theta_a$$

where θ_1, θ_2 and θ_a are, respectively, the angles of the vectors P_1P, P_2P, and P_aP in Fig. 12-19(a).

For very low values of Q_{co}, the pole is quite close to one of the zeros, as Fig. 12-19(a) shows, and their angles essentially cancel each other. Therefore

$$\text{Arg } Y(s) \approx \theta_1$$

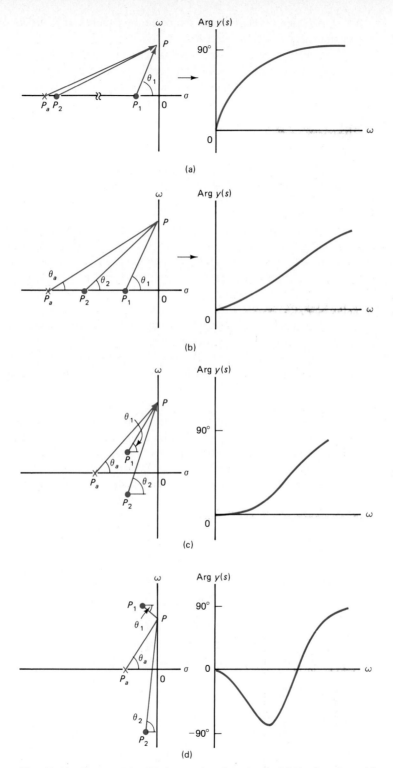

Fig. 12-19 Phase of the admittance function of a tuned circuit estimated from its pole-zero diagram. The Q of the coil starts at a very low value in (a) and increases successively to a very high value in (d).

12-5 Complex Frequency Viewpoint of Resonance

As ω increases (P starts from the origin and moves up), the angle of $Y(s)$ increases monotonically from 0 toward 90°.

If Q_{co} is increased to a somewhat larger value (but still less than 0.5), the situation is as shown in Fig. 12-19(b). Again the angle due to one of the zeros is not much different from the angle due to the pole, and they essentially cancel each other. The angle of $Y(s)$ equals θ_1 and increases from 0 to 90° monotonically as P moves up along the $j\omega$ axis. It can be seen that the angle increases more slowly for higher values of Q_{co} of the coil.

When Q_{co} is increased above 0.5 so as to make the zeros complex, the monotonic increase of the angle of $Y(s)$ persists at first when the zeros are still far removed from the vertical axis, as in Fig. 12-19(c). But as Q_{co} is increased and the zeros move closer to the $j\omega$ axis, Fig. 12-19(d), the situation becomes interesting. The angle θ_2 from the lower zero to point P remains relatively large throughout (80 to 90°). The angle θ_a from the pole starts at 0 and increases continuously toward 90°, rapidly at first and then more slowly. Note that $\theta_2 > \theta_a$ and $(\theta_2 - \theta_a) > 0$ for all positions of P on the vertical axis above the origin. The angle θ_1 from the upper zero starts at a large *negative* value when P is at the origin, increases toward zero (which occurs when P is directly opposite the location of the zero), and then becomes positive. When P is between the origin and the position directly opposite the upper zero, the total angle of $Y(s)$ is *negative*, since the negative value of θ_1 dominates the positive value of $(\theta_2 - \theta_a)$. As P moves above the position directly opposite the upper zero, the total angle of $Y(s)$ becomes positive since θ_1 and $(\theta_2 - \theta_a)$ are both positive.

As point P starts from the origin and moves along the $j\omega$ axis, the angle of $Y(s)$ starts at zero, becomes negative, and decreases at first and then increases. It becomes zero again when P is in the neighborhood of the position directly across the upper zero and then increases toward 90°.

Now consider a very high value of Q_0 for the coil. The zeros of $Y(s)$ will then lie essentially on the $j\omega$ axis at $j\omega_0$, while the pole will be almost at the origin. The angle θ_1 from the upper zero to the point P will be approximately at $-90°$ or $+90°$ depending upon whether P is below or above ω_0. The angle θ_2 from the lower zero to the point P will be essentially a constant at $+90°$ as P moves along the positive portion of the vertical axis. The angle θ_a from the pole to the point P will be essentially a constant at $+90°$ as P moves along the positive portion of the $j\omega$ axis. Thus the angles θ_2 and θ_a cancel each other, and the angle of $Y(s)$ equals θ_1: $+90°$ above ω_0 and $-90°$ below ω_0. (This case is not shown in the figures.)

The diagrams of Fig. 12-19 show the variation of Arg $Y(s)$ for positive frequencies. To have a complete picture it is necessary to add the portion of the curves for negative frequencies, which is obtained in each case by using the fact that the argument is an *odd* function of ω.

The use of the complex frequency plane provides a means of examining the behavior of a circuit's frequency response *qualitatively* and also reveals the interrelationship between the natural response and the sinusoidal steady state response of a circuit.

12-6 SUMMARY OF CHAPTER

The use of a complex function of the form Ke^{st} where K and s are complex quantities enables us to represent different types of forcing and response functions that are constant, exponential, or sinusoidal by an appropriate choice of the exponent s. s is called the complex frequency with a real part σ and an imaginary part ω. When both σ and ω are zero, the function is a constant; when $\omega = 0$ (but not σ), the function is an increasing or decreasing exponential function; and if $\sigma = 0$ (but not ω), the function is sinusoidal. Complex frequency plays a part similar to the quantity $j\omega$ in sinusoidal steady state analysis using phasors.

When a complex forcing function Ke^{st} acts on a circuit, differentiation and integration transform to multiplication and division by s and the differential equations of the circuit reduce to algebraic equations in s. Impedances and admittances are defined in terms of s and the principles of circuit analysis developed in earlier chapters are applicable in the complex frequency domain.

The values of s at which a network function becomes zero are called the zeros of the function, and the values of s at which the function becomes infinite are called the poles of the function. If the poles and zeros of a network function are known, then the function is completely specified except possibly for a constant multiplying factor; hence these are called the critical frequencies of the function.

Bode diagrams are used to sketch the frequency response of networks in which the critical frequencies are all negative real. The amplitude response shows the variation of $|H(j\omega)|_{dB}$ versus Log ω. For a transfer function with a single pole or a single zero, the amplitude response is drawn by sketching the low-frequency and high-frequency asymptotes. Data points are added at and near the break frequency to obtain the actual amplitude response. For a transfer function with two or more poles or zeros, the component plots can be added, since multiplication is transformed to addition by the introduction of the logarithm in the definition of the decibel. The phase response is also obtained by sketching asymptotes at low and high frequencies and adding several data points at and near the break frequency. Bode diagrams are useful in the study of filter networks and amplifiers.

Vector diagrams drawn in the s plane are useful for a qualitative study of the behavior of network functions. The magnitude of the function is related to the product and ratio of the lengths of the vectors, and the phase of the function to the sum and difference of the angles of the vectors. The behavior of the two-branch parallel resonant circuit (tuned circuit) studied in this fashion shows the close relationship of the natural response of the circuit to the sinusoidal steady state response of the circuit. A high Q coil leads to an oscillatory natural response (with slowly decaying oscillations) and a sharp resonance. A very low Q coil leads to a nonoscillatory natural response and a steady state response that does not exhibit resonant characteristics. Qualitative analysis of this type is possible for other circuits.

Answers to Exercises

12-1 $f(t) = 2\sqrt{(a^2 + b^2)}\ e^{\sigma t} \cos{[\omega t + \arctan{(b/a)}]}$.

12-2 (a) $Z(s) = 5(s^3 + 0.267s)/(s^2 + 0.25)$. $Y(s) = 0.2(s^2 + 0.25)/(s^3 + 0.267s)$. $H(s) = L_3s/Z(s) = 0.6(s^2 + 0.25)/(s^2 + 0.267)$. (b) $Z(s) = 5(s^2 + 1)/(s^3 + 4.75s)$. $Y(s) = 0.2(s^3 + 4.75s)/(s^2 + 1)$. $H(s) = s^2/(s^2 + 1)$.

12-3 (a) $Z(s) = 5s(s + j0.517)(s - j0.517)/(s + j0.5)(s - j0.5)$.
(b) $Z(s) = 5(s + j)(s - j)/s(s + j2.18)(s - j2.18)$.

12-4 (a) $F(s) = (s + 1)$. (b) $F(s) = 1/(s + 4)$. (c) $F(s) = (s^2 + 6s + 5)/(s^3 + 10s^2 + 21s)$. (d) $(s^3 + 9s)/(s^4 + 26s^2 + 25)$. (e) $(s^2 + 4s + 13)/(s^3 + 6s^2 + 25s)$.

12-5 (a) $Z_a = 3(s^2 + 4.92s + 2.5)/(s^2 + 1.25s)$. Zeros at -0.576, -4.34. Pole at 0, -1.25.
(b) $Z_b = 3.33(s + 1)$. Zero at -1. No finite poles.
(c) $Z_c = 0.25(s^4 + 220s + 4000)/(s^3 + 110s)$. Poles at 0, $\pm j10.49$. Zeros at $\pm j4.47$, $\pm j14.14$.

12-6 Loss of 6 dB in an octave.

12-7 (a) Low-frequency asymptote: -40 dB. Break at 100 r/s. High-frequency asymptote passes through -40 dB at 100 r/s and -60 dB at 1000 r/s. (b) Low-frequency asymptote: $+20$ dB. Break at 0.1 r/s. High-frequency asymptote passes through $+20$ dB at 0.1 r/s and 0 dB at 1 r/s.

12-8 $C = 1.59$ nF. Shape same as Fig. 12-8(b).

12-9 (a) Asymptotes: $0°$ up to 10 r/s, $-90°$ above 1000 r/s, and a straight line joining $0°$ and $-90°$ between 10 r/s and 1000 r/s. (b) Asymptotes: $0°$ up to 0.01 r/s, $-90°$ above 1 r/s, and a straight line joining $0°$ and $-90°$ between 0.01 r/s and 1 r/s. (c) LPF: $0°$ up to 1000 Hz, $-90°$ above 10^5 Hz, and a straight line joining $0°$ and $-90°$ between 10^3 and 10^5 Hz.

12-10 (a) Upward straight line for all ω. Slope: 20 dB/decade. Passes through 0 dB at 1 r/s. (b) Downward straight line for all ω. Slope -20 dB/decade. Passes through 0 dB at 1 r/s.

12-11 $1/(s + 10)$: Low-frequency asymptote: -20 dB. Break frequency 10 r/s. High-frequency asymptote passes through -20 dB at 10 r/s and -40 dB at 100 r/s. $1/(s + 200)$: Low-frequency asymptote: -46 dB. Break frequency: 200 r/s. High-frequency asymptote passes through -66 dB at 200 r/s and -86 dB at 2000 r/s. Overall: Asymptotes: -66 dB up to 10 r/s; -20 dB/decade from 10 r/s to 200 r/s; -40 dB after 200 r/s. Data points for curve near the break frequencies: -67 dB at 5 r/s, -69 dB at 10 r/s, -73 dB at 20 r/s, -87 dB at 100 r/s, -79 dB at 200 r/s, and -99 dB at 400 r/s.

12-12 (a) $H(s) = (s + 1000)/(s + 2000)$. Asymptotes: Horizontal at -6 dB up to 1000 r/s, upward at $+20$ dB/decade from 1000 to 2000 r/s, horizontal at 0 dB above 2000 r/s. Data points: -5 dB at 500 r/s, -4 dB at 1000 r/s, -2 dB at 2000 r/s, $+1$ dB at 4000 r/s. (b) $H(s) = 1000/(s + 2000)$. Diagram same as a low pass filter: -6 dB at low frequencies, break at 2000 r/s.

12-13 10^4 gives a constant 80 dB. $1/(s + 20)$: low-frequency asymptote at -26 dB, break at 20 r/s, and high-frequency asymptote -20 dB/decade. $1/(s + 20 \times 10^4)$: low-frequency asymptote at -106 dB, break at 20×10^4 r/s, and high-frequency asymptote -20 dB/decade. Overall: low-frequency asymptote -52 dB up to 20 r/s, downward slope -20 dB/decade from 20 r/s to 20×10^4 r/s, and -40 dB/decade after 20×10^4 r/s. Data points: -53 dB at 10 r/s, -55 dB at 20 r/s, -59 dB at 40 r/s, -133 dB at 10×10^4 r/s, -135 dB at 20×10^4 r/s, and -139 dB at 40×10^4 r/s.

12-14 Locus of zeros: Circle of radius 1000 units. Critical damping at $R = 100\ \Omega$.

12-15 $Z(s) = L[s^2 + (R/L)s + 1/LC]/s$. Zeros of $Z(s)$ are exactly the same as the zeros of $Y(s)$ of the tuned circuit discussed in the text. Pole at $s = 0$. Response exactly similar to the tuned circuit for large Q; at low Q there is a very flat resonant behavior.

PROBLEMS

Secs. 12-1, 12-2 Complex Frequency, Impedance, and Admittance

12-1 Express $f(t) = K_1 e^{s_1 t} + K_2 e^{s_2 t}$ in the form of a real time function in each of the following cases (simplify your result as far as possible):
(a) $K_1 = K_2 = 4$. $s_1 = s_2 = 0$.
(b) $K_1 = K_2 = 4$. $s_1 = +3\underline{/0°}$ and $s_2 = 3\underline{/180°}$.
(c) $K_1 = (3 - j4)$. $K_2 = K_1^*$. $s_1 = j6$. $s_2 = -j6$.
(d) $K_1 = (3 - j4)$. $K_2 = K_1^*$. $s_1 = -5 + j6$. $s_2 = s_1^*$.

12-2 Write each of the following functions in the form of a term (or sum of terms) involving Ke^{st}.
(a) $f_1(t) = 10 \cos (100t + 30°)$.   (b) $f_2(t) = 10 e^{-5t} \sin (377t + 75°)$.
(c) $f_3(t) = 10 \cosh 3t$.   (d) $f_4(t) = 10 \sinh 3t$.

12-3 For each of the networks shown in Fig. 12-20, obtain the input impedance $Z(s)$.

Fig. 12-20 Networks for Problem 12-3.

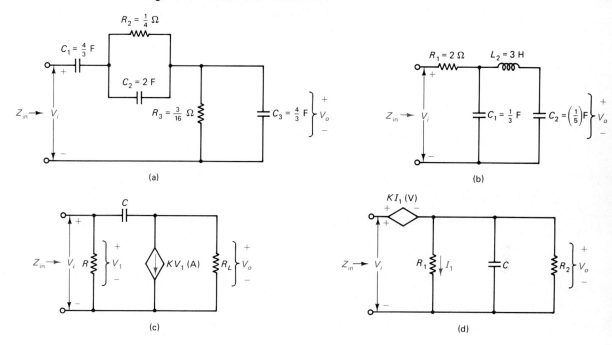

12-4 For each of the networks shown in Fig. 12-21, obtain the input admittance $Y(s)$.

Fig. 12-21 Networks for Problem 12-4.

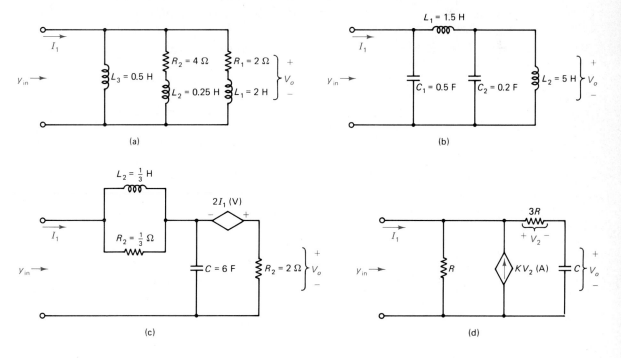

Problems

12-5 For each of the networks of Problem 12-3, obtain the transfer function $H(s) = V_0/V_i$.

12-6 For each of the networks of Problem 12-4, obtain the transfer impedance (V_0/I_1).

Sec. 12-3 Critical Frequencies of Network Functions

12-7 Determine the critical frequencies of each of the impedance functions obtained in Problem 12-3.

12-8 Determine the critical frequencies of each of the admittance functions obtained in Problem 12-4.

12-9 Determine the critical frequencies of each of the transfer functions obtained in Problem 12-5.

12-10 Determine the critical frequencies of each of the transfer impedances obtained in Problem 12-6.

12-11 A coil with an inductance L and resistance R is in parallel with a capacitance C. The input impedance $Z(s)$ has poles at $s = -(a + jb)$ and $-(a - jb)$. Determine the values of L and C in terms of the constants a, b, c, and R.

12-12 Two impedances Z_1 and Z_2 are known to have the following critical frequencies. Z_1: Zeros at $s = 0, -3, -5$, and poles at $s = -1, -4$, and -10. Z_2: Zeros at $s = 0, -1, -3.5, -5$, and poles at $s = -0.5, -3, -4$. What information can you provide about the critical frequencies of $Z(s)$ obtained by (a) connecting the two impedances in series, and (b) connecting the two impedances in parallel?

12-13 It was stated in Chapter 11 that (a) the poles and zeros of the impedance or admittance of a lossless network are all confined to the $j\omega$ axis, and (b) that the poles and zeros alternate. Also $s = 0$ is always a critical frequency. Consider two lossless networks Z_1 and Z_2. Z_1 has zeros at $s = 0$, and $s = \pm j4$ and poles at $s = \pm j2$ and $s = \pm j8$. Z_2 has zeros at $s = \pm j2$, $s = \pm j4$ and poles at $s = 0$, $s = \pm j3$. (a) What is the behavior of Z_1 at infinite frequency? (b) What is the behavior of Z_2 at infinite frequency? (c) Sketch the pole-zero patterns of Z_3 obtained by connecting Z_1 and Z_2 in series. (d) Sketch the pole-zero pattern of Z_4 obtained by connecting Z_1 and Z_2 in parallel. In parts (c) and (d), certain critical frequencies can be determined exactly, only a region of location can be determined for the others.

12-14 Suppose an impedance function $Z(s)$ and the corresponding network with all element values are given. State how each R, each L, and each C will be affected in the following cases. (a) The impedance function is multiplied by a constant K. (b) The frequency for which the given circuit element values are valid is changed by a factor of N.

12-15 Determine the transfer function (V_0/V_s) of the op amp circuit shown in Fig. 12-22. Use the approximate model of the op amp from Chapter 3.

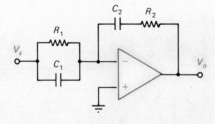

Fig. 12-22 Op amp circuit for Problem 12-15.

12-16 Repeat the previous problem for the op amp circuit of Fig. 12-23.

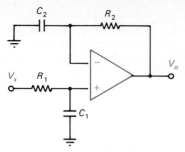

Fig. 12-23 Op amp circuit for Problem 12-16.

12-17 Determine the transfer function of the op amp circuit shown in Fig. 12-24. Show that if the gain A of the op amp is infinitely large, then the given network acts as an integrator: V_o is proportional to (V_i/s).

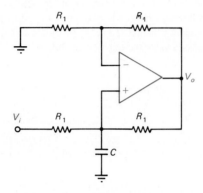

Fig. 12-24 Op amp circuit for Problem 12-17.

Sec 12-3 Frequency Response and Bode Diagrams

12-18 Draw the Bode diagram (amplitude response) of each of the following transfer functions.

(a) $H_a(s) = \dfrac{1}{s + 100}$.

(b) $H_b(s) = \dfrac{s}{s + 100}$.

(c) $H_c(s) = \dfrac{1}{(s + 100)^2}$.

(d) $H_d(s) = \dfrac{s^2}{(s + 100)^2}$.

12-19 Draw the Bode diagram of the phase of each of the transfer functions in Problem 12-15.

12-20 Draw the Bode diagram of the amplitude response of each of the transfer functions given below.

(a) $\dfrac{1000(j\omega + 1)(j100\,\omega + 1)}{(j10\,\omega + 1)(10 + j\omega)}$

(b) $\dfrac{1000/j\omega}{(1 + 10/j\omega)(100 + 1/j\omega)}$

12-21 Three identical amplifier stages, each with a gain of A, are cascaded together so that the overall gain is A^3. If

$$A(s) = \dfrac{10^6 s}{(s + 20)(s + 50 \times 10^3)}$$

Problems

(a) Sketch the Bode diagrams (amplitude response) of a single stage. (b) Sketch the Bode diagram (amplitude response) of the three-stage amplifier. (c) In the three-stage configuration, there is a range of frequencies (called the *midband*) over which the gain remains constant. What is the value of this constant gain? (d) Estimate the two frequencies at which the gain is 3 dB less than the midband gain. The bandwidth of the amplifier is the difference between the two -3 dB frequencies.

12-22 Repeat the previous problem if the gain of each stage of the amplifier is modified to

$$A(s) = \frac{10^8 s}{(s + 2000)(s + 10^4)}$$

12-23 An idealized Bode plot consists of the following straight-line segments: a straight line at -20 dB/decade from $\omega = 0$ to $\omega = 1$ r/s, a horizontal straight line at -20 dB from $\omega = 1$ r/s to $\omega = 30$ r/s, a straight line at -20 dB/decade from $\omega = 30$ r/s to 300 r/s, and a straight line at -40 dB/decade for all $\omega > 300$ r/s. Write the expression for $H(j\omega)$ and obtain $H(s)$.

12-24 An amplifier has the idealized Bode diagram shown in Fig. 12-25(a). (a) Write the transfer function $H_1(s)$. (b) The transfer function $H_1(j\omega)$ is multiplied by a function $H_2(j\omega)$ so that the idealized Bode diagram of the product function $H_3(j\omega)$ is as shown in Fig. 12-25(b). Obtain the idealized Bode diagram of $H_2(j\omega)$ and write the transfer function $H_2(s)$.

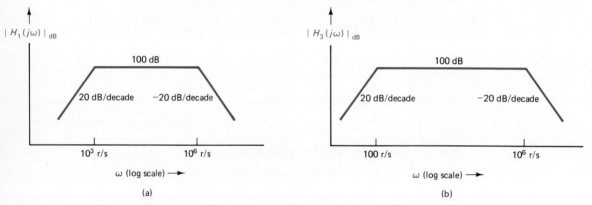

Fig. 12-25 Bode diagram for Problem 12-24.

12-25 Sketch the Bode diagrams (amplitude and phase response) of the following two transfer functions:

$$H_1(s) = \frac{s + a}{s + b} \quad \text{and} \quad H_2(s) = \frac{s + b}{s + a}$$

Assume that $a = 10b$. How are the plots of the two functions related to each other?

12-26 The transfer function of a network is given by

$$H(s) = \frac{10^7}{(s + 100)(s + 3000)}$$

Use Bode diagrams and estimate (a) the phase angle of the function when $|H(j\omega)|_{dB}$ is 0 dB. (b) The value of $|H(j\omega)|_{dB}$ when the phase angle is $-135°$.

12-27 Repeat the previous problem (a) if the given transfer function is modified by adding a zero at $s = 0$, and (b) if the given transfer function is modified by adding a pole at $s = 0$.

12-28 An amplifier has a voltage gain of 10^3 at $\omega = 0$. The poles of the voltage gain are at 250 kHz, 1 MHz, and 2 MHz. There are no finite zeros. (a) Sketch the amplitude and phase response

Bode diagrams. (b) Estimate the phase angle when the amplifier gain is 0 dB. (c) Estimate the gain of the amplifier when the phase angle is $-180°$. (d) The amplifier is cascaded with a network of transfer function $H_1(j\omega)$ so that the resultant transfer function is the product $H(j\omega)H_1(j\omega)$. $H_1(j\omega)$ has a zero at 500 kHz and a magnitude of -20 dB at $\omega = 0$. Repeat parts (a), (b), and (c).

Sec 12-4 Complex Frequency Viewpoint of Resonance

12-29 A conductance $G = 0.5$ S, an inductance $L = 2$ H, and a capacitance $C = 0.25$ F are in parallel. Determine the input admittance $Y(s)$. Sketch the pole-zero diagram on the s plane. Obtain its frequency response characteristics, $|Y(j\omega)|$ and Arg $Y(j\omega)$, as the frequency is varied from 0 to ∞ by using vectors in the s plane.

12-30 The transfer function of a resonant circuit is given by

$$H(j\omega) = \frac{100(s + 0.5)}{(s^2 + 2s + 401)}$$

Use vectors in the s plane to estimate the magnitude and angle of the above transfer function at $\omega = 0$ and $\omega = 20$. Sketch the variation of the magnitude and phase of the transfer function as ω varies from 0 to 200 r/s.

12-31 (a) A series RLC circuit has a maximum impedance of 50 Ω, a resonant frequency of 300 Hz, and a bandwidth of 200 Hz. Obtain its pole-zero pattern. (b) Repeat the problem if the data pertain to a parallel connection of G, L, and C.

12-32 (a) The critical frequencies of the impedance of a series resonant circuit are given by: zeros at $s = -(5 + j50)$, $-(5 - j50)$ and a pole at $s = 0$. Assume $L = 1$H. Determine its resonant frequency, resonant Q_0, bandwidth, the values of R and C. (b) Repeat the problem if the data pertain to the admittance of a parallel GLC circuit.

12-33 The use of vectors in the s plane to study the frequency characteristics of a network is not restricted to resonant circuits. Consider a network with a resistance R in series with an impedance Z_1. Study the variation of the magnitude and phase of $Z_T = (R + Z_1)$ for the following cases: (a) Z_1 is a parallel combination of a resistance R and a capacitance C. (b) Z_1 is a parallel combination of a resistance R and an inductance L.

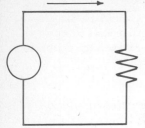

CHAPTER 13
TWO-PORTS

The discussion of nodal and loop analysis in Chapter 9 showed that a knowledge of the nodal admittance or loop (or mesh) impedance matrix of a relaxed network is sufficient for analyzing the network; it is not necessary to know the detailed configuration of the network elements. In the general case, it is possible to provide access to different branches of the network by attaching terminals to a pair of nodes or by cutting into branches that form the links of the network. Each such access forms a *port of entry* into the network. A network with n ports of entry is called an n-port. For example, the determination of Thevenin's and Norton's equivalents pertain to one-ports, since there is one pair of terminals through which we look into the network in such cases. Note that the Thevenin or Norton equivalent is sufficient for determining the voltages and currents in any branch or circuit connected externally to the one-port.

A large number of networks of practical interest and importance have two ports of entry. An amplifier, for example, has a pair of input terminals (the *input port*) and a pair of output terminals (the *output port*). Such networks are called *two-ports*. There are four variables of interest in a two-port; the current and voltage at the input port and the current and voltage at the output port. The response of a two-port is studied by using parameter matrices that express the interdependence of these four variables. The actual configuration of elements is of no interest to anyone making only external connections to the two-port. For example, the user of a given amplifier needs to know only an appropriate parameter matrix in order to analyze the relationship between any external connections at the input and output ports, and does not need to worry about the detailed composition of the amplifier network.

The common occurrence of two-ports in modeling electronic devices, electronic circuits, and communication networks and systems makes it important to study them in some detail. The attractive feature of two-port theory is that the network is viewed simply as a black box described by certain parameter matrices. We can concentrate on the four variables (currents and voltages at the two ports) and any external connections to the two-port.

13-1 BASIC NOTATION AND DEFINITIONS

Consider a general network with four terminals arbitrarily attached to it, as indicated in Fig. 13-1(a). To view the resulting four-terminal network as a two-port, it should be possible to divide the four terminals into two sets of *terminal pairs* in such a way that the current entering one terminal in each pair leaves through the other terminal of that pair. For example, if the current entering terminal P equals the current leaving terminal Q, and the current entering terminal R equals the current leaving terminal S—that is, if

$$i_p + i_q = 0 \quad \text{and} \quad i_r + i_s = 0 \tag{13-1}$$

in Fig. 13-1(a), then the terminal pair PQ form one port and the terminal pair RS form the other port, and the network is a two-port.

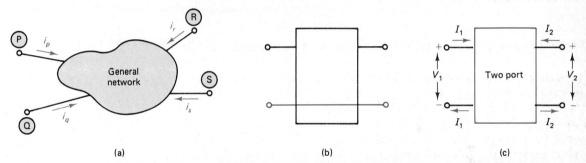

Fig. 13-1 Two-port representation. (a) A four-terminal network. (b) A four-terminal network with two terminals connected together. (c) Notation for voltages and currents in a two-port.

Exercise 13-1 In the diagram of Fig. 13-1(a), what conditions must be satisfied by the four currents in order to treat the network as a two-port with P-R as one port and Q-S as the other?

Note that when conditions of the form in Eq. (13-1) are satisfied, there are only two independent currents. Thus a two-port is characterized by having only two independent currents.

Networks with four terminals, of which two are connected by a short circuit as shown in Fig. 13-1(b), occur frequently in models of electronic circuits. The presence of the common node reduces the number of independent currents to two in such networks, and they are always classified as two-ports.

The block diagram representation of a two-port is shown in Fig. 13-1(c), where V_1 and I_1 denote the voltage and current at the input port, and V_2 and I_2 those at the output port. The network in the black box is assumed to consist of linear components (including linear dependent sources), but no *independent* sources. The standard convention is to choose both the currents I_1 and I_2 *entering* the positive voltage reference terminals at the input and output ports to introduce symmetry in the analysis of two-ports.

The four quantities of interest in a two-port are the input voltage V_1, the input current I_1, the output voltage V_2, and the output current I_2. Note that the only access to the network in the two-port is through the two ports, and the only quantities in the network available for observation are the two currents and two voltages I_1, I_2, V_1, and V_2.

It is possible to express any two of the four variables, V_1, I_1, V_2, and I_2 as *linear functions* of the other two variables. (The validity of this statement will be established in Section 13-2 for one choice of independent variables.) Each choice of two independent variables gives rise to a different two-port parameter matrix useful for analyzing any linear system made up of the two-port and any external connections made to it. The different parameter matrices are related, since all express relationships between and among the same four variables. Therefore, given one parameter matrix, any other parameter matrix can be readily determined.

Four of the possible choices of independent variables and the names of the corresponding sets of two-port parameters are given here.

1. *Open-circuit impedance parameters.* The current I_1 and I_2 are selected as the independent variables, and the voltages V_2 and V_1 are then expressed as functions of I_1 and I_2. The parameters that appear in the resulting equations are called *open-circuit impedance, or z, parameters*.

2. *Short-circuit admittance parameters.* The voltages V_1 and V_2 are chosen as the independent variables, and the currents I_1 and I_2 are then expressed as functions of V_1 and V_2. The parameters appearing in the resulting equations are called *short-circuit admittance, or y, parameters*.

3. *Hybrid parameters.* The input current I_1 and the output voltage V_2 are selected as the independent variables. The output current I_2 and the input voltage V_2 are then expressed as functions of I_2 and V_2. The parameters appearing in the resulting equations are called *hybrid, or h, parameters*. These parameters are particularly important to the analysis of transistor amplifiers.

4. *Transmission parameters.* The output voltage and current V_2 and I_2 are chosen as the independent variables, and the input voltage and current V_1 and I_1 are then expressed as functions of V_2 and I_2. The parameters appearing the resulting equations are called *transmission, or ABCD, parameters*. These parameters are particularly useful in the analysis of general communication networks and transmission lines.

Other choices of independent and dependent variables are available, but the four sets listed here are those most commonly used in network analysis.

13-2 OPEN-CIRCUIT IMPEDANCE PARAMETERS (FORMAL DERIVATION)*

Consider a general linear relaxed network to which two terminal pairs are attached, as shown in Fig. 13-2. Two current sources have been added to excite the network into action. If there are n independent nodes in the network, then the general set of nodal

---

*The analysis presented in Section 13-2 may be skipped by any reader not interested in the formal derivation of the parameters; one may continue with Section 13-3 without loss of continuity.

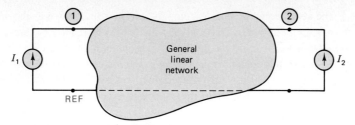

Fig. 13-2 Development of the open-circuit impedance parameters of a two-port.

equations will be of the form

$$[y][V] = [I]$$

where

$[y]$ = the nodal admittance matrix (defined in Chapter 9)

$[V]$ = a column matrix whose elements are the nodal voltages $V_1, V_2, \ldots, V_n$

$[I]$ = a column matrix with I_1 in the first row, I_2 in the second row, and all other elements zero.

The solution of the set of equations (13-2) leads to

$$V_1 = \frac{1}{\|y\|}(A_{11}I_1 + A_{21}I_2) \qquad (13\text{-}2)$$

and

$$V_2 = \frac{1}{\|y\|}(A_{12}I_1 + A_{22}I_2) \qquad (13\text{-}3)$$

where

A_{ij} = the cofactor of the ith row and jth column of $[y]$

and

$\|y\|$ = determinant of $[y]$

Using matrix notation, Eqs. (13-2) and (13-3) are rewritten in the form:

$$\begin{bmatrix} V_1 \\ V_2 \end{bmatrix} = \begin{bmatrix} A_{11}/\|y\| & A_{21}/\|y\| \\ A_{12}/\|y\| & A_{22}/\|y\| \end{bmatrix} \begin{bmatrix} I_1 \\ I_2 \end{bmatrix} \qquad (13\text{-}4)$$

Let

$$z_{ij} = \frac{A_{ji}}{\|y\|} \qquad (i, j = 1, 2) \qquad (13\text{-}5)$$

so that Eq. (13-4) becomes

$$\begin{bmatrix} V_1 \\ V_2 \end{bmatrix} = \begin{bmatrix} z_{11} & z_{12} \\ z_{21} & z_{22} \end{bmatrix} \begin{bmatrix} I_1 \\ I_2 \end{bmatrix} \qquad (13\text{-}6)$$

13-2 Open-Circuit Impedance Parameters (Formal Derivation)

The set of equations represented in Eq. (13-6) is known as the *open-circuit impedance parameter equations*. The four parameters z_{11}, z_{12}, z_{21}, z_{22} are called the *open-circuit impedance parameters* or simply the *z parameters* of the two-port. The matrix consisting of the four *z* parameters is called the *z parameter matrix*. (As we will see in Section 13-3, determination of the *z* parameters involves open-circuit measurements or calculations; hence the name *open-circuit* impedance parameters.)

Example 13-1 Obtain the *z* parameters of the network shown in Fig. 13-3.

Solution The nodal admittance matrix of the network is written by inspection.

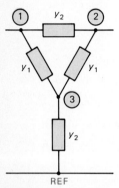

Fig. 13-3 Network for Example 13-1.

$$[y] = \begin{bmatrix} Y_1 + Y_2 & -Y_2 & -Y_1 \\ -Y_2 & Y_1 + Y_2 & -Y_1 \\ -Y_1 & -Y_1 & 2Y_1 + Y_2 \end{bmatrix}$$

The determinant of $[y]$ is given by

$$\|y\| = Y_1 Y_2 (2Y_2 + Y_1)$$

The cofactors A_{ij} are given by

$$A_{11} = A_{22} = Y_1^2 + 3Y_1 Y_2 + Y_2^2$$
$$A_{12} = A_{21} = (Y_1 + Y_2)^2$$

The *z* parameters are obtained by using Eq. (13-5) and the *z* parameter matrix is found to be

$$[z] = \begin{bmatrix} \dfrac{Y_1^2 + 3Y_1 Y_2 + Y_2^2}{Y_1 Y_2 (2Y_2 + Y_1)} & \dfrac{(Y_1 + Y_2)^2}{Y_1 Y_2 (2Y_2 + Y_1)} \\ \dfrac{(Y_1 + Y_2)^2}{Y_1 Y_2 (2Y_2 + Y_1)} & \dfrac{Y_1^2 + 3Y_1 Y_2 + Y_2^2}{Y_1 Y_2 (2Y_2 + Y_1)} \end{bmatrix}$$

In this example $z_{11} = z_{22}$ and $z_{12} = z_{21}$. The equality of z_{12} to z_{21} is a general property of *reciprocal* networks, since the nodal admittance matrix of such a network is symmetrical and $A_{12} = A_{21}$. The equality of z_{11} to z_{22} is not a general property, however, but is due to the *physical symmetry* of the given network, which presents the same configuration when viewed from either port. ∎

Exercise 13-2 Obtain the *z* parameter matrix of the network shown in Fig. 13-4 by using the same procedure as in Example 13-1.

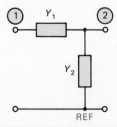

Fig. 13-4 Network for Exercise 13-2.

13-3 OPEN-CIRCUIT IMPEDANCE PARAMETERS OF A TWO-PORT

The open-circuit impedance parameters (or the z parameters) of a two-port are denoted by z_{11}, z_{12}, z_{21}, and z_{22} and are defined by the set of equations:

$$V_1 = z_{11}I_1 + z_{12}I_2 \qquad (13\text{-}7a)$$
$$V_2 = z_{21}I_1 + z_{22}I_2 \qquad (13\text{-}7b)$$

where the voltages and currents are as indicated in Fig. 13-5(a).

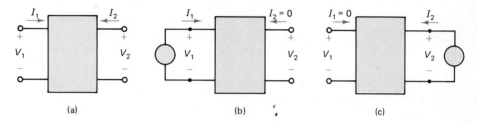

Fig. 13-5 (a)–(c) Determination of the z parameters of a two-port by open-circuit calculations.

One procedure for determining the z parameters is by means of two sets of open-circuit measurements on the network: open circuit at the input port and open circuit at the output port.

Open Circuit at the Output Port

As Fig. 13-5(b) shows, the network is driven by a source at the input port, with the output port open-circuited and $I_2 = 0$. When I_2 is made equal to zero in Eqs. (13-7), they reduce to

$$V_1 = z_{11}I_1 \qquad \text{and} \qquad V_2 = z_{21}I_1$$

The equations above lead to the following relationships:

$$z_{11} = (V_1/I_1) \qquad \text{when } I_2 = 0 \qquad (13\text{-}8a)$$
$$z_{21} = (V_2/I_1) \qquad \text{when } I_2 = 0 \qquad (13\text{-}8b)$$

Eqs. (13-8a,b) lead to the determination of z_{11} and z_{21} by imposing an open circuit (either in actuality or on paper) at the output of the two-port and evaluating the ratios (V_1/I_1) and (V_2/I_1), respectively.

Open Circuit at the Input Port

Repeating the steps with the input port open-circuited, as in Fig. 13-5(c), we obtain the following relationships:

$$z_{12} = (V_1/I_2) \qquad \text{when } I_1 = 0 \qquad (13\text{-}9a)$$
$$z_{22} = (V_2/I_2) \qquad \text{when } I_1 = 0 \qquad (13\text{-}9b)$$

Eqs. (13-9a,b) lead to the determination of z_{12} and z_{22} by imposing an open circuit (either in actuality or on paper) at the input of the two-port and evaluating the ratios (V_1/I_2) and (V_2/I_2), respectively. Note that for *reciprocal* networks, $z_{12} = z_{21}$.

Since the z parameters of a two-port can be determined by means of open-circuit measurements or calculations, they are called *open-circuit* impedance parameters. The four relationships in Eqs. (13-8) and (13-9) also lead to the following specific names of the individual parameters:

- z_{11} = open-circuit *input impedance*
- z_{12} = open-circuit *forward transfer impedance*
- z_{21} = open-circuit *reverse transfer impedance*
- z_{22} = open-circuit *output impedance*

Note that the z parameters determined by open-circuit calculations are valid for use with any input and output connections made to the two-port.

When a given circuit contains no inductances or capacitances, the open-circuit *impedance* parameters become open-circuit *resistance* parameters, and the symbol r may be used with appropriate subscripts.

Example 13-2 Determine the z parameters of the two-port shown in Fig. 13-6(a). Assume that the coils are not mutually coupled.

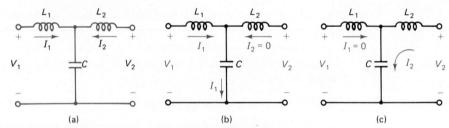

Fig. 13-6 (a) Network for Example 13-2. (b) Calculation of z_{11} and z_{21} by open-circulating the output. (c) Calculation of z_{12} and z_{22} by open-circuiting the input.

Solution   *Open-circuit the Output Port:*   [Fig. 13-6(b)]

$$I_1 = \frac{V_1}{j(\omega L_1 - 1/\omega C)}$$

and

$$V_2 = -j\frac{I_1}{\omega C}$$

Therefore,

$$z_{11} = \frac{V_1}{I_1} \quad \text{when } I_2 = 0$$

$$= j\left(\omega L_1 - \frac{1}{\omega C}\right)$$

and

$$z_{21} = \frac{V_2}{I_1} \quad \text{when } I_2 = 0$$

$$= -j\frac{1}{\omega C}$$

Open-circuit the Input Port: [Fig. 13-6(c)]

$$I_2 = \frac{V_2}{j(\omega L_2 - 1/\omega C)}$$

and

$$V_1 = -j\frac{I_2}{\omega C}$$

Therefore,

$$z_{12} = \frac{V_1}{I_2} \quad \text{when } I_1 = 0$$

$$= -j\frac{1}{\omega C}$$

and

$$z_{22} = \frac{V_2}{I_2} \quad \text{when } I_1 = 0$$

$$= j\left(\omega L_2 - \frac{1}{\omega C}\right)$$

The z parameter matrix of the given two-port is

$$[z] = \begin{bmatrix} j(\omega L_1 - 1/\omega C) & -j(1/\omega C) \\ -j(1/\omega C) & j(\omega L_2 - 1/\omega C) \end{bmatrix}$$

Since the given network is reciprocal, $z_{12} = z_{21}$. ∎

Example 13-3 Determine the z parameters of the two-port shown in Fig. 13-7(a).

Fig. 13-7 (a)–(c) Determination of the z parameters of a network with a dependent source (Example 13-3).

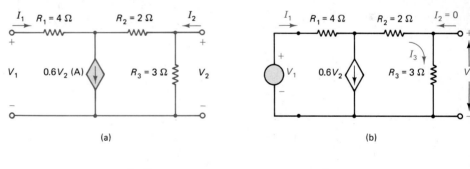

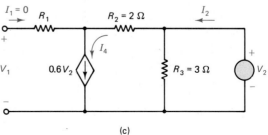

Solution *Open-circuit the Output Port:* [Fig. 13-7(b)] The equations of the circuit in Fig. 13-7(b) are

$$I_1 = I_3 + 0.6V_2$$
$$V_2 = R_3 I_3 = 3I_3$$

and

$$V_1 = R_1 I_1 + (R_2 + R_3)I_3 = 4I_1 + 5I_3$$

These three equations lead to

$$V_1 = 5.79I_1 \quad \text{and} \quad V_2 = 1.07I_1$$

Therefore,

$$z_{11} = 5.79\,\Omega \quad \text{and} \quad z_{21} = 1.07\,\Omega$$

Open-circuit the Input Port: [Fig. 13-7(c)] The equations of the circuit in Fig. 13-7(c) are

$$I_4 = 0.6V_2$$
$$V_2 = R_3(I_2 - I_4) = 3I_2 - 3I_4$$
$$V_1 = -R_2 I_4 + R_3(I_2 - I_4) = 3I_2 - 5I_4$$

The three equations lead to

$$V_1 = -0.214 I_2 \quad \text{and} \quad V_2 = 1.07 I_2$$

Therefore,

$$z_{12} = -0.214\,\Omega \quad \text{and} \quad z_{22} = 1.07\,\Omega$$

The z parameter matrix of the given two-port is

$$[z] = \begin{bmatrix} 5.79 & -0.214 \\ 1.07 & 1.07 \end{bmatrix}$$

∎

Exercise 13-3 Determine the z parameters of the two-ports shown in Fig. 13-8.

Fig. 13-8 Two-ports for Exercise 13-3.

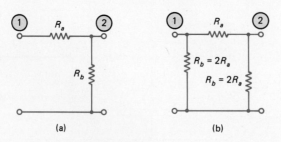

(a) (b)

Exercise 13-4 Determine the z parameters of the two-ports shown in Fig. 13-9.

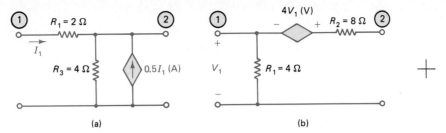

(a)   (b)

Fig. 13-9 Two-ports for Exercise 13-4.

13-4 USE OF THE z PARAMETERS IN THE ANALYSIS OF A TWO-PORT SYSTEM

When a two-port such as an amplifier is used in practice, it is customary to connect a signal source with a series impedance to the input port and a load impedance to the output port, as indicated in Fig. 13-10. The resulting network represents a two-port *system* and is analyzed by using the z parameter equations (13-7) in conjunction with additional constraints imposed on the system by the connections made to the input and output ports.

The connection of the signal source and its impedance at the input port in Fig. 13-10 leads to the following equation:

$$V_1 = V_s - Z_s I_1 \tag{13-10}$$

The presence of the load impedance at the output port leads to the following equation:

$$V_2 = -Z_L I_2 \tag{13-11}$$

The system must satisfy Eqs. (13-7), (13-10), and (13-11). Equations (13-7a) and (13-10) are combined into the single equation:

$$(z_{11} + Z_s)I_1 + z_{12}I_2 = V_s \tag{13-12}$$

and Eqs. (13-7b) and (13-11) are combined into the single equation:

$$z_{21}I_1 + (z_{22} + Z_L)I_2 = 0 \tag{13-13}$$

Solving Eqs. (13-12) and (13-13),

$$I_1 = \frac{(z_{22} + Z_L)V_s}{[(z_{11} + Z_s)(z_{22} + Z_L) - z_{12}z_{21}]} \tag{13-14}$$

$$I_2 = \frac{-z_{21}V_s}{[(z_{11} + Z_s)(z_{22} + Z_L) - z_{12}z_{21}]} \tag{13-15}$$

These expressions form the basis for determining any desired quantities in the two-port system.

Exercise 13-5 Determine the ratio (V_2/V_s) and the input impedance $Z_i = (V_1/I_1)$ in the two-port system of Fig. 13-10.

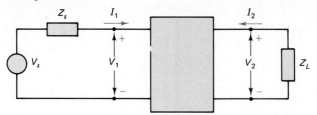

Fig. 13-10 A two-port system.

Exercise 13-6 Connect a load resistance $R_L = 3\Omega$ to the output port in the two-port of Example 13-3 and determine the input resistance $R_i = (V_1/I_1)$ by (a) using the z parameter equations and (b) by using circuit analysis principles directly.

Exercise 13-7 Connect a signal source V_s with a resistance $R_s = 4\Omega$ and a load resistance $R_L = 3\Omega$ to the two-port of Example 13-3. Determine the voltage ratio (V_2/V_s).

Exercise 13-8 Each of the two-ports of Exercise 13-4 is the two-port in the system of Fig. 13-11. For each case, write the system equations, solve them, and determine the ratios (V_2/I_1) and (I_2/I_1).

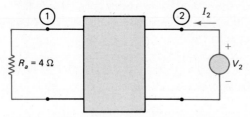

Fig. 13-11 A two-port system for Exercise 13-8.

13-5 MODELS OF A TWO-PORT USING z PARAMETERS

Even though the analysis of a two-port system is performed by using equations of the form (13-12) and (13-13), it is often convenient to use a model or an equivalent circuit, since a model provides a convenient visual aid for remembering the equations. The model of a two-port using z parameters is in the form of a T network, and different models are obtained for reciprocal networks (those without dependent sources) and nonreciprocal networks (those containing dependent sources).

13-5-1 Model of a Reciprocal Two-Port

For a reciprocal two-port,

$$z_{12} = z_{21}$$

so that the z parameter equations become

$$z_{11}I_1 + z_{12}I_2 = V_1 \tag{13-16a}$$
$$z_{12}I_1 + z_{22}I_2 = V_2 \tag{13-16b}$$

Now consider the T network shown in Fig. 13-12(a). The mesh equations of the T network are

$$(Z_1 + Z_2)I_1 + Z_2I_2 = V_1 \tag{13-17}$$
$$Z_2I_1 + (Z_2 + Z_3)I_2 = V_2 \tag{13-18}$$

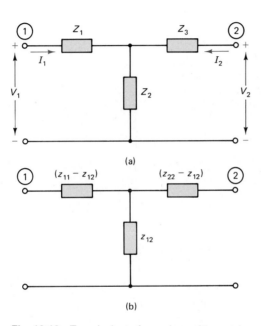

Fig. 13-12 T equivalent of a *reciprocal* two-port.

A comparison of Eq. (13-16) with (13-17) and (13-18) shows that

$$z_{11} = (Z_1 + Z_2) \tag{13-19}$$
$$z_{12} = Z_2 \tag{13-20}$$
$$z_{22} = (Z_2 + Z_3) \tag{13-21}$$

Solving the last three equations, we see that the components in the T equivalent network are related to the z parameter of the two-port through the equations:

$$Z_1 = (z_{11} - z_{12}) \tag{13-22}$$
$$Z_2 = z_{12} \tag{13-23}$$
$$Z_3 = (z_{22} - z_{12}) \tag{13-24}$$

The T equivalent of a reciprocal two-port (described by its z parameters) is of the form shown in Fig. 13-12(b).

13-5 Models of a Two-Port Using z Parameters

Example 13-4 The z parameter matrix of a two-port at $\omega = 500$ r/s is given by

$$[z] = \begin{bmatrix} 5.5 & -1 + j1 \\ -1 + j1 & -j2 \end{bmatrix}$$

Obtain its T equivalent network.

Solution Since $z_{12} = z_{21}$, the two-port may be treated as reciprocal, and Eqs. (13-22), (13-23), and (13-24) can be used for finding the T model.

$$Z_1 = (z_{11} - z_{12}) = (6.5 - j1)\Omega$$
$$Z_2 = z_{12} = (-1 + j1)\Omega$$
$$Z_3 = (z_{22} - z_{12}) = (1 - j3)\Omega$$

The T equivalent network is shown in Fig. 13-13.

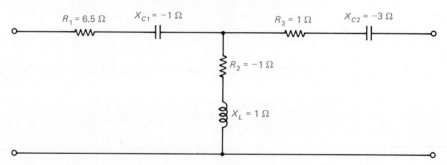

Fig. 13-13 T equivalent of the network for Example 13-4.

Exercise 13-9 Obtain the T equivalents of the two-ports of Exercise 13-3.

Exercise 13-10 Determine the z parameters of the T networks shown in Fig. 13-14.

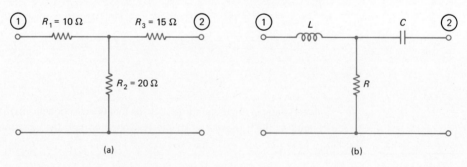

Fig. 13-14 T networks for Exercise 13-10.

13-5-2 Models of a Nonreciprocal Two-Port

For a nonreciprocal network, z_{12} is not equal to z_{21}, and the T equivalent of Fig. 13-12(b) is not valid. It becomes necessary to use dependent sources in the models in order to take into account the inequality between z_{12} and z_{21}. Two different models are possible for a nonreciprocal network: one using two dependent sources, and the other using only one dependent source.

The Two-Source Model

The two-source model is simply a network made up of two subnetworks, each of which is a direct representation of the z parameter equations:

$$z_{11}I_1 + z_{12}I_2 = V_1 \qquad (13\text{-}25)$$
$$z_{21}I_1 + z_{22}I_2 = V_2 \qquad (13\text{-}26)$$

The terms $z_{12}I_2$ and $z_{21}I_1$ in the two equations are accounted for by using dependent voltage sources, as shown in the model of Fig. 13-15.

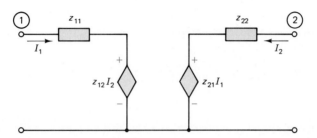

Fig. 13-15 Two-source z parameter model of a nonreciprocal two-port.

It is evident that the two-source model is not particularly imaginative.

The One-Source Model

In order to develop the one-source model of a nonreciprocal two-port, rewrite Eq. (13-25) in the form

$$(z_{11} - z_{12})I_1 + z_{12}(I_1 + I_2) = V_1$$

This equation leads to the partial model shown in Fig. 13-16(a). Note that this is the same as the left side of the model of the *reciprocal* two-port.

The z_{12} branch in the partial mode of Fig. 13-16(a) is common to the two meshes and will give rise to the term $z_{12}(I_1 + I_2)$ in the equation of mesh 2 of that model. Such a term is, however, missing from the z parameter equation for mesh 2 [Eq. (13-26)]. This

Fig. 13-16 Development of the one-source z parameter model of a nonreciprocal two-port. (a) Input side same as for the reciprocal T equivalent. (b) The complete T equivalent of a nonreciprocal two-port.

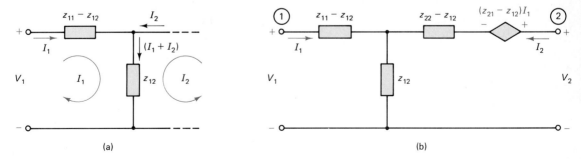

(a)   (b)

deficiency is rectified by adding and subtracting the term $z_{12}(I_1 + I_2)$ from Eq. (13-26), which becomes

$$z_{21}I_1 + z_{22}I_2 + z_{12}(I_1 + I_2) - z_{12}(I_1 + I_2) = V_2$$

or

$$z_{12}(I_1 + I_2) + (z_{22} - z_{12})I_2 + (z_{21} - z_{12})I_1 = V_2$$

The second term in the last equation leads to a branch of impedance $(z_{22} - z_{12})$ in mesh 2 of the model. The third term of the last equation is taken care of by introducing a *dependent voltage source* $(z_{21} - z_{12})I_1$, which is controlled by the current I_1. The complete model of a nonreciprocal two-port using one dependent source is shown in Fig. 13-16(b). It is useful in the analysis of two-port systems.

Example 13-5 The z parameter matrix of a bipolar junction transistor in the common base configuration is given by

$$[r] = \begin{bmatrix} (r_e + r_b) & r_b \\ (\alpha r_c + r_b) & (r_c + r_b) \end{bmatrix}$$

where r_e, r_b, r_c and α are device constants that are specified (or can be determined). (a) Draw the one-source model of the transistor. (b) Connect a signal source V_s with a resistance R_s and a load resistance R_L to the model. Determine the voltage ratio (V_2/V_s).

Solution (a) The one-source model is obtained by referring to Fig. 13-16(b) and is shown in Fig. 13-17(a).
(b) The two-port system is as shown in Fig. 13-17(b). The mesh equations of the circuit in Fig. 13-17(b) are

Mesh 1: $\qquad (R_s + r_e + r_b)I_1 + r_b I_2 = V_s$

Mesh 2: $\qquad (r_b + \alpha r_c)I_1 + (r_b + r_c + R_L)I_2 = 0$

Solving the two mesh equations, we find that

$$I_2 = \frac{(r_b + \alpha r_c)V_s}{(R_s + r_e)(R_L + r_b + r_c) + r_b r_c(1 - \alpha) + r_b R_L}$$

Fig. 13-17 (a) T equivalent of the bipolar transistor for Example 13-5. (b) Model of the amplifier circuit.

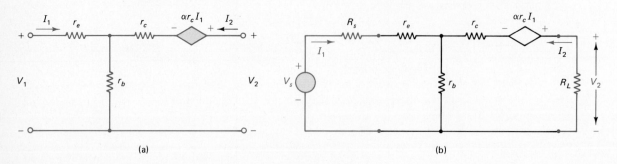

and, since $V_2 = -R_L I_2$, the voltage ratio (V_2/V_s) is given by

$$\frac{V_2}{V_s} = \frac{-R_L(r_b + \alpha r_c)}{(R_s + r_e)(R_L + r_b + r_c) + r_b r_c(1 - \alpha) + r_b R_L}$$

Exercise 13-11 Set up the T equivalent of a two-port whose z parameter matrix is given by

(a) $\begin{bmatrix} 10 & 5 \\ -5 & 9 \end{bmatrix}$ (b) $\begin{bmatrix} (3 + j4) & 0 \\ -j5 & (3 + j4) \end{bmatrix}$

Exercise 13-12 Obtain the T equivalents of the two-ports of Exercise 13-4.

Exercise 13-13 Obtain the one-source z parameter model of the network shown in Fig. 13-18.

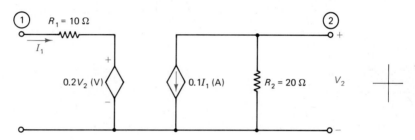

Fig. 13-18 Network for Exercise 13-13.

Exercise 13-14 The two-port of Exercise 13-13 is fed by a signal source V_s in series with a resistance $R_s = 2\Omega$ and drives a load resistance $R_L = 5\Omega$. Determine the ratio (V_2/V_s).

13-6 SHORT-CIRCUIT ADMITTANCE PARAMETERS OF A TWO-PORT

By choosing the voltages V_1 and V_2 in a two-port as the independent variables, it is possible to express the currents I_1 and I_2 as functions of the two voltages through the equations

$$y_{11}V_1 + y_{12}V_2 = I_1 \tag{13-27a}$$

$$y_{21}V_1 + y_{22}V_2 = I_2 \tag{13-27b}$$

The parameters y_{ij} ($i, j = 1, 2$) are called the *short-circuit admittance parameters* or simply the *y parameters* of the two-port. The y parameters represent the dual of the z parameters, and duality is used to modify the discussion of z parameters to apply to the y parameters. For this, replace V by I, I by V, and z by y; and open circuit by short circuit.

Here is a summary of the results obtained.

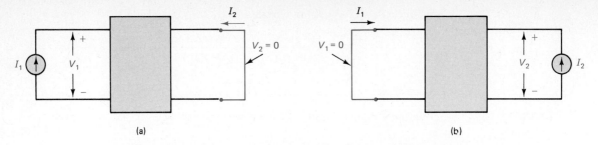

Fig. 13-19 Determination of the y parameters from short-circuit calculations.

Determination of the y Parameters from Short-Circuit Measurements

Short-Circuit the Output Port ($V_2 = 0$): [Fig. 13-19(a)]

$$y_{11} = (I_1/V_1) \quad \text{when } V_2 = 0$$
$$y_{21} = (I_2/V_1) \quad \text{when } V_2 = 0$$

Short-Circuit the Input Port ($V_1 = 0$): [Fig. 13-19(b)]

$$y_{12} = (I_1/V_2) \quad \text{when } V_1 = 0$$
$$y_{22} = (I_2/V_2) \quad \text{when } V_1 = 0$$

For reciprocal networks,

$$y_{12} = y_{21}$$

Example 13-6 Determine the y parameter matrix of the two-port shown in Fig. 13-20(a).

Solution *Short-Circuit the Output Port:* [Fig. 13-20(b)] The *KVL* equations of the circuit in Fig. 13-20(b) are

$$\frac{I_1 + I_2}{G_1} = V_1 \tag{13-28}$$

$$\frac{I_1 + I_2}{G_1} + \frac{I_2}{G_2} = KV_1 \tag{13-29}$$

Solving Eqs. (13-28) and (13-29),

$$I_1 = [G_1 - G_2(K - 1)]V_1$$

and

$$I_2 = (K - 1)G_2 V_1$$

Therefore,

$$y_{11} = (I_1/V_1) \quad \text{when } V_2 = 0$$
$$= [G_1 - G_2(K - 1)]$$
$$y_{21} = (I_2/V_1) \quad \text{when } V_2 = 0$$
$$= G_2(K - 1)$$

Two-Ports

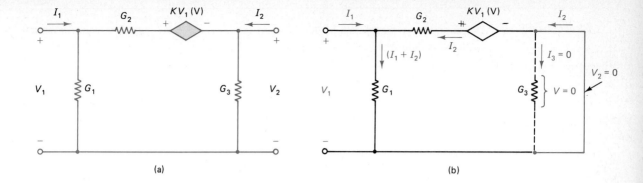

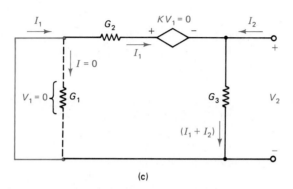

Fig. 13-20 (a) Two-port for Example 13-6. (b) Calculation of y_{11} and y_{21} by short-circuiting the output. (c) Calculation of y_{12} and y_{22} by short-circuiting the input.

Short-Circuit the Input Port: [Fig. 13-20(c)] The dependent source is dead, since the controlling voltage V_1 has been made zero by the short circuit. The *KVL* equations are

$$\frac{I_1 + I_2}{G_3} = V_2$$

and

$$-\frac{I_1}{G_2} = V_2$$

Therefore,

$$y_{12} = (I_1/V_2) \quad \text{when } V_1 = 0$$
$$= -G_2$$
$$y_{22} = (I_2/V_2) \quad \text{when } V_1 = 0$$
$$= (G_2 + G_3)$$

The *y* parameter matrix of the given two-port is

$$[y] = \begin{bmatrix} G_1 - G_2(K-1) & -G_2 \\ G_2(K-1) & G_2 + G_3 \end{bmatrix}$$

13-6 Short-Circuit Admittance Parameters of a Two-Port

Exercise 13-15 Determine the y parameters of the two-ports in Exercises 13-3 and 13-4 by using short-circuit calculations.

Analysis of a Two-Port System Using y Parameters

When a current source I_s with a shunt admittance Y_s is connected to the input port and a load admittance Y_L is connected to the output port, the resulting two-port system (Fig. 13-21) can be analyzed by using the y parameter equations in conjunction with the constraints imposed at the input and output ports by the external connections.

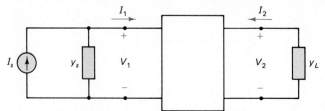

Fig. 13-21 A two-port system.

The connection of the source I_s and admittance Y_s at the input port leads to the following constraint between V_1 and I_1:

$$I_1 = I_s - Y_1 V_1 \tag{13-30}$$

and the connection of the load admittance leads to the following constraint between V_2 and I_2:

$$I_2 = -Y_L V_2 \tag{13-31}$$

Eqs. (13-30) and (13-27a) are combined into a single equation:

$$(y_{11} + Y_s)V_1 + y_{12}V_2 = I_s \tag{13-32}$$

and Eqs. (13-31) and (13-27b) are combined into a single equation:

$$y_{21}V_1 + (y_{22} + Y_L)V_2 = 0 \tag{13-33}$$

Solution of Eqs. (13-32) and (13-33) leads to V_1 and V_2 and any other desired quantities in the two-port system.

Exercise 13-16 Obtain the expression for the transfer function (V_2/I_s) of the two-port system of Fig. 13-21.

13-7 MODELS OF A TWO-PORT USING y PARAMETERS

The y parameter description of a two-port is useful in arriving at an equivalent circuit (or model) of the two-port in a manner exactly like the model obtained by using the z parameters. The equivalent circuit is in the form of a π network, since the π network is the dual of the T network. There are separate models for the reciprocal and nonreciprocal networks, and these are set up by using the principles of duality on the T equivalents already obtained. The resulting models are shown in Fig. 13-22.

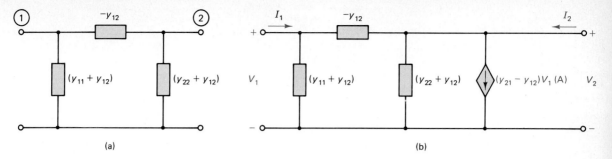

Fig. 13-22 π equivalents of (a) a reciprocal two-port and (b) a nonreciprocal two-port.

Example 13-7 Obtain the π equivalent of the two-port of Example 13-6.

Solution Using the results of Example 13-6 and the diagram of Fig. 13-22(b), the π equivalent assumes the form shown in Fig. 13-23. ∎

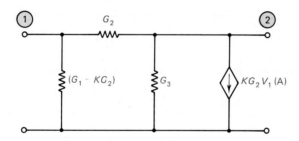

Fig. 13-23 π equivalent of the two-port for Example 13-7.

Example 13-8 The y parameter matrix of a two-port is given by

$$[y] = \begin{bmatrix} 1.5 & -0.6 \\ 0.25 & 2.1 \end{bmatrix}$$

(a) Obtain the π equivalent of the two-port.
(b) A current source I_s in shunt with a resistance of 8Ω is connected to the input port and a load resistance of 4Ω is connected to the output port. Determine the input resistance $R_i = (V_1/I_1)$ and the voltage ratio (V_2/V_1) of the resulting two-port system.

Solution (a) The π equivalent is obtained by referring to Fig. 13-22(b) and is shown in Fig. 13-24(a).
(b) The two-port system is shown in Fig. 13-24(b). The nodal equations of the circuit in Fig. 13-12(b) are

Node 1: $\qquad (Y_s + Y_1 + Y_2)V_1 - Y_2V_2 = I_s$

or

$$1.625V_1 - 0.6V_2 = I_s \qquad (13\text{-}34)$$

Node 2: $\qquad -Y_2V_1 + (Y_2 + Y_3 + Y_L)V_2 = -0.85V_1$

13-7 Models of a Two-Port Using y Parameters

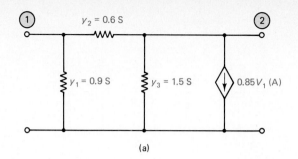

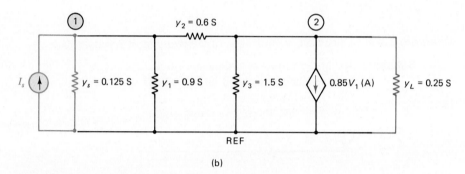

Fig. 13-24 (a) π equivalent of the network for Example 13-8. (b) Two-port system using the equivalent.

or

$$0.25V_1 + 2.35V_2 = 0 \tag{13-35}$$

Solving Eqs. (13-34) and (13-35), we find that

$$V_1 = 0.592I_s \text{ V} \quad \text{and} \quad V_2 = -0.0630I_s \text{ V}$$

The current I_1 is given by

$$I_1 = I_s - Y_s V_1 = 0.926I_s$$

so that the input resistance becomes

$$R_i = (V_1/I_1) = (0.592/0.926) = 0.639\Omega$$

The voltage ratio

$$(V_2/V_1) = (-0.063/0.592) = -0.106$$

(The negative sign means that V_2 is 180° out of phase with V_1. As I_s increases, V_1 increases, but V_2 decreases.) ∎

Exercise 13-17 Set up the π equivalent networks of the two-ports in Exercise 13-15.

Exercise 13-18 Obtain the π equivalent of the two-port shown in Fig. 13-25.

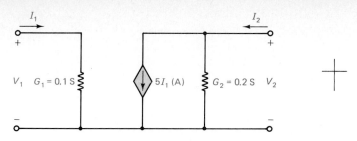

Fig. 13-25 Two-port for Exercise 13-18.

Exercise 13-19 Connect a signal source I_s in shunt with $G_s = 0.4$ S to the input port and a load $G_L = 0.8$ S to the output port of the network in the previous exercise. Determine the current ratio (I_2/I_s) and the input conductance (I_1/V_1).

13-8 RELATIONSHIP BETWEEN z AND y PARAMETERS

Since a given two-port can be described by a z parameter matrix or a y parameter matrix, if one matrix is known, it is possible to obtain the other. The z parameter matrix and y parameter matrix are the *inverses* of each other. To establish this, consider the matrix equation of a two-port in terms of the z parameter matrix [z].

$$[z][I] = [V] \tag{13-36}$$

Premultiplying both sides of the equation by $[z]^{-1}$ of the matrix [z],

$$[z]^{-1}[z][I] = [z]^{-1}[V]$$

which leads to

$$[I] = [z]^{-1}[V] \tag{13-37}$$

Comparing Eq. (13-37) with the y parameter matrix equation of the two-port

$$[y][V] = [I]$$

we see that

$$[y] = [z]^{-1} \tag{13-38}$$

It follows that

$$[z] = [y]^{-1} \tag{13-39}$$

Thus the z parameter matrix and y parameter matrix are mutual inverses, and matrix inversion procedures are used to find one matrix when the other is known. The relationships between the two are summarized here:

Determination of [y] When [z] Is Known:

$$[y] = \begin{bmatrix} \dfrac{z_{22}}{\|z\|} & -\dfrac{z_{12}}{\|z\|} \\ -\dfrac{z_{21}}{\|z\|} & \dfrac{z_{22}}{\|z\|} \end{bmatrix}$$

where $\|z\|$ is the determinant of the matrix [z].

Determination of [z] When [y] Is Known:

$$[z] = \begin{bmatrix} \dfrac{y_{22}}{\|y\|} & -\dfrac{y_{12}}{\|y\|} \\ -\dfrac{y_{21}}{\|y\|} & \dfrac{y_{11}}{\|y\|} \end{bmatrix}$$

where $\|y\|$ is the determinant of the matrix $[y]$.

Exercise 13-20 Obtain the y parameter matrix of each of the z parameter matrices given below:

(a) $\begin{bmatrix} 10 & -6 \\ 2 & 8 \end{bmatrix}$ (b) $\begin{bmatrix} (3 + j4) & -j6 \\ (4 - j6) & j9 \end{bmatrix}$

Exercise 13-21 Determine the y parameter matrix of each of the two-ports in Exercises 13-3 and 13-4 by using matrix inversion.

Exercise 13-22 Determine the z parameter matrix of the network in Exercise 13-18 by matrix inversion.

13-9 THE HYBRID PARAMETERS

Linear models of electronic devices are used in analyzing the response of an electronic circuit to a time-varying signal. The linear model of a device may be obtained by treating it as a two-port and finding a set of two-port parameters (e.g., the z and the y parameters) to describe its behavior. In Example 13-5, the z parameter matrix of a common base transistor was used. Measurement of the z and y parameters of a bipolar junction transistor presents some practical difficulties, however; a set of parameters, called the *hybrid* or *h parameters*, is found to be preferable for modeling the transistor.

The input current I_1 and the output voltage V_2 are selected as the independent variables in the h parameter description of a two-port. The input voltage V_1 and the output current I_2 are then written as functions of I_1 and V_2. The h parameters are denoted by h_{ij} ($i, j = 1, 2$). The h parameter equations of a two-port are

$$h_{11}I_1 + h_{12}V_2 = V_1 \qquad (13\text{-}40\text{a})$$

$$h_{21}I_1 + h_{22}V_2 = I_2 \qquad (13\text{-}40\text{b})$$

Consider the dimensions of the different parameters. To be dimensionally consistent, each term in Eq. (13-40a) must have the dimensions of a voltage:

$h_{11}I_1$ is a voltage → h_{11} is in ohms

$h_{12}V_2$ is a voltage → h_{12} is dimensionless

Similarly, each term in Eq. (13-40a) must have the dimensions of a current:

$h_{21}I_1$ is a current → h_{21} is dimensionless

$h_{22}V_2$ is a current → h_{22} is in siemens

13-9-1 Determination of the *h* Parameters of a Two-Port

The *h* parameters of a two-port may be determined by means of open-circuit and short-circuit measurements or calculations on the network. The *output* is *short*-circuited for finding h_{11} and h_{21}, while the *input* is *open*-circuited for finding h_{12} and h_{22}.

Open-Circuit the Input Port, Making $I_1 = 0$: Apply a source at the output port, as in Fig. 13-26(a).

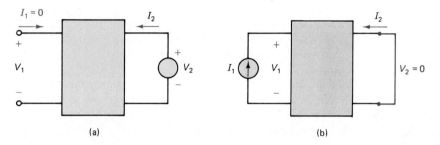

Fig. 13-26 Determination of the *h* parameters: (a) open circuit at the input for finding h_{12} and h_{22}; (b) short circuit at the output for finding h_{11} and h_{21}.

Since $I_1 = 0$, the *h* parameter equations (13-40) reduce to

$$V_1 = h_{12}V_2$$

and

$$I_2 = h_{22}V_2$$

The parameters h_{12} and h_{22} are then given by

$$h_{12} = (V_1/V_2) \quad \text{when } I_1 = 0 \tag{13-41}$$

$$h_{22} = (I_2/V_2) \quad \text{when } I_1 = 0 \tag{13-42}$$

Short-Circuit the Output Port, Making $V_2 = 0$: Apply a source at the input port, as in Fig. 13-26(b).

Since $V_2 = 0$, the *h* parameter equations (13-40) reduce to

$$V_1 = h_{11}I_1$$

and

$$I_2 = h_{21}I_1$$

The parameters h_{11} and h_{21} are then given by

$$h_{11} = (V_1/I_1) \quad \text{when } V_2 = 0 \tag{13-43}$$

$$h_{21} = (I_2/I_1) \quad \text{when } V_2 = 0 \tag{13-44}$$

Thus the h parameters can be determined by means of two tests: open-circuiting the input port and short-circuiting the output port and using Eqs. (13-41) through (13-44).

The relationships for the four h parameters in Eqs. (13-41) through (13-44) lead to the following specific names for the individual h parameters:

- h_{11} = short-circuit input impedance
- h_{12} = open-circuit reverse voltage ratio
- h_{21} = short-circuit forward current gain
- h_{22} = open-circuit output admittance

An alternative set of subscripts is commonly used for h parameters of bipolar transistors: h_i for h_{11}, h_r for h_{12}, h_f for h_{21}, and h_o for h_{22}. These new subscripts arise, respectively, from the terms *input, reverse, forward,* and *output* appearing in the names listed above. A second subscript is also added to indicate the particular connection in which the transistor is used. For example, the symbols h_{ie}, h_{re}, h_{fe}, and h_{oe} are used for the transistor in the common emitter connection.

We will use the symbols for the h parameters with numerical double subscripts in this chapter.

Example 13-9 Determine the h parameters of the two-port in Fig. 13-27(a).

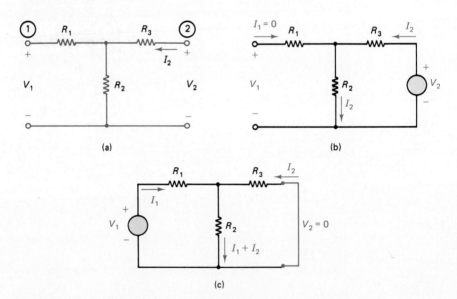

Fig. 13-27 (a)–(c) Determination of the h parameters of the two-port network for Example 13-9.

Solution *Open-Circuit the Input Port ($I_1 = 0$):* [Fig. 13-27(b)] The *KVL* equations are

$$V_1 = R_2 I_2$$

and

$$V_2 = (R_2 + R_3) I_2$$

Therefore,
$$h_{12} = (V_1/V_2) \quad \text{when } I_1 = 0$$
$$= \frac{R_2}{R_2 + R_3}$$
$$h_{22} = (I_2/V_2) \quad \text{when } I_1 = 0$$
$$= \frac{1}{R_2 + R_3}$$

Short-Circuit the Output Port ($V_2 = 0$): [Fig. 13-27(c)] The *KVL* equations are

$$(R_1 + R_2)I_1 + R_2 I_2 = V_1 \tag{13-45}$$

and

$$R_2 I_1 + (R_2 + R_3)I_2 = 0 \tag{13-46}$$

Solving Eqs. (13-45) and (13-46), we find that

$$I_1 = \frac{(R_2 + R_3)V_1}{R_1 R_2 + R_1 R_3 + R_2 R_3}$$

and

$$I_2 = \frac{-R_2 V_1}{R_1 R_2 + R_1 R_3 + R_2 R_3}$$

Therefore,

$$h_{11} = (V_1/I_1) \quad \text{when } V_2 = 0$$
$$= (R_1 R_2 + R_1 R_3 + R_2 R_3)/(R_2 + R_3)$$
$$h_{21} = (I_2/I_1) \quad \text{when } V_2 = 0$$
$$= -R_2/(R_2 + R_3)$$

The *h* parameter matrix of the given two-port is

$$[h] = \begin{bmatrix} \dfrac{R_1 R_2 + R_1 R_3 + R_2 R_3}{R_2 + R_3} & \dfrac{R_2}{R_2 + R_3} \\ \dfrac{-R_2}{R_2 + R_3} & \dfrac{1}{R_2 + R_3} \end{bmatrix}$$

The two-port in the examples above was a reciprocal network, and we can see that

$$h_{12} = -h_{21}$$

This is a general property of reciprocal two-ports, as we will prove in Sec. 13-11. ∎

Example 13-10 Determine the *h* parameters of the two-port shown in Fig. 13-28(a).

Solution *Open-circuit the Input Port* ($I_1 = 0$): [Fig. 13-28(b)] The three nodal equations are:

Node 1: $$(G_1 + G_2)V_1 - G_2 V_3 = 0$$

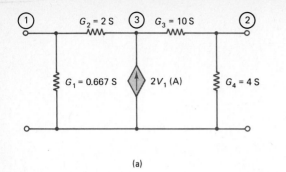
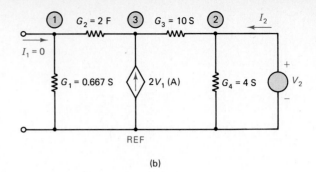

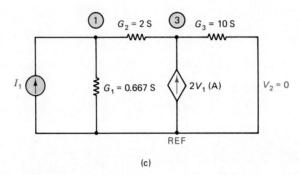

Fig. 13-28 (a)–(c) Determination of the h parameters of the two-port network for Example 13-10.

which becomes
$$2.667V_1 - 2V_3 = 0 \tag{13-47}$$

Node 2:
$$(G_3 + G_4)V_2 - G_3V_3 = I_2$$

which becomes
$$14V_2 - 10V_3 = I_2 \tag{13-48}$$

Node 3:
$$-G_2V_1 - G_3V_2 + (G_2 + G_3)V_3 = 2V_1$$

which becomes
$$-4V_1 - 10V_2 + 12V_3 = 0 \tag{13-49}$$

Solving Eqs. (13-47) through (13-49), we find that
$$V_1 = 0.289I_2 \quad \text{and} \quad V_2 = 0.347I_2$$

Therefore,
$$h_{12} = (V_1/V_2) \quad \text{when } I_1 = 0$$
$$= (0.289/0.347) = 0.833$$
$$h_{22} = (I_2/V_2) \quad \text{when } I_1 = 0$$
$$= 2.89 \text{ S}$$

Output Port Short-circuited ($V_2 = 0$): [Fig. 13-28(c)] The two nodal equations are:

Node 1:
$$(G_1 + G_2)V_1 - G_2V_3 = I_1$$
which becomes
$$2.667V_1 - 2V_3 = I_1 \qquad (13\text{-}50)$$

Node 3:
$$-G_2V_1 + (G_2 + G_3)V_3 = 2V_1$$
which becomes
$$-4V_1 + 12V_3 = 0 \qquad (13\text{-}51)$$

Solving Eqs. (13-50) and (13-51), we find that
$$V_1 = 0.50I_1 \qquad \text{and} \qquad V_3 = 0.167I_1$$

The current I_2 is given by
$$I_2 = -G_3V_3 = -1.67I_1$$

Therefore,
$$h_{21} = (I_2/I_1) \quad \text{when } V_2 = 0$$
$$= -1.67$$
$$h_{11} = (V_1/I_1) \quad \text{when } V_2 = 0$$
$$= 0.50\,\Omega$$

The h parameter matrix of the given two-port is
$$[h] = \begin{bmatrix} 0.375 & 0.833 \\ -1.25 & 2.89 \end{bmatrix}$$
■

Exercise 13-23 Determine the h parameter matrix of each of the two-ports of Exercises 13-3 and 13-4.

Exercise 13-24 Determine the h parameter matrix of each of the two-ports shown in Fig. 13-29.

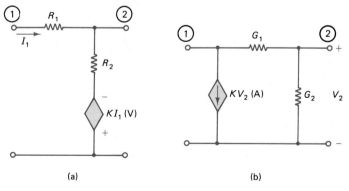

Fig. 13-29 Two-ports for Exercise 13-24.

Exercise 13-25 The equations of a two-port are given by

$$10I_1 - 8I_2 = V_1$$
$$-4I_1 + 12I_2 = V_2$$

Obtain the h parameter equations of the two-port.

13-10 MODEL OF A TWO-PORT USING h PARAMETERS

The h parameter description of a two-port leads to an equivalent circuit for a two-port (similar to the case of models obtained with z and y parameters). The h parameter model is a *literal translation* of the h parameter equations to two subnetworks linked through two dependent sources, as shown in Fig. 13-30(a).

Eq. (13-40a) is in the form of a *KVL* equation:

$$h_{11}I_1 + h_{12}V_2 = V_1$$

The input side of the model is therefore a series connection of a resistance h_{11} (to account for the term $h_{11}I_1$) and a dependent voltage source $h_{12}V_2$.

Equation (13-40b) is in the form of a *KCL* equation:

$$h_{21}I_1 + h_{22}V_2 = I_2$$

The output side of the model is therefore a parallel combination of a conductance h_{22} (to account for the term $h_{22}V_2$) and a dependent current source $h_{21}I_1$.

It is interesting to note that the *h parameter model of a two-port is the same whether the two-port is reciprocal or not*. That is, two dependent sources are necessary in the

Fig. 13-30 (a) h parameter model of a two-port (the same model is used for both reciprocal and nonreciprocal two-ports). (b) A two-port system using the h parameter model.

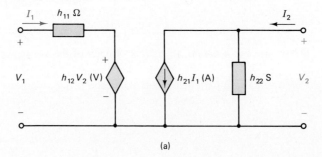

(a)

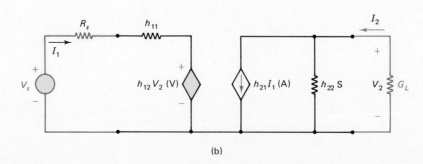

(b)

model even for reciprocal networks. In the case of reciprocal networks, $h_{12} = -h_{21}$, but this does not help simplify the model.

13-10-1 Analysis of a Two-Port System Using h Parameters

A two-port system in which a signal source V_s and an impedance Z_s are connected to the input port and a load impedance Z_L is connected to the output port is shown in Fig. 13-30(b).

The structure of the two portions of the system in Fig. 13-30(b) is such that it is convenient to use a mesh equation for the input side and a nodal equation for the output side.

Input Side:
$$(Z_s + h_{11})I_1 + h_{12}V_2 = V_s \tag{13-52}$$

Output Side:
$$h_{21}I_1 + (h_{22} + Y_L)V_2 = 0 \tag{13-53}$$

Equations (13-52) and (13-53) provide the basis for the analysis of a two-port system using h parameter equations.

Example 13-11 The h parameter matrix of a bipolar junction transistor is given by

$$[h] = \begin{bmatrix} 2500 & 0 \\ 50 & 2 \times 10^{-4} \end{bmatrix}$$

A signal-source V_s with a resistance $R_s = 500\Omega$ is connected to the input port and a load resistance $R_L = 25$ kΩ is connected to the output port. (a) Set up the h parameter model of the system. (b) Determine the ratio (V_2/V_s).

Solution (a) The h parameter model of the system is shown in Fig. 13-31.
(b) *Input Side Mesh Equation:*

$$(R_s + h_{11})I_1 = V_s$$

which leads to

$$3000 I_1 = V_s \tag{13-54}$$

Output Side Nodal Equation:

$$h_{21}I_1 + (h_{22} + G_L)V_2 = 0$$

Fig. 13-31 Analysis of a bipolar transistor amplifier using the h parameters for Example 13-11.

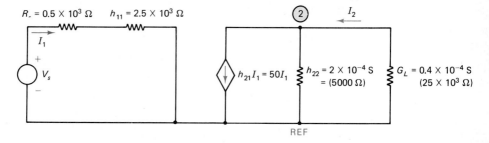

which leads to

$$50I_1 + 2.4 \times 10^{-4}V_2 = 0 \quad (13\text{-}55)$$

Solving Eqs. (13-54) and (13-55), we find

$$V_2 = -69.4V_s$$

and therefore,

$$(V_2/V_s) = -69.4 \qquad \blacksquare$$

Exercise 13-26 Determine the input impedance $Z_i = (V_s/I_1)$ and the transfer function (I_2/I_1) of the two-port system shown in Fig. 13-30.

Exercise 13-27 Calculate the ratio (I_2/I_1) in the system of Example 13-11.

Exercise 13-28 (a) Obtain the h parameter model of the network shown in Fig. 13-32.
(b) Connect a resistance of $R_L = 20\ \Omega$ to the output port and determine the input resistance $R_i = (V_1/I_1)$.

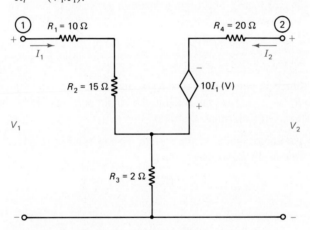

Fig. 13-32 Network for Exercise 13-28.

13-11 RELATIONSHIP BETWEEN h AND z PARAMETERS

If the z or y parameters of a two-port are known, then the h parameters of a two-port can be determined. Conversely, given the h parameters, the z or y parameters can be determined.

Consider the h parameter equations:

$$h_{11}I_1 + h_{12}V_2 = V_1 \quad (13\text{-}56)$$

$$h_{21}I_1 + h_{22}V_2 = I_2 \quad (13\text{-}57)$$

These equations are manipulated to resemble the z parameter equations. Equation (13-57) is easily put in the form of a z parameter equation by a rearrangement of terms:

$$V_2 = -(h_{21}/h_{22})I_1 + (1/h_{22})I_2 \quad (13\text{-}58)$$

Comparing Eq. (13-58) with the z parameter equation,

$$V_2 = z_{21}I_1 + z_{22}I_2$$

we see that

$$z_{21} = -(h_{21}/h_{22}) \tag{13-59}$$

and

$$z_{22} = (1/h_{22}) \tag{13-60}$$

Equation (13-56) is changed to the form of a standard z parameter equation by combining it with Eq. (13-58) and some manipulations. The resulting equation is

$$V_1 = \frac{h_{11}h_{22} - h_{12}h_{21}}{h_{22}}I_1 + \frac{h_{12}}{h_{22}}I_2$$

Comparing the last equation with the z parameter equation

$$V_1 = z_{11}I_1 + z_{12}I_2$$

we see that

$$z_{11} = \frac{h_{11}h_{22} - h_{12}h_{21}}{h_{22}} \tag{13-61}$$

and

$$z_{12} = \frac{h_{12}}{h_{22}} \tag{13-62}$$

The z parameter matrix, in terms of the h parameters, is therefore given by

$$[z] = \begin{bmatrix} \dfrac{\|h\|}{h_{22}} & \dfrac{h_{12}}{h_{22}} \\ -\dfrac{h_{21}}{h_{22}} & \dfrac{1}{h_{22}} \end{bmatrix}$$

where $\|h\|$ is the determinant of the h parameter matrix.

Note that for a *reciprocal network*, $z_{12} = z_{21}$, and when this condition is applied to the matrix above, it leads to

$$h_{12} = -h_{21} \quad \text{(for } reciprocal \text{ networks)}$$

which was found to be the case in Example 13-9.

Exercise 13-29 Obtain the z parameters of the two-ports of Exercise 13-24 by using the conversion procedure outlined above.

Exercise 13-30 Derive the equations needed to determine the y parameters when the h parameters are known.

Exercise 13-31 Obtain the y parameters of the two-ports of Exercise 13-24 by using the conversion procedure of Exercise 13-30.

13-12 TRANSMISSION OR *ABCD* PARAMETERS

In the *cascaded* interconnection of two-ports (where the output of each two-port is connected to the input of the next two-port), as well as in transmission lines, it is convenient to express the input variables as functions of the output variables; that is, V_1 and I_1 are expressed as functions of V_2 and I_2. The parameters in this system of equations are denoted by the symbols A, B, C, D, and the equations are

$$AV_2 - BI_2 = V_1$$
$$CV_2 - DI_2 = I_1$$

or, in matrix form

$$\begin{bmatrix} A & B \\ C & D \end{bmatrix} \begin{bmatrix} V_2 \\ -I_2 \end{bmatrix} = \begin{bmatrix} V_1 \\ I_1 \end{bmatrix}$$

The exercises that follow are intended to provide a discussion of the determination, properties, and uses of the *ABCD* parameters.

Exercise 13-32 The *ABCD* parameters of a two-port can be determined by two tests: open-circuit and short-circuit of the output port. Obtain the relationships for the parameters based on such tests.

Exercise 13-33 Determine the *ABCD* parameters of the two-ports in Exercises 13-3 and 13-4.

Exercise 13-34 Obtain the relationships that lead to a determination of the *ABCD* parameters when the z parameters of a two-port are known.

Exercise 13-35 Use the results of the previous exercise to show that $(AD - BC) = 1$ for all reciprocal networks.

13-13 INTERCONNECTION OF TWO-PORT NETWORKS

When the design of a two-port leads to a rather complex structure, it is advisable to break the design down into simpler two-ports. A suitable interconnection of the simpler two-ports is sought to meet the specifications of the complex two-port. Two commonly used interconnections are the *parallel* connection of two-ports and the *cascaded* connection of two-ports.

13-13-1 Parallel Connection of Two-Ports

The parallel connection of two-ports requires that the voltages V_1 and V_2 of each two-port are, respectively, equal to the voltages V_1 and V_2 of the other two-ports. In addition, an important condition must be observed: the parallel connection must not alter any of the original networks.

An example of *improper* parallel connection of two-ports is shown in Fig. 13-33(a). Note that the bottom branch of the T network is a short circuit, and this shorts out the

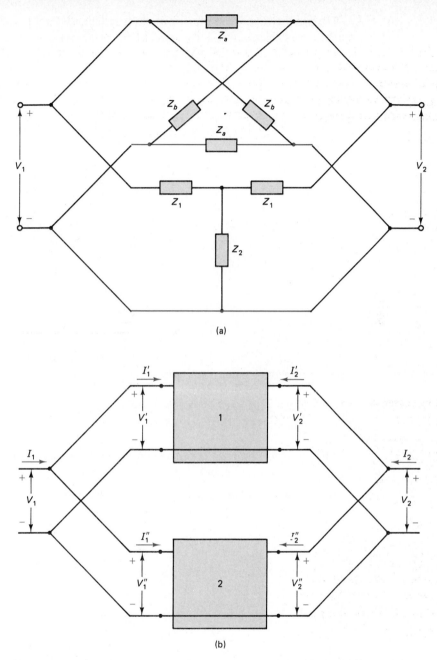

Fig. 13-33 Parallel connection of two-ports: (a) an improper connection; (b) a proper connection.

bottom branch of the lattice. In such cases, an isolating transformer is inserted to provide a proper parallel connection.

The general parallel connection of a pair of two-ports is indicated in Fig. 13-33(b), where the presence of a common ground between input and output for both two-ports ensures a proper parallel connection.

13-13 Interconnection of Two-Port Networks

The parallel connection of two two-ports in Fig. 13-33(b) is equivalent to a two-port whose y parameter matrix is the sum of the y parameter matrices of the individual two-ports. To establish this, let the y parameter equations of two-port 1 be given by

$$I'_1 = y'_{11}V'_1 + y'_{12}V'_2 \qquad (13\text{-}63a)$$

$$I'_2 = y'_{21}V'_1 + y'_{22}V'_2 \qquad (13\text{-}63b)$$

and those of two-port 2 by

$$I''_1 = y''_{11}V''_1 + y''_{12}V''_2 \qquad (13\text{-}64a)$$

$$I''_2 = y''_{21}V''_1 + y''_{22}V''_2 \qquad (13\text{-}64b)$$

The parallel connection of the two-ports imposes the following constraints:

$$V_1 = V'_1 = V''_1 \qquad (13\text{-}65)$$

$$V_2 = V'_2 = V''_2 \qquad (13\text{-}66)$$

$$I_1 = I'_1 + I''_1 \qquad (13\text{-}67)$$

$$I_2 = I'_2 + I''_2 \qquad (13\text{-}68)$$

Equations (13-63) through (13-68) are combined to obtain

$$I_1 = (y'_{11} + y''_{11})V_1 + (y'_{12} + y''_{12})V_2$$

$$I_2 = (y'_{21} + y''_{21})V_2 + (y'_{22} + y''_{22})V_2$$

Therefore, the y parameter matrix of the parallel combination is given by

$$[y] = \begin{bmatrix} (y'_{11} + y''_{11}) & (y'_{12} + y''_{12}) \\ (y'_{21} + y''_{21}) & (y'_{22} + y''_{22}) \end{bmatrix}$$

The result for the parallel connection of two two-ports is readily extended to cover the general case of n two-ports in parallel:

the y *parameter matrix of the* parallel connection is the sum of the y parameter matrices of the individual two-ports.

Example 13-12 The parallel combination of two T networks, known as a *twin-T network*, is shown in Fig. 13-34(a). Obtain its y parameter matrix.

Solution The two component T networks are shown separately in Figs. 13-34(b) and (c). The y parameter matrices of the individual networks are found to be:

$$[y'] = \begin{bmatrix} 0.0667 & -0.0167 \\ -0.0167 & 0.0067 \end{bmatrix} \qquad [y''] = \begin{bmatrix} -j0.0667 & j0.133 \\ j0.133 & -j0.0667 \end{bmatrix}$$

The y parameter matrix of the twin-T network is therefore

$$[y] = \begin{bmatrix} (0.0667 - j0.0667) & (-0.01167 + j0.133) \\ (-0.0167 + j0.133) & (0.0067 - j0.0067) \end{bmatrix} \qquad \blacksquare$$

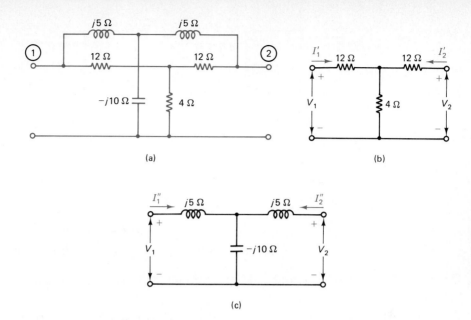

Fig. 13-34 (a), (b) Determination of the y parameter matrix of a twin T network.

Exercise 13-36 The network of Fig. 13-35 can be treated as the parallel combination of a pair of two-ports. Obtain its y parameter matrix.

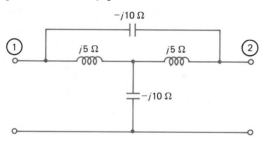

Fig. 13-35 Network for Exercise 13-36.

13-13-2 Cascaded Connection of Two-Ports

When the output port of one network is connected to the input port of the following network, as indicated in Fig. 13-36, the networks are said to be *cascaded*. Cascading of multiple stages occurs often in electronics and communication systems. Since the output variables of one stage become the input variables of the next, the *ABCD* parameters are found to be particularly effective in dealing with cascaded two-ports.

Let the *ABCD* equations of network N' be given by

$$\begin{bmatrix} V_1' \\ I_1' \end{bmatrix} = \begin{bmatrix} A' & B' \\ C' & D' \end{bmatrix} \begin{bmatrix} V_2' \\ -I_2' \end{bmatrix} \qquad (13\text{-}69)$$

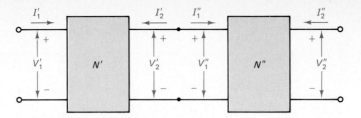

Fig. 13-36 Cascaded two-ports.

and those of N'' by

$$\begin{bmatrix} V_1'' \\ I_1'' \end{bmatrix} = \begin{bmatrix} A'' & B'' \\ C'' & D'' \end{bmatrix} \begin{bmatrix} V_2'' \\ -I_2'' \end{bmatrix} \tag{13-70}$$

The cascaded connection imposes the following constraints:

$$\begin{bmatrix} V_2' \\ -I_2' \end{bmatrix} = \begin{bmatrix} V_1'' \\ I_1'' \end{bmatrix}$$

Therefore, Eqs. (13-69) and (13-70) lead to

$$\begin{bmatrix} V_2' \\ -I_2' \end{bmatrix} = \begin{bmatrix} V_1'' \\ I_1'' \end{bmatrix} = \begin{bmatrix} A'' & B'' \\ C'' & D'' \end{bmatrix} \begin{bmatrix} V_2'' \\ -I_2'' \end{bmatrix} \tag{13-71}$$

and

$$\begin{bmatrix} V_1' \\ I_1' \end{bmatrix} = \begin{bmatrix} A' & B' \\ C' & D' \end{bmatrix} \begin{bmatrix} V_2'' \\ -I_2'' \end{bmatrix}$$

$$= \begin{bmatrix} A' & B' \\ C' & D' \end{bmatrix} \begin{bmatrix} A'' & B'' \\ C'' & D'' \end{bmatrix} \begin{bmatrix} V_2'' \\ -I_2'' \end{bmatrix}$$

Therefore,

the *ABCD* parameter matrix of the cascaded connection of two two-ports is the product of the *ABCD* parameter matrices of the individual two-ports.

The result can readily be extended to the general case of n two-ports connected in cascade.

Example 13-13 The model of an amplifier is of the form shown in Fig. 13-37(a), where R_i = input resistance of the amplifier, R_o = output resistance of the amplifier, and g_m = transconductance of the amplifier device. (a) Determine the *ABCD* parameter matrix of the amplifier. (b) Two amplifier stages are cascaded together. The values of the parameters of stage 1 are: $R_i = 5\text{k}\Omega$, $R_o = 4\text{k}\Omega$, and $g_m = 50$ mS. The values of the parameters of the second stage are: $R_i = 0.1\text{k}\Omega$, $R_o = 8\text{k}\Omega$, and $g_m = 40$ mS. Determine the *ABCD* parameters of the cascaded connection. (c) The cascaded connection is connected to a load resistance of $R_L = 10\text{k}\Omega$. Determine the voltage gain, current gain, and input resistance of the system.

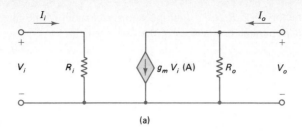

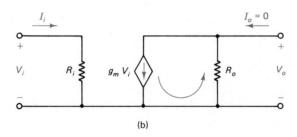

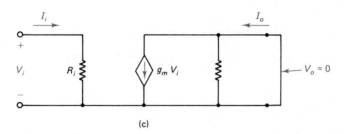

Fig. 13-37 (a) Amplifier for Example 13-13. (b)–(c) Determination of the ABCD parameters.

Solution (a) The ABCD parameters are determined by using the open-circuit ($I_o = 0$) and short-circuit ($V_o = 0$) calculations on the output port. The two cases are illustrated in Fig. 13-37(b) and (c).

From Fig. 13-37(b)

$$V_o = -g_m R_o V_i \quad \text{and} \quad I_i = (V_i/R_i)$$

so that, by using the results of Exercise 13-32

$$A = -(1/g_m R_o) \quad C = -(1/g_m R_o R_i)$$

From Fig. 13-37(c)

$$I_o = g_m V_i \quad \text{and} \quad I_i = (V_i/R_i)$$

so that by using the results of Exercise 13-32

$$B = -(1/g_m) \quad D = -(1/g_m R_i)$$

The ABCD parameter matrix of the amplifier can be written as

$$-(1/g_m R_o R_i) \begin{bmatrix} R_i & R_o R_i \\ 1 & R_o \end{bmatrix}$$

13-13   Interconnection of Two-Port Networks

where the factor $(-g_m R_o R_i)^{-1}$ multiplies every element of the matrix.

(b) The *ABCD* parameter matrices of the two stages are:

Stage 1:

$$(-10^{-6}) \begin{bmatrix} 5 \times 10^3 & 20 \times 10^6 \\ 1 & 4 \times 10^3 \end{bmatrix}$$

Stage 2:

$$(-3.125 \times 10^{-5}) \begin{bmatrix} 100 & 0.8 \times 10^6 \\ 1 & 8 \times 10^3 \end{bmatrix}$$

The *ABCD* parameter matrix of the *cascaded combination* is

$$(10^{-6} \times 3.125 \times 10^{-5}) \begin{bmatrix} 5 \times 10^3 & 20 \times 10^6 \\ 1 & 4 \times 10^3 \end{bmatrix} \begin{bmatrix} 100 & 0.8 \times 10^6 \\ 1 & 8 \times 10^3 \end{bmatrix}$$

$$= \begin{bmatrix} 6.41 \times 10^{-4} & 5.125 \\ 1.28 \times 10^{-7} & 1.025 \times 10^{-3} \end{bmatrix}$$

(13-72)

(c) The addition of the load resistance introduces the constraint

$$I_2 = -(V_2/R_L)$$

which modifies the *ABCD* parameter equations:

$$V_1 = AV_2 - BI_2 = \left(A + \frac{B}{R_L}\right) V_2 \qquad (13\text{-}73)$$

and

$$I_1 = CV_2 - DI_2 = -(CR_L + D)I_2 \qquad (13\text{-}74)$$

Using the values of the parameters from the matrix (13-72) in Eqs. (13-73) and (13-74)

$$V_1 = 1.154 \times 10^{-3} V_2 \quad \text{and} \quad I_1 = 2.305 \times 10^{-3} I_2$$

which leads to

$$(V_2/V_1) = 867$$
$$(I_2/I_1) = 434$$

and

$$R_{in} = (V_1/I_1) = 5 \times 10^3 \; \Omega$$

∎

Exercise 13-37 Interchange the two stages of the previous example and recalculate the answers of part (c).

13-14 BALANCED AND UNBALANCED TWO-PORTS

Two-ports can be classified as balanced or unbalanced on the basis of the following conditions. If a horizontal center line can be drawn through the two-port in such a way as to make the two halves *mirror images* of each other, the two-port is said to be *balanced*. An example of the balanced two-port is the symmetrical lattice (Fig. 13-38).

Any two-port that does not satisfy the condition for a balanced two-port is an *unbalanced two-port*. A sufficient condition for a two-port to be unbalanced is that there be a *common ground between the input and output ports*. All the two-ports we have discussed in this chapter, with the exception of the symmetrical lattice, are unbalanced two-ports. An example of a two-port which is unbalanced but does not have a common ground between the input and output ports is the *unsymmetrical* lattice. Common usage, however, permits the assumption of a common ground between the input and output ports when we refer to unbalanced two-ports.

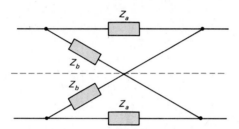

Whether or not a two-port is balanced does not affect the definitions and results derived in this chapter. Also, almost all important networks, especially those used in electronic circuits, are unbalanced two-ports.

13-15 SUMMARY OF CHAPTER

A two-port model of a network views the network as a black box with two terminal pairs, one of which may be considered the input port and the other the output port. The relationships between the voltages and currents are written in terms of different sets of parameter matrices. Two-port parameters of a network permit the analysis of a system without paying attention to the details of the network inside the black box.

In the open-circuit impedance (or z) parameters, the two voltages are expressed as functions of the two currents. One method of determining the z parameters is by means of measurements or calculations when the input port and the output ports are open circuited, one at a time. The z parameters are useful for analyzing a two-port system (which includes a signal source and a load). The z parameter description of a two-port leads to a T network model for a reciprocal two-port (with the property $z_{12} = z_{21}$) and a T network with a dependent voltage source for a nonreciprocal two-port. The models are useful in the analysis of two-ports to which external connections are made.

The short-circuit admittance (or y) parameters are the dual of the z parameters. The two currents are expressed as functions of the two voltages. Short-circuit measurements or calculations permit the determination of the y parameters. The model of a reciprocal network ($y_{12} = y_{21}$) is a π network, while the model of a nonreciprocal network is a π network that includes a dependent current source. The y and z parameter matrices of a two-port are mutual inverses.

Hybrid or h parameters express the input voltage and output current as functions of the input current and output voltage. One of the parameters (h_{11}) is an impedance, another (h_{22}) is an admittance, and the other two are ratios of voltages (h_{12}) or currents (h_{21}). Hybrid parameters are particularly important in the analysis of amplifiers using bipolar junction transistors. An open-circuit test at the input and a short-circuit test at the output permit the determination of the h parameters of a two-port. For a reciprocal network, $h_{21} = -h_{12}$. The h parameter model of a two-port is a literal translation of the equations for a circuit diagram.

The $ABCD$ or transmission parameters express the input voltage and current as functions of the output voltage and current in a network. They are important in communication and power networks and particularly convenient in cascaded networks. The determinant of the $ABCD$ parameter matrix of a reciprocal network has a value of 1.

When several two-ports are interconnected, the overall description of the combination can be expressed in terms of the two-port parameters of the individual network. For a parallel connection of two-ports, the sum of the individual y parameter matrices gives the y parameter matrix of the combination. For a cascaded connection of two-ports, the product of the $ABCD$ parameter matrices gives the $ABCD$ parameter matrix of the combination. Such properties are useful not only in analysis, but also in design, where a complicated two-port can be synthesized by using simpler component two-ports.

Answers to Exercises [*Matrices in answers are given below in the form of a list of elements starting with the first row. NOTE: Several of the exercises depend upon the results of earlier exercises. Save your solutions to all exercises until the chapter is finished.*]

13-1 $i_p + i_r = i_q + i_s = 0$.

13-2 Elements of the nodal admittance matrix are: $Y_1, -Y_1, -Y_1, (Y_1 + Y_2)$. The elements of the z parameter matrix are: $(Y_1 + Y_2)/Y_1Y_2, (1/Y_2), (1/Y_2), (1/Y_2)$.

13-3 (a) $(R_a + R_b), R_b, R_b, R_b$. (b) $1.2R_a, 0.8R_a, 0.8R_a, 1.2R_a$.

13-4 (a) When $I_2 = 0$, $V_1 = 8I_1$ and $V_2 = 6I_1$. When $I_1 = 0$, $V_2 = V_1 = 4I_2$. The z parameter matrix is: 8, 4, 6, 4. (b) When $I_2 = 0$, $V_1 = 4I_1$ and $V_2 = 20\,I_1$. When $I_1 = 0$, $V_1 = 4I_2$ and $V_2 = 28\,I_2$. The z parameter matrix is: 4, 4, 20, 28.

13-5 $(V_2/V_s) = Z_L z_{21}/$(denominator same as in Eqs. (13-14) and (13-15)). $Z_{in} = z_{11} - z_{12}z_{21}/(z_{22} + Z_L)$.

13-6 $R_{in} = 5.85\ \Omega$.

13-7 $V_2/V_s = 0.0801$.

13-8 (a) Equations: $(12I_1 + 4I_2) = 0$; $(6I_1 + 4I_2) = V_2$. $(V_2/I_1) = -6$. $(I_2/I_1) = -3$. (b) Equations: $(8I_1 + 4I_2) = 0$; $(20I_1 + 28I_2) = V_2$. $(V_2/I_1) = -36$; $(I_2/I_1) = -2$.

13-9 (a) Given circuit is already in the form of a T (with one limb zero). (b) Elements of the T are: $0.4R_a, 0.8R_a, 0.4R_a$.

13-10 (a) Elements are: 30, 20, 20, 35 Ω. (b) Elements are: $(R + j\omega L), R, R, (R - j/\omega C)$.

13-11 (a) Elements of the T are: 5 Ω, 5 Ω, $10I_1$ V, and 4 Ω. (b) Elements of the T are: $(3 + j4)\ \Omega, 0, -j5I_1$ V, $(3 + j4)\ \Omega$.

13-12 (a) Elements of the T are: 4 Ω, 4 Ω, $2I_1$ V, 0 (b) 0, 4 Ω, $16I_1$ V, 24 Ω.

13-13 Elements of the z parameter matrix are: $9.6, 4, -2, 20\ \Omega$. Elements of the T are: $5.6\ \Omega, 4\ \Omega, 6I_1$ V, $16\ \Omega$.

13-14 Equations: $(11.6I_1 + 4I_2) = V_s$. $(-2I_1 + 25I_2) = 0$. $V_2/V_s = -0.0336$.

13-15 (a) Elements of the y matrix are: $(1/R_a)$, $-(1/R_a)$, $-(1/R_a)$, $(R_a + R_b)/R_aR_b$. (b) When $V_2 = 0$, $V_1 = (2/3)R_aI_1$ and $I_2 = -(2/3)I_1$. Elements of the y matrix are: $(3/2R_a)$, $-(1/R_a)$, $-(1/R_a)$, $(3/2R_a)$. (c) When $V_2 = 0$, $I_2 = -1.5I_1$, $V_1 = 2I_1$. When $V_1 = 0$, $V_2 = -2I_1$, $V_2 = 6I_1 + 4I_2$. Elements of the y matrix are: $0.5, -0.5, -0.75, 1$ S. (d) When $V_2 = 0$, $V_1 = R_1(I_1 + I_2)$, $-R_1(I_1 + I_2) - R_2I_2 = 4V_1$. When $V_1 = 0$, the dependent source is dead. $I_1 = -I_2$, $V_2 = R_2I_2$. Elements of the y matrix are: $0.875, -0.125, -0.625, 0.125$.

13-16 $(V_2/I_s) = -y_{21}/[(y_{11} + Y_s)(y_{22} + Y_L) - y_{12}y_{21}]$.

13-17 The first two networks of Exercise 13-15 are already π networks. (c) Elements of the π network are: $0, 0.5$ S, 0.5 S, $-0.25V_1$ A. (d) Elements of the π network are: 0.75 S, 0.125 S, 0 S, $-0.5V_1$ A.

13-18 y matrix: $0.1, 0, 0.5, 0.2$ S. π network same as the given circuit.

13-19 $(I_2/I_s) = 0.8$. $(I_1/V_1) = 0.5$ S.

13-20 (a) $0.0870, 0.0652, -0.0217, 0.109$. (b) $\|z\| = j51$. y matrix: $0.176, 0.118$, $(0.118 + j0.0784)$, $(0.0784 - j0.0588)$ S.

13-21 Answers same as in Exercise 13-15(a) and (b).

13-22 z matrix: $10, 0, -25, 5$.

13-23 (a) h matrix: $R_a, 1, -1, (1/R_b)$. (b) h matrix: $(2/3)R_a, (2/3), -(2/3), 0.833/R_a$. (c) When $V_2 = 0$, $I_2 = -1.5I_1$, $V_1 = R_1I_1$. When $I_1 = 0$, $V_2 = V_1 = R_3I_2$. h matrix: $2, 1, -1.5, 0.25$ S. (d) When $V_2 = 0$, $R_1I_1 + R_1I_2 = V_1$. $R_1I_1 + (R_1 + R_2)I_2 = -4V_1$. When $I_1 = 0$, $V_1 = R_1I_2$. $V_2 = (R_1 + R_2)I_2 + 4V_1$. h matrix: 1.14 Ω, $0.143, -0.714, 0.0357$ S.

13-24 (a) When $V_2 = 0$, current in $R_2 = (KI_1/R_2)$. $V_1 = R_1I_1$. When $I_1 = 0$, the dependent source is dead. $V_1 = V_2 = R_2I_2$. h matrix: $R_1, 1, (K - R_2)/R_2, (1/R_2)$. (b) When $V_2 = 0$, the dependent source is dead. $V_1 = I_1/G_1$. $I_2 = -I_1$. When $I_1 = 0$, $V_2 = (I_2 - KV_2)/G_2$, $V_1 = V_2 - (KV_2/G_1)$. h matrix: $(1/G_1), (G_1 - K)/G_1, -1, (G_2 + K)$.

13-25 h matrix: 7.33 Ω, $-0.667, 0.333, 0.0833$ S.

13-26 Equations: $(R_s + h_{11})I_1 + h_{12}V_2 = V_s$. $h_{21}I_1 + (h_{22} + G_L)V_2 = 0$. $(I_2/I_1) = h_{21}G_L/(h_{22} + G_L)$. $(V_s/I_1) = [(R_s + h_{11})(h_{22} + G_L) - h_{12}h_{21}]/(h_{22} + G_L)$.

13-27 8.33.

13-28 (a) When $V_2 = 0$, $(R_1 + R_2 + R_3)I_1 + R_3I_2 = V_1$, $R_3I_1 + (R_3 + R_4)I_2 = 10I_1$. When $I_1 = 0$, $V_1 = R_3I_2$, $V_2 = (R_3 + R_4)I_2$. h matrix: 27.7 Ω, $0.0909, 0.3636, 0.0455$ S. (b) $[R_L \| (1/h_{22})] = 10.48$ Ω. $V_2 = -3.81I_1$. $V_1 = 27.7 I_1 + 0.0909 V_2$. $R_1 = 27.4$ Ω.

13-29 (a) z matrix: $(R_1 + R_2 - K), R_2, (R_2 - K), R_2$. (b) z matrix: $(G_1 + G_2)/[G_1(K + G_2)]$, $(G_1 - K)/[G_1(K + G_2)]$, $1/(K + G_2)$, $1/(K + G_2)$.

13-30 y matrix: $1/h_{11}, -h_{12}/h_{11}, h_{21}/h_{11}, \|h\|/h_{11}$.

13-31 (a) y matrix: $1/R_1, -1/R_1, (K - R_2)/R_1R_2, (R_1 + R_2 - K)/R_1R_2$. (b) y matrix: G_1, $(K - G_1), -G_1, (G_1 + G_2)$.

13-32 $A = (V_1/V_2)|_{I_2 = 0}$. $B = -(V_1/I_2)|_{V_2 = 0}$. $C = (I_1/V_2)|_{I_2 = 0}$. $D = -(I_1/I_2)|_{V_2 = 0}$.

13-33 (a) When $I_2 = 0$, $V_1 = (R_a + R_b)I_1$, $V_2 = R_bI_1$. When $V_2 = 0$, $V_1 = R_aI_1$, $I_2 = -I_1$. ABCD matrix: $(R_a + R_b)/R_b, R_a, (1/R_b), 1$. (b) When $I_2 = 0$, $V_1 = 1.2R_aI_1$, $V_2 = 0.8R_aI_1$. When $V_2 = 0$, $I_2 = -0.667R_aI_1$, $V_1 = 0.667R_aI_1$. ABCD matrix: $1.5, R_a, (1.25/R_a), 1.5$. (c) When $I_2 = 0$, $V_1 = 8I_1$, $V_2 = 6I_1$. When $V_2 = 0$, $I_2 = -1.5I_1$, $V_1 = 2I_1$. ABCD matrix: $1.333, 1.333$ Ω, 0.167 S, 0.667. (d) When $I_2 = 0$, $V_1 = 4I_1$, $V_2 = 20I_1$. When $V_2 = 0$, $I_1 = 0.875V_1$, $I_2 = -0.625V_1$. ABCD matrix: $0.2, 1.6$ Ω, 0.05 S, 1.4.

13-34 *ABCD* matrix: (z_{11}/z_{21}), $(\|z\|/z_{21})$, $(1/z_{21})$, (z_{22}/z_{21}).

13-36 Treat the $-j10$ branch and ground line as two-port 1 and the remaining T as two-port 2. y matrix of 1: $j0.1$, $-j0.1$, $-j0.1$, $j0.1$. y matrix of 2: $-j0.0667$, $j0.133$, $j0.133$, $-j0.0667$. y matrix of combination: $j0.0333$, $-j0.033$, $-j0.033$, $j0.0333$.

13-37 *ABCD* matrix of the combination: 4.06×10^{-5}, 0.162, 4.06×10^{-7}, 1.62×10^{-3}. $(V_1/I_1) = 1.76 \times 10^4$. $(I_2/I_1) = -176$. $R_{in} = 100\ \Omega$.

PROBLEMS

Secs. 13-3, 13-4, 13-5 Open Circuit Impedance Parameters

13-1 Determine the z parameters of each of the circuits shown in Fig. 13-39.

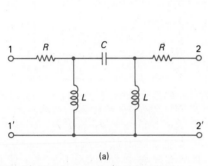
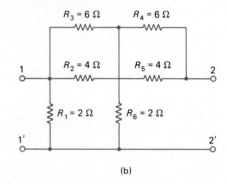

(a)

(b)

Fig. 13-39 Circuits for Problem 13-1.

13-2 Determine the z parameters of each of the circuits shown in Fig. 13-40.

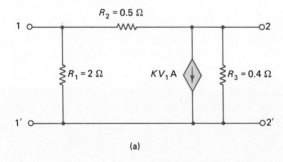

(a)

Fig. 13-40 Circuits for Problem 13-2.

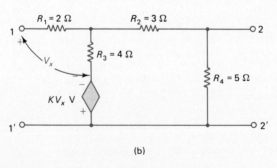

(b)

656   Two-Ports

13-3 Determine the z parameters of the symmetrical lattice network of Fig. 13-38.

13-4 A two-port network is described by the set of equations:
$$V_2 = K_1V_1 + K_2I_1$$
$$I_2 = K_3V_1 + K_4I_1$$

Obtain the z parameters of the network in terms of the Ks.

13-5 A lossless twin T network is shown in Fig. 13-41. Determine its z parameters.

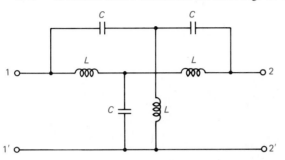

Fig. 13-41 Lossless twin T network for Problem 13-5.

13-6 Obtain the T equivalent of each of the two-ports of Problem 13-1.

13-7 Obtain the T equivalent (one-source model) of each of the two-ports of Problem 13-2.

13-8 Obtain the T equivalent of the symmetrical lattice of Problem 13-4.

13-9 A *gyrator* is a two-port element whose symbol is shown in Fig. 13-42(a), with the governing relationships: $V_1 = KI_2$, and $V_2 = KI_1$ where K is a constant. The gyrator is analogous to the ideal transformer but is nonreciprocal. A gyrator is connected as shown in Fig. 13-42(b). Determine the T equivalent of the connection.

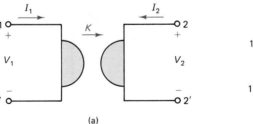

Fig. 13-42 Gyrator connection for Problem 13-9.

13-10 (a) Obtain the T equivalent of the network shown in Fig. 13-43(a). (b) The network is used in the system shown in Fig. 13-43(b). Determine (i) the voltage ratio (V_o/V_s) and the output impedance. (i.e., the Thevenin impedance as seen from the output port).

Fig. 13-43 (a) Network and (b) system for Problem 13-10.

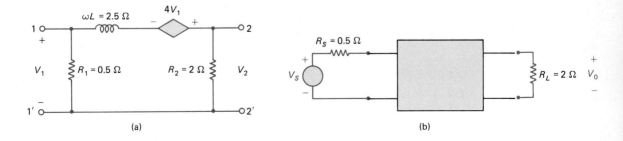

Problems

Secs. 13-6, 13-7, 13-8 Short-Circuit Admittance Parameters

13-11 (a) Determine the y parameters of the two-ports of Problem 13-1 by direct short-circuit calculations. (b) Set up the equivalent π network in each case.

13-12 (a) Determine the y parameters of the two-ports of Problem 13-2 by direct short-circuit calculations. (b) Set up the equivalent π network (one source model) in each case.

13-13 Obtain the π equivalent of the symmetrical lattice.

13-14 Obtain the one-source model of (a) the equivalent T and (b) the equivalent π of the circuit shown in Fig. 13-44.

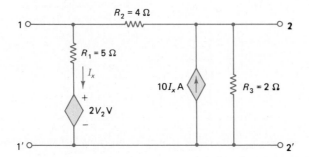

Fig. 13-44 Circuit for Problem 13-14.

13-15 A two-port has the T equivalent and the π equivalent shown in Fig. 13-45. (a) Obtain the relationships for Y_a, Y_b, and Y_c in terms of Z_1, Z_2, and Z_3. (b) Obtain the relationships for Z_1, Z_2, Z_3 in terms of Y_a, Y_b, and Y_c. These two sets of equations are usually referred to as the *delta-wye* and *wye-delta conversion formulas*.

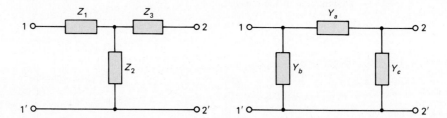

Fig. 13-45 T and π equivalent for Problem 13-15.

13-16 The circuit of Fig. 13-46 is called the hybrid-π model of a bipolar junction transistor. (a) Determine the y parameters of the transistor. (b) Determine the input admittance when a load $R_L = 10$ kΩ is connected to the output terminals, and $g_{bb'} = 0.02$ S, $g_{b'e} = 4 \times 10^{-4}$ S, $g_{b'c} = 2 \times 10^{-5}$ S, $g_{ce} = 2 \times 10^{-5}$ S, and $g_m = 4 \times 10^{-2}$ S.

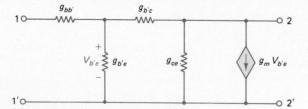

Fig. 13-46 Circuit for Problem 13-16.

Secs. 13-9, 13-10, 13-11 Hybrid Parameters

13-17 Determine the h parameters of the two-ports of Problem 13-1 directly by using appropriate short- and open-circuit calculations.

Two-Ports

13-18 Determine the h parameters of the two-ports of Problem 13-2 directly by using appropriate short- and open-circuit calculations.

13-19 Determine the h parameters of the symmetrical lattice.

13-20 (a) Determine the h parameters of the circuit in Problem 13-14. (b) Draw the h parameters model of the circuit. (c) Connect a current source I_s with a shunt resistance of $R_s = 5\Omega$ at the input port and a load resistance of $R_L = 2\Omega$ at the output port. Calculate the current ratio (I_2/I_1).

13-21 (a) Determine the h parameters of the model of the transistor in Problem 13-16. (b) Connect a load resistance R_L to the output terminals and determine the input impedance of the resulting circuit.

13-22 A bipolar junction transistor has the following specifications: $h_{11} = 2.5$ kΩ, $h_{12} = 0$, $h_{21} = 50$, and $h_{22} = 2 \times 10^{-4}$ S. A signal source V_s with a series resistance of $R_s = 2.5$ kΩ is connected to the input side and a load resistance $R_L = 5$ kΩ is connected to the output side. Calculate (a) the voltage gain (V_2/V_s) and (b) the current gain (I_2/I_1).

13-23 An ideal negative impedance converter (NIC), Fig. 13-47(a), has the property: $Z_{in} = (V_1/I_1) = -Z_L$. (a) Obtain a set of h parameters of the NIC. Verify that the h parameters satisfy the condition $Z_{in} = -Z_L$. (b) Use the set of h parameters to analyze the arrangement in Fig. 13-47(b) and obtain a relationship between Z_s and Z_o.

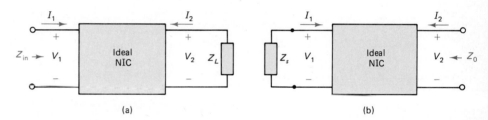

Fig. 13-47 Ideal NIC systems for Problem 13-23.

13-24 Derive the formulas for converting the y parameter matrix of a two-port to its h parameter matrix. Verify your formulas by trying them on the two-ports of Problems 13-12 and 13-18.

13-25 Derive the formulas for converting the z parameter matrix of a two-port to its h parameter matrix. Verify your set of formulas by proving it to be the inverse of the matrix that gives z parameters in terms of the h parameters (derived in the text).

Sec. 13-12 *ABCD* Parameters

13-26 Determine the *ABCD* parameters of the two-ports in Problem 13-1.

13-27 Determine the *ABCD* parameters of the two-ports in Problem 13-2.

13-28 Derive the formulas that permit the (a) determination of the h parameters from *ABCD* parameters and (b) the determination of the *ABCD* parameters from the h parameters.

13-29 A signal source V_s in series with a resistor R_s is connected to the input port of a network, and a load resistance R_L is connected to the output port. If the two-port is described by its *ABCD* parameters, obtain the expressions for the following: (a) the voltage ratio (V_2/V_s); (b) the current ratio (I_2/I_1); (c) the input resistance (V_1/I_1).

13-30 A set of parameters, which will be called the *EFGH* parameters here, can be defined for a two-port by the equations:

$$V_2 = EV_1 - FI_1$$
$$I_2 = GV_1 - HI_1$$

(a) Write a set of relationships to determine the *EFGH* parameters in terms of appropriate open- and short-circuit measurements or calculations.
(b) Determine the relationships that permit the conversion of the *ABCD* parameters to the *EFGH* parameters.
(c) Verify if $(EH - FG) = 1$ for a reciprocal network.
(d) Determine the *EFGH* parameters of the two-port in Problem 13-14.

13-31 Repeat Problem 13-29 by using the *EFGH* parameters defined in Problem 13-30.

Sec. 13-13 Interconnection of Two-Ports

13-32 Figure 13-48 shows two interconnections of two-ports. In each case, (a) discuss which parameters are most convenient to describe the overall system and (b) determine the parameters of the overall network in terms of the selected parameter matrices. Assume that the interconnections do not alter the parameters of the original networks.

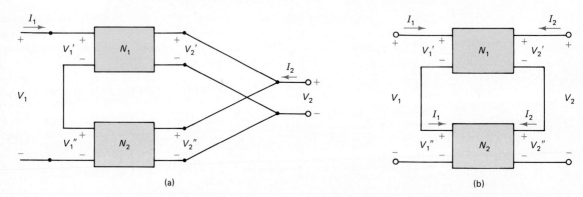

Fig. 13-48 Interconnections of two-ports for Problem 13-32.

13-33 In the case of two cascaded networks, derive the overall *EFGH* parameter matrix in terms of the *EFGH* matrices of the individual two-ports.

13-34 Assume that $h_{12} = h_{22} = 0$ for two networks, A and B. Designate the parameters of network A by an additional subscript a and those of network B by an additional subscript b. The two networks are cascaded, with network A being fed by a voltage source V_s with a series resistance R_s and a load R_L connected to the output of network B.
(a) Obtain the *ABCD* parameters of each of the two networks and use them to determine the voltage gain of the cascaded connection.
(b) Use the *h* parameter models directly and determine the voltage gain of the cascaded connection.

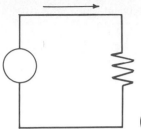

CHAPTER 14
NONSINUSOIDAL SIGNALS– FOURIER METHODS

We have focused on sinusoidal signals up to this point in the analysis of circuits not only because of their common occurrence in practice and mathematical convenience, but also because they serve as building blocks for periodic nonsinuoidal signals. Periodic signals can be decomposed into a sum of sinusoidal components, called *Fourier series*, and the response of the network can be determined for the individual sinusoidal components by using the methods discussed in the preceding chapters. The principle of superposition is then used to find the response of the network to the general periodic signal. *Fourier transforms* (or *Fourier integrals*) permit the extension of the method of Fourier series to nonperiodic signals of finite duration. When the response of networks to arbitrary signals is to be determined, the approach is to slice the signal into rectangular pulses of infinitesimally short duration. The principle of superposition is then applied in the form of an operation called *convolution*. Fourier series, Fourier transforms, and convolution are the subject of this chapter.

14-1 FOURIER SERIES REPRESENTATION OF A PERIODIC SIGNAL

For the purposes of analysis, it is desirable to express an arbitrary signal $f(t)$, defined over an interval (a, b) in the form

$$f(t) = p_1\phi_1(t) + p_2\phi_2(t) + \ldots + p_k\phi_k(t) + \ldots \tag{14-1}$$

where the component functions $\phi_k(t)$ are some familiar functions, and the coefficients p_k are selected so that the series on the right side of Eq. (14-1) *converges* to $f(t)$. The determination of the coefficients p_k is straightforward if the component functions satisfy the *orthogonality condition:*

$$\int_a^b \phi_m(t)\phi_n(t)dt = \begin{cases} 0 \text{ when } m \neq n \\ \text{some nonzero constant when } m = n \end{cases} \quad (14\text{-}2)$$

When $f(t)$ is a periodic signal defined over the interval $(0, T)$, one possible choice of the component function is

$$\phi_k(t) = \cos(2\pi kt/T)$$

since it satisfies the orthogonality condition over the interval $(0, T)$. When $\cos(2\pi kt/T)$ is used for $\phi_k(t)$ in Eq. (14-2)

$$\int_0^T \cos(2\pi mt/T) \cos(2\pi nt/T) dt = \begin{cases} 0 \text{ when } m \neq n \\ (T/2) \text{ when } m = n \end{cases} \quad (14\text{-}3)$$

See Exercise 14–1. A similar statement is valid for the component function $\sin(2\pi kt/T)$.

Exercise 14-1 Verify Eq. (14-3) by using the trigonometric identity

$$(\cos X)(\cos Y) = (1/2)[\cos(X + Y) + \cos(X - Y)]$$

and performing the indicated integration.

Exercise 14-2 Show that the function $\cos[(2\pi kt/T) + \theta)]$, where θ_k is an arbitrary angle, also satisfies the orthogonality condition.

The Fourier series expansion of a periodic signal $f(t)$ with period T uses sines and cosines as the component functions:

$$f(t) = a_0 + \sum_{k=1}^{\infty} [a_k \cos(k\omega t) + b_k \sin(k\omega t)] \quad (14\text{-}4)$$

where a_0, a_k and b_k ($k = 1, 2, \ldots, n, \ldots$) are coefficients to be determined and

$$\omega = (2\pi/T) = 2\pi f \quad (14\text{-}5)$$

where $f = (1/T)$ is called the *fundamental frequency* of the signal.

> The Fourier series expansion is seen to consist of a constant (or dc) term a_0 and a number of sinusoidal components whose frequencies are multiples of the fundamental frequency f. The frequencies $2f, 3f, \ldots$ are called, respectively, the second, third, ... *harmonic* components of the signal. The kth harmonic is given by the sum
>
> $$[a_k \cos(k\omega t) + b_k \sin(k\omega t)]$$
>
> which can also be expressed as a single cosine term with a phase angle by using the procedure discussed in Chapter 7.

Exercise 14-3 Let $d_k \cos(k\omega t + \theta_k)$ represent the kth harmonic component of a periodic signal. Write the equations relating d_k and θ_k to the coefficients a_k and b_k.

Exercise 14-4 A signal is known to have the Fourier series

$$v(t) = 6 + (4.7 \cos 200t - 1.18 \cos 400t + \ldots$$
$$+ (6.47 \sin 200t + 4 \sin 400t + \ldots$$

(a) What is the fundamental frequency of the signal? (b) Write each component (fundamental frequency and second harmonic) in the form of a single cosine term with a phase angle. (c) What are the amplitudes of the fundamental frequency and the second harmonic components?

14-1-1 Determination of the Fourier Series Coefficients

The expressions for the coefficients in Fourier series are obtained by using the orthogonality properties of the sine and cosine functions. The interval $(-T/2, T/2)$ is used in the following discussion, since it will be seen to offer some theoretical advantages over the interval $(0, T)$.

Determination of a_0

Consider the *average value* of $f(t)$ defined by

$$\text{average value of } f(t) = (1/T) \int_{-T/2}^{T/2} f(t) dt$$

Using Eq. (14-4)

$$\text{average value of } f(t) = (1/T) \left[\int_{-T/2}^{T/2} a_0 \, dt + \int_{-T/2}^{T/2} \sum_k [a_k \cos(k\omega t) dt + b_k \sin(k\omega t) dt] \right] \quad (14\text{-}6)$$

Assuming that the series converges uniformly over the interval $(-T/2, T/2)$, the infinite series in Eq. (14-6) is integrated term by term. Each term in the infinite series is periodic and goes through k complete cycles in the interval $(-T/2, T/2)$. Therefore, the integrals due to the sine and cosine terms in the infinite series will all lead to zero, and the only term remaining is due to the constant term a_0:

$$a_0 = \frac{1}{T} \int_{-T/2}^{T/2} f(t) dt \quad (14\text{-}7)$$

Thus the constant term a_0 in the Fourier series expansion equals the average value of $f(t)$ taken over a period, which is given by Eq. (14-7). An alternative expression is

$$a_0 = (1/T) [\text{area under } f(t) \text{ for one period}] \quad (14\text{-}8)$$

since the integral of $f(t)$ represents the area under the signal. If a signal $f(t)$ is observed through a dc ammeter or voltmeter, the meter reading will equal its average value: hence a_0 is called the *dc component* of $f(t)$.

Determination of the Coefficient a_m

The value of a_m, for any integer m ($\neq 0$), is obtained by using the formula

$$a_m = \frac{2}{T} \int_{-T/2}^{T/2} f(t) \cos(m\omega t)\,dt \qquad (14\text{-}9)$$

that is, $f(t)$ is multiplied by $\cos(m\omega t)$, the product integrated over a period, and the result divided by $(T/2)$ in order to determine the coefficient of the cosine term of the mth harmonic.

Proof of Formula in Eq. (14-9):   Consider the integral

$$I_2 = \int_{-T/2}^{T/2} f(t) \cos(m\omega t)\,dt \qquad m \neq 0$$

which becomes after using Eq. (14-4)

$$I_2 = \int_{-T/2}^{T/2} [a_0 \cos(m\omega t)\,dt + \sum_k a_k \cos(k\omega t) \cos(m\omega t)\,dt +$$

$$\sum_k b_k \sin(k\omega t) \cos(m\omega t)\,dt] \qquad (14\text{-}10)$$

Since $m \neq 0$, the term $\int_{-T/2}^{T/2} [a_0 \cos(m\omega t)]\,dt$ in Eq. (14-10) equals zero.

From Eq. (14-3),

$$\sum_k \int_{-T/2}^{T/2} a_k \cos(k\omega t) \cos(m\omega t)\,dt = \begin{cases} 0 \text{ when } k \neq m \\ a_m (T/2) \text{ when } k = m \end{cases}$$

Therefore, in the series of terms

$$\sum_k \int_{-T/2}^{T/2} a_k \cos(k\omega t) \cos(m\omega t)\,dt$$

in Eq. (14-10), the only surviving term will be $a_m (T/2)$. For example, if a_7 is being evaluated, then in the series of terms $\sum_k \int a_k \cos(k\omega t) \cos(7\omega t)\,dt$, all terms except $k = 7$ will be zero, while the $k = 7$ term will result in $a_7(T/2)$.

A procedure similar to the one used in Exercise 14-1 will show that

$$\int_{-T/2}^{T/2} b_k \sin(k\omega t) \cos(m\omega t)\,dt = 0$$

for *all* k (including $k = m$).

Thus, when the integrals in Eq. (14-10) are evaluated for any specific value of m, the result will be

$$I_2 = a_m(T/2)$$

or
$$a_m = (2/T)I_2$$
which proves the formula in Eq. (14-9) for the evaluation of a_m.

Determination of b_m

The value of b_m, for any integer m, is obtained by using the formula

$$b_m = \frac{2}{T} \int_{-T/2}^{T/2} f(t) \sin(m\omega t)\, dt \qquad (14\text{-}11)$$

That is, $f(t)$ is multiplied by $\sin(m\omega t)$, the product integrated over a period, and the result divided by $(T/2)$ in order to determine the coefficient of the sine term of the mth harmonic.

The proof of the formula follows steps similar to those used in connection with a_m.

Exercise 14-5 Use the trigonometric identity

$$[(\sin X)(\cos Y)] = (1/2)[\sin(X+Y) + \sin(X-Y)]$$

and show that

$$\int_{-T/2}^{T/2} b_k \sin(k\omega t) \cos(m\omega t)\, dt = 0$$

for *all* k, including $k = m$

Exercise 14-6 Use the trigonometric identity

$$[(\sin X)(\sin Y)] = (1/2)[\cos(X-Y) - \cos(X+Y)]$$

and the result of Exercise 14-5 to derive the formula for b_m in Eq. (14-11).

Compare the formulas for a_0 and a_m, and note that the multiplying factor is $(1/T)$ in a_0 and $(2/T)$ in a_m. Thus a_0 is not obtained by putting $m = 0$ in the formula for a_m.

A frequently used relationship in the evaluation of Fourier coefficients is

$$\omega T = 2\pi \qquad (14\text{-}12)$$

Example 14-1 The waveform of a half-wave rectified sinusoid (i.e., a sinusoidal signal in which the negative half-cycles have been removed by means of a rectifier) is shown in Fig. 14-1. Obtain its Fourier series.

Solution Describe the function over the interval $(-T/2, T/2)$.

$$f(t) = \begin{cases} V_0 \cos \omega t & -T/4 < t < T/4 \\ 0 \text{ for the remainder of the interval} & (-T/2, T/2) \end{cases}$$

14-1 Fourier Series Representation of a Periodic Signal

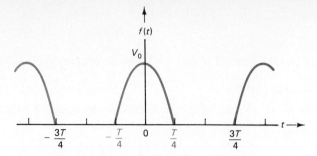

Fig. 14-1 Half-wave rectified sinusoid of Example 14-1.

Determination of a_0: From Eq. (14-7)

$$a_0 = (1/T) \int_{-T/4}^{T/4} V_0 \cos \omega t\, dt$$

$$= (V_0/\omega T)[\sin (\omega T/4) - \sin (-\omega T/4)] = (V_0/\pi)$$

since $\omega T = 2\pi$ and $(\omega T/4) = (\pi/2)$ from Eq. (14-12).

Determination of a_m: From Eq. (14-9)

$$a_m = (2/T) \int_{-T/4}^{T/4} V_o \cos (\omega t) \cos (m\omega t)\, dt$$

Using the trigonometric identity cited in Exercise 14-1

$$a_m = (V_0/2T)\left[\int_{-T/4}^{T/4} \cos (m+1)\omega t\, dt + \int_{-T/4}^{T/4} \cos (m-1)\omega t\, dt \right]$$

$$= \frac{V_0}{\pi} \left[\frac{\sin [(m+1)\pi/2]}{(m+1)} + \frac{\sin [(m-1)\pi/2]}{(m-1)} \right] \qquad (14\text{-}13)$$

If $m = 1$ is used in Eq. (14-13), the second term in it leads to the indeterminate ratio (0/0). Therefore, a_1 cannot be determined from Eq. (14-13) and needs to be determined by returning to Eq. (14-9) with $m = 1$. Deferring that evaluation for the present, let us consider the other values of m.

If m is an *odd* number (but $\neq 1$), both sine terms in Eq. (14-13) become zero. Therefore

$$a_m = 0 \text{ when } m = 3, 5, 7, 9, \ldots$$

If m is an *even* number

$$\sin [(m+1)\pi/2] = \begin{cases} -1 \text{ when } m = 2, 6, 10, \ldots \\ +1 \text{ when } m = 4, 8, 12, \ldots \end{cases}$$

and

$$\sin [(m-1)\pi/2] = \begin{cases} +1 \text{ when } m = 2, 6, 10, \ldots \\ -1 \text{ when } m = 4, 8, 12, \ldots \end{cases}$$

Therefore, after some manipulation of the terms in Eq. (14-13), the expressions for a_m ($m \neq 1$) are found to be

$$a_m = \begin{cases} +\dfrac{2V_0}{\pi(m^2 - 1)} & \text{when } m = 2, 6, 10, \ldots \quad (14\text{-}14) \\[2mm] & \quad (14\text{-}15) \\[2mm] -\dfrac{2V_0}{\pi(m^2 - 1)} & \text{when } m = 4, 8, 12, \ldots \end{cases}$$

Now consider the coefficient a_1. Using $m = 1$ in Eq. (14-9)

$$a_1 = (2/T) \int_{-T/4}^{T/4} V_0 \cos^2(\omega t)\, dt$$

or

$$a_1 = (V_0/2) \tag{14-16}$$

The coefficients for a_m are given by Eqs. (14-14), (14-15), and (14-16).

Determination of b_m: From Eq. (14-11)

$$b_m = (2/T) \int_{-T/4}^{T/4} V_0 \cos(\omega t) \sin(m\omega t)\, dt$$

Using the trigonometric identity cited in Exercise 14-15, the integral leads to

$$b_m = 0 \text{ for all } m$$

The Fourier series of the given $f(t)$ does not contain any sine terms. The Fourier coefficients of the half-wave rectified sinusoid are therefore

$$a_0 = (V_0/\pi)$$
$$a_1 = (V_0/2)$$
$$a_m = \begin{cases} 0 \text{ for } odd \text{ values of } m \text{ (except } m = 1) \\[2mm] +\dfrac{2V_0}{(m^2 - 1)} & \text{when } m = 2, 6, 10, \ldots \\[2mm] -\dfrac{2V_0}{(m^2 - 1)} & \text{when } m = 4, 8, 12, \ldots \end{cases}$$
$$b_m = 0 \text{ for all } m$$

If we write out the first few coefficients of the series,

$$f(t) = (V_0/\pi) + (V_0/2)\cos(\omega t) + (2V_0/\pi)\,[(1/3)\cos(2\omega t)$$
$$- (1/15)\cos(4\omega t) + (1/35)\cos(6\omega t) - (1/63)\cos(8\omega t) + \ldots]$$

■

Exercise 14-7 Obtain the Fourier series of the pulse train shown in Fig. 14-2(a).

Exercise 14-8 Figure 14-2(b) shows a half-wave rectified sinusoid with a different choice of origin from

14-1 Fourier Series Representation of a Periodic Signal

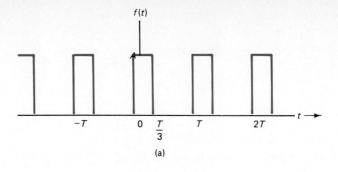

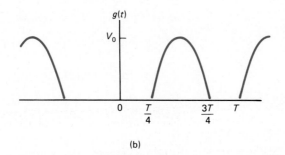

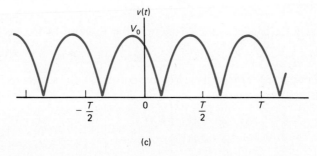

Fig. 14-2 Pulse train and sinusoids for Exercises 14-7, 14-8, and 14-9.

Fig. 14-1. Obtain the series for this waveform by using an appropriate shift in time in the series obtained in Example 14-1.

Exercise 14-9 Figure 14-2(c) shows a full-wave rectified sinusoid. Obtain its Fourier series by combining the results of Example 14-1 and Exercise 14-8.

14-1-2 Effect of Waveform Symmetry on Fourier Series

If the waveform of a signal possesses certain symmetrical features, then some of the coefficients in its Fourier series expansion are found to be zero. For instance, the absence of the sine terms in the half-wave rectified sinusoid in Example 14-1 was due to the symmetry of the waveform (as we will show here). Symmetry properties of waveforms are useful in reducing the amount of computation in evaluating Fourier coefficients.

Even and Odd Symmetry

Even Functions: If t is replaced by $-t$ in the expression for $f(t)$ and does not affect the function, then $f(t)$ is an even function. That is

$$f(t) \text{ is even if } f(t) = f(-t) \tag{14-17}$$

Some examples of even functions are $\cos(m\omega t)$, t^2, $|t|$, and a constant.

Visually, if the portion of the waveform to the left of the vertical axis is a mirror image of the portion to its right, then the waveform has even symmetry.

The half-wave rectified sinusoid of Fig. 14-1 is seen to have even symmetry.

Odd Functions: If t is replaced by $-t$ in the expression for $f(t)$ and the result is a multiplication of the original function by (-1), then $f(t)$ is an odd function. That is

$$f(t) \text{ is odd if } f(-t) = -f(t) \tag{14-18}$$

Some examples of odd functions are $\sin(m\omega t)$ and t^3.

Visually, if the given function is rotated through 180° by using the origin as pivot and the rotated function coincides with the original function, then $f(t)$ has odd symmetry.

The even and odd symmetries lead to the following properties of Fourier coefficients. (The proof of some of the properties will be presented later in this section.)

Even Symmetry: The Fourier coefficients b_m of an even function are zero for all m. All sine terms will be absent from the series.

For an even function, the coefficient a_m is given by the expression [which is an alternative choice to the general formula for a_m in Eq. (14-9)]

$$a_m = 2\left[\frac{2}{T}\int_0^{T/2} f(t) \cos(m\omega t)\, dt\right] \tag{14-19}$$

Therefore, it is sufficient to deal with $f(t)$ over the half period $(0, T/2)$.

Odd Symmetry: The Fourier coefficients a_m of an odd function are zero for all m (including $m = 0$). All cosine terms and the constant term will be absent from the series.

For an odd function, the coefficient b_m is given by the expression [which is an alternative choice to the general formula for b_m in Eq. (14-11)]

$$b_m = 2\left[\frac{2}{T}\int_0^{T/2} f(t) \sin(m\omega t)\, dt\right] \tag{14-20}$$

Therefore, it is sufficient to deal with $f(t)$ over the half period $(0, T/2)$.

Proof That $b_m = 0$ For an Even Function: Consider the integral in Eq. (14-11) for b_m

$$\int_{-T/2}^{T/2} f(t) \sin(m\omega t)\, dt$$

The product $f(t)\sin(m\omega t)$ is an odd function, since $f(t)$ is even and $\sin(m\omega t)$ is an odd function. The integral of an odd function is an even function. That is

$$\int f(t) \sin(m\omega t) dt = \text{an even function}$$

and the value of an even function will be the same at the upper limit $(T/2)$ as at the lower limit $(-T/2)$. Therefore, $b_m = 0$ for all m.

A similar argument is used to establish the property that $a_m = 0$ for the odd functions.

Note that odd symmetry of a function is changed by the presence of the constant term (which itself is an even function). Therefore, it is advisable to check the odd symmetry after removing the dc component from a given function.

Example 14-2 Determine the Fourier series of the periodic signal shown in Fig. 14-3.

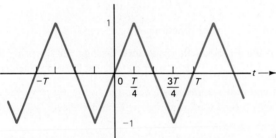

Fig. 14-3 Waveform for Example 14-2.

Solution The waveform has odd symmetry, as can be verified by rotating the signal through 180° about the origin and seeing that it coincides with the original signal. Therefore,

$$a_m = 0 \text{ for all } m \text{ (including } m = 0\text{)}$$

From Eq. (14-11), the description of the function over the half period $(0, T/2)$ is

$$f(t) = \begin{cases} \dfrac{4t}{T} & 0 < t < \dfrac{T}{4} \\ 2 - \dfrac{4t}{T} & \dfrac{T}{4} < t < \dfrac{T}{2} \end{cases}$$

From Eq. (14-20)

$$b_m = \frac{4}{T}\left[\int_0^{T/4} \frac{4t}{T} \sin(m\omega t) dt + \int_{T/4}^{T/2} \left(2 - \frac{4t}{T}\right) \sin(m\omega t) dt\right]$$

Using the integral

$$\int t \sin(m\omega t) dt = -\frac{t}{m\omega} \cos(m\omega t) + \frac{1}{m^2\omega^2} \sin(m\omega t)$$

it is found after a fair amount of manipulation,

$$b_m = \frac{8}{m^2\pi^2} \sin\left(\frac{m\pi}{2}\right)$$

That is

$$b_m = \begin{cases} 0 & \text{when } m \text{ is even} \\ +8/m^2\pi^2 & \text{when } m = 1, 5, 9, \ldots \\ -8/m^2\pi^2 & \text{when } m = 3, 7, 11, \ldots \end{cases}$$

Therefore, the Fourier series of the given signal is

$$f(t) = (8/\pi^2)[\sin(\omega t) - (1/9)\sin(3\omega t) + (1/25)\sin(5\omega t) - \ldots]$$ ∎

Exercise 14-10 Use the symmetry property and obtain the Fourier series of the waveform shown in Fig. 14-4(a).

Exercise 14-11 Use the symmetry property and obtain the Fourier series of the waveform shown in Fig. 14-4(b).

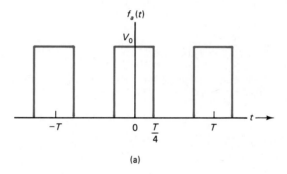

(a)

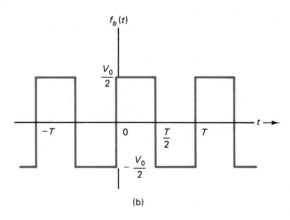

(b)

Fig. 14-4 Waveforms for Exercises 14-10 and 14-11.

Half Wave Symmetry

A signal is said to have *half-wave symmetry* if it satisfies the condition:

$$f\left(t + \frac{T}{2}\right) = -f(t) \tag{14-21}$$

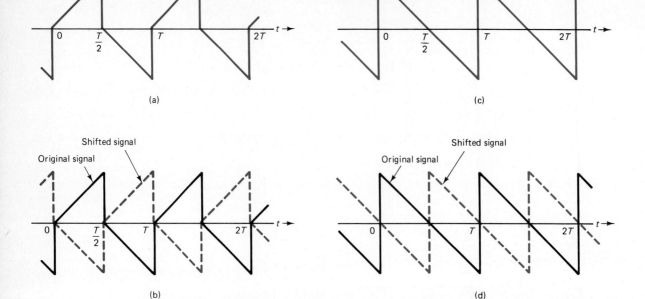

Fig. 14-5 (a) A waveform with half-wave symmetry when shifted by *T*/2 and folded coincides with the original waveform as in (b). (c) A waveform without half-wave symmetry does not satisfy the above property as shown in (d).

That is, if the signal is shifted by *T*/2 in either direcion, the *shifted signal becomes a mirror image* of the original signal about the *horizontal* axis. Figure 14-5 shows examples of signals with and without half-wave symmetry. The waveform of Fig. 14-5(a) has half-wave symmetry, which is made evident by Fig. 14-5(b), while the signal of Fig. 14-5(c) does not have half-wave symmetry, as is made evident by Fig. 14-5(d).

A signal with half-wave symmetry does not contain any even harmonics. That is, a_m and b_m are zero for all *even* values of *m*. To prove this statement, note that

$$\cos[k\omega(t + T/2)] = \cos(k\omega t + k\pi) = \begin{cases} \cos(k\omega t) & \text{when } k \text{ is even} \\ -\cos(k\omega t) & \text{when } k \text{ is odd} \end{cases}$$

$$\sin[k\omega(t + T/2)] = \sin(k\omega t + k\pi) = \begin{cases} \sin(k\omega t) & \text{when } k \text{ is even} \\ -\sin(k\omega t) & \text{when } k \text{ is odd} \end{cases}$$

Write the series for $f(t)$ shifted by *T*/2 and separate the $k =$ even terms and $k =$ odd terms. Then

$$f(t + T/2) = a_0 + \sum_{k \text{ even}} [a_k \cos(k\omega t) + b_k \sin(k\omega t)]$$
$$- \sum_{k \text{ odd}} [a_k \cos(k\omega t) + b_k \sin(k\omega t)] \quad (14\text{-}22)$$

Compare the series in Eq. (14-22) with the series for $[-f(t)]$:

$$-f(t) = -a_0 - \sum_{k=1}^{\infty} [a_k \cos(k\omega t) + b_k \sin(k\omega t)] \quad (14\text{-}23)$$

For a signal having half-wave symmetry,

$$f(t + T/2) = -f(t)$$

and the series in Eqs. (14-22) and (14-23) must be equal to each other, which means that any terms that are different between the two series must be equated to zero. This leads to the conditions: $a_0 = 0$, and $a_k = 0$ and $b_k = 0$ when k is even.

Half-wave symmetry does not lead to any reduction in the computational work needed to find the Fourier series of a signal. It is useful, however, in checking the results obtained for those signals having half-wave symmetry.

One important distinction between half-wave symmetry and the other two symmetries (even or odd) is that if a signal has half-wave symmetry, a shift in the origin of the waveform will not destroy such a symmetry. On the other hand, even and odd symmetry depend upon the location of the origin and can be removed by shifting the origin.

14-2 STEADY STATE RESPONSE OF A CIRCUIT WITH PERIODIC INPUT

The steady state response of a circuit driven by a periodic signal is determined by using the principle of superposition and the Fourier series expansion of the signal. That is, the circuit is analyzed separately for each of the sinusoidal components in the Fourier series: Each individual harmonic component is first converted to a phasor, and the desired response for that harmonic component is found by using phasors, impedance, and admittance. Since phasors of different frequencies are involved, the individual responses are first converted to time functions and added in the time domain to get the total response of the circuit. The procedure is illustrated by Example 14-3.

Example 14-3 The source current i_s in the circuit of Fig. 14-6 has the Fourier series:

$$i_s(t) = 7.5 - 2.03 \cos 100t - 0.225 \cos 300t - 3.18 \sin 100t$$
$$- 1.59 \sin 200t - 1.06 \sin 300t - 0.796 \sin 400t \text{ A}$$

Determining the output voltage $v_0(t)$.

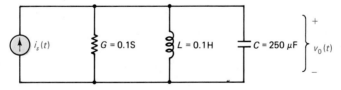

Fig. 14-6 Circuit for Example 14-3.

Solution Since the series contains both sine terms and cosine terms, each harmonic component should first be expressed in the form of a single cosine term with a phase angle.

Using the identity (refer to Chapter 7):

$$a_k \cos (k\omega t) + b_k \sin (k\omega t) = \sqrt{(a_k^2 + b_k^2)} \cos [k\omega t + \arctan (-b_k/a_k)]$$

the given series becomes

$$i_s(t) = 7.5 + 3.77 \cos (100t + 122°) + 1.59 \cos (200t + 90°)$$
$$+ 1.08 \cos (300t + 102°) + 0.796 \cos (400t + 90°) \text{ A}$$

Note that every term is written as a *cosine* to facilitate conversion to a phasor: $A \cos(\omega t + \theta) \rightarrow A\underline{/\theta}$.

For the given circuit, the admittance at any particular frequency is

$$Y = G + j(\omega C - 1/\omega L)$$
$$= 0.1 + j[2.5 \times 10^{-4}\omega - (10/\omega)] \quad (14\text{-}24)$$

and the output voltage at that frequency is given (as a phasor) by

$$\mathbf{V}_o = (\mathbf{I}_s/\mathbf{Y}) \quad (14\text{-}25)$$

Each phasor $\mathbf{V}_o$ is then converted to a sinusoidal time function.
The procedure is quite systematic (though tedious), as shown below.

DC Component of Output: For dc, the inductance acts as a short circuit, and

$$V_o = 0$$

Fundamental Frequency Component of Output: ($\omega = 100$ r/s)

- Input: $i_s = 3.77 \cos(100t + 122°)$
- Phasor $\mathbf{I}_s = 3.77\underline{/122°}$ A
- Admittance: $\mathbf{Y} = 0.1 + j[2.5 \times 10^{-4} \times 100 - (10/100)]$ from Eq. (14-24)
  $= (0.1 - j0.075) = 0.125\underline{/-36.9°}$ S
- Output Phasor: $\mathbf{V}_o = (\mathbf{I}_s/\mathbf{Y})$

$$= \frac{3.77\underline{/122°}}{0.125\underline{/-36.9°}}$$

$$= 30.2\underline{/158.9°} \text{ V}$$

- Output time function: $v_o = 30.2 \cos(100t + 158.9°)$ V

Second Harmonic Component: ($\omega = 200$ r/s)

- $i_s = 1.59 \cos(200t + 90°)$
- $\mathbf{I}_s = 1.59\underline{/90°}$ A
- $\mathbf{Y} = (0.1 + j0) = 0.1\underline{/0°}$ S
- $\mathbf{V}_o = 15.9\underline{/90°}$ V
- $v_o = 15.9 \cos(200t + 90°)$ V

Third Harmonic Component: ($\omega = 300$ r/s)

- $\mathbf{I}_s = 1.08\underline{/102°}$ A
- $\mathbf{Y} = (0.1 + j0.0417) = 0.108\underline{/22.6°}$ S
- $\mathbf{V}_o = 10\underline{/79.4°}$ V
- $v_o = 10 \cos(300t + 79.4°)$ V

Fourth Harmonic Component: ($\omega = 400$ r/s)

- $\mathbf{I}_s = 0.796\underline{/90°}$ A
- $\mathbf{Y} = (0.1 + j0.075) = 0.125\underline{/36.9°}$ S

- $\mathbf{V}_0 = 6.37\underline{/53.1°}$ V
- $v_0 = 6.37 \cos(400t + 53.1°)$ V

Total Output:
$$v_0(t) = 30.2 \cos(100t + 159°) + 15.9 \cos(200t + 90°)$$
$$+ 10 \cos(300t + 79.4°) + 6.37 \cos(400t + 53.1°) \text{ V} \quad\blacksquare$$

The circuit of Example 14-3 is used in certain types of amplifiers to produce an output that is close to a pure sinusoid even though the input is not a sinusoid. By choosing the resonant frequency of the circuit equal to the frequency of the desired sinusoidal output and making the Q of the circuit high, the output is very close to a pure sinusoid with very little distortion. Exercise 14-12 illustrates such an application.

Exercise 14-12 Use $L = 0.1$ H, $C = 1000$ µF, and $G = 0.01$ S in the circuit of Fig. 14-6 and the same current i_s as before. Recalculate the output voltage. Compare the amplitudes of the various harmonics to that of the fundamental frequency component in the input current and the output voltage.

In Section 14-4, the analysis of a circuit for a periodic input will be performed by using the complex exponential form of Fourier series and complex exponential forcing functions. In such cases, the responses for the different harmonic components are in the form of complex exponential time functions and may be added directly. But when complex exponential Fourier series and complex exponential time functions are not used, the component responses are in the form of phasors and must be converted to time functions before being added, as was done in the last example.

14-2-1 Average Power in a Circuit with Periodic Input

Let the voltage and current in a circuit be given by

$$v(t) = V_{dc} + \sum_{k=1}^{\infty} V_k \cos(k\omega t + \theta_k) \quad (14\text{-}26)$$

and

$$i(t) = I_{dc} + \sum_{k=1}^{\infty} I_k \cos(k\omega t + \phi_k) \quad (14\text{-}27)$$

where V_k and I_k are the amplitudes of the kth harmonic components of the voltage and current, and θ_k and ϕ_k the corresponding phase angles. V_{dc} and I_{dc} are the average values or the dc components of the voltage and current.

The instantaneous power received by the circuit is given by

$$p(t) = v(t)i(t) \quad (14\text{-}28)$$

which is also an infinite series. As in the sinusoidal steady state, the *average power* received by the circuit is of practical importance. The average power P_{av} (see Chapters 7 and 8) is defined by

$$P_{av} = \frac{1}{T} \int_{-T/2}^{T/2} p(t)dt = \frac{1}{T} \int_{-T/2}^{T/2} v(t)i(t)dt \qquad (14\text{-}29)$$

When Eqs. (14-26) and (14-27) are substituted in Eq. (14-29), the integrand is the product of two infinite series. The integration of the product of two infinite series might at first appear to be a formidable task. Fortunately, the various product terms in the integrand fit into several categories, each of which is readily integrable. Note that the expression for average power in Eq. (14-29) is made up of terms of the following forms.

$$P_1 = \frac{1}{T} \int_{-T/2}^{T/2} V_{dc} I_{dc} \, dt$$

$$P_2 = \frac{1}{T} \int_{-T/2}^{T/2} V_{dc} I_k \cos(k\omega t + \phi_k) \, dt$$

$$P_3 = \frac{1}{T} \int_{-T/2}^{T/2} I_{dc} V_k \cos(k\omega t + \theta_k) \, dt$$

$$P_4 = \frac{1}{T} \int_{-T/2}^{T/2} V_k I_k \cos(k\omega t + \theta_k) \cos(k\omega t + \phi_k) \, dt$$

$$P_5 = \frac{1}{T} \int_{-T/2}^{T/2} V_m I_n \cos(m\omega t + \theta_m) \cos(n\omega t + \phi_n) \, dt \qquad (m \neq n)$$

In P_2, P_3, and P_4, k can be any integer 1, 2, 3, . . . In P_5, m and n are integers (m, $n = 1, 2, 3, \ldots$), but $m \neq n$.

If the indicated integrations are performed in the above expressions, it is found that:

$$P_1 = V_{dc} I_{dc}$$
$$P_2 = 0 \text{ for all } k$$
$$P_3 = 0 \text{ for all } k$$
$$P_4 = (1/2)V_k I_k \cos(\theta_k - \phi_k) \text{ for each } k$$
$$P_5 = 0 \text{ for all } m \text{ and } n, \; m \neq n$$

Therefore, the only terms that survive in Eq. (14-29) after performing the integrations are those due to P_1 and P_4. Therefore

$$P_{av} = V_{dc} I_{dc} + \sum_{k=1}^{\infty} \frac{V_k I_k}{2} \cos(\theta_k - \phi_k) \qquad (14\text{-}30)$$

The term $V_{dc} I_{dc}$ represents the power due to the dc component of the voltage combining with the dc component of the current. This can be thought of as the average power due to the dc components. For each k, the term

$$\frac{V_k I_k}{2} \cos(\theta_k - \phi_k)$$

represents the average power due to the *k*th harmonic component of the voltage combining with the *same* harmonic component of the current. There is no contribution to average power due to the combination of any harmonic component of voltage with a *different* harmonic component of current. This arises from the orthogonality property of the component functions in Fourier series, mentioned in Section 14-1.

The average power received by a circuit driven by a periodic nonsinusoidal signal is therefore the *sum of the average power contributions from each harmonic component of the voltage (including dc) combining with the same harmonic component of the current (including dc)*.

Example 14-4 The voltage and current associated with a circuit are given by

$$v(t) = 20 + 17.3 \cos 200t + 9.43 \cos(400t + 60°) + 10 \cos 600t \text{ V}$$

and

$$i(t) = 10 + 5.21 \cos(200t - 30°) + 3.16 \cos(400t - 25°) + 10 \cos 600t \text{ A}$$

Determine the average power received by the circuit.

Solution The contribution from each component of voltage and current is calculated first.

DC Component: $P = (20)(10) = 200$ W

Fundamental Frequency Component: ($\omega = 200$ r/s). Amplitude of voltage = 17.3 V, amplitude of current = 5.21 A, and phase angle between voltage and current = 30°.

$$P = (0.5)(17.3)(5.21) \cos 30° = 39.0 \text{ W}$$

Second Harmonic Component: ($\omega = 400$ r/S). Amplitude of voltage = 9.43 V, amplitude of current = 3.16 A, and phase angle between them = (60 + 25) = 85°.

$$P = (0.5)(9.43)(3.16) \cos 85° = 1.30 \text{ W}$$

Third Harmonic Component: ($\omega = 600$ r/s). Amplitude of voltage = 10 V, amplitude of current = 10 A, and phase angle between them = 0°.

$$P = (0.5)(10)(10) = 50 \text{ W}$$

The total average power received by the circuit is

$$P_{av} = (200 + 39.0 + 1.30 + 50) = 290.3 \text{ W}$$ ∎

Exercise 14-13 The voltage and current associated with a network are given by

$$v(t) = 100 + 100 \cos 500t + 30 \cos 1500t \text{ V}$$

and

$$i(t) = 50 + 50 \cos(500t - 45°) + 30 \cos(1500t - 90°) \text{ A}$$

Determine the average power received by the circuit.

Exercise 14-14 Calculate the average power received by the circuit in Example 14-3.

Exercise 14-15 In the circuit of Exercise 14-12, calculate the average power due to each harmonic component as a percentage of the total average power to see the dominant effect of the fundamental frequency component.

Case of a Single Resistance as Load

In communication systems, it is customary to assume the load to be a single resistance (and 1 ohm is often used as the standard load). If a periodic voltage $v(t)$ is applied to a resistor R, then the current $i(t) = [v(t)/R]$. In Eq. (14-30), the phase angles are all zero, $I_{dc} = (V_{dc}/R)$, and $I_k = (V_k/R)$ for all k. The average power fed to a single resistance R by a periodic signal is therefore

$$P_{av} = (1/R)[V_{dc}^2 + \sum_{k=1}^{\infty} (V_k^2/2)] \qquad (14\text{-}31)$$

For the standard load of 1 ohm,

$$P_{av} = V_{dc}^2 + \sum_{k=1}^{\infty} \frac{V_k^2}{2} \qquad (14\text{-}32)$$

In communication systems, the quantity in Eq. (14-32) is often referred to simply as the *power contained in a periodic signal*.

Exercise 14-16 Determine the average power fed to a 1 ohm load by the voltage given in Example 14-4.

14-2-2 RMS Value of a Periodic Signal

The RMS (root mean square) value of a voltage was defined in Chapter 7, and it was found that the average power received by a resistance R is given by

$$P_{av} = \frac{V_{rms}^2}{R} \qquad (14\text{-}33)$$

Comparing Eqs. (14-31) and (14-33), we see that the *rms value of a periodic voltage* $v(t)$ is given by

$$V_{rms} = \sqrt{V_{dc}^2 + \sum_{k=1}^{\infty} \frac{V_k^2}{2}} \qquad (14\text{-}34)$$

Exercise 14-17 Calculate the rms value of the half-wave rectified sinusoid by using its Fourier series obtained in Example 14-1 and a handbook of mathematical tables.

Exercise 14-18 Calculate the rms value of a full-wave rectified sinusoid by using its Fourier series obtained in Exercise 14-9.

14-3 EXPONENTIAL FOURIER SERIES

The form of Fourier series using sine and cosine terms (the form used in the preceding sections) is called the *trigonometric* form of Fourier series. An alternative form uses complex exponentials instead of trigonometric functions, and such a series is referred to as the *exponential* form of Fourier series. The exponential form has certain theoretical advantages, as we will see shortly.

The exponential form of Fourier series is written as

$$f(t) = \sum_{k=-\infty}^{\infty} c_k e^{jk\omega t} \qquad (14\text{-}35)$$

where the coefficients c_k are, in general, complex. The exponential and trigonometric forms of the series are related through Euler's formulas

$$e^{+jk\omega t} = \cos(k\omega t) + j\sin(k\omega t) \qquad (14\text{-}36\text{a})$$

$$\cos(k\omega t) = \frac{1}{2}[e^{jk\omega t} + e^{-jk\omega t}] \qquad (14\text{-}36\text{b})$$

$$\sin(k\omega t) = \frac{1}{2j}[e^{jk\omega t} - e^{-jk\omega t}] \qquad (14\text{-}36\text{c})$$

In the exponential form of the series, the kth harmonic component of the signal is represented by two terms: one involving $e^{jk\omega t}$ and the other involving $e^{-jk\omega t}$.

14-3-1 Determination of the Coefficients c_k

Consider the integral

$$I = \int_{-T/2}^{T/2} f(t) e^{-jm\omega t} dt \qquad (14\text{-}37)$$

where m is any integer (positive, zero, or negative). Using the series for $f(t)$ from Eq. (14-35) in Eq. (14-37)

$$I = \int_{-T/2}^{T/2} [\sum c_k e^{jk\omega t}] e^{-jm\omega t} dt$$

$$= \sum \int_{-T/2}^{T/2} c_k e^{j(k-m)\omega t} dt \qquad (14\text{-}38)$$

where uniform convergence of the series has been assumed.

When $k = m$, the term of interest in Eq. (14-38) is

$$\int_{-T/2}^{T/2} c_m dt = c_m T \tag{14-39}$$

When $k \neq m$, the terms of interest in Eq. (14-38) are of the form:

$$\int_{-T/2}^{T/2} c_k e^{j(k-m)\omega t} \, dt = \frac{1}{j(k-m)\omega}[e^{j(k-m)\omega t}]\Big|_{-T/2}^{T/2}$$

$$= \frac{1}{j(k-m)\omega}[e^{j(k-m)\pi} - e^{-j(k-m)\pi}]$$

since $(\omega T) = 2\pi$. The value of the last expression is zero. Therefore, the integral I in Eq. (14-38) becomes

$$I = c_m T$$

or

$$c_m = \frac{1}{T} \int_{-T/2}^{T/2} f(t) e^{-jm\omega t} dt \tag{14-40}$$

which leads to the determination of the coefficient c_m for *any* value of m (positive, zero, or negative).

Note that there is only *a single formula* for the coefficients of the exponential Fourier series, whereas there were three different formulas for the coefficients of the trigonometric series.

Example 14-5 Obtain the exponential form of Fourier series for the half-wave rectified sinusoid (Fig. 14-1).

Solution The function $f(t)$ is described by

$$f(t) = V_0 \cos(\omega t) \qquad -T/4 < t < T/4$$

It is more convenient to use $f(t)$ in the exponential form

$$f(t) = \frac{V_0}{2}[e^{j\omega t} + e^{-j\omega t}] \qquad -T/4 < t < T/4$$

From Eq. (14-40)

$$c_m = (V_0/2T)\left[\int_{-T/4}^{T/4} e^{j(1-m)\omega t} \, dt + \int_{-T/4}^{T/4} e^{-j(1+m)\omega t} \, dt\right]$$

$$= \frac{V_0}{2T}\left[\frac{e^{j(1-m)\omega t}}{j(1-m)} + \frac{e^{-j(1+m)\omega t}}{j(1+m)}\right]_{-T/4}^{T/4}$$

which reduces (after some manipulations) to

$$c_m = \frac{V_0}{4\pi} \left[\frac{2}{1-m} \sin\frac{(1-m)\pi}{2} + \frac{2}{1+m} \sin\frac{(1+m)\pi}{2} \right]$$

Using the identity $\sin[(1 \pm m)\pi/2] = \cos(m\pi/2)$, the expression becomes

$$c_m = \frac{V_0}{(1-m^2)\pi} \cos\left(\frac{m\pi}{2}\right) \qquad m \neq \pm 1$$

Note that when m is odd, but $\neq \pm 1$, $c_m = 0$. If $m = \pm 1$ is used in the above expression, it becomes $(0/0)$ which is indeterminate. Therefore, Eq. (14-40) is used directly with $m = 1$ to find c_1:

$$c_1 = \frac{V_0}{2T} \left[\int_{-T/4}^{T/4} dt + \int_{-T/4}^{T/4} e^{-j2\omega t} dt \right]$$

which yields

$$c_1 = \frac{V_0}{2T} \left[\frac{T}{2} - \frac{1}{j2\omega}(e^{-j\pi} - e^{j\pi}) \right] = \frac{V_0}{4}$$

Similarly it can be shown that when $m = -1$,

$$c_{-1} = \frac{V_0}{4}$$

Therefore, the exponential Fourier series of $f(t)$ is given by

$$f(t) = \frac{V_0}{4}(e^{j\omega t} + e^{-j\omega t}) + \sum_{\substack{k=-\infty \\ k \neq 1}}^{\infty} \left[\frac{V_0}{(1-k^2)\pi}\right] \cos\frac{k\pi}{2} e^{jk\omega t}$$

The series contains no odd harmonics except $k = 1$. ∎

Exercise 14-19 Obtain the exponential form of Fourier series of the waveform of Exercise 14-7, as shown in Fig. 14-2(a).

Exercise 14-20 Consider the half-wave rectified sinusoid of Fig. 14-2(b). Obtain the coefficients of the exponential series for this waveform by introducing a suitable shift in time t in the series obtained in Example 14-5.

14-3-2 Relationships between Exponential and Trigonometric Forms of Fourier Series

We should expect that the coefficients of the exponential series and those of the trigonometric series must be related. To show this, equate the expression of the mth harmonic of a periodic signal given in the trigonometric form

$$a_m \cos(m\omega t) + b_m \sin(m\omega t) \tag{14-41}$$

with its exponential form:

$$c_m e^{jm\omega t} + c_{-m} e^{-jm\omega t} \qquad (14\text{-}42)$$

Using Euler's identities, Eq. (14-36), we obtain the following relationships:

$$a_m = (c_m + c_{-m}) \qquad (14\text{-}43a)$$
$$b_m = j(c_m - c_{-m}) \qquad (14\text{-}43b)$$

and

$$c_m = (a_m - jb_m)/2 \qquad (14\text{-}44a)$$
$$c_{-m} = (a_m + jb_m)/2 \qquad (14\text{-}44b)$$

The coefficients a_m and b_m in the trigonometric series are real for a real function $f(t)$. Therefore, $(c_m + c_{-m})$ in Eq. (14-43a) must equal a real quantity. Therefore, c_m and c_{-m} must be *complex conjugates*. That is

$$c_m = c_{-m}{}^* \qquad (14\text{-}45a)$$

This means that

$$|c_m| = |c_{-m}| \qquad (14\text{-}45b)$$
$$\text{angle of } c_m = - \text{ angle of } c_{-m} \qquad (14\text{-}45c)$$

Equations (14-43) have the alternative forms:

$$a_m = 2\mathbf{Re}[c_m] \text{ and } b_m = -2\mathbf{Im}[c_m]$$

Exercise 14-21 Verify that Eqs. (14-44) and (14-45) are satisfied by the coefficients obtained in Examples 14-1 and 14-5; and in Exercises 14-7 and 14-19.

The amplitude and phase angle of the *m*th harmonic, using the coefficients c_m and c_{-m}, are obtained by using Eq. (14-44):

$$\text{amplitude of the } m\text{th harmonic} = \sqrt{a_m^2 + b_m^2} = 2|c_m| \qquad (14\text{-}46a)$$
$$\text{phase angle of the } m\text{th harmonic} = \arctan(-b_m/a_m) = \text{angle of } (c_m) \qquad (14\text{-}46b)$$

Exercises 14-22 The Fourier series of a periodic signal is given by

$$f(t) = \sum_{n=-\infty}^{\infty} \frac{0.667}{1 + jn\pi} e^{jn\omega t}$$

(a) Determine the amplitude and phase angle of the *n*th harmonic. Calculate their values for the specific cases: $n = 1, 2,$ and 3. (b) Determine the coefficients a_n and b_n of the trigonometric form of the series.

14-4 STEADY STATE RESPONSE OF A CIRCUIT USING EXPONENTIAL FOURIER SERIES

The procedure for finding the steady state response of a circuit with a periodic input using the trigonometric series was discussed in Section 14-2. A similar procedure is used when the exponential form of the series is available. In fact, the use of the exponential series offers a certain economy in analysis, as will be seen in the following discussion.

Let the input voltage to a linear circuit be

$$v_i(t) = \sum_{k=-\infty}^{\infty} c_k e^{jk\omega t} \qquad (14\text{-}47)$$

and let the desired response be the output voltage $v_0(t)$.

Denote the *transfer function* between the output and input voltages by $H(j\omega)$:

$$H(j\omega) = V_0(j\omega)/V_i(j\omega) \qquad (14\text{-}48)$$

Then, if the input voltage is a single term of the exponential series:

$$v_{ik}(t) = c_k e^{jk\omega t}$$

the corresponding output voltage will be

$$v_{ok}(t) = H(jk\omega)c_k e^{jk\omega t} \qquad (14\text{-}49)$$

Note that v_{ok} in the last equation is a complex exponential function, and not a real function of time. When the input is given by the series in Eq. (14-47), the output will be the sum of terms of the form in Eq. (14-49). That is, the total output of the circuit will be

$$v_0(t) = \sum_{k=-\infty}^{\infty} v_{ok}(t) = \sum_{k=-\infty}^{k=\infty} H(jk\omega)c_k e^{jk\omega t} \qquad (14\text{-}50)$$

Even though each of the terms in the summation is not a real function of time, the total output $v_0(t)$ is a real function, as will be proved shortly. The advantage of using the exponential series is that the output function is obtained directly by using the transfer function and Eq. (14-49), and it is not necessary first to convert each component to a real time function, as was done in connection with the trigonometric series.

Example 14-6 A periodic pulse train is applied to a series RC circuit, as shown in Fig. 14-7. The Fourier series of the pulse train is given by

$$v_s(t) = \sum_{k=-\infty}^{\infty} V_{sk} e^{jk\omega t}$$

where

$$V_{sk} = \frac{Ad}{T} \frac{\sin(k\omega d/2)}{(k\omega d/2)}$$

Determine the output voltage $v_0(t)$.

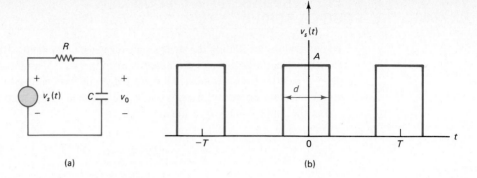

Fig. 14-7 (a) Circuit for Example 14-6. (b) Input to the circuit.

Solution The transfer function $H(j\omega)$ is

$$H(j\omega) = \frac{V_c(j\omega)}{V_s(j\omega)} = \frac{Z_c}{(R + Z_c)} = \frac{1}{1 + j\omega RC}$$

The transfer function at the different harmonics is obtained by using $k\omega$ in the above expression:

$$H(jk\omega) = \frac{1}{1 + jk\omega RC}$$

The output $v_0(t)$ is given by

$$v_0(t) = \sum_{k=-\infty}^{\infty} H(jk\omega) V_{sk} e^{jk\omega t}$$

$$v_c(t) = \sum_{k=-\infty}^{\infty} \frac{1}{(1 + jk\omega RC)} \frac{Ad}{T} \frac{\sin(k\omega d/2)}{(k\omega d/2)} e^{jk\omega t}$$

which is the desired result. ∎

The amplitude and phase angle of any particular harmonic in the output voltage are determined by using Eqs. (14-46).

Exercise 14-23 A voltage $v(t)$ whose Fourier series is the same as that given in Exercise 14-22 is applied to a series RL circuit. Obtain the series for the voltage across the inductance L.

Exercise 14-24 A resistance $R = 2\Omega$ in parallel with a capacitance $C = 0.0159$ F is fed by a current source i_s whose Fourier series is given by

$$i_s(t) = \sum_{k=-\infty}^{\infty} \frac{2}{1 + jk} e^{j20\pi k t} \text{ A}$$

(a) Obtain the series for the current $i_c(t)$ in the capacitor. (b) Evaluate the amplitude and phase of each of the first three harmonics of i_c.

14-5 FREQUENCY SPECTRA OF PERIODIC SIGNALS

The distribution of frequencies in a signal is often the most important piece of information needed for the design of filters, amplifiers, and communication networks. Even though it is possible to observe the waveform of a signal in the laboratory by means of an oscilloscope, it is difficult to make sense out of the time domain displays of most practical signals. More useful are diagrams (or displays) showing the amplitudes, phase angles, and power in the different frequency components of a signal. Such diagrams are known as *frequency spectra*, or more specifically as *amplitude spectra, phase spectra*, and *power spectra*. Since a periodic signal contains components at discrete frequencies (dc, fundamental frequency, and the various harmonics), its frequency spectra are made up of vertical lines at those specific frequencies; that is, they are *line* spectra.

14-5-1 Amplitude and Phase Spectra

The amplitude spectrum of a periodic signal shows the values of the magnitudes of the coefficients of the exponential series, $|c_k|$, and the phase spectrum shows the angles of the coefficients of the exponential series, θ_k, for different values of $k\omega$ where ω is the fundamental frequency.

For example, if the coefficients of the Fourier series of a signal are given by

$$c_k = \frac{1}{1 + jk\pi}$$

then its amplitude spectrum shows the values of

$$|c_k| = \frac{1}{\sqrt{(1 + k^2\pi^2)}}$$

and its phase spectrum shows the values of

$$\theta_k = -\arctan k\pi$$

The two spectra are shown in Fig. 14-8.

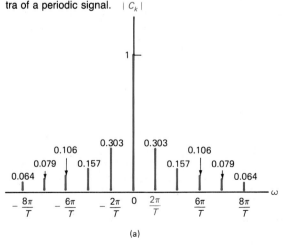
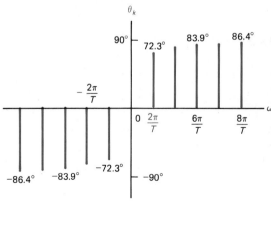

Fig. 14-8 (a)-(b) Amplitude and phase spectra of a periodic signal.

Properties of Amplitude and Phase Spectra

1. Spectral lines extend over positive as well as negative frequencies, for reasons discussed in Chapter 9 in connection with the frequency response of transfer functions.
2. The amplitude spectrum has an *even symmetry;* that is, the lines at $k\omega$ and $-k\omega$ have the same height. This follows from the fact that since c_k and c_{-k} are complex conjugates, their magnitudes must be the same.
3. The phase spectrum has an *odd symmetry;* that is, the line at $k\omega$ has the same length but directed opposite to the line at $-k\omega$. This follows from the fact that since c_k and c_{-k} are complex conjugates, their angles must be equal in value but opposite in sign.
4. The amplitude and phase spectra are useful for calculating the amplitude and phase angle of each of the harmonies. From Eq. (14-46)

$$\text{amplitude of the } k\text{th harmonic} = 2|c_k|$$

Therefore, the amplitude of any harmonic is the *sum* of the heights of the two lines corresponding to that harmonic in the amplitude spectrum.

$$\text{phase of the } k\text{th harmonic} = \text{angle of } c_k$$

Therefore, the phase angle of the *k*th harmonic is given by the line at $k\omega$ in the phase spectrum.

Exercise 14-25 Draw the amplitude and phase spectra of the half-wave rectified sinusoid. Use the spectra to determine the amplitudes and angles of the fundamental component and the fourth harmonic.

Exercise 14-26 Draw the amplitude spectra of the input and output voltages of Example 14-6, when $A = 25$ V, $d = (T/5)$, $\omega = 100$ r/s, and $RC = 0.005$ s.

Exercise 14-27 Repeat the work of the previous exercise on the circuit and input given in Exercise 14-23. Use $\omega = 100$ r/s, $R = 20\Omega$, and $L = 0.1$ H.

14-5-2 Power Spectrum

The average power received by a 1 ohm load when a periodic signal $v(t)$ is applied to it forms the basis of the power spectrum. It was shown in Section 14-2 that the average power in a 1 ohm load is

$$P_{av} = V_{dc}^2 + \sum_{k=1}^{\infty} \frac{V_k^2}{2} \qquad (14\text{-}51)$$

where V_{dc} is the dc component and V_k is the amplitude of the kth harmonic component of the voltage $v(t)$. Using the coefficients of the exponential series

$$\text{amplitude of the } k\text{th harmonic: } V_k = 2|c_k|$$

and

$$V_{dc} = c_0$$

so that Eq. (14-51) becomes

$$P_{av} = c_0^2 + \sum_{k=1}^{\infty} \frac{1}{2}(4|c_k|^2)$$

$$= c_0^2 + \sum_{k=1}^{\infty} 2|c_k|^2$$

or using the property that $|c_k| = |c_{-k}|$, the expression for the average power becomes

$$P_{av} = c_0^2 + \sum_{k=1}^{\infty} \{[|c_k|^2 + |c_{-k}|^2]\}$$

$$= \sum_{k=-\infty}^{\infty} |c_k|^2 \quad (14\text{-}52)$$

The power spectrum consists of lines of height $|c_k|^2$ at all possible integer values of k (positive, zero, and negative).

Figure 14-9 shows the power spectrum of the signal whose amplitude spectrum was shown in Fig. 14-8(a).

From Eq. (14-52), we see that

power in the kth harmonic = sum of the heights of the two lines at $+k\omega$ (14-53a)
and $-k\omega$ in the power spectrum

and

total average power in a signal = sum of the heights of all the lines (14-53b)
in the power spectrum

Fig. 14-9 Power spectrum of a periodic signal.

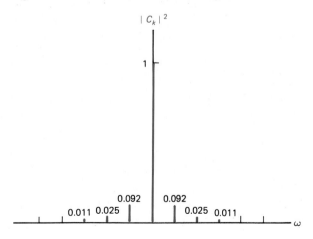

14-5   Frequency Spectra of Periodic Signals

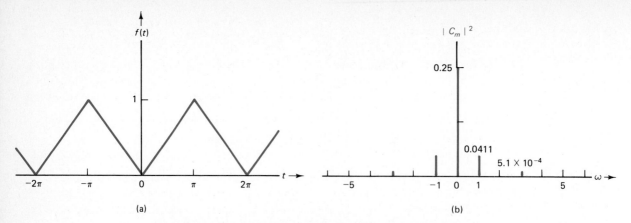

Fig. 14-10 (a) A triangular waveform and (b) its power spectrum.

Example 14-7 The coefficients of the exponential Fourier series of the triangular waveform of Fig. 14-10(a) are given by

$$c_k = \begin{cases} 0.5 & k = 0 \\ -(2/k^2\pi^2) & \text{when } k \text{ is odd} \\ 0 & 0 \text{ when } k \text{ is even} \end{cases}$$

Sketch the power spectrum of the signal. Calculate the average power contained in the dc component and each of the first three odd harmonics.

Solution The power spectrum is shown in Fig. 14-10(b), where the heights of the lines equal $|c_k|^2$: 0.25 at dc, and $(4/k^4\pi^4)$ for all odd values of k. Since the signal does not contain any even harmonics, there are no lines in the power spectrum for those harmonics.

$$\text{power in the dc component} = c_0^2 = 0.25 \text{ W}$$
$$\text{power in the fundamental component} = 2(0.0411) = 0.0822 \text{ W}$$
$$\text{power in the third harmonic} = 2(5.07 \times 10^{-3}) = 1.01 \times 10^{-3} \text{ W}$$
$$\text{power in the fifth harmonic} = 2(6.57 \times 10^{-5}) = 1.31 \times 10^{-4} \text{ W}$$

Exercise 14-28 Draw the power spectrum of the half-wave rectified sinusoid.

Exercise 14-29 Draw the power spectrum of the output of the circuit in Exercise 14-26.

Exercise 14-30 Draw the power spectrum of the output of the circuit in Exercise 14-27.

One of the important applications of the power spectrum is to determine the bandwidth of a system needed to transmit a specified percentage of the total power in a signal. Theoretically, the power spectrum extends to all discrete frequencies from $-\infty$ to $+\infty$, and the system must respond equally to all harmonics. This is an impractical expectation, since the bandwidth of such a system must be infinite. From a practical viewpoint, however, most of the power in a signal is usually concentrated in the first few harmonics. The power spectrum of the triangular wave (in Example 14-7) shows that almost all of its power is concentrated in the first three harmonics, and a bandwidth of three times the fundamental frequency is acceptable for that signal. Therefore, the bandwidth of a communication system is usually determined on the basis of being able to transmit a specified large percentage (90 to 95 percent) of the total power contained in the signal. Then the lines of the power spectrum are added (starting at dc and proceeding in both directions) until the desired percentage is reached, and the number of harmonics included in the computation determines the bandwidth.

14-6 MINIMUM MEAN SQUARE ERROR PROPERTY OF FOURIER SERIES

Theoretically, a periodic signal can be reconstructed by combining the dc component and an infinite number of harmonic components (with the proper amplitudes and phase angles). From a practical viewpoint, it is necessary to limit the number of harmonics in analysis to a finite number N. A signal obtained by using a finite number of harmonics is called the *truncated signal* and denoted by $\tilde{f}(t)$ and the error introduced by truncation is $[f(t) - \tilde{f}(t)]$. It is found that the *mean square error* between $f(t)$ and $\tilde{f}(t)$, defined by

$$\frac{1}{T} \int_{-T/2}^{T/2} [f(t) - \tilde{f}(t)]^2 \, dt$$

is a minimum when the coefficients of the truncated series $\tilde{f}(t)$ are those given by the Fourier series. Therefore, the use of Fourier coefficients ensures that the *mean square error due to truncation is a minimum,* compared with the choice of other series expansions possible for a periodic signal.

The minimum mean square error is not always the most desirable criterion, however. As an example, consider the reconstruction of a rectangular pulse [Fig. 14-11(a)] from its Fourier series. When only a small number of harmonics is used, the reconstructed pulses exhibit oscillations, as indicated in Fig. 14-11(b). As more harmonics are included, the ripples do not disappear; they get together and attenuate more rapidly. Even if all the infinite harmonics are used, there is a vertical line representing a 9 percent overshoot at each discontinuity. This peculiar behavior at a discontinuity of a signal is

Fig. 14-11 (a) A rectangular pulse train. (b) Pulse train reconstructed by using a finite number of its harmonics.

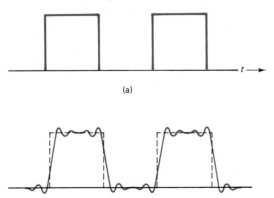

known as the *Gibbs phenomenon*. There are other types of series expansions for a periodic signal which do not exhibit the Gibbs phenomenon; the interested reader is referred to the literature.

14-7 NONPERIODIC SIGNALS: THE FOURIER TRANSFORM

When a signal is not periodic, the Fourier series formulas will not lead to a description of the signal valid for all values of time. An extension of the approach used with Fourier series is possible for signals of finite duration (such as, for example, single pulses or groups of pulses).

Consider a single rectangular pulse of width t_p. It is possible to think of the pulse as being part of a periodic pulse train, except that the period $T = \infty$; that is, the next pulse never appears. The Fourier analysis of the pulse is performed by letting the period T approach infinity in the Fourier series of a periodic pulse train.

For a pulse train of period T and pulse width t_p, as in Fig. 14-12(a), the coefficients

Fig. 14-12 (a) A pulse train and (b) its amplitude spectrum. As T increases, the *envelope* of a plot of $(c_k T)$ is unaffected but the lines get closer together, as shown in (c) and (d).

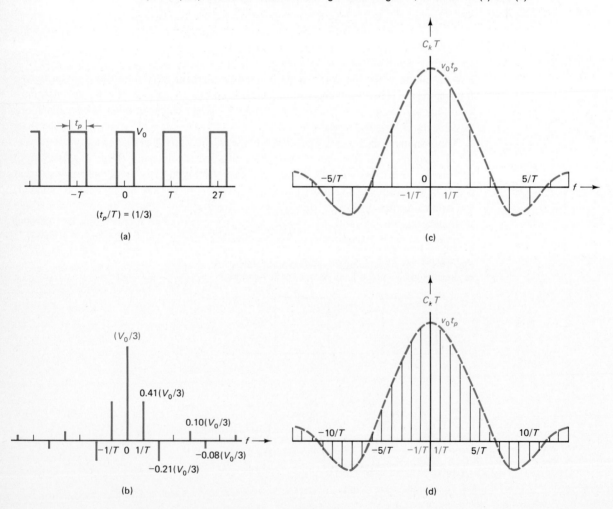

Nonsinusoidal Signals—Fourier Methods

of the exponential Fourier series are given by

$$c_k = \frac{V_0 t_p}{T} \frac{\sin(k\omega t_p/T)}{(k\omega t_p/T)} \tag{14-54}$$

The amplitude spectrum of the pulse train is shown in Fig. 14-12(b). The spacing between the lines equals the fundamental frequency $(1/T)$. Suppose t_p is kept constant while the period T is increased. As T increases, the spectral lines move closer together. At the same time, their heights decrease also, since c_k is proportional to $(1/T)$. If T approaches infinity (so as to simulate a single pulse), the spacing between the lines approaches zero so that the spectrum becomes a continuous curve. But c_k approaches zero so that the spectrum reduces to nothing! To avoid this difficulty, consider a modified spectrum which shows the quantity $(c_k T)$ rather than c_k. As T approaches infinity, $c_k T$ can have a finite value even though c_k itself becomes zero. From Eq. (14-54)

$$c_k T = (V_0 t_p) \left[\frac{\sin(k\omega t_p/T)}{(k\omega t_p/T)} \right] \tag{14-55}$$

When $c_k T$ is plotted as a function of frequency, a line spectrum is again obtained with line spacing $= (1/T)$. The height of the line at dc is $(V_0 t_p)$, independent of T. The envelope of the spectral lines is of the form $[(\sin X)/X]$ where $X = (k\omega t_p/T)$. As T increases, the lines get closer together while the shape of the envelope of the spectrum is unaffected. Figures 14-12(c) and (d) show the spectrum for two different values of T. As $T \to \infty$, the spacing between the lines becomes zero and the spectrum becomes a continuous curve given by the envelope. Therefore the amplitude spectrum of a single rectangular pulse is a continuous curve, with the height of the curve varying in the form $[(\sin X)/X]$.

To put this discussion on a formal basis, let the fundamental angular frequency of a periodic signal be denoted by

$$\Delta\omega_0 = \frac{2\pi}{T} \tag{14-56}$$

The formula for the coefficient c_k is

$$c_k = \frac{1}{T} \int_{-T/2}^{T/2} f(t) e^{-jk\Delta\omega_0 t} \, dt \tag{14-57}$$

where $(k\Delta\omega_0)$ has been used instead of ω to emphasize that only discrete frequencies (multiples of $\Delta\omega_0$) are relevant for periodic signals, while the symbol ω denotes a continuous variable. From Eq. (14-57)

$$(c_k T) = \int_{-T/2}^{T/2} f(t) e^{-jk\Delta\omega_0 t} \, dt$$

Using $T = (2\pi/\Delta\omega_0)$, the last equation becomes

$$\frac{2\pi c_k}{\Delta\omega_0} = \int_{-T/2}^{T/2} f(t) e^{-jk\Delta\omega_0 t} \, dt \tag{14-58}$$

As $T \to \infty$, c_k and $\Delta\omega_0$ both approach zero, but their ratio is finite. The ratio $(c_k/\Delta\omega_0)$ is a function of the angular frequency and will be denoted by $F(j\omega)$. As $T \to \infty$,

$\Delta\omega_0 = 0$ and the product $(k\Delta\omega_0)$ becomes the continuous variable ω. Thus Eq. (14-58) becomes

$$F(j\omega) = \int_{-\infty}^{\infty} f(t)e^{-j\omega t}\, dt \qquad (14\text{-}59)$$

Equation (14-59) is known as the *(forward) Fourier transform of the function $f(t)$*. The quantity $F(j\omega)$ is analogous to the coefficient c_k of a periodic signal, but the units of c_k are volts, while the units of $F(j\omega)$ are volts/(radians/second) when $f(t)$ is a voltage. Therefore a plot of $|F(j\omega)|$ as a function of frequency is called the *amplitude density spectrum*. Note that the limits of integration are $-\infty$ and $+\infty$ in $F(j\omega)$ instead of $-T/2$ and $T/2$. Because of the infinite limits on the integral, the function $f(t)$ must satisfy the condition

$$\int_{-\infty}^{\infty} |f(t)|\, dt < \infty \qquad (14\text{-}60)$$

in order for $F(j\omega)$ to exist.

The following statement summarizes the discussion:

Given a signal $f(t)$ with the property that $\int_{-\infty}^{\infty} |f(t)|\, dt$ is finite, the signal has a Fourier transform $F(j\omega)$ defined by

$$F(j\omega) = \int_{-\infty}^{\infty} f(t)e^{-j\omega t}\, dt \qquad (14\text{-}61)$$

$F(j\omega)$ represents the amplitude density of the signal $f(t)$; that is, a plot of $|F(j\omega)|$ shows the distribution of amplitude per unit of angular frequency as a function of ω in the signal.

Example 14-8 Obtain the Fourier transform of the rectangular pulse shown in Fig. 14-13(a). This is a formal verification of the earlier qualitative discussion of the rectangular pulse.

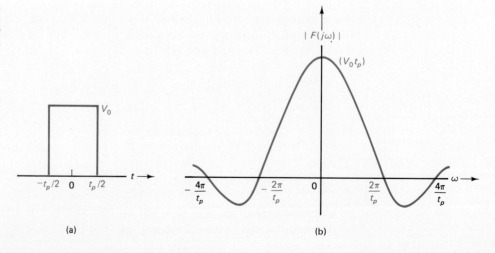

Fig. 14-13 (a) A rectangular pulse and (b) its amplitude density spectrum; Example 14-8.

Solution The signal is described by

$$f(t) = V_0 \quad (-t_p/2 < t < t_p/2)$$

From Eq. (14-61)

$$F(j\omega) = \int_{-t_p/2}^{t_p/2} V_0 e^{-j\omega t}\, dt = -\frac{V_0}{j\omega} e^{-j\omega t}\bigg|_{-t_p/2}^{t_p/2}$$

$$= (2V_0/\omega)\sin(\omega t_p/2)$$

It is customary to introduce the function [(sin X)/X], since it is a well-known (and ubiquitous) function in signal analysis. Therefore, the last expression is written in the form

$$F(j\omega) = (V_0 t_p)\left[\frac{\sin(\omega t_p/2)}{(\omega t_p/2)}\right]$$

The amplitude density spectrum of the rectangular pulse is shown in Fig. 14-12(b).

Exercise 41-31 Given

$$v(t) = \begin{cases} V_0 \sin(100\pi t) & (0 < t < 0.01 \text{ s}) \\ 0 & \text{for all other } t \end{cases}$$

Determine the Fourier transform of $v(t)$. (Using complex exponentials to represent the sine will be helpful.)

14-7-1 Energy Density Spectrum of a Signal

Before proceeding with a detailed discussion of the Fourier transform and introducing the inverse Fourier transform, which transforms $F(j\omega)$ to $f(t)$, it is worth studying the significance of the Fourier transform in signal analysis. That is, what kind of insight can be gained by examining the Fourier transform of a given signal? To address this question, it is more convenient to examine the *energy density spectrum* of a signal, rather than its amplitude density spectrum.

The energy supplied to a 1 ohm resistor by a voltage $v(t)$ is referred to as the *normalized energy* of the signal $v(t)$ and is given by

$$w = \int_{-\infty}^{\infty} [v(t)]^2\, dt \tag{14-62}$$

It is possible to show, by using the convolution theorem presented at the end of this chapter, that

$$\int_{-\infty}^{\infty} [v(t)]^2\, dt = \frac{1}{2\pi}\int_{-\infty}^{\infty} |V(j\omega)|^2\, d\omega \tag{14-63}$$

where $V(j\omega)$ is the Fourier transform of $v(t)$. Therefore, Eq. (14-62) assumes the form

$$w = \int_{-\infty}^{\infty} \frac{1}{2\pi} |V(j\omega)|^2 \, d\omega \qquad (14\text{-}64)$$

The equation states that *if* $|V(j\omega)|^2/2\pi$ *is plotted as a function of* ω, *the total area under the curve gives the normalized energy of the signal* $v(t)$.

For example, the plot of $|V(j\omega)|^2/2\pi$ of a rectangular pulse (from Example 14-8) is shown in Fig. 14-14. The total area under the curve gives the total normalized energy contained in the rectangular pulse.

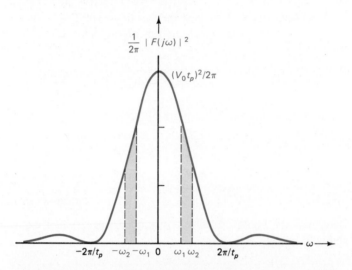

Fig. 14-14 Energy density spectrum of a rectangular pulse. The total shaded area gives the energy content of the pulse in the frequency band ω_1 to ω_2.

Since the area under it represents energy, the function $|V(j\omega)|^2/2\pi$ is called the *energy density function* and a plot of the energy density as a function of frequency is called the *energy density spectrum*.

The area under the energy density spectrum in the two shaded portions $(-\omega_2, -\omega_1)$ and (ω_1, ω_2) of Fig. 14-14 taken together represents the energy contained in the signal in the band of frequencies $\omega_1 < \omega < \omega_2$.

Thus *the square of the magnitude of the Fourier transform of a signal* provides *information about the energy content of the signal* over *different bands of frequencies*. It gives an indication of the distribution of energy in a signal as a function of frequency.

The energy density spectrum is a useful aid in determining the bandwidth needed to transmit a signal (analogous to the use of the power spectrum for a periodic signal). For example, in the transmission of a rectangular pulse, a common rule of thumb is to use the band of frequencies between the first zero crossings, shown shaded in Fig. 14-15(a). Consider two pulses with different widths t_p and t'_p ($t_p < t'_p$), whose energy density spectra are shown in Figs. 14-14(b) and (c). Using the rule of thumb mentioned earlier, we see that

(bandwidth needed for transmitting the narrower pulse)

> (bandwidth needed for transmitting the wider pulse)

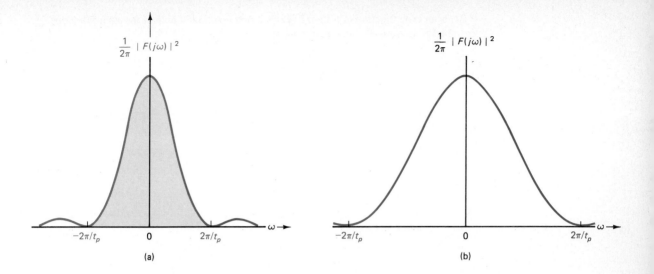
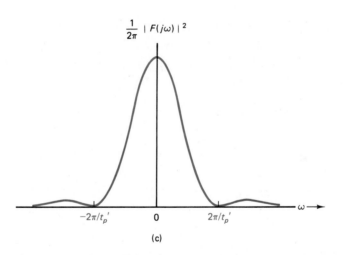

Fig. 14-15 Energy density spectrum of a rectangular pulse. The bandwidth ($2\pi/t_p$), shown by the shaded area in (a), contains a large portion of the total energy in the pulse. (b)-(c) The bandwidth of the pulse is inversely proportional to the pulse duration.

As the pulses to be transmitted in a communication system becomes narrower, the bandwidth of the system must become wider.

This discussion gives a general idea of the importance of Fourier transform as a tool in signal analysis.

14-7-2 Properties of the Fourier Transform

Effects of Even and Odd Symmetry

The Fourier transform of a signal is, in general, a complex quantity, except in the case of signals with even or odd symmetry.

If $f(t)$ is an even function of t, then $F(j\omega)$ is a real function of ω, and for such functions, Eq. (14-61) assumes the alternative form

$$F(j\omega) = 2 \int_0^\infty f(t) \cos \omega t \, dt \qquad (14\text{-}65)$$

If $f(t)$ is an odd function of t, then $F(j\omega)$ is a pure imaginary function of ω and for such functions, Eq. (14-61) assumes the alternative form

$$F(j\omega) = -j2 \int_0^\infty f(t) \sin \omega t \, dt \qquad (14\text{-}66)$$

The proof of the last two statements is based on the properties of even and odd functions and Euler's theorem. Using

$$e^{-j\omega t} = \cos \omega t - j \sin \omega t$$

Eq. (14-61) becomes

$$F(j\omega) = \int_{-\infty}^\infty f(t) \cos \omega t \, dt - j \int_{-\infty}^\infty f(t) \sin \omega t \, dt \qquad (14\text{-}67)$$

If $f(t)$ is even, then the product $f(t) \sin (\omega t)$ is an odd function and the integral of any odd function between $-\infty$ and $+\infty$ equals zero. The product $f(t) \cos (\omega t)$ is an even function and the integral from $-\infty$ to $+\infty$ of an even function equals twice the integral from 0 to ∞. Therefore

$$F(j\omega) = 2 \int_0^\infty f(t) \cos \omega t \, dt \qquad \text{if } f(t) \text{ is even}$$

which establishes Eq. (14-65). A similar argument will show the validity of Eq. (14-66).

When $f(t)$ is neither even nor odd, $F(j\omega)$ is complex, with both a real and an imaginary part.

Example 14-9 Obtain the Fourier transform of the signal shown in Fig. 14-16.

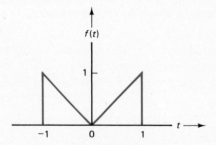

Fig. 14-16 Pulse for Example 14-9.

Solution The function is described by

$$f(t) = |t| \qquad -1 < t < 1$$

Therefore $f(-t) = f(t)$, and it is an even function. Using Eq. (14-65),

$$F(j\omega) = 2\int_0^1 t \cos \omega t \, dt$$

$$= 2\left[\frac{t \sin \omega t}{\omega} + \frac{1}{\omega^2}\cos \omega t\right]_0^1$$

$$= 2\left[\frac{\sin \omega}{\omega} + \frac{1}{\omega^2}(\cos \omega - 1)\right] \qquad \blacksquare$$

Exercise 14-32 In the previous example, show by actual integration that

$$\int_{-1}^{1} f(t) \sin \omega t \, dt = 0$$

Exercise 14-33 Find the Fourier transform of the signal shown in Fig. 14-17.

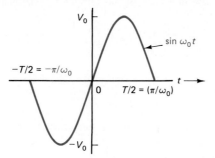

Fig. 14-17 Pulse for Exercise 14-33.

Differentiation and Integration in the Time Domain

The operations of differentiation and integration of a signal in the time domain transform, respectively, to operations of multiplication and division by $j\omega$ in the frequency domain. This is, of course, the same as in the case of complex exponential forcing functions discussed in Chapter 8.

Consider a function $f(t)$ whose Fourier transform is $F(j\omega)$. Then

$$F(j\omega) = \int_{-\infty}^{\infty} f(t) e^{-j\omega t} \, dt \qquad (14\text{-}68)$$

Let $f'(t) = [df(t)/dt]$. The Fourier transform of $f'(t)$ is given by

$$\mathscr{F}[f'(t)] = \int_{-\infty}^{\infty} f'(t) e^{-j\omega t} \, dt \qquad (14\text{-}69)$$

where $\mathscr{F}[\ ]$ denotes Fourier transform of the function in brackets.

Equation (14-69) is evaluated by using integration by parts. Let

$$u = e^{-j\omega t} \quad \text{and} \quad dv = f'(t)dt$$

so that

$$du = -(j\omega)e^{-j\omega t} \quad \text{and} \quad v = f(t)$$

Then Eq. (14-69) becomes

$$\mathcal{F}[f'(t)] = f(t)e^{-j\omega t}\Big|_{-\infty}^{\infty} + \int_{-\infty}^{\infty} f(t)(j\omega)e^{-j\omega t}]dt \qquad (14\text{-}70)$$

If the Fourier transform $F(j\omega)$ exists for the given function $f(t)$, then, from Eq. (14-60),

$$\int_{-\infty}^{\infty} |f(t)|\, dt < \infty$$

which can be satisfied if and only if $|f(t)|$ has decayed to a value of zero at $t = -\infty$ and $t = +\infty$. Therefore, the first term in Eq. (14-70) is zero at both limits and

$$\mathcal{F}[f'(t)] = (j\omega)\int_{-\infty}^{\infty} f(t)\, e^{-j\omega t}\, dt$$

which leads to

$$\mathcal{F}[f'(t)] = (j\omega)F(j\omega) \qquad (14\text{-}71)$$

Thus, *the effect of differentiating $f(t)$ in the time domain is equivalent to multiplying its Fourier transform by $(j\omega)$.*

A similar procedure shows that

$$\mathcal{F}\left[\int_{-\infty}^{t} f(x)dx\right] = \frac{1}{j\omega}F(j\omega) \qquad (14\text{-}72)$$

where x is a dummy variable.

That is, *the effect of integrating $f(t)$ in the time domain is equivalent to dividing its Fourier transform by $(j\omega)$.*

Example 14-10 Determine the Fourier transform of the signal shown in Fig. 14-18(a). Use the result to find the Fourier transform of the signal in Fig. 14-18(b).

Solution By inspection, the function $f_a(t)$ is an even function described by

$$f_a(t) = \begin{cases} 1 - \dfrac{t}{p} & 0 < t < p \\ 1 + \dfrac{t}{p} & -p < t < 0 \end{cases}$$

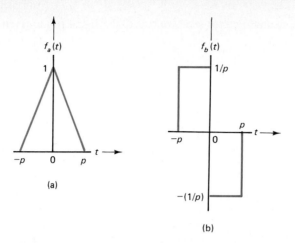

Fig. 14-18 (a) A pulse and (b) its derivative. (Example 14-10).

Using Eq. (14-65)

$$F_a(j\omega) = 2\int_0^p \left[1 - \frac{t}{p}\right]\cos(\omega t)\,dt$$

$$= \frac{2}{\omega}\sin\omega t \Big|_0^p - \frac{2}{p}\left[\frac{t}{\omega}\sin\omega t + \frac{1}{\omega^2}\cos\omega t\right]_0^p$$

$$= \frac{2}{p\omega^2}(1 - \cos p\omega)$$

The function $f_b(t)$ in Fig. 14-18(b) is obtained by differentiating $f_a(t)$. Since

$$f_b(t) = \frac{df_a(t)}{dt}$$

$$F_b(j\omega) = (j\omega)F_a(j\omega) = \frac{j2}{p\omega}(1 - \cos p\omega)$$

■

Exercise 14-34 Verify by direct evaluation the expression obtained for $F_b(j\omega)$ in the previous example.

14-8 THE INVERSE FOURIER TRANSFORM

The transformation of the time domain representation of a signal to its frequency domain representation is performed through the (forward) Fourier transform. The reverse transformation—that is, from the frequency domain representation of a signal to its time domain representation—is accomplished through the *inverse Fourier transform*. The development of the relationship for the inverse Fourier transform starts with a periodic signal and extends its period T to infinity (exactly similar to the development of the forward Fourier transform).

The exponential Fourier series of a periodic signal $f(t)$ is given by

$$f(t) = \sum_{k=-\infty}^{\infty} c_k e^{jk\Delta\omega_0 t} \qquad (14\text{-}73)$$

where c_k is given by the formula

$$c_k = \frac{1}{T} \int_{-T/2}^{T/2} f(t) e^{-jk\Delta\omega_0 t} \, dt \qquad (14\text{-}74)$$

where $\Delta\omega_0 = (2\pi/T)$ as in Eq. (14-56).

The aim is to obtain a formula for $f(t)$ in terms of other quantities. Combining Eqs. (14-73) and (14-74)

$$f(t) = \sum_{k=-\infty}^{\infty} \left[\frac{1}{T} \int_{-T/2}^{T/2} f(u) e^{-jk\Delta\omega_0 u} \, du\right] e^{jk\Delta\omega_0 t} \qquad (14\text{-}75)$$

where u is a dummy variable of integration. Eq. (14-75) leads to

$$f(t) = \sum_{k=-\infty}^{\infty} \frac{\Delta\omega_0}{2\pi} \left[\int_{-T/2}^{T/2} f(u) e^{-jk\Delta\omega_0 u} du\right] e^{jk\Delta\omega_0 t} \qquad (14\text{-}76)$$

Now, let $T \to \infty$. Then $\Delta\omega_0$ becomes the differential $(d\omega)$, $k\Delta\omega_0$ becomes the continuous variable ω, and the summation becomes an integral. Therefore, Eq. (14-76) becomes

$$f(t) = \frac{1}{2\pi} \left[\int_{-\infty}^{\infty} f(u) e^{-j\omega u} du\right] e^{j\omega t} d\omega \qquad (14\text{-}77)$$

The quantity within brackets is recognized as $F(j\omega)$, and Eq. (14-77) becomes

$$f(t) = \frac{1}{2\pi} \int_{-\infty}^{\infty} F(j\omega) e^{j\omega t} \, d\omega \qquad (14\text{-}78)$$

The Fourier transform pair is thus given by the following equations. Time domain to frequency domain (forward transform):

$$F(j\omega) = \int_{-\infty}^{\infty} f(t) e^{-j\omega t} \, dt \qquad (14\text{-}79)$$

Frequency domain to time domain (inverse transform):

$$f(t) = \frac{1}{2\pi} \int_{-\infty}^{\infty} F(j\omega) e^{j\omega t} \, d\omega \qquad (14\text{-}80)$$

The two transform relationships are seen to be quite similar to each other. The differences between them are: (1) the presence of the factor $(1/2\pi)$ in the inverse relationship, and (2) the change in sign of the exponent in the two equations. (The term "Fourier transform" without any qualifying adjective is normally used to denote the forward Fourier transform, while the qualifying adjective "inverse" is always used to refer to the inverse Fourier transform. In the literature, the factor $(1/2\pi)$ is occasionally found associated with the *forward* Fourier transform, in which case it is absent from the equation of the inverse Fourier transform. It is therefore important to check the notation used by a particular author.

The determination of the inverse transform is not always a simple problem, since the integral in Eq. (14-80) cannot always be expressed in closed form.

14-9 RESPONSE OF LINEAR NETWORKS TO NONPERIODIC SIGNALS

The Fourier transform pair is useful in finding the response of a linear network or system to a nonperiodic input signal. Since the operations of differentiation and integration in the time domain become multiplication and division by $j\omega$ in the frequency domain, the differential equations of the network become algebraic equations. The situation is exactly like the sinusoidal steady state analysis under complex exponential forcing functions. Concepts of impedance, admittance, and transfer functions are therefore applicable in connection with nonperiodic inputs as well. The procedure for the solution of a linear network problem consists of (a) replacing the signal $f(t)$ by its Fourier transform; (b) using impedance, admittance, or appropriate transfer functions as needed to find the response in the frequency domain; and (c) using the inverse Fourier transform to find the response as a function of time. As was the case in sinusoidal steady state analysis, the last step is not always carried out. We illustrate the procedure in Example 14-11.

Example 14-11 A rectangular pulse of amplitude 1 V and duration t_p is applied to a low pass filter whose transfer function is given by

$$H(j\omega) = \frac{1}{1 + j(\omega/\omega_0)}$$

where ω_0 is the bandwidth (that is, the frequency at which the transfer function has a magnitude of 0.707). Determine the output of the filter.

Solution The Fourier transform of the rectangular pulse is given by (refer to Example 14-8):

$$V_i(j\omega) = t_p \frac{\sin(\omega t_p/2)}{(\omega t_p/2)} \tag{14-81}$$

The output transform is

$$V_o(j\omega) = H(j\omega)V_i(j\omega)$$
$$= t_p \frac{\sin(\omega t_p/2)}{(\omega t_p/2)} \frac{1}{[1 + j(\omega/\omega_0)]}$$

Since the magnitude of $H(j\omega)$ decreases as frequency increases, the magnitude of $V_o(j\omega)$ decreases as frequency increases also. At $\omega = 0$, $|V_o| = |V_i|$. At $\omega = \omega_0$, the cutoff frequency, $|H(j\omega)| = 0.707$ and $|V_o| = 0.707|V_i|$.

If the time domain representation of the output is desired, then the inverse transform is applied to $V_0(j\omega)$:

$$v_0(t) = \frac{1}{2\pi} \int_{-\infty}^{\infty} V_0(j\omega) e^{j\omega t} \, d\omega \tag{14-82}$$

This is not an easy integral to evaluate (and not given in standard handbooks of integrals). But it is possible to show that

$$v_0(t) = \begin{cases} 1 - e^{-\omega_0 t} & 0 < t < t_p \\ (1 - e^{\omega_0 t}) \, e^{-\omega_0(t - t_p)} & t > t_p \end{cases}$$

The sketch of $v_0(t)$ is shown for different values of the quantity $(\omega_0 t_p)$ in Fig. 14-19.

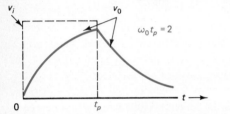
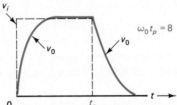
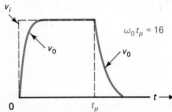

Fig. 14-19 Output of a low pass filter for different values of the product (pulse duration × bandwidth): the larger the product, the more faithful is the output to the input.

Relationship of Filter Bandwidth to Its Time Domain Response

Compare the filter outputs in the last example for the different values of the product: (bandwidth × pulse duration) = $(\omega_0 t_p)$. When $(\omega_0 t_p) \gg 1$, the output pulse is seen to resemble the input very closely. This corresponds to the filter having a large bandwidth for a given pulse duration. When $(\omega_0 t_p) = 1$, the output pulse is a rounded-off version of the input pulse. That is, much of the energy at high frequencies is lost by transmission through the filter. As $(\omega_0 t_p)$ becomes smaller, the rounding off becomes more pronounced. For a given pulse duration, therefore, the bandwidth of the filter must be sufficiently wide to maintain a reasonable fidelity of the output to the input.

Looking at the *speed of response* of the filter (that is, the rate at which the output rises from 0 to 1 at $t = 0$ and drops from 1 to 0 at $t = t_p$), the response is faster when the filter has a large bandwidth for a given pulse duration.

14-10 RESPONSE OF A LINEAR NETWORK TO ARBITRARY SIGNALS

The discussion up to this point has dealt with methods of analysis of linear networks under periodic forcing functions and nonperiodic signals that are Fourier transformable. When the input signal to a system is not Fourier transformable, the Fourier transform method is not adequate and other methods are needed. One of them, called *convolution*, involves computations directly in the time domain and will be discussed in this chapter. Another method involves the use of Laplace transform, which is an extension of the concept of Fourier transform to cover a wider variety of signals; it will be discussed in Chapter 15.

14-10-1 Representation of an Arbitrary Signal by Means of Impulse Functions

The tool available for the analysis of a linear network is, of course, the principle of superposition. In order to use superposition, a signal $x(t)$ must first be expressed as a sum of simpler component functions. And the component functions must be such that the response of a linear system to each of them can be determined. Narrow rectangular pulses are found to be suitable component functions to start with.

Staircase Approximation

It is possible to chop up a signal $x(t)$ into narrow rectangular pulses so that the given signal is approximated by the "staircase" function $\hat{x}(t)$, as in Fig. 14-20. The error due to the approximation approaches zero as the width of each pulse approaches zero.

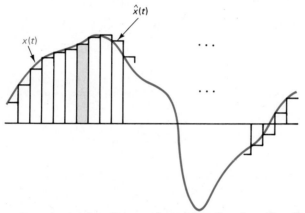

Fig. 14-20 Staircase approximation of a signal.

In order to describe a staircase function formally, first define a *pulse function* $p(t)$, Fig. 14-21(a), whose height and width are mutual reciprocals so as to make the area under the pulse always equal to 1. That is

$$p(t) = \begin{cases} \dfrac{1}{\Delta t_p} & 0 < t < \Delta t_p \\ 0 & \text{for all other } t \end{cases} \quad (14\text{-}83)$$

and

$$p(t)\Delta t_p = 1 \qquad 0 < t < \Delta t_p \quad (14\text{-}84)$$

A *shifted* pulse function starting at $t = t_1$, as in Fig. 14-21(b), is described by

$$p(t - t_1) = \begin{cases} \dfrac{1}{\Delta t_p} & t_1 < t < (t_1 + \Delta t_p) \\ 0 & \text{for all other } t \end{cases} \quad (14\text{-}85)$$

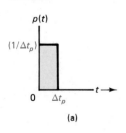
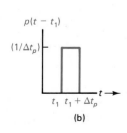

Fig. 14-21 Pulse function occurring at (a) $t = 0$ and (b) $t = t_1$. Height of pulse = reciprocal of its width.

The pulse function (shifted as needed) is used as a narrow window through which we can peek at any portion of a given signal. By moving the window across the signal we get to see the whole signal, a small portion of it at a time.

The whole signal is reconstructed by adding these segmented views, as follows.

Consider an arbitrary signal $x(t)$, as in Fig. 14-22(a). Subdivide the time axis into a large number of narrow intervals of width Δt_p, as in Fig. 14-22(b), set up a rectangular

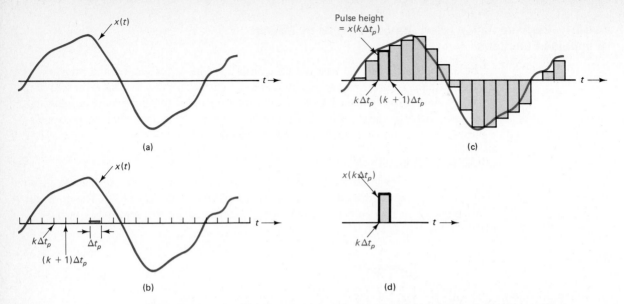

Fig. 14-22 (a) An arbitrary signal. (b) Time axis chopped up into small intervals of width Δt_p. (c) Approximation to the signal by a series of pulse functions. (d) A single pulse from the staircase approximation.

pulse in each interval with *pulse height* equal to the *value of x(t) at the beginning of that interval,* as in Fig. 14-22(c).

A *single* pulse at $t = k\Delta t_p$, as in Fig. 14-22(d), is given by

$$x(k\Delta t_p)\,[p(t - k\Delta t_p)\Delta t_p] \tag{14-86}$$

In Eq. (14-86)

$$p(t - k\Delta t_p) = 0 \quad \text{except where } k\Delta t_p < t < (k + 1)\Delta t_p$$

from Eq. (14-85). Also

$$[p(t - k\Delta t_p)\Delta t_p] = 1 \quad k\Delta t_p < t < (k + 1)\,\Delta t_p$$

from Eq. (14-84).

Therefore, the expression in Eq. (14-86) is equal to $x(k\Delta t_p)$ when $k\Delta t_p < t < (k + 1)\Delta t_p$ and zero for all other t. That is, the expression in Eq. (14-86) covers up the given signal except for the narrow portion between $k\Delta t_p$ and $(k + 1)\Delta t_p$ visible through the pulse window.

Each pulse of the staircase approximation of the given signal is therefore given by an expression of the form in Eq. (14-86), with $k = \ldots, -n, \ldots, 0, \ldots, n, \ldots,$ and the *staircase approximation* $\hat{x}(t)$ is given by

$$\hat{x}(t) = \sum_{k=-\infty}^{\infty} x(k\Delta t_p)p(t - k\Delta t_p)\Delta t_p \tag{14-87}$$

What has been accomplished so far is to synthesize an *arbitrary* signal by the *sum of component pulses* as a first step in the application of superposition to the analysis of a linear system. The error between $x(t)$ and $\hat{x}(t)$ approaches zero as Δt_p approaches zero.

That is, the actual signal $x(t)$ is given by

$$x(t) = \lim_{\Delta t_p \to 0} \sum_{k=-\infty}^{\infty} x(k\Delta t_p) p(t - k\Delta t_p) \Delta t_p \qquad (14\text{-}88)$$

14-10-2 The Impulse Function

When the width of the pulse $p(t)$ defined by Eq. (14-83) approaches zero, it becomes an *impulse function*. A *unit* impulse, denoted by $\delta(t)$, has the rather unusual property of being zero everywhere except at $t = 0$ and having an infinite amplitude at $t = 0$. It is represented symbolically by a vertical arrow of *unit* height at $t = 0$; see Fig. 14-23(a).

Fig. 14-23 (a) A unit impulse $\delta(t)$. (b) Impulse of strength A occurring at $t = t_1$.

That is

$$\delta(t) = \begin{cases} 0 & t \neq 0 \\ \infty & t = 0 \end{cases} \qquad (14\text{-}89)$$

and the area under a unit impulse is defined as being equal to unity. That is

$$\int_{-\infty}^{\infty} \delta(t) dt = 1 \qquad (14\text{-}90)$$

More specifically

$$\int_{-\infty}^{t} \delta(x) dx = \begin{cases} 0 & t < 0 \\ 1 & t > 0 \end{cases} \qquad (14\text{-}91)$$

Eq. (14-91) shows that the *integral of a unit impulse is a unit step function*. Conversely, the *derivative of a unit step function is a unit impulse*.

When an impulse encloses an area of arbitrary value A, it is written as $[A\delta(t)]$. The quantity A is called the *strength* or *weight* of the impulse. Eqs. (14-90) and (14-91) then become:

$$\int_{-\infty}^{\infty} A\delta(t) dt = A \qquad (14\text{-}92)$$

and

$$\int_{-\infty}^{t} A\delta(x) dx = \begin{cases} 0 & t < 0 \\ A & t > 0 \end{cases} \qquad (14\text{-}93)$$

An impulse of strength A occurring at $t = t_1$, as in Fig. 14-23(b), is written as $[A\delta(t - t_1)]$, and has the properties:

$$\int_{-\infty}^{\infty} A\delta(t - t_1)\,dt = A \tag{14-94}$$

and

$$\int_{-\infty}^{t} A\delta(x - t_1)\,dx = \begin{cases} 0 & t < t_1 \\ A & t > t_1 \end{cases} \tag{14-95}$$

The product of a function $f(t)$ and a unit impulse at $t = t_0$ is $[f(t)\,\delta(t - t_0)]$. It is zero except at $t = t_0$. At $t = t_0$, the product is an impulse whose strength equals $f(t_0)$. Therefore

$$\int_{-\infty}^{\infty} f(t)\,\delta(t - t_0)\,dt = f(t_0) \int_{-\infty}^{\infty} \delta(t - t_0)\,dt = f(t_0) \tag{14-96}$$

since $\int_{-\infty}^{\infty} \delta(t - t_0)\,dt = 1$.

Eq. (14-96) is known as the *sifting theorem*.

The value of a signal $f(t)$ at any given instant time t_0 is sifted out by means of a unit impulse occurring at $t = t_0$.

The sifting property of the impulse function is useful in representing a signal $f(t)$ as a summation of impulse functions.

Example 14-12 Evaluate

(a) $\int_{-\infty}^{\infty} f(t)\,\delta(t)\,dt$ and (b) $\int_{-\infty}^{\infty} f(t)\,\delta(t - 2)\,dt$ for each of the following functions.

(i) $f(t) = e^{-j0.5t}$ (ii) $f(t) = 10 \cos(t/5)$.

Solution In each case, the value of the integral is obtained by evaluating the given function at the instant of occurrence of the impulse.

(i) $\int_{-\infty}^{\infty} e^{-j0.5t}\,\delta(t)\,dt = e^{-j0} = 1$.

$\int_{-\infty}^{\infty} e^{-j0.5t}\,\delta(t - 2)\,dt = e^{-j1} = (0.540 - j0.842)$.

(ii) $\int_{-\infty}^{\infty} 10 \cos(t/5)\delta(t)\,dt = 10 \cos 0 = 10$.

$$\int_{-\infty}^{\infty} 10 \cos(t/5)\delta(t-2)\,dt = 10\cos(2/5) = 9.21.$$

Exercise 14-35 Evaluate the integrals

(a) $\int_{-\infty}^{\infty} f(t)\delta(t+1)\,dt$ and (b) $\int_{-\infty}^{\infty} f(t)\delta(t-1)\,dt$ for each of the following functions.

(i) $f(t) = 20\sin(t/2)$. (ii) $f(t) = -10t + 5$.

A unit impulse is obtained as the limiting case of certain specially defined functions. For example, consider the pulse function $p(t)$ defined in the previous section

$$p(t) = \begin{cases} (1/\Delta t_p) & 0 < t < \Delta t_p \\ 0 & \text{for all other } t \end{cases}$$

In the limit $\Delta t_p \to 0$, $p(t)$ becomes zero everywhere except at $t = 0$. At $t = 0$, it has an infinite amplitude, but the total area enclosed by it is still 1. Thus

$$\delta(t) = \left\{ \lim_{t_p \to 0} p(t) \right. \tag{14-97}$$

Similarly, a pulse function occurring at $t = t_1$ leads to

$$\delta(t - t_1) = \left\{ \lim_{t_p \to 0} p(t - t_1) \right. \tag{14-98}$$

Exercise 14-36 Sketch the function

$$f(t) = ae^{-at} u(t) \qquad a > 0$$

for different values of $a > 0$. Show that it becomes a unit impulse in the limit $(1/a) = 0$.

Representation of a Signal x(t) as a Sum of Impulse Functions

Consider the staircase approximation of Eq. (14-88) for a given signal

$$x(t) = \lim_{\Delta t_p \to 0} \sum_{k\Delta t_p = -\infty}^{\infty} [x(k\Delta t_p)p(t - k\Delta t_p)\Delta t_p]$$

where the index of summation is in terms of $(k\Delta t_p)$.

Figure 14-24(a) shows the time axis divided into intervals of width Δt_p. Refer to Fig. 14-24(b). As Δt_p approaches zero

$$k\Delta t_p \to \tau, \text{ a continuous variable}$$
$$\Delta t_p \to d\tau$$
$$p(t - k\Delta t_p) \to \delta(t - \tau)$$

$$\text{(summation over } k\Delta t_p\text{)} \to \text{integral } \int (\ )\, d\tau$$

14-10 Response of a Linear Network to Arbitrary Signals

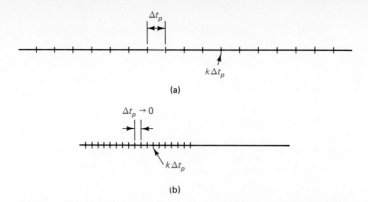

Fig. 14-24 (a)-(b) As the width Δt_p of the interval on the time axis approaches zero, the initial points of the interval ($k\Delta t_p$) become a continuous variable τ and the interval $\Delta t_p = d\tau$.

Therefore

$$x(t) = \int_{-\infty}^{\infty} x(\tau)\delta(t - \tau)d\tau \qquad (14\text{-}99)$$

That is, a signal $x(t)$ can be written as the sum (integral) of a series of impulse functions whose weights are given by the amplitudes of the signal at values of the variable τ.

Since it is possible to represent any signal $x(t)$ as the sum (or integral) of weighted impulse functions, the response of a system to a signal $x(t)$ can be determined by summing (or integrating) its response to weighted impulses. The response of a linear system to a unit impulse is called its *impulse response*. The impulse response is the building block in the time domain analysis of linear systems. The process in which a signal and the impulse response are used for analyzing a system is known as *convolution*.

Let $h_p(t)$ represent the time domain response of a linear system to a pulse input $p(t)$ and let $h(t)$ be the response to a unit impulse at $t = 0$. Then

$$h_p(t) \rightarrow h(t) \quad \text{as} \quad \Delta t_p \rightarrow 0 \qquad (14\text{-}100)$$

Even though an impulse is a theoretical idealization, it is possible to approximate it in practice by means of a very strong, sharp pulse in order to determine the impulse response of a system. For example, the reaction set up by a bullet leaving a gun is an approximation to an impulse in mechanical systems. At a more mundane level, the sharp kick one gives a tire of a used car to test its mechanical fitness is an attempt at studying the impulse response, albeit a crude one.

14-11 THE CONVOLUTION INTEGRAL

When an arbitrary signal is expressed as the weighted sum of impulses, the relationship between output and input is given by the *superposition integral,* usually called the *convolution integral.*

Consider an input $x(t)$ applied to a linear time-invariant system. Let $h_p(t)$ be the response to the pulse function $p(t)$. The diagrams of Fig. 14-25 (a), (b), (c) illustrate the

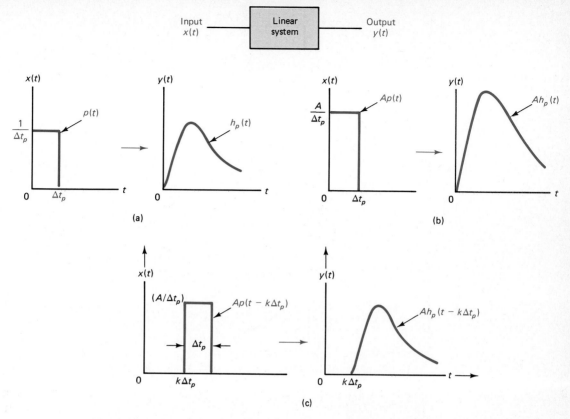

Fig. 14-25 (a) A linear time-invariant system. (b), (c) Effect of changing the amplitude and shifting the input pulse on the response of a linear time invariant system.

effect of changing the amplitude or position of a pulse input to a linear time invariant system, and the diagrams of Fig. 14-26 (a), (b), (c) show the case where $\Delta t_p \to 0$.

$$
\begin{array}{cc}
\text{Input} & \text{Output} \\
p(t) & \to \quad h_p(t) \\
Ap(t) & \to \quad Ah_p(t) \\
Ap(t - k\Delta t_p) & \to \quad Ah_p(t - k\Delta t_p)
\end{array}
\qquad (14\text{-}101)
$$

Using Eq. (14-88), the expression for a signal in terms of pulse functions is

$$x(t) = \lim_{\Delta t_p \to 0} \sum_{k\Delta t_p = -\infty}^{\infty} x(k\Delta t_p) p(t - k\Delta t_p) \Delta t_p$$

which is rewritten in the form

$$x(t) = \lim_{\Delta t_p \to 0} \sum_{k\Delta t_p = -\infty}^{\infty} [x(k\Delta t_p)\Delta t_p] \, p(t - k\Delta t_p) \qquad (14\text{-}102)$$

Each term in the series is a pulse function occurring at $t = k\Delta t_p$ and of amplitude $[x(k\Delta t_p)\Delta t_p]$. From Eq. (14-101), the output of the system to such an input is

14-11 The Convolution Integral

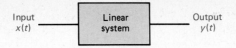

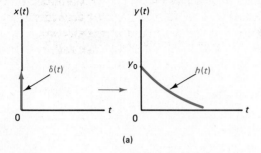

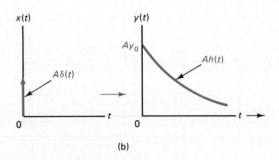

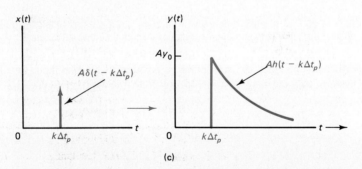

Fig. 14-26 The diagrams here are parallel to the corresponding diagrams of Fig. 14-25 and show the effect of changing the strength and shifting an input impulse on the response of a linear time invariant system.

$$[x(k\Delta t_p)\Delta t_p]h_p(t - k\Delta t_p) \qquad (14\text{-}103)$$

Considering the whole input $x(t)$, the output $y(t)$ is then given by using superposition and Eq. (14-103)

$$y(t) = \lim_{\Delta t_p \to 0} \sum_{k\Delta t_p = -\infty}^{\infty} [x(k\Delta t_p)\Delta t_p]h_p(t - k\Delta t_p) \qquad (14\text{-}104)$$

As discussed in connection with Eq. (14-99), when $t_p \to 0$, $k\Delta t_p \to \tau$, $\Delta t_p \to d\tau$, and $h_p(t - k\Delta t_p) \to h(t - \tau)$ and Eq. (14-104) becomes

$$y(t) = \int_{-\infty}^{\infty} x(\tau)h(t-\tau)d\tau \qquad (14\text{-}105)$$

Equation (14-105) is called the *convolution integral: convolving* means "twisting," and it will be seen that the use of Eq. (14-105) requires a twisting or *folding over* of the impulse response function. The convolution integral is a statement of the superposition principle for a general input $x(t)$ to a linear time-invariant system with an impulse response $h(t)$. The notation $[x(t) \circledast h(t)]$ is used to denote the convolution of $x(t)$ and $h(t)$:

$$y(t) = [x(t) \circledast h(t)] = \int_{-\infty}^{\infty} x(\tau)h(t-\tau)d\tau \qquad (14\text{-}106)$$

The time domain response of a linear time-invariant system to any input signal may be obtained by using convolution, provided the impulse response of the system is known. The output of the system at any given (or specified) value of time t is found by integrating the product $x(\tau)$ and $h(t-\tau)$ over all τ. In the calculations, therefore, t acts as if it were a constant; that is, t is simply a specific point arbitrarily chosen on the τ axis as the time of observation of the output. In order to find the output for all t, it is necessary to consider different values (or ranges of values) of t, and for each value (or range of values) of t, the integral has to be evaluated separately.

Exercise 14-37 Show that the convolution integral can also be written in the form

$$y(t) = \int_{-\infty}^{\infty} x(t-\tau)h(\tau)d\tau \qquad (14\text{-}107)$$

This result shows that *either* of the two functions, $x(t)$ or $h(t)$, can be selected for shifting and folding, while the other remains fixed.

14-11-1 Evaluation of the Convolution Integral

The actual evaluation of the integral in Eq. (14-106) or (14-107) needs a certain amount of care, as will become painfully evident shortly. The saying "A picture is worth a thousand words" is particularly true in the application of convolution; it is important (in fact, *vital*) to draw a series of diagrams for the solution of convolution problems.

Examine the various items in the convolution integral. The variable of integration is τ. This dictates that *all* diagrams *must* use τ as the horizontal coordinate (abscissa). Therefore,

the first step is: draw diagrams of $x(\tau)$ and $h(\tau)$; that is, as functions of τ.

$x(\tau)$ is one of the terms in the integral of Eq. (14-106). The second term in the integral is $h(t-\tau)$. In order to obtain $h(t-\tau)$

Shift $h(\tau)$ by a distance t along the τ axis to get $h(\tau - t)$. Fold the shifted function $h(\tau - t)$ backward about a vertical line at $\tau = t$. The resulting function is $h(t - \tau)$.

Several examples of the shifting and folding operation are shown in Fig. 14-27.

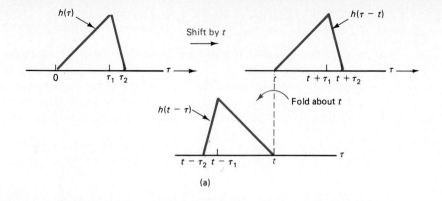

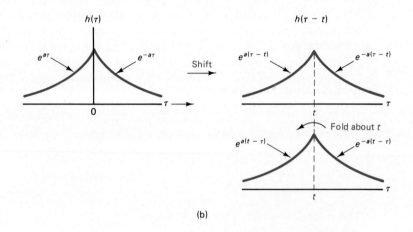

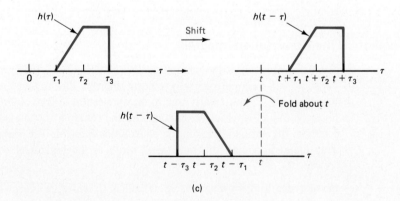

Fig. 14-27 (a)-(c) Examples of shifting and folding a given $h(\tau)$.

Exercise 14-38 For each of the functions $h(t)$ shown in Fig. 14-28, draw a sketch of (a) $h(5 - \tau)$ and (b) $h(-2 - \tau)$, both as functions of τ.

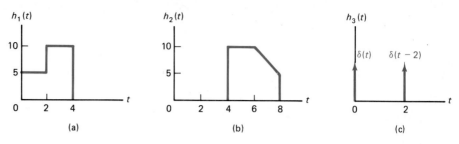

Fig. 14-28 Functions for Exercise 14-38.

In the evaluation of the integral in Eq. (14-106), $x(\tau)$ is defined by the given input signal. As we saw in the last exercise, the term $h(t - \tau)$, however, assumes different positions on the τ axis for different values of t.

That is, $x(\tau)$ remains fixed while $h(t - \tau)$ moves on the τ axis, as illustrated in Fig. 14-29(a)–(g) for a sample set of functions.

Note that the intersection of $x(\tau)$ and $h(t - \tau)$ and the limits of integration for τ vary from one situation to the next.

Fig. 14-29 (a)-(g) In evaluating the convolution integral, one of the functions remains stationary while the other one moves to different positions along the τ axis.

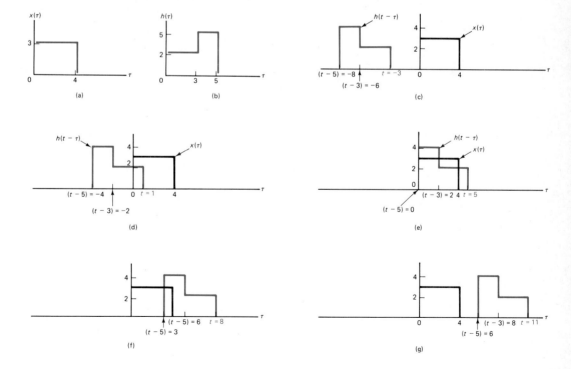

14-11 The Convolution Integral

The intersection of $x(\tau)$ and $h(t - \tau)$ and the limits of integration in Eq. (14-106) for different ranges of values of t are best determined by a series of diagrams similar to those in Fig. 14-29. Trying to evaluate the convolution integral without the help of such diagrams is like touring a strange land without a good map.

Example 14-13 Given the impulse response of a linear time-invariant system

$$h(t) = 5e^{-2t}u(t)$$

The input $x(t)$ is a rectangular pulse of amplitude 3V in the interval (0, 4). Determine the output $y(t)$ of the system.

Solution
1. Draw the input x and the impulse response h as functions of τ [Fig. 14-30(a)].
2. Draw the function $h(t - \tau)$ as a function of τ. This involves shifting $h(\tau)$ by t and then folding the shifted function backward about the vertical at t. The result is shown in Fig. 14-30(b). The shifted and folded function is given by

$$h(t - \tau) = 5e^{-2(t-\tau)} \qquad \tau < t$$

3. Starting with a large negative value of t, move $h(t - \tau)$ and examine how the intersection between that curve and $x(\tau)$ changes. It is seen that there is no inter-

Fig. 14-30 Convolution of (a) a rectangular pulse and an exponential function (Example 14-13). (b)-(d) show the diagrams needed to evaluate the convolution. (e) Output waveform.

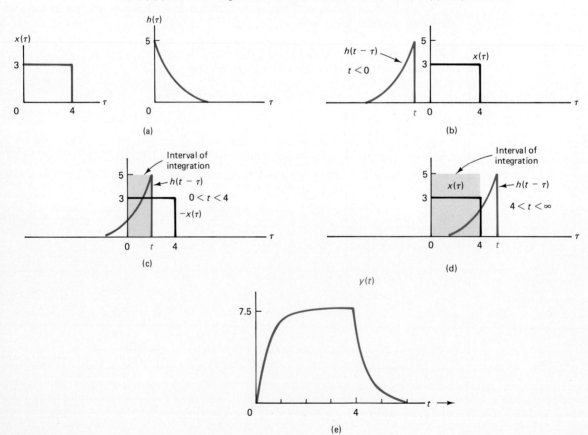

section of $h(t - \tau)$ and $x(\tau)$ when $t < 0$ in this example. This condition persists until $t = 0$.

When $t > 0$, the situation changes at first to that shown in Fig. 14-30(c). The curves intersect over the interval $(0, t)$, which is shaded in the diagram. This situation is valid when $0 < t < 4$.

When $t > 4$, the situation changes, as shown in Fig. 14-30(d). The two curves intersect over the interval $(0, 4)$, which is shaded in the diagram. This situation is valid for all $t > 4$.

Therefore, there are three *regions of values of* t to be considered in the present example: $t < 0$, $0 < t < 4$, and $t > 4$.

4. In each of the regions identified in step 3, form the product of $x(\tau)$ and $h(t - \tau)$ and integrate over the range of values of τ, as defined by the shaded portion of the diagram.

Region 1: $t < 0$

$$x(\tau)h(t - \tau) = 0$$

Therefore, $y(t) = 0$ when $t < 0$.

Region 2: $0 < t < 4$

$$x(\tau)h(t - \tau) = (3)[5e^{-2(t-\tau)}] \qquad 0 < \tau < t$$

$$y(t) = \int_0^t 15e^{-2(t-\tau)}\, d\tau = 15e^{-2t}\int_0^t e^{2\tau}\, d\tau$$

$$= 7.5(1 - e^{-2t}) \qquad 0 < t < 4$$

Region 3: $t > 4$

$$x(\tau)h(t - \tau) = (3)[5e^{-2(t-\tau)}] \qquad 0 < \tau < 4$$

$$y(t) = \int_0^4 15e^{-2(t-\tau)}\, d\tau$$

$$= 15e^{-2t}\int_0^4 e^{2\tau}d\tau$$

$$= 7.5e^{-2t}(e^8 - 1) \qquad t > 4$$

The output of the system is, therefore, described by the following set of equations.

$$y(t) = \begin{cases} 0 & t < 0 \\ 7.5(1 - e^{-2t}) & 0 < t < 4 \\ 7.5e^{-2t}(e^8 - 1) & t > 4 \end{cases}$$

Figure 14-30(e) shows the output. ■

Exercise 14-39 Repeat the calculations of the previous example when the input pulse occurs in the interval $(-1, 3)$ with the same amplitude as before.

Exercise 14-40 A rectangular pulse of amplitude 10 V occurring in the interval (0, 4) is applied to a system with $h(t) = 5e^{2t}$ when $t < 0$, and 0 when $t > 0$.
Determine the output.

Recall that, in Exercise 14-37, we showed that the convolution integral can also be written in the alternative form

$$y(t) = \int_{-\infty}^{\infty} x(t - \tau)h(\tau)d\tau \qquad (14\text{-}107)$$

Therefore, the output of a system may also be determined by shifting and folding the input x while the impulse response $h(\tau)$ remains fixed.

Example 14-14 is a reworking of Example 14-13 by using Eq. (14-107).

Example 14-14 See Example 14-13 for the statement of this problem.

Solution The impulse response:

$$h(\tau) = 5e^{-2\tau} \quad \tau > 0 \quad \text{and} \quad 0 \text{ for } \tau < 0$$

and the input is

$$x(\tau) = 3 \quad 0 < \tau < 4 \quad \text{and} \quad 0 \text{ everywhere else}$$

The diagrams of Fig. 14-31(a)–(c) show $x(t - \tau)$ for different regions of t of interest: $t < 0$; $0 < t < 4$; and $t > 4$.

Region 1: $t < 0$ $\qquad\qquad\qquad y(t) = 0$

Fig. 14-31 (a)-(c) An alternative evaluation of the convolution. $h(\tau)$ remains stationary while $x(t - \tau)$ is moved.

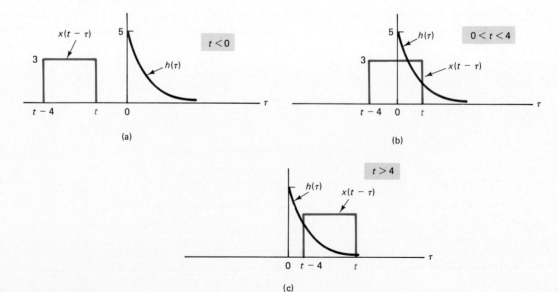

Region 2: $0 < t < 4$

$$y(t) = \int_0^t (3)(5e^{-2\tau})d\tau = 7.5(1 - e^{-2t}) \qquad 0 < t < 4$$

Region 3: $t > 4$

$$y(t) = \int_{t-4}^t (3)(5e^{-2\tau})d\tau$$
$$= -7.5(e^{-2(t-4)} - e^{-2t}) = 7.5e^{-2t}(e^8 - 1)$$

The three expressions for $y(t)$ are fortunately seen to match those obtained in the previous solution to this system problem. ∎

Exercise 14-41 Redo Exercises 14-39 and 14-40 by using Eq. (14-107).

14-11-2 Fourier Transform of Convolution

The convolution integral provides a rigorous basis for using the procedure already employed in the analysis of linear systems when subjected to a Fourier transformable signal. This is a result of the property:

> convolution of two functions in the time domain transforms to a multiplication of their Fourier transforms.

Fourier Transform of an Impulse Response

The Fourier transform of a unit impulse is given by

$$\mathcal{F}[\delta(t)] = \int_{-\infty}^{\infty} \delta(t)e^{-j\omega t}dt = e^{-j0} = 1 \text{ for all } \omega$$

That is,

$$\mathcal{F}[\delta(t)] = 1, \text{ a constant at all frequencies} \qquad (14\text{-}108)$$

Let the input to a linear system be

$$v_1(t) = \delta(t)$$

and the output be denoted by $v_2(t)$. If $h(t)$ is the impulse response of the system, then from convolution

$$v_2(t) = \int_{-\infty}^{\infty} h(\tau)\delta(t - \tau)\, d\tau = h(t) \qquad (14\text{-}109)$$

where the sifting theorem, Eq. (14-96) has been used.

The transfer function of the system is defined by

$$H(j\omega) = \frac{V_2(j\omega)}{V_1(j\omega)}$$

where $V_2(j\omega)$ and $V_1(j\omega)$ are the Fourier transforms of the input and output, respectively. If $v_1(t) = \delta(t)$, then $V_1(j\omega) = 1$ from Eq. (14-108).

The transfer function of a linear system may therefore be determined by driving the system with a unit impulse and observing the output in the frequency domain. When $V_1(j\omega) = 1$, we have

$$H(j\omega) = \frac{V_2(j\omega)}{V_1(j\omega)} = V_2(j\omega) = \mathcal{F}[h(t)]$$

from Eq. (14-109). Therefore,

The transfer function of a linear system is the Fourier transform of its impulse response.

Fourier Transform of Convolution

Let $f(t)$ and $g(t)$ be two functions whose Fourier transforms are given by $F(j\omega)$ and $G(j\omega)$. Then their convolution is given by

$$[f(t) \circledast g(t)] = \int_{-\infty}^{\infty} f(\tau)g(t - \tau)d\tau$$

Taking the Fourier transform of $[f(t) \circledast g(t)]$

$$\mathcal{F}[f(t) \circledast g(t)] = \int_{-\infty}^{\infty} \int_{-\infty}^{\infty} [f(\tau)g(t - \tau)]e^{-j\omega t} dt\, d\tau \qquad (14\text{-}110)$$

Integrating first with respect to t by using the substitution

$$(t - \tau) = x$$

Eq. (14-110) becomes

$$\mathcal{F}[f(t) \circledast g(t)] = \int_{-\infty}^{\infty} f(\tau)\left[\int_{-\infty}^{\infty} g(x)e^{-j\omega x}dx\right]e^{-j\omega \tau}d\tau$$

$$= G(j\omega)\int_{-\infty}^{\infty} f(\tau)e^{-j\omega \tau}d\tau \qquad (14\text{-}111)$$

since $G(j\omega)$ is the Fourier transform of $g(t)$.

Equation (14-111) leads to

$$\mathcal{F}[f(t) \circledast g(t)] = G(j\omega)\left[\int_{-\infty}^{\infty} f(\tau)e^{-j\omega \tau}d\tau\right]$$

$$= G(j\omega)F(j\omega) \qquad (14\text{-}112)$$

since $F(j\omega)$ is the Fourier transform of $f(t)$.

Therefore,

$$\mathcal{F}[f(t) \circledast g(t)] = F(j\omega)G(j\omega) \qquad (14\text{-}113)$$

That is,

Convolution in the time domain → multiplication in the frequency domain.

Example 14-15 Given $f(t) = e^{-at}u(t)$ and $g(t) = e^{-bt}u(t)$, determine the Fourier transform of their convolution: (a) by first evaluating the convolution integral and finding its Fourier transform, and (b) by evaluating the individual Fourier transforms and using Eq. (14-113).

Solution (a) Let

$$c(t) = [f(t) \circledast g(t)]$$

Referring to Fig. 14-32

$$c(t) = 0 \text{ for } t < 0$$

Fig. 14-32 Convolution for Example 14-15.

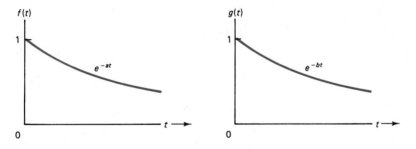

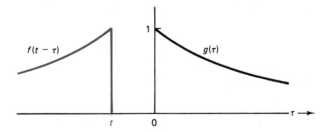

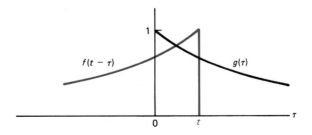

14-11 The Convolution Integral

and, for $t > 0$,

$$c(t) = \int_0^t e^{-a(t-\tau)} e^{-b\tau} \, d\tau$$

$$= \frac{e^{-bt} - e^{-at}}{a - b} \qquad t > 0$$

The Fourier transform of $c(t)$ is obtained by using the identity:

$$\mathcal{F}[e^{-\alpha t}u(t)] = \frac{1}{\alpha + j\omega}$$

Therefore,

$$\mathcal{F}[c(t)] = \frac{1}{(a - b)} \left[\frac{1}{b + j\omega} - \frac{1}{a + j\omega} \right]$$

$$\mathcal{F}[c(t)] = \frac{1}{(a + j\omega)(b + j\omega)}$$

(b) The Fourier transforms of $f(t)$ and $g(t)$ are:

$$\mathcal{F}[a(t)] = \frac{1}{(a + j\omega)} \qquad \mathcal{F}[b(t)] = \frac{1}{(b + j\omega)}$$

and, from Eq. (14-113),

$$\mathcal{F}[a(t)] \, \mathcal{F}[b(t)] = \frac{1}{(a + j\omega)(b + j\omega)}$$

The two results are seen to be the same. ∎

Exercise 14-42 Two rectangular pulses $f(t)$ and $g(t)$ are shown in Fig. 14-33. Evaluate their convolution $c(t)$ and its Fourier transform $C(\omega)$. Show that $C(\omega) = F(\omega)G(\omega)$.

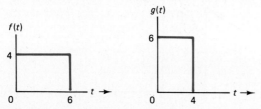

Fig. 14-33 Pulses for Exercise 14-42.

14-12 RESPONSE OF A LINEAR SYSTEM

If $v_1(t)$ is the input to a linear system with an impulse response $h(t)$, then its output in the time domain is given by

$$v_2(t) = [h(t) \circledast v_1(t)]$$

which leads to, from Eq. (14-113)

$$V_2(j\omega) = H(j\omega)V_1(j\omega) \tag{14-114}$$

The Fourier transform of the output of a linear system is the product of the Fourier transforms of the input and the impulse response.

The response of a linear system may therefore be determined in the time domain by using the convolution integral or in the frequency domain by using the transfer function. When the analysis is performed in the frequency domain, the output is obtained in the frequency domain also. If the response in the time domain is desired, it is necessary to use the inverse transform, which may not always be easy.

Whether convolution or transfer function offers the better approach to a problem depends upon a number of factors. The most important consideration is perhaps whether the desired solution is in the time or frequency domain.

14-13 LIMITATIONS OF THE FOURIER TRANSFORM

In this chapter we have shown that the Fourier transform is an effective tool in the analysis of linear systems, provided the input signal has a Fourier transform. The utility of Fourier transform arises from the fact that many signals are in the form of the sum (or integral) of component functions of the form $e^{j\omega t}$. When such a signal is the input to a linear system, the output is determined readily by using Eq. (14-114).

As stated earlier, a signal must satisfy the condition

$$\int_{-\infty}^{\infty} |f(t)|dt < \infty \tag{14-115}$$

in order to have a Fourier transform. When a signal does not satisfy the condition in Eq. (14-115), it is still possible in some cases to obtain a Fourier transform by using an indirect approach. An example is the unit step function, which does not satisfy the condition in Eq. (14-115); however, its Fourier transform can be determined by starting with the function e^{-at} and letting $a \to 0$. Such an indirect approach does not work in all cases. Many signals do not have a Fourier transform, and it becomes necessary to make some suitable modifications to widen the area of applicability. The result is the Laplace transform, discussed in the next chapter.

14-14 SUMMARY OF CHAPTER

A periodic signal can be expanded as an infinite series using component functions. An expansion in which the coefficients of the component functions are selected so as to minimize the mean square error between the series representation and the actual signal is the Fourier series.

The given signal is written as a sum of a constant term (the dc component), a fundamental frequency term, and terms involving all the harmonics of the fundamental frequency. The series involves either sines and cosines (the trigonometric form) or complex exponential functions (the exponential form). The coefficients of the series in either form are determined by the use of suitable integral formulas. The fundamental frequency and each harmonic are represented by two terms in the series: a sine and a cosine term in the trigonometric series, and two exponential terms in the exponential series. Given either form of the series, the other form is obtained by means of suitable conversion formulas.

The exponential form has theoretical advantages over the trigonometric form, and it permits extension to the case of nonperiodic signals.

Symmetry of a signal leads to certain special features of the series expansion. For an even function, the trigonometric form contains no sine terms, while the coefficients of the exponential form are all pure real. For an odd function, the trigonometric form contains no dc component or cosine terms, while the coefficients of the exponential form are all pure imaginary. A signal with half-wave symmetry is found to contain no even harmonics. Symmetry properties are useful in saving computational labor and checking the results.

The response of a circuit to a periodic signal uses the Fourier decomposition and the principle of superposition. When using the trigonometric form, the response of the circuit to each individual harmonic component of the input is determined by employing techniques of ac steady state analysis; the individual responses are converted to sinusoidal time functions and then added. When using the exponential form, the response is obtained by using impedance, admittance, or transfer function and adding all the complex exponential response functions.

A plot of the magnitudes of c_k as a function of frequency is known as the amplitude spectrum. It is a line spectrum and extends to both positive and negative frequencies. Negative frequency is not a realizable parameter in the real world, but it is a mathematical necessity since the complex exponential representation of a real sinusoid requires a combination of $e^{jk\omega t}$ and $e^{-jk\omega t}$. The amplitude of each harmonic component in a signal is given by the sum of the heights of the two lines at that harmonic in the amplitude spectrum. The phase spectrum is a plot of the angle of c_k, and the power spectrum is a plot of $|c_k|^2$ as a function of frequency. The power in any harmonic component is the sum of the two lines for that harmonic in the power spectrum. The total power is the sum of all the lines in the power spectrum.

For nonperiodic signals for which the area contained by their magnitude is finite, the discrete series changes to a continuous function, and the relationship is called the Fourier transform. The Fourier transform represents the amplitude density (that is, volts per radians per second in the case of a voltage signal) of the signal. The energy density spectrum obtained by squaring the magnitude of the Fourier transform (and divided by 2π) displays the energy distribution as a function of frequency in a signal. It is like the power spectrum of a periodic signal in its significance. Symmetry of a signal leads to special features of its Fourier transform. For an even function, the Fourier transform is a pure real function, while for an odd function, the Fourier transform is pure imaginary.

Differentiation and integration of a function in the time domain are equivalent to multiplication and division, respectively, of the Fourier transform by $j\omega$. This property provides the basis for using impedance, admittance, and transfer functions in the determination of the response of a network.

The transformation of the frequency domain representation of a signal to its time domain representation uses the inverse transformation integral. The integral is not always easy to use, however, since it cannot always be evaluated in closed form.

An arbitrary signal may be expressed as the integral of component functions, each of which is an impulse function of strength equal to the amplitude of the given signal. The response of a network to a unit impulse input gives the impulse response. Using superposition with the response of the network to each of the weighted impulses leads to the operation of convolution in the time domain. Convolution permits the determination of the response of a network when the input and the impulse response are available. The Fourier transform of the convolution of two functions equals the product of the Fourier transforms of the two functions. The transfer function of a network is the Fourier transform of its impulse response.

Signals which do not have the property that the area enclosed by their magnitude is finite do not possess a Fourier transform. A modification of the component function $e^{j\omega t}$ leads to the Laplace transform.

Answers to Exercises

14-1 The integral splits into two terms, each of which completes an integral number of complete cycles in the interval $(0, T)$ when $m \neq n$. When $m = n$, one of the terms is a constant. This leads to Eq. (14-3).

14-2 $f(t) = (\cos \theta) \cos (2\pi kt) - (\sin \theta) \sin (2\pi kt)$. Since $\cos (2\pi kt)$ and $\sin (2\pi kt)$ are orthogonal and $\cos \theta$ and $\sin \theta$ are constants, $f(t)$ is also orthogonal.

14-3 From Chapter 7, $d_k = \sqrt{a_k^2 + b_k^2}$. $\theta_k = \arctan(-b_k/a_k)$.

14-4 (a) 200 r/s (or 31.8 Hz). (b) $8 \cos (200t - 54°)$ V, $4.17 \cos (400t - 106°)$V. (c) 8 V, 4.17 V.

14-5 The integral splits into two sine terms, each of which integrates to zero in the interval $(-T/2, T/2)$.

14-6 The integral splits into two cosine terms, each of which integrates to zero when $m \neq n$. When $m = n$, one of the terms is a constant, leading to the formula for b_m.

14-7 $a_0 = (A/3)$. $a_m = (A/m\pi) \sin (2\pi m/3)$. $b_m = (A/m\pi)[1 - \cos (2\pi m/3)]$.
$f(t) = A[0.333 + 0.276 \cos (\omega t) - 0.138 \cos (2\omega t) + 0.0688 \cos (4\omega t) + \ldots + 0.477 \sin (\omega t) + 0.239 \sin (2\omega t) + 0.119 \sin (4\omega t) + \ldots]$. Note that the signal does not contain any harmonics that are multiples of (3ω).

14-8 New signal $g(t) = f(t - T/2)$. The $\cos (m\omega t)$ terms become $\cos (m\omega t - m\pi)$. The only difference between the series for $g(t)$ and $f(t)$ is a change in sign of the coefficient of the *fundamental* frequency component.

14-9 Series $= (2V_0/\pi) + (4V_0/\pi)[(1/3) \cos (2\omega t) - (1/15) \cos (4\omega t) + (1/35) \cos (6\omega t) - (1/63) \cos (8\omega t) + \ldots]$

14-10 Even function. $b_m = 0$ for all m. $a_0 = (V_0/2)$. $a_m = (2V_0/m\pi) \sin (m\pi/2)$. $a_m = 0$ for all even m, $(2V_0/m\pi)$ for $m = 1, 5, 9, \ldots$, and $-(2V_0/m\pi)$ for $m = 3, 7, 11, \ldots$

14-11 Odd function. $a_m = 0$ for all m. $b_m = (V_0/m\pi)(1 - \cos m\pi)$. $b_m = 0$ for all even m. $b_m = (2V_0/m\pi)$ for all odd m.

14-12 $Y = [0.01 + j(10^{-3}\omega - 10/\omega)]$. $Y = 0.01\underline{/0°}$ S at $\omega = 100$ r/s, $0.150\underline{/86.1°}$ at 200 r/s, $0.267\underline{/88°}$ at 300 r/s, and $0.375\underline{/88.5°}$ at 400 r/s. $v_0(t) = 377 \cos (100t + 122°) + 10.6 \cos (200t + 4°) + 4.04 \cos (300t + 14°) + 2.12 \cos (400t + 1.5°)$ V.

14-13 $P_{av} = (100 \times 50) + (1/2)[(100 \times 50)\cos (45°) + (30 \times 30) \cos (90°)] = 6768$ W.

14-14 Average power components are: 45.5 W, 12.6 W, 4.98 W, 2.03 W. Total average power $= 65.1$ W.

14-15 Average power components are 711 W, 0.588 W, 0.0761 W, 0.0221 W. Total $= 711.7$ W. Power in the resonant frequency component is 99.9% of the total power.

14-16 $P_{av} = 20^2 + (1/2)[17.3^2 + 9.43^2 + 10^2] = 644$ W.

14-17 $V_{rms}^2 = (V_0^2/\pi^2) + (V_0^2/8) + (2V_0^2/\pi^2) \sum_{k=1}^{\infty} [1/(4k^2 - 1)]$. $V_{rms} = 0.5V_0$.

14-18 $V_{rms}^2 = (4V_0^2/\pi^2) + (8V_0^2/\pi^2) \sum_{k=1}^{\infty} [1/(4k^2 - 1)]$. $V_{rms} = 0.707V_0$.

14-19 $c_k = -(A/j2\pi k)(e^{-j2\pi k/3} - 1) = (A/k\pi)\sin(k\pi/3)e^{-jk\pi/3}$. Note that $c_0 = A/3$.

14-20 $g(t) = f(t - T/2) \cdot e^{-jk\omega(t - T/2)} = (\cos k\pi)e^{-jk\omega t}$. For even k, the coefficients are the same for both series. When $k = 1$, $c_1 = -V_0/4$ in the present waveform.

14-22 (a) Amplitude $= 1.334/\sqrt{(1 + n^2\pi^2)}$ Angle $= -\arctan(n\pi)$. Amplitudes of the first three harmonics are: 0.405, 0.210, and 0.141. Angles of the first three harmonics are: $-72.3°$, $-81°$, $-84°$. (b) $a_n = 1.334/(1 + n^2\pi^2)$. $b_n = 1.334n\pi/(1 + n^2\pi^2)$.

14-23 $H(jk\omega) = jk\omega L/(R + jk\omega L)$. $v_0(t) = \sum_{k=-\infty}^{\infty}[j0.667k\omega L/(R + jk\omega L)(1 + jk\pi)]e^{jk\omega t}$.

14-24 $\mathbf{I}_c = R\mathbf{I}_s/(R - j1/k\omega C)$. $i_c(t) = \sum_{k=-\infty}^{\infty}[4/(3 + j(2k - 1/k)]e^{jk20\pi t}$ A. Amplitudes of the first three harmonics are: 2.53 A, 1.74 A, 1.25 A. Angles of the first three harmonics are: $-18.4°$, $-49.4°$, $-62.1°$.

14-25 Heights of lines in the amplitude spectrum: $0.318V_0$ at dc and then proceeding in both directions: $0.25V_0$, $0.106V_0$, $0.0212V_0$, $0.00909V_0$, ... Phase spectrum: $0°$ at the first, second, sixth, tenth ... harmonics, and $180°$ at the fourth, eighth, ... harmonics.

14-26 Input spectrum: Line heights are: 5, 4.68, 3.78, 2.52, 1.17, 0, ... Output spectrum: 5, 4.19, 2.67, 1.4, 0.523, 0, ...

14-27 Input spectrum: 0.667, 0.202, 0.105, 0.0704, 0.0529, ... Output spectrum: 0, 0.090, 0.074, 0.0586, 0.0473, ...

14-28 $0.101V_0^2$, $0.0625V_0^2$, $1.12 \times 10^{-2}V_0^2$, $4.49 \times 10^{-4}V_0^2$, ...

14-29 25, 17.6, 7.12, 1.96, 0.274 W.

14-30 0, 0.0082, 5.50×10^{-3}, 3.43×10^{-3}, 2.24×10^{-3} W, 1.55×10^{-3} W.

14-31
$$F(j\omega) = \frac{V_0}{j^2}\left[\frac{e^{-j(\pi - 0.01\omega)} - 1}{j(100\pi - \omega)} + \frac{e^{-j(\pi + 0.01\omega)} - 1}{j(100\pi + \omega)}\right]$$
$$= [(100\pi V_0)/(10^4\pi^2 - \omega^2)](e^{-j0.01\omega} + 1)$$

14-33 Odd function.
$$F(j\omega) = -j\left[\frac{\sin(\pi - \omega/\omega_0)}{(\omega_0 - \omega)} - \frac{\sin(\pi + \omega/\omega_0)}{(\omega_0 + \omega)}\right]$$
$$= -j\left(\frac{2\pi\omega}{\omega_0^2 - \omega^2}\right)\left[\frac{\sin(\pi\omega/\omega_0)}{(\pi\omega/\omega_0)}\right]$$

14-35 (a) $f(-1)$: -9.59, 15. (b) $f(+1)$: 9.59, -5.

14-36 Area under the curve $= 1$. As $a \to \infty$, height $\to \infty$, and width $\to 0$. Hence the function approaches the unit impulse in the limit $a \to \infty$.

14-37 Use $(t - \tau) = u$, $\tau = (t - u)$, $d\tau = -du$. Careful about the two limits. Result follows.

14-38 See Fig. 14-34.

14-39 $t < -1$: $y(t) = 0$. $-1 < t < 3$: $y(t) = \int_{-1}^{t} 15e^{-2(t-\tau)} d\tau = 7.5(1 - e^{-2(t+1)})$.

$t > 3$: $y(t) = \int_{-1}^{3} 15e^{-2(t-\tau)} d\tau = 7.5e^{-2t}(e^6 - e^{-2})$.

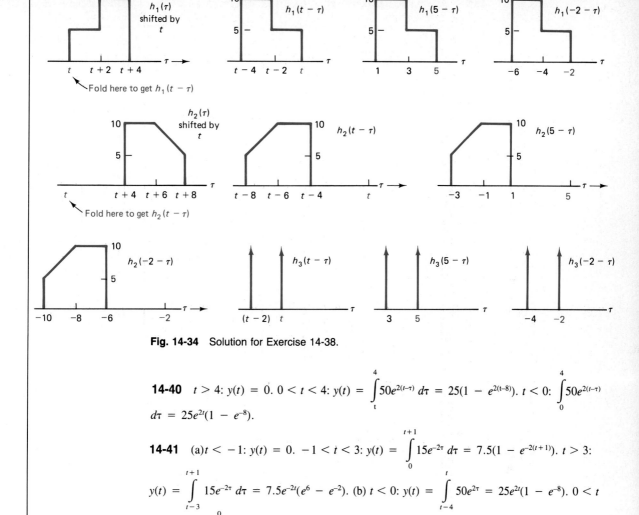

Fig. 14-34 Solution for Exercise 14-38.

14-40 $t > 4$: $y(t) = 0$. $0 < t < 4$: $y(t) = \int_t^4 50e^{2(t-\tau)} d\tau = 25(1 - e^{2(t-8)})$. $t < 0$: $\int_0^4 50e^{2(t-\tau)} d\tau = 25e^{2t}(1 - e^{-8})$.

14-41 (a) $t < -1$: $y(t) = 0$. $-1 < t < 3$: $y(t) = \int_0^{t+1} 15e^{-2\tau} d\tau = 7.5(1 - e^{-2(t+1)})$. $t > 3$: $y(t) = \int_{t-3}^{t+1} 15e^{-2\tau} d\tau = 7.5e^{-2t}(e^6 - e^{-2})$. (b) $t < 0$: $y(t) = \int_{t-4}^t 50e^{2\tau} = 25e^{2t}(1 - e^{-8})$. $0 < t < 4$: $y(t) = \int_{t-4}^0 50e^{2\tau} d\tau = 25(1 - e^{(2t-8)})$. $t > 4$: $y(t) = 0$.

PROBLEMS

Sec. 14-1 Fourier Series Representation of a Periodic Signal

14-1 Determine the coefficients a_0, a_m, and b_m of the Fourier series for the signal shown in Fig. 14-35. Evaluate the coefficients up to $m = 5$. Write out the series showing the numerical values through the fifth harmonic.

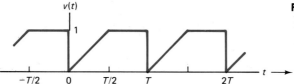

Fig. 14-35 Series for Problem 14-1.

14-2 A periodic function is given by $f(t) = e^{-t}$ $(0 < t < 1)$. $T = 1$ s. Repeat the work of Problem 14-1 for this.

14-3 Obtain the Fourier series of the following functions. (*Hint:* Try using well-known trigonometric formulas rather than formulas for a_m and b_m.)
(a) $f_a(t) = \cos^2 1000t$.
(b) $f_b(t) = 8 \sin 50t \sin 100t \cos 250t$.
(c) $f_c(t) = 10[1 + \cos(2000t)] \cos(15 \times 10^4 t)$.

14-4 The Fourier series of a periodic signal is given by

$$f(t) = \sum_{k=1}^{\infty} \left\{ \frac{-20}{k^2 \pi^2} \cos 200k\pi t - \frac{20}{k\pi} \sin(200k\pi t) \right\}.$$

(a) Calculate the period of the waveform. (b) Determine the amplitude and phase of the kth harmonic. (c) Write the third and fifth harmonics in the form of a single sinusoidal term with a phase angle.

14-5 The output $v_0(t)$ and input $v_i(t)$ of a nonlinear device are related by

$$v_0(t) = v_i + v_i^2$$

If the input to the device is $v_i(t) = 10 \cos 800t + 5 \cos 1600t$, obtain the Fourier series of $v_0(t)$.

14-6 (a) Determine the Fourier series of the signal shown in Fig. 14-36(a). (b) Use the results obtained in part (a) to obtain the series for the signal shown in Fig. 14-36(b). (c) Use the results obtained in the previous parts to obtain the series for the signal shown in Fig. 14-36(c).

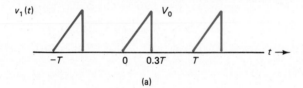

(a)

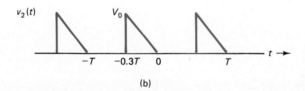

(b)

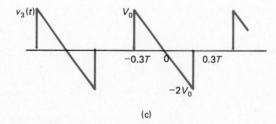

(c)

Fig. 14-36 Signals for Problem 14-6.

14-7 The Fourier series of the waveform shown in Fig. 14-37(a) is given by

$$f(t) = 0.5 + 0.637 \cos t - 0.212 \cos 3t + 0.127 \cos 5t - 0.0909 \cos 7t + \ldots$$

Use the above series and obtain the series for the signal $g(t)$ shown in Fig. 14-37(b). (Do not obtain the series for $g(t)$ by using the formulas of Fourier coefficients directly.)

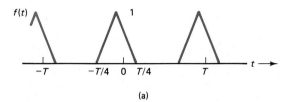

(a)

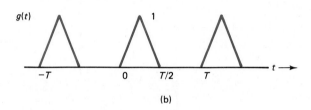

(b)

Fig. 14-37 Waveform for Problem 14-7.

14-8 Figure 14-38 shows a *portion* of a cycle of a periodic function $f(t)$. (a) Sketch the remainder of the cycle so that there is no dc component in the signal. (b) Sketch the remainder of the cycle so that the series expansion does not have any cosine terms. (c) Sketch the remainder of the cycle so that the series does not have any sine terms. (d) Sketch the remainder of the cycle so that the even harmonics are absent from the series expansion.

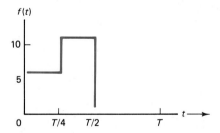

Fig. 14-38 Cycle portion for Problem 14-8.

14-9 For the waveform shown in Fig. 14-39, (a) choose a set of axes so that the function has even symmetry and (b) choose a set of axes so that the function has odd symmetry. In each case, evaluate the appropriate coefficients of the third and fourth harmonics (only).

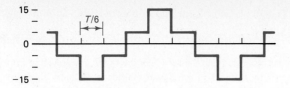

Fig. 14-39 Waveform for Problem 14-9.

Sec. 14-2 Steady State Response of a Circuit with Periodic Input:

14-10 A periodic voltage $v(t)$ whose Fourier series is given by

$$v_s(t) = 15 + 31.8 \cos 500t - 15.9 \cos 1500t + 0.106 \cos 2500t$$

is applied to the circuit shown in Fig. 14-40. Obtain the expression for the voltage $v_o(t)$.

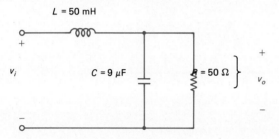

Fig. 14-40 Circuit for Problem 14-10.

14-11 The input to the circuit shown in Fig. 14-41 is a half-wave rectified sinusoid of amplitude 100 V and fundamental frequency 60 Hz (or 377 r/s). Determine the amplitudes of (a) the dc component, (b) the fundamental frequency component, and (c) the second harmonic component of the output voltage $v_o(t)$.

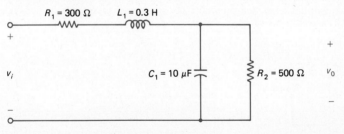

Fig. 14-41 Circuit for Problem 14-11.

14-12 A periodic voltage $v_s(t)$ given by

$$v_s(t) = 50 + 30 \cos 1000t + 15 \cos 2000t + 10 \cos 3000t \text{ V}$$

is applied to a series RLC circuit with $R = 500\Omega$. Determine the values of L and C so that the component of the current in the circuit at 2000 r/s is in phase with the corresponding input voltage component and the amplitude of the voltage across the capacitor at that frequency is 45 V. Also, obtain the expression for $i(t)$ in the circuit.

14-13 A periodic current $i_s(t)$ given by

$$i_s(t) = 10 + 3.5 \cos 500t - 1.75 \cos 1000t + 0.875 \cos 1500t - 5 \sin 500t + 2.5 \sin 1500t - 1.25 \sin 2000t$$

is applied to a parallel combination of $G = 0.1$ S and $C = 100$ μF. Obtain the expression for the voltage across the circuit.

14-14 Determine the average power consumed by the circuit in Problem 14-12.

14-15 Determine the average power consumed by the circuit in Problem 14-13.

14-16 Determine the rms value of each of the functions of Problem 14-3.

14-17 The voltage applied to a circuit is

$$v_s = 100 + 70.7 \cos 400t + 25 \cos 1500t \text{ V}$$

and the resulting current is

$$i_s = 50 \cos (400t - 60°) + 10 \cos (1500t + 25°) \text{ A}.$$

(a) Determine the rms value of the voltage v_s. (b) Determine the rms value of the current i_s. (c) Determine the average power delivered to the circuit.

Sec. 14-3 Exponential Fourier Series:

14-18 Obtain the exponential form of the Fourier series of the signal in Problem 14-1 directly.

14-19 Obtain the exponential series for the signal in Problem 14-2 directly.

14-20 A periodic voltage is given by

$$v(t) = \sum_{k=-\infty}^{\infty} \frac{3}{4 + k^2\pi^2} e^{jk(200t - 0.125\pi)}.$$

(a) Calculate the period of the signal. (b) Calculate the dc component of $v(t)$. (c) Calculate the amplitude and phase angle of the first three harmonics.

14-21 Convert the series of Problem 14-4 to the exponential form.

14-22 A periodic voltage is given by

$$v(t) = 17.8 \cos (1000t + 93.4°) + 25 \cos (3000t + 153.7°) + 9.29 \cos (5000t + 21.8°) \text{ V}.$$

Obtain the exponential series for the signal.

14-23 The coefficient c_k of the exponential Fourier series of a periodic signal is given by

$$c_k = [0.432/(1 + j3.14k)]e^{-jk\pi/4}.$$

Determine a_k and b_k of the trigonometric form of the series.

Secs. 14-4, 14-5 Steady State Response and Frequency Spectra:

14-24 A periodic voltage whose coefficients c_k are those given in Problem 14-23 is applied to the circuit of Problem 14-10. Determine the output voltage $v_0(t)$ in the form of a complex exponential series. Use $\omega = 500$ rls. Calculate the amplitude and phase of the first three harmonics of the output.

14-25 A periodic sawtooth current waveform is described by $i(t) = 5(t/T)$ A in the interval (0, T). Assume a fundamental frequency of 100 r/s. The current is fed to a parallel circuit with $R = 0.25\Omega$ and $L = 1.25$ mH. Obtain the series for the current in the inductance. Calculate the output voltage at the following frequencies: 100 r/s and 300 r/s.

14-26 Sketch the amplitude and phase spectra of the signal in Problem 14-19.

14-27 Sketch the amplitude spectra of the input current and the current in the inductance of the circuit in Problem 14-23.

14-28 The amplitude and phase spectra of a periodic signal are shown in Fig. 14-42. (a) Sketch the power spectrum. (b) Calculate the total power content of the signal (standard load of 1 ohm). (c) Write the trigonometric series for the signal.

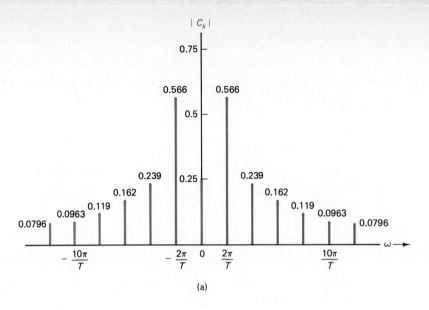

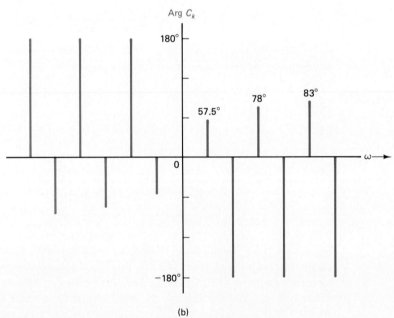

Fig. 14-42 Amplitude and phase spectra for Problem 14-28.

14-29 The coefficients c_k of the Fourier series of a periodic voltage signal are given by

$$c_k = \frac{j2k}{1 - j2k}$$

(a) Sketch the amplitude spectrum, and (b) sketch the power spectrum for a fundamental frequency of 100 r/s. (c) The voltage is fed to a low pass RC filter with cutoff frequency at 300 r/s. Sketch the amplitude and power spectrum of the output voltage of the filter.

14-30 The transfer function of an *ideal* low pass filter is given by

$$H(j\omega) = 10e^{-j(\omega/10^4)} \qquad -10^4 < \omega < 10^4$$

and zero outside the above frequency range. The input $x(t)$ is a pulse train of amplitude 5 V, period $T = 2$ ms, and pulse width $t_p = 0.2$ ms. Assume $x(t)$ is an even function. (a) Obtain the exponential series of the output of the filter. (b) Sketch the amplitude and power spectra of the output. (c) Write $v_0(t)$ as a *real* time function.

14-31 The periodic voltage of Problem 14-1 has Fourier series coefficients given by
$a_0 = 0.75$
$a_k = -(2/k^2\pi^2)$ for odd k and 0 for all even k.
$b_k = -(1/k\pi)$ for all k.
(a) Determine the average power contained in the signal by writing the expression for $f(t)$ of Fig. 14-35 and using

$$P_{av} = (1/T) \int_{-T/2}^{T/2} [f(t)]^2 \, dt.$$

(b) Sketch the power spectrum of the signal using the given coefficients. Starting with the dc component, and proceeding in both directions along the frequency axis, determine the frequency at which the average power equals (a) 80% of the total average power in the signal, and (b) 90% of the total average power in the signal.

Sec. 14-7 Fourier Transform:

14-32 Determine the Fourier transform of the signal

$$f(t) = -\frac{At}{t_p} \qquad 0 < t < t_p$$

and zero everywhere else. Draw the amplitude density and energy density spectra.

14-33 (a) Determine the Fourier transform of a cosine pulse defined by

$$f(t) = \cos t \qquad -(\pi/2) < t < (\pi/2)$$

and zero for all other t.

14-34 Starting with the Fourier transform of a rectangular pulse (Example 14-8), obtain the Fourier transform of each of the signals shown in Fig. 14-43. The functions shown are zero outside the interval covered by the diagrams.

Fig. 14-43 Signals for Problem 14-34.

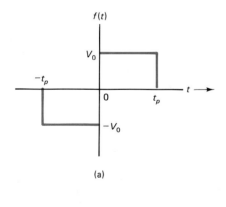

14-35 Determine the Fourier transform of a trianguar pulse with two different choices of origin, as shown in Fig. 14-44. Verify that $|F(j\omega)|$ is the same in both cases.

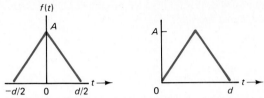

Fig. 14-44 Triangular pulses for Problem 14-35.

14-36 Given a signal $x(t) = 5e^{-10t} u(t)$: (a) determine and sketch its energy density spectrum; (b) calculate the frequency at which the energy density is one-half of its largest value; and (c) calculate the total energy contained in the signal.

14-37 Given that $\mathcal{F}[f(t)] = F(j\omega)$, prove the following theorems:
(a) $\mathcal{F}[f(t)e^{j\omega_0 t}] = F[j(\omega - \omega_0)]$
(b) $\mathcal{F}[f(t)\cos(\omega_0 t)] = (1/2)\{F[j(\omega - \omega_0)] + F[j(\omega - \omega_0)]\}$
(c) The amplitude density spectrum of a signal $f(t)$ is given as being equal to 20 volts/radians/second in the interval $-10 < \omega < 10$ and zero outside that band of frequencies. Sketch the amplitude density spectrum of $f(t) \cos 5t$.

14-38 A function $f(t)$ has the Fourier transform:
$F(j\omega) = 10e^{-[0.01|\omega| - j \text{ arc tan } (0.01\omega)]}$ for all ω.
(a) Obtain an expression for the magnitude $|F(j\omega)|$. (b) Sketch the amplitude density spectrum $|F(j\omega)|$ versus ω. Indicate significant values on your sketch. (c) Obtain an expression for the phase function Arg $[F(j\omega)]$. (d) Sketch the phase spectrum. Indicate significant values on your sketch.

Sec. 14-8 Inverse Fourier Transformation:

14-39 The Fourier transform of a signal $f(t)$ is given by

$$F(j\omega) = \frac{1}{1 + j\omega}(e^{j3\omega} - e^{-j3\omega}).$$

Given that the inverse Fourier transform of $\frac{1}{1 + j\omega}$ is $e^{-t}u(t)$ and the result of Problem 14-37, determine $f(t)$. Simplify it as far as possible.

14-40 The Fourier transform of a signal $f(t)$ is given by

$F(j\omega) = K$, a constant in the intervals $-\omega_b < \omega < -\omega_a$ and $\omega_a < \omega < \omega_b$

and zero for all other ω. Determine $f(t)$ by using Eq. (14-40).

Sec. 14-9 Analysis of Linear Networks:

14-41 A signal $f(t)$ has the Fourier transform

$$F(j\omega) = \frac{1}{4 + j\omega} \text{ for all } \omega.$$

The signal is passed through a low pass filter with a cutoff at 4 r/s. Calculate the energy contained in (a) the input signal, and (b) the output signal.

14-42 A voltage pulse $v(t) = [u(t) - u(t - 1)]$ V is applied to a series RC circuit with $R = 0.5\Omega$ and $C = 1$ F. (a) If the output voltage is taken across the capacitor, determine $V_c(j\omega)$. (b) If the output voltage is taken across the resistor, determine $V_R(j\omega)$. (c) On a single graph sheet, draw neat and informative plots of the amplitude density spectra of the input, the voltage across the capacitor, and the voltage across the resistor, all to the same scale.

Secs. 14-11 Convolution:

14-43 Let $f(t)$ be a single cosine pulse defined by

$$f(t) = \cos \pi t \quad -0.5 < t < 0.5 \text{ s}.$$

Determine the convolution of $f(t)$
(a) with itself
(b) with $g(t) = [u(t + 0.5) - u(t - 0.5)]$ and
(c) with $g(t) = e^{-2t}u(t)$.

14-44 Let $f(t)$ be a single symmetric triangular pulse with a peak value of 5 at $t = 0$ with a slope of $+5$ when $-1 < t < 0$, and a slope of -5 when $0 < t < 1$. $f(t) = 0$ outside the interval $(-1, 1)$. Determine the convolution of $f(t)$ with a periodic train of unit impulses $\sum_{n=-\infty}^{\infty} \delta(t - nT)$ for the following values of the period T: (a) $T = 1.5$ s. (b) $T = 2$ s. (c) $T = 3$ s.

14-45 The impulse response $h(t)$ of a linear network is given by

$$h(t) = -t \quad 0 < t < 1$$

and zero for all other t. The input to the network is a unit step function. Determine the output $v_0(t)$ of the network.

14-46 The impulse response $h(t)$ of a linear network is given by

$$h(t) = \begin{cases} (-2t + 3) & 0 < t < 1 \\ (-0.5t + 1.5) & 1 < t < 3 \\ 0 & \text{for all other } t \end{cases}$$

The input to the network is the *ramp* function $tu(t)$. Determine the output $y(t)$ for the following *specific* values of t: (a) $t = 0.5$ s. (b) $t = 2$ s. (c) $t = 4$ s.

14-47 If

$$f(t) = (-0.5t + 1) \quad 0 < t < 2$$

and zero for all other t, and

$$g(t) = 4\delta(t) + 2\delta(t - 1) + \delta(t - 2),$$

find the convolution $r(t) = [f(t) \circledast g(t)]$. Obtain the result by (a) shifting $f(t)$, and (b) shifting $g(t)$.

14-48 (a) Obtain the transfer function $H(j\omega)$ of a low pass filter with $R = 2\Omega$ and $C = 0.5$ F. (b) Determine the impulse response $h(t)$. (c) If the input to the filter is a rectangular pulse of unit amplitude in the interval $(0, 1)$, obtain the output by using convolution.

14-49 Repeat the work of the previous problem when the output is taken across R in the filter.

14-50 When the input to a linear network is a rectangular pulse $v_i(t) = [u(t - 1) - u(t - 3)]$, the output is found to be $v_0(t) = 10[u(t - 4) - u(t - 8)]$. Determine the transfer function $H(j\omega)$ of the network. Sketch the magnitude and phase of $H(j\omega)$.

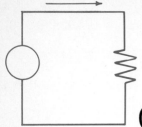

CHAPTER 15
LAPLACE TRANSFORMATION

The methods of network analysis of the preceding chapters permit us to analyze linear networks under various types of forcing functions: constant voltage or current, sinusoidal signals, periodic nonsinusoidal signals, nonperiodic signals that have Fourier transforms, and time domain analysis using convolution in the case of arbitrary signals. Our discussion dealt primarily with the steady state (forced) response of linear networks rather than their complete (natural and forced) response, except in a few simple cases (*RL* and *RC* circuits under constant forcing functions). Convolution yields the complete response, of course, and the response obtained through Fourier transforms also includes both the natural and forced response components.

Even though convolution leads to the complete response in the time domain for arbitrary signals, analysis in the frequency domain offers certain advantages, such as the concepts of transfer functions and critical frequencies, which are powerful tools in the study of networks and their synthesis. The complex frequency introduced informally in Chapter 12 provided the appropriate frequency domain in the case of arbitrary signals.

The discussion of Fourier transforms showed that the condition to be satisfied by a signal (to have a Fourier transform) is the finite area condition: the area enclosed by $|f(t)|$ must be finite. In order to extend the idea of Fourier transform to a wider variety of signals, the finite area condition needs to be modified. The real part σ of the complex frequency $s = (\sigma + j\omega)$ permits the modification of the finite area condition and makes it applicable to a large variety of signals. The transformation obtained by using the complex frequency is called *Laplace transformation*, which converts functions of time to functions of complex frequency. Laplace transformation exists for almost all signals of practical interest.

Important properties and basic applications of Laplace transformation in network analysis will be presented in this chapter. It will also provide a rigorous foundation for the concept of complex frequency introduced in Chapter 12.

15-1 THE LAPLACE TRANSFORM

The Fourier transform description of a signal uses the component function $e^{-j\omega t}$, which has the drawback that it does not converge when $t \to -\infty$ or $t \to +\infty$. Consider the function $e^{-j\omega t}$ multiplied by $e^{-\sigma t}$, where σ is real. The product $(e^{-\sigma t})(e^{-j\omega t})$ may be made to converge when $t \to \infty$ or $t \to -\infty$ by choosing an appropriate range of values of σ. The product $(e^{-\sigma t})(e^{-j\omega t})$ is denoted by e^{-st}, where the variable s is defined by

$$s = (\sigma + j\omega). \tag{15-1a}$$

That is,

$$\sigma = \mathbf{Re}(s) \quad \text{and} \quad \omega = \mathbf{Im}(s) \tag{15-1b}$$

The variable s is called the *complex frequency*. Note that $e^{-j\omega t}$ is a special case of e^{-st} when $s = j\omega$ or $\sigma = 0$. Consider the integral of the forward Fourier transform with $j\omega$ replaced by the new variable s. Then

$$F(s) = \int_{-\infty}^{\infty} f(t) e^{-st}\, dt \tag{15-2}$$

$F(s)$ is called the *Laplace transform* of $f(t)$. The lower limit of the integral in Eq. (15-2) is $-\infty$, since a function $f(t)$, in general, exists for all t ($-\infty < t < \infty$).

In most practical systems, however, the instant at which a system is subjected to a forcing function $f(t)$ is taken as $t = 0$. The system does not respond to $f(t)$ until $f(t)$ is applied to it at $t = 0$. Such systems are called *causal systems* and the signal $f(t)$ is called a *causal signal*. The lower limit in Eq. (15-2) is zero for causal signals. That is,

$$F(s) = \int_0^{\infty} f(t) e^{-st} dt \quad \text{(causal signals)} \tag{15-3}$$

Even though Eq. (15-3) is only a special case of (15-2), it is customary to refer to the relationship in Eq. (15-2) as the *two-sided* (or *bilateral*) Laplace transform, whereas Eq. (15-2) for causal signals is referred to as the *one-sided* (or *unilateral*) Laplace transform. The one-sided Laplace transform is usually called simply the Laplace transform without any qualifying adjective, since causal signals are commonly used in circuit analysis. The existence and nature of the transform function $F(s)$ itself are not uniquely controlled by whether a signal is causal or not. Therefore, the adjective "bilateral" provides just a distinction to the definition of the signal $f(t)$ in the *time domain*.

15-1-1 Region of Convergence for the Laplace Transform

In order for a function $f(t)$ to have a Laplace transform $F(s)$, it is necessary to satisfy the condition

$$\int_{-\infty}^{\infty} |f(t)e^{-st}| dt < \infty \qquad (15\text{-}4)$$

which leads to the condition

$$\int_{-\infty}^{\infty} |f(t)| e^{-\sigma t} dt < \infty \qquad (15\text{-}5)$$

since $|e^{-st}| = e^{-\sigma t}$. Even if $|f(t)|$ itself does not vanish at the two extreme limits $t = \pm\infty$, it is possible to find an appropriate range of values of σ so that the condition in Eq. (15-5) is satisfied for a given signal.

The range of values of σ for which the condition in Eq. (15-5) is satisfied defines a region of convergence in the complex frequency plane over which the Laplace transform of the given $f(t)$ exists.

There are functions that do not have a region of convergence, and such functions do not have a Laplace transform anywhere in the s plane. Fortunately, almost all signals of practical interest do have a Laplace transform.

Example 15-1 Determine the Laplace transform of

$$f(t) = e^{at} u(t)$$

where a is a constant (positive or negative). Identify the region of convergence of the transform.

Solution Using Eq. (15-3), since $f(t) = 0$ for $t < 0$,

$$F(s) = \int_{0}^{\infty} e^{at} u(t) e^{-st} dt$$

$$= \int_{0}^{\infty} e^{(a-s)t} dt \qquad (15\text{-}6)$$

$$= \frac{1}{s - a}$$

The condition to be satisfied for convergence is, from Eq. (15-5):

$$\int_{-\infty}^{\infty} |f(t)| e^{-\sigma t} dt < \infty$$

That is,

$$\int_{-\infty}^{\infty} e^{(a - \sigma)t} dt < \infty \qquad (15\text{-}7)$$

which is satisfied if the exponential function $e^{(a-\sigma)t}$ decays to zero at $t = \infty$. Therefore, the region of convergence is given by

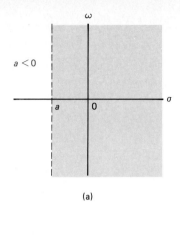
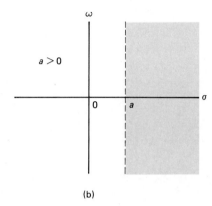

Fig. 15-1 Regions of convergence of the Laplace transform of $e^{at}u(t)$. (a) $a < 0$. (b) $a > 0$.

$$(a - \sigma) < 0 \quad \text{or} \quad \sigma > a$$

The regions of convergence for the two cases $a < 0$, and $a > 0$ are shown in Fig. 15-1 (a) and (b). When $a < 0$, the Laplace transform exists in the entire right half of the s plane as well as a portion of the left half; when $a > 0$, the transform exists only over a portion of the right half of the plane. When $a > 0$, the given function has a Laplace transform, but does not have a *Fourier* transform. ∎

Exercise 15-1 Determine the Laplace transform of a unit step function $u(t)$ and the region of convergence of the transform.

Exercise 15-2 Repeat the previous exercise for a step in the *third* quadrant: $f(t) = -u(-t)$.

Exercise 15-3 Find the Laplace transform of a rectangular pulse of unit amplitude in the interval $(-b, b)$ on the time axis. Identify the region of convergence.

The example and exercises show that the region of convergence for the Laplace transform of a function is, in general, a strip in the s plane whose sides are parallel to the vertical axis. In some cases, the strip occupies the entire s plane; in other cases, it occupies only a portion of the s plane. An example of a function that does not have a region of convergence for its Laplace transform is

$$f(t) = e^{b|t|}$$

where b is a positive constant, as Fig. 15-2 shows.

15-1 The Laplace Transform

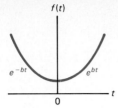

Fig. 15-2 $f(t) = e^{|b|t}$ does not have a region of convergence for its Laplace transform.

15-1-2 Causal Signals and Systems

Causal signals and systems were mentioned earlier in connection with Eq. (15-3). *Physically realizable* systems have the property that they do not produce a response to a signal until that signal is applied; that is, the signal acts as the cause, and the response of the system is the effect. Such systems are known as *causal systems*. Any signal that corresponds to the impulse response of a causal system is defined as a causal signal. Accepted common practice, however, is to refer to *any signal which is identically zero for* t < 0 *as a causal signal.*

It is usually convenient (especially when making observations or measurements) to define the particular instant of starting the observations or measurements as $t = 0$. Even if the system has already been under the influence of some forcing functions before $t = 0$, the effect of the earlier forcing functions is taken into account by means of the energy stored in the inductors and capacitors at $t = 0-$. It is therefore no loss of generality to concentrate on the response of linear systems to causal signals, and Eq. (15-3) is used for finding the Laplace transform. That is,

$$F(s) = \int_0^\infty f(t)e^{-st}dt \qquad (15\text{-}8)$$

In certain circuits, there may be a change of conditions occurring at the instant $t = 0$; for example, a switch may be closed at $t = 0$, activating a source at that instant. In such cases, the *lower limit* of Eq. (15-3) is taken as $t = 0-$.

The region of convergence for the Laplace transform of a causal signal lies to the right of a vertical line in the s plane; that is, $\sigma > \sigma_0$.

A function that satisfies the condition

$$\lim_{t \to \infty} f(t)e^{-\sigma_0 t} = 0 \quad \text{when } \sigma > \sigma_0 \qquad (15\text{-}9)$$

is said to be of *exponential order* σ_0, and for such functions the Laplace transform has a region of convergence given by $\sigma > \sigma_0$.

If $f(t)$ is *not* an exponentially increasing function, then the region of convergence is the *entire right half* of the s plane, and part of the left half of the plane as well, depending upon the actual form of $f(t)$.

If $f(t)$ is an exponentially increasing function of the form e^{at}, where $a > 0$, then the region of convergence is to the right of the vertical line at $\sigma = a$.

There are functions (t^t for example) for which the region of convergence does not exist, but such functions are of only theoretical interest and have no practical significance. It may be safely assumed that, for any causal signal, the Laplace transform has a region of convergence in the right half of the complex frequency plane.

Example 15-2 Determine the Laplace transform of the rectangular pulse

$$f(t) = u(t) - u(t - t_0)$$

where t_0 is a positive constant. Find the region of convergence.

Solution Using Eq. (15-3),

$$F(s) = \int_0^\infty [u(t) - u(t - t_0)]e^{-st}\, dt \qquad (15\text{-}10)$$

Since $u(t - t_0) = 0$ for $t < t_0$, and 1 for $t > t_0$, Eq. (15-10) becomes

$$F(s) = \int_0^\infty e^{-st}\, dt - \int_{t_0}^\infty e^{-st}\, dt$$

$$= -\frac{1}{s}e^{-st}\bigg|_0^\infty + \frac{1}{s}e^{-st}\bigg|_{t_0}^\infty$$

$$= \frac{1}{s}[1 - e^{-st_0}]$$

The condition to be satisfied for convergence is

$$\int_0^\infty [u(t) - u(t - t_0)]e^{-\sigma t}\, dt < \infty$$

Since the term $[u(t) - u(t - t_0)]$ is zero at $t = \infty$, the condition is satisfied for all values of $\sigma > 0$. The region of convergence is the entire right half of the s plane. ∎

Example 15-3 Determine the Laplace transform of $f(t) = (\cos bt)u(t)$, where b is a real constant.

Solution

$$F(s) = \int_0^\infty (\cos bt)e^{-st}\, dt$$

Using Euler's theorem,

$$F(s) = \frac{1}{2}\int_0^\infty [e^{jbt} + e^{-jbt}]e^{-st}\, dt$$

$$= \frac{1}{2}\left[\int_0^\infty e^{-(s-jb)t}\, dt + \int_0^\infty e^{-(s+jb)t}\, dt\right]$$

$$= \frac{1}{2}\left[\frac{1}{(s - jb)} + \frac{1}{(s + jb)}\right]$$

$$= \frac{s}{s^2 + b^2}$$

The region of convergence is again the entire right half of the plane. ∎

Exercise 15-4 Determine the Laplace transform of each of the following functions: (a) $e^{-at}u(t)$, where a is a real positive constant. (b) $(\sin bt)u(t)$, where b is a real constant. (c) $\delta(t - t_0)$.

The Laplace transforms of a number of functions are listed in Table 15-1.

Table 15-1 Laplace Transform Pairs and Properties of Laplace Transforms

$f(t)$	$F(s)$	
$\delta(t)$	1	
$u(t)$	$\dfrac{1}{s}$	
$\dfrac{t^n}{n!}u(t)$	$\dfrac{1}{s^{n+1}}$	
$e^{-at}u(t)$	$\dfrac{1}{s+a}$	
$\dfrac{t^n e^{-at}}{n!}u(t)$	$\dfrac{1}{(s+a)^{n+1}}$	Compare with the transform of $(t^n/n!)$
$[\cos(bt)]u(t)$	$\dfrac{s}{s^2+b^2}$	
$[\sin(bt)]u(t)$	$\dfrac{b}{s^2+b^2}$	
$[e^{-at}\cos(bt)]u(t)$	$\dfrac{s+a}{(s+a)^2+b^2}$	Compare with the transform of $\cos bt$
$[e^{-at}\sin(bt)]u(t)$	$\dfrac{b}{(s+a)^2+b^2}$	Compare with the transform of $\sin bt$
$a_1 f(t) + a_2 g(t)$	$a_1 F(s) + a_2 G(s)$	Linearity
$f(t - t_0)$	$e^{-st_0} F(s)$	Time shift
$f(t)e^{at}$	$F(s - a)$	Effect of multiplying by e^{at}
$f(t/a)$	$aF(as)$	Scaling in time
$f(t) \circledast g(t)$	$F(s)G(s)$	Convolution
$\left(\dfrac{df(t)}{dt}\right)$	$sF(s) - f(0-)$	Differentiation
$\int f(t)dt$	$(1/s)[F(s) + \int f(t)dt\,\|_{0-}]$	Indefinite integral
$\int_0^t f(t)dt$	$(1/s)F(s)$	Definite integral

740 Laplace Transformation

15-2 USE OF LAPLACE TRANSFORMS IN ANALYSIS OF LINEAR NETWORKS

The use of Laplace transforms in network analysis follows the same general steps as that of Fourier transforms. But the method of Laplace transform leads to the complete response of circuits (with and without zero initial conditions) subjected to any signal of practical importance.

Laplace transforms convert ordinary linear differential equations into algebraic equations in the variable s. Concepts of impedance, admittance, and transfer functions are again used in the analysis of a network, except that they are functions of s instead of $j\omega$. The response of a network appears as a function of s and may be left in that form or transformed to the time domain by using the *inverse* Laplace transformation.

15-2-1 Some Properties of the Laplace Transform

To apply the Laplace transform to circuit analysis, it is first necessary to examine some of its important properties.

Multiplication by a Constant

If $\mathcal{L}[f(t)] = F(s)$, then

$$\mathcal{L}[af(t)] = aF(s) \qquad (15\text{-}11)$$

where a is a constant. The notation $\mathcal{L}[\ ]$ denotes "Laplace transform of the function in brackets." The proof of Eq. (15-11) is straightforward and will not be presented here.

Transform of a Derivative

Let $F(s) = \mathcal{L}[f(t)]$. Consider the function $f'(t) = df/dt$. The transform of $f'(t)$ is given by

$$\mathcal{L}[f'(t)] = \int_0^\infty f'(t) e^{-st}\, dt$$

Using integration by parts,

$$\mathcal{L}[f'(t)] = f(t)e^{-st}\Big|_0^\infty + s\int_0^\infty f(t) e^{-st}\, dt$$

The term $f(t)e^{-st}$ becomes $f(0)$ at the lower limit. Since $F(s)$ exists (in the region of convergence), $f(t)e^{-st}$ goes to zero at $t = \infty$. The integral in the second term is $F(s)$. Therefore,

$$\mathcal{L}[f'(t)] = -f(0) + sF(s)$$

where $f(0)$ represents the *initial value* of the function $f(t)$. In the case of functions with a discontinuity at $t = 0$ (for example a unit step, or $[e^{-at}u(t)]$), the initial value is taken at $t = 0-$. In general then

$$[f'(t)] = sF(s) - f(0-) \qquad (15\text{-}12)$$

For the special case $f(0-) = 0$,

$$[f'(t)] = sf(s) \qquad \text{when } f(0-) = 0 \qquad (15\text{-}13)$$

Thus differentiation in the time domain is equivalent to *multiplication* by s in complex frequency domain (except for including the initial value of the function).

Exercise 15-5 Use the result obtained in Eq. (15-12) and determine the Laplace transforms of the second-order derivative of $f(t)$.

Exercise 15-6 Assuming the Laplace transform of $(\cos bt)\, u(t)$ derived in Example 15-3, determine the Laplace transform of $\sin bt\, u(t)$ using Eq. (15-12).

Transform of a Definite Integral

Consider the definite integral $\int_0^t f(u)\, du$.

$$\mathcal{L}\left[\int_0^t f(u)\, du\right] = \int_0^\infty \int_0^t [f(u)\, du]\, e^{-st}\, dt$$

Again, by using integration by parts,

$$\mathcal{L}\left[\int_0^t f(u)\, du\right] = -\frac{1}{s} e^{-st} \int_0^t f(u)\, du \Big|_0^\infty + \frac{1}{s}\int_0^\infty f(t) e^{-st}\, dt \qquad (15\text{-}14)$$

At the lower limit $t = 0$, the integral $\int_0^t f(u)\, du$ in the first term becomes $\int_0^0 f(u)\, du$ and equals zero. Also, since the function $f(t)$ has a Laplace transform (in the region of convergence), the value of the first term in Eq. (15-14) goes to zero at $t = \infty$. Thus the first term disappears, and the second term becomes $[F(s)/s]$.

Therefore,

$$\mathcal{L}\left[\int_0^t f(u)\, du\right] = \frac{F(s)}{s} \qquad (15\text{-}15)$$

For definite integrals, integration in the time domain becomes a division by s in the complex frequency domain.

Exercise 15-7 Assuming the transform of $\cos bt\, u(t)$ determine the transform of $\sin bt\, u(t)$ using Eq. (15-15).

Transform of an Indefinite Integral

The transform of an indefinite integral is given by

$$\mathcal{L}\left[\int f(t)dt\right] = \frac{F(s)}{(s)} + \frac{1}{s}\int f(t)dt\,\big|_{t\,=\,0-} \qquad (15\text{-}16)$$

15-2-2 Current Voltage Relationships in Basic Circuit Components

The relationships of the three basic elements, R, L, and C, in terms of Laplace transforms lead to impedance and admittance as functions of the complex frequency s.

Resistance

The relationship in the time domain is

$$v(t) = Ri(t)$$

Taking the Laplace transform of both sides and using Eq. (15-11)

$$V(s) = RI(s)$$

where $V(s) = \mathcal{L}[v(t)]$ and $I(s) = \mathcal{L}[i(t)]$. For a resistor

$$\frac{V(s)}{I(s)} = R$$

Refer to Fig. 15-3(a).

Inductance

The relationship for an inductance in the time domain is

$$v(t) = L\frac{di}{dt}$$

Taking the transform of both sides and using Eqs. (15-11) and (15-12)

$$V(s) = L[sI(s) - i(0-)]$$
$$= LsI(s) - Li(0-) \qquad (15\text{-}17)$$

Note that the *initial condition* of the inductance is present in the form of the term $Li(0-)$, which is the *flux linkage* in the coil at $t = 0-$. It is convenient to use a model for an inductance, as in Fig. 15-3(b), where a voltage source is used to represent the term $Li(0-)$.

In some cases, it is more convenient to use the equation

$$i(t) = \frac{1}{L}\int v(t)\,dt$$

for an inductance. Taking the transform on both sides and using Eqs. (15-11) and (15-16)

$$I(s) = \frac{V(s)}{Ls} + \frac{i(0-)}{s} \qquad (15\text{-}18)$$

and the corresponding model is shown in Fig. 15-3(c). The two circuits of Figs. 15-3(b) and (c) are seen to be equivalent to each other; one is the Thevenin model, and the other the Norton model.

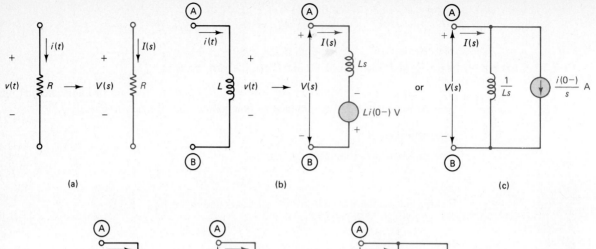

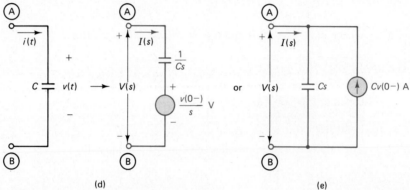

Fig. 15-3 Transform equivalent circuits of single elements. The sources in the models account for the initial energy stored in the element.

In the case of an inductance with no initial stored energy, $i(0-) = 0$, and Eq. (15-17) becomes

$$V(s) = LsI(s)$$

or

$$\frac{V(s)}{I(s)} = Ls \quad \text{(zero initial energy)} \quad (15\text{-}19)$$

The inductance is simply an *impedance* Ls in the s domain, when the initial energy = 0.

Capacitance

By using duality on the relationships obtained for an inductance, the transformed relationships for a capacitance are

$$V(s) = \frac{I(s)}{Cs} + \frac{v(0-)}{s} \quad (15\text{-}20)$$

and

$$I(s) = CsV(s) - Cv(0-) \qquad (15\text{-}21)$$

and the models are shown in Figs. 15-3(d) and (e).

For a capacitor with no initial stored energy

$$V(s) = \frac{1}{Cs}I(s)$$

or

$$\frac{V(s)}{I(s)} = \frac{1}{Cs} \quad \text{(zero initial energy)} \qquad (15\text{-}22)$$

The capacitor is simply as an *impedance* $(1/Cs)$ *in the s domain, when the initial energy* $= 0$.

Exercise 15-8 Show that the voltage current relationship in the time domain for a capacitance leads to Eqs. (15-20) and (15-21). Do each one independently.

When complex frequency was introduced informally in Chapter 12, the impedances of the three basic circuit components were found to be R, Ls, and $(1/Cs)$. (A tacit assumption was made about the initial conditions being zero in the previous treatment.) We now see that those impedance relationships are justified by the properties of Laplace transformation. Other concepts introduced in that chapter, such as poles and zeros of network functions, also have their theoretical basis in Laplace transformation.

15-2-3 Analysis of a Linear Network

Using the previous section, we can replace each of the three basic circuit elements by an equivalent model in the s domain. Now consider a simple series RLC circuit with some arbitrary forcing function $v(t)$ applied at $t = 0$; see Fig. 15-4(a).

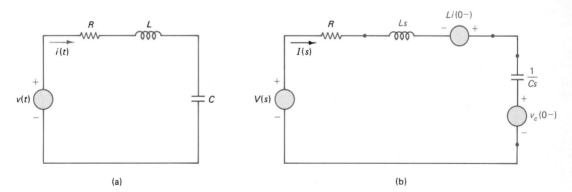

Fig. 15-4 (a) Series RLC circuit under arbitrary excitation and (b) its transform equivalent.

The differential equation of the circuit is

$$L\frac{di}{dt} + Ri + \frac{1}{C}\int i\,dt = v(t)$$

When Laplace transformation is applied to both sides, each term on the left side is replaced by the corresponding expressions in Eqs. (15-17) and (15-20). The differential equation changes to

$$LsI(s) - Li(0-) + RI(s) + \frac{1}{C}\left[\frac{I(s)}{s} + \frac{v(0-)}{s}\right] = V(s) \quad (15\text{-}23)$$

where $V(s) = \mathcal{L}[v(t)]$ and is obtained by referring to Table 15-1. Equation (15-23) leads to

$$I(s) = \frac{V(s) + Li(0-) - v(0-)}{Ls + R + 1/Cs}$$

Instead of writing the differential equation and then replacing each term by the corresponding transform, it is more convenient to replace the elements of the given circuit by their equivalent series or parallel models from Fig. 15-3. The equations of the resulting circuit are then written directly in the s domain by using KVL, KCL, nodal, loop, or mesh analysis. The transform equivalent of the series RLC circuit is as shown in Fig. 15-4(b) and its KVL equation will be exactly the same as Eq. (15-23).

Example 15-4 Determine the current $I(s)$ in the circuit shown in Fig. 15-5(a). Assume that the circuit is in zero initial state.

Solution Replace each element of the circuit by its transform equivalent. Since the initial condition is given as zero, the sources in the models for the inductance and capacitance are absent.

Fig. 15-5 (a) Circuit for Example 15-4 and (b) its transform equivalent.

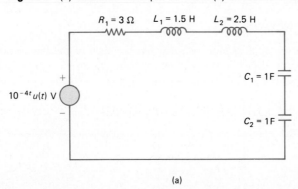

(a)

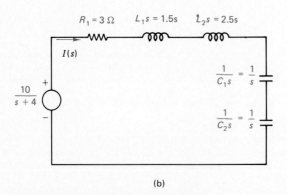

(b)

Laplace Transformation

The forcing function is replaced by its transform, obtained from Table 15-1. The resulting transformed circuit is shown in Fig. 15-5(b). The KVL equation is

$$[R_1 + L_1 s + L_2 s + (1/C_1 s) + (1/C_2 s)]I(s) = \frac{10}{s+4}$$

or

$$(3 + 4s + 2/s)I(s) = \frac{10}{s+4}$$

Therefore

$$I(s) = \frac{10s}{(s+4)(4s^2 + 3s + 2)}$$ ∎

Example 15-5 In the circuit shown in Fig. 15-6(a), assume that $v_c = 15$ V at $t = 0-$. Determine the voltage $V_c(s)$.

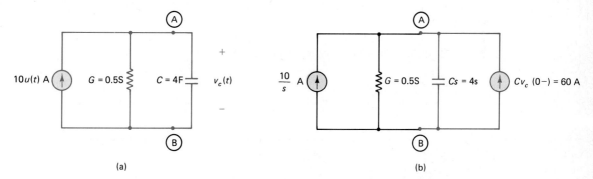

(a)   (b)

Fig. 15-6 (a) Circuit for Example 15-5 and (b) its transform equivalent.

Solution The transformed circuit is shown in Fig. 15-6(b). A parallel model for the capacitance permits analysis through a single KCL equation:

$$(G + Cs)V_c(s) = \frac{10}{s} + Cv_c(0-)$$

That is,

$$(0.5 + 4s)V_c(s) = \frac{10}{s} + 60$$

or

$$V_c(s) = \frac{60s + 10}{s(4s + 0.5)}$$ ∎

Exercise 15-9 Obtain the current $I(s)$ in each of the circuits shown in Fig. 15-7(a) and (b). Assume there is an initial current of 5 A in the inductance. The capacitor voltage is zero at $t = 0-$.

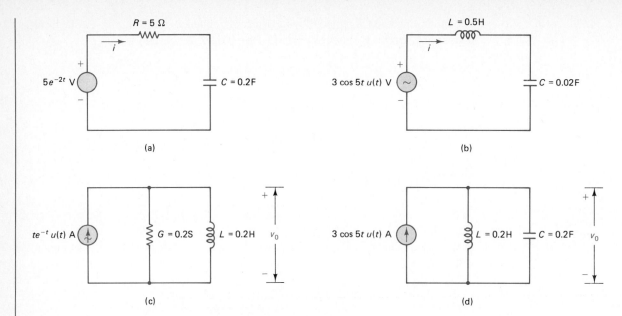

Fig. 15-7 Circuits for Exercises 15-9 and 15-10.

Exercise 15-10 Obtain the voltage $V_0(s)$ in each of the circuits shown in Fig. 15-7(c) and (d). Assume there is an initial current of 2 A flowing downward in the inductance and 4 V in the capacitance, with the polarity indicated.

15-3 DETERMINATION OF THE INVERSE LAPLACE TRANSFORM

The response of a network is obtained as a function of the complex frequency by using the Laplace transform approach. In many cases, the complex frequency description of the response is sufficient for studying the behavior of a network. In other cases, it is necessary to transform the response to the time domain by means of an *inverse Laplace transformation*. The integral formula for the inverse Laplace transform is obtained by replacing $j\omega$ by s in the formula of the inverse Fourier transform of Chapter 14. The evaluation of the resulting integral, however, requires a knowledge of contour integration in the complex plane. Instead of contour integration (which is outside the scope of this book), the method presented here will be to use a table of Laplace transforms. The procedure covers all the functions listed in Table 15-1 and thus includes almost all the important forcing and response functions.

Consider a function $F(s)$ given by

$$F(s) = \frac{P(s)}{Q(s)}$$

where $P(s)$ and $Q(s)$ are polynomials in s and are of the form

$$s^n + a_{n-1}s^{n-1} + \ldots + a_1 s + a_0$$

where $a_{n-1}, a_{n-2}, \ldots, a_1, a_0$ are real constants. A function $F(s)$ that is the *ratio of two polynomials in s* is called a rational function. By factorizing the denominator polynomial $Q(s)$ with each factor of the general form $(s + b)^m$, where m is an integer, it is possible to expand $F(s)$ into a sum of terms of the form

$$\frac{K}{(s + b)^m} \qquad (15\text{-}24)$$

where the values of K are to be determined. Terms of the type in Eq. (15-24) are called *partial fractions*.

For example, consider the function

$$F(s) = \frac{5s + 6}{s^3 + 4s^2 + 5s + 2}$$

The factorized form of the denominator is

$$(s + 1)^2(s + 2)$$

Therefore, the partial fraction expansion of $F(s)$ is given by

$$\frac{5s + 6}{(s + 1)^2(s + 2)} = \frac{K_1}{(s + 1)^2} + \frac{K_2}{(s + 1)} + \frac{K_3}{(s + 2)}$$

where K_1, K_2, and K_3 have to be chosen to satisfy the last equation.

Table 15-1 shows that the expression in Eq. (15-24) is the Laplace transform of the following expression:

$$\frac{K}{(m - 1)!} t^{m-1} e^{-bt} u(t) \qquad (15\text{-}25)$$

Conversely, the expression in Eq. (15-25) is the inverse transform of the partial fraction in Eq. (15-24).

It follows, then, that if a rational function $F(s)$ is expressed as the sum of partial fractions, it is possible to write the inverse transform of $F(s)$ as a sum of the inverse transforms of the partial fraction terms.

Example 15-6 Obtain the inverse Laplace transform of

$$F(s) = \frac{2s + 10}{(s + 4)(s + 6)}$$

Solution The denominator is already factorized, and the given function is found to have the partial fraction expansion

$$F(s) = \frac{1}{s + 4} + \frac{1}{s + 6}$$

(A routine procedure for finding the partial fraction expansion of a rational function will be presented below.)

From Table 15-1

$$\mathcal{L}^{-1} \frac{1}{s + a} = e^{-at} u(t).$$

where $\mathcal{L}^{-1}$ denotes "the inverse Laplace transform of." Therefore, the inverse transform of the given function is

15-3 Determination of the Inverse Laplace Transform

$$f(t) = e^{-4t}u(t) + e^{-6t}u(t) = [e^{-4t} + e^{-6t}]u(t)$$

Example 15-7 Determine the inverse transform of

$$F(s) = \frac{16s + 80}{(s + 2)^2(s + 6)}$$

Solution The partial fraction expansion of $F(s)$ is found to be

$$F(s) = \frac{12}{(s + 2)^2} + \frac{1}{s + 2} - \frac{1}{s + 6}$$

Refer to Table 15-1:

$$f(t) = (12te^{-2t} + e^{-2t} - e^{-6t})u(t)$$

Exercise 15-11 Given the function

$$F(s) = \frac{4s + 16}{(s + 5)(s + 1)}$$

put it in the form

$$F(s) = \frac{K_1}{(s + 5)} + \frac{K_2}{(s + 1)}$$

Evaluate K_1 and K_2 by expanding the second expression and equating it to the first. Write the function $f(t)$.

Exercise 15-12 Given the function

$$F(s) = \frac{4s + 16}{s(s + 2)^2}$$

put it in the form

$$F(s) = \frac{K_1}{s} + \frac{K_2}{(s + 2)^2} + \frac{K_3}{(s + 2)}$$

Evaluate K_1, K_2, and K_3 using the same procedure as in the previous exercise. Write the function $f(t)$.

15-3-1 Procedure for Finding the Partial Fraction Expansion of F(s)

Consider a rational function $F(s)$ given by

$$F(s) = \frac{P(s)}{Q(s)}$$

where $P(s)$ and $Q(s)$ are polynomials, all of whose coefficients are real. Assume that the order of $P(s)$ is *less than* that of $Q(s)$; that is, the highest power of s in the numerator is less than the highest power of s in $Q(s)$. If this is not the case, *the numerator polynomial* is first divided by the denominator polynomial until a ratio $P(s)/Q(s)$ is obtained, with the property that $P(s)$ is of a lower order than $Q(s)$.

In order the expand $F(s)$ into a sum of partial fractions, it is first necessary to find the factors of the denominator polynomial and write that polynomial in the form

$$Q(s) = (s - s_1)(s - s_2) \ldots (s - s_n)$$

where $s_1, s_2, \ldots, s_n$ are the roots of $Q(s)$.

Some or all of the roots of $Q(s)$ may repeat themselves in the factorized form of $Q(s)$; that is, they are of a *multiple order*. If a root s_j appears as a factor $(s - s_j)^k$ in $Q(s)$, where k is an integer, then the root is of order k.

Poles of F(s)

The poles of a function $F(s)$ were defined in Chapter 12 as those values of s for which the function assumes a value of ∞. When $F(s)$ is expressed as the ratio $[P(s)/Q(s)]$, the value of $F(s)$ is infinity whenever $Q(s) = 0$. Therefore, the roots of $Q(s)$ are poles of $F(s)$, and the partial fraction expansion of $F(s)$ is about its poles. A kth order root of $Q(s)$ becomes a kth order pole of $F(s)$. Poles due to roots which are not of multiple order are called *simple poles* of $F(s)$.

Suppose $F(s)$ has some complex poles. A complex pole by itself will lead to complex coefficients in the polynomial $Q(s)$. But all the coefficients of the polynomials in a rational function $F(s)$ must be real. All the coefficients of $Q(s)$ will be real if and only if each of its complex factors is rationalized by the presence of the complex conjugate of that factor, since the product of two complex conjugate factors has only real coefficients. It follows, then, that *any complex poles of $F(s)$ must occur in complex conjugate pairs*.

There are two different cases to be considered for developing the rules of partial fraction expansion: (1) All poles of $F(s)$ are simple, and (2) $F(s)$ has one or more multiple poles.

Case 1: All Poles of F(s) Are Simple

The function $F(s)$ is then of the form

$$F(s) = \frac{P(s)}{(s - s_1)(s - s_2) \ldots (s - s_n)}$$

where each root s_j is distinct from the others.

Then

$$F(s) = \frac{K_1}{(s - s_1)} + \frac{K_2}{(s - s_2)} + \ldots + \frac{K_j}{(s - s_j)} + \ldots + \frac{K_n}{(s - s_n)} \quad (15\text{-}26)$$

The constants $K_1, K_2, \ldots, K_j, \ldots, K_n$ are called the *residues* of the function $F(s)$ at the corresponding poles. If both sides of Eq. (15-26) are multiplied by $(s - s_j)$, then

$$(s - s_j) F(s) = \frac{K_1(s - s_j)}{(s - s_1)} + \ldots + K_j + \ldots + \frac{K_n(s - s_j)}{(s - s_n)} \quad (15\text{-}27)$$

If both sides of Eq. (15-27) are evaluated when $s = s_j$, then, except for the term K_j, all the other terms on the right side disappear. That is

$$(s - s_j)F(s)\big|_{s=s_j} = K_j \quad (15\text{-}28)$$

Therefore, the *residue at any simple pole s_j of $F(s)$ is obtained by finding the value of the product $[(s - s_j)F(s)]$ at that pole.*

Example 15-8   Expand each of the following functions of s into a sum of partial fractions.

(a) $F_1(s) = \dfrac{s + 12}{(s + 4)(s + 6)}$

(b) $F_2(s) = \dfrac{s^2 + 3s + 1}{s^3 + 4s^2 + 3s}$

Solution   (a) The denominator of $F_1(s)$ is already factorized, and the poles are at $s = -4$ and $s = -6$. Therefore, the partial fraction expansion is

$$\frac{s + 12}{(s + 4)(s + 6)} = \frac{K_1}{(s + 4)} + \frac{K_2}{(s + 6)}$$

Equation (15-28) is used to calculate the residues.

Residue at $s = -4$:

$$K_1 = (s + 4)F_1(s)\big|_{s = -4} = \frac{s + 12}{s + 6}\bigg|_{s=-4} = 4$$

Residue at $s = -6$:

$$K_2 = (s + 6)F_1(s)\big|_{s = -6} = \frac{s + 12}{s + 4}\bigg|_{s=-6} = -3$$

Therefore

$$F_1(s) = \frac{4}{(s + 4)} - \frac{3}{(s + 6)}$$

and the inverse transform of $F_1(s)$ is

$$f_1(t) = (4e^{-4t} - 3e^{-6t})u(t)$$

(b) The denominator of $F_2(s)$ is factorized

$$Q(s) = s(s^2 + 4s + 3)$$
$$= s(s + 1)(s + 3)$$

The poles are at $s = 0$, $s = -1$, and $s = -3$. Therefore

$$\frac{s^2 + 3s + 1}{s(s + 1)(s + 3)} = \frac{K_1}{s} + \frac{K_2}{s + 1} + \frac{K_3}{s + 3}$$

The residues are evaluated by using Eq. (15-28).

Residue at $s = 0$:

$$K_1 = sF_2(s)\big|_{s = 0} = \frac{s^2 + 3s + 1}{(s + 1)(s + 3)]}\bigg|_{s=0} = \frac{1}{3}$$

Residue at $s = -1$:

$$K_2 = (s+1)F_2(s)|_{s=-1} = \frac{s^2 + 3s + 1}{s(s+3)}\bigg|_{s=-1} = \frac{1}{2}$$

Residue at $s = -3$:

$$K_3 = (s+3)F_2(s)|_{s=-3} = \frac{s^2 + 3s + 1}{s(s+1)}\bigg|_{s=-3} = \frac{1}{6}$$

Therefore

$$F(s) = \frac{(1/3)}{s} + \frac{(1/2)}{(s+1)} + \frac{(1/6)}{(s+3)}$$

and

$$f(t) = [(1/3) + (1/2)e^{-t} + (1/6)e^{-3t}]u(t) \quad \blacksquare$$

Exercise 15-13 Return to Exercises 15-11 and 15-12 obtain the partial fraction expansions by using Eq. (15-28).

Exercise 15-14 Determine the partial fraction expansion and the inverse transform of each of the following functions:

(a) $\dfrac{2s + 16}{s^2 + 7s + 6}$ (b) $\dfrac{4s}{s^2 - 4}$

Example 15-9 Determine the inverse transform of the function

$$F(s) = \frac{s^2 + 2s + 4}{s^2 + 5s + 6}$$

Solution The numerator is not lower in order than the denominator in the given function. Therefore, first divide the numerator by the denominator to obtain

$$F(s) = 1 + \frac{-3s - 2}{s^2 + 5s + 6}$$

The second term is expanded in partial fractions leading to

$$\frac{-3s - 2}{(s+2)(s+3)} = \frac{K_1}{(s+2)} + \frac{K_2}{(s+3)}$$

Residue at Pole $s = -2$:

$$K_1 = \frac{-3s - 2}{s + 3}\bigg|_{s=-2} = 4$$

Residue at Pole $s = -3$:

$$K_2 = \frac{-3s - 2}{s + 2}\bigg|_{s=-3} = 7$$

15-3 Determination of the Inverse Laplace Transform

Therefore

$$F(s) = 1 + \frac{4}{(s+2)} - \frac{7}{(s+3)}$$

and

$$f(t) = \delta(t) + [4e^{-2t} - 7e^{-3t}]u(t)$$ ∎

Exercise 15-15 Determine the inverse transform of each of the following functions.

(a) $\dfrac{3s^2 - 6}{s^2 + 6s + 8}$ (b) $\dfrac{s^2 + 6s + 9}{s^2 + 3s + 2}$

Complex Conjugate Poles

If $F(s)$ has *complex conjugate* poles that are not of multiple order, the procedure for finding the residues at those poles is exactly the same as for simple real poles. The *residues at a pair of complex conjugate poles* are *complex conjugates* of each other, as shown below.

Consider a function with poles at $s = -(a + jb)$ and $-(a - jb)$ so that

$$F(s) = \frac{P(s)}{(s + a + jb)(s + a - jb)}$$

$$= \frac{K_1}{(s + a + jb)} + \frac{K_2}{(s + a - jb)}$$

The residues are calculated by using Eq. (15-26).

Residue at $s = -(a + jb)$:

$$K_1 = (s + a + jb)F(s)\big|_{s = -(a+jb)}$$

$$= \frac{P(s)}{s + a - jb}\bigg|_{s = -(a+jb)}$$

$$= \frac{P(s)|_{s = -(a+jb)}}{(-j2b)}$$

Residue at $s = -(a - jb)$:

$$K_2 = (s + a - jb)F(s)\big|_{s = -(a-jb)}$$

$$= \frac{P(s)|_{s = -(a-jb)}}{(j2b)}$$

Since $P(s)$ is a polynomial with real coefficients, the values of $P(s)$ at the complex conjugate quantities $s = -(a - jb)$ and $s = -(a + jb)$ will also be complex conjugates. Thus the numerators for K_1 and K_2 are complex conjugates. The denominators are seen to be complex conjugates also. Therefore

$$K_1 = K_2{}^*$$

This property helps reduce the work in the calculation of residues when there are complex conjugate poles.

Example 15-10 Expand the function

$$F(s) = \frac{2s + 7}{(s + 4)(s^2 + 6s + 25)}$$

into partial fractions and find its inverse transform.

Solution The poles of $F(s)$ are at $s = -4, -(3 + j4)$ and $-(3 - j4)$. Therefore

$$F(s) = \frac{K_1}{(s + 4)} + \frac{K_2}{(s + 3 + j4)} + \frac{K_3}{(s + 3 - j4)}$$

where K_2 and K_3 are complex conjugates.

Residue at $s = -4$:

$$K_1 = (s + 4)F(s)|_{s=-4}$$

$$= \frac{2s + 7}{s^2 + 6s + 25}\bigg|_{s=-4}$$

$$= -(1/17) = -0.0588$$

Residue at $s = -(3 + j4)$:

$$K_2 = (s + 3 + j4)F(s)|_{s=-(3+j4)}$$

$$= \frac{2s + 7}{(s + 4)(s + 3 - j4)}\bigg|_{s=-(3+j4)}$$

$$= \frac{1 - j8}{-32 - j8} = 0.0294 + j0.243$$

Residue at $s = -(3 - j4)$: K_3 will be directly evaluated here to verify the value of K_2.

$$K_3 = (s + 3 - j4)F(s)|_{s=-(3-j4)}$$

$$= \frac{2s + 7}{(s + 4)(s + 3 + j4)}\bigg|_{s=-(3-j4)}$$

$$= \frac{1 + j8}{-32 + j8} = 0.0294 - j0.243$$

which verifies the correctness of K_2.

$$F(s) = \frac{0.0588}{(s + 4)} + \frac{0.0294 + j0.243}{(s + 3 + j4)} + \frac{0.0294 - j0.243}{(s + 3 - j4)}$$

and

$$f(t) = [-0.0588e^{-4t} + (0.0294 + j0.243)e^{(-3 - j4)t} + (0.0294 - j0.243)e^{(-3 + j4)t}]u(t)$$

The two complex conjugate terms in the function must be simplified to obtain a *real* function of time. For this, consider the last two terms in $f(t)$:

$$(0.0294 + j0.243)e^{(-3-j4)t} + (0.0294 - j0.243)e^{(-3+j4)t}$$
$$= e^{-3t}(0.2448e^{j83.1°}e^{-j4t} + 0.2448e^{-j83.1°}e^{j4t})$$
$$= e^{-3t}[0.2448e^{-j(4t-83.1°)} + 0.2448e^{j(4t-83.1°)}]$$
$$= e^{-3t}[0.2448 \times 2\cos(4t - 83.1°)]$$

since the sum of two complex conjugate terms equals twice the real part of either term and $\mathbf{Re}[\theta e^{j(\omega t + \theta)}] = \cos(\omega t + \theta)$.
Therefore

$$f(t) = -0.0588e^{-4t} + 0.490e^{-3t}\cos(4t - 83.1°) \qquad \blacksquare$$

Example 15-11 Find the inverse transform of

$$F(s) = \frac{11s^2 - 28s + 27}{(s^2 + 9)(s^2 + 2s + 2)}$$

Solution The poles of $F(s)$ are at $s = -j3, +j3, -(1+j)$, and $-(1-j)$.
Therefore

$$F(s) = \frac{K_1}{(s+j3)} + \frac{K_2}{(s-j3)} + \frac{K_3}{(s+1+j)} + \frac{K_4}{(s+1-j)}$$

The residues K_1 and K_2 will be complex conjugates of each other, and the residues K_3 and K_4 will be complex conjugates of each other.

Residue at $s = -j3$:

$$K_1 = (s + j3)F(s)\big|_{s=-j3}$$
$$= \frac{11s^2 - 28s + 27}{(s - j3)(s^2 + 2s + 2)}\bigg|_{s=-j3}$$
$$= \frac{-72 + j84}{-36 + j42} = 2$$
$$K_2 = K_1^* = 2$$

Residue at $s = -(1+j)$:

$$K_3 = (s + 1 + j)F(s)\big|_{s=-(1+j)}$$
$$= \frac{11s^2 - 28s + 27}{(s^2 + 9)(s + 1 - j)}\bigg|_{s=-(1+j)}$$
$$= \frac{55 + j50}{4 - j18} = -2 + j3.5$$
$$K_4 = K_3^* = -2 - j3.5$$

Therefore

$$F(s) = \frac{2}{(s+j3)} + \frac{2}{(s-j3)} + \frac{(-2+j3.5)}{(s+1+j)} + \frac{(-2-j3.5)}{(s+1-j)}$$

and

$$f(t) = [2e^{-j3t} + 2e^{+j3t} + (-2 + j3.5)e^{(-1-j)t} + (-2 - j3.5)e^{(-1+j)t}]u(t)$$
$$= [2(e^{-j3t} + e^{j3t}) + e^{-t}(4.03e^{j120°}e^{-jt} + 4.03e^{-j120°}e^{jt}]u(t)$$
$$= [4\cos 3t + 8.06e^{-t}\cos(t - 120°)]u(t)$$

∎

Exercise 15-16 Determine the inverse transform of each of the following functions.

(a) $\dfrac{40s^2 + 20}{(s + 2)(s^2 + 2s + 5)}$ (b) $\dfrac{25s^2 + 40s}{(s^2 + 4)(s^2 + 16)}$

Case 2: F(s) Has a Multiple Order Pole

Consider a function $F(s)$ given by the partial fraction expansion

$$F(s) = \frac{A_m}{(s - s_1)^m} + \frac{A_{m-1}}{(s - s_1)^{m-1}} + \ldots + \frac{A_1}{(s - s_1)} \qquad (15\text{-}29)$$

When the m partial fractions are added, there is a single common denominator $(s - s_1)^m$ and the result will be of the form

$$F(s) = \frac{P(s)}{(s - s_1)^m} \qquad (15\text{-}30)$$

where $P(s)$ is a polynomial in s.

Conversely, a function of the form in Eq. (15-30) will, in general, have a partial fraction expansion of the form in Eq. (15-29), with the possibility that one or more of the constants A_j may be zero. Therefore, an mth order pole of $F(s)$ will, in general, give rise to m residues, some of which may be zero.

Multiplication of both the expressions in Eqs. (15-29) and (15-30) by $(s - s_1)^m$ leads to

$$A_m + A_{m-1}(s - s_1) + \ldots + A_1(s - s_1)^{m-1} = P(s) = (s - s_1)^m F(s) \qquad (15\text{-}31)$$

Therefore

$$A_m = P(s)|_{s = s_1} = [(s - s_1)^m F(s)]|_{s = s_1} \qquad (15\text{-}32)$$

which gives the value of the residue A_m.

Next, differentiate Eq. (15-31) once with respect to s:

$$A_{m-1} + 2A_{m-2}(s - s_1) + \ldots + (m - 1)A_1(s - s_1)^{m-2} = \frac{d}{ds}[(s - s_1)^m F(s)]$$

When the last equation is evaluated at $s = s_1$, all terms on the lefthand side, except A_{m-1}, become zero and

$$A_{m-1} = \frac{d}{ds}[(s - s_1)^m F(s)]|_{s = s_1}$$

which give the value of the residue A_{m-1}.

15-3 Determination of the Inverse Laplace Transform

If we proceed in a similar manner, we obtain the following general result:

$$A_{m-r} = \frac{1}{r!} \frac{d^r}{ds^r} [(s - s_1)^m F(s)]|_{s = s_1}$$

where $r = 0, 1, 2, 3, \ldots, m - 1$. ($0! = 1$.)

To summarize the discussion: if a function $F(s)$ has a pole of order m at $s = s_1$, then the partial fraction terms due to such a pole will be

$$F(s) = \frac{A_m}{(s - s_1)^m} + \frac{A_{m-1}}{(s - s_1)^{m-1}} + \cdots + \frac{A_{m-r}}{(s - s_1)^{m-r}} + \cdots + \frac{A_1}{(s - s_1)}$$

A_m is given by the value of $(s - s_1)_m F(s)$ at $s = s_1$. Differentiate the product $(s - s_1)_m F(s)$. After each differentiation, evaluate the derivative at $s = s_1$. The value of the rth derivative at $s = s_1$, divided by $r!$, gives the residue A_{m-r}, where $r = 1, 2, \ldots, (m - 1)$.

The inverse transform of each partial fraction term is then given by (see Table 15-1):

$$\mathscr{L}^{-1}\left[\frac{A_{m-r}}{(s - s_1)^{m-r}}\right] = \frac{A_{m-r} t^{m-r-1} e^{-s_1 t}}{(m - r - 1)!} u(t)$$

Example 15-12 Find the inverse transform of

$$F(s) = \frac{s^2 + 8}{(s + 2)^3}$$

Solution Partial fraction expansion:

$$F(s) = \frac{A_3}{(s + 2)^3} + \frac{A_2}{(s + 2)^2} + \frac{A_1}{(s + 2)}$$

Product function $(s + 2)^3 F(s) = (s^2 + 8)$
Residues:

$$A_3 = (s^2 + 8)|_{s = -2} = 12$$

$$A_2 = \left[\frac{d}{ds}(s^2 + 8)\right]\bigg|_{s = -2} = -4$$

$$A_1 = \left[\frac{1}{2!} \frac{d^2}{ds^2}(s^2 + 8)\right]\bigg|_{s = -2} = 1$$

Therefore,

$$F(s) = \frac{12}{(s + 2)^3} - \frac{4}{(s + 2)^2} + \frac{1}{(s + 2)}$$

The inverse transform of the given function is

$$f(t) = [6t^2 e^{-2t} - 4te^{-2t} + e^{-2t}]u(t) \qquad \blacksquare$$

Example 15-13 Find the inverse transform of

$$F(s) = \frac{s^3 + 2s^2 + 1}{(s + 1)(s + 3)^3}$$

Solution Partial fraction expansion:

$$F(s) = \frac{K_1}{(s+1)} + \frac{A_3}{(s+3)^3} + \frac{A_2}{(s+3)^2} + \frac{A_1}{(s+3)}$$

Residue K_1 at $s = -1$:

$$K_1 = \left.\frac{s^3 + 2s^2 + 1}{(s+3)^3}\right|_{s=-1} = \frac{1}{4} = 0.25$$

The product function needed for the residues at the multiple order pole is

$$(s+3)^3 F(s) = \frac{s^3 + 2s^2 + 1}{s+1}$$

Residues:

$$A_3 = \left.\frac{s^3 + 2s^2 + 1}{s+1}\right|_{s=-3} = 4$$

$$A_2 = \left.\frac{d}{ds}\frac{s^3 + 2s^2 + 1}{s+1}\right|_{s=-3}$$

$$= \left.\frac{2s^3 + 5s^2 + 4s - 1}{(s+1)^2}\right|_{s=-3} = -\frac{22}{4} = -5.5$$

$$A_1 = \left.\frac{1}{2}\frac{d}{ds}\frac{2s^3 + 5s^2 + 4s - 1}{(s+1)^2}\right|_{s=-3} = \frac{1}{2}\left(\frac{3}{2}\right) = 0.75$$

The partial fraction expansion of the given function is therefore

$$F(s) = \frac{0.25}{(s+1)} + \frac{4}{(s+3)^3} - \frac{5.5}{(s+3)^2} + \frac{0.75}{(s+3)}$$

The inverse transform of the given function is

$$f(t) = [0.25e^{-t} + 2t^2 e^{-3t} - 5.5te^{-3t} + 0.75e^{-3t}]u(t)$$ ∎

Exercise 15-17 Determine the inverse transform of each of the following functions.

(a) $\dfrac{2s}{(s+5)^2}$ (b) $\dfrac{2s - 3}{s(s+5)^3}$

15-4 STEP-BY-STEP PROCEDURE FOR ANALYSIS OF NETWORKS USING LAPLACE TRANSFORMS

The discussion in the last section leads to the following step-by-step procedure for using Laplace transform in the analysis of linear networks. (The steps are followed by examples of the procedure.)

Step 1: Replace the independent sources in the network by their transform equivalents by referring to Table 15-1.

voltage source: $v(t)$ replaced by $V(s)$
current source: $i(t)$ replaced by $I(s)$

Dependent sources are essentially left alone except for changing the voltage and current symbols to uppercase letters that are functions of s.

Step 2: Replace each capacitor and inductor by an equivalent transform model from Fig. 15-3. Use a series or parallel equivalent circuit as appropriate for the problem. Initial conditions are either given or have to be determined for finding the models. If the data of the problem are clearly insufficient to determine the initial conditions, they may be taken as zero.

Step 3: Write equations using s as the variable for the network obtained in Step 2. The principles involved in writing equations are exactly the same as in steady state ac circuit analysis. Solve the equations for the desired response quantities. If the aim of the problem is only to find the responses in the complex frequency domain, this step (step 3) completes the solution of the problem.

Step 4: If it is desired to find the responses in the time domain, then expand the response function $F(s) = P(s)/Q(s)$ into a sum of partial fractions. For this, find the poles of $F(s)$ by solving the characteristic equation

$$Q(s) = 0$$

Set up a sum of partial fractions corresponding to the poles of $F(s)$. Remember that a multiple pole of $F(s)$ of order m will initially generate a set of m different partial fractions. Evaluate the residues at the poles of $F(s)$ by using the steps given above. Write out the partial fraction expansion form of $F(s)$.

Step 5: By referring to Table 15-1, write the inverse transform of each partial fraction term obtained in step 5. *In the case of functions* $F(s)$ *with complex conjugate poles, manipulate the inverse transform function until it is simplified to the form of a real time function.*

Example 15-14 A series RC circuit, as in Fig. 15-8(a), is driven by a step input. Assuming zero initial charge on the capacitor, determine the current $i(t)$ and the voltage $v_c(t)$ across the capacitor.

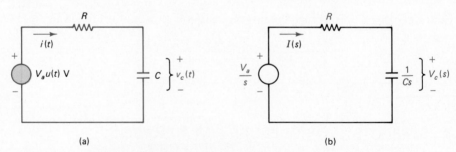

Fig. 15-8 (a) Circuit for Example 15-14 and (b) its transform equivalent.

Solution The circuit is redrawn by using

$$\mathcal{L}\,[V_a u(t)] = \frac{V_a}{s}$$

for the source and the capacitor by an impedance ($1/Cs$) since there is no initial charge. The resulting circuit is shown in Fig. 15-8(b). The KVL equation is

$$\left(R + \frac{1}{Cs}\right)I(s) = \frac{V_a}{s}$$

which leads to

$$I(s) = \frac{(V_a/R)}{s + 1/RC}$$

There is a single pole at $s = -(1/RC)$, and the inverse transform $i(t)$ is obtained from Table 15-1.

$$i(t) = \frac{V_a}{R} e^{-t/RC} u(t)$$

The voltage across the capacitor is found by integrating $i(t)$ and dividing by C, or by means of the relationship

$$V_c(s) = \frac{1}{Cs} I(s) = \frac{V_a}{RC} \frac{1}{s(s + 1/RC)}$$

There are two poles: $s = 0$, and $s = -1/RC$. The partial fraction expansion is found to be

$$V_c(s) = \frac{V_a}{RC} \left[\frac{RC}{s} - \frac{RC}{(s + 1/RC)}\right]$$

Therefore

$$v_c(t) = V_a(1 - e^{-t/RC})u(t)$$

Exercise 15-18 Determine the voltage across the inductance $v_L(t)$ in each of the circuits shown in Fig. 15-9. Assume a zero initial energy storage.

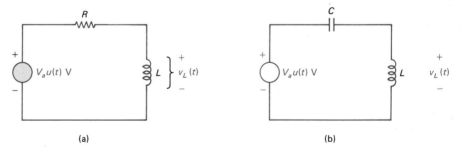

Fig. 15-9 Circuits for Exercise 15-18

Example 15-15 In the RC circuit of Fig. 15-8(a) of Example 15-14, assume that the capacitor has an initial voltage of X volts (with the top terminal positive). Determine the current $i(t)$ and the capacitor voltage $v_c(t)$.

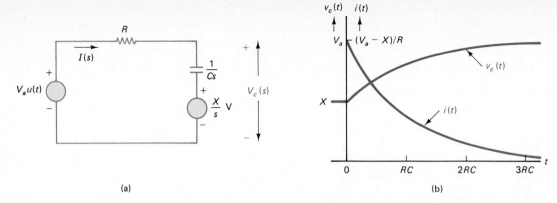

Fig. 15-10 (a) Transform equivalent circuit for Example 15-15. (b) Time-domain response curves.

Solution The circuit is redrawn as shown in Fig. 15-10(a) by replacing the capacitor with a transform model, using Fig. 15-3.
The *KVL* equation is

$$\left(R + \frac{1}{Cs}\right)I(s) = \frac{V_a}{s} - \frac{X}{s}$$

which leads to

$$I(s) = \frac{1}{R}\frac{V_a - X}{s + 1/RC}$$

There is a single pole at $s = -(1/RC)$ and the inverse transform is found directly from Table 15-1.

$$i(t) = \frac{V_a - X}{R}e^{-t/RC}\,u(t)$$

The voltage transform across the capacitor is

$$V_c(s) = \frac{1}{Cs}I(s) + \frac{X}{s}$$
$$= \frac{V_a - X}{RC}\frac{1}{s(s + 1/RC)} + \frac{X}{s}$$
$$= \frac{V_a}{s} - \frac{(V_a - X)}{(s + 1/RC)}$$

Therefore

$$v_c(t) = [V_a - (V_a - X)e^{-t/RC}]u(t)$$
$$= [V_a(1 - e^{-t/RC}) + Xe^{-t/RC}]u(t)$$

The graphs of the current and the capacitor voltage are shown in Fig. 15-10(b). ∎

Exercise 15-19 Repeat the work of Exercise 15-18 if the inductance (in each circuit) had an initial current of $i(0-) = 5$ A. The capacitor has no initial charge.

First-order circuits under constant forcing functions (as in the preceding examples and exercises) were discussed in Chapter 6. The only difference is that instead of solving differential equations in the time domain (as was done in Chapter 6), the present procedure uses algebraic equations in the complex frequency domain and inverse Laplace transform. The method of Laplace transforms is less cumbersome than the time domain solution of differential equations. Even in the case of circuits that could be solved by using phasors or Fourier transforms, the method of Laplace transform has a clear advantage when there is an initial energy storage in the circuit elements and the circuit is subjected to causal signals. This advantage is due to the fact that nonzero initial conditions of circuit elements are systematically included in their transform models. In addition, the Laplace transform method yields expressions for the complete response of networks, rather than just the steady state response, as is the case when phasors are used. The following examples demonstrate the advantages of the method of Laplace transform.

Example 15-16 Determine the voltage $v_c(t)$ in the parallel circuit of Fig. 15-11(a), if the capacitor has an initial voltage $v_c(0-) = V_0$.

Fig. 15-11 (a) Circuit for Example 15-16. (b) Its transform equivalent. (c) Output voltage waveform.

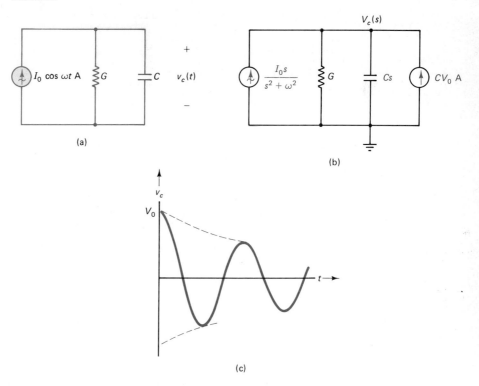

15-4 Step-By-Step Procedure for Analysis of Networks Using Laplace Transforms

Solution From Table 15-1, the Laplace transform of the forcing function is

$$\mathcal{L}\left[I_0 \cos(\omega t)\right] = \frac{I_0 s}{s^2 + \omega^2}$$

The capacitor is replaced by its *parallel* equivalent from Fig. 15-3, since the circuit is best suited for a *nodal* equation. The transformed circuit is shown in Fig. 15-11(b). The KCL equation for $V_c(s)$ is

$$(G + Cs)V_c(s) = CV_0 + \frac{I_0 s}{s^2 + \omega^2}$$

which leads to

$$V_c(s) = \frac{I_0}{C} \cdot \frac{s}{(s + G/C)(s^2 + \omega^2)} + \frac{V_0}{(s + G/C)} \qquad (15\text{-}34)$$

Consider the term

$$T(s) = \frac{s}{(s + G/C)(s^2 + \omega^2)}$$

Since $(s^2 + \omega^2) = 0$ has the roots $s = -j\omega$ and $+j\omega$, $T(s)$ becomes

$$T(s) = \frac{s}{(s + G/C)(s + j\omega)(s - j\omega)}$$

$$= \frac{K_1}{(s + G/C)} + \frac{K_2}{(s + j\omega)} + \frac{K_3}{(s - j\omega)}$$

Residue at $s = -G/C$:

$$K_1 = \left.\frac{s}{(s + j\omega)(s - j\omega)}\right|_{s = -G/C}$$

$$= \frac{-(G/C)}{(G^2/C^2) + \omega^2} = \frac{-G/C}{(G^2 + \omega^2 C^2)}$$

Residue at $s = -j\omega$:

$$K_2 = \left.\frac{s}{(s + G/C)(s - j\omega)}\right|_{s = -j\omega}$$

$$= \frac{-j\omega}{(-j\omega + G/C)(-2j\omega)} = \frac{C}{2(G - j\omega C)}$$

Residue at $s = +j\omega$:

$$K_3 = \left.\frac{s}{(s + G/C)(s + j\omega)}\right|_{s = +j\omega}$$

$$= \frac{+j\omega}{(+j\omega + G/C)(+2j\omega)} = \frac{C}{2(G + j\omega C)}$$

$K_3 = K_2^*$ could have been used instead. Equation (15-34) becomes

$$V_c(s) = \left[\frac{I_0 G}{(G^2 + \omega^2 C^2)} \frac{1}{(s + G/C)} + \frac{I_0}{2} \frac{1}{(G - j\omega C)} \frac{1}{(s + j\omega)}\right.$$
$$\left. + \frac{I_0}{2} \frac{1}{(G + j\omega C)} \frac{1}{(s - j\omega)} + \frac{V_0}{(s + G/C)}\right]$$

Taking the inverse transform

$$v_c(t) = \left[V_0 - \frac{I_0 G}{G^2 + \omega^2 C^2}\right] e^{-(G/C)t}$$
$$+ (I_0/2)\left(\frac{1}{G - j\omega C} e^{-j\omega t} + \frac{1}{G + j\omega C} e^{+j\omega t}\right) \quad (15\text{-}35)$$

The terms involving $e^{j\omega t}$ and $e^{-j\omega t}$ in Eq. (15-35) are combined and simplified to get a real sinusoidal function of time. Let

$$x(t) = \frac{(G + j\omega C)e^{-j\omega t} + (G - j\omega C)e^{+j\omega t}}{G^2 + \omega^2 C^2}$$

$$= \frac{1}{G^2 + \omega^2 C^2}[2G \cos \omega t + j C(-2j \sin \omega t)]$$

$$= \frac{1}{G^2 + \omega^2 C^2}(2G \cos \omega t + 2 C \sin \omega t)$$

$$= \frac{2}{\sqrt{G^2 + \omega^2 C^2}} \cos [\omega t - \arctan (\omega C/G)]$$

The voltage $v_1(t)$ of Eq. (15-35) is therefore

$$v_c(t) = \left[V_0 - \frac{I_0 G}{G^2 + \omega^2 C^2}\right] e^{-(G/C)t} +$$
$$\frac{I_0}{\sqrt{G^2 + \omega^2 C^2}} \cos\left[\omega t - \arctan \frac{\omega C}{G}\right]$$

Check to make sure that the initial condition is satisfied: $v_c(t) = V_0$ at $t = 0$.

The voltage waveform is shown in Fig. 15-11(c). The response consists of an exponentially decaying component and a sinusoidal component. Note that the exponential term is not strictly due to the initial charge on the capacitor; there would be an exponentially decaying term even if $V_0 = 0$. An adjustment of the initial phase angle is necessary to eliminate the exponentially decaying term.

If this problem had been solved using phasors, the initial condition could not have been included. Also, the solution would have yielded only the steady state component (the sinusoidal term) of the response. ∎

Exercise 15-20 A parallel combination of $L = 10$ mH and $G = 0.1$ S is driven by a current source $i_s(t) = 20 \cos 1000t$. Assume that the initial current in the inductance is zero, and determine the voltage $v_0(t)$ across the circuit for $t > 0$.

Example 15-17 Determine the current $i_L(t)$ and the voltage $v_c(t)$ for $t > 0$ in the circuit shown in Fig. 15-12(a). The $20u(-t)$ V source provides 20 V until $t = 0$ and then becomes zero. The $10u(t)$ A source becomes active at $t = 0$.

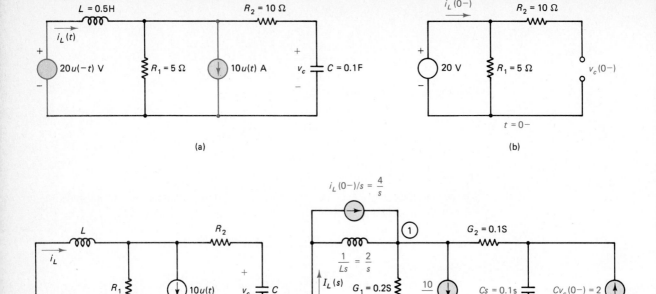

Fig. 15-12 (a) Circuit for Example 15-17. (b) Determination of initial condition. (c) Circuit after $t = 0$, and (d) its transform equivalent.

Solution In order to determine the initial conditions, consider the circuit at $t = 0-$. A steady state dc condition exists: the voltage across the inductance is zero (short circuit) and current in the capacitor is zero (open circuit), as shown in Fig. 15-12(b). Therefore

$$i_L(0-) = 4 \text{ A} \quad \text{and} \quad v_c(0-) = 20 \text{ V}$$

For $t > 0$, the circuit is as shown in Fig. 15-12(c), where the 20 V source has been replaced by a short circuit, since its voltage becomes zero at $t = 0$.

The circuit elements are replaced by their transform equivalents, as indicated in Fig. 15-12(d). Parallel equivalents have been used for the inductor and capacitor in order to use nodal analysis. The nodal equations are

$$\left(G_1 + G_2 + \frac{1}{Ls}\right)V_1(s) - G_2 V_2(s) = \frac{i_L(0-) - 10}{s} \tag{15-36}$$

and

$$-G_2 V_1(s) + (G_2 + Cs)V_2(s) = Cv_c(0-) \tag{15-37}$$

After substituting the numerical values, the last two equations become

$$\left(0.3 + \frac{2}{s}\right)V_1(s) - 0.1V_2(s) = -\frac{6}{s} \tag{15-38}$$

$$-0.1V_1(s) + (0.1 + 0.1s)V_2(s) = 2 \tag{15-39}$$

766   **Laplace Transformation**

Solving Eqs. (15-38) and (15-39), we find that

$$V_1(s) = \frac{-(0.4s + 0.6)}{0.03s^2 + 0.22s + 0.2}$$

$$= \frac{0.4}{0.03} \frac{(s + 1.5)}{s^2 + 7.33s + 6.667}$$

$$= \frac{-13.33(s + 1.5)}{(s + 1.064)(s + 6.270)} \quad (15\text{-}40)$$

which has the partial fraction expansion

$$V_1(s) = \frac{-1.116}{(s + 1.064)} + \frac{-12.21}{(s + 6.270)}$$

Taking the inverse transform

$$v_1(t) = -(1.116e^{-1.064t} + 12.21e^{-6.270t})u(t)$$

The current $i_L(t)$ is given by

$$i_L(t) = -\frac{1}{L} \int v_1(t) dt \quad (15\text{-}41)$$

where the minus sign is due to the relative directions of the current i_L and v_1 in the inductance. Substituting Eq. (15-40) in Eq. (15-41), performing the integration, and using $i_L(0-) = 4$, we find

$$i_L(t) = -2.18e^{-1.064t} - 3.895e^{-6.27t} + 10$$

As $t \to \infty$, $i_L = 10$ A, which checks with the steady state condition in the circuit of Fig. 15-12(c). The expression for $i_L(t)$ could also have been obtained by first getting $I_L(s)$ from the expression for $V_1(s)$ in Eq. (15-40) and then finding $i_L(t)$.

The voltage $V_2(s)$ is, from Eqs. (15-38) and (15-39),

$$V_2(s) = \frac{0.6s + 3.4}{0.03s^2 + 0.22s + 0.2}$$

$$= \frac{0.6}{0.03} \frac{(s + 5.667)}{(s^2 + 7.333s + 6.667)}$$

$$= \frac{20(s + 5.667)}{(s + 1.064)(s + 6.270)}$$

which has the partial fraction expansion

$$V_2(s) = \frac{17.68}{(s + 1.064)} + \frac{2.32}{(s + 6.270)}$$

Therefore

$$v_2(t) = (17.68e^{-1.064t} + 2.32e^{-6.270t})u(t)$$

which is the voltage across the capacitor. The initial condition is verified: $v_2(0) = 20$ V. As $t \to \infty$, the expression for $v_2(t)$ goes to zero, which checks with the steady state condition in Fig. 15-12(c). ∎

Exercise 15-21 In the previous example, $I_L(s)$ can be found from Eq. (15-40) for $V_1(s)$. Apply *KCL* at the leftmost node 1 in Fig. 15-12(c) carefully to obtain an expression for $I_L(s)$. Obtain $i_L(t)$ by using inverse transform on $I_L(s)$.

Exercise 15-22 Mesh analysis can be used in the circuit of the previous example by converting the 10 A source and the resistor R_1 into an equivalent voltage source and series resistor configuration. Do this and determine the current $i_L(t)$ and the voltage $v_c(t)$ for $t > 0$.

15-5 INITIAL VALUE AND FINAL VALUE THEOREMS

It is not always necessary (or even desirable) to invert the response function $F(s)$ of a network and obtain the corresponding $f(t)$ in the time domain. It would be convenient, however, to check the solution by examining its values in the time domain at $t = 0$ and at $t = \infty$, since the values of $f(t)$ at these two extreme instants of time can usually be obtained by inspecting the circuit. Voltage continuity in a capacitor and current continuity in an inductor are used to determine the response of the circuit at $t = 0$. The steady state response of the circuit is used to find the value of the response at $t = \infty$. The values of $f(t)$ at $t = 0+$ and $t = \infty$ may be determined directly from $F(s)$ without using inverse transform by means of two theorems: the *initial value theorem* and the *final value theorem*.

The assumptions in both theorems are:

1. $f(t)$ and its first derivative $f'(t)$ have Laplace transforms.
2. $f(t) = 0$ when $t < 0$.
3. $f(t)$ contains no impulses or higher-order singularities at $t = 0$.

Initial Value Theorem

The value of $f(t)$ at $t = 0+$ is given by

$$f(0+) = \lim_{s \to \infty} [sF(s)] \qquad (15\text{-}42)$$

For example, let $F(s) = 1/(s + a)$ then $[sF(s)] = s/(s + a)$, which equals 1 as $s \to \infty$. This is to be expected, since $f(t) = e^{-at}u(t)$ in this case and $f(0+) = 1$.

Final Value Theorem

The value of $f(t)$ as $t \to \infty$ is given by

$$\lim_{t \to \infty} f(t) = \lim_{s \to 0} [sF(s)] \qquad (15\text{-}43)$$

Again, if $F(s) = 1/(s + a)$, then $[sF(s)] = s/(s + a) = 0$ when $s = 0$, which checks with $f(t) = e^{-at}$ becoming zero as $t \to \infty$.

Exercise 15-23 Consider the expressions of the responses in Examples 15-15, 15-16, and 15-17 in the complex frequency and time domains and verify the initial and final value theorems.

15-6 CONVOLUTION THEOREM

The convolution theorem stated in Chapter 14 in terms of Fourier transformation is readily modified for the case of Laplace transformation by replacing $(j\omega)$ by s. That is, if $\mathcal{L}[f(t)] = F(s)$ and $\mathcal{L}[g(t)] = G(s)$, then

$$\mathcal{L}[f(t) \circledast g(t)] = F(s)G(s)$$

For example, suppose the transfer function

$$H(s) = \frac{V_2(s)}{V_1(s)}$$

of a network is known. The impulse response $h(t)$ is the inverse transform of $H(s)$, and $h(t)$ is found by using the method of partial fraction expansion on $H(s)$. Given an input signal $v_1(t)$, the response $v_2(t)$ is then obtained by using convolution:

$$v_2(t) = h(t) \circledast v_1(t)$$

Thus there are two methods of finding the response of a linear network in the time domain: (1) Determine the response as a function of s using Laplace transform and invert the function to obtain the time domain response. (2) Find the impulse response by using inverse Laplace transform on the transfer function and use convolution.

The principles and applications of convolution in terms of Laplace transform are exactly the same as those discussed at the end of Chapter 14. The only difference is that the determination of the inverse Laplace transform through partial fraction expansion is more methodical for causal systems than finding the inverse Fourier transform.

15-7 ROLE OF CRITICAL FREQUENCIES IN THE RESPONSE OF NETWORKS

Poles and zeros of network functions were discussed in Chapter 12, and we pointed out that if the critical frequencies are known, then the function is completely specified (except for a possible constant multiplicative factor). We also pointed out that the natural response of a circuit (such as a resonant circuit) is dictated by the location of its critical frequencies in the complex frequency domain. Not only does Laplace transform provide the theoretical basis for the concept of critical frequencies, but it also brings some new insight into the behavior of networks by including critical frequencies of forcing functions side by side with those of network functions. One example of the implications of Laplace transform on network theory will be discussed here.

Let the forcing function $F(s)$ in a network and the desired response function $G(s)$ be related by an appropriate network function $H(s)$, which may be an impedance, admittance, or a transfer function. If $F(s)$ and $H(s)$ are known, then

$$G(s) = H(s)F(s)$$

Suppose the poles of $H(s)$ are given by $s_a, s_b, \ldots$, and the poles of $F(s)$ are given by $s_1, s_2, \ldots$. If $g(t)$ is the inverse transform of $G(s)$, then $g(t)$ will be of the form

$$g(t) = K_a e^{s_a t} + K_b e^{s_b t} + \ldots + K_1 e^{s_1 t} + K_2 e^{s_2 t} + \ldots$$

The terms $K_a e^{s_a t}, K_b e^{s_b t}, \ldots$ due to the poles of the network function represent the *natural response* of the network, while the terms $K_1 e^{s_1 t}, K_2 e^{s_2 t}, \ldots$ due to the poles of the forcing function represent the *forced response* of the network. It is important to remember that the *shapes and forms of the natural response components* are completely determined by the *poles of the network function*. Any effect of the forcing function on the natural response components is limited to their amplitudes. Similarly, the *shapes and forms of the forced response components* are completely determined by the *poles of the forcing function,* and any effect of the network function on them is only on their amplitudes.

Now consider two possible network situations: $H_1(s)$ and $F_1(s)$ pertain to network 1 and $H_2(s)$ and $F_2(s)$ pertain to network 2. Let the response $G(s)$ be the same in both cases.

$$G(s) = H_1(s)F_1(s) = H_2(s)F_2(s)$$

Also, let

$$\text{poles of } H_2(s) \text{ of network 1} = \text{poles of } F_2(s) \text{ acting on network 2}$$

and

$$\text{poles of } F_1(s) \text{ of network 1} = \text{poles of } H_2(s) \text{ of network 2}$$

Then the natural response components of network 1 will become the forced response components of network 2, while the forced response components of network 1 will become the forced response components of network 2.

That is, an *exchange of the poles between a network function and a forcing function has no effect on the shape and form of the complete response!*

As an example, consider a series *RL* circuit subjected to a sinusoidal voltage. The complete response of the circuit will contain an exponential term of the form e^{at} ($a < 0$) due to the pole of the admittance of the circuit, and a sinusoidal term due to the complex conjugate poles of the forcing function. Now, the same sort of response is obtained by applying an exponential forcing function of the form e^{at} to a series *LC* circuit.

Therefore, we conclude that it is possible to trade the critical frequencies of circuits and forcing functions without changing the nature of the complete response. The fact that circuits with widely different configurations can be made to produce the same response has important implications in the synthesis of networks.

The insight gained in the preceding discussion was clearly the result of using Laplace transform to determine a circuit's response.

15-8 AN OVERVIEW OF NETWORK ANALYSIS

If we start with a set of two basic laws, Kirchhoff's voltage law and Kirchhoff's current law, and voltage-current relationships of different circuit components, the techniques of circuit analysis require the solution of algebraic equations for purely resistive circuits. The equations become linear differential equations with constant coefficients when the circuits include inductors and capacitors.

For circuits driven by sinusoidal sources of a single frequency, the use of phasors converts the differential equations to linear algebraic equations with complex coefficients. The validity of the notion of phasors rests on the use of complex exponential driving functions. The use of complex exponential functions and phasors leads to concepts of impedance and admittance. Complex driving functions also give rise to complex response functions, but the response due to real sinusoidal functions is obtained by taking the real parts of the complex response functions.

The treatment of networks in the sinusoidal steady state is made more general by addressing the frequency response and transfer functions of networks. The concept of *complex frequency* is found to be useful in dealing with transfer functions. Complex frequency s is defined by $s = (\sigma + j\omega)$. Complex frequency allows us to examine the response of a network through its critical frequencies (poles and zeros). The critical frequencies form the very core of networks: the goal of network design is to realize a specified set of poles and zeros of the appropriate network function.

When the forcing function is nonsinusoidal but periodic, Fourier series is used in the analysis. A periodic forcing function is expanded as an infinite series, each of whose terms is a sinusoidal function of time. The frequencies of the sinusoidal terms are integral multiples of the fundamental frequency of the periodic signal. A combination of Fourier series and the principle of superposition facilitates the analysis of networks driven by periodic forcing functions. The study of Fourier series also leads to the concept of frequency spectra depicting the amplitudes (and phase relationships) of the different harmonics in a signal. The response of a system in the frequency domain is often studied by comparing its output frequency spectrum with the input frequency spectrum.

Single pulses or signals of finite duration may be viewed as the limiting cases of periodic signals with a period of infinity. For such signals, the discrete Fourier series becomes the Fourier integral or transform, and the frequency spectrum becomes a continuous curve. The signal is now expressed as the sum of an infinitely large number of components of the form $e^{-j\omega t}$, where ω can take any value on from $-\infty$ to $+\infty$, instead of only discrete values, as in periodic signals. The response of a system to an input pulse is obtained by using the transfer function of the system and the Fourier transform of the input signal.

The analysis techniques used in sinusoidal steady state and periodic and nonperiodic signals deal almost exclusively with the frequency domain. It is possible, of course, to convert the results to the time domain, but the procedure usually requires extra work. In order to determine the response of a network directly in the time domain, the principle of superposition has to be extended to cover arbitrary signals. Such an extension leads to the operation of convolution: The output of a linear network is found by convolving the input and the impulse response of the network. Convolution of two functions in the time domain is equivalent to multiplication of their Fourier transforms in the frequency domain (and their Laplace transforms in the s domain). This property of the convolution integral also provides the theoretical justification for the technique used in frequency domain analysis of a linear network: the use of the product of the transform of the input and the transfer function to obtain the transform of the output.

The fact that not all signals of practical importance have Fourier transforms makes it necessary to modify the function $e^{-j\omega t}$ used in Fourier integrals to a function of the form e^{-st}, s being the complex variable: $s = \sigma + j\omega$. The complex frequency s introduced earlier in connection with frequency response and transfer functions without any rigorous justification is now found to have a rigorous mathematical basis. The use of the function e^{-st} as the building block of signals leads to the Laplace transformation.

Almost all causal signals of practical interest are found to possess a Laplace transform (within a region of convergence), and a larger class of signals have a Laplace transform than a Fourier transform. This makes the Laplace transform a more versatile tool in network analysis. Laplace transformation permits the conversion of linear differential equations to linear algebraic equations. This is exactly like sinusoidal steady state analysis with phasors, except that Laplace transform is not restricted to sinusoidal signals or steady state response. It permits the determination of the complete response (the sum of the natural and forced responses) of networks under a variety of forcing functions. As the discussion in Section 15-6 indicates, Laplace transform and critical frequencies play a vital role in network theory, and particularly in network synthesis.

The methods of Fourier and Laplace transforms are applicable to all linear systems; that is, any system (electrical as well as nonelectrical) whose behavior is controlled by a set of linear differential equations. Such an extension forms the basis of the study of control and communication systems, as well as of a number of other electrical and mechanical systems.

15-9 SUMMARY OF CHAPTER

The limitations of Fourier transform are due to the use of the component function $e^{-j\omega t}$ that does not converge at $t = -\infty$ or at $t = +\infty$. These limitations are overcome in Laplace transformation, where the component functions are e^{-st}, where s is the complex variable $(\sigma + j\omega)$. The region of convergence of the Laplace transform of a function is a vertical strip in the s plane. The region of convergence for a causal signal is a portion or all of the right half of the s plane. Some functions have a Laplace transform even though they do not have a Fourier transform. Almost all practical causal signals have a Laplace transform.

Differentiation in the time domain transforms to a multiplication by s and integration in the time domain to a division by s in the complex frequency domain (except for the inclusion of initial conditions). The basic elements of electric circuits, R, L, and C, have transform equivalents that are useful in circuit analysis in the s domain. Concepts of impedance, admittance, and transfer function are valid in the s domain.

When the desired response of a circuit is in the time domain, inverse Laplace transform is used. The formal determination of the inverse Laplace transform of a function of s involves contour integration in the s plane. For the vast majority of problems in networks, however, a procedure using partial fraction expansion and a table is sufficient for finding the inverse transform.

The analysis of a circuit is performed by setting up an equivalent circuit with transform models of the elements of the circuit, using the same sort of approach as in sinusoidal steady state analysis (except that s is used instead of $j\omega$) and solving for the desired response. The time domain response (if required) is obtained by partial fraction expansion to find the inverse transform. The time domain response obtained in this manner includes the natural response and the forced response; that is, it is the complete response.

The values of the response at $t = 0$ and $t = \infty$ in the time domain can be calculated by using the initial and final value theorems without finding the response function in the time domain by inverse transform.

Since the Laplace transform approach combines the critical frequencies of network functions and forcing functions, it is useful in gaining some insight into general properties of networks.

Answers to Exercises

15-1 $(1/s)$. Region of convergence: $\sigma > 0$.

15-2 $(1/s)$. Region of convergence: $\sigma < 0$.

15-3 $(e^{bs} - e^{-bs})/s$. Region of convergence: all σ.

15-4 (a) $1/(s + a)$. (b) $b/(s^2 + b^2)$. (c) e^{-st_0}.

15-5 $s^2F(s) - sf(0-) - f'(0-)$.

15-6 Use $\sin bt = -(1/b)(d/dt)(\cos bt)$. Answer same as in Exercise 15-4(b).

15-7 Use $\sin bt = b\int \cos bt \, dt$. Answer same as in Exercise 15-4(b).

15-8 Use $v_c = (1/C)\int i\,dt$ and Eq. (15-16) for the series model. Use $i = C(dv_c/dt)$ and Eq. (15-12) for the parallel model.

15-9

(a) $$I(s) = \frac{5}{s+2} \cdot \frac{1}{(5+5/s)} = \frac{s}{(s+1)(s+2)}$$

(b) $I(s) = [2.5 + 3s/(s^2 + 25)]/(0.5s + 50/s)$
$= 5(s^3 + 1.2s^2 + 25s)/(s^2 + 100)(s^2 + 25)$.

15-10 (a) $V_0(s) = [1/(s+1)^2 - 2/s]/(0.2s + 5/s) = -10(s^2 + 1.5s + 1)/(s+1)^2(s^2 + 25)$. (b) $[3s/(s^2 + 25) - 2/s + 0.8)]/[(5/s) + 0.2s] = 4(s^3 + 1.25s^2 + 25s - 62.5)/(s^2 + 25)^2$.

15-11 $(K_1 + K_2) = 4$; $(K_1 + 5K_2) = 16$. $K_1 = 1$; $K_2 = 3$.
$f(t) = (e^{-5t} + 3e^{-t})u(t)$.

15-12 $(K_1 + K_3) = 0$; $K_1 = 4$; $(4K_1 + K_2 + 2K_3) = 4$. $K_1 = 4$. $K_2 = 4$.
$K_3 = -4$. $f(t) = (4 - 4te^{-2t} - 4e^{-2t})u(t)$.

15-13 Same answers as in (15-11) and (15-12).

15-14 (a) Poles are at $s = -1, -6$. Residue at pole at $s = -1$: 2.8. Residue at pole at $s = -6$: -0.8. $f_a(t) = (2.8e^{-t} - 0.8e^{-6t})u(t)$. (b) Poles are at $s = -2, +2$. Residue $= 2$ at either pole. $f_b(t) = (2e^{-2t} + 2e^{2t})u(t)$.

15-15

(a) $3 - \dfrac{18s + 30}{s^2 + 6s + 8} = 3 - \dfrac{21}{s+4} + \dfrac{3}{s+2}$

$$f_a(t) = 3\delta(t) - 21e^{-4t}u(t) + 3e^{-2t}u(t)$$

(b) $1 + \dfrac{3s + 7}{s^2 + 3s + 2} = 1 - \dfrac{1}{s+2} + \dfrac{4}{s+1}$

$$f_b(t) = \delta(t) - e^{-2t}u(t) + 4e^{-t}u(t)$$

15-16 (a) Residue at pole at $s = -2$: 36. Residue at pole at $s = -(1 + j2)$: $(2 - j21)$. Residue at pole at $s = -(1 - j2)$: $(2 + j21)$. $f(t) = [36e^{-2t} + 42.2e^{-t}\cos(2t + 84.6°)]u(t)$ (b) Residue at pole at $s = -j2$: $(1.6 - j2.08)$. Residue at pole at $s = +j2$: $(1.67 + j2.08)$. Residue at pole at $s = -j4$: $(-1.67 + j4.17)$. Residue at pole at $s = +j4$: $(-1.67 - j4.17)$. $f(t) = [5.33 \cos(2t + 51.2°) + 8.98 \cos(4t - 112°)]u(t)$.

15-17

(a) $-\dfrac{10}{(s+5)^2} + \dfrac{2}{s+5}$

$f(t) = (-10te^{-5t} + 2e^{-5t})u(t)$.

(b) $-\dfrac{0.024}{s} + \dfrac{2.6}{(s+5)^3} + \dfrac{0.12}{(s+5)^2} + \dfrac{0.024}{(s+5)}$

$f(t) = [-0.024 + (1.3t^2 + 0.12t + 0.024)e^{-5t}]u(t)$.

15-18 (a) $V_L(s) = [Ls/(Ls + R)](V_a/s)$. $v_L(t) = V_a e^{-(R/L)t}$.
(b) $V_L(s) = [Ls/(Ls + 1/Cs)](V_a/s)$. $v_L(t) = V_a \cos(t/\sqrt{LC})$.

Answers to Exercises 773

15-19 (a) $V_L(s) = LsI(s) - Li(0-) = (V_a - 5R)/(s + R/L)$. $v_L(t) = (V_a - 5R)e^{-(R/L)t}u(t)$. (b) $V_L(s) = LsI(s) - Li(0-) = (V_a s - I_0/C)/(s^2 + 1/LC)$. $v_L(t) = [V_a \cos t/\sqrt{LC} - 5\sqrt{L/C} \sin t/\sqrt{LC}]u(t)$.

15-20 $Y(s) = G + (1/Ls) = (0.1s + 100)/s$. $V_0(s) = 200s^2/(s^2 + 10^6)(s + 1000)$. Residue at pole at $s = -1000$: 100. Residue at pole at $s = -j1000$: $(50 - j50)$. Residue at pole at $s = +j1000$: $(50 + j50)$. $v_0(t) = [141.4 \cos (1000t + 45°) + 100e^{-1000t}]u(t)$ V.

15-21

$$I_L = (4/s) - (V_1/Ls) = \frac{4s^2 + 56s + 66.67}{s(s + 1.063)(s + 6.27)}$$

$= (10/s) - 2.11/(s + 1.063) - 3.896/(s + 6.27)$, which gives the same expression as Eq. (15-42).

15-22 Mesh equations: $(0.5s + 5)I_1 - 5I_2 = (2s + 50)/s$; and $-5I_1 + [(15s + 10)/s]I_2 = -(70/s)$. Expression for $I_1(s)$ will be the same as in Exercise 15-21. $I_2(s) = -(25s + 4)/(7.5s^2 + 55s + 50)$. Final expressions same as in Example 15-17.

PROBLEMS

Sec. 15-1 Laplace Transformation

[Whenever a function of s [is to be determined, it should be simplified to a ratio of polynomials].

15-1 Starting from the basic relationship of the Laplace transform, (a) determine the Laplace transforms and (b) the regions of convergence of the following functions. Show all steps (do not use Table 15-1).
(a) $f_1(t) = e^{a|t|}$
(b) $f_2(t) = \cosh at\, u(t)$
(c) $f_3(t) = \cos(t - t_0)\, u(t - t_0)$

15-2 Starting from the basic relationship of the Laplace transform, determine the Laplace transforms of the functions shown in Fig. 15-13. The functions exist only in the interval shown.

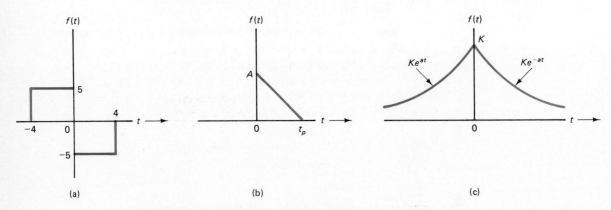

Fig. 15-13 (a)-(c) Functions for Problem 15-2.

15-3 (a) Show that if $\mathcal{L}[f(t)u(t)] = F(s)$, then $\mathcal{L}[f(t - t_0)u(t - t_0)] = e^{-st_0}F(s)$. (b) A staircase function $f(t)$ consists of (an infinie number of) ascending steps, each of which is one unit high and has a width of T seconds. Such a staircase function can be expressed as the sum of unit step functions starting at different instants of time. Write such a sum. Assume the transform of $u(t)$ and the result obtained in part (a) to obtain the Laplace transform of the staircase function. Use a mathematical handbook to sum the series obtained in the solution.

Sec. 15-2 Use of Laplace Transform in Analysis of Linear Circuits

15-4 Obtain the Laplace transforms of the following functions. Use appropriate transforms and theorems from Table 15-1 instead of starting from the basic relationship for Laplace transform in each case.
(a) $f_1(t) = [t \cos(4t)]u(t.$
(b) $f_2(t) = [te^{-3t} \cos(4t)]u(t)$.
(c) $f_3(t) = [\cos b(t-5)]u(t-5)$.
(d) $f_4(t) = \left(\dfrac{d}{dt}\right)[t \cos(4t)]u(t)$.
(e) $f_5(t) = [A \cos(Bt + D)]u(t)$.

15-5 Transform each of the following differential equations to algebraic equations in s and obtain an expression for $Y(s)$.
(a) $3\left(\dfrac{dy}{dt}\right) + y = e^{-t}u(t)$. Initial condition: $y(0) = 1$.
(b) $\left(\dfrac{d^2y}{dt^2}\right) + 2\left(\dfrac{dy}{dt}\right) + y = e^{-3t}\cos 4t$. Initial conditions: $y(0) = -2$, $y'(0) = 4$.

15-6 In the circuit of Fig. 15-14(a), $v_1(t) = \sin 10t$ V, and in Fig. 15-14(b), $v_2(t) = e^{-10t}$. Assume zero initial energy storage. Obtain an expression for the current as a function of s in each circuit. If $C = 0.05$ F, determine the values of R and L so that the currents in the two circuits are equal to each other.

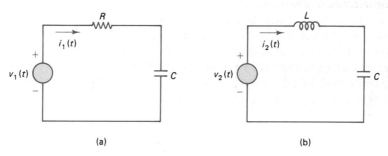

Fig. 15-14 (a), (b) Circuits for Problem 15-6.

15-7 In the circuit of Fig. 15-15, the switch is at position A for $t < 0$ (for a long time). At $t = 0$, it is moved to position B. Obtain an expression for $I(s)$ for the interval $t > 0$.

Fig. 15-15 Circuit for Problem 15-7.

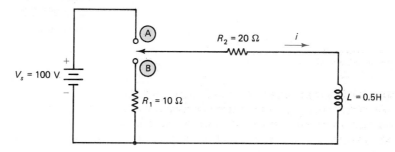

Problems

15-8 Assuming that the initial current is zero in the two inductances, obtain expressions for $I_1(s)$ and $I_2(s)$ in the circuit of Fig. 15-16 for the interval $t > 0$.

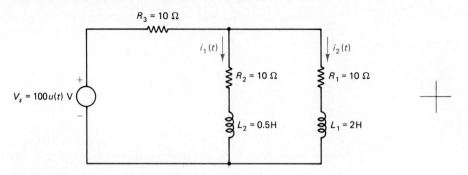

Fig. 15-16 Circuit for Problem 15-8.

15-9 Determine the impedance $Z(s)$ and admittance $Y(s)$ for each of the networks shown in Fig. 15-17.

Fig. 15-17 Networks for Problem 15-9.

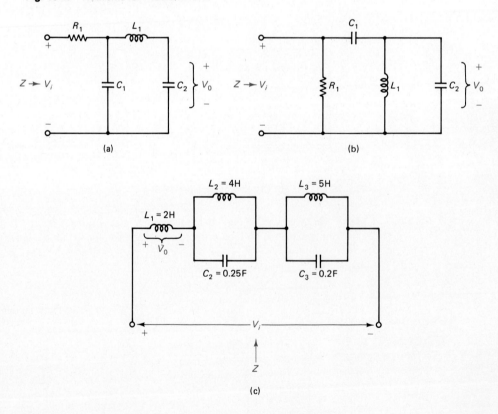

Laplace Transformation

15-10 Determine the transfer function $H(s) = V_o/V_i$ in each of the networks shown in Fig. 15-18 (from the previous problem).

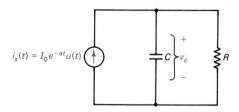

Fig. 15-18 Circuit for Problem 15-18.

Sec. 15-3 Determination of the Inverse Laplace Transform

15-11 Obtain the inverse Laplace transform of each of the following functions.

(a) $\dfrac{(s-2)}{(s+2)}$ (b) $\dfrac{(s+2)}{(s-2)}$ (c) $\dfrac{(s-1)(s-2)}{(s+1)(s+2)}$

15-12 Obtain the inverse Laplace transform of each of the following functions.

(a) $\dfrac{(7s+2)}{(s^3+3s^2+2s)}$ (b) $\dfrac{s^3+3s^2+3s+1}{s(s^2-9)}$

(c) $\dfrac{6s^2+22s+18}{s^3+6s^2+11s+6}$

15-13 Obtain the inverse Laplace transform of each of the following functions.

(a) $\dfrac{s}{(s^2+9)(s^2+16)}$ (b) $\dfrac{(s+3)}{(s^2+4s+9)}$

(c) $\dfrac{s^3+3}{(s+1)(s^2+4)}$

15-14 Obtain the inverse Laplace transform of each of the following functions.

(a) $\dfrac{s^2-s+1}{s^2(s+1)}$ (b) $\dfrac{1}{(s+1)(s+2)^2}$

(c) $\dfrac{1}{s^3(s^2-1)}$ (d) $\dfrac{s^2}{(s^2+1)^2}$

15-15 For the expressions $Y(s)$ obtained for the differential equations in Problem 15-5, find the corresponding time functions.

15-16 Determine the current $i(t)$ in Problem 15-7.

15-17 Determine the currents $i_1(t)$ and $i_2(t)$ in Problem 15-8.

Sec. 15-4 Step-by-Step Procedure of Network Analysis

15-18 Assume that the capacitor is initially uncharged in the circuit of Fig 15-18. Determine $v_c(t)$ for $t > 0$.

15-19 A resistance R and a capacitance C are in a series and driven by a voltage source $v_s(t) = V_m \cos(\omega t + \theta)u(t)$. Assuming the capacitor to be initially uncharged, determine the value of the phase angle θ so as to eliminate the exponentially decaying term (the transient or natural response term) in the response.

15-20 Determine the voltage $v_L(t)$ in the circuit of Fig. 15-19, if the initial current in the inductance is 5 A (in the downward direction).

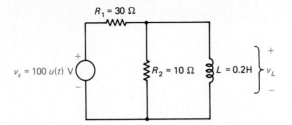

Fig. 15-19 Circuit for Problem 15-20.

15-21 In the circuit of Fig. 15-20, the switch K is open for an indefinite period of time before it is closed at $t = 0$. Determine the expression for the capacitor voltage $v_c(t)$ for $t > 0$.

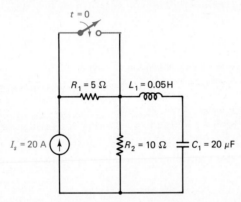

Fig. 15-20 Circuit for Problem 15-21.

15-22 In the circuit of Fig. 15-21, there is an initial voltage of $v_1 = 2$ V on C_1, and the initial voltage on C_2 is zero. The switch is closed at $t = 0$. Determine the voltages $v_1(t)$ and $v_2(t)$.

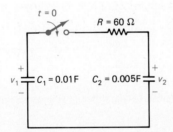

Fig. 15-21 Circuit for Problem 15-22.

15-23 Find the input impedance $Z(s)$ of the circuit shown in Fig. 15-22. For what value of g_m can the circuit be replaced by a resistance? What is the value of the resistance?

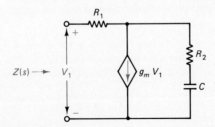

Fig. 15-22 Circuit for Problem 15-23.

Laplace Transformation

15-24 A rectangular pulse $v_1(t) = [u(t - a) - u(t - b)]$ where a and b are constants is applied to a linear network and the output is found to be $v_2(t) = A[u(t - a - t_0) - u(t - b - t_0)]$, where t_0 is a constant. Determine the transfer function $H(s) = V_2(s)/V_1(s)$.

15-25 The response of a linear network to a unit step is found to be $g(t) = (t + 3e^{-5t})u(t)$. Determine the response of the network for an input $f(t) = 10e^{-3t}u(t)$.

15-26 The impulse response $h(t)$ of a network is given by $h(t) = [u(t) - u(t - T)]$ where T is a constant. Determine the output for each of the following inputs.
(a) $x(t) = \delta(t) + \delta(t - T/2) + \delta(t - 2T)$.
(b) $x(t) = \delta(t) + u(t - T)$.
(c) $x(t) = \cos(2\pi t/T)$ in the interval $(0, T)$ and zero everywhere else.

15-27 When the forcing function on a network is a unit impulse voltage $\delta(t)$, the resulting current is found to be $i(t) = 4e^{-2t} + 3e^{-2t}\cos 8t$ A. Determine (a) the impedance $Z(s)$ of the network, (b) its poles and zeros, and (c) the current in the circuit if a unit step voltage is applied to it.

15-28 When a unit step current is fed to the terminals of a network, the resulting voltage is found to be $v(t) = 3 \cos 2t + 2 \sin 3t$ V. Determine (a) the impedance $Z(s)$, and (b) the admittance $Y(s)$ of the network.

Sec. 15-5 Initial and Final Value Theorems

15-29 For each of the functions given below, (a) calculate the initial and final values of $f(t)$ by using the initial and final value theorems; (b) verify your answers by finding $f(t)$ using inverse transform.

(i) $F(s) = \dfrac{s + a}{(s + b)(s + c)}$

(ii) $F(s) = \dfrac{s + 4}{s^2 + 8s + 25}$

Sec. 15-6 Convolution

15-30 Prove the convolution theorem: $\mathcal{L}[f(t) \circledast g(t)] = F(s)G(s)$, where $F(s) = \mathcal{L}[f(t)]$, and $G(s) = \mathcal{L}[g(t)]$.

15-31 (a) Obtain the transfer function $H(s) = V_2(s)/V_1(s)$ of a series RL circuit where the output voltage is measured across the resistor. (b) Determine $h(t)$ by using inverse Laplace transform. (c) Obtain $v_2(t)$ by using convolution when the input is $v_1(t) = [u(t) - u(t - L/R)]$. (d) Verify your result by first finding $V_2(s)$ and then $v_2(t)$ by inverse Laplace transformation.

15-32 Repeat the work of the previous problem when the output is taken across the inductor.

Sec. 15-7 Critical Frequencies in the Response of Networks

15-33 A series RC circuit ($R = 50$ Ω and $C = 150$ μF) is driven by a voltage source $v_s(t) = 10e^{-2t}\cos 500t u(t)$ V. A series RLC circuit which will produce the same response as the given RC circuit is to be designed. Determine the values of R, L, C and the forcing function to be applied to the series RLC circuit.

15-34 Consider two networks. Network 1 is a parallel RL circuit with $L = 50$ mH and driven by a current source $i_1(t) = 5 \sin 4000t u(t)$ A. Network 2 is a parallel LC network with $L = 50$ mH and driven by a current source $i_2(t) = I_0 e^{-20t}$ A. The voltages across the two networks are to be identical. Determine the values of I_0, R, and C and obtain the expression for $v(t)$ across each circuit.

15-35 A series RLC circuit with zero initial energy is driven by a source $v_s(t) = V_m \cos tu(t)$. (a) Obtain the expression for the current $I(s)$ in the circuit. (b) Assuming $L = 1$ H, find R and C so that the critical frequencies of $I(s)$ are: zero at $s = 0$, poles at $-(1 \pm j1)$, $(0 \pm j1)$. (c) Find $i(t)$ for the critical frequencies specified in part (b). (d) Consider a series LC circuit (with $R = 0$)

driven by a voltage source $v_s'(t)$. Determine the capacitance C and $v_s'(t)$ so that the current $I(s)$ has the same critical frequencies as specified in part (b).

15-36 For the network shown in Fig. 15-23, let $H(s) = [V_0(s)/V_1(s)]$. (a) Find the element values so as to have the poles of $H(s)$ at $(-0.383 \pm j0.924)$ and (-0.924 ± 0.383). (b) Determine the impulse response $h(t)$.

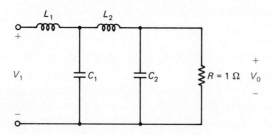

Fig. 15-23 Network for Problem 15-36.

APPENDIX A
COMPUTER-BASED PROBLEMS

[The following is a selection of problems suitable for solution by means of computer programs using FORTRAN.]

A-1 (Simpson's rule for evaluating an integral)

The integral of a function $f(t)$ between two limits t_1 and t_2 represents the area enclosed by $f(t)$ between those limits. Simpson's rule uses the following approach to evaluate the integral: The total interval (t_1, t_2) is divided into N small subintervals of width $2w$. Let t_k denote the beginning of the kth subinterval. The segment of $f(t)$ in each interval is then approximated by an expression of the form

$$f(t) \approx A + Bt + Ct^2$$

where

$$A = f(t_k)$$
$$B = [-3f(t_k) + 4f(t_k + w) - f(t_k + 2w)]2w$$
$$C = [f(t_k) - 2f(t_k + w) + f(t_k + 2w)]/2w^2$$

Then the area of each segment of $f(t)$ in the interval $(t_k, t_k + 2w)$ becomes

$$\text{area of each segment} = (w/3)[f(t_k) + 4f(t_k + w) + f(t_k + 2w)] \quad \text{(A-1)}$$

and the total area under $f(t)$ in the interval (t_1, t_2) is obtained by summation of the expression in Eq. (A-1) over all possible k. A computer program can be written in which the expression in Eq. (A-1) is evaluated for different values of k (by means of a DO loop) and the results added successively.

Consider a component whose voltage-current relationship is given by

$$i = 10^{-6} e^{(38.6v)} \text{ A} \qquad (v > 0)$$

with

$$v = t e^{-t} \text{ V} \qquad 0 < t < 10 \text{ s}$$

Write a program to evaluate the total energy dissipated by the component using Simpson's rule. Evaluate the energy by choosing (a) $N = 50$, and (b) $N = 100$.

A-2 (Gauss's elimination method for solving a set of simultaneous equations)

The solution of a set of simultaneous algebraic equations can be found by a systematic procedure due to Gauss (instead of using Cramer's rule). Consider, for example, the set of three equations:

$$a_{11}x_1 + a_{12}x_2 + a_{13}x_3 = b_1$$
$$a_{21}x_2 + a_{22}x_2 + a_{23}x_3 = b_2$$
$$a_{31}x_1 + a_{32}x_2 + a^{33}x^3 = b^3$$

where a_{jk} and b_j ($j, k = 1, 2, 3$) are all known constants and x_n ($n = 1, 2, 3$) are to be evaluated. Gauss's procedure consists of setting up an *augmented matrix* consisting of the coefficients a_{jk} and a fourth column consisting of b_j:

$$\begin{bmatrix} a_{11} & a_{12} & a_{13} & b_1 \\ a_{21} & a_{22} & a_{23} & b_2 \\ a_{31} & a_{32} & a_{33} & b_3 \end{bmatrix}$$

Then, by multiplying an appropriate row by a suitable constant and adding it to the other rows, the set of equations is reduced to the form:

$$x_1 + a'_{12}x_2 + a'_{13}x_3 = b'_1$$
$$x_2 + a'_{23}x_3 = b'_2$$
$$x_3 = b'_3$$

which is readily solved for the three unknowns:

$$x_1 = b'_1 - a'_{12}(b'_2 - a'_{23}b'_3) - a'_{13}b'_3;$$
$$x_2 = (b'_2 - a'_{23}b'_3);$$
$$x_3 = b'_3$$

The following example illustrates the procedure:
Given the set of equations

$$3x_1 + 5x_2 + 2x_3 = -1$$
$$9x_1 + 3x_2 - x_3 = 5$$
$$2x_1 + x_2 - 4x_3 = -15$$

the augmented matrix is:

$$M_1 = \begin{bmatrix} 3 & 5 & 2 & -1 \\ 9 & 3 & -1 & 5 \\ 2 & 1 & -4 & -15 \end{bmatrix}$$

Divide the first row by 3 so as to make $a_{11} = 1$. This leads to

$$M_2 = \begin{bmatrix} 1 & 1.67 & 0.667 & -0.333 \\ 9 & 3 & -1 & 5 \\ 2 & 1 & -4 & -15 \end{bmatrix}$$

Multiply the elements of the first row of M_2 by $-(9/1)$ and add to the corresponding elements of the second row so as to make $a_{21} = 0$. Multiply the elements of the first row by M_2 by $-(2/1)$ and add to the corresponding elements of the third row so as to make $a_{31} = 0$. This leads to

$$M_3 = \begin{bmatrix} 1 & 1.67 & 0.667 & -0.333 \\ 0 & -12.0 & -7.00 & 8.00 \\ 0 & -2.34 & -5.33 & -14.33 \end{bmatrix}$$

Multiply the elements of the second row of M_3 by $-(-2.34/-12.0)$ and add to the corresponding elements of the third row so as to make $a_{32} = 0$. This leads to

$$M_4 = \begin{bmatrix} 1 & 1.67 & 0.667 & -0.333 \\ 0 & -12.0 & -7.00 & 8.00 \\ 0 & 0 & -3.96 & -15.89 \end{bmatrix}$$

Then,

$$-3.96x_3 = -15.89 \quad \text{or } x_3 = 4.01$$
$$-12x_2 - 7x_3 = 8 \quad \text{or } x_2 = -3.01$$
$$x_1 + 1.67x_2 + 0.667x_3 = -0.333 \quad \text{or } x_1 = 2.02$$

Write a computer program to carry out the necessary operations of Gauss's elimination procedure for (a) a system of three simultaneous equations, and (b) a system of four simultaneous equations. Use your program for the three equations on the example used above. Use your program for the four equations to solve the following set:

$$2x_1 - 4x_2 + 5x_2 + 6x_4 = 21.32$$
$$-4x_1 + 10x_2 + 7x_3 - 8x_4 = -66.04$$
$$5x_1 + 7x_2 + 2x_3 - 10x_4 = -28.55$$
$$6x_1 - 8x_2 - 10x_3 + 12x_4 = 76.37$$

A-3 (Resistive ladder networks)
Ladder networks of the form in Fig. 2-40 (Chapter 2) lend themselves to analysis by a recursive formula, and a computer program can be set up to solve for the branch currents and voltages. Consider a ladder network with n resistors and use the same scheme for numbering the resistors as in Fig. 2-40. Let v_k be the voltage across resistor R_k and i_k the current in R_k. Then the relevant equations are:

$$v_1 = 1. \; i_1 = (v_1/R_1). \; i_2 = i_1. \; v_2 = R_2 i_2. \; v_3 = v_2 + v_1, \; i_3 = (v_3/R_3) \ldots$$

The general forms of the above equations become:

$$i_k = (v_k/R_k) \quad \text{when } k \text{ is odd}$$
$$i_k = i_{k-1} + i_{k-2} \quad \text{when } k \text{ is even}$$
$$v_k = v_{k-1} + v_{k-2} \quad \text{when } k \text{ is odd}$$
$$v_k = R_k i_k \quad \text{when } k \text{ is even}$$

Set up a computer program for evaluating the recursive equations and finding the voltages and currents of a ladder network. Use your program and analyze the circuits of Figs. 2-40, 2-41, and 2-69.

A-4 (Resistive ladder networks)
Modify the program of the previous problem for cases of the type in Fig. 2-42 and analyze that circuit using your program.

A-5 (Circuit with a nonlinear element)
Consider a circuit with a voltage source v_s, a resistor R_1, and a nonlinear device NL series. Let the

nonlinear device have the voltage-current relationship

$$i = I_0(e^{bv} - 1) \text{ A}$$

where v is the voltage across, i is the current in NL (using associated reference polarities), and I_0 and b are device constants. The operating point of the nonlinear device (that is, the value of v and i that will satisfy the circuit constraints and the characteristic of the device) can be obtained by means of a trial and error procedure. Set up a program for such a procedure. Use your program to analyze a circuit in which $I_0 = 10^{-6}$ A, $b = 38.6$ volt^{-1}, $R_1 = 100\Omega$, and v_s assuming values from -5 to $+5$V in increments of 0.2 V.

A-6 (Ladder network with impedances)
Modify the ladder network program of Problem A-3 to cover the case when the branches are impedances (instead of resistances). Use your program on the circuits of Figs. 9-5, 9-57, and 9-59.

A-7 (Pulse response of an RC circuit)
Consider a series RC circuit driven by a rectangular voltage pulse: $v_s = V_0$ $(0 < t < t_p)$. Write a program to determine and plot (a) the voltage across C, and (b) the voltage across R. Use your program on a circuit with $V_0 = 10$ V and the ratio (RC/t_p) starting at 0.02 and having *geometric* increments of 5 (that is, 0.02, 0.1, 0.5, . . .) to 62.5.

A-8 (Pulse response of an RC circuit)
Modify the program of Problem A-7 for the case of the input being a group of pulses (each of width t_p and with an interval of t_p between successive pulses): $v_s = V_0$ $(0 < t < t_p)$ and $(2t_p < t < 3t_p)$ . . . and zero everywhere else.

A-9 (Fourier series)
The coefficients of the Fourier series of a periodic signal can be determined by using a summation in place of the integral relationships for the coefficients. For this, chop up each period T into N subintervals of width (T/N). Let a periodic function $f(t)$ be approximately represented by pulses of height $f(kT/N)$, $k = 0, 1, 2, \ldots$ Then the coefficients of the Fourier series are given (approximately) by

$$a_o = (1/T) \sum_{k=0}^{N-1} f(kT/n)$$

$$a_m = (2/T) \sum_{k=0}^{N-1} f(kT/n) \cos(2\pi mk/n)$$

$$b_m = (2/T) \sum_{k=0}^{N-1} f(kT/n) \sin(2\pi mk/n)$$

Set up a program to compute the coefficients using these formulas. Apply the program to determine the Fourier series for the periodic function described by

$$V(t) = \begin{cases} V_m \sin \omega t & 0.1T < t < 0.3T \\ 0.951 V_m e^{-(t-0.3T)/1.66T} & 0.3T < t < 1.1T \end{cases}$$

A-10 (Gibbs phenomenon)
Write a program to plot the truncated Fourier series made up of the dc component and the first N harmonics of a periodic rectangular pulse train: $v(t) = V_m$ $(-T/2 < t < T/2)$ and zero for the rest of each period. Plot the series for $N = 5, 10$, and 50.

A-11 Write a computer program to evaluate the convolution of two functions $f(t)$ and $g(t)$. The integral itself is to be evaluated by using Simpson's rule (see problem A-1). Make sure that the program includes the establishment of the lower and upper limits of τ and the proper ranges of values of t. Use your program for the cases: (a) $f(t) = e^{-atu(t)}$; $g(t) = e^{-btu(t)}$, and (b) $f(t) = [u(t) - u(t - 2)]$ and $g(t) = \cos \pi t$ $(0 < t < 2)$, and zero everywhere else.

APPENDIX B
DETERMINANTS AND MATRICES

This appendix presents a summary of the basic principles and relationships needed for using determinants and matrices in circuit analysis. The discussion is informal and only those items needed directly for the material discussed in several chapters of this textbook are presented here. More formal and complete treatment can be found in the literature.

B-1 DETERMINANTS

Even though determinants are usually thought of as a tool for solving numerical equations, they have more important applications in circuit theory. The theoretical properties of certain aspects of determinant theory lead to the development of some general results in circuit analysis, and this is the primary reason for our interest in the theory of determinants in circuit analysis.

Determinants of the Second Order

A second-order determinant contains two rows and two columns and written in the form

$$\|D_1\| = \begin{vmatrix} a_{11} & a_{12} \\ a_{21} & a_{22} \end{vmatrix} \tag{B-1}$$

The value of the determinant in Eq. (B-1) is

$$\|D_1\| = (a_{11}a_{22} - a_{12}a_{21}) \tag{B-1a}$$

For example,

$$\begin{vmatrix} 2 & -3 \\ -4 & -5 \end{vmatrix} = [2(-5) - (-3)(-4)] = -22$$

B-2 MINORS AND COFACTORS OF A DETERMINANT

For determinants of order higher than 2, evaluation is systematically done by using their cofactors. Consider a determinant of order n:

$$\|D_2\| = \begin{vmatrix} a_{11} & a_{12} & \cdots & a_{1k} & \cdots & a_{1n} \\ a_{21} & a_{22} & \cdots & a_{2k} & \cdots & a_{2n} \\ \cdot & \cdot & & \cdot & & \cdot \\ a_{j1} & a_{j2} & \cdots & a_{jk} & \cdots & a_{kn} \\ \cdot & \cdot & & \cdot & & \cdot \\ a_{n1} & a_{n2} & \cdots & a_{nk} & \cdots & a_{nn} \end{vmatrix} \tag{B-2}$$

where a_{jk} is the element in the jth row and kth column.

The *minor* M_{jk} of any element a_{jk} is defined as the determinant remaining after the jth row and kth column are deleted from the original determinant. For example, M_{32} is the minor of the element a_{32} and is obtained by deleting the third row and the second column of the original determinant.

The *cofactor* A_{jk} of any element a_{jk} of the determinant in Eq. (B-2) is defined by

$$A_{jk} = (-1)^{j+k} M_{jk} \tag{B-3}$$

That is, the *cofactor of any element is the minor of that element multiplied by a sign given by* $(-1)^{j+k}$.

The cofactor of the element a_{32}, for example, is

$$A_{32} = -M_{32}$$

since $(-1)^{3+2} = -1$, while the cofactor of the element a_{13} is

$$A_{13} = M_{13}$$

since $(-1)^{1+3} = +1$.

As an example, consider the determinant

$$D_3 = \begin{vmatrix} 10 & -4 & -9 \\ -8 & 7 & 6 \\ 2 & 5 & 13 \end{vmatrix}$$

Various minors and cofactors of the determinant D_3 are listed below.

$$M_{21} = \begin{vmatrix} -4 & -9 \\ 5 & 13 \end{vmatrix} \qquad A_{21} = (-1) \begin{vmatrix} -4 & -9 \\ 5 & 13 \end{vmatrix}$$

$$M_{33} = \begin{vmatrix} 10 & -4 \\ -8 & 7 \end{vmatrix} \qquad A_{33} = (+1) \begin{vmatrix} 10 & -4 \\ -8 & 7 \end{vmatrix}$$

$$M_{13} = \begin{vmatrix} -8 & 7 \\ 2 & 5 \end{vmatrix} \qquad A_{13} = (+1) \begin{vmatrix} -8 & 7 \\ 2 & 5 \end{vmatrix}$$

Each of the cofactors is evaluated by using the formula in (B-1a):

$$A_{21} = 7 \qquad A_{33} = 38 \qquad A_{13} = -54$$

B-3 LAPLACE'S EXPANSION OF A DETERMINANT

The value of a determinant may be found by using *Laplace's expansion,* which states that the determinant equals the sum of the products of the elements of any row (or column) and their cofactors.

That is, the determinant D_2 in Eq. (A-2) can be expanded about its *j*th row and written as

$$\|D_2\| = a_{j1}A_{j1} + a_{j2}A_{j2} + \ldots + a_{jk}A_{jk} + \ldots + a_{jn}A_{jn}$$

or it can be expanded about its *k*th column and written as

$$\|D_2\| = a_{1k}A_{1k} + a_{2k}A_{2k} + \ldots + a_{jk}A_{jk} + \ldots + a_{nk}A_{nk}.$$

For example, the determinant

$$\|D_3\| = \begin{vmatrix} 10 & -4 & -9 \\ -8 & 7 & 6 \\ 2 & 5 & 13 \end{vmatrix}$$

can be expanded about any row or any column. Some possible expansions are shown below.

Expansion about the First Row:

$$\|D_3\| = 10(+1)\begin{vmatrix} 7 & 6 \\ 5 & 13 \end{vmatrix} - 4(-1)\begin{vmatrix} -8 & 6 \\ 2 & 13 \end{vmatrix} - 9(+1)\begin{vmatrix} -8 & 7 \\ 2 & 5 \end{vmatrix}$$
$$= 10(91 - 30) + 4(-104 - 12) - 9(-40 - 14) = 632$$

Expansion about the Third Column

$$\|D_3\| = -9(+1)\begin{vmatrix} -8 & 7 \\ 2 & 5 \end{vmatrix} + 6(-1)\begin{vmatrix} 10 & -4 \\ 2 & 5 \end{vmatrix} + 13(+1)\begin{vmatrix} 10 & -4 \\ -8 & 7 \end{vmatrix}$$
$$= -9(-40 - 14) - 6(50 + 8) + 13(70 - 32) = 632$$

The computational advantage of Laplace's expansion is that the value of a given determinant is found through the evaluation of lower order determinants. Laplace's expansion is used frequently in circuit analysis for writing a determinant in the form of a series of terms, each of which is the product of an element of the determinant and the cofactor of that element.

B-4 CRAMER'S RULE

Consider a system of *n* equations in *n* unknowns $x_1, x_2, \ldots, x_n$:

$$\begin{aligned} a_{11}x_1 + a_{12}x_2 + \ldots + a_{1n}x_n &= y_1 \\ a_{21}x_1 + a_{22}x_2 + \ldots + a_{2n}x_n &= y_2 \\ &\vdots \\ a_{n1}x_1 + a_{n2}x_2 + \ldots + a_{nn}x_n &= y_n \end{aligned} \quad \text{(B-4)}$$

The solution of the equations is given by *Cramer's rule*. Let $\|a\|$ be the determinant of the coefficients on the lefthand sides of the equations:

$$\|a\| = \begin{vmatrix} a_{11} & a_{12} & \cdots & a_{1k} & \cdots & a_{1n} \\ a_{21} & a_{22} & \cdots & a_{2k} & \cdots & a_{2n} \\ \cdot & \cdot & & \cdot & & \cdot \\ a_{j1} & a_{j2} & \cdots & a_{jk} & \cdots & a_{jn} \\ \cdot & \cdot & & \cdot & & \cdot \\ a_{n1} & a_{n2} & \cdots & a_{nk} & \cdots & a_{nn} \end{vmatrix}$$

The value of the unknown x_k in Eq. (B-4) is then given by the following procedure. Replace the coefficients in the kth column of the determinant $\|a\|$ by the quantities $y_1, y_2, \ldots, y_n$ from the righthand side. Calling the newly formed determinant the "modified determinant," Cramer's rule states that

$$x_k = \frac{\text{(modified determinant)}}{\|a\|} \tag{B-5}$$

Laplace's expansion of the numerator of Eq. (B-5) leads to

$$x_k = \frac{y_1 A_{1k} + y_2 A_{2k} + \ldots + y_n A_{nk}}{\|a\|}$$

or

$$x_k = \frac{1}{\|a\|} \sum_{j=1}^{n} A_{jk} y_j \tag{B-6}$$

where $A_{1k}, A_{2k}, \ldots, A_{nk}$ are the cofactors of the kth column of the determinant. That is, in order to determine a particular unknown x_k, the cofactors of the kth column are multiplied by the corresponding quantities y from the righthand side, and the product divided by the determinant $\|a\|$.

Equation (B-6) is used frequently in several of the chapters in this text.

As an example, consider the set of equations:

$$\begin{align} g_{11}v_1 + g_{12}v_2 + g_{13}v_3 &= I_1 \\ g_{21}v_1 + g_{22}v_2 + g_{23}v_3 &= I_2 \\ g_{31}v_1 + g_{32}v_2 + g_{33}v_3 &= I_3 \end{align} \tag{B-7}$$

where the gs are known coefficients, I_1, I_2, and I_3 are known quantities, and the unknowns are v_1, v_2, v_3.

Then

$$v_1 = \frac{1}{\|g\|} \sum_{j=1}^{3} A_{j1} y_j$$

$$= \frac{1}{\|g\|} [A_{11}I_1 + A_{21}I_2 + A_{31}I_3]$$

where $\|g\|$ is the determinant of the coefficients of the lefthand side of Eqs. (B-7) and A_{j1} ($j = 1, 2, 3$) are the cofactors of the elements of the first *column* of the determinant $\|g\|$.

Similarly

$$v_2 = \frac{1}{\|g\|} \sum_{j=1}^{3} A_{j2} y_j$$

$$v_3 = \frac{1}{\|g\|} \sum_{j=1}^{3} A_{j3} y_j$$

B-5 MATRICES

A matrix is an array of elements arranged in the form of m rows and n columns, in general. The matrix notation and definitions permit the representation of equations and their solutions in a methodical form.

An $n \times n$ *matrix* contains n rows and n columns. The matrix of the coefficients of Eq. (B-4) is written as

$$[a] = \begin{bmatrix} a_{11} & a_{12} & \cdots & a_{1k} & \cdots & a_{1n} \\ a_{21} & a_{22} & \cdots & a_{2k} & \cdots & a_{2n} \\ \vdots & \vdots & & \vdots & & \vdots \\ a_{j1} & a_{j2} & \cdots & a_{jk} & \cdots & a_{jn} \\ \vdots & \vdots & & \vdots & & \vdots \\ a_{n1} & a_{n2} & \cdots & a_{nk} & \cdots & a_{nn} \end{bmatrix}$$

A *column matrix* contains a single column. For example, the quantities y in Eqs. (B-4) have the column matrix representation

$$\begin{bmatrix} y_1 \\ y_2 \\ \vdots \\ y_j \\ \vdots \\ y_n \end{bmatrix}$$

Sum of Two Matrices

The sum of two $n \times n$ matrices is defined as an $n \times n$ matrix each of whose elements is the sum of the corresponding elements of the two given matrices.

$$\begin{bmatrix} a_{11} & a_{12} \\ a_{21} & a_{22} \end{bmatrix} + \begin{bmatrix} b_{11} & b_{12} \\ b_{21} & b_{22} \end{bmatrix} = \begin{bmatrix} (a_{11} + b_{11}) & (a_{12} + b_{12}) \\ (a_{21} + b_{21}) & (a_{22} + b_{22}) \end{bmatrix}$$

Product of Two Matrices

The product of two matrices $[A][B]$ is defined when the number of columns in $[A]$ is equal to the number of rows in $[B]$. The *order* of multiplication is important: $[A][B]$ is not in general equal to $[B][A]$. If

$$[A][B] = [C]$$

then the element c_{jk} of the product matrix is obtained by multiplying the elements of jth row of $[A]$ by the corresponding elements of $[B]$ and adding these products.

$$c_{jk} = a_{j1}b_{1k} + a_{j2}b_{2k} + \ldots + a_{jn}b_{nk}$$

For example

$$\begin{bmatrix} a_{11} & a_{12} \\ a_{21} & a_{22} \end{bmatrix} \times \begin{bmatrix} b_{11} & b_{12} \\ b_{21} & b_{22} \end{bmatrix} = \begin{bmatrix} a_{11}b_{11} + a_{12}b_{21} & a_{11}b_{12} + a_{12}b_{22} \\ a_{21}b_{11} + a_{22}b_{21} & a_{21}b_{12} + a_{22}b_{22} \end{bmatrix}$$

A system of equations as in Eq. (B-4) has the matrix representation

$$[A][X] = [Y]$$

where $[A]$ is an $n \times n$ matrix whose elements are the coefficients a_{jk}, (j, k = 1, 2, 3, . . . , n), $[X]$ is a column matrix whose elements are $x_1, x_2, \ldots, x_n$, and $[Y]$ is a column matrix whose elements are $y_1, y_2, \ldots, y_n$.

Inverse of a Matrix

Given a matrix $[A]$, the inverse of the matrix (if it exists) is denoted by $[A]^{-1}$ and has the property

$$[A][A]^{-1} = [A]^{-1}[A] = [I]$$

where $[I]$ is called the unit matrix. The elements on the principal diagonal of the unit matrix $[I]$ are all equal to 1, while all the other elements in it are zero.

In order for the inverse to exist, the given matrix must have a *nonzero determinant*. Since only second-order matrices are used in Chapter 13, the following formula for inverting a second order matrix is sufficient for our purposes:

$$[A] = \begin{bmatrix} a_{11} & a_{12} \\ a_{21} & a_{22} \end{bmatrix} \quad [A]^{-1} \begin{bmatrix} a_{22}/\|a\| & a_{12}/\|a\| \\ a_{21}/\|a\| & a_{11}/\|a\| \end{bmatrix}$$

The procedure for finding the inverse of a higher-order matrix can be found in the literature.

APPENDIX C
COMPLEX NUMBERS AND COMPLEX ALGEBRA

An informal discussion of some of the important properties and rules of operation involving complex numbers is presented here. A more formal and complete treatment of complex variables can be found in the literature.

A complex number **Z** is an ordered pair of real numbers A and B which is commonly written in the form

$$\mathbf{Z} = (A + jB) \tag{C-1}$$

where A is called the *real part* of **Z** and B is called the *imaginary part* of **Z**. The symbol j is used to designate the imaginary part* of a complex number. The quantity j satisfies the relationship

$$j^2 = -1 \tag{C-2}$$

The form of the complex number in Eq. (C-1) is called its *rectangular form*. Note that the symbol j is *not* part of the definition of the imaginary part, but only an indicator of the imaginary part. The notation **Re**[ ] is used to denote "take the real part of the quantity within the brackets" and **Im**[ ] to denote "take the imaginary part of the quantity within the brackets." In Eq. (C-1)

$$\mathbf{Re}[\mathbf{Z}] = A \text{ and } \mathbf{Im}[\mathbf{Z}] = B$$

Addition and Subtraction of Two Complex Numbers

The sum of two complex numbers $(A + jB)$ and $(C + jD)$ is defined by

$$(A + jB) + (C + jD) = (A + C) + j(B + D) \tag{C-3}$$

---

*The symbol i is used in mathematics and physics to designate the imaginary part of a complex quantity. It is more convenient to use the symbol j in electrical engineering, however, since the symbol i normally denotes current in an electric circuit.

Note that the

$$\text{real part of the sum} = \text{sum of the real parts}$$

and the

$$\text{imaginary part of the sum} = \text{sum of the imaginary parts}$$

Examples:

$$(3 + j4) + (-5 + j23) = -2 + j27$$
$$(3 + j4) - (-5 + j23) = (3 + j4) + (-1)(-5 + j23)$$
$$= (3 + j4) + (-5 - j23) = 8 - j19$$

Product of Two Complex Numbers

The product of two complex numbers $(A + jB)$ and $(C + jD)$ is defined by

$$[(A + jB)(C + jD)] = (AC - BD) + j(AD + BC)$$

where the relationship $j^2 = -1$ of Eq. (C-2) has been used. That is, multiplication of two complex quantities in rectangular form is done in the same manner as in the algebra of real numbers, with the exception that j^2 is replaced by -1. Note that

$$\text{Re}[(A + jB)(C + jD)] = (AC - BD)$$

and

$$\text{Im}[(A + jB)(C + jD] = (AD + BC)$$

Examples:

$$(3 + j4)(-5 + j23) = -107 + j49$$
$$(3 + j4)(3 - j4) = 25 + j0$$
$$(3 + j4)(4 + j3) = 0 + j25$$

Multiplication is often more convenient when the complex numbers are in *exponential* (or *polar*) form. The discussion of division will be deferred until after the exponential form has been introduced.

Graphic Representation of a Complex Number

The ordered pair (A, B) representing a complex number is represented by a point on a plane, called the *complex plane*, where the horizontal axis is chosen as the "real part axis" and the vertical axis as the "imaginary part axis." The real part of the complex number is the abcissa and the imaginary part of the complex number is the ordinate. With such a representation, it is possible to go one step further and identify the complex number with a *vector emanating from the origin and terminating at the point* (A, B). Graphic representations of several complex numbers are shown in Fig. C-1.

The length of the vector is called the *magnitude* (or *modulus*) and the angle made by the vector with the real axis is called the *angle* (or *argument*) of the complex number.

Given a complex number

$$\mathbf{Z} = A + jB$$

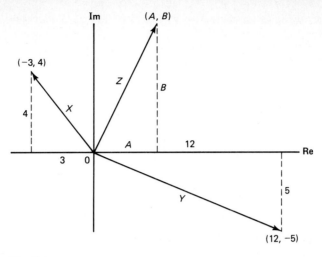

Fig. C-1

the relationships between its magnitude, argument, real part, and imaginary part are given by

$$|Z| = \sqrt{A^2 + B^2} \tag{C-4a}$$
$$\Theta = \text{Arg } Z = \text{arc tan } (B/A) \tag{C-4b}$$
$$A = |Z| \cos \Theta \tag{C-5a}$$
$$B = |Z| \sin \Theta \tag{C-5b}$$

Examples: Given $\mathbf{X} = -3 + j4$. Referring to Fig. C-1,
$$|X| = \sqrt{(-3)^2 + (4)^2} = 5$$
$$\text{Arg } X = \text{arc tan } (4/-3) = 127°$$

Given $Y = 12 - j5$,
$$|Y| = \sqrt{12^2 + (-5)^2} = 13.$$
$$\text{Arg } Y = \text{arc tan } (-5/12) = -22.6°$$

Exponential and Polar Forms of a Complex Number

The exponential function e^{jx} is a complex number whose real part and imaginary part are given by Euler's identity:

$$e^{jx} = \cos x + j \sin x \tag{C-6}$$

If R is some real number, then Eq. (C-6) leads to

$$Re^{jx} = R \cos x + jR \sin x \tag{C-7}$$

Equation (C-7) shows that the complex number $\mathbf{Z}$ of Eqs. (C-4) and (C-5) can also be written as

$$|Z|e^{j\Theta} = |Z| \cos \Theta + j|Z| \sin \Theta = \mathbf{Z} \tag{C-8}$$

Complex Numbers and Complex Algebra

Thus we have an alternative way, called the *exponential* form, of writing a complex number:

$$\mathbf{Z} = |Z|e^{j\theta} \qquad \text{(C-9)}$$

In electric circuit analysis, a notation called the *polar form* is convenient for computational purposes:

$$\mathbf{Z} = |Z|\underline{/\theta}. \qquad \text{(C-10)}$$

where $|Z|$ is the magnitude and θ the angle of the complex number Z. That is

$$|Z|\underline{/\theta} = |Z|e^{j\theta} \qquad \text{(C-11)}$$

Examples:

$$(3 + j4) = 5\underline{/53.1°}$$
$$(-5 + j12) = 13\underline{/113°}$$
$$(-10 - j10) = 14.14\underline{/-135°}$$

The conversion of one form of a complex number to another is a frequent task in ac circuit analysis. Many calculators have the ability to do this routinely. Some calculators, however, do not give the angle in the correct quadrant; that is, arc tan $(-3/4)$ and arc tan $(3/-4)$ will lead to the same result in such calculators. It is always advisable (and frequently essential) to draw a sketch of the complex number in each case to make sure to place the angle of a complex number in the correct quadrant.

The Unit Circle

The locus of complex numbers of *unit* magnitude is a circle of unit radius centered at the origin. Conversely, any point on a unit circle in the complex plane is of the form

$$1\underline{/\theta} = e^{j\theta} = \cos\theta + j\sin\theta$$

Consider, for example, the point P on the imaginary axis and the unit circle: Then

$$\mathbf{P} = 1\underline{/90°} = +j$$

Thus

$$j = 1\underline{/90°} = 1\underline{/\pi/2} = e^{j(\pi/2)}$$

Therefore

$$(1/j) = 1/e^{j(\pi/2)} = e^{-j(\pi/2)} = -j$$

or $(1/j) = -j$. Also

$$\sqrt{j} = \sqrt{e^{j\pi/2}} = e^{j(\pi/4)} = \cos(\pi/4) + j\sin(\pi/4)$$
$$= 0.707 + j0.707$$

Multiplication of Complex Numbers in Polar Form

Given two complex numbers

$$\mathbf{X} = |X|e^{j\theta} \quad \text{and} \quad \mathbf{Y} = |Y|e^{j\Phi}$$

their product is

$$XY = |X||Y|e^{j(\Theta+\Phi)} \tag{C-12a}$$

When polar form is used, the above equation becomes

$$XY = |X||Y|\underline{/(\Theta + \Phi)} \tag{C-12b}$$

Equation (C-12) states:

$$\text{magnitude of the product} = \text{product of the magnitudes}$$

and

$$\text{angle of the product} = \text{sum of the angles}$$

Examples:

$$(3.25 - j13.1)(-4.32 + j2.46) = (13.5\underline{/76.1°})(4.97\underline{/150°}) = 67.1\underline{/226.1°}$$

(It is generally advisable to keep angles in the range of $-180°$ to $+180°$ for computational convenience. So add or subtract $360°$ as needed. The answer above is then written as: $67.1\underline{/-134°}$.)

$$(0.125 + j0.321)(-0.342 - j0.419)$$
$$= (0.344\underline{/68.7°})(0.541\underline{/-129°})$$
$$= 0.186\underline{/-60.3°}$$

Division of Complex Numbers in Polar Form

Given two complex numbers

$$\mathbf{X} = |X|e^{j\Theta} \quad \text{and} \quad \mathbf{Y} = |Y|e^{j\Phi}$$

then

$$\frac{\mathbf{X}}{\mathbf{Y}} = \frac{|X|}{|Y|}e^{j(\Theta-\Phi)} \tag{C-13a}$$

When polar form is used, the above equation becomes

$$\frac{\mathbf{X}}{\mathbf{Y}} = \frac{|X|}{|Y|}\underline{/(\Theta - \Phi)} \tag{C-13b}$$

Equation (C-13) states:

$$\text{magnitude of the ratio} = \text{ratio of the magnitudes}$$

and

$$\text{angle of the ratio} = (\text{numerator angle}) - (\text{denominator angle})$$

Examples:

$$(3.25 - j13.1)/(-4.32 + j2.46) = (13.5\underline{/76.1°})/(4.97\underline{/150°}) = 2.72\underline{/-73.9°}$$
$$(0.125 + j0.321)/(-0.342 - j0.419) = (0.344\underline{/68.7°})/(0.541\underline{/-129°})$$
$$= 0.636\underline{/198°} = 0.636\underline{/162°}$$

Complex Conjugates

Two complex numbers that differ only in the *sign* of their imaginary parts are called *complex conjugates*. For example, $(3 + j4)$ and $(3 - j4)$ are complex conjugates; and $(-5 - j12)$ and $(-5 + j12)$ are complex conjugates.

The symbol $\mathbf{A}^*$ denotes the complex conjugate of $\mathbf{A}$. That is, if

$$\mathbf{A} = (q + jr) \quad \text{then} \quad \mathbf{A}^* = (q - jr)$$

Eq. (C-4) for the magnitude and angle of a complex number show that:

>the magnitudes of a pair of complex conjugates are equal to each other

and that

>the angles of a pair of complex conjugates are opposite in sign

That is

$$(|A|\underline{/\Theta})^* = |A|\underline{/-\Theta}$$

Important Relationships Involving Complex Conjugates

$$\mathbf{A}\mathbf{A}^* = |A|^2 \qquad (\text{C-14})$$

That is, the product of a pair of complex conjugates equals square of the magnitude of either number

$$(\mathbf{A} + \mathbf{A}^*) = 2\mathbf{Re}(\mathbf{A}) \qquad (\text{C-15})$$

That is, the sum of a pair of complex conjugates equals twice the *real* part of either number.

$$(\mathbf{A} - \mathbf{A}^*) = j[2\mathbf{Im}(\mathbf{A})] \qquad (\text{C-16})$$

That is, the difference between a pair of complex conjugates equals j times twice the imaginary part of the minuend.

APPENDIX D
SPICE PROGRAM FOR CIRCUIT ANALYSIS

With the proliferation of computers, large and small, and especially personal computers, a wide choice of computer software is available for computer analysis. Almost all of this software is based directly or indirectly on the program called SPICE. The brief discussion of the SPICE program presented here is not intended to be complete, and it will be necessary for the student to refer to more detailed manuals and instructions for the particular computer being used.

The *MicroSim Corporation* of Laguna Hills, CA, has developed a modification of SPICE called *PSpice* that can be used directly on the personal computer. Since many students and professors have personal computers, they should find PSpice attractive. Prentice Hall also publishes *A Guide to Circuit Simulation and Analysis Using PSpice*™, a tutorial by Paul Tuinenga. The hardcover edition includes a student version of the PSpice program. A paperback edition of the text only is also available.

D-1 INTRODUCTION

SPICE was developed at the Department of Electrical and Computer Engineering of the University of California, Berkeley, by Dr. Lawrence Nagel, and modified extensively by Dr. Ellis Cohen. Dr. Richard Dowell and Dr. Sally Liu have contributed to the development of the present version. SPICE stands for *Simulation Program with Integrated Circuit Emphasis*. It is a general-purpose circuit simulation program for nonlinear dc, nonlinear transient, and linear ac analyses. It covers circuits containing resistors, capacitors, inductors, coupled coils, sources (independent and dependent), and transmission lines, and it has built-in models of semiconductor devices.

D-2 TYPES OF ANALYSIS

D-2-1 DC Analysis

The *dc analysis* portion of SPICE determines the dc operating point of a circuit with inductors replaced by short circuits and capacitors by open circuits. A dc analysis is performed automatically prior to a transient analysis to determine the initial conditions. It is also performed to determine the linear, small-signal models of nonlinear devices. Other capabilities of dc analysis are: dc small-signal value of transfer functions, input and output resistances, transfer characteristics, and dc small-signal sensitivities. The dc analysis options are specified in these control statements: .DC, .TF, .OP, and .SENS.

D-2-2 AC Small-Signal Analysis

The *ac small-signal* portion of SPICE calculates that ac output variables as a function of frequency. The program first determines the linear, small-signal models of all the nonlinear devices in a circuit and then analyzes the equivalent circuit so obtained over the specified range of frequencies. The desired output is usually a transfer function. In the case of circuits with a single ac input, it is convenient to make it $= 1\underline{/0°}$ so that it serves as the reference.

D-2-3 Transient Analysis

The *transient analysis* of SPICE determines the output variables as a function of time over a specified time interval. The initial conditions are automatically calculated by a dc analysis. Sources that do not vary with time are set to their dc values.

D-3 INPUT FORMAT

The input for SPICE is of the free format type. Fields on a statement line are separated by one or more blanks, a comma, an equality sign (=), or a parenthesis mark; extra spaces are ignored.

A *name field* must begin with a letter (A to Z) and cannot contain any delimiters. Only the first eight characters of a name are used.

A *number field* may be an integer field (e.g., 12, −44), a floating point field (3.1416), either an integer or floating point number followed by an integer exponent (e.g., 1E-13, 2.65E6), or an integer or a floating point number followed, without a space, by one of the following *scale factors:*

T = 1E12	G = 1E9	MEG = 1E6	K = 1E3
MIL = 25.4E-6	M = 1E-3	U = 1E-6	N = 1E-9
P = 1E-12	F = 1E-15		

Note that M does not denote meg (10^6). Letters immediately following a number that are not scale factors are ignored, and letters immediately following a scale factor are ignored. That is, 10, 10V, 10 volts, and 10 Hz all represent the same numbers. Similarly, M, MA, and MSEC all represent the same scale factors. Note also that 1000, 1000.0, 1000 Hz, 1E3, 1.0E3, 1KHz, and 1K all represent the same number.

D-4 CIRCUIT DESCRIPTION

A set of descriptive statements defining the circuit topology and values of the components and a set of control statements defining the model parameters and run controls are used to describe a circuit in SPICE. The first statement line in the input file must be a title and the last line must be .END (note the preceding period). The order of the remaining statements is arbitrary.

Each element in the circuit is specified by an *element statement* which contains the name of the element, the circuit nodes to which it is connected, and the values of the parameters that determine the electrical characteristics of the element. The first letter of the element name specifies the element type; e.g., a resistor name must begin with the letter R. The name may have up to eight arbitrary (alphanumeric) characters to identify specific elements; e.g., R1, RSE, R3AC2ZY are valid names for resistors.

Data fields that are optional are enclosed in LT ("less than") and GT ("greater than") signs: < and >. SPICE uniformly uses the *associated reference polarity* convention for all branches: that is, current flows from reference positive to reference negative of the voltage.

Nodes must be non-negative integers (but need not be numbered sequentially). The *datum node* must be numbered *zero*. The circuit must not contain a loop of voltage sources or a loop of inductors or a cut set of current sources or a cut set of capacitors. Each node in the circuit must have a *dc path to ground*. Every node must have at least two connections.

D-5 TITLE, COMMENT, AND .END STATEMENTS

The title statement must be the first in the input file. Its contents are printed verbatim as the heading of each section of the output.

EXAMPLES OF TITLE STATEMENTS:

Power amplifier circuit

AC analysis of a low pass filter

An .END statement must always be the last line in the input file.

Example: .END

Note that the period is an integral part of the statement.

The general form of a comment statement is: * <*any comment*>. The leading asterisk indicates that this is a comment. Comments may be placed anywhere in the circuit description.

EXAMPLES OF COMMENT STATEMENTS:

* RF = 1K Gain Should be 100
* This circuit is a low-pass filter

D-6 ELEMENT STATEMENTS

Resistors

The general form is

  RXXXXXXX N1 N2 value

where N1 and N2 are the nodes between which the resistor is connected. "Value" is the resistance in ohms and may be positive or negative (but not zero). In this statement and those that follow, XXXXXXX, YYYYYY, etc. represent identifying descriptions used for the different elements in the circuit.

EXAMPLES:

  R1 1 2 100
  RC1 12 17 1K

Capacitors and Inductors

The general form for a capacitor and inductor are, respectively,

  CXXXXXXX N+ N− value

and

  LYYYYYYY N+ N− value

where N+ and N− are, respectively, the positive and negative nodes of the element. "Value" is the capacitance in farads or inductance in henries.

EXAMPLES:

  C1 13 0 1UF
  COSC 17 23 10U
  L2 42 69 1UH

Coupled Coils

The general form is

  KXXXXXXX LYYYYYY LZZZZZZZ value

where LYYYYYY and LZZZZZZZ are the names of the two coupled coils and "value" is the value of the coefficient of coupling k ($0 < k < 1$).

  In terms of the *dot convention* of coupled coils, SPICE assumes that the *first specified node* (N+) of each of the two coupled coils is the *dotted node*.

EXAMPLES:

  K43 LAA LBB 0.999
  KXFRMR L1 L2 0.36

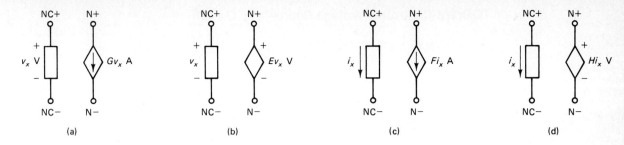

Fig. D-1

D-7 LINEAR DEPENDENT SOURCES

There are four possible linear dependent sources: voltage-controlled current sources, voltage-controlled voltage sources, current-controlled current sources, and current-controlled voltage sources.

For a linear *voltage-controlled current source,* as in Fig. D-1(a), the general form of specification is

GXXXXXXX N+ N− NC+ NC− value

where G is the transconductance, N+ and N− are, respectively, the positive and negative nodes of the source, and NC+ and NC− are, respectively, the positive and negative *controlling nodes.* "Value" is the value of the transconductance in siemens. Current flow in the source is taken as from N+ to N−.

EXAMPLE:

G1 2 0 5 0 0.1 mS.

The specifications of the other types of linear dependent sources are similar to that for the voltage-controlled current source and are given below:

Voltage-Controlled Voltage Source, Fig. D-1(b)

EXXXXXXX N+ N− NC+ NC− value

where E is the voltage gain and "value" is the value of the voltage gain.

Current-Controlled Current Source, Fig. D-1(c)

FXXXXXXX N+ N− VNAM value

where F is the current gain and "value" is the value of the current gain. VNAM is the name of the component, defined elsewhere in the program, through which the controlling current flows (from N+ to N−).

Current-Controlled Voltage Source, Fig. D-1(d)

$$HXXXXXXX \ N+ \ N- \ VNAM \ value$$

where H is the transfer impedance and "value" is the value of the transfer resistance (in ohms).

Independent Sources

The general form is

$$Name \ N+ \ N- \ \underbrace{Type \ DC/TRAN \ value}_{Optional}$$

where $N+$ and $N-$ are the positive and negative nodes of the source. Positive current is assumed to flow from $N+$ to $N-$ through the source. DC/TRAN is the dc and transient analysis values of the source. This may be omitted if the value of the dc and the transient components are zero for the source.

EXAMPLES:

$$VCC \quad 10 \quad 0 \quad DC \quad 6$$

represents a dc voltage source of 6 V connected between nodes 10 and 0.

$$IS \quad 23 \quad 21 \quad AC \quad \underset{ACMAG}{\underset{\uparrow}{0.33}} \quad \underset{ACPHASE}{\underset{\uparrow}{0.45}}$$

represents an ac current source connected between nodes 23 and 21 with a current of $0.33\underline{/0.45}$ A. ACMAG and ACPHASE denote the magnitude and phase of an ac source. The default values are 1 for ACMAG and 0 for ACPHASE.

D-8 INDEPENDENT SOURCE FUNCTIONS

There are five *independent source functions* in the program: pulse, exponential, sinusoidal, piecewise linear, and single frequency FM. The first three are discussed below.

Pulse Functions, Fig. D-2(a)

The general form is

$$NAME \ N+ \ N- \ PULSE(V1 \ V2 \ TD \ TR \ TF \ PW \ PER)$$

where V1 is the initial value, V2 is the pulsed value, TD is the delay time, TR is the rise time, TF is the fall time, PW is the pulse width, and PER is the period.

EXAMPLE:

$$VIN \ 3 \ 0 \ PULSE(-1 \ 1 \ 2NS \ 2NS \ 50NS \ 100NS)$$

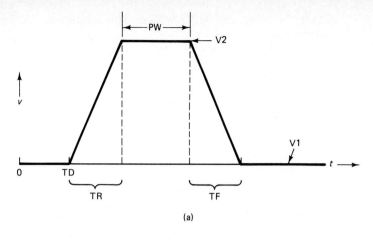

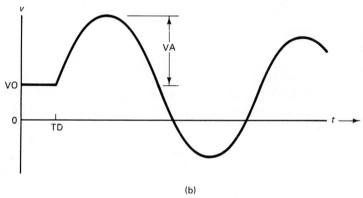

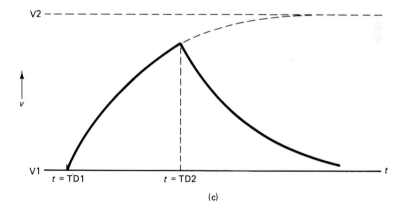

Fig. D-2

Sinusoidal Functions, Fig. D-2(b)

The general form is

NAME N+ N− SIN(VO VA FREQ TD THETA)

where VO is the dc offset value, VA is the amplitude, FREQ is the frequency (in Hz),

TD is the delay, and THETA is the damping factor. The function corresponds to the following expressions:

$$V(t) = \begin{cases} VO & (0 < t < TD) \\ VO + (VA)e^{-THETA(t-TD)} \sin[2\pi FREQ(t - TD)] & (t > TD) \end{cases}$$

EXAMPLE:

VIN 3 0 SIN(0 1 100 MEG 1NS 1E10)

represents the function

$$V(t) = \begin{cases} 0 & (t < 1 \text{ ns}) \\ e^{-(t-10^{-9})10^{10}} \sin 2\pi[100 \times 10^6(t - 10^{-9})] \end{cases}$$

Exponential Functions, Fig. D-2(c)

The general form of the function is

Name N+ N− EXP(V1 V2 TD1 TAU1 TD2 TAU2)

where V1 is the initial value, V2 is the pulsed value, TD1 is the rise delay time, TAU1 is the rise time constant, TD2 is the fall delay time, and TAU2 is the fall time constant. The function corresponds to the following expressions:

$$V(t) = \begin{cases} 0 & t < TD1 \\ V1 + (V2 - V1)[1 - e^{-(t-TD1)/TAU1}] & TD1 < t < TD2 \\ V1 + (V2 - V1)[e^{-(t-TD2)/TAU2} - e^{-(t-TD1)/TAU2}] & t > TD2 \end{cases}$$

EXAMPLE:

VIN 3 0 EXP(−4 −1 2ns 30ns 60ns 40ns)

represents the function

$$V(t) = \begin{cases} 0 & (t < 2 \text{ ns}) \\ -4 + 3[1 - e^{-(t-2\times 10^{-9})/30\times 10^{-9}}] & 2 \text{ ns} < t < 60 \text{ ns} \\ -4 + 3[e^{-(t-60\times 10^{-9})/40\times 10^{-9}} - e^{-(t-2\times 10^{-9})/40\times 10^{-9}}] & t > 60 \text{ ns} \end{cases}$$

D-9 CONTROL STATEMENTS

All control statements must be *preceded* by a period.

D-9-1 Output Width

For proper display on the VAX computer terminal, an 80 column format must be specified with the statement:

.WIDTH OUT = 80

D-9-2 Options

The options statement permits the user to modify program control for specific simulation purposes. The general form is

$$.\text{OPTIONS OPT1 OPT2} \ldots \underbrace{\text{OPTN (OPT = OPTVAL)}}_{\text{alternative form}}$$

combination of the following options may be included in any order. This is only a partial list of available questions. Wherever X occurs below, it denotes a positive number.

OPTION	EFFECT
ACCT	Causes accounting and run time statistics to be printed.
LIST	Causes a summary listing of the input data to be printed.
NOMOD	Suppresses printout of the model parameters.
NOPAGE	Suppresses page ejects.
NODE	Causes printing of the Node Table.
OPTS	Causes the option values to be printed.
GMIN=X	Resets the value of GMIN, the minimum conductance allowed by the program. The default value is 1.0E-12.
RELTOL=X	Resets the relative error tolerance of the program. The default value is 0.001 (0.1 percent).
ABSTOL=X	Resets the absolute current error tolerance of the program. The default value is 1 picoamp.
VNTOL=X	Resets the absolute voltage error tolerance of the program. The default value is 1 microvolt.

D-9-3 .OP Statement

The inclusion of the statement

.OP

in the input file will force SPICE to determine the dc operating point of the circuit with the inductors replaced by short circuits and capacitors by open circuits. (Note that a dc analysis is automatically performed prior to a transient analysis to find the initial conditions and prior to a small-signal analysis to determine the linear small-signal models for non-linear devices.)

D-9-4 .AC Statement

The inclusion of the .AC statement makes SPICE perform an ac analysis of the circuit over a specified frequency range. The general forms of the statement are:

```
.AC  DEC  ND  FSTART  FSTOP
.AC  OCT  NO  FSTART  FSTOP
.AC  LIN  NP  FSTART  FSTOP
```

where DEC denotes decade variation, with ND being the number of points per decade;

OCT denotes octave variation, with NO the number of points per octave; and LIN a linear variation, with NP being the number of points. FSTART is the starting frequency and FSTOP is the stopping frequency. Note that at least one source must have a specified ac value for getting meaningful results with the .AC statement.

EXAMPLES:

.AC DEC 10 1 10K
.AC DEC 10 1K 100MEG
.AC LIN 100 1 100Hz
.AC LIN 1 1K 1K

The last statement produces a single sinusoid of 1 kHz.

D-9-5 .TRAN Statement

The .TRAN statement must be included for transient analysis and has the general form

.TRAN TSTEP TSTOP <TSTART <TMAX>>
                   optional

where TSTEP is the printing or plotting increment for line printer output, TSTOP is the final time, and TSTART is the start time. If TSTART is omitted, it is assumed to be zero. The transient analysis always begins at $t = 0$. In the interval (0, TSTART), the circuit is analyzed, but no outputs are stored. In the interval (TSTART, TSTOP), the circuit is analyzed, and the outputs are stored. TMAX is the maximum step size, and it is specified when we wish to guarantee a computing interval smaller than the printer increment, TSTEP. If TMAX is not specified, the program chooses either TSTEP or [(TSTOP-TSTART)/50], whichever is smaller.

EXAMPLES:

.TRAN 1NS 100NS
.TRAN 1NS 1000NS 500NS

D-9-6 .IC Statement

The .IC statement is used for setting transient initial conditions and has the form

.IC V(NODENUM) = VAL .... V(NODENUM) = VAL

sets the initial node voltages (at up to eight nodes).

EXAMPLE:

.IC V(11) = 5   V(4) = −5   V(2) = 2.2

D-9-7 .PRINT Statement

The .PRINT statement defines the contents of a tabular listing of one to eight output variables and is of the form

.PRINT PRTYPE OV1 .... OV8

where PRTYPE is the type of analysis (DC, AC, TRAN) for which the specified outputs are desired.

EXAMPLES:

.PRINT TRAN V(4) I(VIN)
.PRINT AC VM(4, 2) VR(7) VP(8, 3)
.PRINT DC V(2) I(VRSC) V(23, 27)

where

V(N1, N2) specifies the voltage between nodes N1 and N2
VR = the real part
VI = the imaginary part
VM = magnitude
VOP = phase
VDB = 20 Log(magnitude)
I(VXXXXXXX) = current in the independent source named VXXXXXX

D-9-8 .PLOT Statements

The .PLOT statement defines the contents of a single plot of one to eight output variables and is of the form

.PLOT PLTYPE OV1 <(PLO1, PHI1> <OV2 <(PLO2, PHI2)>... OV8>

optional

where PLTYPE is the type of analysis (DC, AC, TRAN) corresponding to the specified outputs. The syntax for the OV statement is the same as that for the .PRINT statement. PLO and PHI are optional plot limits, and all outputs to the left of a pair of plot limits will be plotted using the same upper and lower bounds. If plot limits are not specified, SPICE determines the minimum and maximum values of all output variables to be plotted and scales the plot to fit.

EXAMPLES:

.PLOT DC V(4) V(5) V(1)
.PLOT TRAN V(17, 5) (2, 5) I(VIN) V(17) (1, 9)

D-10 EXAMPLES OF SPICE APPLICATIONS

Example D-1

Determine the ac response of the *RLC* circuit shown in Fig. D-3. Plot the real and imaginary parts of the source current, the phase of the current, the real and imaginary parts of the output voltage across R2, and its magnitude and phase.

The data file for this problem is as follows.

```
AC CIRCUIT
L1 1 2 250MH
R1 2 0 470              ── No Space
C1 2 3 .22UF
R2 3 0 1K               ── If omitted, VS is a DC voltage
VS 1 0 AC 10
.AC LIN 1 100 100
         │ │  │
         │ │  └── End frequency (Hz)
         │ └───── Start frequency (Hz)
         └─────── Number of steps
    Linear frequency scan

.PRINT AC IR(VS) II(VS) IP(VS)
         │   │      │      │
         │   │      │      └── Phase of current in VS relative to VS
         │   │      └───────── Imaginary part of current in VS
         │   └──────────────── Real part of current in VS
         └──────────────────── Type of analysis

.PRINT AC VM(2) VR(2) VI(2) VP(2)
            │     │     │     │
            │     │     │     └── Phase of V(2,0) relative to VS
            │     │     └──────── Imaginary part of voltage V(2,0)
            │     └────────────── Real part of V(2,0)
            └──────────────────── Magnitude of V(2,0)

.WIDTH OUT=80
.END
```

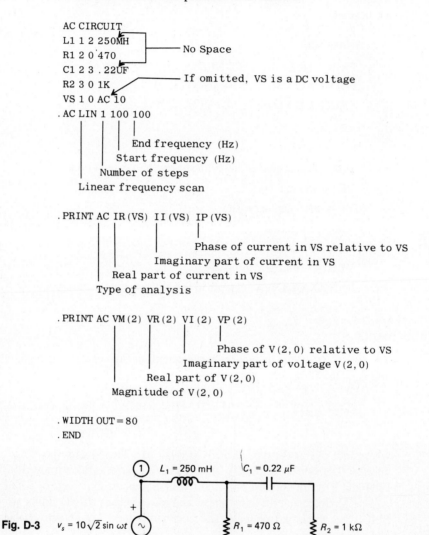

Fig. D-3 $v_s = 10\sqrt{2} \sin \omega t$, $L_1 = 250$ mH, $C_1 = 0.22$ μF, $R_1 = 470$ Ω, $R_2 = 1$ kΩ

Example D-2

Determine the transient response of the *RC* circuit shown in Fig. D-4. Assume that the initial voltage on the capacitor is 5V.

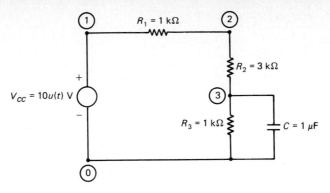

Fig. D-4

The data file for this problem is as follows.

```
RC TRANSIENT ANALYSIS
*INITIAL VOLTAGE ACROSS CAPACITOR IS 5V

VCC 1 0 10V
R1 1 2 1K
R2 2 3 3K
R3 3 0 1K
 C 3 0 1UF
.IC V(3) =5V
.TRAN .1MS 2MS
.PLOT TRAN V(2)
.WIDTH OUT=80
.END
```

BIBLIOGRAPHY

This bibliography lists a selection of books that provide more information on topics not covered in depth in the present book and those that extend the subject areas covered in it. The list is not exhaustive, but is dictated by my own preferences.

Physics

HALLIDAY, D., and RESNICK, R. *Physics for Students of Science and Engineering.* New York: Wiley, 1960.

TILLEY, D. E. *University Physics for Science and Engineering.* New York: Cummings Publishing Co., 1976.

These two books are useful sources of the physics background needed for circuit analysis.

Mathematics

LEPAGE, WILBUR. *Complex Variables and the Laplace Transform for Engineers.* New York: McGraw-Hill, 1961. A rigorous, theoretical, and detailed treatment (at an advanced level) of complex variables and Laplace transforms.

PALIOURAS, J. D. *Complex Variables for Scientists and Engineers.* New York: Macmillan, 1975. A good reference for complex numbers and complex algebra, as well as the more advanced topics in complex variables.

WYLIE, C. R. *Advanced Engineering Mathematics.* New York: McGraw-Hill, 1960. A comprehensive but concise treatment of determinants, differential equations, complex variables, and other mathematical topics.

General Circuit Analysis Topics

BRENNER, EGON, and JAVID, MANSOUR. *Analysis of Electric Circuits.* New York: McGraw-Hill, 1967. Contains a detailed treatment of the response of first- and second-order circuits, power considerations in the sinusoidal steady state, resonance, three-phase circuits and magnetically coupled circuits.

CLOSE, CHARLES M. *The Analysis of Linear Circuits*. New York: Harcourt, Brace and World, 1966. A good reference for Bode diagrams, Fourier series and integrals, and Laplace transforms.

DESOER, CHARLES A., and KUH, ERNEST S. *Basic Circuit Theory*. New York: McGraw-Hill, 1969. Noteworthy for its rigorous definitions and treatment of the topics in circuit analysis.

GUILLEMIN, ERNST A. *Introductory Circuit Theory*. New York: Wiley, 1953. An excellent reference for many of the topics in circuit analysis (especially topology), particularly for insight into certain areas of network response.

VAN VALKENBURG, M. E. *Network Analysis*. Englewood Cliffs, NJ: Prentice-Hall, 1958. Probably the most lucid treatment of network functions, critical frequencies, and their importance in the response of networks.

Operational Amplifiers

MILLMAN, JACOB. *Microelectronics*. New York: McGraw-Hill, 1979. An almost encyclopedic volume on electronics with an extensive discussion of operational amplifiers and Bode diagrams.

Signals, Convolution, and Transform Methods

LATHI, B. P. *Signals, Systems and Communication*. New York: Wiley, 1965. A good (and lucid) reference for exponential signals, Fourier series, Fourier transforms, convolution, and Laplace transforms.

OPPENHEIM, ALAN V., and WILLSKY, ALAN S. *Signals and Systems*. Englewood Cliffs, NJ: Prentice-Hall, 1983. A good reference for a rigorous and detailed treatment of convolution, Fourier transforms, and Laplace transforms.

STREMLER, FERREL G. *Introduction to Communication Systems*, 2nd ed. Boston: Addison Wesley, 1982. A detailed discussion of signal representation, Fourier transform and applications, and power and energy spectral density.

TAUB, HERBERT, and SCHILLING, DONALD. *Principles of Communication Systems*, 2nd ed. New York: McGraw-Hill, 1986. Contains a detailed treatment of spectral analysis.

Magnetic Circuits

GOURISHANKAR, VEMBU. *Electromechanical Energy Conversion*. Scranton, PA: International Textbook Co., 1965. Contains several chapters on magnetic circuits and their applications.

Two-Ports and Generalized Networks

MURDOCH, JOSEPH. *Network Theory*. New York: McGraw-Hill, 1970. Contains a detailed discussion of network parameters and matrices, topological properties of networks, and transform methods.

WEINBERG, LOUIS. *Network Analysis and Synthesis*. New York: McGraw-Hill, 1962. A classic text with a detailed discussion of two-ports and filter design.

Computers in Circuit Analysis

GUPTA, S., BAYLESS J. W., and PEIKARI, B. *Circuit Analysis with Computer Applications to Problem Solving*. Scranton, PA: Intext Educational Publishers, 1972. Provides examples of FORTRAN-based programs to problems in a wide variety of topics in circuit analysis.

LEY, B. J., *Computer Aided Analysis and Design for Electrical Engineers*. New York: Holt, Rinehart and Winston, 1970. Provides FORTRAN examples for a variety of topics in circuit analysis, as well as other topics in electrical engineering.

ANSWERS TO EVEN-NUMBERED PROBLEMS

Chapter 1

1-2 Decreases by 4×10^{-17} J.

1-4 2.5 A ($0 < t < 6$); -5 A ($6 < t < 9$). Function is repetitive every 9 seconds.

1-6 $p(t) = 180 - 180 \sin 2\pi t$ W. 147 J.

1-8 $P_o = 400$ kW. $P_{av} = 20$ kW.

1-10 (a) $v_1 = v_A - v_B$, $v_2 = v_B - v_c$. (b) $v_B = -19$ V, $v_1 = 34$ V.

1-12 $v_P = v_1$; $v_2 = (v_1 - v_2)$; $v_R = v_1 - v_2 + v_3$; $v_S = v_1 - v_2 + v_3 - v_4$.

1-14 (a) Delivers $+400$ W. (b) Delivers $+100$ W. (c) Receives $+550$ W. (d) Receives $+200$ W.

1-16 No. 1 receives -700 W. No. 2 receives -650 W. No. 3 delivers $+975$ W. No. 4 delivers -1575 W. No. 5 delivers -750 W.

1-18 75 MHz.

1-20 Nonlinear.

1-22 (a) $[26.4 + 10 \cos 10t - 25 \sin 10t]$ A. (b) $(-16 + 4 \cos 10t - 10 \sin 10t)$ A. (c) $(-22 - 5.12 \cos 10t + 12.8 \sin 10t)$ A.

1-24 $v(t) = 0$ when $t < -4$ s; $-24e^{-3(t+4)}$ A when $t > -4$ s.

1-26 9.02×10^{-17} J.

1-28 $v < -8$: 10.7 V, 2.67 Ω. $-8 < v < -4$: 4V, 0.571 Ω. $-4 < v < 0$: open circuit. $0 < v < 5$: 0.625 Ω. $v > 5$: 6.43 V, 1.43 Ω.

1-30 (a) $v_x = 10$ V, $v_y = 50$ V. (b) $v_x = -97$ V, $v_y = -5$ V.

1-32 i_a source: 141 V, 707 W. i_b source: -189 V, -567 W. i_c source: -211 V, -2321 W.

1-34 (a) 0.5 mA. (b) 57 A. (c) -22 mA. (d) 11A.

1-36 v_1 source: 29A, 493 W. v_2 source: 0 A, 0 W. v_3 source: -13A, -104 W.

1-38 (a) 500 V (positive at Q), 25 Ω. (b) 720 V (positive at P), 6 Ω.
(c) 4 A (out at P), 20 Ω. (d) 0.15 A (out at Q), 200 Ω.

Chapter 2

2-2 $R_x = 5.3$ Ω. Powers: 11.25 W, 15.75 W, 6.75 W, 29.25 W, 11.92 W.

2-4 21.5 kΩ, 1.31 kΩ.

2-6 (a) 1%. (b) 20%. (c) 50%. (d) 0.05%.

2-8 $R_T = (R_1 + R_2 + R_3)$; $v_T = (v_a + v_b + v_c)$; $i = v_T/R_T$. $v_1 = (R_1/R_T)v_T$; $v_2 = (R_2/R_T)v_T$; $v_3 = (R_3/R_T)v_T$.

2-10 134.3 V.

2-12 $G_x = 6.83$ S. Powers: 4.5 W, 0.9 W, 61.5 W, 8.1 W.

2-14 2.22 kΩ, 6.67 kΩ.

2-16 (a) 1%. (b) 20%. (c) 0.05%.

2-18 $G_T = (G_1 + G_2)$; $i_T = (-i_a + i_b - i_c - i_d)$. $v = (i_T/G_T)$. $i_1 = (G_1/G_T)i_T$. $i_2 = (G_2/G_T)i_T$.

2-20 142.5 A.

2-22 (a) $v_1 = -58.1$ V. i_a delivers 1046 W; i_b delivers 988 W; v_c delivers 2896 W.
(b) $v_1 = -203$ V. i_a delivers 2640 W; i_b delivers 3452 W; v_c delivers 2424 W.

2-24 (a) $v_a = (i_a + i_b)/(G_a + G_b + K)$. Power $= -Kv_a^2$.
(b) $v_a = -i_s/[G_1(1 + K) + G_2]$. Power $= -KG_1v_a^2$.

2-26 $v_1 = [i_a(1 + KG_b) + i_b]/(G_a + G_b + KG_aG_b)$.

2-28 (a) $R_T = 10.3$ Ω. $i_s = 9.69$ A. (b) $R_T = 8.29$ Ω. $i_s = 12.1$ A.

2-30 Using the same subscripts as the resistors: (a) $i_1 = 9.69$ A, $i_2 = 5.61$ A, $i_3 = 4.08$ A, $i_4 = 1.71$ A, $i_5 = 1.28$ A, $i_6 = i_7 = 1.09$ A. (b) $i_1 = i_2 = 1.72$ A; $i_3 = i_4 = 3.45$ A; $i_5 = i_6 = 3.45$ A; $i_7 = 6.90$ A; $i_8 = 12.1$ A; $i_9 = 5.17$ A; $i_{10} = 4.31$ A; $i_{11} = i_{12} = 0.862$ A.

2-32 Using the same subscripts as the resistors: $i_1 = 1.12$ A; $i_2 = 0.875$ A; $i_3 = 0.854$ A; $i_4 = 0.250$ A; $i_5 = 0.521$ A; $i_6 = 0.625$ A; $i_7 = i_9 = 0.333$ A; $i_8 = 0.146$ A.

2-34 Percent error $= 100[R_LR_s/(R_LR_v + R_LR_s + R_vR_s)]$.
For 10% error: $R_v = 9R_LR_s/(R_L + R_s)$. $R_v = 54$ kΩ.

Chapter 3

3-2 (b) (i) A. (ii) E, H, and G.

3-4 (a) $-v_D$. (b) $-v_J + v_c - v_D + v_L$.
(c) $v_K - v_D$.

3-6 $v_1 = v_K - v_D$; $v_2 = v_K - v_C + v_B$; $v_3 = v_K - v_C$; $v_4 = v_K$; $v_5 = v_K - v_D + v_L$; $v_6 = v_K - v_C + v_B - v_I$; $v_7 = v_K - v_C + v_J$.

3-8 (a) $(G_2 + G_1)(v_1 - v_2) + G_3(v_1 - v_6) + G_4(v_1 - v_5) + (G_5 + G_6)(v_1 - v_4) + G_7v_1 = -i_a$.
(b) $(G_1 + G_2 + G_3)(v_1 - v_2) + (G_4 + G_5 + G_6)(v_1 - v_3) = -i_a + i_b - i_c - i_d - i_e$.

3-10 $v_1 = -1.45$ V; $v_2 = 1.39$ V; $v_3 = 0.866$ V. Starting with G_a and going in alphabetical order of the subscripts, the branch currents and power dissipated in the conductances are: 5.68 A (16.1 W), 8.52 A (24.2 W), 5.8 A (8.41 W), 0.261 A (0.136 W), 0.174 A (0.150 W), 0.0866 A (0.0750 W), 5.56 A (7.73 W). Power delivered by the three current sources: 34.8 W, 11.6 W, 10.4 W.

3-12 New voltages: $v_1 = 11.6$ V, $v_2 = 14.4$ V, $v_3 = 13.87$ V, $v_{REF} = 13$ V. Other quantities not affected.

3-14 $G_2/[(G_1G_2 + G_1G_3 + G_1G_4 + G_2G_3 + G_2G_4)]$.

3-16 Starting with G_1 and going in numerical order of the subscripts, the branch currents are: 6.67 A, 13.3 A, 0 A, 5.77 A, 19.2 A.

3-18 31.3 A.

3-20 v_a source: 1276 W; v_b source: 1253 W; i_c source: 666 W.

3-22 Same as in (3-20) above.

3-24 $v_1 = 6.97$ V; $v_2 = 4.76$ V; $v_3 = 6.18$ V. Starting with G_a and proceeding in alphabetical order of the subscripts, the branch currents are: 5.58 A, 4.42 A, 0.618 A, 5.68 A, 3.81 A.

3-26 (a) $(v_2/v_1) = (KG_2 + G_3)/(G_3 + G_4)$. Let $\Delta = G_1G_3 + G_1G_4 + G_3G_4 + G_2G_3 + G_2G_4 (1 + K)$. (b) $(i_2/i_s) = -G_4(KG_2 + G_3)/\Delta$. (c) $(v_1/i_s) = (G_3 + G_4)/\Delta$.
(d) $(v_2/i_s) = [G_1 + G_2 (1 + K) + G_3]/\Delta$.

3-28 Matrix elements are: (a) $(G_1 + G_2 + G_6 + G_8 + G_9), -(G_1 + G_2), -(G_8 + G_9), -(G_1 + G_2), (G_1 + G_2 + G_3 + G_4 + G_5), -(G_4 + G_5), -(G_8 + G_9), -(G_4 + G_5), (G_5 + G_4 + G_7 + G_8 + G_9)$. (b) $(G_a + G_b + G_c), -G_a, -G_b, -G_a, (G_a + G_b + G_c), -G_c, -G_b, -G_c, (G_a + G_b + G_c)$.

3-30 (a) $v_1 = 2.94$ V; $v_2 = 3.31$ V; $v_3 = 3.06$ V. (b) $v_1 = 10/(G_1 + G_2)$; $v_2 = 15(G_3 + G_5)/\Delta$; $v_3 = 15G_3/\Delta$,
where $\Delta = (G_1 + G_2)[(G_3 + G_4)(G_4 + G_5) - G_3^2]$.

3-34 (a) Matrix elements are: $2G_1, 0, -G_1, -G_1, 0, 2G_1, -G_1, -G_1, -G_1, -G_1, (2G_1 + G_2), 0, -G_1, -G_1, 0, (2G_1 + G_2)$. (b) $v_2 = i_1/(2G_2)$. (c) $(v_1/i_1) = (G_1 + G_2)/2G_1G_2$.

3-36 (a) $(v_1/i_2) = G_2/\Delta$; $(v_2/i_1) = (G_2 - K)/\Delta$, where $\Delta = G_1(G_2 + G_3) + G_2(K + G_3)$.
(b) $(v_1/i_2) = 0.121$ Ω, $(v_2/i_1) = 0.0213$ Ω.

3-38 (a) 0.163 Ω, 0.163 Ω, 0.463 Ω. (b) $1/(G_1 + G_2)$, $(G_3 + G_5)/\Delta$, $(G_3 + G_4)/\Delta$,
where $\Delta = [(G_3 + G_4)(G_3 + G_5) - G_3^2]$.

3-40 (a) $R_{in} = (G_1 + G_2)[G_3(\beta + 1) + G_4] + G_3G_4/G_1G_2[G_3(\beta + 1) + G_4] + G_1G_3G_4$.
$R_o = (G_2 + G_3)/G_2[G_3 (\beta + 1) + G_4] + G_3G_4$.
(b) $R_{in} = (1.9 + 0.2$ K$)$ Ω. $R_o = 1.6$ Ω.

3-42 $R_o = (G_1 + G_i + G_F)/[(G_1 + G_i)(G_o + G_F) + (1 + A) G_FG_o]$.

3-44 (a) 1000 (b) 2.

3-46 $R_F = 124$ MΩ.

3-48 $v_o = A[(G_1 + G_F)v_b - G_1v_a]/[G_1 + G_F(1 + A)]$.

3-50 Use Eq. (3-92) with $v_s = (R_2R_3v_{i1} + R_1R_3v_{i2} + R_1R_2v_{i3})/(R_1R_2 + R_1R_3 + R_2R_3)$ and G_4 instead of G_1.

Answers to Even-Numbered Problems

Chapter 4

4-2 (a) $i_1 = -0.106$ A; $i_2 = 0.417$ A; $i_3 = -1.298$ A. Currents in the other branches are: 1.19 A in R_2, 1.61 A in R_5, and 1.72 A in v_a. 20.6 W delivered by v_a, 21.4 W delivered by v_b. $v_{AB} = 18.8$ V. (b) $i_a = 0.636$ A, $i_b = i_c = 0.454$ A. Currents in the other branches are: 0.182A in R_L, 1.09 A in v_a, and 0.636 in R_a. 10.9 W delivered by v_a. $v_{AB} = 3.62$ V.

4-4 $(v_D/v_{be}) = [1 + (R_2/R_1)]$.

4-6 Same result as Eq. (3-93) in Chapter 3.

4-8 v_a delivers 1084 W; i_b delivers -288 W. 520 W in R_1, 22.7 W in R_2, 163 W in R_3, and 90.8 W in R_4.

4-10 i_a delivers $(v_b i_a)$ W; v_b delivers $v_b[(v_b/R_L) - i_a]$ W; R_L consumes (v_b^2/R_L)W. v_b receives positive power (or delivers negative power) when $R_L > (v_b/i_a)$.

4-12

	i_a	i_b	i_c	i_d	i_e	i_f	i_g	i_h
i_1	+1	+1	0	0	-1	0	+1	0
i_2	0	+1	+1	0	0	-1	0	+1
i_3	0	0	+1	+1	+1	0	-1	0
i_4	+1	0	0	+1	0	+1	0	-1

4-14 $R_1 R_x = R_2 R_3$. R_s and R_m have no effect on the condition for balance.

4-16 Open circuit: $i_s = 2v_s/(R_a + R_b)$; $v_{AB} = v_s(R_b - R_a)/(R_b + R_a)$. Short circuit: $i_s = v_s(R_a + R_b)/2R_a R_b$. $i_{sc} = v_s(R_b - R_a)/2R_a R_b$.

4-20 $R_1 = R_a R_b/(R_a + R_b + R_c)$; $R_2 = R_a R_c/(R_a + R_b + R_c)$;
$R_3 = R_b R_c/(R_a + R_b + R_c)$.
$R_a = (R_1 R_2 + R_1 R_3 + R_2 R_3)/R_3$; $R_b = (R_1 R_2 + R_1 R_3 + R_2 R_3)/R_2$;
$R_b = (R_1 R_2 + R_1 R_3 + R_2 R_3)/R_1$.

4-22 $v_3 = -K_2 R_2 R_3 v_s/\{(R_1 + K_2 R_1)(R_1 + R_2 + R_3) - (R_1 + K_1 R_3)[R_1 + K_2(R_1 + R_2)]\}$.

4-24 v_a delivers $v_a i_c$ W. i_c delivers $(R_b i_c - v_a) i_c$ W; R_b consumes $(R_b i_c^2)$W. i_c receives positive power when $R_b < (v_a/i_c)$.

4-26 $v_o = R[(2/3) i_a + i_b + (4/3) i_c]$.

4-32 Elements of the matrix are: $(R_1 + R_2 + R_3 + R_4)$, $-R_4$, $-(R_2 + R_3)$, $-R_4$, $(R_4 + R_6 + R_7 + R_8)$, $-(R_6 + R_7)$, $-(R_2 + R_3)$, $-(R_6 + R_7)$, $(R_2 + R_3 + R_6 + R_7 + R_5)$. $i_1 = 4.57$ A, $i_2 = 2.57$ A, $i_3 = 2.86$ A.

4-34 Elements of the matrix are: $(R_1 + R_2 + R_3)$, $-(R_3 + KR_5)$, KR_5, $-R_3$, $[R_3 + R_4 + R_5(1 + K)]$, $-R_5(1 + K)$, 0, $-R_5$, $(R_5 + R_6)$.

4-36 4.57×10^{-2} S, 2.57×10^{-2} S, 2.86×10^{-2} S.

4-38 Elements of the matrix are: $(R_1 + R_2)$, $-R_2$, $(K - R_2)$, $(R_2 + R_3)$.

Chapter 5

5-2 $V_{Th} = v_s(R_2 - R_1)/(R_2 + R_1)$. $R_{Th} = 2R_1 R_2/(R_1 + R_2)$.

5-4 $R_{in} = (20 + 32K)/(7 + 8K)$.

5-6 (a) $i_N = 5.74$ A. (b) $i_N = -12.7$A. (c) $i_N = 6.64$ A. (d) $i_N = (R_2 - R_1)v_s/2R_1 R_2$. (e) $i_N = (R_1 - KR_2)v_s/(R_s + R_1) R_2 + R_s R_1$.

5-8 $v_{Th} = 4.8$ V, $I_N = 2$ A, $R_{Th} = 2.4$ Ω.

5-10 $v_{Th} = 10$ V, $R_{Th} = 6$ Ω. $i_L = 1.37$ A, $v_L = 1.88$ V.

Answers to Even-Numbered Problems

5-12 (a) $R = 0$. (b) $R_b = \infty$. (c) $R_a = \infty$. (d) $R_b = 0$.

5-14 $(P_L/P_s) = R_L/(R_L + R_s)$. $(P_L/P_s) = 0$ when $R_L = 0$ and approaches 1 as $R_L \to \infty$. P_s is maximum when $R_L = 0$; P_L is maximum when $R_L = R_s$; (P_L/P_s) is maximum when $R_L = \infty$.

5-16 $i_{sc} = i_a + (i_c/2) + (v_b/R)$.

5-18 $v_b = -Ri_a - 2Ri_c$.

5-20 Both ratios $= (R_1 + R_2)/8R_1R_2$.

5-22 Passive network: $r_{11} = 102\ \Omega$; $r_{12} = r_{21} = 107\ \Omega$; $r_{13} = r_{31} = -37.9\ \Omega$; $r_{22} = 128\ \Omega$; $r_{23} = r_{32} = -51.7\ \Omega$; $r_{33} = 34.5\ \Omega$. Data insufficient for active network.

Chapter 6

6-2 $i(t) = CV_m\{(1 + m\cos At)\ B\cos Bt - mA\sin At\sin Bt\}$.

6-4 (a) $v_c(t) = (1/b^2C)[1 - (1 + bt)e^{-bt}]$. (b) $w_c(t) = (1/2b^4C)[1 - (1 + bt)e^{-bt}]^2$. $p_c(t) = (1/b^2C)[1 - (1 + bt)e^{-bt}]te^{-bt}$.

6-6 (a) $v_L = -0.125$ V $(0 < t < 4)$; 0.25 V $(4 < t < 6)$. Repetitive.
(b) $w_L = 0.025(-2.5t + 10)^2\ 0 < t < 4$; $0.025(5t - 20)^2\ (4 < t < 6)$.
(c) $p(t) = (0.3125t - 1.25)$ W $(0 < t < 4)$; $(1.25t - 5)$ W $(4 < t < 6)$.

6-8 (a) $i(t) = (-0.1t^2 + 2t)$ A $(0 < t < 10)$; $(-1.33t + 23.3)$ A $(10 < t < 20)$; $(4.44 \times 10^{-2}t^2 - 3.11t + 41.3)$ A $(20 < t < 35)$; -13.2 A $(t > 35)$.
(b) $i(t) = (-0.1t^2 + 2t + 3)$A $(0 < t < 10)$; $(-1.33t + 26.3)$ A $(10 < t < 20)$; $(4.44 \times 10^{-2}t^2 - 3.11t + 44.3)$ A $(20 < t < 35)$; -10.2 A $(t > 35)$.

6-10 $x(t) = 75e^{-10t} + 25$.

6-12 (a) $i(t) = I_o e^{-At}$. (b) $i(t) = (B/A)(1 - e^{-At})$.

6-14 $R = 10.5$ kΩ $C = 22$ μF.

6-16 $i_1(t) = -0.018e^{-2.4t}$ A $(t > 0)$. $i_2(t) = (1.67 - 15e^{-2.4t})$mA $(t > 0)$.

6-18 $v_c(t) = [V_o(1 - k)/2][1 - e^{-t/\tau}]$ where $\tau = (K + 3)RC/2$.

6-20 (a) v_c increases to 9.93 V at $t = 50$ ms, decays to zero and increases again to 9.93 V at $t = 200$ ms and decays to zero. (b) v_c increases to 3.94 V at $t = 50$ ms, decays to 1.45 V at $t = 150$ ms, increases to 4.81 V at $t = 200$ ms, and decays to zero.

6-22 $v_{c1} = 80e^{-5t}$ V. $v_{c2} = 100 + 20e^{-5t}$ V.

6-24 $R = 2$ kΩ. $L = 39.2$ mH.

6-26 $i_L(t) = -170$ mA $(t < 0)$; $(-42.6 - 127\ e^{-1.08 \times 10^4 t})$ mA $(t > 0)$.

6-28 $i_L = -7$ A $(t < 0)$; $(-2.5e^{-14.3t} - 4.50)$ A $(t > 0)$. $i_1 = -4$ A $(t < 0)$; $(-2 - 1.43e^{-14.3t})$ $(t > 0)$.

6-30 $i_L = (V_m/R)(1 - e^{-t/\tau})$ $(0 < t < t_P)$; $(V_m/R)(1 - e^{-t_P/\tau})e^{-(t-t_P)/\tau}$ $(t > t_P)$.

6-32 $R = 4$ kΩ. $L = 111$ H.

6-34 (a) $v_c(t) = 25e^{-1.33t}$ V; $i_c(t) = -0.0167e^{-1.33t}$ A. (b) $v_c(t) = 45(1 - e^{-1.33t})$ V; $i_c(t) = 0.03e^{-1.33t}$ A. (c) $v_c(t) = (45 - 20e^{-1.33t})$ V; $i_c(t) = 0.0133e^{-1.33t}$ A.

6-36 (a) $i_L(t) = 0.03e^{-3.13 \times 10^4 t}$ A; $v_L(t) = -14.1e^{-3.13 \times 10^4 t}$ V.
(b) $i_L(t) = (0.2 - 0.2e^{-3.13 \times 10^4 t})$ A; $v_L(t) = 94e^{-3.13 \times 10^4 t}$ V.
(c) $i_L = (0.2 - 0.17e^{-3.13 \times 10^4 t})$ A; $v_L(t) = 79.9e^{-3.13 \times 10^4 t}$ V.

6-40 $R = 7.36\ \Omega$; $L = 92$ mH; $C = 6793$ μF.

Chapter 7

7-2 (a) 477 HZ. (b) $i(t) = 1.5 \cos(3000t - 90°)$ A.

7-4 $i_1(t) = 20 \cos 524t$ A; $i_2(t) = 16 \cos(524t + 120°)$ A.

7-6 (a) $(-104 \cos 100t + 60 \sin 100t)$ V. (b) $(26.0 \cos 450t + 15 \sin 450t)$ A. (c) $(-7.24 \cos 50t + 49.0 \sin 50t)$ V.

7-8 (a) $36.0 \cos(100t + 56.3°)$ V. (b) $28.0 \cos(200t + 105°)$ V. (c) $10 \cos 300t$ V.

7-10 (a) Amplitude $= 2A \cos(\phi/2)$. (b) Phase $= (\phi/2)$. (c) Maximum when $\phi = 0$ and minimum when $\phi = 180°$.

7-12 $i(t) = 4.47 \cos 105t$ A. $v(t) = 112 \cos 105t$. V. $P_{av} = 250$ W.

7-14 $i(t) = -7.2 \times 10^{-2} \sin(5 \times 10^6 t)$ A. $p(t) = -0.432 \sin 10^7 t$ W.

7-16 $v(t) = 17.8 \cos(1.26 \times 10^4 t)$ V. $i(t) = -5.61 \sin(1.26 \times 10^4 t)$ A. $p(t) = -50 \sin(2.52 \times 10^4 t)$ W.

7-18 $v(t) = -30 \sin 500t$ V. $p(t) = -60 \sin 1000t$ W.

7-20 $i(t) = 20 \cos 1000t$ A. $v(t) = -50 \sin 1000t$ V. $L = 2.5$ mH. $p(t) = -500 \sin 2000t$ W.

7-22 (a) Peak value increases by a factor K_1^2 and frequency is not affected for all three elements. Constant component for the resistor changes by a factor of K_1^2, but no change for the inductor or capacitor. (b) Peak value unaffected for the resistor, decreases by K_2 for the inductor, and increases by K_2 for the capacitor. Frequency increases by a factor K_2 for all three elements. Constant component not affected for all three elements. (c) No effect on any of the items.

7-24 $i(t) = 10 \cos(2000t - 2.86°)$ A.

7-26 $v(t) = [I_m/\sqrt{G^2 + 1/\omega^2 L^2}] \cos(\omega t + \arctan 1/\omega LG)$.

7-28 $I_m = V_m/\sqrt{R^2 + \omega^2 L^2}$. $\tan \phi = -(\omega L/R)$.

7-30 $G = 0.313$ S. $C = 1.56 \times 10^{-2}$ F.

7-32 (a) 1168 r/s. (b) 316 r/s. (c) 184 r/s.

7-34 $p(t) = [19.9 + 42.3 \cos(5000t - 61.9°)]$ W. $P_{ave} = 19.9$ W.

7-36 (a) $I_m = 4.62$ A. $p(t) = [500 + 577 \cos(2\omega t - 30°)]$W. (b) $I_m = 4.62$ A. $p(t) = 500 + 577 \cos(2\omega t + 30°)$ W. (c) $I_m = 4$ A. $p(t) = 500 + 500 \cos 2\omega t$.

7-38 (a) $0.577 V_m$. (b) $0.745 I_m$.

Chapter 8

8-2 (a) $v_1(t) = 10 \cos(200t - 135°)$ V; (b) $v_2(t) = 10 \sin(200t - 75°)$ V; (c) $v_3(t) = 10 \cos(200t - 180°)$ V.

8-4 (a) $f_1(t) = 2\sqrt{p^2 + q^2} \cos[bt + \arctan(q/p)]$. (b) $f_2(t) = 2\sqrt{p^2 + q^2} \cos bt$. (c) $f_3(t) = 2\sqrt{p^2 + q^2} \cos[bt + r + \arctan(q/p)]$. (d) $f_4(t) = 2\sqrt{p^2 + q^2} \cos bt$.

8-6 (a) $h_1(t) = \text{Im}[22.4e^{j(500t + 63.4°)}]$. (b) $h_2(t) = \text{Im}[14.1e^{j(500t + 98.1°)}]$. (c) $h_3(t) = \text{Im}[18e^{j(500t - 146°)}]$.

8-8 (a) $10\underline{/-30°}$ V. (b) $24.8\underline{/-13°}$ A.

8-10 (a) $5 \cos(100t + 36.9°)$ A. (b) $8.06 \cos(100t - 60.2°)$ A. (c) $3 \cos(100t - 90°)$ A.

8-12 (a) $2e^{j(3t - 3°)}$ A. (b) $16 \sin(3t - 12°)$ V.

8-14 (a) $47i + 8333 \int i\, dt + 14 \times 10^{-3}(di/dt) = 160 \cos(377t)$. (c) $\mathbf{i}(t) = 3.21 e^{j(377t + 20°)}$ A. (d) $\mathbf{i}^*(t) = 3.21 e^{-j(377t + 20°)}$ A. (e) $i(t) = 3.21 \cos(377t + 20°)$ A.

8-16 $i(t) = \cos(2t - 97°)$ A.

8-18 (a) $(v/R) + C(dv/dt) = I_o \cos \omega t$. (c) $\mathbf{V} = R\mathbf{I}_o/(1 + j\omega RC)$. (d) $R = 0.463 \, \Omega$, $C = 0.125$ F. (e) $v(t) = 0.756 \cos(50t - 70.9°)$ V.

8-20 $i_L + LC(d^2i/dt^2) = I_m \cos(t/2\sqrt{LC})$. $i_L(t) = (4/3)I_m \cos(t/2\sqrt{LC})$.

8-22 (a) $166\underline{/-73.6°} \, \Omega$, $47 \, \Omega$, $-160 \, \Omega$. (b) $48.9\underline{/-16°} \, \Omega$, $47 \, \Omega$, $-13.5 \, \Omega$. (c) $52.5\underline{/26.5°} \, \Omega$, $47 \, \Omega$, $23.4 \, \Omega$. (d) $254\underline{/79.3°} \, \Omega$, $47 \, \Omega$, $250 \, \Omega$.

8-24 (a) $300\underline{/0°}$ V, $4000\underline{/-90°}$ V, $2000\underline{/90°}$ V, $1000\underline{/-90°}$ V, $3041\underline{/-80.5°}$ V. (b) 790 r/s. (c) 856 r/s. (d) 730 r/s.

8-26 (a) $\omega = (1/RC)$. $\mathbf{I} = (0.707\underline{/45°})(V_s/R)$. (b) $\omega = 0.378/RC$. $\mathbf{I} = (0.353\underline{/69.3°})(V_s/R)$.

8-28 $C = 100 \, \mu\text{F}$. $\mathbf{Z}_x = 22.4\underline{/26.4°} \, \Omega$. $20.1 \, \Omega$ in series with 10 mH.

8-30 $\mathbf{Z}_1 = 100\underline{/61°} \, \Omega$. $48.5 \, \Omega$ in series with 0.875 H.

8-32 (a) $\mathbf{Y} = 0.125\underline{/-80°}$ S. (b) $\mathbf{I} = 5 \times 10^3\underline{/-113°}$ A. (c) $\mathbf{V} = 0.962\underline{/-30.5°}$ V.

8-34 (a) 110 µF. (b) 90 µF. (c) 140 µF. (d) 60 µF.

8-36 (a) $R = 10 \, \Omega$, $X = 4.12 \, \Omega$, $G = 0.0856$ S, $B = -0.0353$ S. (b) $R = 5.95 \, \Omega$, $X = -4.91 \, \Omega$, $G = 0.1$ S, $B = 0.0824$ S. (c) $R = 0.684 \, \Omega$, $X = 1.880 \, \Omega$, $G = 0.171$ S, $B = -0.470$ S. (d) $R = 0.12 \, \Omega$, $X = -0.16 \, \Omega$, $G = 3$ S, $B = 4$ S.

8-38 (a) $\omega = (G/C)$. $\mathbf{V} = (0.707\underline{/-45°})(I_s/G)$ (b) $2.64(G/C)$. $\mathbf{V} = (0.353\underline{/-69.3°})(I_s/G)$.

8-40 (a) $C = 250 \, \mu\text{F}$. (b) $\mathbf{Y}_x = j1.75$ S. 1750 µF.

8-42 $C = 0.188 \, \mu\text{F}$.

8-44 $G = R/[R^2 + (\omega L - 1/\omega C)^2]$. $B = -(\omega L - 1/\omega C)/[R^2 + (\omega L - 1/\omega C)^2]$.

8-46 (a) 6.67 r/s. (b) 3.33 r/s. (c) $G = 0.002$ S, $B = 0.004$ S.

8-50 $R = 18.7 \, \Omega$, $L = 0.054$ H.

8-52 $\mathbf{Z}_1 = 7.07\underline{/-45°} \, \Omega$. $L = 50$ mH.

8-54 (a) 25.2 mH or 19.4 mH. (b) 83.1 µF or 155 µF. (c) 1.04 mH or 0.487 mH.

8-56 133 kW, 117 kVAR.

8-58 (a) $2\underline{/-60°}$ kVA in Z_1. $2.66\underline{/41.4°}$ kVA in Z_2. (b) 3000 W, 28 VAR, 3000 VA. pf = 1. (c) $2.12 \, \Omega$ in series with 0.0586 mH.

Chapter 9

9-2 (a) $-j4.11 \, \Omega$. (b) $10.0\underline{/45°} \, \Omega$. (c) $3.60\underline{/-55.6°} \, \Omega$.

9-4 (a) $4.83\underline{/10.2°} \, \Omega$. (b) Multiply each R and each L by 25 and divide each C by 25. (c) R remains unaffected. Multiply each L and C by 0.2.

9-6 $(\omega L) = (R/3)$.

9-8 $\mathbf{V}_s = 1592\underline{/15.3°}$ V. $\mathbf{Z}_{in} = 9.42\underline{/60°} \, \Omega$. $P_{av} = 67.3$ kW. pf = 0.5 (current lagging).

9-10 $\mathbf{Z}_x = 8\underline{/67.6°} \, \Omega$.

9-12 $\begin{bmatrix} (3.5 + j2.5) & -(0.5 + j2.5) \\ -(0.5 + j2.5) & (2.5 + j1.5) \end{bmatrix}$

9-14 $\mathbf{V}_1 = (\omega L I_a)/[\omega LG + j(\omega^2 LC + K - 1)]$. $\mathbf{V}_2 = \mathbf{V}_1(1 - K)$.

9-16 V_a delivers 1000 W. V_B delivers -500 W. V_c delivers 500 W.

9-18 $\mathbf{Y}_{in} = [G_1(\mathbf{Y}_F + \mathbf{Y}_L) + \mathbf{Y}_F(g_m + \mathbf{Y}_L)]/(\mathbf{Y}_F + \mathbf{Y}_L)$.

9-20 $\begin{bmatrix} (8 + j2.5) & -(5 + j2.5) \\ -(5 + j2.5) & (7 + j3.5) \end{bmatrix}$

9-22 $\mathbf{I}_1 = 2.84\underline{/55.4°}$ A. $\mathbf{I}_2 = 5.68\underline{/125°}$ A.

9-24 I_a delivers 470 W. I_b delivers 2243 W. I_c delivers 1431 W.

9-26 $Z_{in} = [R_1(Z_1 + Z_L) + Z_1(Z_L - KR_1)]/(Z_1 + Z_L)$.

9-28 $V_c = 2.59/\underline{-129°}$ V.

9-30 $V_a = 348/\underline{-57.4°}$ V.

9-32 $(I_o/V_s) = \omega^2 C^2 R/[\omega^2 C^2 R^2 - 3j\omega RC - 1]$.

9-34 (a) $i_o(t) = 3.85 e^{j(2t+46.1°)} + 4.62 e^{j(4t+25.7°)}$ A.
(b) $i_o(t) = 3.85 e^{-j(2t+46.1°)} + 4.62 e^{-j(4t+25.7°)}$ A.
(c) $i_o(t) = 3.85 \cos(2t + 46.1°) + 4.62 \cos(4t + 25.7°)$ A.

9-36 (a) $|H(j\omega)| = \sqrt{(\omega^4 + 61\omega^2 + 100)/(\omega^4 + 5\omega^2 + 4)}$. $\theta(j\omega) = \arctan[9\omega/(10 - \omega^2)] - \arctan[3\omega/(2 - \omega^2)]$. (b) $|H(j\omega)| = 10^3 \sqrt{(1 + \omega^2)(10^6 \omega^2 + 1)/(\omega^2 + 10^6)}$. $\theta(j\omega) = \arctan(\omega) + \arctan(1/10^3\omega) - \arctan(10^3/\omega)$. (c) $|H(j\omega)| = \sqrt{9(\omega^2 + 1)/(\omega^4 + 9\omega^2)}$. $\theta(j\omega) = \pi + \arctan\omega + \arctan(3/\omega)$

9-38 $(V_2/V_1) = 1/[(1 - 0.125\omega^2 + j0.125\omega]$. Amplitude response starts at 1 at $\omega = 0$, reaches a peak value of 2.86 at 2.74 r/s, and approaches zero as $\omega \to \infty$. Phase response starts at 0° at $\omega = 0$, becomes $-45°$ at 2.37 r/s, and approaches zero as $\omega \to \infty$.

9-42 $(V_2/V_1) = (1 - K)/(2 + j\omega RC)$. $0.186(1 - K)/\underline{-68.2°}$ at $\omega = (5/RC)$; $0.354(1 - K)/\underline{-45°}$ at $\omega = (2/RC)$; $0.447(1 - K)/\underline{-26.6°}$ at $\omega = RC$; $0.485(1 - K)/\underline{-14°}$ at $\omega = (1/2RC)$; $0.498(1 - K)/\underline{-5.7°}$ at $\omega = (1/5RC)$.

9-44 $(V_m/I_j) = A_{jm}/\|y\|$. $(V_2/I_1) = (V_1/I_2) = -j(1/\omega L_2)/\|y\|$, where $\|y\| = [G_2 + j(\omega C - 1/\omega L_2)][G_1 - j(1/\omega L_1 + 1/\omega L_2)] + (1/\omega^2 L_1 L_2)$.

9-46 (a) $V_{Th} = 28.5/\underline{52°}$ V; $Z_{Th} = 1.46/\underline{31°}$ Ω. (b) $V_{Th} = 10.9/\underline{-35.2°}$ V; $Z_{Th} = 0.752/\underline{76.3°}$ Ω.

9-48 (a) AA': $V_{Th} = [(0.91 + 0.182K) - j0.364]$ V; $Z_{Th} = (3.09 - 0.454K) + j0.182(K - 1)$ Ω. BB': $V_{Th} = (4 - 2K)/[(8 - K) + j2]$ V; $Z_{Th} = 2(6 + j2)/[(8 - K) + j2]$ Ω. (b) AA': $V_{Th} = -10K/[3 + j(0.8K - 0.2)]$ V. $Z_{Th} = 0$. BB': $V_{Th} = (3 + j1)(1 + K)$ V; $Z_{Th} = (0.3 + j0.1)(1 + K)$ Ω.

9-50 $Z_{Th} = j\omega L R_1/[(R_1 - K + j\omega L]$.

9-52 (a) AA': $I_N = [(10 + 2K - j4]/[(34 - 5K) + j(2K - 2)]$ A. BB': $I_N = (2 - K)/(6 + j2)$ A. (b) BB': $I_N = 10$ A.

9-54 $V_{Th} = 100/\underline{-30°}$ V. $Z_{Th} = 50.1/\underline{53.1°}$ Ω.

9-56 (a) $R = 0.6$ Ω, $X = 0.2$ Ω. $P_{max} = 209$ W. (b) $R = 0.565$ Ω, $X = 0.283$ Ω. $P_{max} = 208$ W.

9-58 $i(t) = (2/3)\sqrt{C/L}\, V_1 \cos(t/2\sqrt{LC} + 90°) + V_2 \cos(2t/\sqrt{LC} - 90°)$ A.

9-60 $i_c(t) = 0.447 I_1 \cos(t/2RC + 63.4°) + 0.707 I_2 \cos(t/RC + 45°) + 0.892 I_3 \cos(2t/RC + 26.6°)$ A. $i_R(t) = I_o + 0.894 I_1 \cos(t/2RC - 26.6°) + 0.707 I_2 \cos(t/RC - 45°) + 0.446 I_3 \cos(2t/RC - 63.4°)$ A.

9-62 $v_o(t) = 10^4 + 53.3 \cos(500t - 90.1°) + 66.7 \cos(1000t + 92°)$ V.

9-64 $v_c(t) = 415 \cos(1000t + 90°) + 104 \cos(2000t - 90°)$ V.

Chapter 10

10-2 (a) 4.74 mWb. (b) 2.53×10^5.

10-4 (a) 1.28×10^{-5} Wb. (b) 0.938×10^8.

10-6 1.43 A.

10-8 3.55 mWb.

10-10 (a) *Mesh 1:* $[R_1 + j(\omega L_1 - 1/\omega C)]I_1 - j(\omega M - 1/\omega C)I_2 = V_s$. *Mesh 2:* $-j(\omega M - 1/\omega C)(I_1 + [R_2 + R_3 + j(\omega L_2 - 1/\omega C)]I_2 = 0$. (b) *Loop 1:* $[R_1 + j\omega(L_1 + L_2 - 2M)]I_1 + [R_1 + j\omega(L_1 - M)]I_2 = V_s$. *Loop 2:* $[R_1 + j\omega(L_1 - M)]I_1 + [R_1 + R_2 + R_3 + j(\omega L_1 - 1/\omega C)]I_2 = V_s$.

10-12 (a) $\mathbf{V}_{Th} = 88.5\underline{/-81.9°}$ V. $\mathbf{Z}_{Th} = 66.8\underline{/63°}$ Ω.
(b) $\mathbf{V}_{Th} = 75.6\underline{/6.34°}$ V. $\mathbf{Z}_{Th} = 20.8\underline{/70°}$ Ω.

10-14 (a) 12.3 mJ. (b) 9.42 mJ.

10-16 See answers to Problem 10-12.

10-18 (a) $1.92\underline{/-50.3°}$ A; 61.3 W. (b) $0.625\underline{/-14.5°}$ A; 30.2 W.

10-20 $Z_{in} = R(1 - a)^2/(1 + ja^2\omega RC)$, where $a = (N_1/N_2)$.

10-22 (a) $\mathbf{V}_{ab} = 398\underline{/30°}$ V. (b) $\mathbf{V}_{ac} = 398\underline{/-30°}$ V. (c) $\mathbf{V}_{bn} = 230\underline{/-120°}$ V.
(d) $\mathbf{V}_{ca} = 398\underline{/150°}$ V. (e) $\mathbf{V}_{cb} = 398\underline{/90°}$ V.

10-24 (a) $\mathbf{I}_a = 5.02\underline{/41.6°}$ A. $\mathbf{I}_b = 5.02\underline{/-78.4°}$ A. $\mathbf{I}_c = 5.02\underline{/162°}$ A.
(b) 252 W. (c) 756 W.

10-26 $\mathbf{Z}_2 = 3\mathbf{Z}_1$.

10-28 (a) $\mathbf{I}_a = 3.26\underline{/68.8°}$ A. $\mathbf{I}_b = 11.4\underline{/-75°}$ A. $\mathbf{I}_c = 8.97\underline{/117°}$ A.
(b) Z_{ab} consumes 722 W, Z_{bc} 1083 W, and Z_{ca} 468 W. (c) 2273 W.

10-30 *Phase sequence a-b-c:* (a) $\mathbf{I}_1 = 13.9\underline{/18.4°}$ A; $\mathbf{I}_2 = 10.4\underline{/-75°}$ A; $\mathbf{I}_3 = 13.9\underline{/-168°}$ A.
(b) $\mathbf{I}_a = 27.8\underline{/15.2°}$ A; $\mathbf{I}_b = 17.8\underline{/-126°}$ A; $\mathbf{I}_c = 17.8\underline{/156°}$ A. (c) $\mathbf{Z}_1$: 483 W. $\mathbf{Z}_2$: 809 W. $\mathbf{Z}_3$: 1451 W. (d) 2741 W. *Phase sequence a-c-b:* (a) $\mathbf{I}_1 = 13.9\underline{/18.4°}$ A. $\mathbf{I}_2 = 10.4\underline{/165°}$ A. $\mathbf{I}_3 = 13.9\underline{/-48.4°}$ A. (b) $\mathbf{I}_a = 15.3\underline{/75°}$ A. $\mathbf{I}_b = 23.3\underline{/-176°}$ A. $\mathbf{I}_c = 23.2\underline{/-34.3°}$ A.
Answers to (c) and (d) are not affected by the phase sequence.

10-32 (a) $\mathbf{Y} = 4.17 \times 10^{-2}\underline{/-60°}$ S. 48.1 Ω in parallel with 88.2 mH. (b) $C = 56.7$ μF.
(c) Same average power as before. Apparent power = 20 kVa.

10-34 (a) Reading of $W_1 = (3\sqrt{3}V_m{}^2/2|Z_L|) \cos(\theta - 30°)$. Reading of $W_2 = (3\sqrt{3}V_m{}^2/2|Z_L|) \cos(\theta + 30°)$. (c) $|\theta| > 60°$. (d) 0°.

10-36 W_1 reads 20 kW. W_2 reads 25 kW. Total average power = 45 kW. Total reactive power = 5 kVAR (current lagging). Total apparent power = 45.3 kVA.

Chapter 11

11-2 (a) $L = 0.507$ μH. (b) $\mathbf{I}_R = 10\underline{/0°}$ A. $\mathbf{I}_L = 62.8\underline{/-90°}$ A. $\mathbf{I}_C = 62.8\underline{/90°}$A.

11-4 $i_G = 25 \cos 1000t + 20 \cos 2000t + 15 \cos 3000t$ mA. $i_L = 50 \cos(1000t - 90°) + 20 \cos(2000t - 90°) + 10 \cos(3000t - 90°)$ mA. $i_C = 12.5 \cos(1000t + 90°) + 20 \cos(2000t + 90°) + 22.5 \cos(3000t + 90°)$ mA. $i_s = 45.1 \cos(1000t - 56.3°) + 20 \cos(2000t) + 19.5 \cos(3000t + 39.8°)$ mA.

11-6 (a) $G = 0.01$ S. $C = 50$ μF. $L = 0.2$ mH.
(b) $w_c = 50 \times 10^{-6}[1 + \cos(2 \times 10^4 t)]$ J. $w_L = 50 \times 10^{-6}[1 - \cos(2 \times 10^4 t)]$ J.
(c) $w_{max} = 10^{-4}$ J.

11-8 (b) $\omega = 1/\sqrt{(LC - G^2L^2/2)}$.

11-10 (a) $C_o = 1/\omega_o{}^2 L$. (b) $C = (1/\omega_o{}^2 L) \pm (\sqrt{3}G/\omega_o)$.

11-12 (a) $G = 0.0223$ S. $L = 4.56$ μH. $C = 88.9$ nF. (b) $i_G = 3.34 \cos(1.57 \times 10^6 t)$ A. $i_L = 20.9 \cos(1.57 \times 10^6 t - 90°)$ A. $i_C = 20.9 \cos(1.57 \times 10^6 t + 90°)$ A.
(c) $w_L = 5 \times 10^{-4}[1 - \cos(3.14 \times 10^6 t)]$ J. $w_c = 5 \times 10^{-4}[1 + \cos(3.14 \times 10^6 t)]$ J.
(d) $w_R = 10^{-3}$ J/cycle.

11-16 (a) $BW = 200$ r/s. Half-power at 124 r/s and 324 r/s. (b) $Y = 0.01$ S. (c) $0.0141\underline{/45°}$ S.

11-18 (a) $G = 2 \times 10^{-4}$ S. $L = 0.319$ mH. $C = 1.27$ nF. (b) Half-power at (1.495×10^6) r/s and (1.652×10^6) r/s. (c) $w_T = 3.96 \times 10^{-9}$ J. (d) $w_R = 2.5 \times 10^{-9}$ J/cycle.

11-22 (a) $BW = 1.25 \times 10^5$ r/s. Half-power at 6.262×10^6 r/s and 6.387×10^6 r/s. (b) $R = 1$ MΩ.

11-24 (a) 6324 r/s. (b) $G = 2 \times 10^{-4}$ S; $L = 50$ mH; $C = 500$ nF. (c) $\mathbf{I}_c = 790\underline{/90°}$ mA. $\mathbf{I}_{\text{coil}} = 792\underline{/-86.4°}$ mA. (d) $BW = 400$ r/s. Half-power at: 6128 r/s and 6528 r/s.

11-26 $R_L = 0.305$ Ω. $L = 4.86$ μH. $C = 3.18$ nF.

11-32 (a) $L = 0.507$ μH. (b) $\mathbf{V}_R = 10\underline{/0°}$ V. $\mathbf{V}_L = 1.59\underline{/90°}$ V. $\mathbf{V}_C = 1.59\underline{/-90°}$ V.

11-34 (a) $R = 100$ Ω. $L = 0.5$ H. $C = 20$ nF. (b) $w_L = 1.25 \times 10^{-3}[1 + \cos(2 \times 10^4 t)]$ J. $w_c = 1.25 \times 10^{-3}[1 - \cos(2 \times 10^4 t)]$ J. (c) $w_T = 2.5 \times 10^{-3}$ J.

11-36 (b) $\omega = \sqrt{1/LC - R^2/2L^2}$

11-38 (a) $R = 44.7$ Ω. $L = 0.178$ mH. $C = 2.28$ nF. (b) $\mathbf{V}_R = 150\underline{/0°}$ V. $\mathbf{V}_L = 938\underline{/90°}$ V. $\mathbf{V}_c = 938\underline{/-90°}$ V. (c) $w_L = 5 \times 10^{-4}[1 + \cos(1.57 \times 10^6 t)]$ J. $w_c = 5 \times 10^{-4}[1 - \cos(1.57 \times 10^6 t)]$ J. (d) $w_R = 1.01 \times 10^{-3}$ J/cycle.

11-40 (a) $BW = 100$ r/s. Half-power at: 1951 r/s and 2051 r/s. (b) $Z_{\min} = 50$ Ω. (c) $70.7\underline{/\pm 45°}$ Ω.

11-42 (a) $R = 5$ kΩ. $L = 31.8$ mH. $C = 12.7$ pF. (b) Half-power at: 1.494×10^6 r/s and 1.651×10^4 r/s. (c) $w_T = 6.36 \times 10^{-8}$ J. (d) $w_R = 4 \times 10^{-8}$ J/cycle.

Chapter 12

12-2 (a) $f_1(t) = 5e^{j30°}e^{j100t} + 5e^{-j30°}e^{-j100t}$. (b) $f_2(t) = 5e^{j15°}e^{(-5-j377)t} + 5e^{-j15°}e^{(-5+j377t)}$. (c) $f_3(t) = 5e^{3t} + 5e^{-3t}$. (d) $f_4(t) = 5e^{3t} - 5e^{-3t}$.

12-4 (a) $Y_{\text{in}} = 6.5(s^2 + 7.08s + 4.92)/(s^3 + 17s^2 + 16s)$. (b) $Y_{\text{in}} = 0.5(s^4 + 5.67s^2 + 1.33)/(s^3 + 4.33s)$. (c) $Y_{\text{in}} = 3(s + 1)/s$. (d) $Y_{\text{in}} = [(4 - 3KR)/3R][s + 1/RC(4 - 3KR)]/(s + 1/3RC)$.

12-6 (a) $0.154(s^3 + 17s^2 + 16s)/(s^2 + 7.08s + 4.92)$. (b) $6.67s/(s^4 + 5.67s^2 + 1.33)$. (c) 2. (d) $1/[C(4 - 3KR)][s + 1/(4 - 3KR)RC]$.

12-8 (a) Zeros at $s = -0.781, -6.30$; poles at $s = 0, -1, -16$. (b) Zeros at $s = \pm j0.495, \pm j2.33$; poles at $s = 0, \pm j2.08$. (c) Zero at $s = -1$; pole at $s = 0$. (d) Zero at $s = -1/RC(4 - 3KR)$; pole at $s = -1/3RC$.

12-10 (a) Zeros at $s = 0, -1, -16$; poles at $s = -0.781, -6.30$. (b) Zero at $s = 0$; poles at $s = \pm j0.495, \pm j2.33$. (c) no critical frequencies. (d) pole at $s = -1/(4 - 3KR)RC$.

12-12 *Series:* Any pole of Z_1 or Z_2 is a pole of the total impedance. Any zero common to both Z_1 and Z_2 is a zero of the total impedance. *Parallel:* Any zero of Z_1 or Z_2 is a zero of the total impedance. Any pole common to both Z_1 and Z_2 is a pole of the total impedance.

12-16 $(V_o/V_s) = (A/R_1C_1)(s + 1/R_2C_2)/(s + 1/R_1C_1)[s + (1 + A)/R_2C_2]$.

12-18 The asymptotic plots of the functions are as follows. (a) Horizontal at -40 dB up to 100 r/s and downward at -20 dB/decade after that. (b) Upward at a slope of $+20$ dB/decade from $\omega = 0$ to 100 r/s and levels off at a constant value of 0 dB after that. (c) Horizontal at -80 dB up to 100 r/s and downward at -40 dB/decade after that. (d) Upward at a slope of $+40$ dB/decade up to 100 r/s and horizontal at 0 dB after that.

12-20 The asymptotic plots of the functions are as follows. (a) Horizontal at 40 dB to 0.01 r/s, upward at 20 dB/decade from 0.01 to 0.1 r/s, horizontal at 60 dB from 0.1 r/s to 1 r/s, and upward at 20 dB/decade from 1 to 10 r/s, and horizontal at 80 db after that. (b) Upward at 20 dB/decade to 10 r/s, horizontal at 20 dB from 10 r/s to 100 r/s, and downward at -20 dB/decade after that.

12-22 (a) Single-stage asymptotic plot: Upward at 20 dB/decade to 2000 r/s, horizontal at 80 dB from 2000 r/s to 10^4 r/s, and downward at -20 dB/decade after that. (b) For the three-stage amplifier, upward at 60 dB/decade to 2000 r/s, horizontal at 240 dB from 2000 to 10^4 r/s, and downward at -60 dB/decade after that. (c) 240 dB. (d) Bandwidth = 1000 r/s (approx).

12-24 (a) $H_1(s) = 10^{11}s/(s + 10^3)(s + 10^6)$. (b) $H_2(s) = (s + 10^3)/(s + 100)$.

12-26 (a) $-128°$. (b) -2.95 dB.

12-28 (b) $-246°$. (c) 36.5 dB. (d) $-164°$ at 0 dB; -40 dB at $-180°$.

12-32 (a) $\omega_o = 50.2$ r/s. $Q_o = 5.02$. $BW = 10$ r/s. $R = 10\ \Omega$. $C = 396\ \mu$F. (b) Same as in (a) except that $G = 3.96 \times 10^{-3}$S.

Chapter 13

13-2 Elements of the z matrix are: (a) $1.8/(2.9 + 0.8K)$, $0.8/(2.9 + 0.8K)$, $(0.8 - 0.4K)/(2.9 + 0.8K)$, $1/(2.9 + 0.8K)$. (b) $14(1 - K)/(3 - K)$, $5(1 - K)/(3 - K)$, $(5 - 7.5K)/(3 - K)$, $(8.75 - 5K)/(3 - K)$.

13-4 Elements of the z matrix are: $-K_4/K_3$, $1/K_3$, $(K_2K_3 - K_1K_4)/K_3$, K_1/K_3.

13-10 (a) Elements of the z matrix: $0.311\underline{/22.3°}\ \Omega$, $0.194\underline{/-29.2°}\ \Omega$, $0.969\underline{/-29°}\ \Omega$, $1.38\underline{/15.8°}\ \Omega$. (b) $V_o/V_s = 0.745\underline{/-47.3°}$. $Z_o = 1.36\underline{/25.9°}$.

13-12 Elements of the y matrix are: (a) $(R_1 + R_2)/R_1R_2$, $-1/R_2$, $(K - 2)$, $(R_2 + R_3)/R_2R_3$. (b) $(4K - 7)/26(K - 1)$, -0.154, $(4 - 6K)/26(K - 1)$, 0.431.

13-14 T model elements: $6.08\ \Omega$, $0.963\ \Omega$, $-0.296\ \Omega$, $2.37I_1$. Pi model elements: -0.20 S, 0.65 S, 4.10 S, $-1.6V_1$.

13-16 (a) Elements of the y matrix: $g'_{bb}(g'_{be} + g'_{bc})/(g'_{bb} + g'_{be} + g'_{bc})$, $-(g'_{bb}g'_{bc})/(g'_{bb} + g'_{be} + g'_{bc})$, $(g_m - g'_{bc})g'_{bb}/(g'_{bb} + g'_{be} + g'_{bc})$, $g_{ce} + [g'_{bc}(g_m + g'_{bb} + g'_{be})]/(g'_{bb} + g'_{be} + g'_{bc})$. (b) 4.69×10^{-3}S.

13-18 Elements of the h matrix: (a) $0.4\ \Omega$, 0.8, $(0.4K - 0.8)$, $(2.9 + 0.8K)$ S. (b) $26(K - 1)/(4K - 7)$, $4(1 - K)/(7 - 4K)$, $(4 - 6K)/(4K - 7)$, $(2.4 - 0.8K)/(7 - 4K)$.

13-20 Elements of the h matrix: $2.22\ \Omega$, 1.44, -5.00, 1.5 S. (b) $(I_2/I_s) = -1.25$.

13-22 (a) -25. (b) 25.

13-24 $h_{11} = 1/y_{11}$. $h_{12} = -y_{12}/y_{11}$. $h_{21} = y_{21}/y_{11}$. $h_{22} = (y_{11}y_{22} - y_{12}y_{21})/y_{11}$.

13-26 (a) $A = (L^2Cs^3 + 2RLCs^2 + Ls + R)/L^2Cs^3$. $B = [2RL^2Cs^3 + (L^2 + 2R^2LC)s^2 + 2RLs + R^2]/L^2Cs^3$. $C = (2LCs^2 + 1)/L^2Cs^3$. $D = (L^2Cs^3 + 2RLCs^2 + Ls + R)/L^2Cs^3$. (b) $A = 1.63\ B = 6.30\ \Omega$, $C = 1.08$ S, $D = 4.79$.

13-28 (a) $h_{11} = B/D$. $h_{12} = (AD - BC)/D$. $h_{21} = -1/D$. $h_{22} = C/D$. (b) $A = -(h_{11}h_{22} - h_{12}h_{21})/h_{21}$. $B = -h_{11}/h_{21}$. $C = -h_{22}/h_{21}$. $D = -1/h_{21}$.

13-30 (a) With $I_1 = 0$, $E = (V_2/V_1)$, and $G = (I_2/V_1)$. With $V_1 = 0$, $F = -(V_2/I_1)$, and $H = -(I_2/I_1)$. (b) $E = D/(AD - BC)$, $F = B/(AD - BC)$, $G = C/(AD - BC)$. $H = A/(AD - BC)$. (d) EFGH parameters are: 0.694, $1.54\ \Omega$, 1.04 S, 7.31.

13-32 h parameters. Overall h matrix = sum of the individual h matrices.

13-34 (a) $A = 0$, $B = -h_{11}/h_{21}$. $C = 0$. $D = -1/h_{21}$. (b) $V_2/V_s = R_Lh_{21a}h_{21b}/R_s + h_{11a}$.

Chapter 14

14-2 $a_o = 0.632$. $a_m = 1.264/(1 + 4\pi^2m^2)$. $a_1 = 0.0312$; $a_2 = 7.95 \times 10^{-3}$; $a_3 = 3.55 \times 10^{-3}$; $a_4 = 2 \times 10^{-3}$; $a_5 = 1.28 \times 10^{-3}$. $b_m = 2.53\pi m/(1 + 4\pi^2m^2)$. $b_1 = 0.196$; $b_2 = 0.100$; $b_3 = 0.0669$; $b_4 = 0.0503$; $b_5 = 0.0402$.

14-4 (a) $T = 10$ ms. (b) Amplitude = $(20/k\pi)\sqrt{(1 + k^2\pi^2) + 1}$. Phase = $(\pi - \arctan k\pi)$. (c) Third harmonic: $2.13 \cos (600\pi t + 96°)$. Fifth harmonic: $1.28 \cos (1000\pi t + 93.6°)$.

14-6 (a) $v_1(t) = 0.15 + 0.271 \cos(\omega t - 72.6°) + 0.199 \cos(2\omega t - 148°) + 0.117 \cos(3\omega t + 124°) + 0.0699 \cos(4\omega t + 12°) + \ldots$
(b) $0.15 + 0.271 \cos(\omega t + 72.6°) + 0.199 \cos(2\omega t + 148°) + 0.117 \cos(3\omega t - 124°) + 0.0699 \cos(4\omega t - 12°) + \ldots$ (c) $v_3(t) = 0.517 \sin(\omega t + 180°) + 0.211 \sin(2\omega t + 180°) + 0.194 \sin(3\omega t) + 0.0291 \sin(4\omega t) + \ldots$

14-10 $v_o(t) = 15 + 31.2 \cos(500t - 29.4°) + 10.6 \cos(1500t + 89.6°) + 0.0343 \cos(2500t - 126°)$ V.

14-12 $L = 0.75$ H. $C = 0.333$ μF. $i(t) = 13.0 \cos(1000t + 77.5°) + 30.0 \cos(2000t) + 7.43 \cos(3000t - 68.2°)$ mA.

14-14 131 mW.

14-16 (a) 0.612 V. (b) 2.83 V. (c) 8.66 V.

14-18 $c_o = 0.75$. $c_m = (1/2m^2\pi^2)[(1 + jm\pi)(-1)^m - 1] - (1/j2\pi m)[1 - (-1)^m]$

14-20 (a) $T = 31.4$ ms. (b) dc component $= 0.75$ V. (c) $0.433 \cos(200t - 22.5°)$ V, $0.138 \cos(400t - 45°)$ V, $0.0646 \cos(600t - 67.5°)$ V.

14-22 $v(t) = 4.64e^{-j(5000t+21.8°)} + 12.5e^{-j(3000t+153.7°)} + 8.9e^{-(1000t+93.4°)} + 8.9e^{j(1000t+93.4°)} + 12.5e^{j(3000t+153.7°)} + 4.64e^{(5000t+21.8)}$.

14-24 $v_o(t) = 0.0596e^{-j(1000t-232°)} + 0.130e^{-j(500t-146°)} + 0.432 + 0.130e^{j(500t-146°)} + 0.0596e^{j(1000t-232°)}$.

14-28 (b) 1.43 W. (c) $f(t) = 0.75 + 1.13 \cos(\omega t + 57.5°) + 0.478 \cos(2\omega t - 180°) + 0.324 \cos(3\omega t + 78°) + 0.238 \cos(4\omega t - 180°) + 0.193 \cos(5\omega t + 83°) + 0.154 \cos(6\omega t - 180°)$.

14-30 (a) $v_o(t) = \sum_{k=-3}^{3} (50/k\pi)\sin(k\pi/10)e^{j(1000k\pi t - 0.1k\pi)}$ (c) $v_o(t) = 5 + 9.84 \cos(1000\pi t - 18°) + 9.36 \cos(2000\pi t - 36°) + 8.58 \cos(3000\pi t - 54°)$ V.

14-32 $F(j\omega) = (A/\omega^2 t_p)[(1 - \cos \omega t_p - \omega t_p \sin \omega t_p) + j(\sin \omega t_p - \omega t_p \cos \omega t_p)]$.
$|F(j\omega)| = (A/\omega^2 t_p)\sqrt{[2 + \omega^2 t_p^2 - 2\cos(\omega t_p) - 2\omega t_p \sin(\omega t_p)]}$.

14-34 (a) $(4V_o/j\omega) \sin^2(\omega t_p)$. (b) $(V_o/j\omega)[2\cos(\omega t_p) - 2\cos(2\omega t_p)]$.

14-36 (a) $25/2\pi(\omega^2 + 100)$. (b) 10 r/s. (c) 1.25 J.

14-38 (a) $|F(j\omega)| = 10e^{-0.01|\omega|}$. (c) Arg $F(j\omega) = -\arctan(0.01\omega)$.

14-40 $(K/\pi t)[\sin(\omega_b t) - \sin(\omega_a t)]$.

14-42 (a) $V_c(j\omega) = (1 - e^{-j\omega})/[j\omega(1 + j0.5\omega)]$. (b) $V_R(j\omega) = 0.5(1 - e^{j\omega})/(1 + j0.5\omega)$.
(c) $|V_c(j\omega)| = [(2/\omega) \sin(\omega/2)]/\sqrt{1 + 0.25\omega^2}$. $|V_R(j\omega)| = \sin(\omega/2)/\sqrt{1 + 0.25\omega^2}$.

14-46 (a) 0.333. (b) 3.58. (c) 9.50.

14-48 (a) $H(j\omega) = 1/(j\omega + 1)$. (b) $h(t) = e^{-t}u(t)$. $y(t) = 0$ $(t < 0)$; $(1 - e^{-t})$ $(0 < t < 1)$; $e^{-t}(e - 1)$ $(t > 1)$.

14-50 $H(j\omega) = (20 \cos \omega)e^{-j4\omega}$.

Chapter 15

15-2 (a) $(5/s)(e^{4s} + e^{-4s} - 2)$. (b) $(A/s^2 t_p)(e^{-st_p} - 1 + st_p)$. (c) $2aK/(a^2 - s^2)$.

15-4 (a) $(s^2 - 16)/(s^2 + 16)^2$. (b) $(s^2 + 6s - 7)/(s^2 + 6s + 25)^2$. (c) $se^{-5s}/(s^2 + b^2)$.
(d) $s(s^2 - 16)/(s^2 + 16)^2$. (e) $[(A \cos D)s - (A \sin D)B]/(s^2 + B^2)$.

15-6 $I_1(s) = (10/R)[s/(s^2 + 100)(s + 1/RC)]$. $I_2(s) = (1/L)[s/(s + 10)(s^2 + 1/LC)]$. $R = 2$ Ω. $L = 0.2$ H.

15-8 $I_1(s) = 200(s + 5)/[s(s^2 + 50s + 300)]$. $I_2(s) = 50(s + 20)/[s(s^2 + 50s + 300)]$.

15-10 (a) $H(s) = 1/[R_1L_1C_1C_2s^3 + L_1C_2s^2 + R_1(C_1 + C_2)s + 1]$.
(b) $H(s) = L_1C_1s^2/[L_1(C_1 + C_2)s^2 + 1]$. (c) $H(s) = (s^2 + 1)/(s^2 + 5.5)$.

15-12 (a) $f(t) = (1 + 5e^{-t} - 6e^{-2t})u(t)$. (b) $f(t) = \delta(t) - 0.111u(t) - 0.444e^{-3t}u(t) + 3.56e^{3t}u(t)$. (c) $f(t) = (e^{-t} + 2e^{-2t} + 3e^{-3t})u(t)$.

15.14 (a) $f(t) = (t + 3e^{-t} - 2)u(t)$. (b) $f(t) = (e^{-t} - te^{-2t} - e^{-2t})u(t)$.
(c) $f(t) = (-0.5t^2 - 1 + 0.5 e^{-t} + 0.5e^t)u(t)$. (d) $0.5(t \cos t + \sin t)$.

15-16 $i(t) = 5e^{-60t}u(t)$ A.

15-18 $v_c(t) = [I_oR/(1 - aRC)](e^{-at} - e^{-t/RC})u(t)$.

15-20 $v_L(t) = -12.5e^{-37.5t}u(t)$ V.

15-22 $v_1(t) = (0.667e^{-5t} + 1.33)u(t)$ V.

15-24 $H(s) = Ae^{-st_0}$.

15-26 (a) $y(t) = [u(t) - u(t - T) + u(t - T/2) - u(t - 3T/2) + u(t - 2T) + u(t - 3T)]$. (b) $y(t) = [u(t) + (t - T)u(t - T) + (t - 2T)u(t - 2T)]$.
(c) $y(t) = [(T/2\pi)\sin(2\pi t/T)][u(t) - 2u(t - T) + u(t - 2T)]$.

15-28 $Z(s) = (3s^4 + 6s^3 + 27s^2 + 24s)/(s^4 + 13s^2 + 36)$.

15-32 (a) $H(s) = s/(s + R/L)$. (b) $h(t) = [\delta(t) - (R/L)e^{-Rt/L}u(t)]$.
(c) $v_2(t) = [e^{-Rt/L}u(t) - e^{(1-Rt/L)}u(t - L/R)]$.

15-34 $R = 10 \, \Omega$. $C = 1.25 \, \mu\text{F}$. $I_o = 0.25$ A.
$v(t) = [-2.49e^{-200t} + 50 \cos(4000t - 87.1°)]$V.

15-36 (a) $L_1 = 1.53$ H. $L_2 = 1.08$ H. $C_1 = 1.58$ F. $C_2 = 0.383$ F. (b) $h(t) = [-e^{-0.383t}\cos(0.924t - 22.5°) + 2.41e^{-0.924t}\cos(0.383t - 67.5°)]$.

INDEX

A

Active circuit, 132, 137
Admittance, 369–372
   conductance and susceptance, 370
   relationship to impedance components, 377
   triangle, 375
Alternating current (*see* sinusoidal signals)
Ampere's law (*see* magnetic circuits)
Amplifiers (*see* op amp)
Analog computer circuit, 317
Angular frequency, 319
Associated reference polarities, 10
Automatic exposure control, 285

B

Bode diagrams, 582–598
   amplitude response, 583–588, 590–591
   asymptotes, 583, 585, 590
   break (corner) frequency, 585
   high-pass filter, 592
   low-pass filter, 588
   phase response, 589–590
   procedure for, 586
Branch currents, Method of, 168–179
   circuits with current sources, 177
   summary of method, 179
Bridge, *see* Wheatstone's bridge.

C

Capacitor, 264
   ac applied, 326, 362, 371
   in complex frequency, 576, 744
Causal signals, 738
Characteristic equation, 298, 574
Circuit, classification of, 13
   distributed, 14
   linear, 15
   lumped, 14
   time-invariant, 17
Circuit analysis overview, 42, 770
Closed path, 24
Cofactor, 134, 434, 786

Complex exponential functions, 347–350
   differentiation and integration in time domain, 357
   real part convention, 352
   Euler's theorem, 348
Complex frequency, 573, 735
   analysis of circuits, 743–748
   differentiation and integration in time domain, 576
   impedance and admittance, 576–577
Complex numbers, 791–796
   complex conjugate, 796
   exponential and polar forms, 793
   multiplication and division, 794
Complex power, 389–391
Conductance, 65
   in ac circuit, 370
   driving point and transfer, 213–215
Conservative of charge, 28
Conservative property, 24
Controlled sources (*see* dependent sources)
Convolution, 708–720
   Fourier transform of, 718
   integral, 711
   Laplace transform of, 769
Coulomb, 2
Coupled coils (*see* magnetically coupled coils)
Cramer's rule, 133, 788
Critical frequencies, 561, 578–582
   role in network response, 769
Current (*see also* loop currents)
   direction, 3, 10
   ideal current source, 10
Current continuity condition, 269
Current divider, 68–71
   approximations, 71
Current gain, 126

D

D'Arsonval movement, 60, 70
   analog dc ammeter, 70
      multiple range, 96
   analog dc voltmeter, 60
      multiple range, 94
Dead network (*see* relaxed network)
Decibels, 582
Delta-wye transformation, 226, 658

Dependent sources (*see* sources)
Determinants, 785–790
   cofactor, 786
   Laplace's expansion, 787
   minor, 786
Direct current (dc), 7
Driving point conductance, 214
   resistance, 143–144
Duality, 74, 203–205, 268, 372
   table of dual pairs, 74, 205
   use in mesh analysis, 333
   use in resonant circuits, 551–552
   use in two-port networks, 629

E

Electromagnetic induction, 268
Electromotive force (emf), 2
Energy, electrical, 6
   in a capacitor, 267
   in coupled coils, 494–496
   in an inductor, 269
Error, allowable margin of, 61
Euler's theorem, 348

F

Farad, 265
Filters:
   low-pass, 429, 588–89
   high-pass, 592
Forced response, 276
Fourier series, 661–690
   amplitude and phase of m-th harmonic, 682
   amplitude spectrum, 685
   average power, 675–677
   dc component, 663
   even and odd symmetry, 669
   exponential series, 679–684
   Gibbs phenomenon, 690, 784
   half-wave symmetry, 671
   minimum mean square error, 689–690
   phase spectrum, 685
   power spectrum, 686
   relationship between trigonometric and exponential series, 681
   rms value, 678
   steady state response, 673–675, 683–684

trigonometric series, 663–677, 784
truncated signal, 689
Fourier transform, 692–702
   amplitude density spectrum, 692
   of convolution, 718–720
   differentiation and integration in time domain, 697
   energy density spectrum, 693–695
   even and odd symmetry, 695
   filter bandwidth, 702
   of impulse response, 717
   inverse Fourier transform, 699–700
   limitations, 721
   normalized energy, 693
   response of a linear system, 701–702
Frequency, 319
   negative, 430
Frequency response diagrams (*see* Bode diagrams; transfer functions).

G

Gauss's elimination method, 782
Graph of a network, 104, 181
Gyrator, 657

I

Impedance, 360–363
   relationship to admittance components, 377
   resistance and reactance, 361
   of single elements, 362
   triangle, 367
Impulse function, 705
   sifting theorem, 706
   in signal representation, 707–710
Impulse response, 708
   Fourier transform of, 717
Inductor, 268
   ac applied, 328, 362, 371
   in complex frequency, 576, 743

K

Kilowatt-hour (kWh), 7
Kirchhoff's laws, 23–24, 28–29
   sign convention for KCL, 29, 65, 108
   sign convention for KVL, 25, 57, 170

L

Ladder network procedure, 82–88, 406–407, 783–784
Laplace transform, 735–768
   analysis of linear networks, 745–748, 759–768
   differentiation and integration in time domain, 741
   final value theorem, 768
   initial value theorem, 768
   inverse Laplace transform, 748–759
   properties, 741
   region of convergence, 736
   step by step procedure for networks, 759
   table of Laplace transform pairs, 740
   transforms of circuit elements, 744
Lenz's law, 268
Linearity, 14
   additive property, 15
   scaling property, 14
Loop analysis, 181–188, 417–426
   circuits with current sources, 197–200
   circuits with dependent sources, 195–196, 420–421
   loop resistance matrix, 211–212
   sinusoidal steady state, 417–426
   standard form of loop equations, 191–193

M

Maclaurin's series, 347
Magnetically coupled coils, 485–504
   circuits with coupled coils, 491–499
   coefficient of coupling, 487
   dot polarity convention, 489
   energy stored, 494–497
   ideal transformer, 499–504
   impedance transformation, 504
   polarity of induced voltage, 486
   T equivalent, 497
Magnetic circuits and fields, 473–484
   Ampere's law, 474
   basic relationships, 474
   magnetomotive force (*mmf*), 475
   mmf and flux calculation, 479–484
   reluctance, 477
Matrix, 789–790
   inverse, 790
Maximum power transfer, 244–246, 443–446
   matching, 246

Maximum power transfer (*Cont.*)
  sinusoidal steady state, 443–446
Mesh analysis, 189–196, 201–202, 422
  circuits with current sources, 201–202, 422
  circuits with dependent sources, 209, 422
  Cramer's rule in, 207–208
  mesh impedance matrix, 433–436
  mesh resistance matrix, 205–210
  sinusoidal steady state, 417–422
  standard form of equations, 191–193
Mho, 65
Model of amplifier, 76
  bipolar junction transistor, 41
  conversion of source models, 38
  nonideal current source, 34
  nonideal voltage source, 34
  Norton, 34
  Thevenin, 34
Mutual induction (*see* Magnetically coupled coils)

N

Natural response, 276
Negative frequency, 430
Negative impedance converter, 659
Neper frequency, 575
Network functions, 573–582
  critical frequencies, 578–581
  magnitude and angle in the s plane, 600
  poles and zeros, 579–580
Nodal analysis, 103–144, 408–415
  algebraic discussion, 130–144
  circuits with dependent sources, 125, 136, 410
  circuits with voltage sources, 118, 125, 412–414
  source conversion procedure, 162, 412–414
  general procedure, 110
  nodal conductance matrix, 130–132
  sinusoidal steady state, 408–415
  standard form of equations, 113–115
  summary of procedure, 125
Node, 29
  reference (datum), 106, 107
Node voltage, 106
  independence of, 65, 106
  node-pair voltage, 105

Nonlinear series circuit, graphical solution, 88–90
  use of Thevenin equivalent, 244
Nonohmic device, 22
Norton's equivalent, 238–244, 440–443
  conversion to Thevenin, 242
  procedure, 239
  sinusoidal steady state, 440–443

O

Ohm's law, 20
  in ac circuit, 361
  in nodal analysis, 107
Op amp (operational amplifier), 144–153
  analog computer circuit, 317
  integrating circuit, 307–308
  inverting circuit, 146, 176
  model of op amp, 145
    approximate model, 148
  noninverting circuit, 149
  summing circuit, 151
  valid approximations, 148
Open circuit, 26
Orthogonality, 662

P

Parallel circuits, 65–73, 372–376
  current divider, 68
  equivalent in ac circuits, 378
  with multiple sources, 71
Partial fraction expansion, 749–759
Passive circuit, 132
Permeability, 474
Phase angle, 320
  lag and lead, 322
Phasor, 350–354
  use in circuit analysis, 358
  diagrams, 368, 376
  different forms of, 354
Polarities, associated reference, 10
Poles, 561, 579–580
Potential difference, 2
Power, 4
  delivered or received, 11
  dissipated in a resistor, 21
  negative values of, 12

Power factor, 382–383
  modification, 386
Power in the sinusoidal steady state, 334–337, 381–390
  apparent power, 384–385
  average power, 335–337, 382–383, 675–678
  complex power, 389–391
  content of periodic signal, 678
  instantaneous, 335
  reactive power, 384
  in single elements, 326
PSpice, 797
Pulse function, 703

R

Rational function, 748
RC circuit, 273–287, 331–332
  capacitor discharge, 280
  change of conditions, 283
  pulse response, 280–282, 784
  sinusoidal steady state, 331–332
  step response, 273–286
  time constant and rise time, 277
  voltage buildup, 275
Reactance, 361
Reactance/susceptance curves of lossless networks, 561–567
  properties of, 561
  rules for sketching, 564
Reciprocity theorem, 250
  proof, 256
Rectifying device, 22
Relaxed network, 139, 213, 232, 433
Residues, 751–759
  at complex conjugate poles, 754–756
  at multiple poles, 757–759
  at simple poles, 751–754
Resistance, 20
  driving point, 143–144
  feedback, 145
  of impedance, 361
  input, 76, 139, 145
  internal, 34
  negative, 132, 139, 233
  output, 76, 139, 145
  Thevenin, 139
  transfer resistance, 140–142

Resistor, 20
  ac applied, 326, 362, 371
  in complex frequency, 576, 743
  current direction, 20
  linear time-invariant, 20
  piecewise linear, 22
Resonance in parallel GLC circuit, 529–543
  admittance triangles at half-power frequencies, 541
  bandwidth, 541–543
  complex frequency viewpoint, 598
  energy relationships, 534–538
  half-power frequencies, 540
  properties at resonance, 532
  Q, 537
  summary of relationships, 543
Resonance in series RLC circuit, 551–560
  bandwidth, 557–558
  energy relationships, 554–556
  half-power frequencies, 558
  Q, 555
  summary of relationships, 558
  use of duality, 551–552
Resonance in two-branch parallel RLC circuit, 544–551
  application, 549–550
  bandwidth calculations, 547–548
  equivalent GLC circuit, 545
  harmonic generator, 551
  Q of a coil, 544
Rise time, 277
RL circuit, 288–295, 354–356
  complex exponential excitation, 354–356
  current buildup, 289
  sinusoidal steady state, 329–330
  step response, 288–295
  time constant, 290
Root mean square (rms) value, 338–340
  of periodic signal, 678
  of a sinusoid, 340

S

Second order circuit, step response 297–307
  characteristic equation, 298
  critically damped, 301
  overdamped, 302
  underdamped (oscillatory), 304

Second order circuit, step response (*Cont.*)
  zero input response, 298
  zero state response, 305
Series circuit, 56–64, 363–369
  equivalent in ac circuits, 378
  with multiple sources, 62
  voltage divider, 59, 365
Series-parallel reduction, 78–82, 403–405
Short circuit, 31
Shunt resistor, 70
Siemens, 65
Sifting theorem, 706
Simpson's rule, 781
Sinusoidal signals, 318–328, 350–353
  combination, 324
  decomposition, 322
  definitions, 319
  half-wave rectified, 665
  phase angle, 320
  phasor representation, 350
  real part convention, 352
Sinusoidal steady state, 326–333, 356–359, 402–425
  ladder network method, 406–407
  loop and mesh analysis, 417–427
  maximum power transfer, 443–446
  nodal analysis, 408–415
  Norton equivalent, 440–443
  reciprocity theorem, 450–452
  series parallel reduction, 403–405
  single elements, 326–328
  superposition theorem, 447–449
  Thevenin equivalent, 437–440
  time domain solution, 331–333
  transfer functions, 426–436
Sources, dependent, 41–42
  conversion of models, 38–40, 162, 412–413
  ideal sources, 9–10
  nonideal source models, 34–40
SPICE program, 797–809
Staircase approximation of a signal, 703–705
Step function, 272
Superposition Principle, 17
  theorem, 247–250, 447–449
  proof of theorem, 255

Susceptance, 370
Synchronous motor in power factor correction, 388–389

T

Tellegen's theorem, 48
Temperature effect on resistance, 23
Thevenin's equivalent, 232–238, 437–440
  conversion to Norton, 242
  procedure, 232, 437
  proof of theorem, 253
  sinusoidal steady state, 437–440
  Thevenin impedance, 437
  Thevenin resistance, 139, 233
Three phase circuits, 505–520
  balanced load, 509–514
  balanced source, 507
  double subscript notation, 505
  line to line voltage, 508
  phase sequence, 508
  phase voltage, 508
  power, 510, 515–520
  voltage triangle, 508, 513
  wattmeters in measuring power, 517–520
Tie set schedule, 223
Time constant, 277
Topology of networks, 104–106, 181–183
  graph of a network, 181
  links in a graph, 181
Transfer conductance, 214
  resistance, 139
Transfer functions, 426–436
  frequency response, 428–432
  from mesh impedance 433–435
Transformers (*see* Magnetically coupled coils)
Transistor, bipolar junction, 41
  analysis using h parameters, 659
  dc analysis, 173–177
Tree, 104
Two-ports, 614–656
  ABCD (transmission) parameters, 646, 649–652
  analysis of a two-port system, 623–624, 632, 643–644
  balanced and unbalanced, 653
  h parameter model 642–643
  hybrid (*h*) parameters, 636–645
  interconnection of two-ports, 646–652

pi network models, 632–634
relationships between parameters, 635, 644
T network models, 624–628
y parameters, 629–636
z parameters, 616–629

V

Voltage, 2
   average value, 7
   reference polarities, 8
   root-mean-square (rms), 7
Voltage continuity condition, 267
Voltage divider, 59
   approximations, 61
Voltage gain, 77, 126, 146
Volt-ampere (*VA*), 384
Volt-ampere-reactive (*VAR*), 384

W

Wheatstone's bridge, 225, 380–381
Wye-delta transformation, 226, 658

Z

Zero input response, 295–296, 298–304
Zero state response, 295–296, 305–306
Zeros, 561, 579